Behavioral Genetics in the Postgenomic Era

Behavioral Genetics in the Postgenomic Era

EDITED BY

ROBERT PLOMIN, JOHN C. DEFRIES, IAN W. CRAIG, AND PETER MCGUFFIN

AMERICAN PSYCHOLOGICAL ASSOCIATION, WASHINGTON, DC

Published by
American Psychological Association
750 First Street, NE
Washington, DC 20002
www.apa.org

To order
APA Order Department
P.O. Box 92984
Washington, DC 20090-2984

Tel: (800) 374-2721; Direct: (202) 336-5510
Fax: (202) 336-5502; TDD/TTY: (202) 336-6123
Online: www.apa.org/books/
Email: order@apa.org

In the U.K., Europe, Africa, and the Middle East, copies may be ordered from
American Psychological Association
3 Henrietta Street
Covent Garden, London
WC2E 8LU England

Typeset in Century Schoolbook by EPS Group Inc., Easton, MD

Printer: Port City Press, Baltimore, MD
Cover Designer: Anne Masters, Washington, DC
Technical/Production Editor: Casey Ann Reever

Library of Congress Cataloging-in-Publication Data
Behavioral genetics in the postgenomic era / edited by Robert Plomin . . . [et al.].
p. cm.
Includes bibliographical references and index.
ISBN 1-55798-926-5 (alk. paper)
1. Behavior genetics. I. Plomin, Robert, 1948–

QH457 .B427 2002
155.7—dc21

2002018681

British Library Cataloguing-in-Publication Data
A CIP record is available from the British Library.

Printed in the United States of America
First Edition

Contents

Contributors

Katherine J. Aitchison, BMBCh, MRC Psych, Clinical Senior Lecturer in Adult Psychiatry, Section Clinical Neuropharmacology and Social, Genetic, and Developmental Psychiatry Research Centre, Institute of Psychiatry, London

Seth A. Balogh, PhD, Research Associate, Institute for Behavioral Genetics, University of Colorado, Boulder

Robert H. Belmaker, MD, Division of Psychiatry, Faculty of Health Sciences, Ben Gurion University of the Negev, Beersheva, Israel

Jonathan Benjamin, MD, Chairman, Division of Psychiatry, Faculty of Health Sciences, Ben Gurion University of the Negev, Beersheva, Israel

Dorret I. Boomsma, PhD, Department of Biological Psychology, Vrije University, Amsterdam, The Netherlands

Kathleen K. Bucholz, PhD, Missouri Alcoholism Research Center, Department of Psychiatry, Washington University School of Medicine, St. Louis

Lon R. Cardon, PhD, Wellcome Trust Centre for Human Genetics, University of Oxford, Oxford, England

John C. Crabbe, PhD, Portland Alcohol Research Center, Department of Behavioral Neuroscience, Oregon Health and Science University and Veterans Affairs Medical Center, Portland

Ian W. Craig, PhD, Head of Molecular Genetics Group, Social, Genetic, and Developmental Psychology Research Centre, Institute of Psychiatry, London

John C. DeFries, PhD, Institute for Behavioral Genetics, University of Colorado, Boulder, CO

Eco J. C. de Geus, PhD, Department of Biological Psychology, Vrije University, Amsterdam, The Netherlands

Alan Doyle, PhD, Scientific Programme Manager, The Wellcome Trust Centre, London

Richard P. Ebstein, PhD, Professor and Director of Research Laboratory, S. Herzog Memorial Hospital, Jerusalem, Israel

Simon E. Fisher, DPhil, Royal Society Research Fellow, Wellcome Trust Centre for Human Genetics, Oxford, England

Jonathan Flint, PhD, Wellcome Trust Centre for Human Genetics, Oxford, England

Gina M. Geffen, PhD, Cognitive Psychophysiology Laboratory, University of Queensland, St. Lucia, Australia

Laurie B. Geffen, PhD, Cognitive Psychophysiology Laboratory, University of Queensland, St. Lucia, Australia

Michael Gill, MD, MRCPsych, FTCD, Associate Professor of Psychiatry, Trinity Centre for Health Sciences, St. James' Hospital, Dublin, Ireland

Irving I. Gottesman, PhD, FRCPsych, Departments of Psychiatry and Psychology, University of Minnesota, Minneapolis

Seth G. N. Grant, MB, Department of Neuroscience, University of Edinburgh, Scotland

Peter Greenaway, PhD, FRCPath, Assistant Director of Research and Development, Department of Health, London

Elena L. Grigorenko, PhD, Center for the Psychology of Abilities, Competencies, and Expertise and Child Study Center, Yale University, New Haven, CT

Andrew C. Heath, DPhil, Missouri Alcoholism Research Center, Department of Psychiatry, Washington University School of Medicine, St. Louis, MO

Jerome Kagan, PhD, Research Professor of Psychology, Harvard University, Cambridge, MA

Sridevi Kalidindi, MBBS, MRCPsych, Institute of Psychiatry, King's College, London

K. Peter Lesch, MD, Department of Psychiatry and Psychotherapy, University of Wuerzburg, Wuerzburg, Germany

Michelle Luciano, PhD, Queensland Institute of Medical Research and Cognitive Psychophysiology Laboratory, University of Queensland, Brisbane, Queensland, Australia

Pamela A. F. Madden, PhD, Missouri Alcoholism Research Center, Department of Psychiatry, Washington University School of Medicine, St. Louis, MO

Nicholas G. Martin, PhD, Genetic Epidemiology Section, Queensland Institute of Medical Research, Brisbane, Queensland, Australia

Joseph McClay, BSc, Molecular Genetics Group, Institute of Psychiatry, King's College, London

Peter McGuffin, MB, PhD, Institute of Psychiatry, King's College, London

Tom Meade, DM, FRCP, FRS, School of Hygiene and Tropical Medicine, MRC Epidemiology and Medical Care Unit, London

Elliot C. Nelson, PhD, Missouri Alcoholism Research Center, Department of Psychiatry, Washington University School of Medicine, St. Louis

Michael C. O'Donovan, PhD, FRCPsych, Professor of Psychiatric Genetics, Department of Psychological Medicine, University of Wales College of Medicine, Cardiff, UK

Richard K. Olson, PhD, Department of Psychology, University of Colorado, Boulder

Michael J. Owen, PhD, FRCPsych, FMedSci, Head, Department of Psychological Medicine, University of Wales College of Medicine, Neuropsychiatric Genetics Unit, Cardiff

Bruce F. Pennington, PhD, Department of Psychology, University of Denver, Denver, CO

Margaret A. Pericak-Vance, PhD, James B. Duke Professor of Medicine, Director, Center for Human Genetics, Duke University Medical Center, Durham, NC

Robert Plomin, PhD, Institute of Psychiatry, King's College, London
Daniëlle Posthuma, PhD, Department of Biological Psychology, Vrije University, Amsterdam, The Netherlands
Rumi Kato Price, PhD, MPE, Missouri Alcoholism Research Center, Department of Psychiatry, Washington University School of Medicine, St. Louis
Frances Rawle, PhD, Strategy and Liaison Manager, Medical Research Council, London
David C. Rowe, PhD, Division of Family Studies and Human Development, School of Family and Consumer Sciences, University of Arizona, Tucson
Pak Sham, PhD, MRCPsych, Institute of Psychiatry, King's College, London
Glen A. Smith, PhD, Department of Psychology, University of Queensland, Brisbane, Queensland, Australia
Shelley D. Smith, PhD, Center for Human Molecular Genetics, Munroe Meyer Institute, University of Nebraska Medical Center, Omaha
Anita Thapar, MBBch, MRCPsych, PhD, Professor of Child and Adolescent Psychiatry, Department of Psychological Medicine, University of Wales College of Medicine, Cardiff
Alexandre Todorov, PhD, Missouri Alcoholism Research Center, Department of Psychiatry, Washington University School of Medicine, St. Louis, MO
James D. Watson, PhD, President, Cold Spring Harbor Laboratory, Cold Spring Harbor, NY
Jeanne M. Wehner, PhD, Department of Psychology, Institute for Behavioral Genetics, University of Colorado, Boulder
John B. Whitfield, PhD, Royal Prince Alfred Hospital, Sydney, Australia and Genetic Epidemiology Section, Queensland Institute of Medical Research, Brisbane, Queensland, Australia
Erik G. Willcutt, PhD, Department of Psychology, Institute for Behavioral Genetics, University of Colorado, Boulder
Julie Williams, PhD, Department of Psychological Medicine, University of Wales College of Medicine, Cardiff
Margaret J. Wright, PhD, Queensland Institute of Medical Research, Brisbane, Queensland, Australia

A Behavioral Genetics Perspective

Irving I. Gottesman

I am personally grateful to the editors and authors of this volume because it constitutes an advanced base camp in the assault on Mt. Ignorance, a camp so far up the slope that it permits glimpses, through the fog, of the summit where enlightenment about the nature of individual differences and the factors that impact on their plasticity will prevail. The summit may be only 20 years away, with exciting victories on the way. The fact that the American Psychological Association (APA) hereby seconds their vote of confidence in the "authenticity" of behavioral genetics as a worthy facet of the many that define the behavioral sciences making up the broader discipline of psychology may well mark the end of a long cold war.

In 1993, the APA published *Nature, Nurture, and Psychology,* edited by Robert Plomin and Gerald E. McClearn, which was timed to help celebrate the 100th anniversary of the APA. The 29 spirited "units" in that volume showed (off) behavioral genetics, tempered by informed critics, in its various guises and roles for the preceding century wherein genetics came of age but had been largely resisted by establishment psychology. And, in 2001, the APA presented me with the Distinguished Scientific Contributions Award for my career in behavioral genetics.

The contributors to the present volume, confident but not smug, eager to get on with the task at hand—to actually take the field of behavioral genetics into the postgenomic era—do just that. You will find here a mature collection of 26 advanced treatments of the topics that are at the core of informed psychological research and concerns, informed from the vantage point that defines psychology as a natural science, but knowing full well that it is also codefined as a social science. The book ends with a bold agenda and strategies that use quantitative trait loci in the context of a neurodevelopmental systems approach as the weapons of choice for the behavioral neurosciences, but I will not spoil it for you by premature revelations. It seems paradoxical that none of the dozens of APA divisions represents behavioral genetics—a consequence of historical origins and science politics.

Perhaps the heterogeneity of professional roots saved the domain of behavioral genetics from becoming overly organized. When the 20 founding fathers (no women attended, although some were invited) of the Behavior Genetics Association (BGA) convened on a wintry March 30, 1970, at the University of Illinois, their professional affiliations included physical anthropology, statistics-psychometrics, biology, zoology, psychology, psychiatry, agriculture, and medicine. The groundwork had been laid over 3 years earlier at a 1967 conference at Princeton University (F. Osborn presiding) underwritten by the Society for the Study of Social Biology. The active participants included V. E. Anderson, R. B. Cattell, T. Dobzhansky (who would become the first elected BGA president), B. Eckland, I. I. Gottesman, J. Hirsch, J. Loehlin, R. H. Osborne, S. C. Reed, and J. P. Scott.

In brief, the momentum leading to organization was supplied by J. L. Fuller and W. R. Thompson's (1960) text *Behavior Genetics*, J. Shields's (1962) *Monozygotic Twins: Brought Up Apart and Brought Up Together*, S. G. Vandenberg's (1965) edited *Methods and Goals in Human Behavior Genetics*, J. P. Scott and J. L. Fuller's (1965) *Genetics and the Social Behavior of the Dog*, J. Hirsch's (1967) edited *Behavior-Genetic Analysis*, D. C. Glass's (1968) edited *Biology and Behavior: Genetics*, S. G. Vandenberg's (1968) edited *Progress in Human Behavior Genetics*, D. Rosenthal's (1970) *Genetic Theory and Abnormal Behavior*, and E. Slater and V. Cowie's (1971) *The Genetics of Mental Disorder*. The preceding list is idiosyncratically illustrative and far from exhaustive. S. G. Vandenberg and J. C. DeFries founded the journal *Behavior Genetics* in 1970, in part to provide an outlet for research that was underappreciated by the editors and referees of other journals in the behavioral sciences. Authors of texts dealing in human genetics have provided a solid and receptive setting for the subject matter of behavioral genetics, with a few exceptions, going back to the first modern text *Principles of Human Genetics* by Curt Stern in 1949; extensive and informed treatments could be found in I. M. Lerner, T. Dobzhansky, S. Wright, L. S. Penrose, J. A. F. Roberts, L. L. Cavalli-Sforza, F. Vogel, and A. Motulsky. Only J. V. Neel and W. J. Schull (1954), skeptical of twin methodologies, held back their endorsement.

Human genetics itself, as an organized scientific discipline, has a rather short history. Planning for the first professional society to focus on human genetics was initiated in late 1947, with the first meeting held in September 1948. The first issue of *American Journal of Human Genetics* appeared one year later. The many scientists and physicians who were members of the American Society of Human Genetics had all received their higher degrees in some other area because degrees in genetics or departments so-named did not yet exist.

The intellectual pedigrees of both disciplines (behavioral genetics and human genetics) have in common an ancestral concept from the middle of the 19th century, when Belgian statistician and astronomer L. A. J. Quetelet demonstrated (in 1846), using the heights of French soldiers arranged along the abscissa with frequency on the ordinate, the so-called normal curve. The concept of course was variation or individual differences. The implications informed and inspired the research programs of Charles Darwin, Sir Francis Galton, and Karl Pearson so that by 1895, Alfred Binet and Victor Henri in Paris and William Stern (1900) in Germany were actively pursuing focused attempts to understand the nature and extent of individual differences in complex psychological processes among children. Differential psychology was "born" as the nearest relative to what would become behavioral genetics. The rapid rise and success of the testing movement for intellectual abilities in school children, following on the formal test developed by Binet and Théodore Simon in 1905 so as to match abilities to classroom resources, permitted the rapid implementation of such devices for the massive assessment and placement of men for World War I duties.

Appreciation and enthusiasm for research programs that focused on the psychology of individual differences between 1920 and 1960 was scattered. Edward C. Tolman had demonstrated as early as 1924 that maze learning in rats was under genetic control, and his student Robert C. Tryon added greatly to the elegant proof of that proposition between 1934 and 1940, all at the University of California, Berkeley. William T. Heron had conducted similar independent studies at the University of Minnesota in 1935. Working together with B. F. Skinner while both were at Minnesota, they published an obscure paper in 1940 titled, "The Rate of Extinction in Maze-Bright and Maze-Dull Rats." If Skinner had only pursued those interests, the history and attitudes of the discipline of psychology toward the conceptual development of behavior genetics would have been entirely different. Leona E. Tyler's *The Psychology of Human Differences* (1947) and Anastasi and Foley's *Differential Psychology* (1949), influential textbooks on differential psychology, fostered and chronicled the manifold research projects that we would call the province of behavioral genetics today, setting the stage for those milestones mentioned above, and culminating in the remarkable chapters to follow.

After this brief retrospective look at our intellectual pedigree, and afforded the look into our future through the present efforts of Plomin, DeFries, Craig, and McGuffin, I would offer the advice attributed to John Maynard Keynes (1883–1946), the British economist, "When the facts change, I change my mind."

References

Anastasi, A., & Foley, J. P., Jr. (1949) *Differential psychology: Individual and group differences in behavior.*

Binet, A., & Henri, V. (1895). La Mémoire des phrases. *L'Année Psychologique, 1*, 24–59.

Binet, A., & Simon, T. (1905). New methods for the diagnosis of intellectual level of abnormal people. *Psychologic Year, 11*, 191–244.

Cavalli-Sforza, L. L., & Bodmer, W. F. (1971). *The genetics of human populations.* San Francisco: W. H. Freeman.

Dobzhansky, T. (1962). *Mankind evolving.* New Haven, CT: Yale University Press.

Fuller, J. L., & Thompson, W. R. (1960) *Behavior genetics.* New York: Wiley.

Glass, D. C. (1968). *Biology and behavior: Genetics.* New York: Rockefeller Press.

Hirsch, J. (1967). *Behavior-genetic analysis.* New York: McGraw-Hill.

Lerner, I. M. (1968). *Heredity, evolution, and society.* San Francisco: W. H. Freeman.

Neel, J. V., & Schull, W. J. (1954). *Human heredity.* Chicago: University of Chicago Press.

Penrose, L. S. (1949). *The biology of mental defect.* New York: Grune & Stratton Inc.

Quetelet, L. A. J. (1946/1849). (Trans. O. G. Downes). *Letters on probabilities.* London: Layton.

Roberts, J. A. F. (1952). The genetics of mental deficiency. *Eugenics Review, 44*, 71–83.

Rosenthal, D. (1970). *Genetic theory and abnormal behavior.* New York: McGraw-Hill.

Scott, J. P., & Fuller, J. L. (1965). *Genetics and the social behavior of the dog.* Chicago: University of Chicago Press.

Shields, J. (1962). *Monozygotic twins: Brought up apart and brought up together.* London: Oxford University Press.

Skinner, B. F., & Heron, W. T. (1940). The rate of extinction in maze-bright and maze-dull rats. *Psychological Record, 4,* 11–18.

Slater, E., & Cowie, V. (1971). *The genetics of mental disorder*. London: Oxford University Press.

Stern, W. (1900). *On the psychology of individual differences.* Liepzig: Barth.

Stern, C. (1949). *Principles of human genetics.*.

Tyler, L. E. (1947). *The psychology of human differences.* New York: Appleton-Century-Crofts.

Vandenberg, S. G. (1965). *Methods and goals in human behavior genetics.* New York: Academic Press.

Vandenberg, S. G. (1968), *Progress in human behavior genetics*. Baltimore: Johns Hopkins University Press.

Vogel, F., & Motulsky, A. G. (1979). *Human genetics.* New York: Springer-Verlag.

Wright, S. (1934). An analysis of variability in the number of digits in an inbred strain of guinea pigs. *Genetics, 19,* 506–536.

A Behavioral Science Perspective

Jerome Kagan

The sharp feeling of surprise provoked in readers perusing a paper in 1950 on the inheritance of human traits has become almost extinct as biologists continue to publish evidence supporting a rational scaffolding for such claims. The number of social scientists who resist the suggestion that a person's genes make some contribution to variation in intellectual, behavioral, and emotional traits dwindles each day. Even though our understanding of the nature of the contribution of genetic variation to human psychological features is in a very early phase, the enterprise is worth pursuing because the deep premise that motivates the work is true. A behavior as complex as whether or not a strain of vole will pair bond depends, as neurobiologist Tom Insel has shown, on a small bit of promoter material contiguous to the gene for the vasopressin receptor. Some 4-month-old infants who react to unfamiliar visual and auditory stimuli with vigorous thrashing of limbs and crying because of a lower threshold in the limbic system retain a coherent set of physiological features through 10 years of age.

However, important theoretical problems must be resolved as this domain of inquiry makes its inevitable progress. One critical issue is whether we should conceive of the basis for a psychological trait as a continuum, with its origin in the additive influences of many genes or as a category defined by a very particular combination of genes and their biological products. The existing evidence permits a defense of either stance.

A second puzzle stems from the fact that a person's temperament and gender can operate as background conditions to affect the behavioral phenotype emerging from a particular set of alleles. Even though high levels of infant motor activity and crying to stimulation appear to reflect an inherent excitability of the amygdala, the bed nucleus, and their projections, high reactive boys and girls do not develop identical behavior or biological phenotypes in late childhood. High reactive boys tend to show extreme left frontal activation in the resting EEG; most high reactive girls develop right frontal activation. Moreover, if a bias to be right frontal active is under some genetic control, the evidence suggests that different psychological phenotypes are correlated with that particular feature.

It is likely that profiles of measurements, rather than single variables, will become the referent for the next set of constructs. To illustrate, a particular IQ score combined with high energy level and a vulnerability to uncertainty defines a biological category different from one defined by an equivalent IQ score combined with low energy level and low vulnerability to uncertainty.

The most critical requirement is to assess directly the interactions between experience and genes. It is not sufficient to infer the magnitude of environmental variance from the heritability equations alone. The next set of investigators will have to evaluate environmental contributions directly by gathering historical data in longitudinal designs.

It will also be useful to evaluate physiological profiles that are partial functions of inheritance to understand the mechanisms that underlie a phenotype. At present, the relations between the behavioral variables and brain processes are weak. The emergent nature of psychological events is analogous to the temperature and pressure of a closed container of gas. Pressure and temperature describe the consequences of large numbers of molecular collisions and are inappropriate terms for the components. Similarly, anxiety is a property of a person and not of the neurons and circuits that contribute to this emotion. All of nature cannot be described with one vocabulary because brains have qualitatively different structures than cognitive configurations like schemata and semantic networks.

A class of brain states, not any particular one, can be associated with a particular psychological phenomenon. If an investigator could remove one spine from a dendrite that is a regular component of a brain state, it is unlikely that that change would alter the person's psychological state in any measurable way. An individual's pattern of possible psychological representations for an event is analogous to the physical concept of a phase space. A collection of gas molecules in a vessel can assume a very large number of states, only one of which can be measured at a given time to a given incentive. Thus, the pattern of neural activation in a person lying in an FMRI (functional magnetic resonance imaging) scanner looking at pictures of poisonous snakes is not to be regarded as the true or only neural configuration that these stimuli could provoke. These pictures would probably create a different brain state if the person were looking at them on a television screen in his or her bedroom.

Perhaps the most difficult problem in predicting the future profile from knowledge of an infant's genome is that the future is unknowable, a state of affairs analogous to the inability to predict which new species might arise because no scientist can know future ecological changes. Had biologists been present 200 million years ago, they could not have known that Australia and Tasmania would break away from the Antarctic land mass and that the separate land masses now called North and South America would be connected by the isthmus of Panama. These events changed world climate in dramatic ways, including the establishment of several ice ages that led to altered profiles of flora and fauna. Analogously, two girls born with the genes that represent a diathesis for panic attacks will have different symptom profiles at age 20 if one is born to an insecure, chaotic family while the other is born to a nurturing, economically secure pair of parents.

The social class of rearing in North America and Europe, where most genetic research is conducted, is one of the most important determinants of the adult psychological phenotype—this condition may represent the most important determinant. Unfortunately, behavioral geneticists rarely parse their samples by class of rearing. If they did so, they would discover different heritability quotients for a particular trait for poor, uneducated adults compared with affluent, educated ones.

It may be helpful to think of some genetic factors as constraining certain outcomes rather than determining a particular trait. Although relations between variables are always continuous, the tightness of any link can vary over a broad range of probabilities from the near-perfect relation between the force with which a stone is thrown and the distance it travels to the far-from-perfect relation between an infant's Apgar score and grade point average in the third grade. When the probability of one event following a prior one is relatively high, for example the probability is greater than 0.6 or 0.7, it is fair to conclude that the prior event influenced the latter one directly. However, when the probability is low, for example less than 0.4, it is likely that the prior event affected the consequent indirectly or in combination with other factors. In this case, it is more accurate to use the verb *constrain* rather than *determine*. For example, consistent nurturance of young children during the first 5 years does not predict, at a high level of probability, quality of marriage, amount of education, or level of professional achievement. But consistent nurturance probably constrains seriously the likelihood that the adults who enjoyed that experience will become criminals or homeless people. Less than 20% of the school-age children who were high reactive infants became consistently fearful throughout childhood, but not one high reactive infant developed a complementary profile of extreme sociability and fearlessness. Because more than 80% of high reactive infants did not become consistently anxious, timid children, it is misleading to suggest that a high reactive temperament determines a consistently inhibited profile. It is more accurate to conclude that a high reactive temperament constrains the probability of becoming a consistently sociable, uninhibited child. Replacing the word *determine* with *constrain* is not idle word play, for the connotations surrounding the two words are different. Determine implies a particular consequence from a prior cause, whereas constrain implies a restriction on a set of outcomes. It is probably useful to regard many genetic factors as imposing constraints on the probability of developing a particular family of psychological profiles rather than assuming that a set of genes determines the development of a particular trait.

Finally, we must invent psychological constructs that specify incentive and context. For example, we must replace a construct like intelligence with a set that specifies the cognitive process in more detail; for instance, speed of perceptual inference with visual information or the capacity to acquire a semantic network. Constructs such as anxiety or fear should also be regarded as a family of constructs with particular and separate estimates of vulnerability to acquiring phobias, excessive worry about the future, panic attacks, or concern over evaluation by others.

It appears that the amygdala is necessary for acquiring a conditioned reaction of freezing or heart rate acceleration to a conditioned stimulus but is not necessary for the unconditioned response of freezing or heart rate acceleration to the unconditioned stimulus. A critical role of the amygdala in the conditioning paradigm is to respond to unexpected events, whether or not they create a fear state. The neurons that mediate the motor circuit for freezing are fed by sites other than the basal amygdala, for example, the bed nucleus, and perhaps other structures. This fact implies that the role of the amygdala in the conditioning paradigm is to respond to unexpected events, and not necessarily because it creates a fear state. The major principle is that every response, in this case freezing, is ambiguous as to its origin, and, therefore, with respect to its meaning. The more complex a behavior, the more ambiguous the meaning of its display. Most behavioral variables are not homogeneous in their etiology, and we must combine the behaviors with both physiology and the context of display to arrive at a more accurate construct for behavioral geneticists to study.

The journey, therefore, has just begun. But there has to be a beginning to every mission, and the chapters in this volume present interested readers who do not follow this literature with a wonderful opportunity to assimilate the current state of understanding. Much progress will occur because bright minds are attracted to these problems. The current cohort of scientists are scouting the territory, searching for the best places to establish a permanent settlement. Tycho Brahe had to precede the contemporary study of black holes; Gregor Mendel had to sort peas before scientists could discover the HOX genes. Although the facts the future promises will be very different from those we read about today, like the ship that had each of its parts replaced during a journey, the goals of the mission remain unchanged.

A Molecular Genetics Perspective

James D. Watson

My first interest in science involved animal behavior. For Christmas 1937, when I was 7, my aunt and uncle gave me a children's book on bird migration. They knew my father was a keen birder and wanted me to follow in his steps. As I grew into adolescence, I was increasingly dominated by bird spotting, first on Sundays with my father, and later by myself when I went to the marshes on Chicago's far south side to look for migrating shore birds soon to fly off to Arctic tundra or lower South America. When and where to migrate were encoded as instincts within the bird's brain. By the time I finished my undergraduate years as a zoology major at the University of Chicago, I saw bird migration as too complicated to be understood by the science of those years. So my attention turned to the molecular essence of the gene, an objective just becoming reachable through advances in chemistry and physics.

At the start of the 20th century, ascribing genetic bases to human behavior was at best imprudent. My predecessor as director of the Cold Spring Harbor Laboratory, Charles Davenport, proposed genetic causes for all sorts of family traits. In his 1911 book, *Heredity in Relation to Eugenics*, he said members of the Banning family had "determination and will power almost to the point of stubbornness" (p. 243). In contrast, Brinckerhoff family members were not only "magnetic and generous, but usually could whistle a tune or sing a song without apparent efforts" (p. 243). Behind these even then nutty thoughts were not only racial prejudices but also social Darwinism, which ascribed human successes to good genes and failures to bad genes. Davenport and his fellow eugenicists wanted to improve human society through curtailing the passing of bad genes to subsequent generations. Out of their thinking came sterilization programs for the mentally ill and the Immigration Act of 1924 that effectively blocked further Italian and Jewish immigration to the United States. Less than a decade later, the Nazis came to power soon to sterilize some 500,000 people they characterized as "lives not worthy of being lived."

In the immediate post–World War II days, even thinking about human behavior as having a genetic basis was a no-no. The realization of how the Nazis had used bogus science to justify the genocide of Jews and Gypsies gave a stench to much of human genetics that would long persist. Past ideas that much of human nature had an instinctive genetic basis went out the window. The human brain at birth became regarded by most psychologists and many biologists as a blank slate. Given the right environmental inputs, society no longer needed to be bedeviled with members unable to read, to be mentally disturbed, or to have criminal personality traits. Through the right combinations of money and good intention, all human children would have equal potentials for meaningful adult lives.

With our future fates not falsely attributed to genes, there should be no limit to human progress.

If we were indeed born with blank slates, then human nature was totally different from that of the animals about us. Conceivably, human evolution has led to the loss of most of the instinctive behavior that permits a dog or cat to survive on its own. Ed Wilson, however, challenged this belief with his seminal 1975 book *Sociobiology*. Those in the blank slate camp, and particularly its left-wing contingent, so disliked its implications that they once saw the need to pour water over Ed's head when he lectured to an audience of the American Association for the Advancement of Science in Washington, DC. Though human twin studies by then had provided incontrovertible evidence for genetic involvement in personality and intelligence differences, those on the radical left continued to shout "not in your genes." Sadly, many of their students enthusiastically accepted their antigenetics diatribes, wanting futures determined by free will as opposed to genes.

The older one gets, however, the more most of us conclude that children come into the world with fixed personalities that are hard to ascribe to specific home or school environmental influences. Particularly happy children almost seem to be born that way. Now with the arrival of the human DNA genome sequence and its attendant list of human genes, the experimental procedures will soon be on hand to finally settle the long contentious nature–nurture arguments. Useful soon will be looks at polymorphisms within brain expressing genes. Sequence differences within the genes coding for endorphin receptors might, say, correlate with contented or discontented personalities. Polymorphism within the oxytocin gene likewise might explain differences in degree of human maternal affection. To be sure, over the near future, most such initial claims will prove wrong—the apparent correlations due to statistics working against us. Firmer links between precise genes and much of human behavior may take many decades, if not most of the coming century, to clearly establish.

The finding of important human behavioral genes will, it is hoped, lead over the long term to effective remedies for socially inappropriate behaviors. But a world in the near future without any mental illness or ugly aggression is too much to ask for. However, even a few lives, so saved by new medicines from the horrors of the deranged mind, more than give us reason to find the genes we know are there.

References

Davenport, C. B. (1911). *Heredity in relation to eugenics.* New York: Henry Holt.

Wilson, E. O. (1975). *Sociobiology: The new synthesis.* Cambridge MA: Harvard University Press.

Acknowledgments

The editors are grateful to The Wellcome Trust for funding a conference on the theme of behavioral genetics in the postgenomic era, which led to this book. The meeting to which 45 of the top people in the relevant fields were invited was held March 27–29, 2001, just one month after the publication of the working draft of the human genome sequence, the single most important event on the road to the postgenomic era. The meeting was magical because it was held at the Wellcome Trust's Hinxton Hall Conference Center on the campus of the Sanger Centre near Cambridge, England, which contributed so much to the Human Genome Project. The exciting possibilities for behavioral genetics in the postgenomic era discussed at the meeting compelled the organizers of the conference, who are the editors of this book, to make chapters informed by the conference proceedings available to a wider audience.

We are grateful to Irv Gottesman, Jerry Kagan, and Jim Watson for writing forewords to the book. Although it might seem a bit over the top to have three forewords, the authors are leaders in the three target areas of the book—behavioral genetics, behavioral sciences, and molecular genetics, respectively. We are indebted to the excellent staff at APA Books: Lansing Hays, Senior Acquisitions Editor, whose idea it was to create a book based on the Hinxton Hall conference; Judy Nemes, Development Editor, for carefully processing and reviewing the manuscript; and Casey Ann Reever, Production Editor, for her excellent management of the book's production in the face of a tight schedule. Finally, a special thanks to Tara McKearney at the Institute of Psychiatry in London, who spent many months working with the editors and authors to obtain chapters, reviews, and revisions and to deliver the manuscript on schedule.

Part I

Introduction

1

Behavioral Genetics

Robert Plomin, John C. DeFries, Ian W. Craig, and Peter McGuffin

The 20th century began with the rediscovery of Mendel's laws of heredity. The word *genetics* was first coined in 1903. Fifty years later the double-helix structure of deoxyribonucleic acid (DNA) was discovered. The genetic code was cracked in 1966: The four-letter alphabet (G, A, T, C, for guanine, adenine, thymine, and cytosine, respectively) of DNA is read as three-letter words that code for the 20 amino acids that are the building blocks of proteins. The crowning glory of the century and the beginning of the new millennium is the Human Genome Project, which has provided a working draft of the sequence of the 3 billion letters of DNA in the human genome.

Some curmudgeons grumble that the four nucleotide letters that constitute the DNA alphabet are not GATC, but HYPE. They complain that the DNA sequence by itself does not tell us how we can begin life as a single fertilized cell and end up with trillions of cells containing the same DNA but expressing different genes. They argue that knowing the DNA sequence does not by itself cure any diseases. Although it is true that sequencing the human genome is just a first step toward understanding how genes work, it is a giant step forward that ought to be celebrated. The Human Genome Project and the new techniques that are emerging from it will greatly speed up the pace of discovery in genetics, even though the current pace is amazing.

This book appears at a time when we are at the threshold of a post-genomic era in which all genes and all DNA variation will be known. The major beneficiary of this new era will be research on complex traits that are influenced by multiple genes as well as multiple environmental influences. This is where behavior comes in. Behavioral dimensions and disorders are the most complex traits of all and are already becoming a focus for DNA analysis. The genetic analysis of behavior includes both quantitative genetic and molecular genetic strategies, the two worlds of genetics that are rapidly coming together in the study of complex traits such as behavioral dimensions and disorders. The goal of this book is to assess where we are and where we are going in genetic research on behavior.

Nearly a decade ago, an edited volume summarized human behavioral genetic research (Plomin & McClearn, 1993). At that time, quantitative genetic research such as twin and adoption studies of human behavior and strain and selection studies in other species had made the case that nearly all behavioral dimensions and disorders show at least moderate genetic influence. What was new was that quantitative genetic research was going beyond estimating heritability to ask questions about developmental change and continuity in genetic influence, about multivariate issues such as comorbidity and heterogeneity, and about the interface between genes and environment. A few chapters in the volume mentioned the potential of molecular genetics to identify some of the many genes that are responsible for the ubiquitous heritability of behavioral dimensions and disorders. However, few confirmed linkages and associations for complex behavioral traits were known at that time. With the emergence of a draft sequence of the human genome and progress toward identifying polymorphisms in the DNA sequence, the time seemed right to review progress and promise in quantitative and molecular genetic analysis of behavior. Integration of these approaches is especially valuable because quantitative genetic findings can inform the search for genes, and, conversely, identifying genes holds great promise for advancing the understanding of post-heritability issues.

Although some of these issues are relevant to all complex traits, the book focuses on behavioral dimensions and disorders, specifically those that have received the most attention in genetic research: cognitive abilities and disabilities, personality, psychopathology, and psychopharmacology. We also emphasize substantive behavioral issues rather than quantitative or molecular genetic methodologies. As genes associated with behavior are identified, the next step is to understand the mechanisms by which genes have their effect (functional genomics). However, this book focuses on quantitative and molecular genetic issues rather than tackling the much larger topic of functional genomics. Its intended audience is not just psychologists and psychiatrists but anyone in the behavioral, biomedical, and biological sciences interested in the genetics of behavior.

The Very Standard Deviation

It is important to begin with a discussion of the different perspectives or levels of analysis used to investigate behavior because so much follows conceptually as well as methodologically from these differences in perspective (see Figure 1.1). Research in behavioral genetics has focused on within-species interindividual differences, for example, why some children have reading disabilities and others do not. In contrast, many areas of psychology and neuroscience seldom mention individual differences and concentrate instead on species-universal or species-typical (normative) phenomena. For example, leading textbooks in cognitive neuroscience (i.e., Gazzaniga, 2000; Thompson, 2000) do not mention individual differences. Neuroscience has focused on understanding how the brain works on average, for example, which bits of the brain light up under neuroimaging for particular tasks. Until now, genetics has entered neuroscience primar-

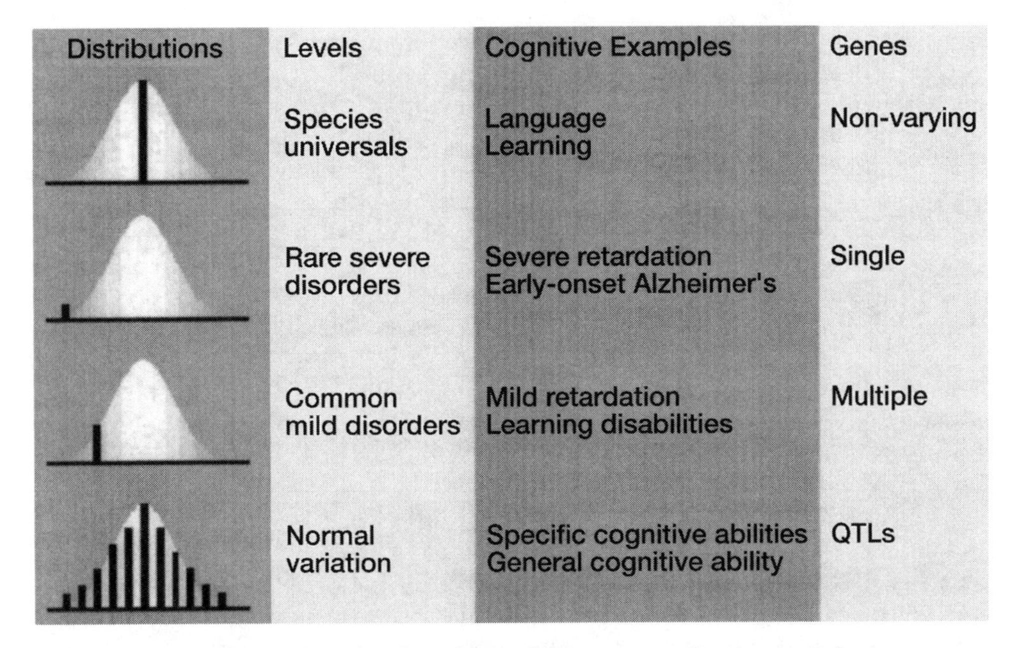

Figure 1.1. Levels of analysis. QTLs = quantitative trait loci.

ily in relation to gene targeting in mice in which mutations are induced that interfere with normal brain processes. In humans, rare single-gene mutations have been the center of attention. This approach tends to treat all members of the species as if they were genetically the same except for a few rogue mutations that disrupt normal processes. In this sense, the species-typical perspective of neuroscience assumes that mental illness is a broken brain. In contrast, the individual differences perspective considers variation as normal—the very standard deviation. Common mental illness is thought to be the quantitative extreme of the normal distribution.

Although perspectives are not right or wrong—just more or less useful for particular purposes—the species-typical perspective and the individual differences perspective can arrive at different answers because they ask different questions. The distinction between the two perspectives is in essence the difference between means and variances. There is no necessary connection between means and variances, either descriptively or etiologically. Despite its name, *analysis of variance (ANOVA)*, the most widely used statistical analysis in the life sciences, is actually an analysis of mean effects in which individual differences are literally called the *error term*. Instead of treating differences between individuals as error and averaging individuals across groups as in ANOVA, individual differences research focuses on these interindividual differences. Variation is distributed continuously, often in the shape of the familiar bell curve, and is indexed by variance (the sum of squared deviations from the mean) or the square of variance, which is called the *standard deviation*.

The two perspectives also differ methodologically. Most species-typical research is experimental in the sense that subjects are randomly assigned to conditions that consist of manipulating something such as genes, lesions, drugs, or tasks. The dependent variable is the average effect of the manipulation on outcome measures such as single-cell recordings of synaptic plasticity, activation of brain regions assessed by neuroimaging, or performance on cognitive tests. Such experiments ask whether manipulations *can* have an effect on average in a species. For example, a gene knock-out study investigates whether an experimental group of mice which inherit a gene that has been made dysfunctional differs, for example, in learning or memory from a control group with a normal copy of the gene. A less obvious example can be seen in recent experimental research that manipulated tasks and found that average blood flow assessed by positron emission tomography (PET) in the human species is greater in the prefrontal cortex for high-intelligence tasks than for low-intelligence tasks (Duncan et al., 2000).

In contrast, rather than creating differences between experimental and control groups through manipulations, the individual differences perspective focuses on naturally occurring differences between individuals. One of the factors that makes individuals different is genetics. The individual differences perspective is the foundation for quantitative genetics, which focuses on naturally occurring genetic variation, the essence of heredity. Although 99.9% of the human DNA sequence is identical for all human beings, the 0.1% that differs—3 million base pairs—is ultimately responsible for the ubiquitous genetic influence found for complex traits including behavioral dimensions and disorders (Plomin, DeFries, McClearn, & McGuffin, 2001). Individual differences research is correlational in the sense that it investigates factors that *do* have an effect in the world outside the laboratory. Continuing with the previous examples, an individual differences approach would ask whether naturally occurring genetic variation in mice is associated with individual differences in mouse learning and memory. The brain is like the engine of an automobile with thousands of parts that need to work for the engine to run properly. In trying to diagnose the defect in a faulty engine, we may test the effect of removing any of these parts. The engine may not run properly as a consequence, but that part is not likely to be the reason why your automobile is not running properly. In the same way, genes can be knocked out and shown to have major effects on learning and memory, but this does not imply that the gene has anything to do with the naturally occurring genetic variation that is responsible for hereditary transmission of individual differences in performance on learning and memory tasks. The PET experiment that compared average performance on high- and low-intelligence tasks could use an individual differences perspective by comparing cortical blood flow in high- and low-intelligence individuals rather than comparing average performance on tasks.

Other perspectives lie in between these two extremes of species universals and normal variation. The effects of rare severe disorders caused by single gene defects can be dramatic. For example, mutations in the gene that codes for the enzyme phenylalanine hydroxylase if untreated cause phenylketonuria (PKU) that is associated with a severe form of mental retardation. This inherited condition occurs in 1 in 10,000 births. At least 100 other rare single-gene disorders include mental retardation as part of the syndrome (Wahlström, 1990). Such rare single-gene disorders are viewed from the species-universal perspective as aberrations from the species type, exceptions to the species rule. However, common disorders, such as mild mental retardation and learning disabilities, seldom show any sign of single-gene effects and appear to be caused by multiple genes as well as by multiple environmental factors. Indeed, quantitative genetic research suggests that such common disorders are usually the quantitative extreme of the same genes responsible for variation throughout the distribution. Genes in such multiple-gene (polygenic) systems are called quantitative trait loci (QTLs) because they are likely to result in dimensions (quantitative continua) rather than disorders (qualitative dichotomies). In other words, in terms of the genetic etiology of common disorders, there may be no disorders, just dimensions. Other than simple and rare single-gene or chromosomal disorders, mental illness may represent the extreme of normal variation.

In summary, the individual differences perspective views variation as normal and distributed continuously; common disorders are viewed as the extremes of these continuous distributions. As indicated at the outset, perspectives are not right or wrong. But they are different, and the proper interpretation of genetic research depends on understanding these differences. The perspectives are complementary in the sense that a full understanding of behavior requires integration across all levels of analysis.

The Two Worlds of Genetics

The species-universal perspective and the individual-differences perspective contributed to the drifting apart of the two worlds of quantitative genetics and molecular genetics since their origins a century ago. Quantitative genetics focused on naturally occurring individual differences in complex phenotypes that are commercially important in agriculture and animal husbandry or socially important in the human species. Genetic tools were developed to assess the extent to which such traits were inherited. In nonhuman animals, these tools included experimental crosses, inbred strains, and artificial selection. In the human species, twin and adoption methods were developed in the 1920s. The essence of the theoretical foundation for quantitative genetics is the extension of Mendel's laws of single-gene inheritance to complex multifactorial traits influenced by multiple genes as well as multiple environmental factors (Fisher, 1918). If multiple genes affect a trait, the trait is likely to be distributed quantitatively as each gene adds its effects to the mix. The goal of quantitative genetics is to decompose observed variation in a trait and covariation between traits into genetic and environmental sources of variation. The conclusion gradually emerged from quantitative genetic research that genetic factors contribute to nearly all complex traits including behavior. However, the polygenic nature of complex traits made it seem unlikely that we would ever identify more than a few of the specific genes responsible for the inheritance of complex traits.

In contrast to the focus of quantitative genetics on naturally occurring sources of variation in complex traits, the origins of molecular genetics can be traced back to species-universal research early in the century that created new genetic variation by mutating genes chemically or through X-irradiation. This experimental orientation continues today with large-scale mutation screening research programs and targeted mutations such as deleting key DNA sequences that prevent a particular gene from being transcribed ("knock outs") or alter its regulation to underexpress or over-express the gene (Capecchi, 1994). The focus here is on the gene rather than the phenotype and creating new genetic variation rather than studying the phenotypic effects of naturally occurring genetic variation. As a result, these approaches concentrate on single genes rather than multiple genes and on simple phenotypes to assess the average effects of mutations on the species rather than individual differences in complex phenotypes that are of interest in their own right.

These conceptual and methodological differences caused the two worlds of genetics to drift apart until the 1980s when a new generation of DNA markers made it possible to identify genes for complex quantitative traits influenced by multiple genes (QTLs) as well as by multiple environmental factors. Until the 1980s, relatively few genetic markers were available that were the products of single genes, such as the red blood cell proteins that define the blood groups. The new genetic markers were polymorphisms in DNA itself rather than in a gene product. Because millions of DNA base sequences are polymorphic, millions of such DNA markers became potentially available to hunt for QTLs responsible for the inheritance of complex traits.

Unlike single-gene effects that are necessary and sufficient for the development of a disorder, QTLs contribute incrementally, analogous to probabilistic risk factors. The resulting trait is distributed quantitatively as a dimension rather than qualitatively as a disorder; this was the essence of Fisher's classic 1918 paper on quantitative genetics. The QTL perspective is the molecular genetic version of the quantitative genetic perspective that assumes that genetic variance on complex traits is due to many genes of varying effect size. As mentioned earlier, from a QTL perspective, common disorders may be just the extremes of quantitative traits caused by the same genetic and environmental factors responsible for variation throughout the dimension. In other words, the QTL perspective predicts that genes found to be associated with complex disorders will also be associated with normal variation on the same dimension and vice versa (Deater-Deckard, Reiss, Hetherington, & Plomin, 1997; Plomin, Owen, & McGuffin, 1994).

In contrast to the species-universal perspective in which we are all thought to be the same genetically except for a few rogue mutations that lead to disorders, the QTL perspective suggests that genetic variation is normal. Many genes affect most complex traits and, together with environmental variation, these QTLs are responsible for normal variation as well as for the abnormal extremes of these quantitative traits. This QTL perspective has implications for thinking about mental illness because it blurs the etiological boundaries between the normal and the abnormal. That is, we all have many alleles that contribute to mental illness but some of us are unluckier in the hand that we drew at conception from our parents' genomes. A more subtle conceptual advantage of a QTL perspective is that it frees us to think about both ends of the normal distribution —the positive end as well as the problem end, abilities as well as disabilities, and resilience as well as vulnerability. It has been proposed that we move away from an exclusive focus on pathology toward considering positive traits that improve the quality of life and perhaps prevent pathology (Seligman & Csikszentmihalyi, 2000).

The initial goal of QTL research is not to find *the* gene for a complex trait, but rather those genes that make contributions of varying effect sizes to the variance of the trait. Perhaps 1 gene will be found that accounts for 5% of the variance, 5 other genes might each account for 2% of the variance, and 10 other genes might each account for 1% of the variance. If the effects of these QTLs are independent, together these QTLs would account for 25% of the trait's variance. All of the genes that contribute to the heritability of a complex trait are unlikely to be identified because some of their effects may be too small to detect. The problem is that we do not know the distribution of effect sizes of QTLs for any complex trait in plant, animal, or human species. Not long ago, a 10% effect size was thought to be small, at least from the single-gene perspective in which the effect size was essentially 100%. However, for behavioral disorders and dimensions, a 10% effect size may turn out to be a very large effect.

If effect sizes are as small as 1%, this would explain the slow progress to date in identifying genes associated with behavior, because research so far is woefully underpowered to detect and replicate QTLs of such small effect size (Cardon & Bell, 2001). Although the tremendous advances during the past few years from the Human Genome Project warrant optimism for the future, progress in identifying genes associated with behavioral traits has been slower than expected. Using schizophrenia and bipolar affective disorder as examples, no clear-cut associations with schizophrenia and bipolar affective disorder have been identified despite a huge effort during the past decade, although there are several promising leads (Baron, 2001). Part of the reason for this slow progress may be that because these were the first areas in which molecular genetic approaches were applied, they happened at a time in the 1980s when large pedigree linkage designs were in vogue. It is now generally accepted that such designs are only able to detect genes of major effect size.

Recent research has been more successful for finding QTLs for complex traits because different designs have been used that can detect genes of much smaller effect size. QTL linkage designs use many small families (usually siblings) rather than a few large families, and they are able to detect genes that account for about 10% of the variance. Association studies such as case control comparisons make it possible to detect genes that account for much smaller amounts of variance (Risch, 2000; Risch & Merikangas, 1996). A daunting target for molecular genetic research on complex traits such as behavior is to design research powerful enough to detect QTLs that account for 1% of the variance while providing protection against false-positive results in genome scans using thousands of markers. To break the 1% QTL barrier (which no study has yet done), samples of many thousands of individuals are needed for research on disorders (comparing cases and controls) and on dimensions (assessing individual differences in a representative sample).

It should be noted that DNA variation has a unique causal status in explaining behavior. When behavior is correlated with anything else, the old adage applies that correlation does not imply causation. For example, when parenting is shown to be correlated with children's behavioral outcomes, this does not imply that the parenting caused the outcome environmentally. Behavioral genetic research has shown that parenting behavior to some extent reflects genetic influences on children's behavior. When it comes to interpreting correlations between biology and behavior, such correlations are often mistakenly interpreted as if biology causes behavior. For example, correlations between neurotransmitter physiology and behavior or between neuroimaging indices of brain activation and behavior are often interpreted as if brain differences cause behavioral differences. However, these correlations do not necessarily imply causation because behavioral differences can cause brain differences. In contrast, in the case of correlations between DNA variants and behavior, the behavior of individuals does not change their genome. Expression of genes can be altered, but the DNA sequence itself does not change. For this reason, correlations between DNA differences and behavioral differences can be interpreted causally: DNA differences can cause the behavioral differences but not the other way around.

Quantitative Genetics: Beyond Heritability

Most of what is known today about the genetics of behavior comes from quantitative genetic research. During the past three decades, the behavioral sciences have emerged from an era of strict environmental explanations for differences in behavior to a more balanced view that recognizes the importance of nature (genetics) as well as nurture (environment). This shift occurred first for behavioral disorders, including rare disorders such as autism (.001 incidence), more common disorders such as schizophrenia (.01), and very common disorders such as reading disability (.05). More recently, it has become increasingly accepted that genetic variation contributes importantly to differences among individuals in the normal range of variability as well as for abnormal behavior (Plomin, DeFries, et al., 2001). The most well-studied domains of behavior are cognitive abilities and disabilities, psychopharmacology, personality, and psychopathology. These domains are the topics of the major sections of this book.

Although twin and adoption studies of individual differences in previously neglected behavioral domains such as learning might yield important new insights into their etiology, asking only whether and how much genetic factors influence such domains would not fully capitalize on the strengths of quantitative genetic analysis. New quantitative genetic techniques make it possible to go beyond these rudimentary questions to investigate how genes and environment affect developmental change and continuity; *comorbidity* and *heterogeneity*, the links between disorders and normal variation; and interactions and correlations between genetic and environmental influences. Many of the chapters in this book emphasize these issues. Using these techniques, quantitative genetic research can lead to better diagnoses based in part on etiology rather than solely on symptomatology. They also can chart the course for molecular genetic studies by identifying the most heritable components and constellations of disorders as they develop and as genetic vulnerabilities correlate and interact with the environment. In this sense, quantitative genetics will be even more important in a postgenomic era. The future of behavioral genetic research lies in identifying specific genes responsible for heritability, which will make it possible to address these same questions with much greater precision.

Although most chapters in this book concentrate on genetics, it should be mentioned at the outset that quantitative genetic research is at least as informative about nurture as it is about nature. In the first instance, it provides the best available evidence for the importance of the environment in that the heritability of complex traits is seldom greater than 50%. In other words, about half of the variance cannot be explained by genetic factors. In addition, two of the most important findings about environmental influences on behavior have come from genetic research. The first finding is that, contrary to socialization theories from Freud onward, environmental influences operate to make children growing up in the same family as different as children growing up in different families, which is called *nonshared environment* (reviewed by Plomin, Asbury, & Dunn, 2001). The second finding, called the *nature of nurture,* is that genetic factors influence the way we experience our environments (reviewed by Plomin, 1994). For this reason, most measures of the environment used in behavioral research show genetic influence. For the same reason, associations between environmental measures and behavioral outcome measures are often substantially mediated genetically. The way forward in research is to bring together genetic and environmental strategies, for example, using environmental measures in genetically sensitive designs to investigate interactions and correlations between nature and nurture.

The present book's focus on genetics is not intended to denigrate the importance of environmental influences or to imply biological determinism. In many ways, it is more difficult to study the environment than genes. There is no Human Environome Project—indeed, environmental research is pre-Mendelian in the sense that the laws of environmental transmission and even the units of transmission are unknown.

Molecular Genomics

Although the publication of the working draft of the human genome sequence in February 2001 (Venter et al., 2001) was a stunning achievement, much work still needs to be done to complete the sequence and to identify all of the genes. For behavioral genetics, the most important next step is the identification of the DNA sequences that make us different from each other. There is no single human genome sequence—we each have a unique genome. Most of the DNA letters are the same for all human genomes, and many of these are the same for other primates and other mammals and even insects. Nevertheless, about one in every thousand DNA letters differs among us, around 3 million DNA variations in total, enough to make us each differ for almost every gene. Most of these DNA differences involve a substitution of a single base pair, called *single nucleotide polymorphisms* (SNPs, pronounced "snips"). These DNA differences are responsible for the widespread heritability of behavioral disorders and dimensions. That is, when we say that a trait is heritable, we mean that variations in DNA exist that cause differences in behavior. Particularly useful are SNPs in coding regions (cSNPs) and other SNPs that are potentially functional, such as SNPs in DNA regions that regulate the transcription of genes.

When the working draft of the human genome sequence was published, much publicity was given to the finding that there appear to be fewer than half as many genes in the human genome as expected—about 30,000 to 40,000 genes, similar to the estimates for mice and worms. A bizarre spin in the media was that having fewer genes somehow implies that nurture must be more important than we thought because there are not enough genes to go around. There is, however, an important implication that follows from the finding that the human species does not have more genes than many other species: The number of genes is not responsible for the greater complexity of the human species. In part, the greater complexity of the human species occurs because during the process of decoding genes into proteins, human genes more than the genes of other species are spliced in alternative ways to create a greater variety of proteins. Such subtle variations in genes rather than the number of genes may be responsible for differences between mice and men. If subtle DNA differences are responsible for the differences between mice and men, even more subtle differences are likely to be responsible for individual differences within the species.

Another relevant finding from the Human Genome Project is that less than 2% of the 3 billion letters in our DNA code involves genes in the traditional sense, that is, genes that code for amino acid sequences. This 2% figure is similar in other mammals that have been studied. On an evolutionary time scale, mutations are quickly weeded out from these bits of DNA by natural selection because their code is so crucial for development. When mutations are not weeded out, they can cause one of the thousands of severe but rare single-gene disorders. However, it seems increasingly unlikely that the other 98% of DNA is just along for the ride.

For example, some variations in this other 98% of the DNA are known to regulate the activity of the 2% of the DNA that codes for amino acid sequences. For this reason, the other 98% of DNA might be a good place to look for genes associated with quantitative rather than qualitative effects on behavioral traits.

Behavioral Genomics

Despite the slower-than-expected progress to date in finding genes associated with behavior, the substantial heritability of behavioral dimensions and disorders means that DNA polymorphisms exist that affect behavior. We are confident that some of the genes responsible for this heritability will be found. Although attention is now focused on finding specific genes associated with complex traits, the greatest impact for behavioral science will come after genes have been identified. Few behavioral scientists are likely to join the hunt for genes because it is difficult and expensive, but once genes are found, it is relatively easy and inexpensive to use them. DNA can be obtained painlessly and inexpensively from cheek swabs—blood is not necessary. Cheek swabs yield enough DNA to genotype thousands of genes, and the cost of genotyping is surprisingly inexpensive.

It is crucial that the behavioral sciences be prepared to use DNA in research and eventually in clinics. What has happened in the area of dementia in the elderly population will be played out in many other areas of the behavioral sciences. The only known risk factor for late-onset Alzheimer's dementia (LOAD) is a gene, apolipoprotein E, involved in cholesterol transport. A form of the gene called allele 4 quadruples the risk for LOAD but is neither necessary nor sufficient to produce the disorder; hence, it is a QTL. Although the association between allele 4 and LOAD was reported less than a decade ago (Corder et al., 1993), it has already become *de rigueur* in research on dementia to genotype subjects for apolipoprotein E in order to ascertain whether the results differ for individuals with and without this genetic risk factor. Genotyping apolipoprotein E will become routine clinically if this genetic risk factor is found to predict differential response to interventions or preventative treatments.

In terms of clinical work, DNA might eventually contribute to gene-based diagnoses and treatment programs. The most exciting potential for DNA research is to use DNA as an early warning system that facilitates the development of primary interventions that prevent or ameliorate disorders before they occur. These interventions for behavioral disorders, and even for single-gene disorders, are likely to be behavioral rather than biological, involving environmental rather than genetic engineering. For example, PKU, a metabolic disorder that results postnatally in severe mental retardation, is caused by a single gene on chromosome 12. This form of mental retardation has been largely prevented, not by high-tech solutions such as correcting the mutant DNA or by eugenic programs or by drugs, but rather by a change in diet that prevents the mutant DNA from having its damaging effects. Because this environmental intervention is so cost-effective, newborns have been screened for decades for PKU to identify those with the disorder so that their diet can be changed. The example of PKU serves as an antidote to the mistaken notion that genetics implies therapeutic nihilism, even for a single-gene disorder. This point is even more relevant to complex disorders that are influenced by many genes and by many environmental factors as well. With behavior-based interventions, psychotherapists eventually will be in the business of preventing the consequences of gene expression.

The search for genes involved in behavior has led to a number of ethical concerns. For example, there are fears that the results will be used to justify social inequality, to select individuals for education or employment, or to enable parents to pick and choose among their fetuses. These concerns are largely based on misunderstandings about how genes affect complex traits (Rutter & Plomin, 1997), but it is important that behavioral scientists knowledgeable about DNA continue to be involved in this debate. Students in the behavioral sciences must be taught about genetics to prepare them for this postgenomic future.

As the advances from the Human Genome Project begin to be absorbed in behavioral genetic research, optimism is warranted about finding more QTLs associated with behavioral dimensions and disorders. The future of genetic research will involve moving from finding genes (genomics) to finding out how genes work (functional genomics). Functional genomics is generally viewed as a bottom-up strategy in which the gene product is identified by its DNA sequence and the function of the gene product is traced through cells and then cell systems and eventually the brain. This view is captured by the term *proteomics*, which focuses on the function of gene products (proteins). But there are other levels of analysis at which we can understand how genes work. At the other end of the continuum is a top-down level of analysis that considers the behavior of the whole organism. For example, we can ask how the effects of specific genes unfold in behavioral development, how they interact and correlate with experience (Plomin & Rutter, 1998). This top-down behavioral level of analysis is likely to pay off more quickly in prediction, diagnosis, and intervention than the slow buildup of knowledge through cell systems. The phrase *behavioral genomics* has been suggested to emphasize the potential contribution of top-down levels of analysis toward understanding how genes work (Plomin & Crabbe, 2000). Bottom-up and top-down levels of analysis of gene-behavior pathways will eventually meet in the brain. The grandest implication for science is that DNA will serve as an integrating force across diverse disciplines.

We hope that this book contributes to this integration, as well as to the integration of quantitative genetics and molecular genetics, normal and abnormal behavior, human and animal behavior, and nature and nurture.

References

Baron, M. (2001). Genetics of schizophrenia and the new millennium: Progress and pitfalls. *American Journal of Human Genetics, 68,* 299–312.

Capecchi, M. R. (1994). Targeted gene replacement. *Scientific American,* 52–59.

Cardon, L. R., & Bell, J. (2001). Association study designs for complex diseases. *Nature Genetics, 2,* 91–99.

Corder, E. H., Saunders, A. M., Strittmatter, W. J., Schmechel, D. E., Gaskell, P. C., Small, G., et al. (1993). Gene dose of apolipoprotein E type 4 allele and the risk of Alzheimer's disease in late onset families. *Science, 261,* 921–923.

Deater-Deckard, K., Reiss, D., Hetherington, E. M., & Plomin, R. (1997). Dimensions and disorders of adolescent adjustment: A quantitative genetic analysis of unselected samples and selected extremes. *Journal of Child Psychology and Psychiatry, 38,* 515–525.

Duncan, J., Seitz, R. J., Kolodny, J., Bor, D., Herzog, H., Ahmed, A., et al. (2000). A neural basis for general intelligence. *Science, 289,* 457–460.

Fisher, R. A. (1918). The correlation between relatives on the supposition of Mendelian inheritance. *Transactions of the Royal Society of Edinburgh, 52,* 399–433.

Gazzaniga, M. S. (Ed.). (2000). *Cognitive neuroscience: A reader.* Oxford, England: Blackwell.

Plomin, R. (1994). *Genetics and experience: The interplay between nature and nurture.* London: Sage.

Plomin, R., Asbury, K., & Dunn, J. (2001). Why are children in the same family so different? Nonshared environment a decade later. *Canadian Journal of Psychiatry, 46,* 225–233.

Plomin, R., & Crabbe, J. C. (2000). DNA. *Psychological Bulletin, 126,* 806–828.

Plomin, R., DeFries, J. C., McClearn, G. E., & McGuffin, P. (2001). *Behavioral genetics* (4th ed.). New York: Worth.

Plomin, R., & McClearn, G. E. (Eds.). (1993). *Nature, nurture, and psychology.* Washington, DC: American Psychological Association.

Plomin, R., Owen, M. J., & McGuffin, P. (1994). The genetic basis of complex human behaviors. *Science, 264,* 1733–1739.

Plomin, R., & Rutter, M. (1998). Child development, molecular genetics, and what to do with genes once they are found. *Child Development, 69,* 1223–1242.

Risch, N. J. (2000). Searching for genetic determinants in the new millennium. *Nature, 405,* 847–856.

Risch, N., & Merikangas, K. R. (1996). The future of genetic studies of complex human diseases. *Science, 273,* 1516–1517.

Rutter, M., & Plomin, R. (1997). Opportunities for psychiatry from genetic findings. *British Journal of Psychiatry, 171,* 209–219.

Seligman, M. E. P., & Csikszentmihalyi, M. (2000). Positive psychology: An introduction. *American Psychologist, 55,* 5–14.

Thompson, R. F. (2000). *The brain: A neuroscience primer* (3rd ed.). New York: Worth.

Wahlström, J. (1990). Gene map of mental retardation. *Journal of Mental Deficiency Research, 34,* 11–27.

Venter, J. C., Adams, M. D., Myers, E. W., Li, P. W., Mural, R. J., Sutton, G. G., et al. (2001). The sequence of the human genome. *Science, 291,* 1304–1351.

Part II

Research Strategies

2

The Role of Molecular Genetics in the Postgenomic Era

Ian W. Craig and Joseph McClay

The interface between the 20th and 21st centuries has heralded the completion of the initial phase of the Human Genome Project, and sequencing projects of other mammals including the mouse are following in its wake. Concomitantly, an array of new techniques for mapping, identifying, and analyzing the 1.5% of the 3 billion base pairs, which constitute our genes, has been developed. Although these advances are of major significance in many areas, it is in behavioral genetics, a field previously generally refractory to the impact of molecular genetics, where we can hope for a particularly timely impact.

The reasons for this are not hard to find. As emphasized in the Introduction (chapter 1), it has become obvious that most behaviors are both dimensional and multifactorial. Although the genetic contribution to many of these ranges from 40% to 60%, the input of each of the many individual genes involved, so-called quantitative trait loci (QTLs), to the population variation may only achieve a few percent. Searching for QTLs across the genome through linkage studies with simple sequence repeats (SSRs) spaced at approximately 10 cM (300–400 markers) can detect genes of large effect with high scanning efficiency (i.e., relatively few markers per genome scan); however, we should anticipate the existence of only relatively few of these major players. The alternative approach of association studies using case control, or selected extremes, designs can detect QTLs of small effect but is restricted to finding only those QTLs within close range of the markers used (see Sham, chapter 3, and Cardon, chapter 4, this volume). Given sufficiently large samples, allelic association can be powerful (Risch, 2000); nevertheless, the efficiency of a scan is limited, among other factors, by the extent of linkage disequilibrium between the QTL and the marker. This parameter is likely to be variable throughout the genome, and its scale currently provides a topic for vigorous debate. The issue will be settled only by empirical studies of a wide range of marker types across the genome. For our purposes, it can be taken that the density of markers required for an efficient genome scan is at intervals of between 10 and 100 kilobase (kb; i.e., a very low efficiency with up to 300,000 markers being required for total genome coverage). It is particularly for this reason that a second seam of major genomic research has been opened alongside that of sequencing, by the search for and discovery of millions of new markers. In the aftermath of the publication of the first draft of the human genome, the next major challenge is to devise optimal strategies for the analysis of these markers in the very high numbers that will be required. Progress to this end is reviewed later in this chapter.

An alternative to linkage and whole genome association scanning is provided by a strategy concentrating only on the coding or control regions of genes, for it is there that variation of functional significance is most likely to exist. Traditionally, this has been achieved by the *candidate gene* approach, in which a locus presumed to be implicated in a disorder on the basis of physiology or neurochemistry is selected for detailed investigation. The studies of the dopamine receptors and dopamine transporter genes in attention deficit hyperactivity disorder are well documented and apparently successful examples of this approach (Curran et al., 2001; Faraone, Doyle, Mick, & Biederman, 2001). On the deficit side of this strategy, an obvious problem is how to decide on the criteria for the selection of the candidate genes. One possible solution to this could be achieved by the development of marker sets specifically targeted to all of the 32,000–45,000 genes now predicted (e.g., see Lander et al., 2001), and in this way, all genes could be considered as candidates. This would significantly reduce the total number of markers compared with using a random genome distribution of polymorphic sites for association studies. We argue that an efficient midterm strategy should be based on a *systems genotype* approach in which, instead of picking on individual genes, variation in all of the loci encoding the relevant steps in well-documented brain pathways should be examined together.

To provide a basis for evaluating the type of genotyping approaches available to the field, we offer an overview of contemporary markers and technologies below. Although this section may be daunting for the uninitiated, we have tried to present the material as simply as possible, and the range of approaches discussed is by no means exhaustive. A recent review of current technology is also available (see Chicurel, 2001).

The Nature of DNA Markers

Although single base pair substitutions (now referred to as *single nucleotide polymorphisms*, or SNPs) provided the first easily detected variation at the DNA level, because they could produce, or destroy, recognition sites for restriction enzymes, it was the discovery of SSRs such as the repeated dinucleotide, ACAC . . . AC, that revolutionized the DNA approach to genetic mapping (Weber, 1990; Weber & May, 1989; see Exhibit 2.1). The existence of high-frequency variation in copy number at such loci coupled with the development of the polymerase chain reaction (PCR) for their amplification led to their use as markers of choice in linkage mapping. Very quickly, marker sets were produced that covered the genome at 10 cM intervals—sufficient for complete genome scans for linkage studies in pedigrees, or by sib pair analysis (Todd & Farrall, 1996).

Exhibit 2.1. Properties of Genetic Markers

Single Base Pair Substitutions, or Single Nucleotide Polymorphisms (SNPs)

- The basis of the traditional restriction fragment length polymorphisms (RFLPs).
- Generally biallelic with maximum heterozygosity, $[2p(1 - p)]$ therefore = 0.5.
- Probably stable and ancient.
- One allele usually identical to that found in primates.
- Occurrence: every few hundred base pairs (i.e., >million/genome).

Simple Sequence Repeats (SSRs)

- Includes mini- and microsatellites (with motifs in the range about 10–50 and 1–10 base pairs, respectively).
- The most commonly used are dinucleotide repeats, for example, (AC)n, but tri- and tetranucleotides also useful.
- They can attain high heterozygosity values (>90%).
- Some alleles quite recent and relatively high mutation rates.
- Recent estimates indicate their occurrence in about 1 per 2 Kb, however, only a minority probably of sufficient variability to be useful as genetic markers.

As emphasized in chapter 1, the DNA sequence from any two chromosomes selected at random from the population will differ on average at intervals of 1,000 base pairs, providing the potential for 3 million SNPs across the genome. Clearly, these variants provide an exciting resource for gene hunters, and new generations of strategies are being developed to exploit them for applications in genomewide association scans.

Relative Merits of SSR and SNP Markers in Genome Scans

Although current attention is focused on the potential of SNPs, SSR markers still have much to contribute. Many SSRs have the advantage of multiple alleles, conferring high information content, which is important in both linkage analysis and association studies. Although alleles calling for SSRs are hindered in some cases by PCR "artifacts" such as stutter and the variable addition of an A residue, these can be largely avoided and are less of a problem with tri- and tetranucleotides. The recent evidence from the genome programs suggests that SSRs are distributed in sufficient density to be of more use than is generally appreciated, as illustrated in the recent data taken from the initial draft of the human genome sequence (Lander et al., 2001); see Table 2.1. There is approximately one SSR per 2 kb, which approaches the observed density of SNPs. In both cases, this distribution considerably overestimates the density of useful polymorphic markers; nevertheless the resource of SSRs should not be overlooked. An additional feature in their favor is that evidence is accumulating that SSRs located within or close to genes may have a regulatory function—precisely the sort of variant that is of interest in the context of behavioral QTLs.

Table 2.1. Simple Sequence Repeat (SSR) Content of the Human Genome

Length of repeat unit	Average bases per Mb	Average no. of SSR elements per Mb
1	1,660	36.7
2	5,046	43.1
3	1,013	11.8
4	3,383	32.5
5	2,686	17.6
6	1,376	15.2

Nevertheless, in spite of the undoubted value of SSRs, the vast number of SNPs available, even given their lower information content, means that they will ultimately provide the backbone of reference points for genome mapping. Useful information on the proportion of SNPs identified that are of practical value as markers in genomic screens has recently been made available (Kruglyak & Nickerson, 2001; see Table 2.2). Furthermore, a large survey investigating the distribution of SNPs showed that about a third of those examined were reasonably polymorphic (the minor allele having a frequency > 20%) in three major population groups (Marth et al., 2001).

Methods for Genotype Analysis

Until the advent of high-throughput semiautomated machines, DNA genotypes were routinely established by detecting variation based either on SSRs or on single base substitutions that caused an alteration to a restriction enzyme cleavage site. Both can be translated into alleles of different DNA fragment lengths, whose detection following size fractionation on agarose gels can be achieved by Southern blotting and hybridization with locus-specific radioactive probes. Figure 2.1 illustrates this approach to detect alleles generated by the presence or absence of a restriction enzyme site at a locus.

In the case of SSRs, restriction enzymes can be used to digest DNA at sites flanking the tandem repeats. The sizes of the resulting fragments will depend on the copy number of the repeat present, and separation and

Table 2.2. Allele Frequency and Distribution of Human Single Nucleotide Polymorphisms (SNPs)

Minimum allele frequency	Estimated no.	Expected frequency	% in collection examined
1%	11 million	290 bp	11–12
10%	5.3 million	600 bp	18–20
20%	3.3 million	960 bp	21–25
40%	0.97 million	3,200 bp	24–28

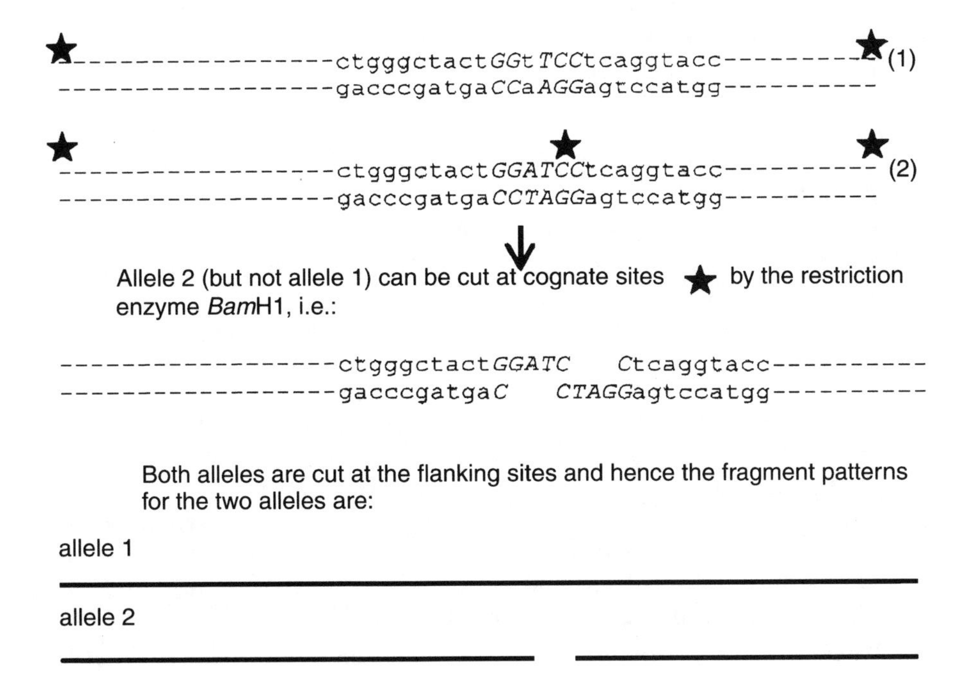

Digested DNA is separated by electrophoresis on agarose gels. Following size fractionation, the fragments are denatured into single strands, and a "print" obtained by blotting these onto an appropriate membrane. The positions of the bands are localized using a "probe" made up from short stretches of radioactively-labeled, single-stranded DNA, complementary to the region of the genome under investigation.

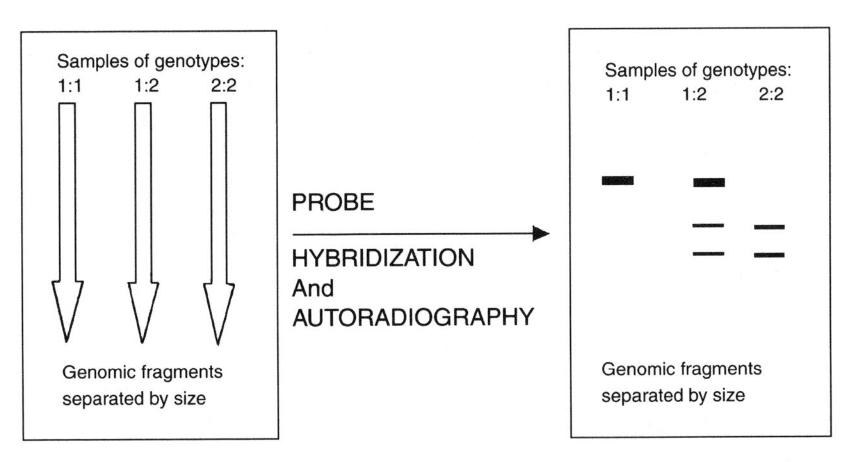

Figure 2.1. Illustration of simple detection system for single nucleotide polymorphisms (SNPs) resulting in alterations to a restriction enzyme cleavage site.

detection can be achieved as for restriction fragment length polymorphisms (RFLPs). For the prevalent di-, tri-, and tetranucleotide repeats, the size differences conferred on the fragments by changes in copy number are relatively small and difficult to resolve on agarose gels. With the increasing availability of PCR technology, however, it became routine to amplify small DNA segments with primers embracing the repeat segment of choice with the resulting products being separated on acrylamide gels to achieve adequate resolution (see Figure 2.2).

Although current methods for analyzing SSRs are derived directly from the above approach, a new generation of techniques is being introduced to improve the efficiency of SNP analysis. These are intended to broaden the range of target sites and to achieve much higher throughput. Current technologies, discussed in more detail below, depend on one of three basic principles, and many of them require the coupling of fluorescent labels to primers enabling the differential tagging of PCR fragments.

In the first approach, the base compositional difference conferred by a SNP, within a PCR-amplified fragment, is exploited to create fragments of differing size. This can be achieved in several ways, including traditional restriction enzyme cleavage, but also by the introduction of nonstandard nucleotides, creating novel sites following further treatment. Size variation can be detected by conventional electrophoresis on gels or capillary systems, or by denaturing high-pressure liquid chromatography (DHPLC).

The second generic technique is based on the extension of primers close to or abutting the variable SNP site and the use of chain-terminating reactions incorporating specific di-deoxyribonucleotides to detect the polymorphism. The extension will be halted at different positions depending on the nature of the SNP variant. Size differences can be distinguished by conventional approaches as above; but this technology also lends itself to applications based on mass spectrometry, as discussed later.

In the third generic strategy, allele-specific hybridization is used to discriminate between the local sequence embracing the dichotomy created by the SNP. As indicated later, several methods are available, including use of allele-specific primers, solid-phase microarrays, molecular beacons, and “invader” assays.

It is not yet clear which of the many variations on the above strategies will become predominant, and it is certainly impossible to cover them all with sufficient detail in this chapter to do justice to the ingenuity inherent in their development. A few approaches illustrating the range of technologies available are described below.

Strategy 1: Creation of Fragment Size Differences by Cleavage

PCR fragments containing SNPs can be differentiated by cleavage, either with restriction enzymes or by introduction of novel bases, which provide targets for breaks introduced by endonucleases or alkali. In one version of the latter approach, primers are used to amplify the template (which

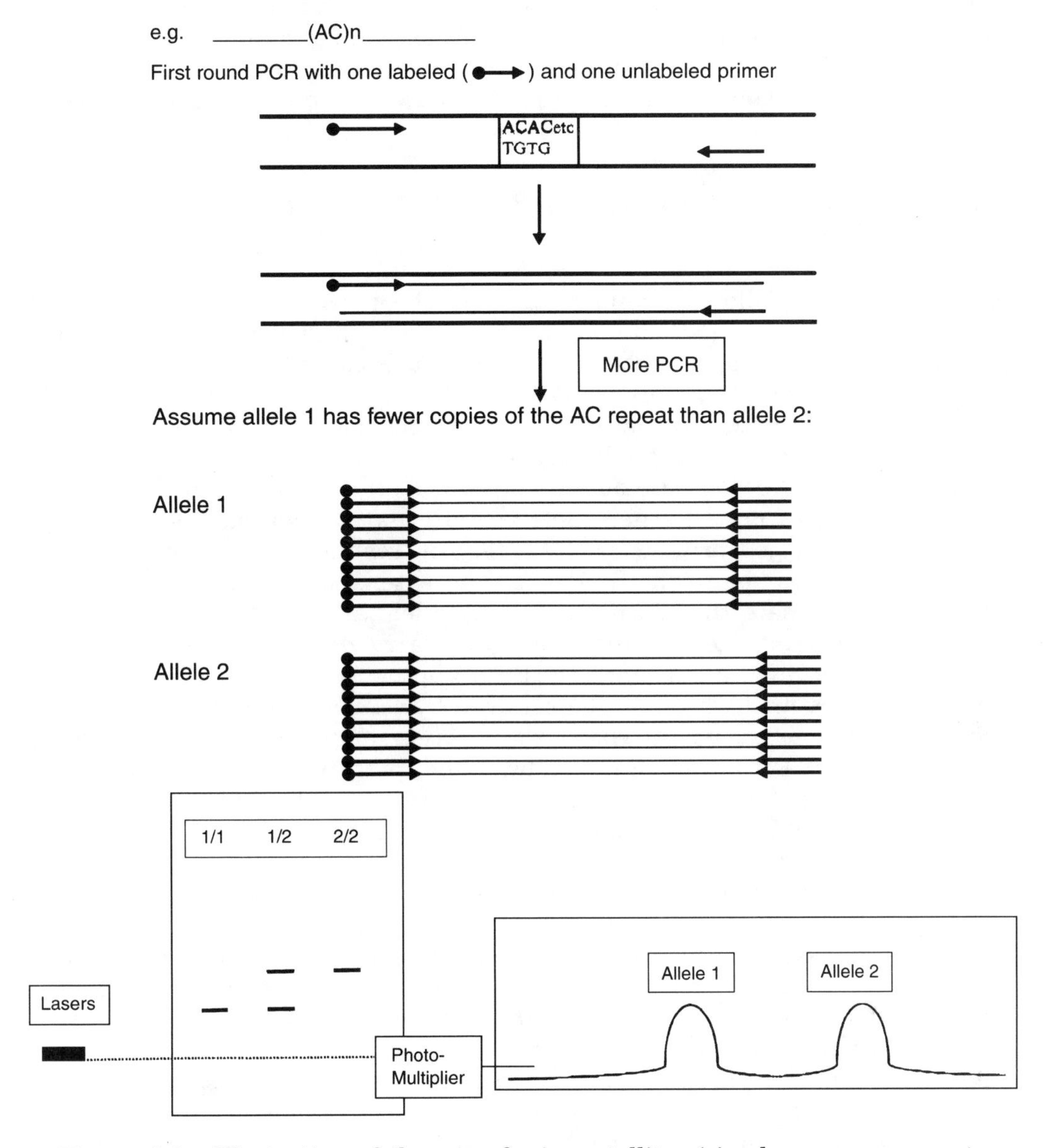

Figure 2.2. Illustration of the use of microsatellites (simple sequence repeats; SSRs) as genetic markers. The variable repeat section is amplified by locus-specific polymerase chain reaction (PCR) and the size differences of the alleles differentiated by electrophoresis. Fluorescent tags added to the 5′ end of one of the primer pairs allow the labeled fragments to be observed via laser excitation and detection with a charge-coupled device (CCD) camera. Several independent loci can be examined in a single gel lane or capillary by using different dyes to label the primers.

can be genomic DNA) across the SNP, but with deoxyuridine triphosphate (dUTP) replacing its deoxythymidine triphosphate (dTTP) in the amplification. The product is then treated with uracil DNA-glycosylase to modify incorporated U, providing a site for subsequent cutting with NaOH, or by endonuclease IV. One of the primers is tagged with a fluorescent label (see Figure 2.3).

Strategy 2: Creation of Fragment Size Differences by Primer Extension PCR

In this method, following amplification of the target region, a common primer abutting the SNP site is used to extend the PCR template across the SNP with an appropriate reaction mixture containing differentially color-labeled terminating di-deoxyribonucleotide triphosphates (ddNTPs), which will terminate the chain at the SNP (see Figure 2.4). The method can be adopted to produce products of different lengths, with the extension terminating at the SNP site on one allele or at the next base of the same type encountered on the other allele. An alternative approach is to use a fluorescently tagged primer for the extension reaction, which terminates one or few bases before the SNP site. The extension reaction mix is designed to terminate the extension for one allele at the SNP site. On the second allele template, the extension will carry until a base similar to allele 1 of the SNP is encountered. Both approaches result in fragments of different lengths, which can be resolved by conventional means.

The primer extension, chain termination approach has also been adapted for DHPLC, providing a robust and quantitative analytical procedure for those laboratories already equipped with the appropriate machines (e.g., WAVE Transgenomics; see Hoogendoorn et al., 2000). A strategy related to that of primer extension is that of "pyrosequencing SNP detection." Here the extension reaction is coupled to a base-specific light emission. Following the first amplification of the target region containing the SNP, an extension primer (or sequencing primer) abutting the SNP is added in a specifically formulated reaction mixture, and the four possible deoxyribonucleotide triphosphates (dNTPs) are added sequentially one at a time. When the one provided is complementary to the next base of the template, it initiates a series of reactions, resulting in the emission of light that is detected by a charge-coupled device (CCD) camera and analyzed (Nyren, Karamohamed, & Ronaghi, 1997).

Strategy 3: Strategies Depending on Differential Hybridization of Primers or Probes

Allele-specific PCR. One of the simplest approaches is to use a common primer in combination with either of a pair of allele-specific primers whose last (3′) base is complementary to one, or other, of the SNP base substitutions. In the absence of 3′ exonuclease activity of the DNA polymerase, the specificity of annealing to the terminal residue of the primer provides

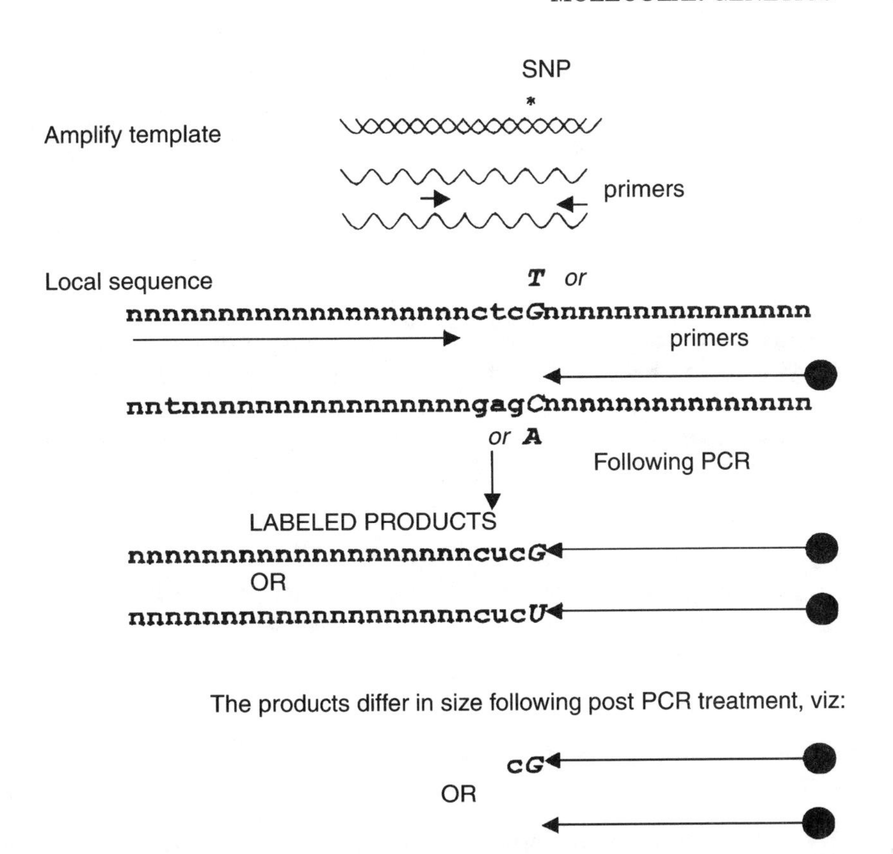

Figure 2.3. Single nucleotide polymorphism (SNP) detection following incorporation of a novel base at the variant site. The incorporation of uracil rather than thymidine at the SNP site for one allele introduces an additional cleavage point in the post-polymerase chain reaction (PCR) treatment, leading to products of differing size. The post-PCR treatment includes digestion with exonuclease 1 (optional), Uracil DNA glycosylase—30 min (to modify the base and sensitize it for digestion), and finally cleavage with endonuclease IV or NaOH. The products differ in size (in this example, by 2 base pairs) and can be resolved by gel or capillary electrophoresis. Because the size ranges of the products can be varied by judicious choice of primers, a range of SNPs can be analyzed on a single lane by multiple staggered loading (up to 10 per lane/capillary). The above is based on a procedure described by Vaughan and McCarthy (1998) and referred to as the SNaPit process.

sufficient power to discriminate between the alleles in two separate reactions. Once again, the alleles can be distinguished by use of different fluorescent tags at the 5′ end of the allele-specific primers. Recently, this approach has been modified such that both allele-specific primers are combined in a single competition reaction (Craig, McClay, Plomin, & Freeman, 2000; McClay, Sugden, Koch, Higuchi, & Craig, 2002). This so-called "SNiPtag" method is illustrated below (Figure 2.5).

Allele-specific primers can also be adapted for use in real-time, or "kinetic," PCR—particularly in the context of analyzing pooled DNA samples (Germer, Holland, & Higuchi, 2000). Real-time PCR follows the course of

Amplify template:

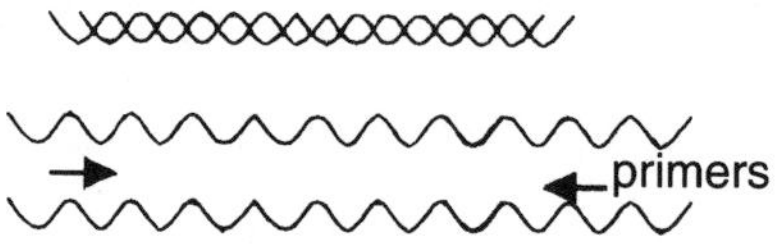

primers

Denature and anneal unlabeled new primer:

C	alternate base
nnagcatgctcaatcgaatccaTnnnn	template strand allele 1
tcgtacgagttagcttaggt	(unlabeled primer)
nntcgtacgagttagcttaggtAnnnnn	complementary strand allele 1

Extend with ddNTP/dNTP mix ↓

nnagcatgctcaatcgaatccaTnnnn	
tcgtacgagttagcttaggtA	primer tagged with appropriate color-labeled ddNTP

OR

nnagcatgctcaatcgaatccaCnnnn
tcgtacgagttagcttaggtG

Figure 2.4. Primer extension methodology using chain termination to detect single nucleotide polymorphisms (SNPs). The amplified templates are first purified by removing deoxyribonucleotide triphosphates (dNTPs) and primers using proprietary columns, or by digestion with shrimp alkaline phosphatase and exonuclease I. After the unlabeled new primer is bound and the extension reaction carried out, unincorporated ddNTPs are removed by digestion with shrimp alkaline phosphatase. The products are denatured and separated, usually on an automated gel- or capillary-based system such as the ABI 310, 377, or 3100 "SNaPshot" or the MegaBace (Amersham) "SNuPe". The products differ in composition of the final base and in the color of the tag, enabling their easy distinction on conventional electrophoresis hardware with laser detection and a charge-coupled device camera. The postextension treatment is required to remove the 5′ phosphoryl group of the fluorescent ddNTP, otherwise unincorporated fluorescent ddNTPs will comigrate with the fragment of interest.

a PCR amplification by monitoring the number of copies made from a given template. There are several approaches available. One in general use employs a TaqMan strategy (Applied Biosystems 7700). In this, a fluorescent dye is released from a probe that has bound to the amplified products during a PCR cycle. During the next cycle, the probe is degraded by the procession of the polymerase, which separates a quencher residue from the fluorochrome of the probe. This generates a signal proportional to the concentration of PCR products after a given number of cycles (see Figure 2.6).

Using competing discriminating primers (labeled with different fluorescent tags) with common primer

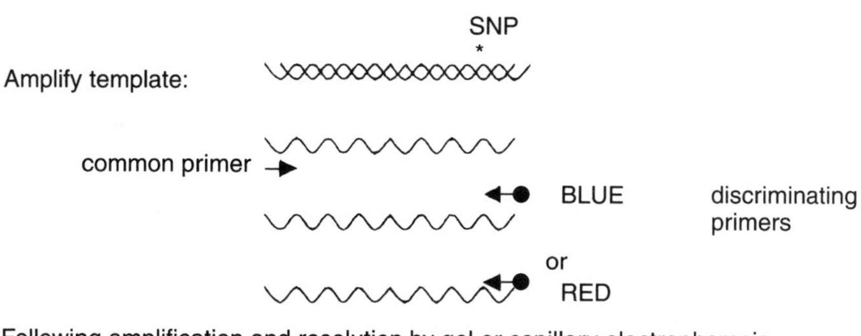

Following amplification and resolution by gel or capillary electrophoresis:

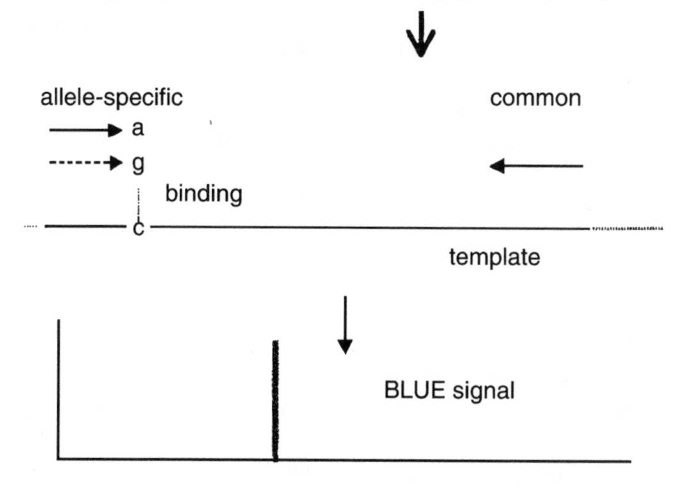

Figure 2.5. Single nucleotide polymorphism (SNP) analysis by competitive allele-specific polymerase chain reaction (PCR), SNiPtag. Simplified schematic shows 3′ guanine primer binding to cytosine of the template, to the exclusion of the competing homologous primer with 3′ adenine. The situation would be reversed with thymine in the template. As the 3′ guanine primer is labeled with blue fluorescent dye (shown dotted arrow in figure), a blue signal is detected by the automated sequencer. Hence, template homozygous for cytosine would be blue, homozygous for thymine would be red (shown black in illustration), and the heterozygote a combination of both. The PCR products are specific for template genotype and can be detected on an automated fluorescent DNA sequencer (e.g., PE Applied Biosystems 377 or 310). The color that the band fluoresces, detected in different channels, indicates the genotype.

Molecular beacons. This ingenious strategy uses a probe, again containing a fluorochrome and a quencher residue. The fluorescence of the "beacon" probe is only activated on binding to the exactly matching template (see Figure 2.7). The technology can accommodate multiple dye signals to be followed in each assay (Tyagi & Kramer, 1996).

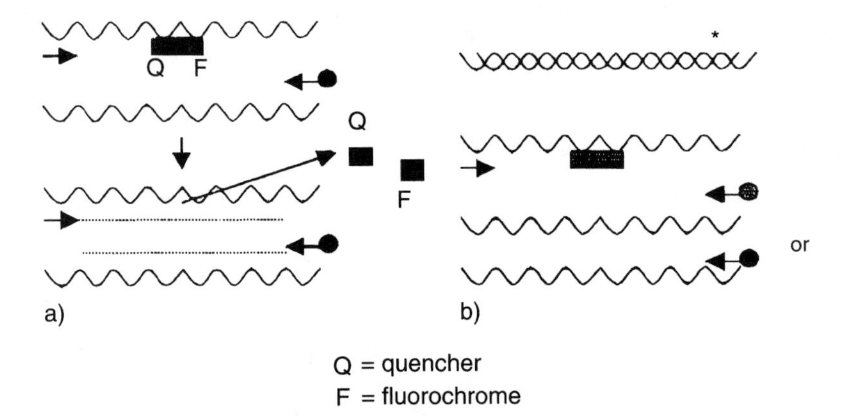

Figure 2.6. Strategy for monitoring polymerase chain reaction (PCR) in real time and detection of single nucleotide polymorphism (SNP) alleles. (a) A specifically designed probe carrying both a quench and a fluorescent moiety binds to the template strand at the beginning of each cycle. As the polymerase progresses along the strand, the probe is degraded, separating the quencher from the fluorochrome. This results in a doubling of fluorescent output per PCR cycle. (b) This can be adapted to distinguish the ratio of SNP alleles by using allele-specific primers in separate reactions.

"Invader" assays. The Invader technology uses an enzyme for allele discrimination whose cleavage specificity is dependent on detecting an absolute match between a signal probe and the template. Successful cleavage drives a universal, secondary reaction with a "reporter probe" in which a fluorochrome is released from close proximity to a quench residue (Mein et al., 2000). The signal is detected at an end point with a conventional fluorescence microtitre plate reader (see Figure 2.8).

Other high-throughput technologies: microarrays. Microarray technology for SNP detection is now becoming more routinely available, both commercially and within the laboratory environment. The technique involves the fixing of oligonucleotides of known sequence to a solid support (usually treated glass), and then probing these with fluorescently labeled samples of unknown sequence. If the unknown sample contains the complementary sequence to the fixed oligonucleotide, it binds and will give a positive signal when scanned by a laser or ultraviolet source. Gridding of the oligonucleotides onto the solid support can be achieved in a number of ways. Current technology would enable all the known variations of several thousand different polymorphic SNP markers to be arrayed onto a 2 cm^2 area of a glass slide. Such arrays can then be probed with multiple PCR products from an individual's DNA and the allelic forms of the various polymorphisms present determined by the patterns of fluorescent dots revealed by scanning the array (see Figure 2.9). Unfortunately, commercially available arrays are at present expensive. However, the great attraction of this approach lies in its ease of automation and in the ability to "read"

Typical Probe

<u>5'FAM</u>-gcgaggAAAGCATC*C*GGGATGGCcctcgc*Quench*

This will normally adopt a hair-pin configuration:

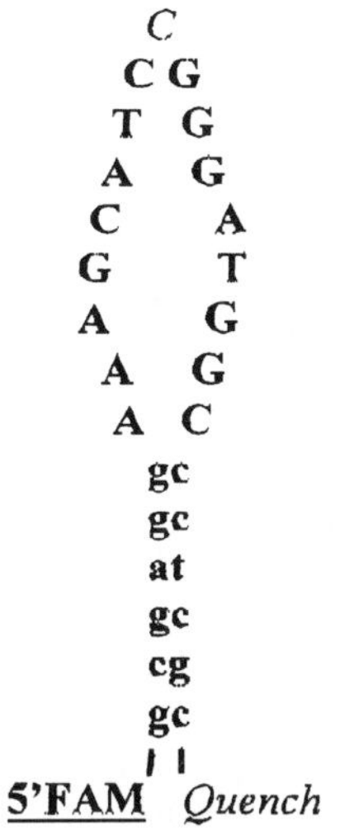

On binding to appropriate template:

Figure 2.7. Single nucleotide polymorphism (SNP) detection with molecular beacons. The hairpin configuration brings the quench and fluorescent moieties into close proximity. It is only when the probe binds to the exactly cognate template that the two active residues are separated, resulting in a fluorescent signal proportional to the template concentration. Use of different dye allows two (or more) polymorphic sequences to be monitored in the appropriate equipment.

thousands of genotypes on a single chip. New developments should soon enable chip technology to be applied to pooled DNA samples, thereby conferring the advantages of both multiple genotyping and high sample throughput.

Other Strategies for Size Fractionating Alleles Generated by SNPs: MALDI TOF

Most procedures that generate fragments of differing size, representing the two alleles of a SNP, should be suitable for genotyping by matrix-

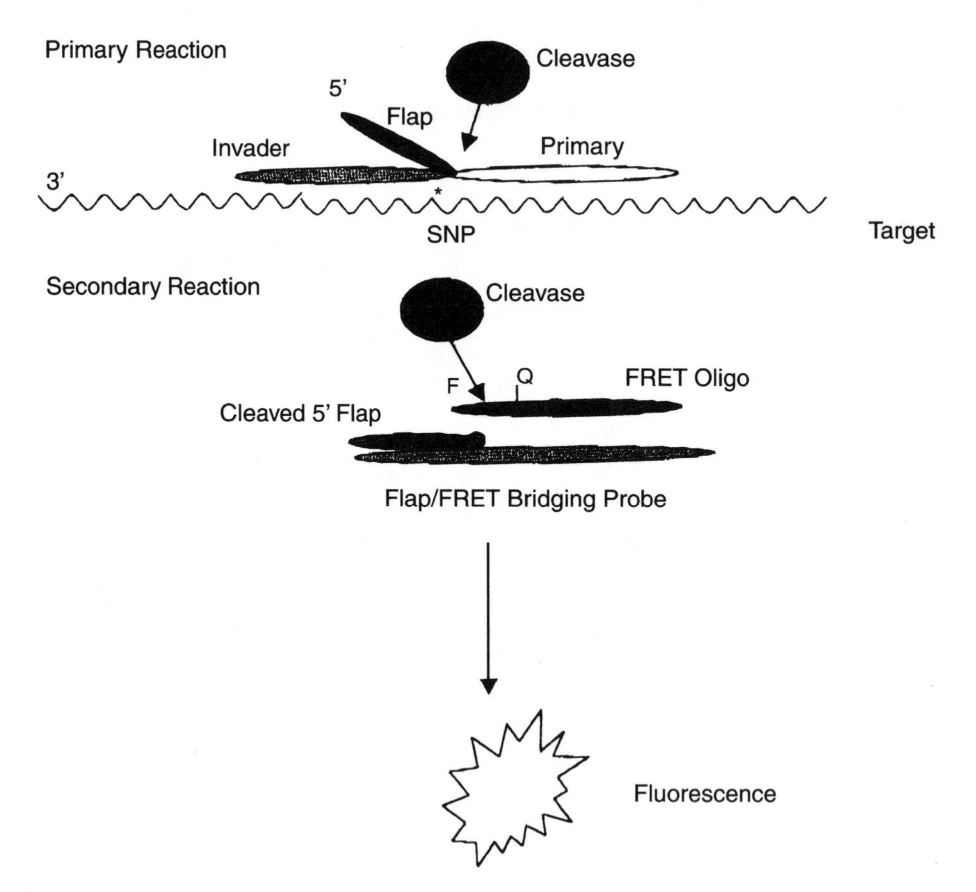

Figure 2.8. Single nucleotide polymorphism (SNP) detection via the "Invader" assay. In the primary reaction, two probes bind to the target DNA; only if the sequence matches precisely that of the allele being followed will the "flap" of the detection probe be displaced and cleaved. The released flap will initiate a secondary cleavage of a signal fragment cobinding to a "bridging" template probe. This releases a dye from its conjunction with a quench moiety resulting in fluorescent output, detected at an end point after several cycles. FRET = Fluorescence Resonance Energy Transfer.

assisted laser desorption ionization–time of flight mass spectrometry (MALDI TOF), a technique with enormous potential for high-throughput analysis. When applied to SNP analysis, this technique involves vaporizing a PCR sample using a laser. The mobilized fragments are then accelerated through a vacuum using an electric field, eventually impacting on a detector. The time taken for the fragments to travel the distance from the plate to the detector depends on the charge to mass ratio of the molecule involved. Two DNA fragments of equal length, but containing a single base change, will have a different charge to mass ratio and therefore will have a varying time of flight; however, the final resolution depends

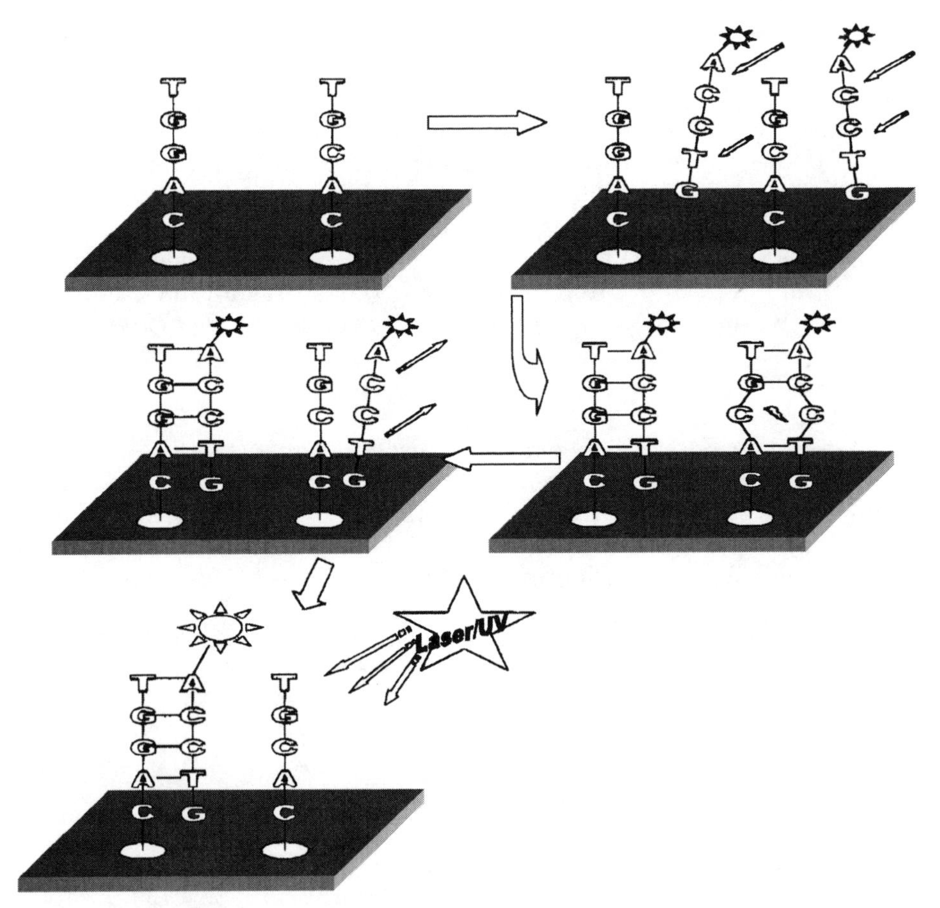

Figure 2.9. Diagrammatic illustration of allele discrimination by oligonucleotide arrays on chips. Microarray containing two spots, representing two versions of a biallelic single nucleotide polymorphism (SNP) detection. The array is incubated with a fluorescently labeled, SNP polymerase chain reaction (PCR) product under appropriate conditions for hybridization. Unknown sequence binds to one spot perfectly and the other spot imperfectly. Array illuminated by laser. Spot with bound tag will "fluoresce."

on the size range of the fragments. PCR products are spotted on a platten (many samples at a time) that is inserted in the machine. Once the process is initiated, the results are available after seconds, making it one of the fastest methods for detecting mutations. Throughput has been estimated at 2,500 samples per day, and multiplexing may allow a 10-fold increase on this. The method still generally depends on PCR for the initial amplification step, and the mass discrimination is frequently based on primer extension (e.g., Li et al., 1999) or ligation-based techniques (Jurinke et al., 1996). Mir and Southern (2000) reviewed the whole field of large-scale sequence variation analysis, including requirements for high throughput PCR amplification.

How to Achieve High Throughput: DNA Pooling

As emphasized in chapter 1, trawling the genome for QTLs of small effect will require sample sizes of 1,000 or more. An association study with markers at 100-kb intervals will require information on 30 million genotypes! Clearly, even the best of the highly sophisticated methodology described above will still fall short of accomplishing this feat in a realistic time frame. In addition, although some of the emerging technology lends itself to high throughput and automation, particularly the chip-based and mass spectrometry approaches, there are many laboratories that have invested in less sophisticated platforms using electrophoretic separation of PCR products. Here, particularly, alternative strategies to cope with the genotyping load are needed.

A possible solution has emerged through the development of DNA pooling in which allele frequencies are estimated on combined samples of multiple cases or controls. Pooling applied to the examination of SSRs has already made considerable advances (Barcellos et al., 1997; Daniels et al., 1998; Daniels, Williams, Plomin, McGuffin, & Owen, 1995; Fisher et al., 1999; Hill et al., 1999), and the first stage of a total genome screen at intervals approaching 2 Mb has been completed (Plomin et al., 2001). Allele frequencies of combined cases, or of combined controls, can be estimated from the genotyping SSR profile obtained from conventional equipment following corrections made for stutter and differential amplification. The robustness of such approaches can be enhanced by undertaking examination of pools in a stepwise manner, with replication of positive results being required before moving on to a final analysis. This, ideally, should be based on a strategy that avoids any possible risks of population stratification, such as the transmission disequilibrium test (TDT), if suitable trios of parents and an affected offspring are available. Other approaches to detect and adjust for stratification errors are available, including that of using a range of unlinked polymorphic markers throughout the genome, whose frequencies should be similar in both cases and controls, to test for homogeneity (see Pritchard & Rosenberg, 1999).

Pooling strategies for SNP analysis are also emerging rapidly. In principle, most single SNP detection systems can be developed for pooling methods. In practice, however, some appear to be more adaptable than others. Both the SNaPit (see Figure 2.3) and SNaPshot (see Figure 2.4) systems described above have been exploited in this regard. An example of the SNaPit technology applied to pooled DNA samples is illustrated in Figure 2.10.

Real-time (kinetic) PCR approaches also appear to offer a robust solution to the problem. In theory, only one allele of a heterozygote would be amplified in each reaction with its appropriate allele-specific primer. For mixed DNA samples, the ratios of alleles are determined by difference in cycle number to reach an arbitrary threshold. In practice, it appears that there may be some amplification of the mismatched allele—but this is likely to be delayed by at least 10 cycles, and the actual allele frequencies have to be adjusted for differential amplification of the two alleles.

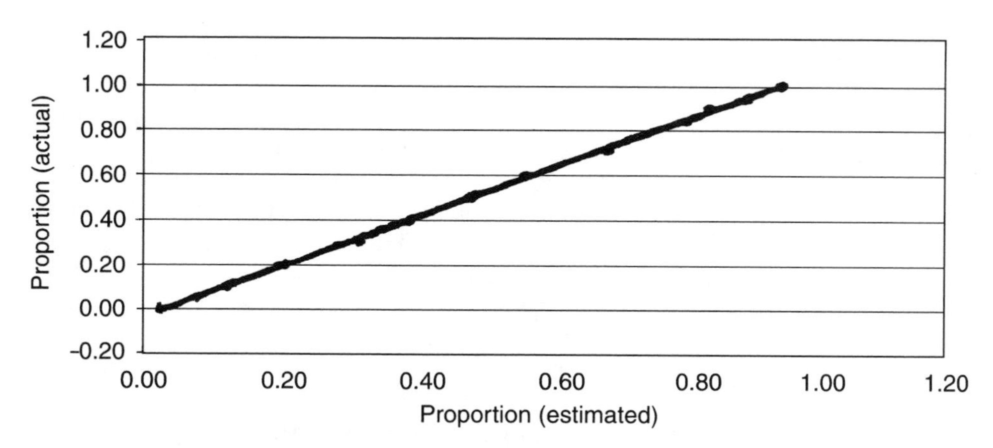

Figure 2.10. Allele frequency estimates for SNaPit genotyping applied to pooled DNA samples. Artificial DNA pools of varying, but known, allele frequencies were prepared and genotyped employing SNaPit methodology. From the ABI 3100 output, the ratios of signals for the two alleles were measured and converted into estimated frequencies. The actual and estimated frequencies correlated well across the whole range of allele frequencies.

The appropriate correction factor can be determined by calibration on a known heterozygote or by mixtures of standard known genotypes (e.g., see Germer et al., 2000). Similar correction problems and calibration will be a requirement for most other pooling approaches to SNP genotyping, in which the efficiency of amplification or extension reactions can differ in an allele-specific manner.

Considerable progress also has been made in the application of pooling approaches with other types of SNP detection system, such as single strand conformation polymorphism (SSCP) analysis (Sasaki et al., 2001), generic high-density oligonucleotide arrays on chips (Fan et al., 2000), and DHPLC (Hoogendoorn et al., 2000). Whatever the approach, ultimately the success of pooling depends on the quality (template availability) and on the accurate estimation of concentrations of the contributing DNA. Nevertheless, for those without access to genotyping factories, pooling provides the only practical solution to the application of association studies to genome scans with sufficient power and marker density to detect a majority of behavioral QTLs.

The Way Forward

Perceptions of the field of behavioral genetics have changed radically over the past decades. From a prevailing mindset that believed that most psychiatric and behavioral disorders were a result of developmental difficulties such as brain damage or lack of parenting skills, we moved on to one that embraced—possibly overenthusiastically—the concept that inheritance makes a major contribution. It also assumed that the major genes implicated would soon be identified by the application of the new genetic technology. In the dawn of the postgenomic era, however, a new realism is emerging, one which accepts that, for many conditions, any single gene will only contribute slightly to the overall variance. The challenge is, therefore, to have enough power to tease out the individual QTLs of small effect and to investigate their interactions. The power to do so by genome screening depends not only on the number of cases and controls available for study and the validity of the diagnosis but also, critically, on the proximity of any genetic marker to the functional locus.

An Alternative to Whole Genome Scans for QTLs

Given the uncertainties concerning the distribution of linkage disequilibrium across the genome and its likely variation in different populations, does the information emerging from genome studies provide a way forward to more efficient screening for QTLs. One obvious improvement would be to concentrate on the approximately 2% of the genome that is coding rather than the 98% remainder of presumed low information content. Indeed much effort is being directed toward the identification of SNPs within coding regions, so-called cSNPs. Even if the functional variant of a locus is not one of the single base pair polymorphisms examined, any significant mutation of an average gene could be tracked by 4 or so SNPs distributed across the region, providing a haplotype for that locus. In the context of behavioral genetics, it would make sense to target those loci expressed in the brain (i.e., about half of the 32,000–45,000 genes estimated for humans). Further rationalization could be provided by examining as a priority only nonhousekeeping loci, thereby further reducing the number of potential targets by approximately 40%. This would suggest a total number of cSNPs to be included in a comprehensive survey to be $4 \times 40{,}000 \times 0.5 \times 0.4 = 32{,}000$—certainly a feasible project for the high-throughput approaches discussed above.

Candidate Genes and "Pathway/Systems Analysis"

Until recently, because of the daunting prospect of hunting through the entire genome for behavioral QTLs, most efforts have concentrated on genes likely to be implicated in behavior because of their known involvement in brain function. This has mostly been done on an ad hoc basis. We have proposed that a more systematic approach could be achieved by coordinated genotyping approaches to examine variants of the related members of neurotransmitter pathways. Such pathways would include the synthetic, degradation, receptor, and transporter functions of the entire network of metabolic processes for a given neurotransmitter and, as such, could be targeted at specific systems such as the dopaminergic or serotonergic pathways (see also Grant, chapter 8, this volume). Indeed, we have developed multiplex PCR analysis to analyze SSRs associated with some of the key players in both of these (e.g., see Figure 2.10) and have complemented this by adapting our allele competition SNP methodology for single-track analysis of several additional loci of the dopamine pathway (see Figures 2.11 and 2.12).

This is an exciting time to be a molecular geneticist. Although the task of identifying even a proportion of the loci implicated in behavior may seem more daunting than some may previously have assumed, we now know that the tools are rapidly becoming available. The promise of mining

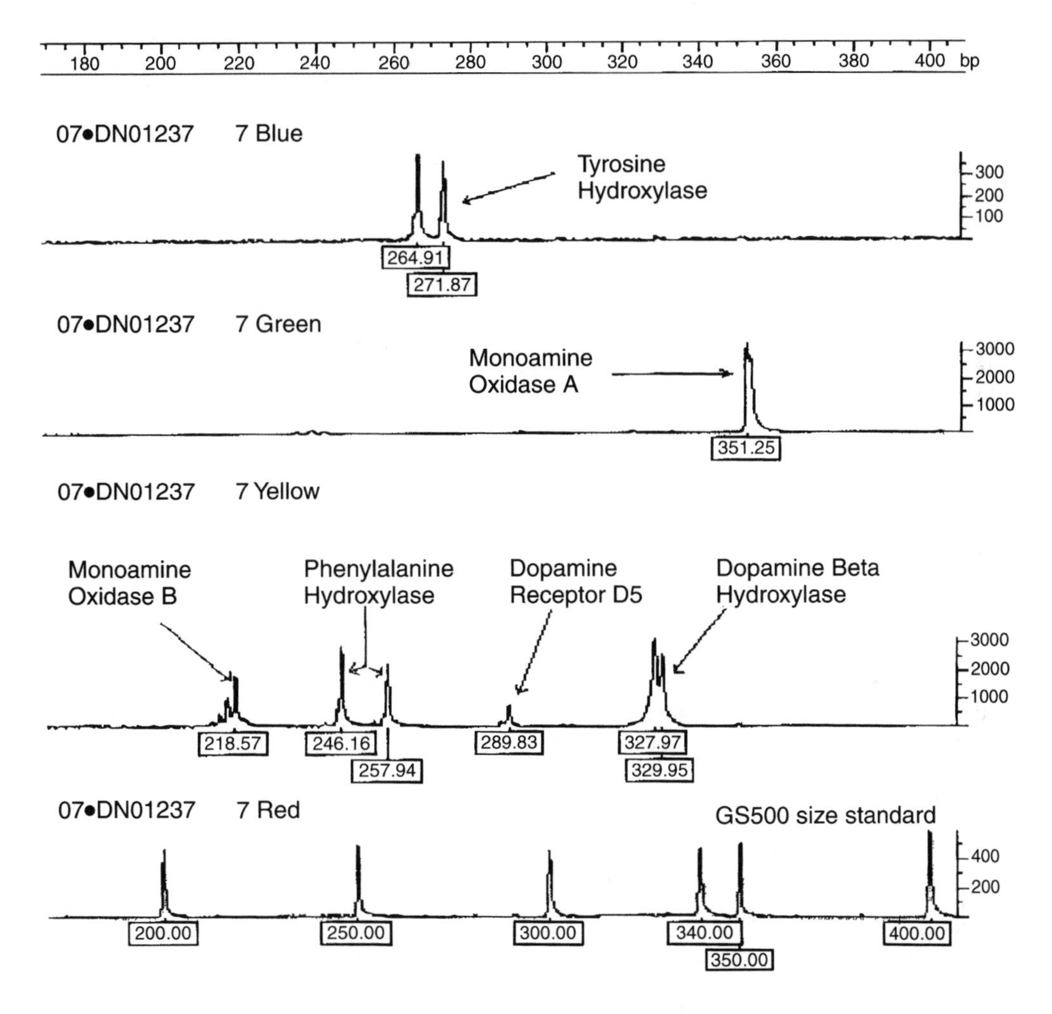

Figure 2.11. The products of a multiplex single polymerase chain reaction (PCR) detecting single sequence repeats (SSRs) at six steps in the dopamine pathway. Use of different fluorescent labels and fragment sizes enable all of the alleles to be detected in a single capillary run (ABI 310).

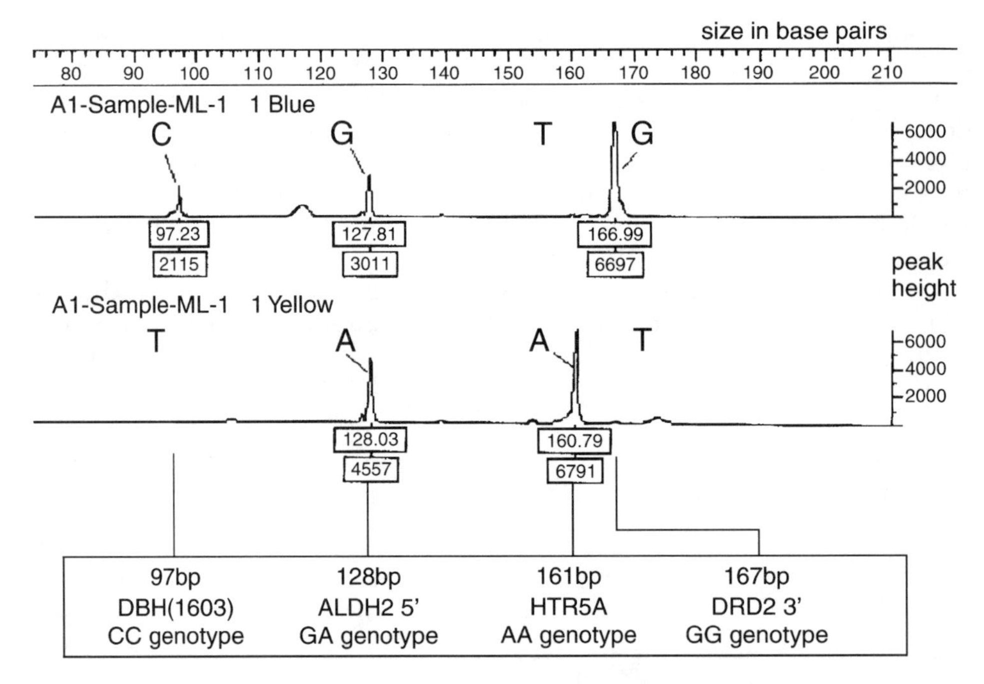

Figure 2.12. Combined single nucleotide polymorphism (SNP) markers for the serotonin and dopamine pathways. The products of several independent competitive, allele-specific SNP reactions at four loci implicated in behavioral disorders are combined and run on the same capillary (ABI 3100). Judicious use of different fragment size ranges generated by the SNiPtag approach should enable up to 20 different reactions to be examined in the same lane or capillary.

the genome for at least some of the more significant QTLs contributing to the major behavioral disorders afflicting humankind is within reach.

References

Barcellos, L. F., Klitz, W., Field, L. L., Tobias, R., Bowcock, A. M., Wilson, R., et al. (1997). Association mapping of disease loci, by use of a pooled DNA genomic screen. *American Journal of Human Genetics, 61,* 734–747.

Chicurel, M. (2001). Faster, better, cheaper genotyping. *Nature, 412,* 580–582.

Craig, I. W., McClay, J., Plomin, R., & Freeman, B. (2000). Chasing behaviour genes into the next millennium. *Trends in Biotechnology, 18,* 22–26.

Curran, S., Mill, J., Tahir, E., Kent, L., Richards, S., Gould, A., et al. (2001). Association study of a dopamine transporter polymorphism and attention deficit hyperactivity disorder in UK and Turkish samples. *Molecular Psychiatry, 6,* 425–428.

Daniels, J., Holmans, P., Williams, N., Turic, D., McGuffin, P., Plomin, R., & Owen, M. J. (1998). A simple method for analyzing microsatellite allele image patterns generated from DNA pools and its application to allelic association studies. *American Journal of Human Genetics, 62,* 1189–1197.

Daniels, J., Williams, N. M., Plomin, R., McGuffin, P., & Owen, M. J. (1995). Selective DNA pooling using automated technology: A rapid technique for allelic association studies in quantitative traits and complex diseases. *Psychiatry Genetics, 5*(Suppl. 1), 50.

Fan, J. B., Chen, X., Halushka, M. K., Berno, A., Huang, X., Ryder, T., et al. (2000). Parallel genotyping of human SNPs using generic high-density oligonucleotide tag arrays. *Genome Research, 10,* 853–860.

Faraone, S. V., Doyle, A. E., Mick, E., & Biederman, J. (2001). Meta-analysis of the association between the 7-repeat allele of the dopamine d(4) receptor gene and attention deficit hyperactivity disorder. *American Journal of Psychiatry, 158,* 1052–1057.

Fisher, P. J., Turic, D., Williams, N. M., McGuffin, P., Asherson, P., Ball, D., et al. (1999). DNA pooling identifies QTLs on chromosome 4 for general cognitive ability in children. *Human Molecular Genetics, 8,* 915–922.

Germer, S., Holland, M. J., & Higuchi, R. (2000). High-throughput SNP allele-frequency determination in pooled DNA samples by kinetic PCR. *Genome Research, 10,* 258–266.

Hill, L., Craig, I. W., Asherson, P., Ball, D., Eley, T., Ninomiya, T., et al. (1999). DNA pooling and dense marker maps: A systematic search for genes for cognitive ability. *Neuroreport, 10,* 843–848.

Hoogendoorn, B., Norton, N., Kirov, G., Williams, N., Hamshere, M. L., Spurlock, G., et al. (2000). Cheap, accurate and rapid allele frequency estimation of single nucleotide polymorphisms by primer extension and DHPLC in DNA pools. *Human Genetics, 107,* 488–493.

Jurinke, C., van den Boom, D., Jacob, A., Tang, K., Worl, R., & Koster, H. (1996). Analysis of ligase chain reaction products via matrix-assisted laser desorption/ionization time-of-flight-mass spectrometry. *Analytical Biochemistry, 237,* 174–181.

Kruglyak, L., & Nickerson, D. A. (2001). Variation is the spice of life. *Nature Genetics, 27,* 234–236.

Lander, E. S., Linton, L. M., Birren, B., Nusbaum, C., Zody, M. C., Baldwin, J., et al. (2001). The International Human Genome Sequencing Consortium. *Nature, 409,* 860–921.

Li, J., Butler, J. M., Tan, Y., Lin, H., Royer, S., Ohler, L., et al. (1999). Single nucleotide polymorphism determination using primer extension and time-of-flight mass spectrometry. *Electrophoresis, 20,* 1258–1265.

Marth, G., Yeh, R., Minton, M., Donaldson, R., Li, Q., Duan, S., et al. (2001). Single-nucleotide polymorphisms in the public domain: How useful are they? *Nature Genetics, 27,* 371–372.

McClay, J., Sugden, K., Koch, H. G., Higuchi, S., & Craig, I. W. (2002). High through-put single nucleotide polymorphism genotyping by fluorescent competitive allele-specific polymerase chain reaction (SNifTag). *Analytical Biochemistry, 301,* 200–206.

Mein, C. A., Barratt, B. J., Dunn, M. G., Siegmund, T., Smith, A. N., Esposito, L., et al. (2000). Evaluation of single nucleotide polymorphism typing with invader on PCR amplicons and its automation. *Genome Research, 10,* 330–343.

Mir, K. U., & Southern, E. M. (2000). Sequence variation in genes and genomic DNA: Methods for large-scale analysis. *Annual Review of Genomics and Human Genetics, 1,* 329–360.

Nyren, P., Karamohamed, S., & Ronaghi, M. (1997). Detection of single-base changes using a bioluminometric primer extension assay. *Analytical Biochemistry, 244,* 367–373.

Plomin, R., Hill, L., Craig, I., McGuffin, P., Purcell, S., Sham, P., et al. (2001). A genome-wide scan of 1842 DNA markers for allelic associations with general cognitive ability: A five-stage design using DNA pooling and extreme selected groups. *Behavior Genetics, 31,* 497–509.

Pritchard, J. K., & Rosenberg, N. A. (1999). Use of unlinked genetic markers to detect population stratification in association studies. *American Journal of Human Genetics, 65,* 220–228.

Risch, N. J. (2000). Searching for genetic determinants in the new millennium. *Nature, 405,* 847–856.

Sasaki, T., Tahira, T., Suzuki, A., Higasa, K., Kukita, Y., Baba, S., & Hayashi, K. (2001). Precise estimation of allele frequencies of single-nucleotide polymorphisms by a quantitative SSCP analysis of pooled DNA. *American Journal of Human Genetics, 68,* 214–218.

Todd, J. A., & Farrall, M. (1996). Panning for gold: Genome-wide scanning for linkage in Type 1 diabetes. *Human Molecular Genetics, 5,* 1443–1448.

Tyagi, S., & Kramer, F. R. (1996). Molecular beacons: Probes that fluoresce upon hybridization. *Nature Biotechnology, 14,* 303–308.

Vaughan, P., & McCarthy, T. V. (1998). A novel process for mutation detection using uracil DNA-glycosylase. *Nucleic Acids Research, 26,* 810–815.

Weber, J. L. (1990). Informativeness of human (dC-dA)n.(dG-dT)n polymorphisms. *Genomics, 7,* 524–530.

Weber, J. L., & May, P. E. (1989). Abundant class of human DNA polymorphisms which can be typed using the polymerase chain reaction. *American Journal of Human Genetics, 44,* 388–396.

3

Recent Developments in Quantitative Trait Loci Analysis

Pak Sham

Behavioral genetics is traditionally concerned with the partitioning of phenotypic variance into various genetic and environmental components. With the advent of molecular genetics and the completion of the Human Genome Project (International Human Genome Sequencing Consortium, 2001; Peltonen & McKusick, 2001; Venter et al., 2001), behavioral geneticists are now ready to attempt the dissection of the genetic component into the contributions of individual genes. Naturally occurring genetic polymorphisms that contribute to quantitative individual differences are called *quantitative trait loci* (QTLs). The identification and subsequent characterization of QTLs promise to bring our understanding of the determinants of human behavior to a deeper, mechanistic level. For behaviors that are related to health or disease, the identification and characterization of QTLs may suggest new approaches to the promotion of health and the prevention and treatment of disease.

Linkage Analysis

Classical Parametric and Nonparametric Linkage Analyses

Until recently, the principal method for the genetic mapping of diseases or other phenotypic traits has been based on linkage. Linkage analysis typically uses phenotypic data on families to infer the presence and extent of nonindependent segregation between a trait and one or more genetic markers, to establish the location of a trait-influencing genetic variant. Linkage analysis has two attractive features: It requires no knowledge of pathophysiological mechanisms, and its range extends over long genomic distances (up to 20 cM). For these reasons, linkage analysis is ideally suited for a systematic screen of the genome for disease-predisposing loci.

Classical linkage analysis, developed by Haldane and Smith (1947), Morton (1955), Elston and Stewart (1971), Ott (1974), and others, is referred to as being parametric or model-based because an explicit model of the disease locus is assumed. Model-based linkage analysis is the method of choice for the genetic mapping of single-gene diseases. With sufficient size or number of segregating pedigrees, convincing log of the odds (LOD) scores (>3) can be generated and replicated, and the cosegregating segment progressively narrowed down until a causal mutation is identified. This positional cloning approach has been successful in mapping hundreds of rare Mendelian diseases (Collins, 1995; Peltonen & McKusick, 2001). This approach also has been successfully applied to Mendelian forms of complex diseases, such as early onset familial Alzheimer's disease and early onset familial breast cancer. It has had much less success with the majority of cases of common disorders, which are non-Mendelian and often only moderately familial. The usual pattern of findings is multiple broad chromosomal regions of only moderately high LOD scores, as well as inconsistent results from attempted replication studies. Some replication studies show small positive LOD scores at various distances from the original finding, and others fail to replicate altogether (for a review, see Bishop & Sham, 2000).

A separate tradition of linkage analysis, referred to as being nonparametric or model-free, was initiated by Penrose (1935) and further developed by Haseman and Elston (1972), Suarez, Rice, and Reich (1978), Risch (1990), Holmans (1993), and others. Unlike parametric linkage, no explicit model of the disease is required. Instead, evidence for linkage between disease and marker is based simply on phenotypic similarity (i.e., concordance or discordance for disease phenotype) and allele sharing at the putative locus, measured technically by the proportion of alleles identical by descent (IBD) at the locus. Nonparametric methods were originally developed for sib pairs but have been extended to general pedigrees (Kong & Cox, 1997; Kruglyak, Daly, Reeve-Daly, & Lander, 1996; Weeks & Lange, 1988). The apparent failure of the parametric linkage to produce consistent findings, and the relative ease of ascertaining affected sib pairs as compared with large multiplex pedigrees, led to the increasing popularity of the nonparametric paradigm (Thomson, 1994). However, to date, these studies also have been largely disappointing (for a review, see Bishop & Sham, 2000).

Variance Components Linkage Analysis

The lack of success of both parametric and nonparametric linkage to genetically map common forms of complex diseases has stimulated the reassessment of how the phenotype should be defined. Evidence for a substantial genetic contribution, in terms of the sibling recurrence ratio (λ_S) for all-or-none traits and heritability (h^2) or sibling correlation (r) for quantitative traits, is now considered a prerequisite for embarking on gene-mapping studies. The fact that all-or-none traits offer poor precision in the measurement of the underlying genetic liability, and the potential gain in power by considering quantitative "intermediate phenotypes," are increasingly being recognized. Thus, instead of attempting to map genes for hypertension, stroke, and depression directly, one might initially focus on finding QTLs for blood pressure, carotid artery thickness, and neuroticism, respectively. Such an approach has been successful in identifying a locus on chromosome 6 for reading disability (Cardon et al., 1994), which has been subsequently replicated by several other groups (Fisher et al., 1999; Gayan et al., 1999; Grigorenko et al., 1997; Grigorenko, Wood, Meyer, & Pauls, 2000; Morris et al., 1999). The shift in paradigm to quantitative traits has been further accelerated by the demonstration that extreme discordant sibpairs have greater linkage information than extreme concordant sibpairs (Risch & Zhang, 1995).

For an additive diallelic QTL, the size of effect on a quantitative trait can be measured by the difference in average trait value between the two homozygous genotypes. The contribution of the QTL to the population phenotypic variance is called *QTL heritability*, and it is proportional to the square of the effect size for any given allele frequencies. Factors that determine the linkage information content of sibpairs are QTL heritability, the variance in IBD sharing (which depends on the informativeness and density of markers), the overall correlation between the sibpairs, and the trait values of the sibpair (Rijsdijk, Hewitt, & Sham, 2001; Sham, Cherny, Purcell, & Hewitt, 2000). In a sibship or pedigree, the overall linkage information content is given approximately by the sum of all pairwise contributions (Rijsdijk et al., 2001), so that large sibships and large pedigrees are highly informative. If r is the overall sib correlation, $\hat{\pi}$ is the estimated proportion of IBD sharing, and Q is the proportion of phenotypic variance explained by the QTL, then the noncentrality parameter (NCP) for linkage contributed by a sibpair is given by:

$$NCP = \frac{1 + r^2}{(1 - r^2)^2} Var(\hat{\pi}) Q^2$$

It is important to note that the information for detecting a QTL by linkage analysis is proportional to the square of QTL heritability (Sham, Cherny, et al., 2000). In other words, halving the effect size of a QTL will increase the required sample size by 16-fold for linkage analysis. This extreme reduction of power with decreasing effect size emphasizes the need to maximize efficiency by selective genotyping, the recruitment of large sibships, and the optimal definition of the phenotype.

For a common, additive QTL of small effect, the information content of a sib pair is proportional to a weighted sum of the squared sum and squared difference of trait values, where the weights are determined by the correlation between the sibs (Purcell, Cherny, Hewitt, & Sham, 2001; Sham & Purcell, 2001). Plotting information content against sibpair trait values gives a surface indicating extreme discordant and extreme concordant sibpairs to be most informative (see Figure 3.1). Selecting the most informative 20% of sibpairs for genotyping retains approximately 80% of the linkage information (see Figure 3.2). The same procedure applies to larger sibships, but the gain in efficiency is less.

Variance components models, which were developed for the partitioning of phenotypic variation into genetic and environmental components

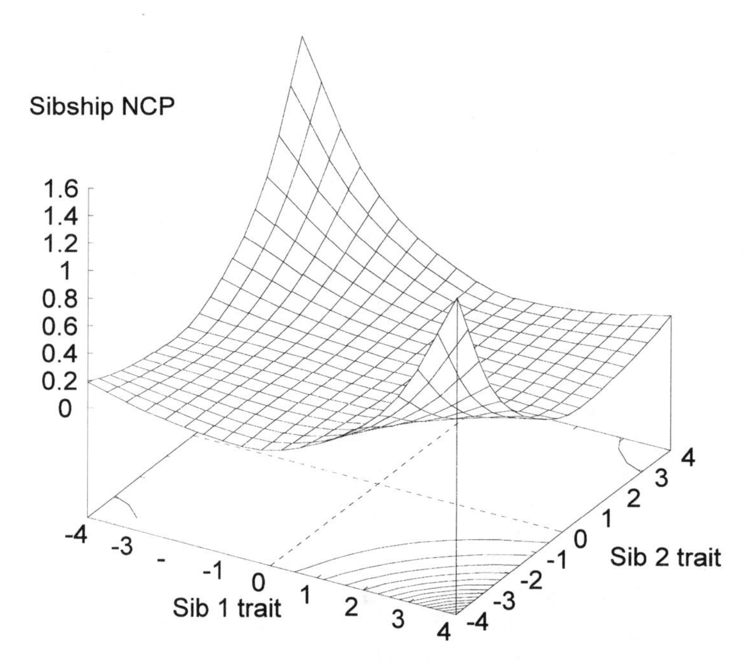

Figure 3.1. Information content for linkage of a sibpair plotted against sibpair trait values. The quantitative trait loci (QTLs) are additive with equal allele frequencies, and the overall sib correlation is .25. NCP = noncentrality parameter.

from correlational data from pairs of relatives (R. A. Fisher, 1918), have been extended for QTL analysis (Almasy & Blangero, 1998; Amos, 1994; Eaves, Neale, & Maes, 1996; Fulker & Cherny, 1996; Schork, 1993). Linkage is modeled in the covariance matrix, as a variance component whose correlation between a pair of relatives is proportional to the degree of IBD sharing at the QTL between the relatives. The dependency of the correlation on locus-specific IBD sharing rather than on overall genetic relationship enables the QTL component to be untangled from the remainder of the genetic component.

One problem in the use of variance components models is the assumption of multivariate normality. While obvious outliers can be readily detected and then either removed or adjusted, more subtle deviations from multivariate normality can inflate the test statistics and lead to false-positive linkage results (Allison, Neale, et al., 1999). Similarly, when applied to the analysis of samples selected for phenotypic extremes, standard variance components models have been shown to produce inflated test statistics and an increase in false-positive linkage findings (Dolan, Boomsma, & Neale, 1999). Several approaches have been proposed to overcome these problems, including variance components modeling conditional on trait values (Sham, Zhao, Cherny, & Hewitt, 2000) and a modified form of Haseman–Elston regression (Sham & Purcell, 2001). For proband-ascertained data, the augmented DeFries–Fulker method is also a simple and effective method of QTL linkage analysis (Fulker et al., 1991).

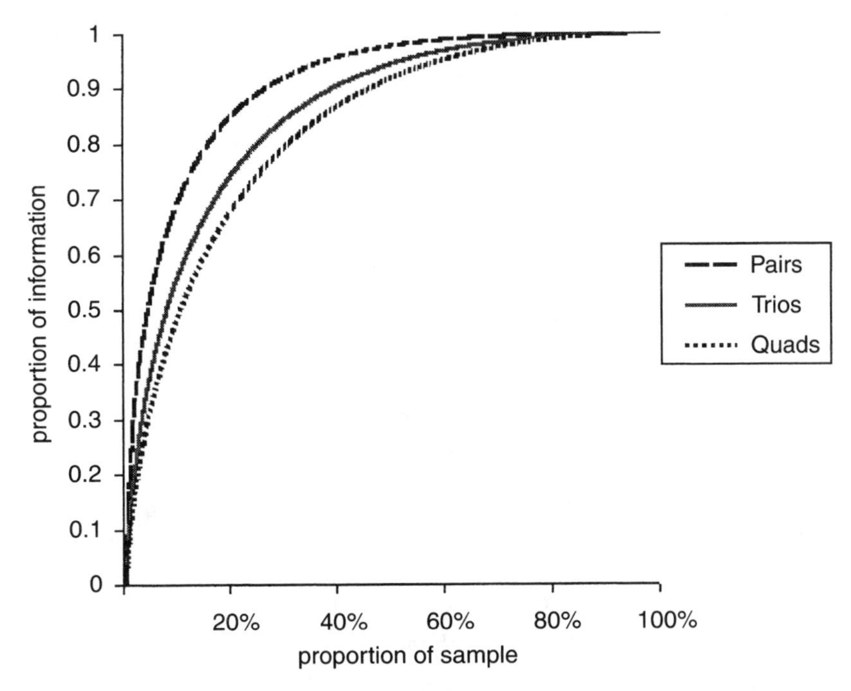

Figure 3.2. Proportion of linkage information retained by selecting a certain proportion of most informative sibships, for sibships of size 2, 3, and 4. The quantitative trait loci (QTLs) are additive with equal allele frequencies, and the overall sib correlation is .25.

Epistasis and Gene × Environment Interactions

The variance components framework, as well as the Haseman–Elston and Defries–Fulker methods, can easily incorporate both epistasis and Gene × Environment (G × E) interactions. Whereas the sib correlation for the main effect of a QTL is proportional to the IBD sharing at the QTL, the correlation for the epistatic effects of two QTLs is related to the product of the IBD sharing at the two QTLs. Epistasis can therefore be incorporated as extra components of variance in a variance components model. However, because the product of IBD sharing at the two QTLs is correlated with each of the IBD sharing of the two QTLs, there will be a degree of confounding between the epistatic and main effect components. As a result, when the epistatic component is omitted, much of its variance will be "absorbed" into the main effect components of the two QTLs. There are two important consequences. First, the power to detect epistasis will be usually limited. Second, analyses that omit epistatic components will often have inflated estimates of QTL variance components.

G × E interactions also can be incorporated into a variance components framework. Instead of having a constant effect on the trait, the QTL can be specified to have an effect that is linearly related to a measured modifier variable. The variance of each sibling, as well as the covariance between two siblings, is therefore dependent on the values of the modifier variables. In addition, the covariance will be determined by the IBD sharing at the QTL. The power of detecting G × E interactions in a variance components framework has not been examined in detail but is likely to be superior to that of detecting epistasis, at least in the absence of GE correlation.

Association Analysis

Transmission / Disequilibrium Test

The greater power of association analysis to detect and localize low-penetrance genes has been long recognized (Bodmer, 1986). However, it was Risch and Merikangas (1996) who emphasized the inherent limitations of linkage studies, whether parametric or nonparametric, in terms of power to detect low-penetrance genes. This has led to the increasing popularity of association studies in recent years. A disadvantage of association, however, is that it may occur between alleles at unlinked loci, as a result of hidden population stratification. The popularity of association analysis was boosted by the introduction by Spielman, McGinnis, and Ewens (1993) of the transmission/disequilibrium test (TDT). Part of the reason for the rapid acceptance of the TDT was that it was introduced as a test of linkage rather than association. The TDT examines heterozygous parents of affected individuals for evidence of preferential transmission of certain alleles over others. Spielman et al. showed that unequal transmission probabilities necessarily implicated linkage. However, for samples consisting of parent–offspring triads or small pedigrees, the power of the TDT depends on the presence of association. Therefore, except for its robustness to population stratification, the properties of the TDT are much more similar to other methods of association analysis than to linkage analysis. It is therefore perhaps more appropriate to regard the TDT as a method of association analysis that is robust to population stratification than a method of linkage analysis.

The original TDT was applicable to a single diallelic locus in parent–offspring triads but has been subsequently extended to more complex designs. These extensions include multiallele loci (Schaid, 1996; Sham & Curtis, 1995), multilocus haplotypes (Zhao, Zhang, et al., 2000), discordant sibships (Curtis, 1997; Horvath & Laird, 1998; Knapp, 1999), general pedigrees (Martin, Monks, Warren, & Kaplan, 2000; Sinsheimer, Blangero, & Lange, 2000), and quantitative traits (Allison, Heo, Kaplan, & Martin, 1999; Rabinowitz, 1997; Waldman, Robinson, & Rowe, 1999). These and other developments of the TDT have been summarized by H. Zhao (2001).

Variance Components Association Analysis

Within a variance components framework, association is modeled either in the covariance matrix (as random effects) or in the mean vector (as fixed effects). Fulker, Cherny, Sham, and Hewitt (1999) proposed a partitioning of the association effects into between- and within-sibship components, with the latter providing a test that is robust to population stratification. This model has been extended to general pedigrees, using the parental average for the between-sibship component when parental genotypes are present and the sibship average when parental genotypes are not available (Abecasis, Cardon, & Cookson, 2000). When applied to parent–offspring triads, the within-sibship component provides a test that can be viewed as a quantitative-trait equivalent of the TDT.

The information content for association is simply proportional to the QTL heritability (Sham, Lin, Zhao, & Curtis, 2000), so that halving the effect size will increase the required sample size by only 4-fold (as compared with 16-fold for linkage). Association analysis is therefore potentially more powerful than linkage for the detecting of QTL with small effects. Analytic expressions for the NCPs for the between-sibship and within-sibship tests have been derived under the assumptions of a standard biometrical genetic model. The parameters V_A, V_D, and r represent the proportions of phenotypic variance explained by additive effects and dominance at the QTL and trait correlation between siblings, respectively. For a randomly selected sibship of size s, the NCPs of the between-sibship association test and the within-sibship association test are as follows:

$$NCP_B \approx \frac{[(s + 1)/2]V_A + [(s + 3)/4]V_D}{V_N + sV_S}$$

$$NCP_W \approx (s - 1)\,\frac{(1.2)V_A + (3/4)V_D}{V_N}.$$

When parental genotype data are included, the NCPs of the between- and within-sibship association tests become

$$NCP_B \approx \frac{(s/2)V_A + (s/4)V_D}{V_N + sV_S}$$

$$NCP_W \approx (s - 1)\,\frac{(1/2)V_A + (3/4)V_D}{V_N} + \frac{(1/2)V_A + (3/4)V_D}{V_N + sV_S}.$$

The effect of having parental genotypes is simply to move part of the NCP from the between-sibship test to the within-sibship test. Thus, singleton families are not informative for the robust within-sibship test unless parental genotypes are available. With increasing sibship size, it becomes less important to have parental genotypes, as was shown in previous simulation studies (Abecasis et al., 2000). From a practical point of view, it is worth emphasizing that, because NCP_W becomes approximately proportional to $(s - 1)$ with or without parental genotypes as sibship size increases, a collection of large sibships would be a valuable resource for robust association analyses of quantitative phenotypes.

Adjustment for Population Stratification Using Genomic Control

The robustness of the TDT and its quantitative-trait equivalents comes with the inherent cost of obtaining genotypes from parents. This can seriously increase the time and effort required for recruiting a sample of sufficient size, especially for adult-onset diseases. Furthermore, some experts have questioned whether hidden population stratification is likely to be a serious problem in actual human populations (Morton & Collins, 1998). The combination of these factors has led to another paradigm shift in favor of case-control studies and other designs requiring only unrelated individuals.

With the development of high-throughput genotyping, the use of random, unlinked markers to detect and to adjust for hidden population stratification has become feasible. Early methods of genomic control, based on estimating an inflation factor from marker data and then using it to adjust association test statistics, were introduced by Pritchard and Rosenberg (1999) and Devlin and Roeder (1999). More recently, latent class models have been proposed for extracting population substructure and for constructing a robust test for association (Pritchard, Stephens, Rosenberg, & Donnelly, 2000; Satten, Flanders, & Yang, 2001; Zhang, Kidd, & Zhao, in press). At present, there is limited experience with the use of these approaches for the detection of and adjustment for population stratification in actual populations. Nevertheless, it seems likely that the refinement and application of these approaches will favor the increasing use of case control and other designs that do not require familial controls.

Systematic Screening for Association Using Pooled DNA

The association between an allele and a phenotype is strongest when the allele has a direct causal relationship with the phenotype. However, because association often occurs between alleles at tightly linked loci, association between an allele and a phenotype also may occur indirectly through their mutual association with the causal allele. The existence of linkage disequilibrium (LD) between tightly linked loci raises the possibility of conducting systematic genomewide scans by association analysis. However, the number of markers required for such an endeavor is likely to be large, perhaps several tens of thousands (Abecasis et al., 2001). This is because of both the rapid decline of average LD as a function of increasing genetic distance and the large variance in LD for a given genetic distance. Thus, at a distance of less than 10 kb, many pairs of markers are at complete LD, whereas others show nearly no LD. The current development of LD maps and of marker sets optimal for LD mapping is therefore of crucial importance for the feasibility of the indirect, LD approach to gene mapping.

The cost of association analysis may be reduced drastically by performing genotyping not on individual DNA samples but on pools made up of DNA from multiple individuals. Thus, the allele frequencies in a sample of 200 cases and 200 controls can be measured from 2 pooled samples rather than 400 individual samples. A genome-wide association scan of general cognitive ability has already been conducted using this approach (Plomin et al., 2001).

There are many problems in the use of pooled DNA for association analysis. For quantitative traits, pooling will result in loss of information concerning the variation among individuals of the same pool. Significant covariates will need to be balanced in the design of the pools, because of the inability for individual adjustment in the analysis. Similarly, there will be limited ability to delineate hidden population stratification and to adjust for this. Analyses will be restricted to the main effects of alleles; it will be impossible to examine for multilocus haplotype effects or interactions. Finally, allele frequency estimation from pooled DNA is subject to both bias and random measurement errors.

Some of these problems are now being addressed. An interesting finding is that, for the detection of a common QTL underlying a quantitative trait, the optimal design is to form the two pools using the lower and upper 27% of trait distribution, assuming no measurement error in allele frequency estimation (Bader, Bansal, & Sham, 2001). This design retains approximately 75% of the information obtainable from individual genotyping. The optimal pooling fraction is reduced in the presence of measurement error, so that for most studies a pooling fraction of 15%–25% is probably reasonable.

Haplotype Analysis

Several theoretical studies have shown that highly polymorphic markers are likely to be more powerful than less polymorphic markers for association analysis (Sham, Zhao, & Curtis, 2000). The main reason for this appears to be the potential for a greater allele frequency difference between cases and controls for a low-frequency allele than one that is moderately common. The same argument would suggest that the analysis of haplotypes might provide greater power for association analysis, because the presence of even a small number of extremely rare haplotypes among cases with a disorder would be highly significant.

The simplest form of haplotype association analysis is the comparison of haplotype frequencies in cases and controls. Although efficient programs have been developed for this purpose (e.g., Zhao, Curtis, & Sham, 2000), they typically ignore the problems of covariates and population stratification and do not handle quantitative phenotypes. Furthermore, the individuals are typically assumed to be unrelated. There is a need for family-based association methods to be extended to haplotype data.

Another fundamental problem in haplotype analysis is the potentially very large number of distinct haplotypes that may exist when multiple highly polymorphic loci are jointly considered (Bader, 2001). Haplotypes are, however, related to one another through being descended from common ancestral chromosomes but differentially modified by the processes of crossing over and mutation. An association between a specific haplotype and a phenotype may therefore be accompanied by similar associations between closely related haplotypes and the same phenotype (Hoehe et al., 2000). A method that attempts to integrate the evidence for association over a class of ancestrally related haplotypes, called the *trimmed haplotype method,* has been proposed (MacLean et al., 2000). Other similar methods also have been and are being developed (for a review, see Lazzeroni, 2001).

Conclusion

The integration of quantitative and molecular genetics promises an exciting future for behavioral genetics, where the determinants of sophisticatedly defined phenotypes will be analyzed to the level of specific genetic variants and environmental influences. Some of the statistical methodology for accomplishing this goal, such as a combined variance components model for linkage and association, is now in place. Substantial progress also has been made in the optimal design of QTL studies, particularly with regard to the selection of maximally informative families.

The success of linkage studies will depend on identifying quantitative phenotypes that are relevant to clinical disorder and have high heritability, with at least one QTL contributing a substantial proportion (10% or more) of phenotypic variance. Several current studies of anxiety and depression are following this approach using quantitative questionnaire measures (e.g., Sham et al., 2001). With increasingly sophisticated methods of measurements using, for example, electrophysiology and magnetic resonance imaging, it may be possible to identify suitable quantitative "endophenotypes" for genetic mapping.

An alternative view is that the degree of clinical heterogeneity and etiological complexity may be such that phenotypic dissection before genetic dissection is not possible. However, even if this is not the case and suitable "endophenotypes" can be found, linkage will still lack the resolution to localize a QTL to a region smaller than about 20 Mb. Therefore, in any case, association analysis on a large scale will certainly become routine in future. It is likely that both indirect association using LD and direct association targeting entire systems of functional polymorphisms will be systematically used. DNA pooling may form an efficient tool for initial screening, followed by individual genotyping for more detailed analysis of putative positive loci.

Increasingly, quantitative and molecular genetic methodology will be applied simultaneously to patients and controls, sibpairs, twins, and so on to study the effects of genes in different environments or the impact of different therapeutic interventions in different genotypic groups. The availability of genetically informative samples and sophisticated statistical methodology, combined with the increasing level of detailed information about the human genome, promises an exciting decade for behavioral genetics.

References

Abecasis, G. R., Cardon, L. R., & Cookson, W. O. (2000). A general test of association for quantitative traits in nuclear families. *American Journal of Human Genetics, 66,* 279–292.

Abecasis, G. R., Noguchi, E., Heinzmann, A., Traherne, J. A., Bhattacharya, S., Leaves, N. I., et al. (2001). Extent and distribution of linkage disequilibrium in three genomic regions. *American Journal of Human Genetics, 68,* 191–198.

Allison, D. B., Heo, M., Kaplan, N., & Martin, E. R. (1999). Sibling-based tests of linkage and association for quantitative traits. *American Journal of Human Genetics, 64,* 1754–1763.

Allison, D. B., Neale, M. C., Zannolli, R., Schork, N. J., Amos, C. I., & Blangero, J. (1999). Testing the robustness of the likelihood ratio test in a variance component quantitative trait loci mapping procedure. *American Journal of Human Genetics, 65,* 531–544.

Almasy, L., & Blangero, J. (1998). Multipoint quantitative-trait linkage analysis in general pedigrees. *American Journal of Human Genetics, 62,* 1198–1211.

Amos, C. I. (1994). Robust variance-components approach for assessing genetic linkage in pedigrees. *American Journal of Human Genetics, 54,* 535–543.

Bader, J. S. (2001). The relative power of SNPs and haplotypes as genetic markers for association tests. *Pharmacogenomics, 2,* 11–24.

Bader, J. S., Bansal, A., & Sham, P. C. (2001). Efficient SNP-based tests of association for quantitative phenotypes using pooled DNA. *Genescreen, 1,* 143–150.

Bishop, T., & Sham, P. C. (2000). *Analysis of multifactorial disease*. London: Bios.

Bodmer, W. F. (1986). Human genetics: The molecular challenge. *Cold Spring Harbour Symposium of Qualitative Biology, 51,* 1–13.

Cardon, L. R., Smith, S. D., Fulker, D. W., Kimberling, W. J., Pennington, B., & DeFries, J. C. (1994). Quantitative trait locus for reading disability on chromosome 6. *Science, 226,* 276–279.

Collins, F. S. (1995). Positional cloning moves from perditional to traditional. *Nature Genetics, 9,* 347–350.

Curtis, D. (1997). Use of siblings as controls in case-control association studies. *Annals of Human Genetics, 61,* 319–333.

Devlin, B., & Roeder, K. (1999). Genomic control for association studies. *Biometrics 55,* 997–1004.

Dolan, C. V., Boomsma, D. I., & Neale, M. C. (1999). A simulation study of the effects of assignment of prior identity-by-descent probabilities to unselected sib pairs, in covariance-structure mapping of a quantitative trait locus. *American Journal of Human Genetics, 64,* 268–280.

Eaves, L. J., Neale, M. C. H., & Maes, H. (1996). Multivariate multipoint linkage analysis of quantitative trait loci. *Behavior Genetics, 26,* 519–525.

Elston, R. C., & Stewart, J. (1971). A general model for the analysis of pedigree data. *Human Heredity, 21,* 523–542.

Fisher, R. A. (1918). The correlation between relatives on the supposition of Mendelian inheritance. *Transactions of the Royal Society of Edinburgh, 52,* 399–433.

Fisher, S. E., Marlow, A. J., Lamb, J., Maestrini, E., Williams, D. F., Richardson, A. J., et al. (1999). A quantitative trait locus on chromosome 6p influences different aspects of developmental dyslexia. *American Journal of Human Genetics, 64,* 146–156.

Fulker, D. W., Cardon, L. R., DeFries, J. C., Kimberling, W. J., Pennington, B. F., & Smith, S. D. (1991). Multiple regression analysis of sib-pair data on reading to detect quantitative trait loci. *Reading and Writing: An Interdisciplinary Journal, 3,* 299–313.

Fulker, D. W., & Cherny, S. S. (1996). An improved multipoint sib-pair analysis of quantitative traits. *Behavior Genetics, 26,* 527–532.

Fulker, D. W., Cherny, S. S., Sham, P. C., & Hewitt, J. K. (1999). Combined linkage and association analysis for quantitative traits. *American Journal of Human Genetics, 64,* 259–267.

Gayan, J., Smith, S. D., Cherny, S. S., Cardon, L. R., Fulker, D. W., Bower, A. M., et al. (1999). Quantitative-trait locus for specific language and reading deficits on chromosome 6p. *American Journal of Human Genetics, 64,* 157–164.

Grigorenko, E. L., Wood, F. B., Meyer, M. S., Hart, L. A., Speed, W. C., Shuster, A., et al. (1997). Susceptibility loci for distinct components of developmental dyslexia on chromosomes 6 and 15. *American Journal of Human Genetics, 60*, 27–39.

Grigorenko, E. L., Wood, F. B., Meyer, M. S., & Pauls, D. L. (2000). Chromosome 6p influences on different dyslexia-related cognitive processes: Further confirmation. *American Journal of Human Genetics, 66*, 715–723.

Haldane, J. B. S., & Smith, C. A. B. (1947). A new estimate of the linkage between the genes for colour-blindness and haemophilia in man. *Annals of Eugenics, 14*, 10–31.

Haseman, J. K., & Elston, R. C. (1972). The investigation of linkage between a quantitative trait and a marker locus. *Behavior Genetics, 2*, 3–19.

Hoehe, M. R., Kopke, K., Wendek, B., Rohde, K., Flachmeier, C., Kidd, K. K., et al. (2000). Sequence variability and candidate gene analysis in complex disease: Association of mu opioid receptor gene variation with substance dependence. *Human Molecular Genetics, 9*, 2895–2908.

Holmans, P. (1993). Asymptotic properties of affected-sib-pair linkage analysis. *American Journal of Human Genetics, 52*, 362–374.

Horvath, S., & Laird, N. M. (1998). A discordant sibship test for disequilibrium and linkage: No need for parental data. *American Journal of Human Genetics, 63*, 1886–1897.

International Human Genome Sequencing Consortium. (2001). Initial sequencing and analysis of the human genome. *Nature, 409*, 860–921.

Knapp, M. (1999). The transmission/disequilibrium test and parental-genotype reconstruction: The reconstruction-combined transmission/disequilibrium test. *American Journal of Human Genetics, 64*, 861–870.

Kong, A., & Cox, N. J. (1997). Allele-sharing models: LOD scores and accurate linkage tests. *American Journal of Human Genetics, 61*, 1179–1188.

Kruglyak, L., Daly, M. J., Reeve-Daly, M. P., & Lander, L. S. (1996). Parametric and non-parametric linkage analysis: A unified multipoint approach. *American Journal of Human Genetics, 58*, 1347–1363.

Lazzeroni, L. C. (2001). A chronology of fine-scale gene mapping by linkage disequilibrium. *Statistical Methods in Medical Research, 10*, 57–76.

MacLean, C. J., Martin, R. B., Sham, P. C., Wang, H., Straub, R. E., & Kendler, K. S. (2000). The trimmed-haplotype test for linkage disequilibrium. *American Journal of Human Genetics, 66*, 1062–1075.

Martin, E. R., Monks, S. A., Warren, L. L., & Kaplan, N. L. (2000). A test for linkage and association in general pedigrees: The pedigree disequilibrium test. *American Journal of Human Genetics, 67*, 146–154.

Morris, D. W., Turic, D., Robinson, L., Duke, M., Webb, V., Easton, M., et al. (1999). Linkage disequilibrium mapping in reading disability. *American Journal of Human Genetics, 65*, A462.

Morton, N. E. (1955). Sequential tests for the detection of linkage. *American Journal of Human Genetics, 7*, 277–318.

Morton, N. E., & Collins, A. (1998). Tests and estimates of allelic association in complex inheritance. *Proceedings of the National Academy of Sciences USA, 95*, 11389–11393.

Ott, J. (1974). Estimation of the recombination fraction in human pedigrees: Efficient computation of the likelihood for human linkage studies. *American Journal of Human Genetics, 26*, 588–597.

Peltonen, L., & McKusick, V. (2001). Dissecting human disease in the postgenomic era. *Science, 291*, 1224–1229.

Penrose, L. S. (1935). The detection of autosomal linkage in data which consist of pairs of brothers and sisters of unspecified parentage. *Annals of Eugenics, 6*, 133–138.

Plomin, R., Hill, L., Craig, I., McGuffin, P., Purcell, S., Sham, P., et al. (2001). A genome-wide scan of 1842 DNA markers for allelic associations with general cognitive ability: A five-stage design using DNA pooling. *Behavior Genetics, 31*, 497–509.

Pritchard, J. K., & Rosenberg, N. A. (1999). Use of unlinked genetic markers to detect population stratification in association studies. *American Journal of Human Genetics, 65*, 220–228.

Pritchard, J. K., Stephens, M., Rosenberg, N. A., & Donnelly, P. (2000). Association mapping in structured populations. *American Journal of Human Genetics, 67*, 170–181.

Purcell, S., Cherny, S. S., Hewitt, J. K., & Sham, P. C. (2001). Optimal sibship selection for genotyping in quantitative trait locus linkage analysis. *Human Heredity, 52*, 1–13.

Rabinowitz, D. (1997). A transmission disequilibrium test for quantitative trait loci. *Human Heredity, 47*, 342–350.

Rijsdijk, F. V., Hewitt, J. K., & Sham, P. C. (2001). Analytic power calculation for QTL linkage analysis of small pedigrees. *European Journal of Human Genetics, 9*, 335–340.

Risch, N. (1990). Linkage strategies for genetically complex traits: III. The effect of marker polymorphism on analysis of affected relative pairs. *American Journal of Human Genetics, 46*, 242–253.

Risch, N., & Merikangas, K. (1996). The future of genetic studies of complex human diseases. *Science, 273*, 1516–1517.

Risch, N., & Zhang, H. (1995). Extreme discordant sib pairs for mapping quantitative trait loci in humans. *Science, 268*, 1584–1589.

Satten, G. A., Flanders, W. D., & Yang, Q. (2001). Accounting for unmeasured population substructure in case-control studies of genetic association using a novel latent-class model. *American Journal of Human Genetics, 68*, 466–477.

Schaid, D. J. (1996). General score tests for association of genetic markers with disease using cases and their parents. *Genetic Epidemiology, 13*, 423–429.

Schork, N. J. (1993). Extended multipoint identity-by-descent analysis of human quantitative traits: Efficiency, power, and modeling considerations. *American Journal of Human Genetics, 55*, 1306–1319.

Sham, P. C., & Curtis, D. (1995). An extended transmission/disequilibrium test (TDT) for multiallele marker loci. *Annals of Human Genetics, 59*, 323–336.

Sham, P. C., Cherny, S. S., Purcell, S., & Hewitt, J. K. (2000). Power of linkage versus association analysis of quantitative traits, by use of variance-components models, for sibship data. *American Journal of Human Genetics, 66*, 1616–1630.

Sham, P. C., Lin, M. W., Zhao, J. H., & Curtis, D. (2000). Power comparison of parametric and non-parametric linkage tests in small pedigrees. *American Journal of Human Genetics, 66*, 1661–1668.

Sham, P. C., & Purcell, S. (2001). Equivalence between Haseman–Elston and variance-components linkage analyses for sib pairs. *American Journal of Human Genetics, 68*, 1527–1532.

Sham, P. C., Sterne, A., Purcell, S., Cherny, S., Webster, M., Rijsdijk, F., et al. (2001). Creating a composite index of the vulnerability to anxiety and depression in a community sample of siblings. *Twin Research, 3*, 316–322.

Sham, P. C., Zhao, J. H., Cherny, S. S., & Hewitt, J. K. (2000). Variance components QTL linkage analysis of selected and non-normal samples: Conditioning on trait values. *Genetic Epidemiology, 19*, S22–28.

Sham, P. C., Zhao, J. H., & Curtis, D. (2000). The effect of marker characteristics on the power to detect linkage disequilibrium due to single or multiple ancestral mutations. *Annals of Human Genetics, 64*, 161–169.

Sinsheimer, J. S., Blangero, J., & Lange, K. (2000). Gamete competition models. *American Journal of Human Genetics, 66*, 1168–1172.

Spielman, R. S., McGinnis, R. E., & Ewens, W. J. (1993). Transmission test for linkage disequilibrium: The insulin gene region and insulin-dependent diabetes mellitus (IDDM). *American Journal of Human Genetics, 52*, 506–516.

Suarez, B. K., Rice, J. P., & Reich, T. (1978). The generalised sib-pair IBD distribution: Its use in the detection of linkage. *Annals of Human Genetics, 42*, 87–94.

Thomson, G. (1994). Identifying complex disease genes: Progress and paradigms. *Nature Genetics, 8*, 108–110.

Venter, C., Adams, M. D., Myers, E. W., Li, P. W., Mural, R. J., Sutton, G. G., et al. (2001). The sequence of the human genome. *Science, 291*, 1304–1351.

Waldman, I. D., Robinson, B. F., & Rowe, D. C. (1999). A logistic regression based extension of the TDT for continuous and categorical traits. *Annals of Human Genetics, 63*, 329–340.

Weeks, D. E., & Lange, K. (1988). The affected-pedigree-member method of linkage analysis. *American Journal of Human Genetics, 42*, 315–326.

Zhang, S., Kidd, K. K., & Zhao, H. (in press). Detecting genetic association in case-control studies using similarity-based association tests. *Statistica Sinica.*

Zhao, H. (2001). Family-based association studies. *Statistical Methods in Medical Research, 9,* 563–587.

Zhao, H., Zhang, S., Merikangas, K. R., Wildenaur, D., Sun, F., & Kidd, K. K. (2000). Transmission/disequilibrium tests for multiple tightly linked markers. *American Journal of Human Genetics, 67,* 936–946.

Zhao, J. H., Curtis, D., & Sham, P. C. (2000). Model-free analysis and permutation tests for allelic associations. *Human Heredity, 50,* 133–139.

4

Practical Barriers to Identifying Complex Trait Loci

Lon R. Cardon

Attempts to identify specific genetic variants for complex traits have not met with their original anticipated success. In the majority of cases, one of two approaches have been undertaken in these attempts: (a) candidate gene association studies, wherein particular polymorphisms, genes, or genomic regions are targeted on the basis of a prior hypothesis; and (b) positional cloning studies, in which the entire genome is screened by linkage analysis with the aim of highlighting broad tracts of the genome for further resolution by high-density genotyping and association analysis. Both of these designs originally held great promise for complex traits, following a superb track record for single-gene disorders (Collins, Guyer, & Chakravarti, 1997). Only a few years ago, it was predicted that by the present time, a large number of multifactorial trait loci would already have been detected by positional cloning, and we would now be well-entrenched in position-based candidate gene studies (Collins, 1995). Unfortunately, the data have not borne out this optimistic prediction: Despite thousands of candidate gene association studies and hundreds of costly genomewide linkage scans, the number of known, verifiable, etiologic loci for complex traits is vanishingly small (for a review, see Cardon & Bell, 2001).

The limited success of genetic studies of multifactorial traits has resulted in widespread criticism, raising questions about the complexity of the phenotypes and the feasibility of the study designs used. Weiss and Terwilliger (2000) reviewed the difficulties, highlighting the indirect nature of genotype–phenotype correlations, the likelihood of extensive inter- and intragenic variation, and almost certain genotype–environment correlations and interactions that are uniformly ignored in nearly all statistical models. These arguments, most of which appear reasonable and compelling, emphasize the clear and largely consistent outcome of complex trait genetic studies: failure to detect genuine contributory loci (see Sham, chapter 3, this volume, for a further perspective on this outcome).

I thank Gonçalo Abecasis for many helpful discussions and suggestions during the course of this work. I appreciate the careful reviews of Robert Plomin, John DeFries, and Pak Sham. This work was supported in part by the Wellcome Trust and by National Institutes of Health Grant EY-12562.

In the context of the broad focus of human genetics studies over the past decade, the outcome of "failure to detect" presents a conundrum, because family-based research has been intensely focused on minimizing false-positive outcomes at the expense of missing true findings. Nearly all association studies have moved from efficient and inexpensive case control models to inefficient and costly family-based designs that protect against spurious results due to population stratification but have greater genotyping costs and effectively throw away much of the available information (Spielman & Ewens, 1993, 1996; Spielman, McGinnis, & Ewens, 1994). The rationale for such widespread use of family-based association is not entirely clear, as the number of false-positive association results due to population stratification does not seem exceptionally high (Cardon & Bell, 2001). In addition, strict guidelines have been proposed for reporting linkage results, thereby standardizing the nomenclature to reserve the use of such terms as *significant* or *replication* for only the most unusual findings (Lander & Kruglyak, 1995; Witte, Elston, & Schork, 1996). Unfortunately, most genome scans for linkage have been too underpowered to ever achieve such statistical levels for realistic effect sizes; thus, the most obvious effect of the guidelines is to minimize reports of real findings detected at the level of a study's power. Moreover, as these guidelines include stringent penalties for screening many markers or phenotypes, they serve to inhibit analyses that might aim to explore the same phenotypic and genetic complexity highlighted by Weiss and Terwilliger (2000).

Although protection against spurious association is worthwhile and appropriate, as is establishment of clear terms for scientific discourse, their ubiquitous endorsement has almost certainly contributed to the "failure to detect" complex trait loci. Interestingly, these factors are unrelated to the phenotypic and genetic/genomic complexity underlying complex traits. That is, they are under the control of the investigator rather than the hand of nature. This is a promising outcome, indicating that there are straightforward and relatively simple means to elevate the potential for identifying specific variants: (a) revisit study designs to increase statistical power and (b) study samples of the appropriate composition and size to detect modest genetic effects. At present, it is inappropriate to say that positional cloning or candidate gene studies have performed poorly, as few studies have been conducted with reasonable *a priori* chances of success. Recent planned and ongoing linkage and association studies have been expressly designed to increase these chances (see Sham, chapter 3, this volume).

The power and study design issues just described provide obvious pathways for more careful evaluation of the etiology of complex traits. It is clear, however, that they are not the only features that hinder detection of linked or associated loci. Here I describe the effects of genotyping error and sample stratification on reducing statistical power to detect real effects. I show that even modest genotype error or stratification can cause otherwise detectable genetic effects to be seriously masked or entirely hidden in linkage and association studies.

Genotyping Error

The past decade has witnessed an explosion in genotyping technology and throughput. The first genome scan for linkage to a complex trait, occurring only 8 years ago, involved extensive marker development, repeated assays, and genetic map construction and validation by a large dedicated group (Davies et al., 1994). Now, there are commercial marker panels available for widespread use, genome scans of hundreds or thousands of samples can be completed by one or two individuals, and cost is not even a great hindrance, reaching fully burdened levels of less than U.S. $0.30 per genotype (Weber & Broman, 2001). For microsatellite applications, one of the largest hurdles has been the manual process of assigning or reviewing genotype calls (Ghosh et al., 1997; Idury & Cardon, 1997; Perlin, Lancia, & Ng, 1995), but advances in genotyping technology (e.g., capillaries) and better software are increasing the automation of the system.

Although microsatellite genome scans now can be completed efficiently and inexpensively, the crucial issue of accuracy is not fully resolved. What is a “good” or even “acceptable” rate of error in genotyping? At first glance, 1%–2% might seem quite adequate given the study of small volumes of a complex biological material (DNA) under a range of chemical reactions, human interactions, and engineering protocols. Indeed, this may approximate the rate in many published studies to date (Gordon, Leal, Heath, & Ott, 2000). The background error rates and their consequences are of obvious importance for genotype–phenotype analysis, and they also are crucial for the genotyping specialist who runs the polymerase chain reaction (PCR) and generates the data before analysis. This person must decide whether to assign or call as many genotypes as possible to maximize power by minimizing missing data, or whether to be selective in calling genotypes to minimize error at the expense of missing data and reduced power.

The effects of genotyping error have been carefully studied in the context of parametric linkage analysis and inferences of genetic map order and distance, yielding various models to facilitate error detection (Buetow, 1991; Lincoln & Lander, 1992; Shields, Collins, Buetow, & Morton, 1991). However, the effects of genotyping error on nonparametric linkage analysis, of the form commonly used in the study of complex traits, have only recently been explored. Following on the initial work of Douglas, Boehnke, and Lange (2000), my colleagues and I compared the power to detect real genetic effects under models of increasing genotype error (see Abecasis, Cherny, & Cardon, 2001). We varied the study design (affected sibling pairs [ASPs] vs. unselected siblings) and examined both discrete and quantitative traits.

The results from 1,000 simulations of 1,000 ASPs are shown in Figure 4.1 for a locus of modest effect size ($\lambda s = 1.25$). This effect corresponds to what might be expected for “typical” complex trait loci in the behavioral or common disease area. Successive descending curves represent no genotyping error, 0.5%, 1%, 2%, and 5% genotyping error for a hypothetical 10 cM chromosome map of 4-allele markers. Thus, comparison of any curve

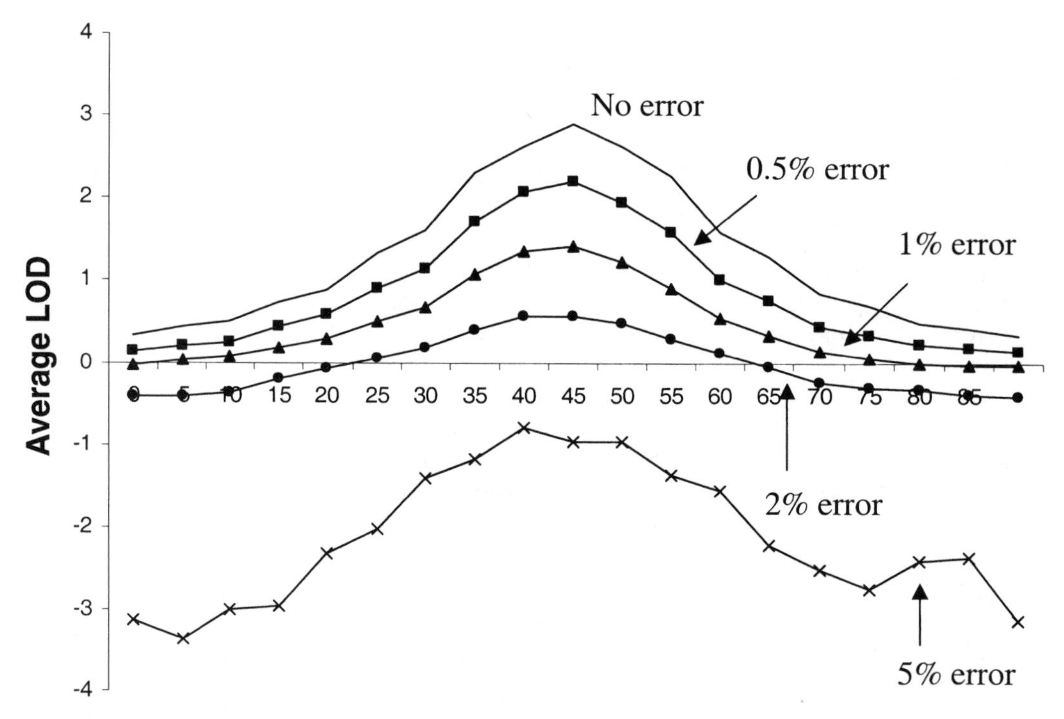

Figure 4.1. The effects of genotyping error in a linkage study of 1,000 affected sibling pairs. Ten markers with 4-equifrequent alleles were simulated at 10 cM intervals along a hypothetical chromosome. A trait locus of $\lambda_s = 1.25$ is situated in the middle (45 cM) of the chromosome (λ_s is the sibling recurrence ratio, a measure of disease familiality). The (descending) successive curves represent the average Kong and Cox (1997) LOD scores according to genotype error rates of 0%, 0.5% 1%, 2%, and 5%. Details concerning simulation conditions are provided in Abecasis, Cardon, and Cookson (2000).

with the top curve provides an indication of reduction in log of the odds (LOD) score due to genotyping error. Note that the families simulated comprise two siblings with no parental genotypes, so that all possible genotypic combinations are permissible under Mendelian transmission. The Kong and Cox (1997) procedure was used to calculate the linkage evidence shown, thereby allowing negative values.

The results in Figure 4.1 reveal some striking effects of genotyping error. In this scenario, the trait locus is positioned at 45 cM. In this study of 1,000 sibling pairs, this locus yields an average LOD score of approximately 3.0. This is slightly higher than, but similar to, the maximum levels often seen in actual genome screens of ASPs. The curves indicate that with only 0.5% error, the LOD score drops by about one third, and with 1% error, the LOD score drops by more than half. At 2% error, the real genetic locus would almost certainly escape all notice (max LOD < 0.5). Genotyping error rates of 5% or greater would lead to exclusion of the correct locus. In ASP linkage studies, small amounts of error can obliterate real results.

Figure 4.2 shows a similar plot for a quantitative trait locus that explains 2% of the phenotypic variance in unselected sibling pairs. In this case, high levels of genotyping error (e.g., 5% or greater) still yield dramatic reductions in LOD scores, but lower error rates do not have the same degree of impact as in studies of ASPs. An error rate of 2% results in a 10% loss, and error of 1% or less produces only slight decrements from the perfect-typing situation.

The differences between ASPs and quantitative trait loci (QTLs) studies highlight an interaction between study design and genotyping error. When a real gene is present in an ASP study (i.e., under the alternative hypothesis of linkage), sibpairs share more alleles than by chance alone. In this case, many sibpairs will share both alleles identical by descent (IBD) from their parents. To have this sharing state, they also will have identical genotypes. Any genotyping error in either sibling will cause their genotypes to differ and thereby reduce the number of alleles shared. This has the effect of diminishing the test statistic and reducing power. In contrast, when a real gene is present in a QTL study, sibling pairs who share either more or fewer alleles than expected by chance will contribute to the

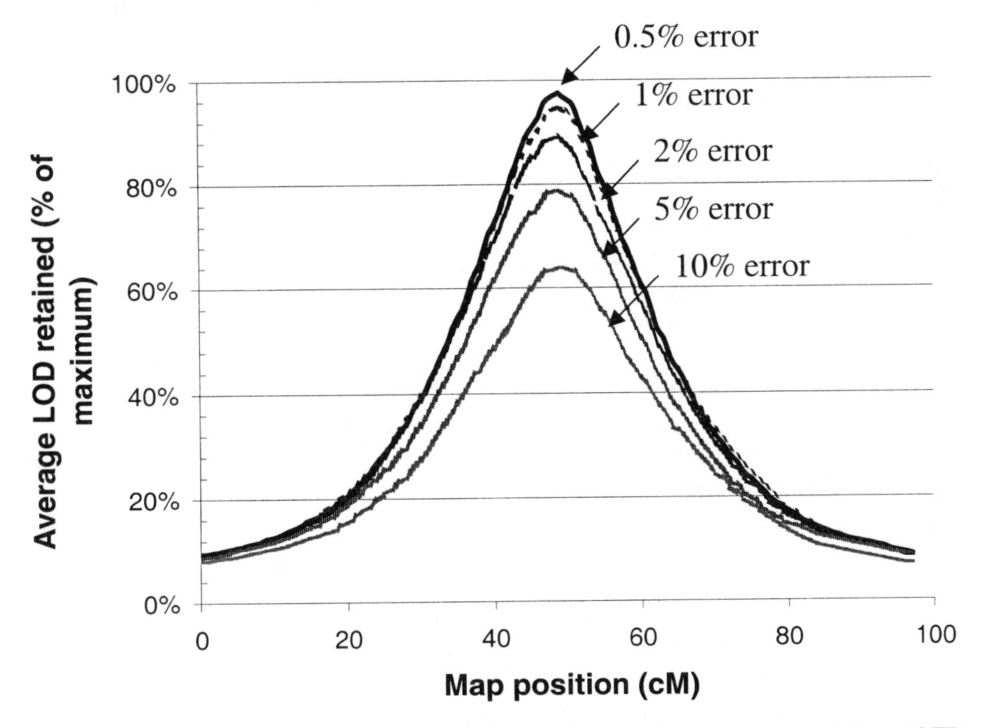

Figure 4.2. The effects of genotyping error in a quantitative trait loci (QTL) linkage study of 1,000 unselected sibling pairs. Ten markers with 4-equifrequent alleles were simulated at 10 cM intervals along a hypothetical chromosome. A QTL accounting for 2% of the phenotypic variance was positioned in the center of the chromosome. The successive curves, from top to bottom, describe the proportion of the peak average LOD (from QTDT; Abecasis, Cardon, & Cookson, 2000) for genotype error rates of 0.5%, 1%, 2%, 5%, and 10%.

test statistic. That is, sibpairs with concordant trait values (either high or low) should share more alleles IBD, whereas pairs with discordant values will share fewer alleles. Genotyping error has the effect of pushing both of these toward the midpoint. For the QTL study, the effect of error is similar to just adding noise (and reducing power) to the signal, rather than the unidirectional decrease that accompanies studies of affected pairs.

These effects are surprisingly concordant with the results from genome screens, which often show too few high LOD scores. That is, even if one conducted a genome screen for a nonheritable trait, a certain number of "significant" LOD scores would appear by chance alone, given a large sample size. As hundreds of genome screens for linkage have been conducted in the past few years alone, there should be quite a large number of high LOD scores in the literature, at least some of which are due to chance. Often, however, the results of genome screens do not yield enough significant results to even approximate randomness, producing maximum LOD scores of only about 2.0 across the entire genome. These diminished LOD scores are entirely consistent with a larger effect (either real or spurious) that has been dampened by genotyping error. It is worth emphasizing that these low statistical levels have nothing to do with complicated issues of phenotypic measurement or the genetic and biochemical pathway from etiological source to trait expression, but simply reflect a moderate amount of genotyping error on top of otherwise detectable loci.

Because genotyping error can devastate a linkage study, the problem of error detection is paramount. In the simulations described above, Abecasis et al. (2001) found that it was nearly as efficient to detect and eliminate errors as it was to detect and correct errors; that is, it is not so important that errors are corrected as it is that they at least are found and excluded from the dataset. However, detecting genotyping errors is nontrivial. The design used above, which is typical of late-onset disorders, does not provide Mendelian information to help detect genotyping error. In addition, while genetic map information can be incorporated to aid detection (Douglas et al., 2000), both multipoint and two-point analyses suffer from errors in genotyping. Moreover, many errors will remain undetected even with parents and with knowledge of the genetic map (Abecasis et al., 2001). Expanding the family size (e.g., including parents or additional sibs) does not fully alleviate the detection problem. Gordon et al. (2000) showed that only about 40%–43% of genotyping errors are detectable in sibling pairs even if parents are available. Family data are helpful, but not sufficient, to identify most genotyping errors.

Earlier in this chapter, it was suggested that genotyping specialists face two options: Liberally call all genotypes to retain power or conservatively call only the best signals to minimize error. The results clearly support the latter strategy, because the loss of power due to genotyping error can drastically exceed the loss of power due to a proportionally smaller sample size. Ironically, the "liberal" strategy actually is more conservative in terms of outcome: If all genotypes are called, a higher error rate will ensue, and with a higher error rate linkage evidence is diminished, thereby reducing false claims of linkage. Unfortunately, this also reduces true claims of linkage.

With current advances in high-throughput genotyping of single nucleotide polymorphisms and the eventual completion of the finished human genome sequence, the scope of genetic applications is poised to explode. Studies involving pooled DNA samples, genotypic correlations between treatment efficacy or adverse drug response, and genomewide case–control association analysis are under way or currently planned. All of these applications involve unprecedented levels of genotyping throughput. Unfortunately, none of these applications even have families to assist in genotyping error detection. Thus, genotype errors must be detected at the source, during the genotyping process itself, for these studies to be fully appreciated. One possibility is to conduct all assays in duplicate, thereby providing a means to carefully scrutinize experimental procedures and minimize random variation. Still, even this costly endeavor is not certain to detect all, or even most errors, as many may be reproducible because of experimental conditions (e.g., primer design, DNA preparation, PCR protocol). Thus, it is incumbent on bench scientists and technology providers to focus on developing new, precise measures of genotype accuracy. By the time the resulting calls are analyzed in the context of the full study, it is too late to see, let alone solve, the problems.

Population Stratification

In a simple case–control study of allelic association, the primary objective is to identify specific alleles that have different frequencies among individuals with different traits (i.e., disease vs. healthy). Similarly, for quantitative measures, the objective is to identify alleles that delineate mean differences in trait scores among a sample of individuals. It is obvious, however, that differences in allele frequencies between groups do not occur exclusively because of differences in disease or trait susceptibility. Different groups of individuals have different allele frequencies because of differences in ancestral histories; mating patterns; migration trends; and a host of other factors, including simple sampling variability (see Cavalli-Sforza, Menozzi, & Piazza, 1994). Whatever their origin, when allele frequency differences are coupled with differences in traits, evidence for association may be obtained. When this occurs and the allele is thought to be relevant to the trait, the finding is taken as evidence for genuine trait association. When this occurs and the allele is thought to reflect some other ethnic, geographic, or sampling issue, the finding (with respect to the trait) is referred to as "spurious" due to population stratification.

Fear of spurious results arising from population stratification has had a demonstrable and ongoing influence on human genetics studies of complex traits over the past decade, driving research away from population-based cohorts to family-based models (Cardon & Bell, 2001). The move toward family-based controls was spurred by the development of a suite of association designs and statistical methods that use the genetic composition of the family (within which there is no stratification) as the background for comparing allele frequencies, instead of comparing frequencies across unrelated individuals. The most popular of these approaches, the transmission disequilibrium test (TDT), minimally involves parents and one offspring. Only heterozygous parents are informative in this design, so whenever parents are homozygous, the genotype data are effectively useless for association testing. A number of quantitative-trait extensions of the TDT have been developed, broadening the discrete models by explicitly partitioning linkage and association effects in random samples (for a review, see Abecasis, Cookson, & Cardon, 2000) and in phenotypically selected sib pairs (Allison, 1997; Cardon, 2000).

It has been clearly shown that the TDT and its variants are unbiased in the presence of population stratification; that is, when no disease or trait locus is present, population stratification will not yield a spurious outcome (see also Sham, chapter 3, this volume). Here I describe a variation on this issue. I am interested in the situation in which stratification inflates false-positive rates but where a gene is also associated with the trait. That is, I explore the situation in which association and stratification occur simultaneously. What are the effects on false-positive rates and power to detect association in this case?

As an extreme example, consider an allele that causes a disease with complete penetrance in all populations. Although one assesses a single collection of individuals, in a subset of the sample, the allele is common, and thus so is the disease. In another subset the allele is rare, and thus, so is the disease. As expected, we now have within-population association, in the form of the original causal effect; however, we also have between-populations association, in the form of population stratification owing to the high versus low differences in allele and trait frequencies. In this example, these effects are indistinguishable because we only study a single sample of individuals that happens to exhibit extensive variability. The ultimate outcome of this study is predictable; the between and within effects combine to yield even higher association evidence than that obtainable from either of the subsamples alone or with a homogeneous group. Admixture mapping (McKeigue, 1997, 1998) capitalizes on precisely these combined effects.

As a less extreme and more realistic example, consider an oligogenic locus, which is again uniformly associated with a quantitative trait. Instead of complete penetrance and causality, the allele accounts for only a small proportion of variance, and trait scores are additionally influenced by other unknown genetic and environmental factors. Again, two population subsets emerge, one with high trait allele frequency and the other with low trait allele frequency. In this case, however, because of the small proportional effect of this locus, the sample with the high allele frequency is not accompanied by a concomitant increase in trait scores. Rather, the trait scores actually go in the opposite direction and are decreased in this group, owing to some other uncontrolled factor, for example, environmental effects, other loci, epistasis, Gene × Environment (G × E) interaction, and so on. Similarly, the low-frequency group does not exhibit decreased trait means, but instead, the trait scores increase. In this case, within each subsample, the allele remains associated with the trait. In addition, between-groups effects still yield association information, because there are trait and allele frequency differences between samples. However, the within- and between-groups effects are in opposite direction. What is the effect of analyzing the overall sample in this case? What is the expected outcome when both population stratification and genuine allelic association are present but not in phase?

Simple Model of Stratification Effects

This question can be addressed more systematically by drawing on some simple biometrical principles (Falconer, 1981). Consider two sample strata ($i = 1, 2$) with equal sample sizes, and a QTL, Q, with allele frequencies p_i and $q_i = 1 - p_i$. Assume that mean differences between strata are not due to any one marker or allele, but from a constant background mean factor, μ_i in each stratum, that results from other genetic loci, environmental effects, G $\times$ E interaction, epistasis, and all other uncontrolled factors. For simplicity, consider only the symmetric case, where the allele frequencies and background means are in exactly opposite direction or sign; that is, $p_2 = q_1$, $q_2 = p_1$, and $\mu_2 = -\mu_1$.

Let $W_F = \sigma_p^2 + \sigma_e^2$: the within-sample background variance comprising polygenic (p) and environmental (e) components; $W_Q = 2p_1q_1a^2 = 2p_2q_2a^2$: the within-sample QTL variance with additive genetic effect, a, under the simplifying assumption of no dominance variance; $B_Q = [a(p_1 - q_1) - a(p_2 - q_2)]^2/4 = (2ap_1 - 2aq_1)^2/4 = a^2(p_1 - q_1)^2 = a^2(p_2 - q_2)^2$: the between-samples variance due to the QTL; and $B_S = (\mu_1 - \mu_2)^2/4 = (2\mu_1)^2/4 = \mu_1^2 = \mu_2^2$: the between-samples variance due to extraneous (i.e., non-QTL) stratification. Then, because the within-sample variances are equal in each stratum, the total variance of the (measured) combined sample is $V_T = W_F + W_Q + B_Q + B_S$. These variance components may be easily rearranged to consider two sources of variation, within sample, $V_W = W_Q + W_F$, and between samples, $V_B = B_Q + B_S = \mu_1^2 + a^2(p_1 - q_1)^2$. Similarly, the total QTL variance, V_Q, can be partitioned into that arising from within-sample effects and that due to group mean differences, $V_Q = W_Q + B_Q = 2p_1q_1a^2 + a^2(p_1 - q_1)^2 = a^2(1 - 2p_1q_1) = a^2(1 - 2p_2q_2)$.

Stratification Simulations

To explore the pattern of stratification effects in this simple model, I conducted a simulation study, varying the values of the parameters: p_1, σ_p^2, B_S, V_Q, V_T, and deriving the other necessary quantities as shown above, and additionally, as $\sigma_e^2 = V_T - (V_Q + B_S + \sigma_p^2)$, $|\mu_1| = |\mu_2| = \sqrt{B_S}$, and $a = \sqrt{V_Q/(1 - 2p_1q_1)} = \sqrt{V_Q/(1 - 2p_2q_2)}$. Genotypes and phenotypes were simulated for 100 sib pairs, and average likelihood ratio chi-square statistics were computed from 1,000 runs of each parameter combination. In all cases, two subsamples with means $\mu_i + a(p_i - q_i)$ and variances V_W were simulated according to the selected parameter values. They were then combined as one sample for the analyses. The simulation procedures and association tests were as described by Abecasis, Cardon, and Cookson (2000).

The results of the simulation analyses are presented in Figure 4.3. For these outcomes, unless noted otherwise, a QTL of 5% total heritability was modeled ($V_Q/V_T = 0.05$), with additional polygenic influences accounting for 50% of the phenotypic variation ($\sigma_p^2 = 0.50 \cdot V_T$). The trait variance was arbitrarily set at 100. Population stratification accounted for 5% of the total variance ($B_S/V_T = 0.05$) unless noted otherwise.

The three curves shown in Figure 4.3 represent different sampling scenarios. First, consider the case of population stratification only (dashed line with filled triangles). In this case, no QTL effects were included, so

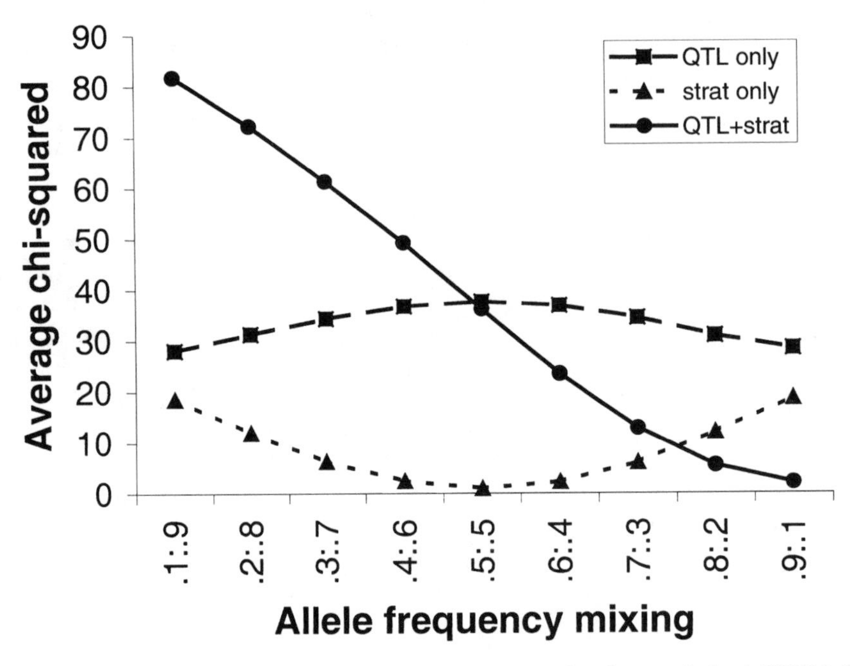

Figure 4.3. The effects of sample mixing on quantitative trait loci (QTL) detection in the presence of stratification. The effects of population stratification accounting for 5% of the phenotypic variance are shown as filled triangles (strat only). The effects of a QTL accounting for 5% of the phenotypic variance are shown in filled squares (QTL only). The combined effects of a 5% QTL and 5% population stratification are shown in the filled circles (QTL+strat). Allele frequency mixing, shown on the x-axis, refers to the allele frequencies in two subsamples of the overall population analyzed, Sample 1:Sample 2. The average chi-square is the result of 1,000 simulations using the QTDT program (Abecasis, Cardon, & Cookson, 2000).

any evidence for association was "spurious," arising from the allele frequency and trait mean differences. The shape of this curve is fully expected for stratification, having a minimum (i.e., chi-square of 1.0 for 1 *df*) when the allele frequencies are identical (.50:.50) but quickly providing spurious evidence as the subsample frequencies differ. The curve is symmetric because of the model imposed; that is, the allele frequencies are identical but for the opposite alleles, and the sample means are opposite in sign.

Second, consider the case of a QTL effect only (long dashed line with filled squares). In this case, no stratification was modeled ($\mu_1 = \mu_2$), so all evidence for association derives from genuine QTL effects. This pattern also is expected, showing maximal power when the subsamples are identical in composition and the allele frequencies are equal. Note that, despite a constant QTL effect (5%) over the entire range of allele frequencies, the average chi-square decreases as the subsample frequencies become more disparate. This pattern arises because, while the total QTL variance, V_Q, is constant over all frequencies, its within- and between-samples components differ. These proportional differences are shown in Figure 4.4. At 50:50 allele frequencies, all association information comes from the within-sample effects; that is, $B_Q = 0$ and $V_Q = W_Q$. However, the between-groups component of variance increases with frequency differences and, accordingly, the within-component must decrease to keep the sum constant. As the frequency mixing tends to 1.0:0.0, the within-sample effects will disappear, and all information will come from between-samples effects. The reduction in average chi-square apparent in Figure 4.3 therefore derives from the greater proportion of between-samples QTL variance, for which there is less power to detect than within-sample variance.

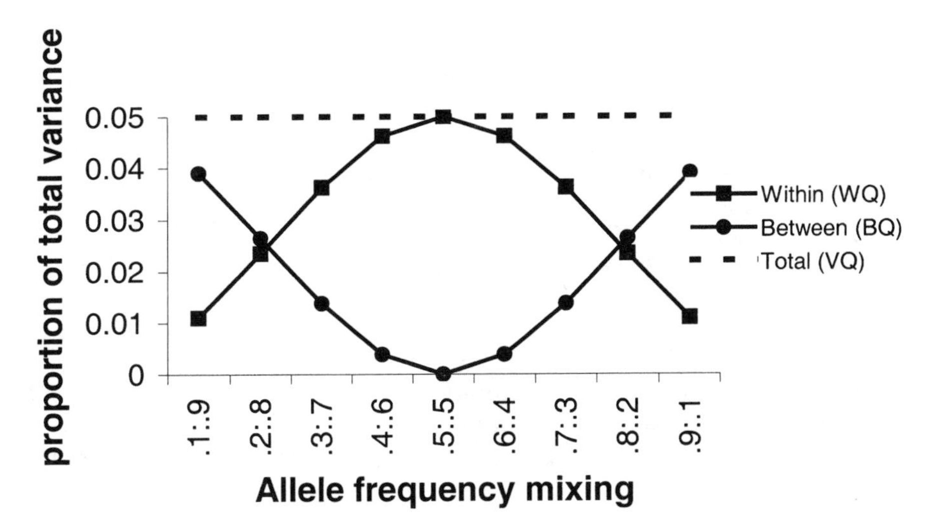

Figure 4.4. Variation in quantitative trait loci (QTL) variance as a function of allele frequency. Variance attributable to within-group (squares) and between-groups (circles) effects are partitioned from a constant level of QTL variance (dashed line).

Finally, the joint effects of a real QTL, coupled with "spurious" population stratification, are shown in the continuous line with solid circles in Figure 4.3. This trend differs dramatically from that for the QTL or stratification alone. At the leftmost end, the QTL and stratification effects are in the same orientation, so that the sample with the higher QTL allele frequency also has the higher mean background effect. Clearly, these two features do not add linearly to the test statistic but in fact offer a substantial increase in power over the joint effects of the individual data. This power increase (for all values to the left of .50:.50 mixing) is one of the main attractions of admixture mapping. At 50:50 mixing, stratification is absent, so all information derives from the QTL, as expected. The most interesting features occur at mixing values to the right of .50:.50. In these cases, the sample with the higher allele frequency has the lower background mean effect. These two pieces of information tend to cancel, rapidly driving down the power to detect the genuine underlying QTL. In these situations, there is a real QTL association, plus real population stratification, yet the QTL goes undetected.

The potential for background factors to mask QTLs has been observed in murine models of behavioral traits (Mott, Talbot, Turri, Collins, & Flint, 2000; Talbot et al., 1999), although it is not clear how frequently the phenomenon might occur in humans. Still, these patterns indicate that population stratification does not exclusively lead to false positives; it can also lead to missed gene identification. In many cases, the latter will be more damaging than the former, because false positives are often at least followed by additional replication studies, which can highlight erroneous conclusions, whereas false negatives may not be pursued at all, thereby leaving important etiological loci undetected.

I noted above that family-based association designs such as the TDT, developed to minimize false positives, were inefficient in their use of genotypic information and thus may have contributed to underdetection of complex trait loci due to reduced power. This point is described further by Sham in chapter 3, this volume. Ironically, the results in Figure 4.3 suggest that TDT is useful for the opposite reason behind its popularity: prevention of false-negative findings. Still, this does not seem an optimal study design because of its inherent inefficiency. Recent theoretical developments in the use of unlinked anonymous markers distributed across the genome to detect and correct for population stratification offer promising tools that do not require family-based samples but instead may be used in population-based cohorts (Bacanu, Devlin, & Roeder, 2000; Devlin & Roeder, 1999; Pritchard & Rosenberg, 1999; Pritchard, Stephens, & Donnelly, 2000). By genotyping a relatively small number of additional markers, the benefits of large sample cohorts may be realized without sacrificing power or suffering from the bidirectional effects of population stratification.

Conclusion

Although previous linkage and association studies of multifactorial human characters have detected relatively few bona fide complex trait loci, future studies that use improved genotyping accuracy and better methods for detecting and controlling for population stratification should yield positive results. This chapter focused on two factors, genotyping error and population stratification, because they can be controlled by the experimenter. This lies in contrast to the fundamentally unmanageable phenotypic and genetic complexity inherent in multifactorial traits. Genotyping error rates can be assessed and addressed explicitly, with full knowledge of the implications on power. The design of new studies can take this into account when determining appropriate sample sizes. Similarly, there are interesting methods for detecting and controlling population stratification in association studies. Family-based controls offer one mechanism, but they are not the most efficient in their use of genotyping data. Genomic control, using background markers in the genome, offers another means to detect stratification. This approach carries an advantage of efficiency and flexibility, as it can be incorporated into almost any type of sampling strategy in which genotypes are assessed at the level of the individual (i.e., not pooled).

Advances in genotyping technology, completion of the human genome sequence, development of the "-omics" disciplines (e.g., proteomics, metabolomics), and the emergence of very large population-based cohorts provide novel and exciting tools for disentangling the etiologies of human characters. However, better tools and larger samples offer no substitute for poor study design and execution. Multifactorial traits, and behavioral traits in particular, are already difficult to study, owing to several layers of phenotypic and biological complexity. Thus far, factors within our control, such as genotyping error and population stratification, have provided yet another layer of complexity that has dampened our ability to isolate specific trait loci. Controlling these practical issues is a mandatory first step to progress into the postgenomic era.

References

Abecasis, G. R., Cardon, L. R., & Cookson, W. O. (2000). A general test of association for quantitative traits in nuclear families. *American Journal of Human Genetics, 66*, 279–292.

Abecasis, G. R., Cherny, S. S., & Cardon, L. R. (2001). The effect of genotype error on family-based analysis of quantitative traits. *European Journal of Human Genetics, 9*, 130–134.

Abecasis, G. R., Cookson, W. O., & Cardon, L. R. (2000). Pedigree tests of transmission disequilibrium. *European Journal of Human Genetics, 8*, 545–551.

Allison, D. B. (1997). Transmission-disequilibrium tests for quantitative traits. *American Journal of Human Genetics, 60*, 676–690.

Bacanu, S. A., Devlin, B., & Roeder, K. (2000). The power of genomic control. *American Journal of Human Genetics, 66*, 1933–1944.

Buetow, K. H. (1991). Influence of aberrant observations on high-resolution linkage analysis outcomes. *American Journal of Human Genetics, 49*, 985–994.

Cardon, L. R. (2000). A family-based regression model of linkage disequilibrium for quantitative traits. *Human Heredity, 50*, 350–358.

Cardon, L. R., & Bell, J. I. (2001). Association study designs for complex traits. *Nature Reviews Genetics, 2*, 91–99.

Cavalli-Sforza, L. L., Menozzi, P., & Piazza, A. (1994). *History and geography of human genes*. Princeton, NJ: Princeton University Press.

Collins, F. (1995). Positional cloning moves from perditional to traditional. *Nature Genetics, 9*, 347–350.

Collins, F. S., Guyer, M. S., & Chakravarti, A. (1997). Variations on a theme: Cataloging human DNA sequence variation. *Science, 278*, 1580–1581.

Davies, J. L., Kawaguchi, Y., Bennett, S. T., Copeman, J. B., Cordell, H. J., Pritchard, L. E., et al. (1994). A genome-wide search for human Type 1 diabetes susceptibility genes. *Nature, 371*, 130–136.

Devlin, B., & Roeder, K. (1999). Genomic control for association studies. *Biometrics, 55*, 997–1004.

Douglas, J. A., Boehnke, M., & Lange, K. (2000). A multipoint method for detecting genotyping errors and mutations in sibling-pair linkage data. *American Journal of Human Genetics, 66*, 1287–1297.

Falconer, D. S. (1981). *Introduction to quantitative genetics*. Harlow, England: Longman.

Ghosh, S., Karanjawala, Z. E., Hauser, E. R., Ally, D., Knapp, J. I., Rayman, J. B., et al. (1997). Methods for precise sizing, automated binning of alleles, and reduction of error rates in large-scale genotyping using fluorescently labeled dinucleotide markers: FUSION (Finland–U.S. Investigation of NIDDM Genetics) Study Group. *Genome Research, 7*, 165–178.

Gordon, D., Leal, S. M., Heath, S. C., & Ott, J. (2000). An analytic solution to single nucleotide polymorphism error-detection rates in nuclear families: Implications for study design. *Pacific Symposium on Biocomputing*, 663–674.

Idury, R. M., & Cardon, L. R. (1997). A simple method for automated allele binning in microsatellite markers. *Genome Research, 7*, 1104–1109.

Kong, A., & Cox, N. J. (1997). Allele-sharing models: LOD scores and accurate linkage tests. *American Journal of Human Genetics, 61*, 1179–1188.

Lander, E., & Kruglyak, L. (1995). Genetic dissection of complex traits: Guidelines for interpreting and reporting linkage results. *Nature Genetics, 11*, 241–247.

Lincoln, S. E., & Lander, E. S. (1992). Systematic detection of errors in genetic linkage data. *Genomics, 14*, 604–610.

McKeigue, P. M. (1997). Mapping genes underlying ethnic differences in disease risk by linkage disequilibrium in recently admixed populations. *American Journal of Human Genetics, 60*, 188–196.

McKeigue, P. M. (1998). Mapping genes that underlie ethnic differences in disease risk: Methods for detecting linkage in admixed populations, by conditioning on parental admixture. *American Journal of Human Genetics, 63*, 241–251.

Mott, R., Talbot, C. J., Turri, M. G., Collins, A. C., & Flint, J. (2000). A method for fine mapping quantitative trait loci in outbred animal stocks. *Proceedings of the National Academy of Sciences USA, 97*(23), 12649–12654.

Perlin, M. W., Lancia, G., & Ng, S. K. (1995). Toward fully automated genotyping: Genotyping microsatellite markers by deconvolution. *American Journal of Human Genetics, 57*, 1199–1210.

Pritchard, J. K., & Rosenberg, N. A. (1999). Use of unlinked genetic markers to detect population stratification in association studies. *American Journal of Human Genetics, 65*, 220–228.

Pritchard, J. K., Stephens, M., & Donnelly, P. (2000). Inference of population structure using multilocus genotype data. *Genetics, 155*, 945–959.

Shields, D. C., Collins, A., Buetow, K. H., & Morton, N. E. (1991). Error filtration, interference, and the human linkage map. *Proceedings of the National Academy of Sciences USA, 88*(15), 6501–6505.

Spielman, R. S., & Ewens, W. J. (1993). Transmission disequilibrium test (TDT) for linkage and linkage disequilibrium between disease and marker. *American Journal of Human Genetics, 53*(3 SS), 863.

Spielman, R. S., & Ewens, W. J. (1996). The TDT and other family-based tests for linkage disequilibrium and association. *American Journal of Human Genetics, 59*, 983–989.

Spielman, R. S., McGinnis, R. E., & Ewens, W. J. (1994). The transmission/disequilibrium test detects cosegregation and linkage. *American Journal of Human Genetics, 54*, 559–560.

Talbot, C. J., Nicod, A., Cherny, S. S., Fulker, D. W., Collins, A. C., & Flint, J. (1999). High-resolution mapping of quantitative trait loci in outbred mice. *Nature Genetics, 21*, 305–308.

Weber, J. L., & Broman, K. W. (2001). Genotyping for human whole-genome scans: Past, present, and future. *Advanced Genetics, 42*, 77–96.

Weiss, K. M., & Terwilliger, J. D. (2000). How many diseases does it take to map a gene with SNPs? *Nature Genetics, 26*, 151–157.

Witte, J. S., Elston, R. C., & Schork, N. J. (1996). Genetic dissection of complex traits. *Nature Genetics, 12*, 355–356.

5

Assessing Genotype–Environment Interactions and Correlations in the Postgenomic Era

David C. Rowe

The goal of this chapter is to consider how knowledge of molecular genetics can advance our understanding of GE interactions and correlations. G stands for *genotype* and E for the *environment*. The term *interaction* can be used in many ways to refer to phenomena that can be quite dissimilar. Interaction is sometimes used to mean how two elements combine to produce something completely new; for instance, hydrogen and oxygen combine to form water, which has properties completely different from a gas (Kagan, 1984, p. 10). In this chapter, another meaning of interaction is used: a statistical sense in which phenotypic mean differences can be divided among main effects and interactions (Plomin, DeFries, McClearn, & McGuffin, 2000, pp. 314–317). For example, if one type of plant grows twice as fast as another on good soil but the slower growing plant does better on poor soil, statistically, this is a GE interaction. To know how well the plant will grow, knowledge is needed about both the plant type and growing conditions.

A genotype to environment (GE) active correlation refers to the tendency of organisms with particular genotypes to seek out, create, and thrive in particular environments. Nonhuman animals show GE correlations. The beaver drastically modifies its environment, cutting down trees and building dams. Yet this environment-modifying behavior is instinctive. Among people, individual differences in characteristics lead them to seek out and prefer different environments. A child with good chess skills gravitates toward the chess club; one with natural assertiveness and strength toward team sports. The chapter's thesis is that, although not without some shortcomings, the use of molecular genetic markers should enhance our understanding of various GE interactions and correlations.

GE Interactions

Dramatic cases of GE interactions can be found in medicine. The fava bean, which is perfectly harmless to most Mediterranean people, is nonetheless a poison to some. Those affected individuals carry an X-linked susceptibility genotype that causes symptoms after eating fava beans, ranging from headaches, dizziness, and nausea to severe abdominal pain and fever, and even to death from hemolytic anemia (i.e., an inability of the red blood cells to carry oxygen). In a more controversial case, a drug against high blood pressure is in a clinical trial for African Americans, who may benefit from it, although it offers no particular benefit to European-descended White Americans. This drug may be a harbinger of future medicines that target particular genotypes, that is, genetic markers detected through molecular genetic technologies. These so-called "designer drugs" would be taking advantage of GE interactions.

In behavioral science, a GE interaction occurs when specific genotypes react differently to the environment than other genotypes do; they show a greater or lesser environmental sensitivity. In a classical analysis of variance, both genotypes—for example those of inbred species of animals—and environments are varied and the resulting phenotypic outcomes are examined. This experimental design can be illustrated by one classic study of GE interaction: a study of maze dull and bright rats (Cooper & Zubek, 1958).

In this study, maze bright rats were selected through a breeding program to perform well in a maze learning task; maze dull rats were bred to perform poorly. Thus, these rats were born with different genetic predispositions attributable to their divergent breeding histories. An interaction effect was shown by placing them into different environments. The enriched environment offered playthings appealing to rats and room for running. A normal condition was ordinary laboratory rearing (a third condition in the experiment, restricted rearing, is not shown in Figure 5.1).

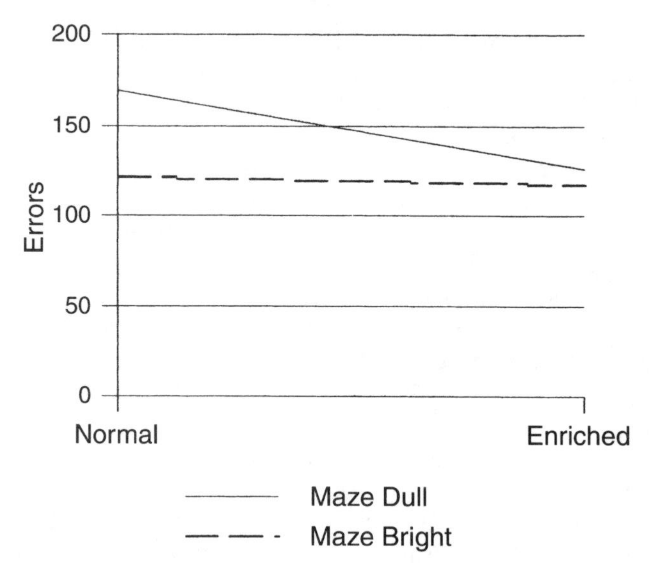

Figure 5.1. Mean learning errors in maze dull and maze bright rats by environmental enrichment.

In Figure 5.1, maze bright rats made fewer learning errors when reared in both normal and enriched environments; maze dull rats learned well under only enriched rearing. This effect constitutes an interaction because the effect of rearing environment depends on the particular line of inbred rats, that is, on their genotypic differences.

The Difficulty of Detecting Interactions

Interaction effects are notoriously difficult to detect because of their low statistical power (McClelland & Judd, 1993). This limitation can be understood intuitively by examining Figure 5.2, in which the lines drawn to connect group means are parallel. These parallel lines show a purely additive effect because the best performing strain's learning mean score is a sum of an effect of being maze bright plus an effect of being reared in an enriched environment. Moving the lines to be nonparallel creates an interaction effect (i.e., Figure 5.1). However, a statistical test must be able to detect a departure of a line from parallelism to find an interaction effect, whereas to find a main effect, it is only necessary that maze bright rats do somewhat better than maze dull rats and that enriched rearing be beneficial. With regard to statistical power, the standard error of the difference between two means (i.e., a test of a main effect) is always less than the standard error of the differences between two differences (i.e., a test of interaction). In behavioral science, one ubiquitous result is that main effects have proved more robust to replication than interactions have. Nonetheless, GE interactions have been detected in some behavioral genetic (BG) studies.

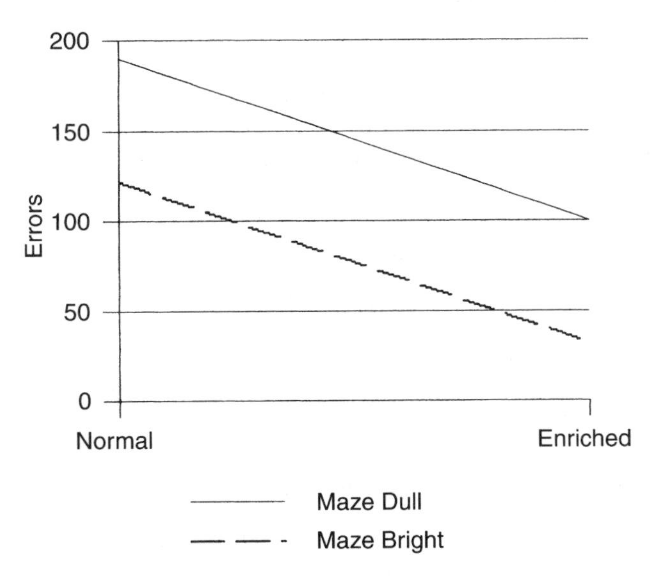

Figure 5.2. Hypothetical outcome of maze experiment with only main effects.

Detection of GE Interactions in BG Studies

GE interactions can be detected in many BG research designs, and a complete survey of these approaches is beyond the scope of this chapter. Instead, two examples are presented that yielded statistically significant interaction effects that can be interpreted. In the National Longitudinal Study of Adolescent Health ("Add Health Study"),[1] a study of verbal intelligence, my colleagues and I (Rowe, Jacobson, & Van den Oord, 1999) took advantage of different kinship levels, from identical twins to biologically unrelated siblings. The other example is from a full adoption design applied to aggression in adolescence.

In our study of verbal IQ, we predicted an interaction between socioeconomic status (i.e., parental educational level) and the heritability of verbal IQ. We expected that in economically more disadvantaged families, genetic influence on verbal IQ would be less and family environmental influences would be greater. In a BG study, these influences can be expressed as variance components. Heritability is the genetic variance divided by the phenotypic variance ($h^2 = V_G/V_P$). Shared environmental influences that operate to make siblings alike behaviorally are estimated as the shared environmental variance divided by the phenotypic variance ($c^2 = V_S/V_P$). In our study, we found that estimates of h^2 declined as social class decreased and estimates of c^2 became greater.

A DeFries–Fulker regression analysis that treated socioeconomic status as a continuous variable was used in the analysis of our data (Rodgers, Rowe, & Li, 1994). The support for our findings, however, can be illustrated by grouping kinship pairs according to their level of parental education: More than a high school education versus a high school education or less. Low-related kinship pairs combined half-siblings, cousins (who were in the unusual situation of living in the same household), and unrelated siblings. Moderately related pairs included fraternal twins and full siblings; high-related pairs included identical twins only. The verbal IQ correlations of each kinship group appear in Table 5.1. The level of parental education partly determines these kinship correlations. In low-related pairs, siblings with less well-educated parents correlated more strongly than those with better-educated parents. The former association constitutes a stronger shared environmental effect among the offspring of less-educated parents. In the other two categories, a stronger rearing effect is in a smaller difference between the kinship correlations; that is, if the identical twin correlation equaled that of the moderately related kinship groups, then shared environment is the main variance component. This difference is smaller with poorly educated parents (.23) than with well-educated parents (.38). Heritability was estimated at 74% in families

[1]The National Longitudinal Study of Adolescent Health ("Add Health Study") was designed by J. Richard Udry and Peter Bearman and funded by Grant P01-HD31921 from the National Institute of Child Health and Human Development. Datasets can be obtained by contacting the Carolina Population Center, 123 West Franklin St., Chapel Hill, NC 27516-3997; addhealth@unc.edu.

Table 5.1. Sibling Correlations by Level of Parental Education

Variable	Low education		High education	
	No. of pairs	*r*	No. of pairs	*r*
High relatedness	52	.55*	124	.75*
Moderate relatedness	424	.33*	718	.37*
Low relatedness	277	.32*	314	.10†

†$p < .10$. *$p < .001$.

with more-educated parents and 24% in families of less-educated parents.

We interpreted these results as indicating that among the offspring of poorly educated parents, some *but not all* family environments must be harmful to intellectual growth. Harmful family environments may include abusive and unconcerned parents, as well as other psychosocial and physical (i.e., lead exposure) influences. Such influences may produce a shared environmental effect of a greater magnitude. In families with unconcerned but better-educated parents, bright children may find other opportunities with which to accelerate their own intellectual growth. If their parents do not push them toward achievement, they may find equal intellectual challenges in schools or peer groups. Brighter children also may extract information from their environments more rapidly, handle complex information better, and expose themselves to information at a greater rate than less-able children do (Carroll, 1997; Gottfredson, 1997). Although interpretations of our finding are possible, they illustrate a gain from having an environmental variable in a BG study.

The next example is a BG study of aggression, an adoption study by Cadoret and his colleagues (Cadoret, Yates, Troughton, Woodword, & Stewart, 1995). Biological parents of adopted-away children were assumed to determine a genetic predisposition toward aggression if the adoptees' biological parents had antisocial personality disorder or an alcohol abuse history. If the adoptive child's biological parents were free of psychiatric disorder, the child was placed in the control group. All of the adoptive children were separated from their biological parents at birth and then placed with an adoptive family—a family that constituted the environment of their upbringing. An environmental risk score was created from a count of problems that existed in the adoptive families; that is, adoptive parents who were divorced, separated, or had various psychiatric disorders were assumed to create a more adverse environment than those adoptive parents with fewer problems. As adolescents, the adoptive children who had a biological predisposition had a greater sensitivity to the adoptive environment than did those children without a predisposition. To state this finding another way, the increase in aggressivity with family–environmental risk was greater in criminally predisposed adoptive children than in the control adoptive children.

Using Measured Genotypes to Detect Interactions

In the two BG studies reviewed above, no use was made of measured genotypes. By measured genotype, I refer to polymorphisms assessed at the level of the DNA sequence. The media has informed nearly everyone that genes are composed of a series of DNA letters (i.e., chemical bases of adenine, A; thymine, T; cytosine, C; and guanine, G) that mainly code for proteins. A short segment of a gene might read: TTA CCA GAC ATT GAC GGG GGG CGT TTA CAT (spaces shown for visual clarity).

Genetic polymorphisms are changes in the letter sequence. A single nucleotide polymorphism (SNP) occurs when a single letter is substituted for another. Thus, in some chromosomes, a G might replace the T as the first letter in the sequence. This change would be a SNP. Other changes involve insertions or deletions of DNA letters that alter the length of a piece of DNA. For instance, in some people, the second series of three Gs in "GGG GGG" may be absent. Many such changes in DNA sequence can have physiological effects, but describing them is best left to a textbook in molecular biology.

Such measured genotypes offer several advantages for detecting GE interactions. One advantage is that any person can be used in a study of GE interaction, not only adoptive children or another special kinship group, which can be expensive to recruit into a study. A second advantage is that the association between genotype and phenotype within a person is potentially stronger than that between a phenotype on one kin and that on another. For instance, in Cadoret et al.'s (1995) adoption study, a child's genetic risk was estimated from his or her biological parent's phenotype. However, because the parent–child coefficient of genetic relatedness is 0.50, not every offspring of a parent with antisocial personality disorder would inherit genes conferring a high risk of disorder. Thus, by this method, some children are bound to be classified incorrectly. Now, if we use a measured genotype to indicate risk, we could know whether a child possesses the high-risk genes. Indeed, some children of phenotypically normal parents may be classified as "at risk" on the basis of their own genotypes. A third advantage is that a measured genotype may help to establish a direction of causation between variables.

What would a measured GE interaction look like? Madrid, MacMurray, Lee, Anderson, & Comings (2001) reported one such interaction in adults genotyped for a polymorphism in the DRD2 gene. This polymorphism has a controversial history with regard to its association with alcoholism (Blum et al., 1995; Gelernter, Goldman, & Risch, 1993). I am not going to detail this history here. Rather, Madrid's finding is presented to illustrate the GE interaction concept. Their environmental risk variable was a questionnaire measure of socioeconomic stress. Their outcome was an alcoholism screening test that classified people from having a low tendency toward alcoholism to being alcoholic. To illustrate the form of the interaction, findings from the two common DRD2 genotypes, A_1A_2 and A_2A_2, are presented in Figure 5.3. The A_1A_2 heterozygote disposed toward

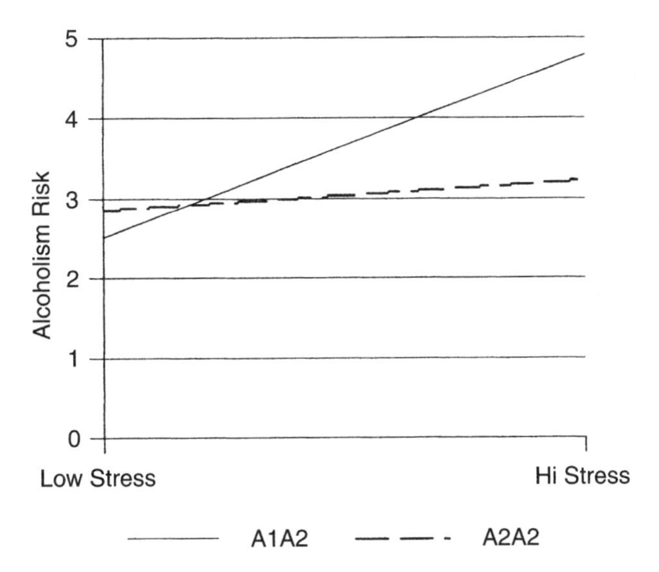

Figure 5.3. Mean alcoholism risk by genotype and environmental stress.

alcoholism, but only when socioeconomic stress was present. Without these stressors of poverty, alcoholism risk was always low.

Direction of Causation in GE Interaction

This section is an important digression from my main theme. One advantage of molecular genetic markers, in my view, is that a direction of causation is most plausibly from a marker to a phenotype and not the reverse. Consider the causal status of the DRD2 genotype. In this design, it is known that socioeconomic stress did not cause the gene sequence difference that created a Taq I restriction site (i.e., a place in the gene sequence where DNA can be cut by an enzyme) that reveals the A_1/A_2 polymorphism.

For an example of the contrary situation, consider an interaction between socioeconomic stress and a parental phenotype such as depression. Assume that the highest level of alcoholism tendency exists in depressed parents who are also socioeconomically disadvantaged. A causal analysis of this interaction would be highly ambiguous. Perhaps their depression caused the parents to fall in social status and become poor. Or perhaps the hardship of living in poverty made the parents depressed. If DNA sequences associated with depression are used, then the causal ambiguity disappears. In an interaction of gene sequence by low social status, we know that neither the low social status nor the outcome, alcoholism, causes a gene sequence. Thus, DNA-based depression risk must causally relate to alcoholism, whereas a phenotypic score on depression may not.

Gottlieb's Position

Gottlieb (1991) is one scholar who takes what superficially appears to be a different intellectual position by arguing that environmental influences may have causal effects on genes. What he really means, however, is that environmental influences can change *gene expression* levels.

Short-term changes in gene expression are common in response to environmental changes. Genes in the gut of a mosquito become active in response to a blood meal. A plant, exposed to heat, can activate "heat shock" proteins that operate to prevent tissue damage. In people, sunlight causes damage to the skin, which leads to a protective physiological response of tanning, as favored in some societies as an ideal of beauty. Turtle eggs may develop into either the male or female sex depending on the temperature of the sand in which they were laid. These examples could be multiplied because gene expression does depend on environmental stimuli.

Gottlieb (1991) presented these ideas in a figure showing a bidirectional causation between genes and environment, a simplified version of which appears in Figure 5.4a. Genes and the environment are shown interacting in an endless cascade. In his research, Gottlieb was most concerned with embryological development because features of an embryo's cytoplasm partly regulate gene expression.

Gottlieb's diagram, however, may be subtly misleading because it ignores the limits on development set by gene sequence; for example, we know that, regardless of early environmental exposures, a chimpanzee's

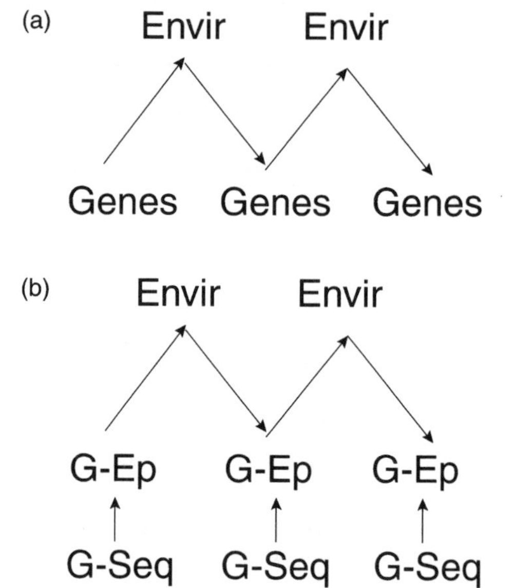

Figure 5.4. (a) Simplified version of Gottlieb's (1991, p. 6) figure showing reciprocal influence of genes and environment (Envir). (b) Modification of Figure 5.4a distinguishing level of gene expression (G–Ep) from gene sequence (G–Seq).

embryo will not become a *Homo sapien's* embryo, and vice versa. Environment does not alter gene sequence. Even in the case of genomic imprinting, sequence is unchanged, although either the maternal or paternal copy of a gene is silenced (i.e., not expressed as the protein or RNA molecule that it codes for; Ferguson-Smith & Surani, 2001). Genomic imprinting is also an inflexible process because whether the maternal or paternal gene is silenced is obligatory and not altered by the environment. Thus, Figure 5.4b is offered as a modification to Gottlieb's original figure, to demonstrate that a bidirectionality of causation exists between gene expression and environment but not between gene sequence and environment.

The uniqueness and protected status of gene sequence is also a rationale against *Lamarckian inheritance*, that is, the inheritance of acquired characteristics. For example, a father using steroids to bulk up muscle carries no advantage to the weight-lifting capacity of his offspring because gene sequence is unchanged by exercise. Another implication of Figure 5.4b is that, for an interaction effect to be repeated in two generations, the environmental events that trigger it must be precisely replicated. On the other hand, gene sequence effects automatically carry over because, except for rare mutations, gene sequences are copied faithfully from one generation to the next. In summary, research designs using molecular genetic markers also help to establish causal direction because of an absence of Lamarckian inheritance.

GE Correlation

Behavioral geneticists distinguish several types of GE correlations: passive, evocative, and active (Plomin et al., 2001). In the passive form, children receive genotypes correlated with their family environment. An example is a child diagnosed with conduct disorder growing up in a family in which the parents also are aggressive—parental and child behavior are correlated. In an evocative GE correlation, children evoke particular social reactions on the basis of their genotypes. For instance, physically attractive children typically encounter more favorable reactions than do less attractive children. And parents of children with a genetic predisposition toward childhood psychiatric disorders may adopt less optimal parenting practices than may parents of children without a genetic risk (Ge et al., 1996). An active GE correlation occurs when a particular genotype is associated with the selection or creation of a particular environmental circumstance. This chapter focuses on this last form of correlation.

The similarity of friends can be used to illustrate the idea of an active GE correlation. Delinquent boys befriend other delinquent boys; smart teenagers join with other smart teenagers; and jocks may seek their company among other athletes. High school cliques often form around common interests and characteristics.

In many cases, the characteristics that led to assortment existed before the assortment took place. In 1978, Kandel completed a classic study of friends by following friendship formation and dissolution during a school year. Her study focused on four characteristics, including minor trivial delinquency. She found that before a friendship formed during the school year, prospective friends were already similar to one another in their levels of trivial delinquency. On trivial delinquency, the correlation of "friends to be" at the start of the school year was .25. The correlation of friends who remained friends throughout the school year was .29. Thus, to count the full .29 correlation as entirely "influence" would be to misinterpret its origin because some part of the .29 association existed before friendship formation (i.e., a selection process). Kandel also found that friendships that dissolved during the school year tended to be those friendships between the most behaviorally dissimilar friends. She also observed a second phenomenon: friends becoming more similar during the duration of their friendships. Using a complex statistical analysis, Kandel determined that about half of friends' similarity was due to selection and half was due to influence (Kandel, 1978, 1996).

Another study examined an even earlier precursor to adolescents' behavior: infants' temper tantrums (Caspi, 2000). In children all born in the same year (i.e., a birth cohort), 3-year-old children were distinguished according to whether they had temper tantrums. Anger and temper in the 3-year-old children predicted their criminal behavior, antisocial personality disorder, suicide attempts, and alcohol dependence at 21 years. Surely, selecting delinquent peer companions cannot have caused the explosive temperament of the 3-year-olds.

GE Correlation in BG Research Designs

It is well known to behavioral geneticists that in kinship studies, the effect of an active GE correlation is methodologically confounded with genotype. That is, whatever variance is contributed by an active GE correlation, it is included in the heritability estimate.

As a way of examining an active GE correlation, my colleagues and I have extended the kinship design to include the *friends* of kinship members (Gilson, Hunt, & Rowe, in press). In our study, we used only adolescent full siblings and their friends drawn from the National Longitudinal Study of Adolescent Health (see Footnote 1). The design is more powerful when kinship groups of different levels of genetic relatedness are used, but sample sizes in the Add Health Study were too small to conduct such analyses. The dependent variable was a self-reported delinquency score on each sibling and an average of the delinquency reports of the siblings' first chosen male and first chosen female friends. Uniquely in this dataset, the friends' reports of delinquency were made by the friends during interviews in their homes. The Add Health Study identified friends from the siblings' nominations on school rosters and used ID numbers to create data-links between them and their friends.

Figure 5.5 presents the structural equation model applied to siblings and their friends. The circle represents all familial influences, genetic and environmental, that affect variation in siblings' self-reported delinquency

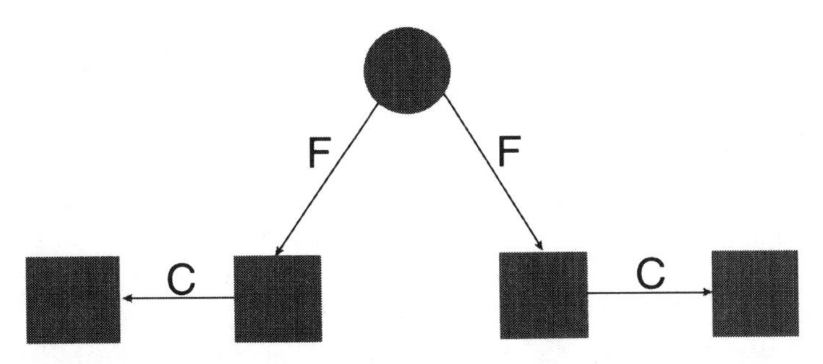

Figure 5.5. Structural equation model of delinquency (from left to right, Friends 1 delinquency, Sibling 1 delinquency, Sibling 2 delinquency, Friends 2 delinquency). Variation in siblings' self-reported delinquency is through the path coefficient F; friends are selected or influenced through the path coefficient C.

through the path coefficient F.[2] The friends are selected or influenced through the path coefficient C. Expectations for correlations are easily derived by following Wright's rules (Loehlin, 1987) for reading path diagrams. The sibling correlation (F^2) should be greater than the one between Friends 1 and Friends 2 ($C \times F \times F \times C$). The expected correlation of Friends 1 with Sibling 2 also follows directly from the structural equation model.

As expected under the model in Figure 5.5, statistically significant correlations were found between Sibling 1 and Sibling 1's friends' delinquency ($r = .23$, $p < .001$), as well as between Sibling 2 and Sibling 2's friends' delinquency ($r = .19$, $p < .01$). This correlation between friends' delinquency is considerably smaller than the .40 to .60 value often reported because the friends' own reports of their delinquency were used instead of children reporting on themselves and on their friends. In criminological studies, this association has been inflated by having a single individual, a target child, report on both his own and his friends' delinquency. Also statistically significant was the sibling correlation ($r = .19$, $p < .05$). The delinquency of the Friends 1 (of Sibling 1) and Friends 2 (of Sibling 2) did not correlate much.

The program Mx was used to fit this structural equation model to a 4×4 covariance matrix. As estimates of standardized path coefficients, the model yielded F = .44 and C = .22. Both parameters proved necessary, and the model had an excellent statistical fit. Thus, in accord with the observed correlations, the expected sibling correlation was .19, the expected correlation between friends was .22, and the expected correlation of Friends 1 and 2 was .01. Although this model fit delinquency successfully, a more complex one was required for verbal intelligence (see Gilson, Hunt, and Rowe, in press), which is not detailed in this chapter.

With multiple kinship groups, the model in Figure 5.5 can be extended

[2]In Gilson, Hunt, and Rowe (in press), the "e" was used for "F" in the structural equation diagram. I made the switch of symbols because "e" may imply environmental influences to some readers, whereas F better captures the idea of all familial influences, including genes.

to estimate genetic and shared environmental influences on each sibling's phenotype and their indirect effect on friends' phenotypes.

GE Correlation Using Measured Genotypes

Measured genotypes permit an analysis of GE correlation using a sample of individuals and correlating variation in their genes with their environmental exposures. The causal status of allelic variation is unique because the environmental variation could not have caused it. Instead, either directly or through a mediated pathway, the allelic variation may have produced the environmental variation.

As an example, consider that people divorce and remarry. In studies of children, divorce is considered to be an environmental variable. Children of married and divorced parents are compared to determine whether divorce can cause psychological harm. Parental life histories may differ depending on whether a divorce or remarriage had occurred. We can ask, Is there a GE correlation such that individuals with a particular genotype would be more likely to be divorced and to remarry than those of different genotypes? One twin study, done in Minnesota with a large sample of twins, suggested that divorce is heritable (Jocklin, McGue, & Lykken, 1996); but a twin study uses the phenotype of a biological relative to estimate genotype.

In contrast, a measured genotype allows for any group of individuals to be used in a study of GE correlation. To illustrate a GE correlation, consider the results on the association of the D4 receptor gene with the number of marriages (see Rowe et al., 2001, for a description of the sample and genotyping methods).

The D4 polymorphism is a number of repeat units in the coding region of the D4 gene. The D4 gene's protein product binds dopamine from the synaptic cleft and then elicits a signaling pathway within the cell. Within the cell's cytoplasm, there is a region of the D4 protein that varies in length (repeats), with common repeat numbers in Caucasians being 2, 4, and 7. The repeat is 48 letters of DNA long, and each repeat codes for 16 amino acids that enter into the D4 protein. The 7-repeat allele is a risk allele for attention deficit hyperactivity disorder and has been associated with the personality disposition of novelty-seeking (Ebstein et al., 1996; Faraone, Doyle, Mick, & Biederman, 2001; see Thapar, chapter 22, this volume, on attention deficit hyperactivity disorder).

In our study, each mother reported on her number of marriages. The frequency distribution of marriages was as follows: one, 73.2%; two, 17.9%; three, 6%; and four or more marriages, 3%; N = 168 mothers.

An analysis of variance was used to compare those mothers possessing a 7-repeat allele with mothers who possessed any other genotype. The mean number of marriages was greater in the 7-repeat carriers (M = 1.5 marriages) versus all other mothers (M = 1.3 marriages), $F(1, 140) = 3.34$, $p < .04$, one tailed. Note that there is evidence of a causal direction established because a greater number of marriages cannot cause a woman's D4 gene to change from 4 to 7 repeats. Instead, a change of repeat number must have increased the divorce/remarriage risk. We can regard this association as a GE correlation to the extent that these mothers evoke conflict from their husbands or select husbands who are psychologically unstable and less able to maintain a marriage.

Conclusions

In this chapter, examples of genes possibly involved in GE interactions and correlations have been presented. In a GE interaction, a polymorphism in the dopamine D2 receptor gene was associated with greater alcoholism, but only when socioeconomic stress was present. The risk gene was the DRD2 A1 variant that has been shown in some studies to increase the risk of alcohol abuse, although this finding is highly controversial (Blum et al., 1995; Gelernter, Goldman, & Risch, 1993). In a GE correlation, the 7-repeat D4 gene variant of the dopamine receptor D4 gene was associated with a tendency toward divorce and multiple marriages. Both associations should be regarded as preliminary; they are in need of replication before they are accepted as established facts. As illustrative examples, however, they demonstrate a potential of genotypes to detect sensitively the presence of GE interactions and correlations.

Molecular genetic markers offer several advantages. A major one is that any group of individuals can be genotyped and used in research, making locating and recruiting rarer kinship types (e.g., twins, adoptive siblings) unnecessary. Sampling costs are reduced by the ability to sample any population, but genotyping costs are added. They are, however, dropping quickly, and eventually commercial labs will perform DNA extraction and genotyping at a low service charge.

A second advantage is that causality is less ambiguous than in a typical research design for examining either GE interactions or correlations. The reason is that the social environment cannot change gene sequence, although it may alter the rate of gene expression. When subgroups are formed around different genotypes, the direction of causation must be from the genotypic classification to the behavioral outcome. For instance, we know that getting married a third time did not change a mother's genotype from a 2-repeat gene to a 7-repeat gene. The genes must correlate with the number of marriages because possessing a 7-repeat variant increases the risk of divorce and later remarriage. Of course, mediator processes may exist between the genotype and the outcome. For instance, if the 7-repeat allele may increase a desire for novelty, it may weaken marriage bonds. A mediating process is necessary because a single gene cannot act directly on a demographic variable such as the number of marriages.

In this regard, it is important to caution that the phrase "gene for trait X" is only a convenient shorthand. Genes are most often expressed by coding for proteins that provide the structural material of the body or that set reaction rates in different biochemical pathways. Some genes are expressed by coding for RNA molecules, which possess a variety of functions. Genes are likely to have pleiotropic effects such that their products influence more than one bodily system; for example, genetic polymorphism related to alcoholism also could influence some seemingly unrelated medical trait, such as blood pressure. Indeed, one argument against trying to rid the population of seemingly harmful genes is that some genes may have pleiotropic effects that are beneficial in another physiological system. A gene for trait X then is really a gene that probabilistically affects a risk of certain physical or behavioral outcomes; it explains variance in them or changes their odds in comparison with prevalence in the general population.

However, the measured genotype approach is not without its pitfalls. The greatest current limitation is that thus far the identified genes account for only a small part of trait variation, with the exception of an association between the aldehyde dehydrogenase gene (ALDH2) and alcohol consumption in Asian populations (the relevant gene variant is rare in other ethnic and racial groups; Wall & Ehlers, 1995). Thus, with a measured genotype, one may be dealing with only 1%–4% of the total variance, and <10% of the genetic variance, in a phenotype. Weak genetic effects imply that GE interactions and correlations can explain only a relatively small part of total trait variation, although explained variance in an interaction is increased through its second independent variable.

Spurious associations also can be obtained with the measured genotype approach due to population stratification (Hamer & Sirota, 2000). The first condition for this methodological confound occurs when a gene is more frequent in one population than in another. A second, necessary condition is that the two populations also differ in their phenotypic means. If, unbeknownst to the investigator, these two populations are combined into one sample, an association may occur between a gene and a phenotype that is noncausal, merely because the gene is more frequent in the population with the higher or lower phenotypic mean. A number of remedies for population stratification have been used, including comparing siblings within families or comparing the genes of affected children with those genes carried by their parents (e.g., Spielman & Ewens, 1996).

A solution to small amounts of explained variance may be to use multiple genetic markers that are combined toward explaining variance. For instance, a multiple regression equation could be used to weight contributions from several genes to trait variation. This method, however, assumes that genes combine additively in fostering trait variation. More complex, configural schemes for weighting genes may be required if Gene × Gene interactions are to be captured (see Grigorenko, chapter 14, this volume).

In an effort to address the complexities of Gene × Gene interaction, Ritchie et al. (2001) developed a statistical method called multifactorial-dimensional reduction that reduces the number of genotypic predictors from n dimensions (number of genotypes) to one dimension. This procedure avoids the problem of empty cells, or cells with low sample sizes, that occur frequently when multiple genes are used. For instance, just four genes with three genotypes each yields 3^4 cells to fill, too many cells except in extremely large samples. The new, one-dimensional variable is then used to predict disease status through permutation testing and cross-validation. Ritchie et al. applied this method to breast cancer, focusing on five genes involved in the metabolism of estrogen because some types of breast cancer are sensitive to estrogen levels. An interaction was found among genotypes at four genes that was related to sporadic breast cancer cases. Such methods also may be applied to behavioral traits in which specific metabolic pathways are thought to underlie risk, for example, as in the case of attention deficit hyperactivity and metabolism of the neurotransmitter dopamine.

In summary, when people are genotyped for genes predisposing toward particular behaviors, the search for GE interactions and correlations becomes both more statistically sensitive and more widely applicable. The rapidly advancing DNA technologies may lead to a discovery of environments that are indeed "best" for particular genotypes.

References

Blum, K., Sheridan, P. J., Wood, R. C., Braverman, E. R., Chen, T. J. H., & Comings, D. E. (1995). Dopamine D2 receptor gene variants: Association and linkage studies in impulsive–addictive–compulsive behavior. *Pharmacogenetics, 5*, 131–141.

Cadoret, R. J., Yates, W. R., Troughton, E., Woodword, G., & Stewart, M. A. (1995). Genetic–environmental interaction in the genesis of aggressivity and conduct disorders. *Archives of General Psychiatry, 52*, 916–924.

Carroll, J. B. (1997). Psychometrics, intelligence, and public perception. *Intelligence, 24*, 25–52.

Caspi, A. (2000). The child is father of the man: Personality continuities from childhood to adulthood. *Journal of Personality and Social Psychology, 78*, 158–172.

Cooper, R. M., & Zubek, J. P. (1958). Effects of enriched and restricted early environments on the learning ability of bright and dull rats. *Canadian Journal of Psychology, 12*, 159–164.

Ebstein, R. P., Novick, O., Umansky, R., Priel, B., Osher, Y., Blaine, D., et al. (1996). Dopamine D4 receptor (D4DR) exon III polymorphism associated with the human personality trait of novelty seeking. *Nature Genetics, 12*, 78–80.

Faraone, S. V., Doyle, A. E., Mick, E., & Biederman, J. (2001). Meta-analysis of the association between the 7-repeat allele of the dopamine D_4 receptor gene and attention deficit hyperactivity disorder. *American Journal of Psychiatry, 158*, 1052–1057.

Ferguson-Smith, A. C., & Surani, M. A. (2001). Imprinting and the epigenetic asymmetry between parental genomes. *Science, 293,* 1086–1089.

Ge, X., Conger, R. D., Cadoret, R. J., Neiderehiser, J. M., Yates, W., Troughton, E., et al. (1996). The developmental interface between nature and nurture: A mutual influence model of child antisocial behavior and parenting. *Developmental Psychology, 52*, 574–589.

Gelernter, J., Goldman, D., & Risch, N. (1993). The A1 allele at the D_2 dopamine receptor gene and alcoholism. *Journal of the American Medical Association, 269*, 1673–1677.

Gilson, M. S., Hunt, C. G., & Rowe, D. C. (in press). The friends of siblings: A test of social homogamy vs. peer selection and influence. *Marriage and Family Review*.

Gottfredson, L. S. (1997). Why g matters: The complexity of everyday life. *Intelligence, 24*, 79–132.

Gottlieb, G. (1991). Experiential canalization of behavioral development: Theory. *Developmental Psychology, 27*, 4–13.

Hamer, D., & Sirota, L. (2000). Beware the chopsticks gene. *Molecular Psychiatry, 5*, 11–13.

Jocklin, V., McGue, M., & Lykken, D. T. (1996). Personality and divorce: A genetic analysis. *Journal of Personality and Social Psychology, 71*, 288–299.

Kagan, J. (1984). *The nature of the child.* New York: Basic Books.

Kandel, D. B. (1978). Homophily, selection and socialization in adolescent friendships. *American Journal of Sociology, 84*, 427–436.

Kandel, D. B. (1996). The parental and peer context of adolescent deviance: An algebra of interpersonal influences. *Journal of Drug Issues, 26*, 289–315.

Loehlin, J. C. (1987). *Latent variable models.* Hillsdale, NJ: Erlbaum.

Madrid, G. A., MacMurray, J., Lee, J. W., Anderson, B. A., & Comings, D. E. (2001). Stress as a mediating factor in the association between the DRD2 Taq I polymorphism and alcoholism. *Alcoholism, 23*, 117–122.

McClelland, G. H., & Judd, C. M. (1993). Statistical difficulties of detecting interaction and moderator effects. *Psychological Bulletin, 114*, 376–390.

Plomin, R., DeFries, J. C., McClearn, G. E., & McGuffin, P. (2001). *Behavioral genetics* (4th ed.). New York: Worth.

Ritchie, M. D., Hahn, L. W., Roodi, N., Bailey, L. R., Dupont, W. D., Parl, F. F., et al. (2001). Multifactor-dimensionality reduction reveals high-order interactions among estrogen-metabolism genes in sporadic breast cancer. *American Journal of Human Genetics, 69*, 138–147.

Rodgers, J. L., Rowe, D. C., & Li, C. (1994). Beyond nature vs. nurture: DF analysis of nonshared influences on problem behaviors. *Developmental Psychology, 30*, 374–384.

Rowe, D. C., Jacobson, K. C., & Van den Oord, E. J. C. G. (1999). Genetic and environmental influences on vocabulary IQ: Parental education level as moderator. *Child Development, 70*, 1151–1162.

Rowe, D. C., Stever, C., Chase, D., Sherman, S., Abramowitz, A., & Waldman, I. D. (2001). Two dopamine genes related to reports of childhood retrospective inattention and conduct disorder. *Molecular Psychiatry, 6*, 429–433.

Spielman, R. S., & Ewens, W. J. (1996). The TDT and other family-based tests for linkage disequilibrium. *American Journal of Human Genetics, 59*, 983–989.

Wall, T. L., & Ehlers, C. L. (1995). Genetic influences affecting alcohol use among Asians. *Alcohol Health and Research World, 19*, 184–189.

6

A U.K. Population Study of Lifestyle and Genetic Influences in Common Disorders of Adult Life

Tom Meade, Frances Rawle, Alan Doyle, and Peter Greenaway

This chapter summarizes the background to a large cohort study, a major objective of the work being to establish the combined effects of lifestyle and genetic influences on many of the leading chronic disorders of middle and older age, the provisional plans for the study and its design, and the ways in which the ethical issues it raises would be dealt with. The rapid and far-reaching developments in genetics over the past decade or so are well known. With this progress in mind, and anticipating the early announcement of the sequencing of the human genome, the Medical Research Council (MRC) and Wellcome Trust of the United Kingdom held a joint meeting in May 1999 to consider the implications for further research at the population level.

As a result of the U.K. Government's Comprehensive Spending Review in 1998, the MRC received additional funds to set up nationally available DNA collections for the genetic epidemiology research that would become increasingly productive with the completion of the Human Genome Project. The focus was on multifactorial diseases of significant public health importance. The Wellcome Trust had simultaneously been considering the possibility of funding a large population collection for studies of genetic and lifestyle risk factors. The workshop brought together experts in the field to consider, in the light of collections already available in the United Kingdom, whether it would be valuable to set up one or more large new collections, and if so, what form it or they should take.

The workshop emphasized the importance of large case control studies for rapid investigations of low penetrance genes associated with increased risk of multifactorial diseases. Although they are well suited for these purposes, case control studies have well-known limitations when the influences of lifestyle, environmental factors, and physiological characteristics are also to be considered. (For simplicity, these three features are henceforth summarized as *lifestyle.*) Cases are included only after they have developed disease, which may have resulted in modifications of lifestyle; when DNA is needed, they can—to all practical intents and purposes—only be drawn from survivor populations so that any differences between those who die and those who survive cannot usually be studied. Retrospective collection of data on lifestyle is also notoriously unreliable anyway, and recall bias introduced by patients who may be more likely than control participants to record an event or exposure is another widely recognized difficulty. Case control studies may not be able to establish whether genetic variations have multiple effects.

All these drawbacks make case control studies unsuitable for the study of the combined effects of genes and lifestyle characteristics, as this requires the collection of high-quality information on risk factors prior to the development of disease. The workshop therefore concluded that, although case control studies were essential to identify genetic polymorphisms associated with common multifactorial diseases, it would be desirable to set up a large prospective study, which would probably need to include between 250,000 and 1,000,000 participants, to investigate genetic and lifestyle influences on a wide range of health outcomes. The workshop therefore recommended that a small expert working group should be established to develop a blueprint for such a study. In making this recommendation, the workshop emphasized that funding for one or more large new population studies should not be at the expense of funding DNA collections from case series and case control studies. In fact, several such collections are now under way as a result of recent calls for proposals, including two on psychiatric disorders, unipolar depression, and late-onset Alzheimer's disease.

Outline Proposal

The expert group reported to the MRC and the Wellcome Trust early in 2000, and both accepted the recommendations in principle and as a basis for proceeding to detailed planning and protocol development. The Department of Health has also accepted the recommendations, having recognized both the importance of the National Health Service (NHS) contribution and the potential value of the study's results for improving health and social care. The main objective of the large new prospective study would be to identify genetic and lifestyle risk factors and, in particular, their combined effects in common multifactorial diseases of adult life. There are already examples with significant practical implications in which the risks conferred by genetic and lifestyle characteristics are greater than the sum of the risks due to either separately. One example is the large combined effect of oral contraceptive use and the factor V-Leiden polymorphism affecting blood coagulability on the risk of venous thromboembolism. This observation raises the question of a similar effect due to hormone replacement therapy in the older women who would be recruited into the study, because this treatment is now also known to increase thromboembolic risk.

First, a cohort study of adequate size would lead to improvements in the prediction of those at risk of various diseases so that preventive and therapeutic measures might be more precisely directed toward those who need them and, of perhaps equal importance, avoided in those who do not. Second, as new gene products that contribute to the development of disease are identified, knowledge about disordered pathways in disease would improve, which should, third, facilitate the earlier and increasingly specific development of new treatments. Alongside these developments, it might be possible to use the cohort for pharmacogenetic studies, which would allow the identification of individuals who are likely to respond well or to have adverse reactions to different medicines. These topics have recently been discussed by Roses (2000).

The study would be concerned with the common causes of morbidity and mortality in adult life, for two main reasons. First, these conditions affect the quality of life of millions of older people and make very heavy demands on health services. Second, the high incidence of many of them means that valuable results could be anticipated within 10–15 years. Consequently, the main disease groupings to be studied would be the following:

1. Cardiovascular disease, particularly coronary heart disease and stroke. The study may also be sufficiently large to confirm or identify associations with abdominal aortic aneurysm.
2. The leading cancers, for example, lung, breast, colorectal, and prostate.
3. Metabolic disorders, principally Type II diabetes.
4. Musculoskeletal conditions, particularly osteoporosis.
5. Respiratory and infectious diseases, including asthma.
6. Psychiatric and neurological disorders, for example, depression, Alzheimer's disease, and Parkinson's disease.

In summary, the study would therefore be concentrating on chronic or long-term disorders of middle and older age that are multifactorial in origin and, as far as genetic influences are concerned, those of low penetrance and in many cases due to more than one gene.

Decisions about the size and duration of the study depend on the incidence, mortality, and morbidity of these conditions at various ages; the ability to detect combined effects for those which, although of interest, occur less frequently than others; allowance for multiple significance testing; the prevalence of different genetic traits; and the obvious desirability of ensuring the likelihood of useful results within a reasonable time period. There also is the trade-off to be considered between the amount of lifestyle information collected at baseline and the numbers that could be recruited for a given research cost; in general, the less baseline information, the larger the number of participants and vice versa. A central consideration is that the proposed study depends as much on information about lifestyle as it does on having DNA available. DNA obtained will not be inexhaustible, but it could be used on several occasions for different purposes. By contrast, there would only be a single opportunity for collecting the baseline lifestyle data on all participants, that is, at entry to the study. If characteristics that turned out to be relevant were not recorded at that stage, the opportunity for defining their importance in combination with genetic markers would be lost. Thus, although studies of the kind proposed are often referred to as DNA, gene, or genetic databases, it is essential not to lose sight of the equal importance of the lifestyle information necessary.

The standardized baseline data collected across the whole cohort would probably include the following:

1. Demographic information, including ethnic origin.
2. Lifestyle information, including socioeconomic data and information on smoking, exercise, diet, and alcohol consumption.
3. Information on existing medical conditions, current medication, and family history of disease.
4. Height, weight, blood pressure, lung function, and urinalysis.
5. Storage of serum and plasma for analysis of biochemical and hematological markers of disease susceptibility.
6. The storage of frozen blood cells for subsequent DNA extraction.

In addition to the baseline data, further information would if possible be collected from samples or from the whole study population, for three reasons. First, repeat samples at some stage would be necessary to allow for the regression dilution effect (MacMahon et al., 1990), which, if uncorrected, results in underestimates of the true strength of associations between risk factors and events. Second, further information on a subset of the cohort, perhaps 5 years after recruitment, would give an idea of whether for example, there have been secular changes over time in the proportions of men and women smoking, in blood pressure levels, and in other characteristics that might have altered. Third, it would be of great value to record information about medicines taken by study participants and any adverse events that might be attributed to them. The study would then make an important contribution to pharmacogenetics, that is, to understanding why patients respond well or badly to various treatments and thus to progress toward "tailor-made" treatments for individuals as distinct from relying—as at present—on generalizations.

Taking account of all these considerations, it is currently proposed that the study should include about 500,000 men and women between ages 45 and 69 years. Although it is hoped that useful results would begin to emerge after about 10 years (possibly less time if pharmacogenetic work is possible), a study of this size and design would go on producing results for many years to come. It should therefore be planned as a resource of long-term value. The incidence of the diseases of interest are all lower at younger ages, and a considerably larger sample would be required if they were to be included, which would significantly increase costs. It is true that although events generally occur more frequently in older age groups, the contribution of genetic and lifestyle characteristics to risk may be lower with increasing age of onset so that an older cohort might give less information on factors influencing morbidity and mortality. The age range suggested represents a balance between these competing considerations. In addition to core information collected on all participants, intensive phenotyping for continuously distributed variables would be carried out on a subsample (because much smaller numbers would be needed; e.g., detailed lipid profiles and to characterize aspects of glucose metabolism), although this subsample would also be used for similar detailed information relevant to the other major disease categories and also perhaps to those of particular interest to some of the regional "spokes" (see later).

Half a million people represents approximately 4% of the total U.K. population in the age range suggested. It might be possible to include some U.K. cohorts already established, provided they have or could obtain data compatible with those to be collected in the new cohort. None of the existing U.K. cohorts are large enough, singly or combined, for the purposes now intended.

Recruitment

Almost all of the U.K. population is registered with general practitioners, so that recruitment through general practice is likely to be an effective way of obtaining representative samples. General practice also provides an important and a practically useful link between those responsible for the research, on the one hand, and the participants themselves, on the other. Recruitment should reflect the heterogeneity of the U.K. population and include members of all ethnic groups, although numbers will only be sufficient for analyses by ethnicity in the larger groups. Recruitment should be according to a common protocol by several centers with epidemiological expertise, each of which would be responsible for managing its own subcohort. All data would be returned to the main coordinating center, responsible for collating results and managing the overall database. This multicenter approach would allow individual centers to undertake additional data collection over and above the core protocol, according to their own interests and expertise. As well as giving the potential for significantly increased richness of the dataset, the multicenter approach should promote involvement of the best available U.K. scientific expertise (e.g., epidemiological, genetic, biochemical) and an increased sense of ownership both by the scientific community and by the U.K. population. It also would significantly expand the potential for training younger scientists in genetic epidemiology and related disciplines.

Blood and DNA Samples

Blood samples would be processed as buffy coat for DNA extraction and as serum and plasma for the later analysis of biochemical and hematological variables of potential interest—for example, cholesterol, fibrinogen, and ascorbic acid—either on the entire cohort or on smaller representative groups. Consideration is also being given to enabling work on proteomics, provided the necessary inhibition of enzyme activity can be achieved.

It has been suggested that initial genotyping should be done on a selected subset of around 10,000 from the cohort for a small number of single nucleotide polymorphisms (SNPs), distributed across each chromosome. Such a pilot genotyping exercise would allow quality control of DNA extraction and storage for the whole cohort. It also would ensure that validated general procedures such as robotic liquid handling, bar-coding of storage and assay plastics, integration of workflow with data generation, and data could be put in place at an early stage. The typing should include human leukocyte antigen markers, particularly those of established medical relevance. This would enable the frequencies of important alleles in the cohort population to be established, the degree of genetic variation to be investigated, and information about patterns of linkage disequilibrium in the study population to be obtained. This pilot genotyping activity could also be the starting point for building up a genotype database on a control panel from within the cohort. The sampling strategy for such a panel would have to be considered carefully, so that there would be representation of the various sectors of the U.K. population to be able to support studies drawing on selected geographic or ethnic populations or both. Individually matched controls for particular case series from either within or outside the cohort could also be selected from the panel. The control panel would require sizable amounts of DNA to allow the samples to be genotyped for large numbers of SNPs. So Epstein–Barr Virus (EBV)-transformed B-lymphoblastoid cell lines might need to be established to secure samples for future use. Genotype data would accumulate rapidly and would be made available to other users. The proposed control panel genotype database should be viewed in the same way as the human genome sequence database (i.e., viewable by all and updated regularly), with allele frequencies, evaluation of Hardy–Weinberg equilibria, and linkage disequilibrium data with neighboring SNPs. Within 5 years, the samples would be genotyped extensively and would serve as a ready-to-use control database for comparison with the genotype of incident cases emerging in the main cohort, and possibly other studies in this age range.

Outcome and Follow-Up

Outcome information on mortality and most cancer incidence can be obtained through the NHS Central Register and local cancer registries (subject to the satisfactory resolution of questions on consent and confidentiality currently under consideration). One of the advantages of carrying out the study through general practice would be that practice records could be used to identify those who might have had events of interest. How this preliminary identification of outcomes would be followed up depends very much on developments in electronic data capture and record linkage, including the extent to which confirmatory evidence from hospitals may be available through practice records. It is difficult to predict how rapidly comprehensive systems of this sort will become available for use in the study, but if and when they do, they would be adopted. Meanwhile, experience in large multipractice studies and trials in the United Kingdom has shown that incident events can be identified not only through close contact of research nurses with general practitioners but also through annual searches of the notes of patients taking part in a study. However, these steps—and indeed notification of events through electronic systems—are only a precursor to then obtaining further information from hospitals and, where necessary, from coroners so that full clinical histories can be assembled and then adjudicated according to standard criteria on whether or not they represent true events.

Ethical Issues

The study does raise a number of ethical issues. These already have been largely considered in the new MRC guidelines, "Human Tissue and Biological Samples for Use in Research" and "Personal Information in Medical Research" (www.mrc.ac.uk). These need to be translated into a practical management strategy for the proposed cohort. A key to both the opportunities and the difficulties is illustrated in Figure 6.1. This figure summarizes the results of a study commissioned in 1999 by the MRC but carried out independently by Opinion Leader Research Ltd. In this survey, participants spontaneously mentioned, first, what they saw as the most exciting developments in medical research and, second, what they saw as the most worrying aspects. At just over 50%, genomic approaches clearly outstripped all the other possible benefits of medical research but, at the same time, genetic research easily raised the most concern, outweighing all the other spontaneously mentioned difficulties combined. The question for the study is, therefore, how to allay the anxieties that were clearly expressed to achieve the benefits that were perceived equally clearly.

The two principal ethical considerations are obtaining meaningful consent to participation and assurances about confidentiality. A third question is whether and, if so, to what extent individual results should be fed back to study participants and their doctors.

In summary, the main point to be settled about obtaining meaningful consent is over the use of DNA samples for purposes not anticipated at the outset. Thus, while it is possible to give potential entrants to the study an account of how the information and samples they provide may be used in studying the main diseases of interest specified, there are bound to be other questions that could be answered but not be anticipated at the outset. Subject to approval by the Multicentre Research Ethics Committee (MREC), to which the study proposal would be submitted, it may be possible to ask participants to give consent of a fairly general nature to cover unanticipated uses, particularly if they are told that these will first have to be approved by the MREC and the independent monitoring authority to be established (see later). Participation would be entirely voluntary in

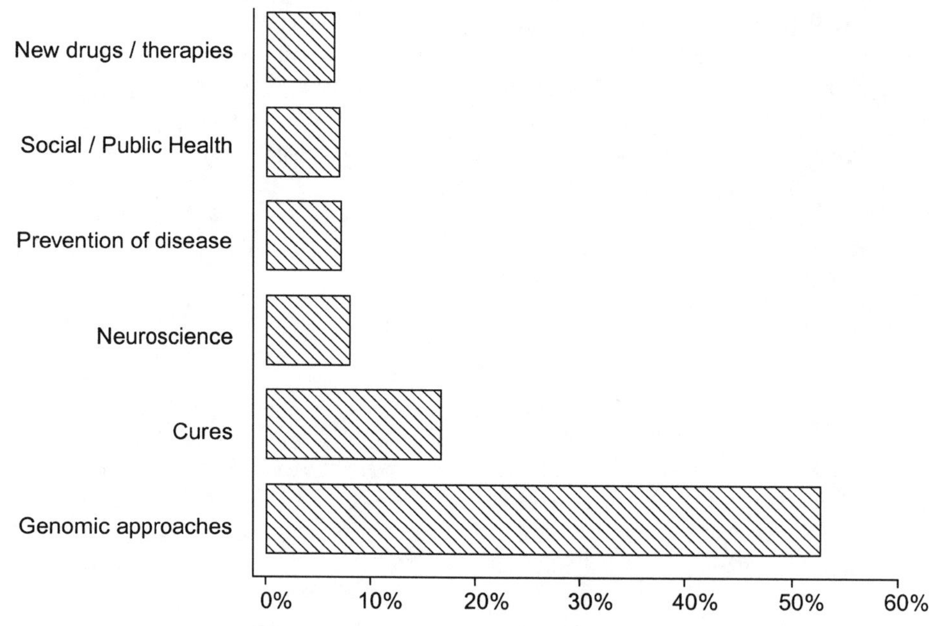

Most worrying development over the next 5 years
(mentioned spontaneously)

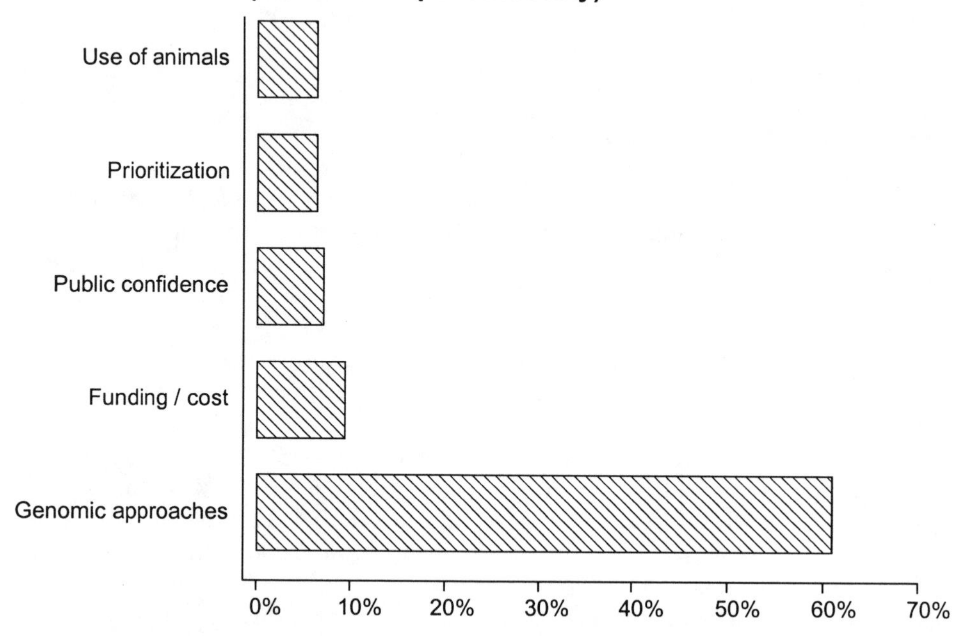

Figure 6.1. Most exciting (top) and most worrying (bottom) developments over the next 5 years (mentioned spontaneously by participants).

all respects, on an opt-in basis (rather than having to opt out if some information could be obtained without the individual's knowledge). Entrants would be given the opportunity of stating that they did not want their information to be used for purposes other than those explained to them to begin with, as well as be given the right to opt out of the study at any time. These wishes would be respected, although naturally it is hoped that this would not occur frequently enough to introduce possible biases.

The question of confidentiality has been raised on numerous occasions in both lay and professional media, sometimes in an unjustifiably alarmist manner. In principle, the importance of maintaining confidentiality in the proposed study does not differ from the importance attributed to this aspect in all medical research involving information about individuals. In fact, the medical research community has a good record in this respect, and it is hard to identify occasions on which there have been serious breaches. Nevertheless, the general concern about confidentiality has to be recognized, and in one sense there certainly is a specific point about genetic research in that information from a DNA sample applies only to that individual. A major concern is that information about genetically determined disease risk might be passed on to insurance companies, to those considering mortgage applications, and to potential employers. Fortunately, the Association of British Insurers has recently stated that participation in genetic research will not be a consideration in making decisions about applications; this will go a long way toward being able to reassure potential study participants.

There have been strikingly polarized views about the requirement or otherwise for feeding back information to participants and their doctors, particularly genetic information. One view is that participants would be giving their time, information about themselves, and blood samples for work on common diseases and that they could therefore expect to be told what the findings are and what the implications might be. The other view is that no feedback of individual results should be made. This is partly because the full significance of genetic variations measured for future health would be unclear to start with and knowledge about their predictive value would be likely to change as time progressed. Feedback of individual information would involve subjective decisions about levels of risk that ought to be notified, as well as a huge counseling burden in explaining the uncertain and changing implications of the information. In addition, quality control for genotyping on this scale would be geared to ensuring the validity of the data for research and not for providing clinical advice, and occasionally there would be errors in the findings on individuals. Also, the value of giving information when it is not clear whether the risk implied can be modified or managed successfully is questionable.

A middle way is to provide regular information through newsletters to participants and their general practitioners on the overall results of the study as they begin to emerge and for participants and their doctors then to decide whether it seems advisable for them to be tested through NHS channels. There are workload implications for general practice in this proposal, but as results where further action might be indicated would not be available for many years, there is time to consider these and other implications before they actually arise. It would be necessary to notify participants and their doctors about some results at baseline that clearly call for intervention according to current standards of medical practice—for example, significantly raised blood pressure and raised blood cholesterol levels, although even in these cases the consent of the participant to informing his or her doctor would be needed.

Pharmaceutical Industry Applications of the Proposed Resource

The proposed cohort study is likely to provide a unique resource to maintain and enhance the world-class reputation of British pharmaceutical research. The ability to link genetic profiles with disease susceptibility and subsequent health care use would facilitate genome-based efforts to identify new medicines and preventive interventions. Understanding the genetics of disease susceptibility would inform the development of new interventions that more fully reflect underlying pathophysiology and (see earlier) the treatment or prevention of clinically detectable illness in smaller, more homogeneous groups of individuals. Together with opportunities for work on pharmacogenetics, the study should represent a unique and valuable resource for the pharmaceutical industry. However, there is no question that the whole database would simply be licensed to an industrial concern.

Ownership and Accessibility

The study would be set up as a national resource, accessible to bona fide researchers wishing to use the data for high-quality research projects, regardless of their primary source of research support, subject to agreed, transparent criteria for assigning priorities to requests for access. Funders and the independent oversight body should have joint custody of the samples and datasets. Researchers generating data would probably have an agreed period of preferential access before being required to place their data in the common dataset. Formal agreements would be required for access by all users, to ensure that data and samples were used only in accordance with the consent given and that confidentiality was protected. These agreements would also specify arrangements for handling intellectual property rights and exploitation of discoveries. Samples or data would never be made available for any purposes other than scientific research (e.g., checking paternity or suitability for insurance or employment), and linkage of samples and data to any identifying information would be strictly controlled so as to maintain confidentiality.

Public Consultation and Acceptability

The study would depend on the cooperation of the general public, so their acceptance, together with an understanding of the potential benefits of this type of research, is essential. The MRC and the Wellcome Trust commissioned a survey by Cragg Ross Dawson on public perceptions about the study (www.wellcome.ac.uk and www.mrc.ac.uk), and, in summary, the findings reflected positively on the proposal. The purpose of the study—to increase understanding of disease and to provide information to combat it—was considered admirable, particularly by those with experience of illness. Although the full range of potential implications was not immediately apparent to most respondents, once these had been raised, discussed, and addressed through factual information, positive views of the collection tended to be restored. The majority also considered that, provided such essential conditions as informed consent and anonymity were integral features of the project, they would be prepared to donate samples to the study. The recent House of Lords Select Committee on Science and Technology Report on Human Genetic Databases (2001) strongly endorsed the importance of the proposed study.

There have also been consultations with general practice through the Royal College of General Practitioners and in focus groups with primary health care professionals from the Trent Focus Collaborative Research Network. These contacts have also been encouraging, while at the same time it is recognized that there are many questions, including workload implications, that need to be considered in detail. The recently established Protocol Development Group (see below) includes two members of the MRC's Consumer Liaison Group, thus ensuring lay representation in the discussions about the detailed planning of the study.

One issue that needs further consideration to be addressed in public consultation is whether guaranteeing an exclusive focus on health and disease (i.e., not on the genetics of intelligence) might increase acceptability. A clear distinction might need to be drawn in the consent process between disease-orientated use and other possible uses, although this should not, for example, be to the detriment of useful information related to the onset and progression of psychiatric disorders.

Current Position

On the scientific side, a Protocol Development Group is preparing a detailed case for the overall design of the study, sample size and aspects of recruitment (e.g., whether the study population should be identified quite randomly through general practice and other lists or whether oversampling of ethnic groups should be attempted), baseline information to be collected, and methods for doing so (e.g., face-to-face interview, self-completed interviews, and electronic methods). Procedures for blood sampling and processing of samples for biochemical, hematological, DNA, and other purposes are being detailed, together with arrangements to enable regression dilution effects and secular changes in lifestyle characteristics to be taken into account (see earlier) and for possible pharmacogenetic studies and for follow-up for outcome events. It is probable that (as indicated earlier) recruitment will be through about six regional centers or "spokes," each collecting core baseline information on all participants but also additional information on topics in which they are themselves especially interested. This information would be relayed to the central "hub" for storage and analysis, although with provision for some analyses to be carried out by the regional spokes where, again, these enable them to study topics of particular interest. A scientific steering group would be set up to provide advice and peer review as work proceeds.

On the wider, organizational side, there would be the high-level independent authority already referred to with responsibility for monitoring all the activities and progress of the study, agreeing policies on access to the data, and requests for use of the samples and for safeguarding the interests of study participants. Again, though, all these plans and arrangements are currently under active consideration. Although the general principles set out here are likely to hold, there may be a number of new or different ways of achieving them. It may be possible to start a pilot study late in 2002 or early in 2003 and a full study soon after feasibility has been demonstrated and methods fully validated.

Conclusion

The proposed study would be of adequate size to establish the combined effects of lifestyle and genetic influences on the major chronic diseases of middle and older age. The United Kingdom is well-suited to this purpose because, first, registration of virtually the whole population with general practitioners provides a sampling frame from which representative populations could readily be drawn. Second, the diversity of the U.K. population provides the opportunity of seeing how far associations differ between ethnic groups, for example. It would also be possible to identify spouse pairs and sibling groupings taking part in the study. The intensive phenotyping of a subsample at recruitment would enable associations between lifestyle and genetic influences with continuously distributed variables to be established both cross-sectionally and prospectively. Results would enable the increasingly precise definition of those who were or who were not at risk of various conditions so that preventive and management measures could be directed with greater accuracy. New and improved knowledge about disordered pathways would suggest new pharmaceutical treatments. Pharmacogenetic information would also help in this way. The study would provide a unique resource for many purposes and for many years and also outstanding training opportunities for young scientists in a range of disciplines, not only epidemiological and genetic but also clinical, pharmacological, in the social sciences, and in other areas.

References

House of Lords Select Committee on Science and Technology. (2001). *Human genetic databases: Challenges and opportunities* (No. HL 57). London: Her Majesty's Stationery Office.

MacMahon, S., Peto, R., Cutler, J., Collins, R., Sorlie, P., Neaton, J., et al. (1990). Blood pressure, stroke, and coronary heart disease: Part 1. Prolonged differences in blood pressure: Prospective observational studies corrected for the regression dilution bias. *Lancet, 335,* 765–774.

Roses, A. D. (2000). Pharmacogenetics and future drug development and delivery. *Lancet, 355,* 1358–1361.

Part III

Learning and Memory in Mice

7

Genetic Studies of Learning and Memory in Mouse Models

Jeanne M. Wehner and Seth A. Balogh

The mouse has become a valuable resource for modeling the critical processes regulating learning and memory in humans. Continued progress toward a finished mouse genomic sequence for several inbred strains holds promise for the elucidation of genes that regulate brain development and function. However, genomic sequence alone will not reveal how genes function in integrated brain circuits or how these circuits regulate behavior. Therefore, there is a need now and in the future for sound phenotypic assessments of learning and memory, better functional genomics, and a better understanding of epistatic (gene–gene) interactions and Gene × Environment interactions. This chapter discusses current strategies used to elucidate genes regulating learning and memory processes, commonly used paradigms to study learning in mice, and possible future directions for research of the genetic regulation of learning and memory.

Genetic Strategies

Currently, three main strategies (quantitative trait loci analysis, transgenics, and chemically induced mutations) are being used to examine the genetic regulation of learning and memory processes in the mouse.

Learning and memory processes, like all complex behavioral traits, are regulated by polygenic systems. Consistent with this, all behavioral studies of inbred and recombinant inbred strains, to date, have revealed a continuous distribution of behavioral scores for all measures of learning examined. Consequently, one strategy for examining the genetic regulation of learning, the quantitative trait loci (QTL) method, attempts to identify the genes regulating this continuous distribution in a defined genetic cross (Gora-Maslak et al., 1991; Lander & Botstein, 1989). This strategy allows the examination of individual differences in a phenotype in a genetically defined mouse population. Importantly, this method has been used exten-

This work was supported by grants from the National Institutes of Health (MH-53668), a Research Career Award to Jeanne M. Wehner (AA-00141), and a postdoctoral traineeship to Seth A. Balogh (MH-16880).

sively to model in mice a wide variety of human disorders including asthma, alcohol and drug abuse, seizure susceptibility, and to a lesser extent for learning and memory (Wehner, Radcliffe, & Bowers, 2001).

In contrast to the polygenic QTL approach, single-gene approaches also are used to identify genes whose expression is essential for adequate performance on a learning and memory task. These methods attempt to identify genes that, when overexpressed or inactivated, impact learning and memory. Several methods are used to manipulate single genes. Increased expression of a single gene can be achieved by introducing additional copies (transgenes) of that gene into the genome of a mouse. Transgenes can be derived from mice or from another species, including humans. Transgenic mice have been particularly useful in creating animal models for diseases such as Alzheimer's disease (Duff, 1998), Down syndrome (Hyde, Crnic, Pollock, & Bickford, 2001), and *c*orpus callosum hypoplasia, *r*etardation, *a*dducted thumbs, *s*pasticity, and *h*ydrocephalus (CRASH) syndrome (Yamasaki, Thompson, & Lemmon, 1997).

Single-gene mutants also can be created by gene-targeting strategies in which a null mutation or a site-specific alteration is made in an important coding region (exon) of a gene of interest. The resulting null mutants (knock outs) or site-specific mutants (knock ins) have the advantage of being created by homologous recombination such that the mutated gene replaces the normal gene. In this design, the essential nature of a gene for learning and memory processes is inferred by how the absence of that gene influences an animal's phenotype. In 1992, two laboratories successfully identified null mutations that caused impairments in learning and memory processes in mice (Grant et al., 1992; Silva, Paylor, Wehner, & Tonegawa, 1992). Although criticisms of the analyses of null mutants to identify genes influencing learning and memory have been made (Routtenberg, 1995), these early studies of α-calcium/calmodulin kinase II and the Fyn tyrosine kinase launched a new era of behavioral research and introduced molecular biologists to behavioral research and psychologists to the power of molecular genetics.

In contrast to gene-targeting approaches in which a known gene is manipulated to elucidate its biological role in behavior, mutants also can be created by random mutagenesis without any knowledge of the genes themselves. This approach has been used extensively in invertebrates to isolate and characterize new genes and allelic series of those genes (Waddell & Quinn, 2001). Random mutagenesis is often performed by chemical methods in which the highly effective mutagen, ethyl-nitroso urea, is injected into male mice to produce mutagenized sperm (see Tecott & Wehner, 2001, for a review of technique). After mutagenesis, each sperm may carry multiple mutations and, when bred with normal females, large populations of mice can be produced. Behavioral screening is then used to identify putative mutants. Breeding of the putative mutants to verify the mutation, mapping of the mutation, and identifying the DNA sequence alteration in a gene are then carried out. Recently large-scale mutagenesis projects that include behavioral screens have been initiated, but the impact of this strategy on studies of the genetics of learning and memory cannot be evaluated as yet and is not discussed here. This chapter describes the progress made using these techniques to uncover the genetic underpinnings of learning and memory processes.

Behavioral Methods

Rodent behavioral models provide excellent methods to study learning and memory. Fortunately, there are many procedures used to assess cognitive function in mice that retain many of the essential components that are tested in humans. For example, the development of spatial learning ability has been assessed in a human version of the Morris task and the radial-arm maze in young children (Overman, Pate, Moore, & Peuster, 1996). Fear conditioning can be measured across a number of species including mice and humans (LeDoux, 2000). These tasks, as well as a few other popular protocols, are reviewed here with emphasis on those that have been used in genetic studies of mutants and QTL analyses.

Several associative–learning and memory–behavioral protocols are used to assess the ability of a mouse to learn about and use relationships between environmental cues and reinforcement. These procedures vary considerably in their level of complexity, type of reinforcement and motivation, and length of training. Some take several days to learn and require an animal to use multiple cues to help it obtain appetitive reinforcement, for example, whereas others depend on only a single exposure to a relatively simple stimulus–response contingency to avoid an aversive stimulus. Another critical factor that distinguishes these tasks is their differential recruitment of disparate memory systems (long term vs. trial specific) and solution strategies (spatial vs. nonspatial). Because most use a mixture of these, they are usually amenable to slight changes in protocol that allow researchers to specifically assess them within the same apparatus. This is a useful commodity because it eliminates the confound of comparing learning ability in entirely different apparatuses.

A great wealth of literature exists delineating the role of the hippocampus and related brain structures in learning and recall in humans, nonhuman primates, and rats (Squire, 1992). As a consequence, the most widely used procedures for evaluating learning and memory in mice are those that have been shown to depend primarily on the functional integrity of the hippocampus in mice. Although the nature of this brain structure's involvement has been debated and the processes it supports referred to by different names (hippocampal-dependent, spatial, or configural associative learning), it is clear that these tasks depend on the integration of multiple environmental cues and are disrupted by the inactivation of the hippocampal formation (Eichenbaum, 1996; Morris, Garrud, Rawlins, & O'Keefe, 1982; O'Keefe & Nadel, 1979; Olton, Becker, & Handelmann, 1979; Rudy & Sutherland, 1995). The most frequently used of these *hippocampus-dependent* tasks are the Morris water maze, the radial-arm maze, the Barnes circular maze, and contextual fear conditioning.

Hippocampus-dependent reference (*spatial*) memory is assessed in the

Morris (1984) maze by placing an animal into a large pool of water from different starting points and recording the time (latency) and distance traveled to locate a hidden escape platform. The pool itself is devoid of cues to guide the animal. Therefore, because the platform remains in the same location throughout testing, an animal must learn to use extramaze cues to find it and expedite its removal from the water (i.e., decrease latency and distance traveled). Spatial learning also is evaluated in this task by using either a probe trial or a reversal test on the final day of training. This consists of removing the platform from the training quadrant or moving the platform to a different location, respectively. Increased time spent searching in the training quadrant during the probe trial or increased latency to find the hidden platform in the reversal test are both taken to indicate the use of a spatial strategy to solve the maze. Importantly, mice with hippocampal lesions or impaired hippocampal function have been shown to perform poorly on this task (Logue, Paylor, & Wehner, 1997).

Further, the Morris maze has been adapted in several instances to allow investigators to examine nonspatial reference memory or spatial working memory. In the nonspatial Morris maze, the location of the escape platform is made visible by either marking it with a flag or raising it above the surface of the water (Paylor, Johnson, Papaioannou, Spiegelman, & Wehner, 1994). Data obtained in several studies show that mice with and without spatial impairments learn this task quite well (Morgan, Holcomb, Saad, Gordon, & Mahin, 1998; Paylor, Baskall-Baldini, Yuva, & Wehner, 1996; Sarnyai et al., 2000). In the working-memory task, a variant of delayed-matching-to-sample, an animal receives multiple problem sets (Morris, 1984; Waters, Sherman, Galaburda, & Denenberg, 1997). As in the spatial version, an animal escapes from the water by swimming to a hidden platform, and the first trial establishes the location of the platform. Thereafter, however, the location of the platform remains the same for only 1–3 additional trials. That is, after every 2nd or 3rd trial, the platform is moved to a new place in the maze, and the animal must relearn its location on Trial 1. Effective spatial working memory is demonstrated if an animal learns to consistently decrease its latency to the platform during the trials following the 1st trial (information trial).

The radial-arm maze also is used to study hippocampal-dependent learning processes (Olton & Samuelson, 1976). Typically, this land maze contains eight arms that radiate from a center platform like the spokes on a wheel. The usual protocol involves food or water deprivation and consequently appetitive reinforcement. Testing consists of placing the animal in the center of the maze and allowing it to make as many choices as necessary to locate all of the food or water that has been placed at the distal end of all or a subset of the arms. The most common measure taken of an animal's performance in this maze is the number of repeat entries (errors) into previously visited arms. Like the Morris maze, mice must rely on extramaze cues in the testing environment to decrease their error scores. The benefit of using this maze is that, when only a subset of arms is baited, a researcher can evaluate both reference memory errors (repeat entries into arms that have never contained a reinforcer) and working-memory errors (repeat entries into arms that previously contained a reinforcer). An unfortunate caveat of this maze is that some mouse strains adapt "chaining" responses or successive choices into adjacent arms. While clearly successful in that the mouse will eventually find the reward, this reflects the use of a motoric rather than a spatial strategy. Several elegant studies have shown that performance in the radial-arm maze is negatively affected by hippocampal damage (Jarrard, 1983, 1986, 1993). It is important to note that several water maze versions of this maze and variations with more and fewer arms have been used successfully with mice as well (Ammassari-Teule, Hoffmann, & Rossi-Arnaud, 1993; Hyde, Hoplight, & Denenberg, 1998; Hyde, Sherman, & Denenberg, 2000). As would be expected, the training procedures in the water maze vary slightly from those used in the land maze.

The Barnes (1979) circular maze is also a land maze used to evaluate hippocampal learning. This apparatus consists of a large circular platform with several holes along its perimeter. Like the aforementioned mazes, a mouse must use extramaze cues to find reinforcement. In this case, reinforcement consists of the hole with an escape tunnel. Like the Morris maze, measures of learning include latency to locate the escape tunnel, searching errors, and the search path of the animal. Senescent animals have been shown to have difficulty solving this maze, presumably because of deficits in hippocampal processing (Barnes, 1979).

Contextual fear conditioning is a testing procedure that, in most instances, is designed to assess long-term hippocampal reference memory. Fear-conditioning tasks have the advantage that they produce robust learning even after a single training trial and the memory produced is long lasting. Being a form of emotional learning, it also has been used as a model in which to explore some psychopathologies, including posttraumatic stress syndrome. Contextual fear, as its name implies, involves the pairing of a neutral context with a fearful stimulus (Blanchard & Blanchard, 1969; Fanselow, 1980). Initially, a mouse explores a small, enclosed environment for several minutes. Following this, a conditioned stimulus (CS tone or clicker) is presented and terminates with the presentation of a brief shock. This is repeated, and the mouse is returned to its home cage. The next day the animal is returned to the same environment, and the percentage of its time spent freezing, a species-specific fear response, is recorded. The animal is then placed into a novel environment. The amount of time spent freezing assessed before and after being reexposed to the CS is measured to control for generalized learning impairments. Mice with intact hippocampal function will not freeze in the novel context. However, they will exhibit pronounced freezing in the context and to the CS. Conversely, animals with hippocampal lesions or otherwise altered hippocampal function have been shown not to exhibit enhanced freezing in the context but normal responses to both the novel context and CS (Anagnostaras, Maren, & Fanselow, 1999; Kim & Fanselow, 1992; Logue et al., 1997; Phillips & LeDoux, 1992).

Most hippocampal-dependent forms of learning are examined using tasks that require the animal to use visual information and other complex cues. For this reason, it is important to determine whether an animal can learn simple associations in behavioral tasks that do not depend on the hippocampus but do require good visual acuity. Black–white or brightness discrimination and horizontal–vertical discrimination T-mazes have proved useful tools with which to answer this question in mutant mice (Balogh, Sherman, Hyde, & Denenberg, 1998; Denenberg, Sherman, Schrott, Rosen, & Galaburda, 1991; Wimer & Fuller, 1966). Discrimination T-mazes have two equidistant arms that connect perpendicularly to the distal end of a stem area in the shape of a capital letter T. Before testing, each animal is randomly assigned a positive stimulus that will contain an escape ladder (i.e., black or white arm, horizontal or vertical striped arm). A different semirandom sequence is then used each day to specify the left–right location of the positive stimulus, so that each stimulus occurs equally often on the left and the right. To solve this maze, an animal must learn to associate a specific intramaze cue, rather than spatial information, with removal from the water. The measure of learning in these mazes is the number of correct choices a mouse makes (swimming down the stem of the T and choosing the correct arm without entering the incorrect arm). Several mouse strains have been shown to learn black–white discrimination very quickly. In contrast, horizontal–vertical discrimination is more difficult and takes longer to learn (Balogh, McDowell, Kwon, & Denenberg, 2001; Reuter, 1987).

Using both visual discrimination and hippocampal-dependent tasks, such as the Morris task or the Barnes maze, more complete assessments can be made as to the nature of deficits in mice. For example, animals that perform well on a simple visual discrimination task but are impaired in a hippocampal-dependent task such as the Barnes maze or Morris task are likely to have a specific genetic or biochemical defect in the hippocampus. In contrast, animals that perform poorly on both the visual discrimination and hippocampal-dependent tasks may have sensory deficits or a very generalized learning deficit.

In addition to the tests described above, active and passive avoidance are also two of the most commonly used tests in mice to assess relatively straightforward "hippocampus-independent" stimulus–response learning. Both involve cued onset of shock. They differ insofar as active avoidance (one-way and two-way) requires an animal to actively evade shock, whereas passive avoidance (step-through and step-down) involves an animal inhibiting its behavior to avoid shock. For active avoidance, a mouse must learn that the onset of a light signals imminent shock. An animal learns this relationship and subsequently that it can "shuttle" to the other side to avoid the shock. Before it learns this relationship, it usually spends the majority of a training session escaping to the other side once it has been shocked. The primary measure of learning in this task, therefore, is an increase in avoidance responses and a decrease in escape responses. The fundamental difference between one- and two-way active avoidance is that training (light/shock pairings) occurs on both sides in two-way avoidance and on only one side in one-way avoidance. Modifications can be made to perform step-through and step-down passive avoidance (e.g., different shock levels, test–retest interval), and these have been used effectively to study biochemical changes that are thought to underlie short-term-memory and consolidation processes. They have been used less for genetic studies but could be easily applied to any genetic strategy.

QTL Analyses of Learning and Memory

Behavioral geneticists have long recognized that learning and memory processes are influenced by both genetic and environmental factors. Selection studies, inbred strain surveys, recombinant inbred strains, and classical Mendelian crosses have all been used to document and model genetic influences on learning using a wide variety of behavioral tasks. However, none of these types of studies allow the identification of genes underlying polygenic differences. The advent of QTL analyses offered the promise of identifying chromosomal regions containing genes that regulate variation in a polygenically regulated phenotype. For a review of the basic technique used for QTL analyses, see Tecott and Wehner (2001).

The most common form of learning that has been examined using QTL analyses is contextual and cued fear conditioning. As described earlier, this form of learning is particularly amenable to QTL analyses because it is a form of one-trial learning that allows sufficiently large numbers of mice to be tested to provide adequate power for QTL mapping. Four studies have localized QTLs for contextual and cued fear conditioning.

In two studies, C57BL/6 (B6) and DBA/2J (D2) mice were used as the parental strains because of the extensive documentation that D2 mice are impaired on many hippocampal-dependent learning tasks (Paylor, Baskall, & Wehner, 1993; Paylor, Tracy, Wehner, & Rudy, 1994; Upchurch & Wehner, 1988). Owen, Christensen, Paylor, and Wehner (1997) localized several putative QTLs for contextual and cued conditioning in B6 X D2 (BXD) recombinant inbred strains, but the 22 strains examined provided insufficient power to place great certainty in these QTLs. Wehner et al. (1997) later performed full genome scans and localized five QTLs in a B6XD2 F2 segregating cross of 479 mice. In this study, QTLs on chromosomes 1, 2, 3, 10, and 16 were identified. The QTLs on chromosomes 2, 3, and perhaps 16 appeared to be specifically regulating contextual learning. As would be expected from the fact that the genetic correlation between cued and contextual conditioning is high, the QTLs on chromosomes 1 and 10 appeared to map variation in both contextual and cued learning. These data suggested that some common genes might regulate processes important for both the amygdala-dependent fear component and the hippocampus-dependent contextual learning component.

In a third study, Calderone et al. (1997) performed a QTL study of fear conditioning in a B6 X C3H backcross segregating population. The C3H strain, like the D2 strain, demonstrates inferior contextual learning compared with the B6 strain. Because adult C3H mice are blind due to their homozygous retinal degeneration (*rd*) gene, Caldarone et al. (1997) used the backcross design so that the *rd* gene can only occur in the heterozygous state. As a result, visual impairments would not interfere with the interpretation of the contextual learning that requires an animal to integrate visual, tactile, and olfactory cues. QTLs were detected on chromosomes 1 and 3, within the same confidence intervals as those detected by Wehner et al. (1997), suggesting that there are common QTLs regulating fear conditioning that generalize across some strains of mice. Additional QTLs were detected, but confirmation studies have not been conducted. Likewise, a QTL for contextual learning also has been detected on chromosome 8 by Valentinuzzi et al. (1998) in a B6 X BALB/c genetic cross, but it too has not been confirmed.

Because QTL studies provide statistical associations, it is important to confirm the QTLs in an independent population and usually an independent breeding strategy. By using procedures similar to those outlined in Buck, Metten, Belknap, and Crabbe (1997), a fourth study confirmed the B6/D2 QTLs for contextual and cued fear conditioning was sought using a short-term phenotypic selection in which contextual fear conditioning was examined in an independent population of F2 mice (Radcliffe, Lowe, & Wehner, 2000). In the confirmation study, cued fear conditioning also was examined. By examining the frequencies of the D2-like and B6-like QTLs in the high versus low lines of mice, QTLs for contextual learning were confirmed on chromosomes 2, 3, and 16. The QTL on chromosome 1 did not show a significant divergence in B6/D2-like QTL frequencies as a function of selection for contextual learning. However, as is described below, most converging evidence suggests that the chromosome 1 QTL may regulate the fear component rather than the contextual component of this form of learning. Moreover, when a DNA marker-assisted breeding strategy was undertaken to create an interval-specific congenic line in which the chromosome 1 D2-like QTL was introgressed into a B6 background, confirmation of the chromosome 1 QTL has been obtained (Wehner & Radcliffe, 2002).

While QTLs have been detected for at least one form of learning, the labor-intensive process of narrowing the chromosomal regions and gene identification within the QTL regions is ongoing. One of the most interesting outcomes of the QTL analyses of fear conditioning is the detection of the chromosome 1 QTL and its potential overlap with the chromosome 1 QTL detected in studies of emotionality in the DeFries high and low open-field lines (DeFries, Gervais, & Thomas, 1978; Flint et al., 1995) and in genetic crosses of B6XA/J mice (Gershenfeld & Paul, 1997). Because the QTL regions for both anxiety and fear have not been sufficiently narrowed, it is impossible to know whether a single polymorphic gene is important for both anxiety- and fear-related behaviors or whether there are multiple QTLs in this region representing numerous genetic polymorphisms.

It is noteworthy that the QTL method has not been widely used for learning and memory studies. This may be for several reasons. First, QTL studies are labor intensive. Some studies have examined 500–1,600 mice to provide sufficient statistical power to detect multiple QTLs of small effect sizes that regulate variation in behavioral traits. Because of this demand, complicated learning tasks that require several days to weeks to train each subject are not particularly attractive.

The second reason why progress is slow is that the ultimate gene discovery phase of QTL analysis is difficult. The polymorphic changes in genes that underlie QTLs may be of two types. A polymorphism in a coding region of a gene may produce a change in the amino acid sequence that alters protein function, whereas a polymorphism in a promoter or regulatory region would produce a quantitative difference in the amount of the protein. Because QTLs are mapped initially to large chromosomal intervals containing hundreds of genes, it is a high-risk proposition to pick candidate genes for sequencing and identify polymorphisms in those genes. Most often, strategies are used to narrow QTL regions to more reasonable sizes that could eventually be useful for positional cloning of genes or candidate gene sequencing.

Narrowing of QTL regions for B6/D2 differences for contextual and cued fear conditioning currently is being done using the interval-specific congenic strategy outlined by Darvasi (1998). In this breeding method, overlapping segments within the large QTL region containing the D2-like form of the QTL were identified in an F2 population or a specific recombinant inbred strain, and the mice were bred with B6 mice in a typical backcross design. The offspring are genotyped, and those mice that are heterozygotes for the required chromosomal segment are backcrossed to B6 again. This procedure is repeated for 10 generations, and then homozygous mice are ultimately created. The resulting product is multiple congenic strains, each carrying a specific chromosomal region. Behavioral testing to identify which of the overlapping intervals carries the QTL for contextual or cued fear conditioning is then undertaken. Further narrowing of the QTL interval is then accomplished by breeding the appropriate interval-specific congenics. This process may require 2–3 years. However, once stable congenic strains have been created that capture the behavioral effect of a QTL, the resulting congenics can be used to examine multiple phenotypes that are suspected to be regulated by the same QTL. For example, chromosome 1 congenics could be tested on contextual and cued fear conditioning and on anxiety tests.

Congenics also may be useful for functional genomic studies in which gene expression is compared between brain tissue derived from a congenic carrying a particular strain-specific QTL and the other parental background strain (Karp et al., 2000). DNA microarray techniques are currently the method of choice because they allow one to screen for differential expression of thousands of genes simultaneously.

Congenics may not be the fastest route to narrowing QTLs. Mapping QTLs to narrower chromosomal intervals can be achieved using genetic stocks of mice that have undergone extensive recombination. This can be accomplished by examining very large segregating populations beyond the F2 generation to identify rare recombinants using advanced intercross lines (Silver, 1995). Another attractive method is to use an existing outbred heterogeneous stock (HS) of mice. At the Institute for Behavioral Genetics (IBG) in Boulder, Colorado, HS mice, the result of systematic breeding of

8 inbred mouse strains for more than 60 generations, has allowed for a great deal of recombination to occur. In this stock, for example, Talbot et al. (1999) showed that they could detect a QTL on chromosome 1 for emotionality in an interval of less than 1 cM.

Studies of contextual learning in HS mice are ongoing, so it is premature to say whether this strategy will be useful for all complex learning tasks. That is, for the original crosses, the IBG HS used a number of the strains either that were albino or that carried the *rd* gene and as a result had potentially compromised vision. However, in the absence of a method to enhance rates of recombination in mammals during meiosis, the HS strategy appears to be the most promising method available for mapping QTLs to smaller intervals.

Regardless of the method used to perform QTL analyses, there is no question that this strategy will become easier in the future. Although the time required for phenotypic analyses will still be dependent on the behavioral task, the time required to map QTLs will be dramatically reduced, and ultimate gene identification will be easier. Mapping using single-nucleotide polymorphisms (SNPs) in combination with current DNA chip technologies will allow the initial chromosomal mapping and identification of QTLs to be done in a few days (Grupe et al., 2001). Finished genomic sequences for multiple inbred mouse strains also will facilitate the identification of polymorphic genes underlying QTLs.

The ultimate proof that a candidate gene is the gene that actually regulates the phenotypic differences between mouse strains has not been well established. However, behavioral differences between one of the parental strains and a genetically engineered mouse that carries a particular site-specific polymorphism (a knock in) in the gene of interest may serve as the ultimate proof that a particular gene underlies a QTL.

Studies of Learning and Memory in Single-Gene Mutant Mice

In general, much of the human behavioral genetic literature in cognition has focused on individual differences in populations. Thus, QTL studies in the mouse are perhaps of greater interest to most behavioral geneticists because it is hoped that polymorphisms in particular genes in the mouse that regulate variation will translate to humans. At this time, the test of this assumption cannot be made because specific genes regulating variation in learning and memory using QTL methods have yet to be identified in the mouse. There is, however, much to be learned about essential neurochemical processes that regulate learning and memory in model organisms that will lead to a better understanding of human cognition. Thus, it would be unwise to dismiss the accumulating literature gained from single-gene mutants that are identifying genes regulating proteins essential for the acquisition and storage of information in mice. The primary technique that has provided insights into the genes regulating behavior in the mouse is the use of transgenic technologies, especially null mutants (knock-out mice). Importantly, a wider range of behavioral protocols has been used to evaluate learning and memory in null mutant studies than has been used in QTL studies. This relates to the fact that the complexity of these protocols can be greater given that smaller numbers of animals can be evaluated. Many criticisms have been generated concerning the use of null mutants. Issues relating to behavioral and neuroanatomical abnormalities in the 129 strains, which serve as the primary source of embryonic stem cells, have been discussed quite thoroughly (Balogh, McDowell, Stavnezer, & Denenberg, 1999; Montkowski, Poettig, Mederer, & Holsboer, 1977; Simpson et al., 1997; Wolfer & Lipp, 2000). Breeding strategies to maintain proper genetic control of experiments also have been addressed (Wehner & Silva, 1996). The possible confounds and compensatory processes that might lead to misinterpretation of primary versus secondary effects of single-gene deletions are clearly a concern, and new strategies to create conditional mutations that are restricted to specific brain regions or inducible mutations that allow careful timing of the expression will improve our ability to interpret the influence of single-gene manipulations on learning and memory processes (Mayford, Mansuy, Muller, & Kandel, 1997). These issues have all been raised previously and are not discussed here. Rather we have chosen to focus on gaps in our understanding of genetics of learning and memory that might be addressed in single-gene mutants, as well as new ways to use the information gained from single-gene mutation studies.

Hundreds of transgenic and null mutants are available, many of which have been analyzed for behavioral, biochemical, and pharmacological alterations. Bolivar, Cook, and Flaherty (2000) reviewed the transgenic and null mutant mouse literature and compiled a list of single-gene manipulated mouse lines that have behavioral abnormalities. Included are those lines that show differences in learning and memory processes. Most frequently, differences between wildtype and mutant mice are revealed in contextual fear conditioning and the Morris water task. Table 7.1 shows a summary of many of the genes that have been implicated in altered performance using freezing in the context or probe trial as measures of spatial learning. They are included here to illustrate several things about single-gene mutations. First, the early fear of behaviorists that every null mutant will show disrupted learning is not supported. Of the 61 mutants that received a Morris maze probe trial, 31 showed abnormalities, whereas 2 showed improvements. Similarly, of 24 that were given contextual fear conditioning, 13 were impaired and 4 were enhanced on contextual learning. Interestingly, of the 9 instances in which both the Morris task and contextual fear were administered, 7 resulted in consistent spatial learning profiles, whereas 2 did not (Bolivar et al., 2000; see also http://ibgwww.Colorado.EDU/~smitham/L&M_genes.html). Examination of this list of candidate genes is quite revealing in that almost all of the genes listed have been implicated in processes related to synaptic plasticity by either pharmacological or electrophysiological experiments using in vitro methods. One might question then if studying single-gene mutants is a useful process. We would argue that it is a worthwhile endeavor for several reasons. First, the transgenics outlined by Bolivar et al. (2000) begin to

Table 7.1. Summary of Genes That Have Been Shown to Alter Spatial Learning in Contextual Fear Conditioning and the Morris Water Maze

Gene name	Transgene	Morris probe trial	Contextual fear context
Acetylcholine receptor β2	KO	Normal	Normal
Adenylate cyclase 1	KO	Lower than controls	NT
Apolipoprotein E	KO	Mixed	NT
Amyloid precursor protein	Tg/KO	Mixed	NT
Brain-derived neurotropic factor	KO	Normal	NT
Calmodulin-kinase II-α subunit	Tg*/KO**	Lower than controls	Lower than controls
CCAAT enhancer binding protein-δ	KO	Normal	Higher than controls
cAMP responsive element binding protein 1	KO	Lower than controls	Lower than controls
Dopamine β hydrolyase	KO	Lower than controls	NT
Discs, large homolog 4	KO	Lower than controls	NT
Dopamine D1A receptor	KO	Lower than controls	NT
Disheveled	KO	Normal	NT
Fragile X-linked mental retardation syndrome	KO	Normal	NT
FosB	KO	Normal	NT
Fyn protooncogene	KO	Conflicting	NT
GABA-A receptor β3	KO	NT	Lower than controls
Glial fibrillary acidic protein with L1	Tg	Normal	NT
Growth hormone (bovine)	Tg	NT	Lower than controls
Ionotropic glutamate receptor, AMPA2	KO	Lower than controls	NT
Ionotropic glutamate receptor, AMPA1	KO	Lower than controls	NT
Ionotropic glutamate receptor, NMDA2A	KO	Lower than controls	NT
Ionotropic glutamate receptor, NMDA2B	Tg	Higher than controls	Higher than controls
Ionotropic glutamate receptor, NMDA2D	Tg	Normal	NT
Glucocorticoid receptor	KO	Lower than controls	NT
Metabotropic glutamate receptor 1	KO	NT	Lower than controls
Metabotropic glutamate receptor 2	KO	Normal	NT
Metabotropic glutamate receptor 4	KO	Normal	NT
Metabotropic glutamate receptor 5	KO	Lower than controls	Lower than controls

Metabotropic glutamate receptor 7	KO	NT	Lower than controls
Huntington's disease	KO	Normal	NT
Heme oxygenase	Tg	Lower than controls	NT
5-HT1b receptor	KO	Normal	Normal
5-HT2c receptor	KO	Lower than controls	Normal
Interleukin 2	KO	Lower than controls	NT
L1 cell adhesion molecule	KO	Lower than controls	NT
Monoamine oxidase A	KO	NT	Higher than controls
MAS oncogene	KO	Normal	NT
Metallothionen III	KO	Normal	NT
Neural cell adhesion molecule	KO	Lower than controls	NT
Neurofibromatosis Type 1	KO	Lower than controls	NT
Neurofilament medium polypeptide	Tg	Lower than controls	NT
Neuropeptide nociceptin I	KO	Higher than controls	NT
Neuropeptide Y	KO	Normal	NT
Protein kinase, cAMP dependent, catalytic-1β	KO	NT	Normal
Protein kinase C-γ	KO	Lower than controls	Lower than controls
Plasminogen activator	KO	Normal	Higher than controls
Plasminogen activator, urokinase	Tg	Lower than controls	NT
Prion protein	KO	Normal	NT
Protein kinase, cAMP dependent regulatory-βI	Tg/KO	Lower than controls/Normal	Lower than controls/Normal
RAS guanine nucleotide releasing factor 1	KO	Normal	Lower than controls
S100 protein-β	Tg	Lower/Conflicting	NT
Spinocerebellar ataxia 1	KO	Lower than controls	NT
Synapsin I	KO	Normal	Normal
Synapsin II	KO	NT	Lower than controls
Thyroid hormone receptor-β	KO	Normal	Normal
Thymus cell antigen I	KO	Normal	NT
Ubiquitin conjugating enzyme-E3A	KO	NT	Lower than controls

Note. Adapted from "List of Transgenic and Knockout Mice: Behavioral Profiles," by V. Bolivar, M. Cook, and L. Flaherty, 2000, *Mammalian Genome, 11,* pp. 260–274. KO = knockout; Tg = transgenic; NT = not tested; GABA-A = gamma-aminobutyric acid; AMPA = alpha-amino-3-hydroxy-5-methylisoxazole-4-propionic acid; NMDA = *N*-methyl-D-aspartate; cAMP = 3′5′ cyclic adenosine monophosphate.

highlight the complexity of the genetic systems that underlie learning and memory processes (i.e., genes in one spatial task are not necessarily involved in another spatial task). The use of transgenic and other molecular genetic techniques will undoubtedly aid in the clarification of these systems. Second, it allows researchers to investigate behavior within a living animal interacting with its environment. Finally, science is built on converging lines of evidence. The techniques discussed throughout this chapter add favorably to the already vast, but incomplete, knowledge base on the biochemical underpinnings of learning and memory storage.

There is no question that interesting mutants will continue to be created using transgenic strategies. However, there are serious gaps in the literature in several areas. An examination of the list in Table 7.1 suggests that most of the genes listed are implicated in early (probably acquisition) processes and few genes have been identified, perhaps with the exception of cAMP-responsive-element-binding protein (CREB), protein kinase A (PKA), and the mitogen-activated protein (MAP) kinases, that may be involved in consolidation or long-term memory formation (Bourtchouladze et al., 1998; Bourtchouladze et al., 1994; Schafe et al., 2000; but see Selcher, Nekrasova, Paylor, Landreth, & Sweatt, 2001). Additionally, most current studies have focused on the learning and long-term potentiation within the hippocampus at the expense of other forms of learning and other brain structures. Lesion studies suggest that the hippocampus has a time-limited role in spatial and contextual learning (Anagnostaras et al., 1999; Kim & Fanselow, 1992; Maren, Aharonov, & Fanselow, 1997) and, therefore, does not appear to be the storage site of permanent memories. Cortical plasticity (i.e., structural and biochemical changes) has been thought to be essential for the establishment of more permanent memory traces, but genes regulating long-term memory have not been described. For example, what new genes are turned on during the protein synthesis dependent stage of learning that influence storage of information? Frankland, O'Brien, Ohno, Kirkwood, and Silva (2001) took an interesting approach to find genes regulating permanent memory formation. Previously, Silva et al. (1992) had demonstrated that homozygous mice lacking α-calcium/calmodulin kinase II were profoundly impaired on spatial learning. Heterozygotes did not show deficits at early times after training. However, by 50 days after training, Frankland et al. (2001) revealed a deficit in long-term memory. In addition, impairments in cortical, but not hippocampal, long-term potentiation were observed in the heterozygous mice. These studies provide an example of how to unravel the molecular events that may lead to permanent memory storage. They also demonstrate that more extensive time courses should be done in learning and memory studies because it is now clear that a widely expressed enzyme such as α-calcium/calmodulin kinase II may have important roles in multiple brain regions and in multiple phases of learning. Moreover, a recent study by Zirlinger, Kreiman, and Anderson (2001) has shown, by way of microarray technology, that gene expression within the amygdala can be captured in a subnuclei specific fashion. That is, genes can be shown to be differentially expressed depending on the amygdalar nuclei being studied. Although no concomitant behavioral assays were undertaken, this suggests that additional genes responsible for mediating fear conditioning may soon be discerned and their expression patterns revealed. Several laboratories are now examining changes in gene expression as a function of learning using the microarray technology.

Conclusion

Considerable progress has been made in using genetic strategies to discover genes regulating learning and memory processes. New strategies for QTL analyses have been proposed that may allow rapid identification of QTLs across multiple inbred strains using SNP maps (Grupe et al., 2001). Technology for gene identification and the application of DNA microarrays will become more prominent as the cost of this technology is reduced. More sophisticated methods for making single-gene mutants using embryonic stem cell methods continue to evolve, including the use of conditional and inducible gene-targeting strategies (Mayford et al., 1997). Random mutagenesis may be greatly facilitated with the recent developments of gene-trapping strategies that use marker transgenes that will only function when introduced into an active gene. Such insertional mutations disrupt the normal gene producing a null mutation but also tag the gene, allowing rapid identification of that gene (Jackson, 2001).

These technological advances will not however completely solve complex behavioral problems. Although many single-gene mutants have been shown to be impaired in learning and memory processes, little is known about whether the identified deficits are true learning deficits. Noncognitive deficits in attention, motivational processes, or baseline anxiety also must be considered using more thorough phenotypic analyses. Cognizant of the increased need for better methods of phenotyping mutant mice, investigators have begun introducing additional behavioral methods into the mouse genetic literature. Many of these are adaptations from the rich behavioral literature in the rat. Minor modifications of operational parameters that relate to intrinsic species differences that exist between mouse and rat are often made. For example, mice are highly explorative animals and will easily nose poke in a designated hole to receive a reward, so that nose poking replaces lever pressing in several operant behaviors. There is a continued need for new behavioral tasks to be adapted to the mouse that will allow examination of complex issues relating to learning.

It is also the case that those interested in genetic variation (i.e., gene polymorphisms) may be missing an important opportunity by ignoring the research on single-gene mutants. Other areas of behavioral genetics that focus on drug abuse or psychopathology have used the candidate gene approach productively. For example, if one would ask a behavioral geneticist who works in the area of drug abuse, "Why are you looking for polymorphisms in the dopamine transporter?" they would be quick to tell you that neuropharmacological studies support that the dopamine system is clearly important in the brain's reward circuit and some psychostimulants like cocaine clearly inhibit the dopamine transporter. The link between the dopamine system and drug abuse was obtained initially from animal studies, and now human geneticists apply this knowledge in the search for polymorphisms in "candidate genes." However, in the field of human behavior genetics of cognition, the trail of information from animal studies to the search for human polymorphisms is very weak.

One possible application of the wealth of knowledge that is being obtained from the studies of single-gene null mutants would be to address whether genes that we believe are essential for acquisition of new information are polymorphic in humans. There is much converging evidence for a role of several genes in information processing, including the NMDA (*N*-methyl-D-aspartate) receptor gene family, the AMPA (alpha-amino-3-hydroxy-5-methylisoxazole-4-propionic acid) receptor gene family, α-calcium/calmodulin kinase II, protein kinase C gene family, *Fyn* kinases, and the cAMP-dependent protein kinase gene family, to name a few potential targets of exploration in human genetics. Such studies might be a fruitful way to merge the growing understanding of learning and memory in animal models with the complex genetic regulation of human cognition. Presumably, the emerging databases from the human and mouse genome projects will allow the identification of SNPs in genes that have been identified in the mouse.

References

Ammassari-Teule, M., Hoffmann, H. J., & Rossi-Arnaud, C. (1993). Learning in inbred mice: Strain-specific abilities across three radial maze problems. *Behavior Genetics, 23,* 405–412.

Anagnostaras, S. G., Maren, S., & Fanselow, M. S. (1999). Temporally graded retrograde amnesia of contextual fear after hippocampal damage in rats: Within-subjects examination. *Journal of Neuroscience, 19,* 1106–1114.

Balogh, S. A., McDowell, C. S., Kwon, Y. T., & Denenberg, V. H. (2001). Facilitated stimulus–response associative learning and long-term memory in mice lacking the NTAN1 amidase of the N-end rule pathway. *Brain Research, 892,* 336–343.

Balogh, S. A., McDowell, C. S., Stavnezer, A. J., & Denenberg, V. H. (1999). A behavioral and neuroanatomical assessment of an inbred substrain of 129 mice with behavioral comparisons to C57BL/6J mice. *Brain Research, 836,* 38–48.

Balogh, S. A., Sherman, G. F., Hyde, L. A., & Denenberg, V. H. (1998). Effects of neocortical ectopias upon the acquisition and retention of a non-spatial reference memory task in BXSB mice. *Developmental Brain Research, 111,* 291–293.

Barnes, C. A. (1979). Memory deficits associated with senescence: A neurophysiological and behavioral study in the rat. *Journal of Comparative and Physiological Psychology, 93,* 74–104.

Blanchard, R. J., & Blanchard, D. C. (1969). Crouching as an index of fear. *Journal of Comparative and Physiological Psychology, 67,* 370–375.

Bolivar, V., Cook, M., & Flaherty, L. (2000). List of transgenic and knockout mice: Behavioral profiles. *Mammalian Genome, 11,* 260–274.

Bourtchouladze, R., Abel, T., Berman, N., Gordon, R., Lapidus, K., & Kandel, E. R. (1998). Different training procedures recruit either one or two critical periods of contextual memory consolidation, each of which requires protein synthesis and PKA. *Learning and Memory, 5,* 365–374.

Bourtchouladze, R., Frenguelli, B., Blendy, J., Cioffi, D., Schutz, G., & Silva, A. J. (1994).

Deficient long-term memory in mice with targeted mutation of the cAMP-responsive element-binding protein. *Cell, 79,* 59–68.

Buck, K. J., Metten, P., Belknap, J. K., & Crabbe, J. C. (1997). Quantitative trait loci involved in genetic predisposition to acute alcohol withdrawal in mice. *Journal of Neuroscience, 17,* 3946–3955.

Caldarone, B. J., Saavedra, C., Tartaglia, K., Wehner, J. M., Dudak, B. C., & Flaherty, L. (1997). Mapping of genes affecting contextual memory in mice. *Nature Genetics, 17,* 335–337.

Darvasi, A. (1998). Experimental strategies for the genetic dissection of complex traits in animal models. *Nature Genetics, 18,* 19–24.

DeFries, J. C., Gervais, M. C., & Thomas, E. A. (1978). Response to 30 generations of selection for open-field activity in laboratory mice. *Behavior Genetics, 8,* 3–13.

Denenberg, V. H., Sherman, G. F., Schrott, L. M., Rosen, G. D., & Galaburda, A. M. (1991). Spatial learning, discrimination learning, paw preference and neocortical ectopias in two autoimmune strains of mice. *Brain Research, 562,* 98–104.

Duff, K. (1998). Recent work on Alzheimer's disease transgenics. *Current Opinion in Biotechnology, 9,* 561–564.

Eichenbaum, H. (1996). Is the rodent hippocampus just for "place"? *Current Opinion in Neurobiology, 6,* 187–195.

Fanselow, M. S. (1980). Conditional and unconditional components of post-shock freezing. *Pavlovian Journal of Biological Science, 15,* 177–182.

Flint, J., Corley, R., DeFries, J. C., Fulker, D. W., Gray, J. A., Miller, S., & Collins, A. C. (1995). A simple genetic basis for a complex psychological trait in laboratory mice. *Science, 269,* 1432–1435.

Frankland, P. W., O'Brien, C., Ohno, M., Kirkwood, A., & Silva, A. J. (2001). α-CaMKII-dependent plasticity in the cortex is required for permanent memory. *Nature, 411,* 309–313.

Gershenfeld, H. K., & Paul, S. M. (1997). Mapping quantitative trait loci for fear-like behaviors in mice. *Genomics, 46,* 1–8.

Gora-Maslak, G., McClearn, G. E., Crabbe, J. C., Phillips, T. J., Belknap, J. K., & Plomin, R. (1991). Use of recombinant inbred strains to identify quantitative trait loci in psychopharmacology. *Psychopharmacology, 104,* 413–424.

Grant, S. G. N., O'Dell, T. J., Karl, K. A., Stein, P. L., Soriano, P., & Kandel, E. R. (1992). Impaired long-term potentiation, spatial learning, and hippocampal development in *fyn* mutant mice. *Science, 258,* 1903–1910.

Grupe, A., Germer, S., Usuka, J., Aud, D., Belknap, J. K., Klein, R. F., et. al. (2001). In silico mapping of complex disease-related traits in mice. *Science, 292,* 1915–1918.

Hyde, L. A., Crnic, L. S., Pollock, A., & Bickford, P. C. (2001). Motor learning in Ts65Dn mice, a model for Down syndrome. *Developmental Psychobiology, 38,* 33–45.

Hyde, L. A., Hoplight, B. J., & Denenberg, V. H. (1998). Water version of the radial-arm maze: Learning in three inbred strains of mice. *Brain Research, 785,* 236–244.

Hyde, L. A., Sherman, G. F., & Denenberg, V. H. (2000). Non-spatial water radial-arm maze learning in mice. *Brain Research, 863,* 151–159.

Jackson, I. J. (2001). Mouse mutagenesis on target. *Nature Genetics, 28,* 198–199.

Jarrard, L. (1983). Selective hippocampal lesions and behavior: Effects of kainic acid lesions on performance of place and cue tasks. *Behavioral Neuroscience, 97,* 873–889.

Jarrard, L. (1986). Selective hippocampal lesions and behavior. In R. L. I. K. H. Pribram (Ed.), *The hippocampus* (Vol. 4, pp. 93–126). New York: Plenum Press.

Jarrard, L. (1993). On the role of hippocampus in learning and memory in the rat. *Behavioural Brain Research, 60,* 9–26.

Karp, C. L., Grupe, A., Schadt, E., Ewart, S. L., Keane-Moore, M., Cuomo, P. J., et al. (2000). Identification of complement factor 5 as a susceptibility locus for experimental allergic asthma. *Nature Immunology, 1,* 221–226.

Kim, J. J., & Fanselow, M. S. (1992). Modality-specific amnesia of fear. *Science, 256,* 675–677.

Lander, E. S., & Botstein, D. (1989). Mapping Mendelian factors underlying quantitative traits using RFLP linkage maps. *Genetics, 121,* 185–199.

LeDoux, J. E. (2000). Emotion circuits in the brain. *Annual Review of Neuroscience, 23*, 155–184.
Logue, S. F., Paylor, R., & Wehner, J. M. (1997). Hippocampal lesions cause learning deficits in inbred mice in the Morris water maze and conditioned-fear task. *Behavioral Neuroscience, 111*, 104–113.
Maren, S., Aharonov, G., & Fanselow, M. S. (1997). Neurotoxic lesions of the dorsal hippocampus and Pavlovian fear conditioning in rats. *Behavioural Brain Research, 88*, 261–274.
Mayford, M., Mansuy, I. M., Muller, R. U., & Kandel, E. R. (1997). Memory and behavior: A second generation of genetically modified mice. *Current Biology, 7*, R580–R589.
Montkowski, A., Poettig, M., Mederer, A., & Holsboer, F. (1977). Behavioural performance in three substrains of mouse strain 129. *Brain Research, 762*, 12–18.
Morgan, D., Holcomb, L., Saad, I., Gordon, M., & Mahin, M. (1998). Impaired spatial navigation learning in transgenic mice over-expressing heme oxygenase-1. *Brain Research, 808*, 110–112.
Morris, R. G. M. (1984). Developments of a water-maze procedure for studying spatial learning in the rat. *Journal of Neuroscience Methods, 11*, 47–60.
Morris, R. G. M., Garrud, P., Rawlins, J. N. P., & O'Keefe, J. (1982). Place navigation impaired in rats with hippocampal lesions. *Nature, 297*, 681–683.
O'Keefe, J., & Nadel, L. (1979). Precis of O'Keefe and Nadel's *The hippocampus as a cognitive map*. *Behavioral and Brain Sciences, 2*, 487–533.
Olton, D. S., Becker, J. T., & Handelmann, G. E. (1979). Hippocampus, space, and memory. *Behavioral and Brain Sciences, 2*, 313–365.
Olton, D. S., & Samuelson, R. J. (1976). Remembrance of places passed: Spatial memory in rats. *Journal of Experimental Psychology: Animal Behavior Processes, 2*, 97–115.
Overman, W. H., Pate, B. J., Moore, K., & Peuster, A. (1996). Ontogeny of place learning in children as measured in the radial arm maze, Morris search task, and open field task. *Behavioral Neuroscience, 110*, 1205–1228.
Owen, E. H., Christensen, S. C., Paylor, R., & Wehner, J. M. (1997). Identification of quantitative trait loci involved in contextual and auditory-cued fear conditioning in BXD recombinant inbred strains. *Behavioral Neuroscience, 111*, 292–300.
Paylor, R., Baskall, L., & Wehner, J. M. (1993). Behavioral dissociations between C57BL/6 and DBA/2 mice on learning and memory tasks: A hippocampal-dysfunction hypothesis. *Psychobiology, 21*, 11–26.
Paylor, R., Baskall-Baldini, L., Yuva, L., & Wehner, J. M. (1996). Developmental differences in place-learning performance between C57BL/6 and DBA/2 mice parallel the ontogeny of hippocampal protein kinase C. *Behavioral Neuroscience, 110*, 1415–1425.
Paylor, R., Johnson, R. S., Papaioannou, V., Spiegelman, B. M., & Wehner, J. M. (1994). Behavioral assessment of c-fos mutant mice. *Brain Research, 651*, 275–282.
Paylor, R., Tracy, R., Wehner, J. M., & Rudy, J. W. (1994). DBA/2 and C57BL/6 mice differ in contextual fear conditioning but not auditory fear conditioning. *Behavioral Neuroscience, 108*, 810–817.
Phillips, R. G., & LeDoux, J. E. (1992). Differential contribution of amygdala and hippocampus to cued and contextual fear conditioning. *Behavioral Neuroscience, 106*, 274–285.
Radcliffe, R. A., Lowe, M. V., & Wehner, J. M. (2000). Confirmation of contextual fear conditioning QTLs by short-term selection. *Behavior Genetics, 30*, 183–191.
Reuter, J. H. (1987). Tilt discrimination in the mouse. *Behavioural Brain Research, 24*, 81–84.
Routtenberg, A. (1995). Knockout mouse fault lines. *Nature, 374*, 314–315.
Rudy, J. W., & Sutherland, R. J. (1995). Configural association theory and the hippocampal formation: An appraisal and reconfiguration. *Hippocampus, 5*, 375–389.
Sarnyai, Z., Sibille, E. L., Pavlides, C., Fenster, R. J., McEwen, B. S., & Toth, M. (2000). Impaired hippocampal-dependent learning and functional abnormalities in the hippocampus in mice lacking serotonin (1A) receptors. *Proceedings of the National Academy of Sciences, 97*, 14731–14736.
Schafe, G. E., Atkins, C. M., Swank, M. W., Bauer, E. P., Sweatt, J. D., & LeDoux, J. E. (2000). Activation of ERK/MAP kinase in the amygdala is required for memory consolidation of Pavlovian fear conditioning. *Journal of Neuroscience, 20*, 8177–8187.

Selcher, J. C., Nekrasova, T., Paylor, R., Landreth, G. E., & Sweatt, J. D. (2001). Mice lacking the ERK1 isoform of MAP kinase are unimpaired in emotional learning. *Learning and Memory, 8*, 11–9.

Silva, A. J., Paylor, R., Wehner, J. M., & Tonegawa, S. (1992). Impaired spatial learning in α-calcium-calmodulin kinase II mutant mice. *Science, 527*, 206–211.

Silver, L. M. (1995). *Mouse genetics: Concepts and applications*. New York: Oxford University Press.

Simpson, E. M., Linder, C. C., Sargent, E. E., Davisson, M. T., Mobraaten, L. E., & Sharp, J. J. (1997). Genetic variation among 129 substrains and its importance for targeted mutagenesis in mice. *Nature Genetics, 16*, 19–27.

Squire, L. R. (1992). Memory and the hippocampus: A synthesis from findings with rats, monkeys, and humans. *Psychological Review, 99*, 195–231.

Talbot, C. J., Nicod, A., Cherny, S. S., Fulker, D. W., Collins, A. C., & Flint, J. (1999). High-resolution mapping of quantitative trait loci in outbred mice. *Nature Genetics, 21*, 305–308.

Tecott, L. H., & Wehner, J. M. (2001). Mouse molecular genetic technologies: Promise for psychiatric research. *Archives of General Psychiatry, 58,* 995–1004.

Upchurch, M., & Wehner, J. M. (1988). DBA/2Ibg mice are incapable of cholinergically-based learning in the Morris water task. *Pharmacology Biochemistry and Behavior, 29*, 325–329.

Valentinuzzi, V. S., Kolker, D. E., Vitaterna, M. H., Shimomura, K., Whiteley, A., Low-Zeddies, S., et al. (1998). Automated measurement of mouse freezing behavior and its use for quantitative trait locus analysis of contextual fear conditioning in (BALB/cJ × C57BL/6J)F2 mice. *Learning and Memory 5*, 391–403.

Waddell, S., & Quinn, W. G. (2001). Flies, genes and learning. *Annual Review of Neuroscience, 24*, 1283–1309.

Waters, N. S., Sherman, G. F., Galaburda, A. M., & Denenberg, V. H. (1997). Effects of cortical ectopias on spatial delayed-matching-to-sample performance in BXSB mice. *Behavioural Brain Research, 84*, 23–29.

Wehner, J. M., & Radcliffe, R. A. (2002). Interval specific congenic strains localize chromasome 1 QTL for fear conditioning [unpublished data].

Wehner, J. M., Radcliffe, R. A., & Bowers, B. J. (2001). Quantitative genetics and mouse behavior. *Annual Review of Neuroscience, 24*, 845–867.

Wehner, J. M., Radcliffe, R. A., Rosmann, S. T., Christensen, S. C., Rasmussen, D. L., Fulker, D. W., et al. (1997). Quantitative trait locus analysis of contextual fear conditioning in mice. *Nature Genetics, 17*, 331–334.

Wehner, J. M., & Silva, S. (1996). Importance of strain differences in evaluations of learning and memory processes in null mutants. *Mental Retardation and Developmental Disabilities Research Reviews, 2,* 243–248.

Wimer, R. E., & Fuller, J. L. (1966). Patterns of behavior. In E. L. Green (Ed.), *Biology of the laboratory mouse* (pp. 629–653). New York: McGraw-Hill.

Wolfer, D. P., & Lipp, H. P. (2000). Dissecting the behaviour of transgenic mice: Is it the mutation, the genetic background, or the environment? *Experimental Physiology, 85*, 627–634.

Yamasaki, M., Thompson, P., & Lemmon, V. (1997). CRASH syndrome: Mutations in L1CAM correlate with severity of the disease. *Neuropediatrics, 28*, 175–178.

Zirlinger, M., Kreiman, G., & Anderson, D. J. (2001). Amygdala-enriched genes identified by microarray technology are restricted to specific amygdaloid subnuclei. *Proceedings of the National Academy of Sciences, 98*, 5270–5275.

8

An Integrative Neuroscience Program Linking Mouse Genes to Cognition and Disease

Seth G. N. Grant

The year 2001 marked the arrival of almost complete genome sequence for four organisms with nervous systems, which are also the subject of intense genetic studies: humans, mice, flies, and worms. In the same way that the advent of molecular biology or electrophysiology opened grand new insights into the function of the nervous system, the genome era offers yet another exciting platform for discovery. As with any new set of tools, we want to see them result in improved forms of health care and in the discovery of new biological processes. This chapter does not aim to predict either the future of this "postgenomic revolution" or all of the ways genome information can be applied to understanding behavior. Here I outline a structure for a program of study, referred to as the Genes to Cognition program (G2C), which takes advantage of genomic information and combines it with a diverse set of methodologies. The G2C is driven by studies in basic genetic organisms with the aim of using this information to understand mechanisms of behavior and diseases of the human nervous system.

The potential of the G2C is illustrated using the biology of learning and memory. Learning is a fundamental cognitive process that has been at the center of mechanistic studies of neural function for more than a century. In the past decade, studies in rodents have led to the identification of a large number of genes involved with learning, which far outstrips those known in any other area of cognitive science. This knowledge can now be applied to humans, where it is likely to be relevant to the pathologies of learning impairment in children, dementias, schizophrenia, and brain injury.

By the very nature of the quest to link genes with behavior, it is necessary to construct a broadly integrative program that encompasses many distinct methods apart from genetics and psychology. These other areas include cell biology, electrophysiology, biochemistry, proteomics, microar-

I thank the following colleagues and collaborators for stimulating discussions and input: David Bentley, Douglas Blackwood, Robert Plomin, and Jane Rogers.

rays, brain imaging, and more. This raises a new and fascinating problem—of how to construct information networks that facilitate linking of these areas within a framework that not only provides rapid and simple access to information but also leads to the generation of new hypotheses and insights.

Despite the logic of constructing a framework linking large datasets ranging from molecular biology to psychology, and the inevitability of information accumulating and being organized in this manner, there needs to be a more purposeful drive to this goal. A potentially powerful focus of organization is based on the general recognition that biological functions are performed by sets of proteins (or genes) working together in pathways or as macromolecular machines (Alberts, 1998; Brent, 2000; Tjian, 1995). This logic can be adapted and applied to the study of behavior. Below are some guiding principles that underpin the G2C strategy:

1. Basic aspects of behavior and brain function are evolutionarily conserved.
2. Core biochemical processes (e.g., signaling pathways and protein complexes) conduct these functions.
3. Sets of genes encoding the core processes can be defined.
4. Core processes are disrupted by mutations and produce phenotypes in humans.
5. Multiple mutations or alleles cooperate to disrupt the core processes.
6. Behaviors and some psychiatric diseases are polygenic; this nature may reflect multiple genes encoding core processes.

The following sections outline aspects of a general strategy for the assembly of a knowledge base that encompasses a biological spectrum from gene to behavior.

General Strategy and Outline of the G2C

A key feature of this strategy is the linking of mouse and human genetics. Genes involved with learning can be discovered in mice (Grant et al., 1992; Grant & Silva, 1994; Silva, Paylor, Wehner, & Tonegawa, 1992). Because of the homology between mice and humans at the levels of the gene, the protein, the synapse, the brain region, and the behavior, there is a high probability that a gene involved with learning in a mouse will also be important in humans. This notion has been substantially supported by many studies.

An alternative approach to finding the genes involved with learning, or other behaviors, is to score for variation in the phenotype between individuals and then seek the genetic differences that correlate with these changes (see Plomin, chapter 11, this volume). This approach has been widely used and particularly successfully for finding large-effect genes underpinning some disorders (e.g., Huntington's disease).

Gene-targeting technology and the use of embryonic stem (ES) cells allow the experimentalist to modify the structure of any given gene or chromosome in the mouse in a controlled manner (Bradley, Zheng, & Liu, 1998). The widespread use of this technology has led to many hundreds of mouse genes being disrupted or modified. These mutant mice are routinely examined in a wide variety of assays, many of which are aimed at exploring the dysfunction of the nervous system. In this way, lists of genes are being developed in which the named genes are known to be required for the normal physiological function in question. These lists can be used to design experiments in humans, in which one asks, Does the behavioral abnormality in humans correspond to an altered structure of the corresponding human gene? To illustrate how this might work, let us consider the molecular mechanisms of learning as revealed by studies in the mouse and ask if this is informative for studies in humans. First, I provide a brief overview of the history of the molecular biology of learning.

Connecting the Molecular Mechanisms of Learning Between Mouse and Human

The recognition that sensory information is encoded in patterns of action potentials and transmitted into the brain (Adrian, 1928) led to the prediction that there must be some "metabolic" mechanism in neurons that is capable of detecting specific patterns of activity and converting them into some "structural" changes (Hebb, 1949). This hypothesis was made experimentally tractable when electrophysiologists found that synapses from the hippocampus, a region of the brain involved with learning, could be stimulated with different patterns of action potentials, and these patterns would induce increases or decreases in the efficiency of communication between two neurons (Bliss & Lomo, 1973). This system allowed pharmacological studies (Collingridge, Kehl, & McLennan, 1983) and mouse genetic studies to be used to identify molecules previously unknown in this process of synaptic plasticity (Grant et al., 1992). By using these methods, a large body of data accumulated during the 1990s, which essentially implicated in excess of 100 proteins in this biology without providing any unifying scheme or molecular hypothesis (Sanes & Lichtman, 1999).

Within this dataset, it was well established that a receptor-ion channel, known as the *N*-methyl-D-aspartate receptor (NR), was an essential component. The NR is a receptor for the excitatory neurotransmitter glutamate and on activation allows Ca^{2+} influx via its central pore. As is inappropriately illustrated in many textbooks and reviews, it would appear that this receptor simply sits in the membrane at the postsynaptic side of the synapse, where it injects Ca^{2+} into the dendrite, which then diffuses to activate a variety of enzymes that seem to float freely in the cytoplasm. These enzymes then drive various poorly understood signaling pathways that control neuronal properties. The first evidence that the NR and signaling proteins do not function in this way came when transgenic mice carrying a mutation in the Post Synaptic Density 95 protein (PSD-95), which normally binds the NR, were found to produce striking changes in the properties of synaptic plasticity and learning (Migaud et al., 1998). This work predicted that there are multiprotein signaling complexes comprised of NR and PSD-95 with other proteins, which control learning.

This genetic evidence for a multiprotein complex was used to justify a proteomic analysis: biochemical isolation of the protein complexes from brain and identification of proteins using mass spectrometry and immunoblotting (Husi & Grant, 2001a; Husi, Ward, Choudhary, Blackstock, & Grant, 2000). These methods, which have general applicability to other receptor complexes (Husi & Grant, 2001b), showed that the NR PSD-95 complexes were approximately 2,000–3,000 kDa, which is several-fold more than would be expected if it was simply the NR subunits alone. A picture emerged of 75 or more proteins that could be broadly categorized into five classes: neurotransmitter receptors, cell adhesion molecules, adaptors, signaling enzymes, and cytoskeletal proteins (for more details, see Husi et al., 2000). A major surprise in this study was that at least 27 proteins from the complexes were known to be required for synaptic plasticity and 18 for learning in rodents and were from each of the five classes of complex components. Thus the organization of these proteins into these multiprotein complexes suggests that they work together in a large "machine," not unlike many other multiprotein molecular machines. The importance of this concept is that it removes the focus of interest away from the individual molecules onto the function of the overall machine. My colleagues and I (Migaud et al., 1998) have proposed that these complexes are a "device" for detecting patterns of synaptic activity and for converting this information into intracellular signals that store the information in the cell. In this way, electrical information can be translated into cellular memory.

These properties were at the basis of Hebb's postulate, and these complexes have been described as *Hebbosomes,* multiprotein complexes that convert patterns of neuronal activity into cellular changes underlying learning. It is beyond the scope of this chapter to broadly discuss Hebbosomes, except to indicate that there are families of such complexes, with different molecular composition, which confer specific properties to different synapses.

The characterization of Hebbosomes has significant implications for human genetics. Three genes previously known in humans to be involved with cognitive deficits were also found to encode proteins found in the complexes. These include two signal transduction enzymes: neurofibromin (also known as NF-1 and mutant in the neurofibromatosis syndrome) and RSK-2 (mutant in the Coffin Lowry syndrome) and the adhesion protein L1 (mutant in CRASH syndrome). These observations open the exciting possibility that other human cognitive disorders that have a genetic component may involve genes encoding the proteins in these complexes. In this way, the named genes from the mouse studies can be used as candidate genes in a human association study.

In the simplest setting, knowing that a mouse gene is important for behavior is a reasonable starting point for a human study. There are a number of potential weaknesses in this setting. For example, the human gene may not be as important to the human as it is to the mouse. A stronger starting point is not to rely on a single gene but to use the information about that gene to build up a set of genes. As described in the example above, one could consider that PSD-95 was a starting point, because the mouse knock out had severe learning deficits (Migaud et al., 1998). By isolating the PSD-95 containing complexes and using proteomic tools, it became clear that at least 75 proteins could now be considered candidates for association studies. Thus a nongenetic strategy, such as proteomics, can be used in conjunction with the genetics to identify molecules involved with learning.

A Multilayer Organization

The G2C can be organized into four layers (see Figure 8.1). These layers are briefly summarized here and discussed in more detail later.

The entry point for this strategy (Layer 1) is molecular information derived from basic science studies. Strong emphasis is placed on the value of genetically modifiable organisms with nervous systems (invertebrates: fruit fly, *Drosophila*; worm, *Caenorhabditis elegans*; vertebrates: mouse, *Mus musculus*; zebra fish, *Danio rerio*). Through the use of genetic screens and mutations, these organisms have generated lists of proteins that are involved with various phenotypes. Compiling the set of genes that are involved in a common phenotype (e.g., learning) or involved in a multiprotein complex, or some other ways of classifying sets, produces useful information for a human genotyping study. A prototype for this set is that derived from the molecular studies of the multiprotein complexes (Hebbosomes) underlying acquisition of learning (Husi et al., 2000).

Layer 2 of the G2C takes forward the candidate genes from Layer 1 into human genotyping. Using genome sequencing technology, human single-nucleotide polymorphisms (SNPs) can be determined for all genes in the set and DNAs from relevant humans genotyped. Given the rapid pace of the SNP identification and characterization, information covering the first phase of this should be available in the public domain in the very near future.

Layer 3 of the G2C is aimed at validating the biological significance of variant alleles found in humans. Here functional assays are required, and mouse ES cell technology again is used to provide several complementary in vivo and in vitro approaches. One could assemble a wide range of molecular and neuroscience methods in a highly integrative research program. These neurobiological studies can be linked to human neurobiological studies, thus providing a broad framework of connections at many levels of analysis.

There will be an important role for informatics at all stages of the G2C, and Layer 4 is the platform for this technology. This will include access to existing databases as well as generating new databases. These

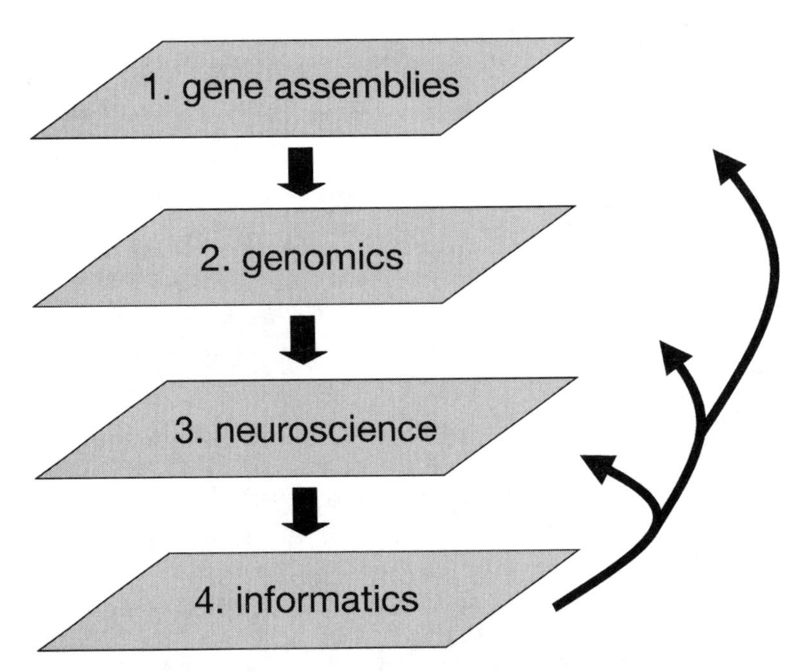

Figure 8.1. Overview of organizational layers. The four main layers of the Genes to Cognition program (G2C) are shown as flat planes. The entry point to the program is primarily via Layer 1 (see Figure 8.2 for detail). In Layer 1, a set of genes is defined according to various criteria and driven by basic neuroscience studies. In Layer 2 (see Figure 8.3 for detail), the genes in Layer 1 are used in human genotyping assays to seek putative functional variants. These variants are examined in biological assays in Layer 3 (see Figure 8.4 for detail). Underpinning and central to the integration of the data in all layers is bioinformatic tools (Layer 4). This information is transferred between layers and is used to modify the experiments as the G2C develops. The arrows indicate the simplest flow of information between layers.

databases and links should generate a novel and valuable resource for the scientific and medical community.

Layer 1: Identification of Genes Encoding Assemblies

Sets of genes are defined using several sources of information (see Figure 8.2). This layer requires bioinformatics and expertise of scientists within the area of basic biology. Types of molecular information that will be used to select genes include the following: (a) mutant phenotypes of mice and other genetic organisms, (b) knowledge of molecular pathways, (c) protein interaction networks obtained from proteomic and yeast 2-hybrid screens, and (d) gene families, chromosomal organization, and syntenic regions between human and mouse. This prioritization of genes will provide the information for Layer 2.

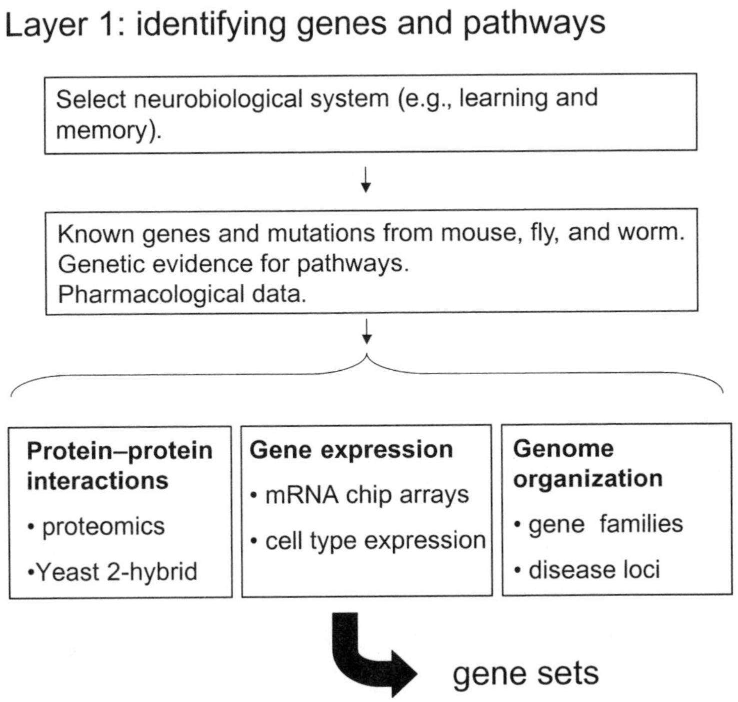

Layer 1: identifying genes and pathways

Select neurobiological system (e.g., learning and memory).

Known genes and mutations from mouse, fly, and worm.
Genetic evidence for pathways.
Pharmacological data.

Protein–protein interactions
• proteomics
•Yeast 2-hybrid

Gene expression
• mRNA chip arrays
• cell type expression

Genome organization
• gene families
• disease loci

gene sets

Figure 8.2. Schematic representation of Layer 1. Layer 1 of the Genes to Cognition program (G2C) is aimed at defining sets of genes encoding "assemblies" or functional sets of genes. The first step is choice of neurobiological problem for which there is considerable molecular data from basic neuroscience experiments, such as learning and memory. Using all available molecular data, with an emphasis on the value of genetic organisms, as well as other data (illustrated in boxes), sets of candidate genes for future human studies are chosen. These genes are used in Layer 2.

Layer 2: Genomics

The overall goal of Layer 2 is to identify variant structures in specific human genes, which are candidates for detailed functional testing (see Figure 8.3). The basic gene structure for those loci that have been selected and prioritized according to Layer 1 of the G2C will be determined for human and mouse using available finished sequence from the various genome sequencing projects (see www.sanger.ac.uk for information on genome projects). The comparative gene structure of mouse and human serves several purposes. First, it provides a basis for comparing gene structure and assigning intron/exon and other regulatory features to the sequence (Wiehe, Guigo, & Miller, 2000). This information is useful in designing genotyping strategies, including those involving SNP detection. A second reason for obtaining mouse sequence is that in Layer 3 of the G2C this information is useful as a guide for construction of gene-targeting vectors for engineering specific mutations into the mouse.

A major collaborative international effort is under way to identify

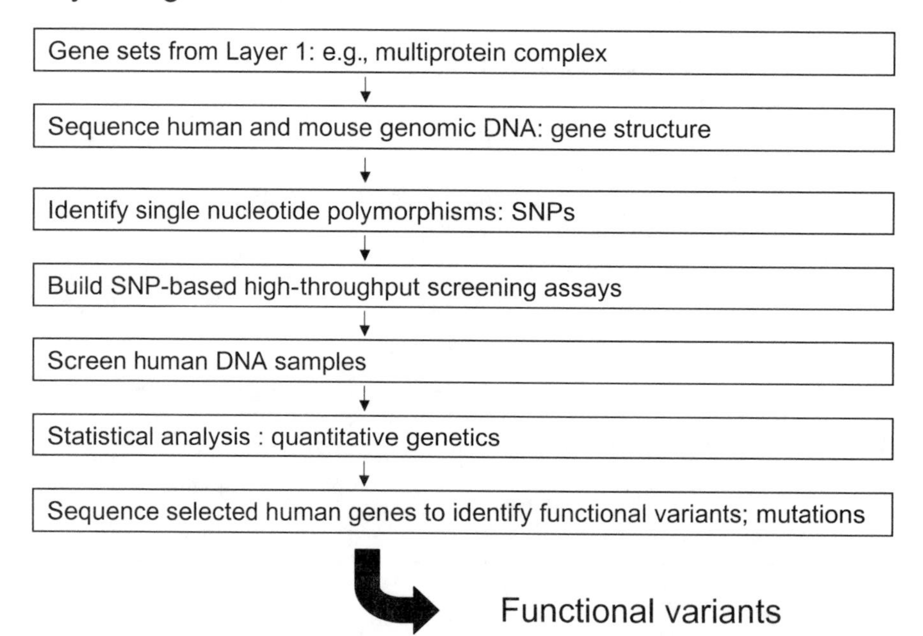

Figure 8.3. Schematic representation of Layer 2. Layer 2 of the Genes to Cognition program (G2C) sets out a scheme for screening DNA from humans for polymorphisms in genes identified in Layer 1. Details relevant to the individual boxes are described in the body of the text. The output from this layer is information on putative functional variants that can be tested in Layer 3 (see Figure 8.4).

SNPs in the human genome. This SNP Consortium (Altshuler et al., 2000; Isaksson et al., 2000; Masood, 1999; Mullikin et al., 2000; Sachidanandam et al., 2001) aims to generate sufficient numbers of SNPs that can then be used in high-throughput genotyping assays (Fors, Lieder, Vavra, & Kwiatkowski, 2000; Kokoris et al., 2000; Kwok, 2000). Statistical analysis (Bader, 2001; Niu, Struk, & Lindpaintner, 2001) of SNP frequency in populations is used to implicate a gene in the phenotype relevant to the human DNAs. The identification of statistical association will motivate resequencing of the alleles in affected individuals to identify potential functional variants. Sequence information may predict the nature of the functional impairment, such as premature termination codons, and these putative functional variants will be tested in Layer 3.

There are potentially interesting features of a genotyping strategy based on genes encoding proteins known to be components of pathways. As has been shown in model genetic organisms, construction of compound mutations allows one to examine the functional relationship between the two genes. *Epistatic* interactions between genes (traditionally defined as the presence of one allele at one locus preventing the expression of an allele at a different locus) is a feature of genes encoding proteins in common pathways. By extension, it may be that some diseases manifest symptoms only if a pathway is debilitated, and this may require the presence of two affected genes. Thus, statistical analysis of the set of genes in Layer 1 may show that sets of SNPs identifying particular variant genes will detect these genes. The effects of these variant genes alone may not give statistical significance association with the disease, although the subsets of genes may do so.

Layer 3: Functional Genomics–Experimental Neuroscience

An output from Layer 2 will be variants in the sequence of a human gene. In addition to the statistical analysis used to make a case that a variant gene may be at the basis of some altered phenotype in humans, there is a need to generate biological data showing this variant has function consequences. The simplest way forward may be to use some kind of specific in vitro assay that is sensitive to the function of the protein involved. The G2C includes this aspect; however, it proposes to use a wider, integrative program of study where the variant is tested in sets of assays relevant to the cells on one hand and the cognitive processes on the other—in other words, many assays at the molecular, cellular, and animal level (see Figure 8.4).

Studying gene function in the nervous system requires general tools applicable to neurons and glia. This is in contrast to some areas of cell biology, such as DNA replication or growth control, which can be studied in generic cells. Moreover, in the context of heritable differences in gene structure and the implications for behavior, it is ultimately necessary to study the gene in the context of the whole animal. Gene targeting in mouse provides an ideal way to bridge the gap between cell biology in cultured neurons and biology of the whole animal. This is because of the pluripotential nature of ES cells and thus the ability to derive cells and animals from the same genetically modified cell. Layer 3 outlines some of the applications of mouse gene targeting and the analysis of the mice.

Gene targeting in mouse ES cells (Box 2) is ideally suited for studies of human gene function in complex organs such as the brain because almost any type of gene or chromosomal engineering is feasible in ES cells. The following are some of the relevant technologies:

1. Gene knock out (complete disruption of expression; Cheah & Behringer, 2000; DeChiara, 2001), including gene traps (Brown & Nolan, 1998; Medico, Gambarotta, Gentile, Comoglio, & Soriano, 2001).
2. Point mutation and other fine mutations (Brown & Nolan, 1998); this may be particularly useful for introducing SNPs into mouse genes.
3. Larger sequence modification, including "humanization" or substitution of human wild-type or mutant genes for mouse genes; this uses techniques of chromosomal engineering (Mills & Bradley, 2001).

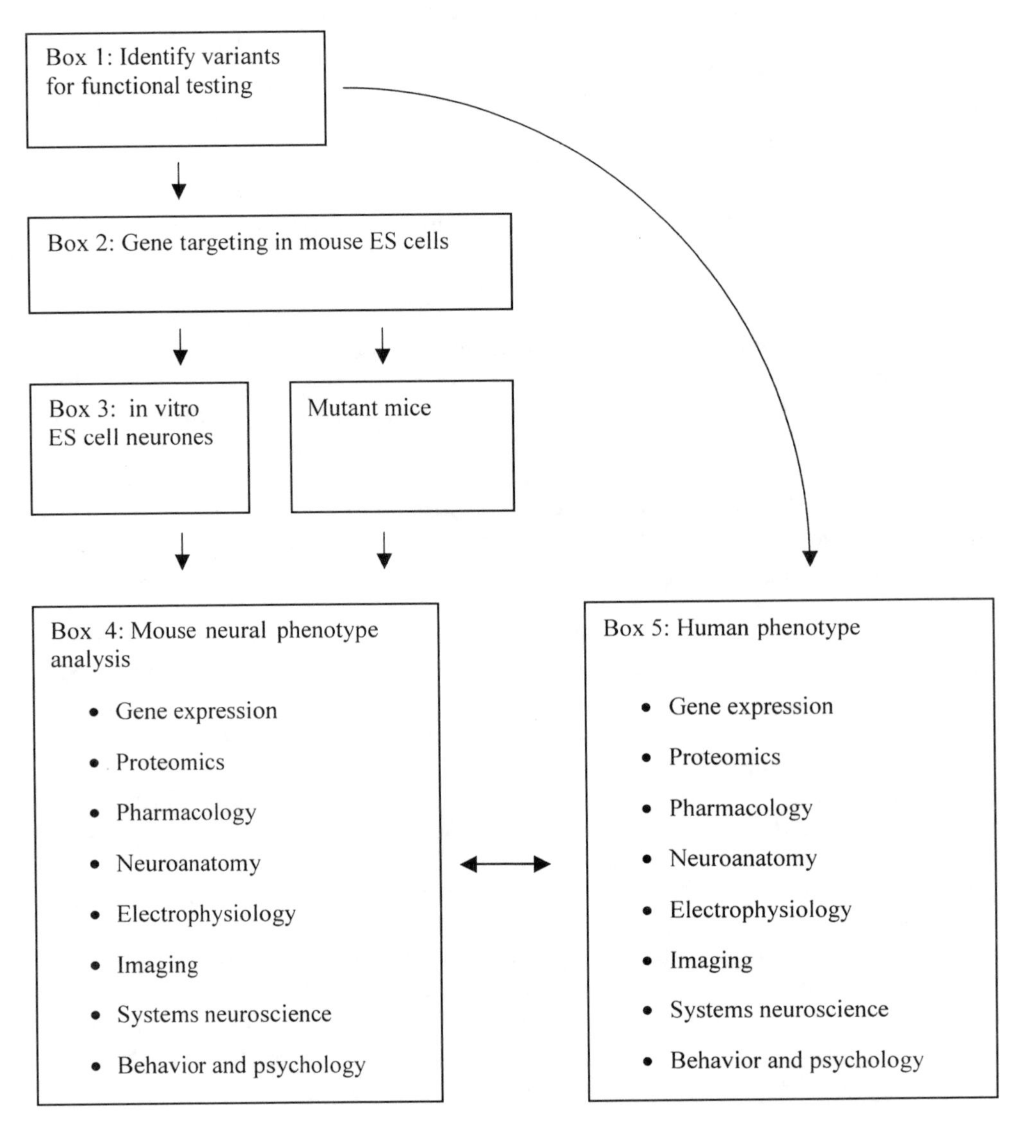

Figure 8.4. Schematic representation of Layer 3. Layer 3 shows a simplified flow-chart beginning with the sequence of putative human variant gene and an outline of assays based on mouse embryonic stem (ES) cell technology. These ES cell experiments are based on generating mutant cell lines that can be used to generate mutant mice (Box 4) as well as in vitro differentiation into neurons (Box 3; see Figure 8.5 for more detail). A list of phenotypes that can be scored are shown in Box 4 for mice, and a comparable set of human phenotype data can also be drawn from the profiling of the subjects used for the DNA used in Layer 2. The arrow from Box 1 to Box 5 is meant to indicate that the human phenotypes listed are those relevant to the DNAs that were used in the genotype screen.

4. Conditional gene modification; these methods allow the desired genetic modification to be "active" or "inactive" in a desired cell (neurone or specific neuronal population) at a specific time during the lifespan of the animal (Le & Sauer, 2000; Mansuy & Bujard, 2000). For example, a gene that regulates synaptic plasticity, which may be encoded in almost all neurons, can be inactivated in a set of neurons in a selected brain region (e.g., hippocampus) and the effects on cognitive functions assessed.
5. Rescue of knock-out allele with mutant or variant genes (Kojima et al., 1997).
6. Insertion of reporter constructs to monitor gene expression (Migaud et al., 1998) and subcellular localization of proteins (e.g., Green Fluorescent Protein technology).

Although mutant mice are useful, there are some neuronal phenotypes that can be studied in neurons grown in culture (see Figure 8.4, Box 4). A major limitation to the study of synapse function has been the lack of clonal cell lines that form synapses with the properties of central nervous system synapses. Very recently, it was found that totipotent murine ES cells can be induced to differentiate in culture into neurones (*e*mbryonic *s*tem cell *n*eurons; ESNs) comparable with those prepared from neonatal cortex (Bain & Gottlieb, 1998). Importantly, the ESNs display the ability to form functional synapses. Combining gene targeting with ESN technology allows the creation of mutant neurons in vitro. This opens the possibility toward various in vitro screens in mutant neurons (see Figure 8.5).

The phenotype of the cells, animal tissues, and whole animal can be systematically studied in a variety of studies ranging from the molecular to the psychological (see example list in Figure 8.4, Box 5). It is unnecessary here to break this list into further detail but rather to draw attention to the value of multiple lines of experimental analysis. The first advantage of testing a variant allele in multiple assays is that it makes it more likely that a phenotype can be identified. A greater challenge is to understand why a variant allele may be involved with the human phenotype. Here it is necessary to have some information on the brain at many levels. For example, if one were to only examine synapse function, one may overlook some other critical role in, say, glial function. The advantage of the mouse is that it is possible to explore many levels using ethically acceptable approaches, unlike humans, for which it is not possible to perform similarly invasive procedures. Thus, it is necessary to compare and contrast at those levels where it is possible—the phenotype of mouse and human (see Figure 8.4).

Comparison of mouse and human phenotypes can be pursued on two levels: (a) comparing the mouse and human where each carries a mutation in the same gene and (b) comparing similar phenotypes where the genetic basis in humans is unknown. As illustrated in Boxes 4 and 5 of Figure 8.4, it would be important to have detailed annotation of phenotype information, assembled in appropriate databases, so that genotype information could be used to ascribe gene function to a phenotype. Developing "neu-

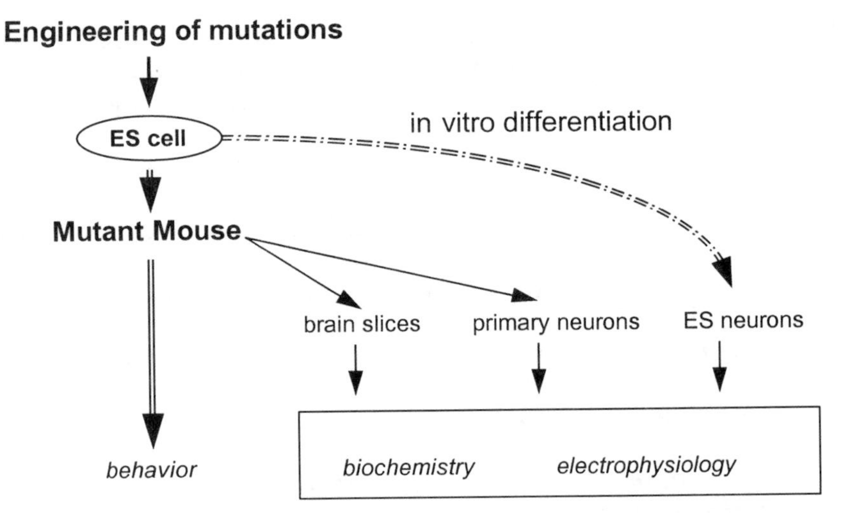

Figure 8.5. Using mutant mouse embryonic stem (ES) cells to generate both mice and neurons in vitro. Mutations or chromosomal alterations are engineered into mouse ES cells. These cells can be used to create mutant mice via blastocyst injection or differentiated in vitro to produce neurons (ES neurons). The mutant mice can be the source of a variety of neural tissue preparations including brain slice and primary neuronal cultures. Thus neurons carrying mutations can be derived from a variety of methods and used in many types of assays, ranging from behavior of the intact animal to electrophysiology and biochemistry.

roscience phenotyping" assays for comparison of humans and mouse is an area that needs further development. Many tests have been developed for rats and are readily transferable to mice. This program of research could promote further efforts to improve and find new ways to examine neurological phenotypes in mice.

Layer 4: Informatics

This broadly integrated program places emphasis on the need for user mobility between datasets as well as storage and recall of information. Linking the datasets together is perhaps one of the most difficult challenges, and a fluid interface would be extremely important. Here are some examples of questions that a well-designed informatic interface may be able to handle:

- List all genes that are encoded in a particular region of chromosome 6 and expressed in the hippocampus. Further sort those genes into those that are known to be important for development or synaptic plasticity of the hippocampus.
- Identify the regions of the human brain that are altered in functional magnetic resonance imaging studies in various genetic diseases, and contrast the regions with the known gene expression profiles and the biochemical function of these genes.

- Catalog the multiprotein complexes involved with synaptic signaling, and list the corresponding human syndromes involving those genes.
- List the genotyping assays that could be used to differentially diagnose chosen psychiatric disorders.
- List the human polymorphisms that result in altered expression of neuronal membrane proteins, and link this to drugs known to modulate those proteins.

Although some aspects of these questions could be answered today, the amount of labor involved with the current data mining tools is enormous. In principle, it should be possible to have answers to these questions in just hours with appropriately designed databases and search engines.

Structural Issues for a Large Multidisciplinary Program

The G2C approach is well-suited to be organized as a national or international consortium. The structure can be broken down into component areas, such as the four layers, where these areas are managed by experts. A relatively fixed framework with considerable flexibility would be desirable during the phase of construction. The initial phase of construction could proceed around the existing framework that has developed from the study of learning. This framework, which could be described as the vertical component because it links genes, proteins, cells, circuits, and behavior, presents the major challenges in terms of constructing the new databases, the general management, and coordination. Once this was established, the G2C could be expanded using several strategies. First, using the entry point at Layer 1, new systems or assemblies could be identified and processed. For example, there are areas of developmental neuroscience in which sets of genes and pathways have been identified. Here, experts in these areas could meet and discuss Layer 1, whereupon Layer 2 could be performed by other genotyping experts. In the broadest sense, the consortium could be seen as a platform for scientists in specialized areas to join a program that connects their work with those at other points in the vertical organization.

A second driving force is the collection of human DNA samples. In the area of clinical neuroscience (psychiatry, neurology, and neurosurgery), there has been substantial effort placed on the collection of DNAs from individuals and families with heritable conditions. It is outside the scope of this chapter to review this in detail; however, it is worth noting that there is a considerable need for further collection, including for conditions that have not gained as much attention as some of the major disorders. The availability of these DNAs for Layer 2 of the G2C would influence the choice of gene sets from Layer 1.

A third way to expand the G2C is around the cell biology of neurons and glia. For example, my colleagues and I have initiated the G2C by focusing on a particular synaptic multiprotein complex. The next step could be to include other synaptic protein complexes and move toward genotyping for all synaptic proteins. The number of synaptic proteins is probably around 2,000, and many of these are now known. These could also be used to constitute a set for Layer 1. Thus, one could expand this approach to include dendritic, axonal, and other sets of neuronal proteins. Similarly, this type of categorization could be applied to glial cells. The use of microarray technology for monitoring messenger RNA expression, will assist in generating these data, especially as arrays encoding all known transcripts are available. In the broadest sense, it would be possible to extend the G2C to examine all genes expressed in the brain.

Conclusion

In the same way that one now understands biochemical pathways underlying various basic biological function including metabolism, cell proliferation, and differentiation, the core molecular pathways utilized in neurons for the generation of behaviors will be found. As is already clear, aberrations in these pathways arising from genetic variation will provide a molecular basis for altered behavior. These molecular insights will provide not only useful diagnostic and therapeutic avenues but also potentially profound insights into human cognition. The development of tools used to link areas of neuroscience through vertical layers ranging from psychology to gene structure will be a valuable resource for new areas of curiosity-led research as well as industrial application.

Perhaps the greatest biological challenge of the 21st century is to understand the mechanisms of human behavior. This intellectual challenge is of enormous practical significance given the burden of neuropsychiatric disease. Here a general discovery framework for understanding behavior at the molecular, cellular, and systems neuroscience levels in humans is outlined. The key hypothesis driving this program is that molecular assemblies are core components of neurobiology and behavior and are the fundamental defective unit in diseases of the brain. A new approach to human studies of cognition and brain disease can be driven by basic science in model genetic organisms and combined with an integrative program of genome sequencing and neuroscience.

References

Adrian. E. (1928). *The basis of sensation, the action of the sense organs*. New York: Norton.

Alberts, B. (1998). The cell as a collection of protein machines: Preparing the next generation of molecular biologists. *Cell, 92,* 291–294.

Altshuler, D., Pollara, V. J., Cowles, C. R., Van Etten, W. J., Baldwin, J., Linton, L., et al. (2000). An SNP map of the human genome generated by reduced representation shotgun sequencing. *Nature, 407*, 513–516.

Bader, J. S. (2001). The relative power of SNPs and haplotype as genetic markers for association tests. *Pharmacogenomics, 2,* 11–24.

Bain, G., & Gottlieb, D. (1998). I. Neural cells derived by in vitro differentiation of P19 and embryonic stem cells. *Perspective Developmental Neurobiology, 5,* 175–178.

Bliss, T. V., & Lomo, T. (1973). Long-lasting potentiation of synaptic transmission in the dentate area of the anaesthetized rabbit following stimulation of the perforant path. *Journal of Physiology, 232,* 331–356.

Bradley, A., Zheng, B., & Liu, P. (1998). Thirteen years of manipulating the mouse genome: A personal history. *International Journal of Developmental Biology, 42,* 943–950.

Brent, R. (2000). Genomic biology. *Cell, 100,* 169–183.

Brown, S. D., & Nolan, P. M. (1998). Mouse mutagenesis-systematic studies of mammalian gene function. *Human Molecular Genetics, 7,* 1627–1633.

Cheah, S. S., & Behringer, R. R. (2000). Gene-targeting strategies. *Methods in Molecular Biology, 136,* 455–463.

Collingridge, G. L., Kehl, S. J., & McLennan, H. (1983). Excitatory amino acids in synaptic transmission in the Schaffer collateral–commissural pathway of the rat hippocampus. *Journal of Physiology (London), 334,* 33–46.

DeChiara, T. M. (2001). Gene targeting in ES cells. *Methods in Molecular Biology, 158,* 19–45.

Fors, L., Lieder, K. W., Vavra, S. H., & Kwiatkowski, R. W. (2000). Large-scale SNP scoring from unamplified genomic DNA. *Pharmacogenomics, 1,* 219–229.

Grant, S. G., O'Dell, T. J., Karl, K. A., Stein, P. L., Soriano, P., & Kandel, E. R. (1992). Impaired long-term potentiation, spatial learning, and hippocampal development in fyn mutant mice. *Science, 258,* 1903–1910.

Grant, S. G., & Silva, A. J. (1994). Targeting learning. *Trends in Neuroscience, 17,* 71–75.

Hebb, D. O. (1949). *The organization of behavior: A neuropsychological theory.* New York: Wiley.

Husi, H., & Grant, S. G. (2001a). Isolation of 2000-kDa complexes of *N*-methyl-D-aspartate receptor and postsynaptic density 95 from mouse brain. *Journal of Neurochemistry, 77,* 281–291.

Husi, H., & Grant, S. G. (2001b). Proteomics of the nervous system. *Trends in Neuroscience, 24,* 259–266.

Husi, H., Ward, M. A., Choudhary, J. S., Blackstock, W. P., & Grant, S. G. (2000). Proteomic analysis of NMDA receptor-adhesion protein signaling complexes. *Nature Neuroscience, 3,* 661–669.

Isaksson, A., Landegren, U., Syvanen, A. C., Bork, P., Stein, C., Ortiago, F., et al. (2000). Discovery, scoring and utilization of human single nucleotide polymorphisms: A multidisciplinary problem. *European Journal of Human Genetics, 8,* 154–156.

Kojima, N., Wang, J., Mansuy, I. M., Grant, S. G., Mayford, M., & Kandel, E. R. (1997). Rescuing impairment of long-term potentiation in fyn-deficient mice by introducing Fyn transgene. *Proceedings of the National Academy of Sciences USA, 94,* 4761–4765.

Kokoris, M., Dix, K., Moynihan, K., Mathis, J., Erwin, B., Grass, P., et al. (2000). High-throughput SNP genotyping with the Masscode system. *Molecular Diagnosis, 5,* 329–340.

Kwok, P. Y. (2000). High-throughput genotyping assay approaches. *Pharmacogenomics, 1,* 95–100.

Le, Y., & Sauer, B. (2000). Conditional gene knockout using cre recombinase. *Methods Molecular Biology, 136,* 477–485.

Mansuy, I. M., & Bujard, H. (2000). Tetracycline-regulated gene expression in the brain. *Current Opinion on Neurobiology, 10,* 593–596.

Masood, E. (1999). As consortium plans free SNP map of human genome. *Nature, 398,* 545–546.

Medico, E., Gambarotta, G., Gentile, A., Comoglio, P. M., & Soriano, P. (2001). A gene trap vector system for identifying transcriptionally responsive genes. *Natural Biotechnology, 19,* 579–582.

Migaud, M., Charlesworth, P., Dempster, M., Webster, L. C., Watabe, A. M., Makhinson, M., et al. (1998). Enhanced long-term potentiation and impaired learning in mice with mutant postsynaptic density-95 protein. *Nature, 396,* 433–439.

Mills, A. A., & Bradley, A. (2001). From mouse to man: Generating megabase chromosome rearrangements. *Trends in Genetics, 17,* 331–339.

Mullikin, J. C., Hunt, S. E., Cole, C. G., Mortimore, B. J., Rice, C. M., Burton, J., et al. (2000). An SNP map of human chromosome 22. *Nature, 407,* 516–520.

Niu, T., Struk, B., & Lindpaintner, K. (2001). Statistical considerations for genome-wide scans: Design and application of a novel software package polymorphism. *Human Heredity, 52*, 102–109.

Sachidanandam, R., Weissman, D., Schmidt, S. C., Kakol, J. M., Stein, L. D., Marth, G., et al. (2001). A map of human genome sequence variation containing 1.42 million single nucleotide polymorphisms. *Nature, 409*, 928–933.

Sanes, J. R., & Lichtman, J. W. (1999). Can molecules explain long-term potentiation? *Nature Neuroscience, 2,* 597–604.

Silva, A. J., Paylor, R., Wehner, J. M., & Tonegawa, S. (1992). Impaired spatial learning in alpha-calcium-calmodulin kinase II mutant mice. *Science, 257*, 206–211.

Tjian, R. (1995). Molecular machines that control genes. *Scientific American, 272*, 54–61.

Wiehe, T., Guigo, R., & Miller, W. (2000). Genome sequence comparisons: Hurdles in the fast lane to functional genomics. *Brief Bioinform, 1,* 381–388.

Part IV

Cognitive Abilities

9

Genetic Contributions to Anatomical, Behavioral, and Neurophysiological Indices of Cognition

Daniëlle Posthuma, Eco J. C. de Geus, and Dorret I. Boomsma

The large genetic contribution to individual differences in cognitive abilities is well established (Bouchard & McGue, 1981; Plomin, Owen, & McGuffin, 1994). From childhood to early adulthood, the relative impact of genetic factors on cognitive abilities increases (Boomsma & van Baal, 1998; Cherny & Cardon, 1994) and becomes even higher from middle adulthood (Posthuma, de Geus, & Boomsma, 2001; Posthuma, Neale, Boomsma, & de Geus, 2001) to late adulthood (Plomin, Pedersen, Lichtenstein, & McClearn, 1994). Data from four large twin studies from the Dutch Twin Registry (see Table 9.1; Figure 9.1), which are partly longitudinal and partly cross-sectional, reflect this increasing heritability of cognitive abilities with age (Bartels, Rietveld, van Baal, & Boomsma, in press; Boomsma & van Baal, 1998; Posthuma, de Geus, et al., 2001; Rietveld, Dolan, van Baal, & Boomsma, in press; Rijsdijk & Boomsma, 1997; Rijsdijk, Boomsma, & Vernon, 1995). Shared environmental influences play a role only before adolescence and are of relatively low importance between ages 7 and 16. This pattern of the relative impact of genetic and environmental influences on cognitive abilities corresponds to that found in many other countries (Plomin, Chipuer, & Neiderhiser, 1994; Plomin, DeFries, & McClearn, 1990).

In spite of the overwhelming evidence for the existence of "genes for cognition," actual identification of such genes is limited to neurological mutations with rather severe cognitive effects (e.g., Pick's disease, X-linked mental retardation, and Huntington's disease), as reviewed by Flint (1999). Like the many rare diseases and disorders listed in online *Mendelian Inheritance in Man* (OMIM; McKusick, 1998), these genetic defects of cognition are largely Mendelian in nature. True polygenes (or quantitative trait locis; QTLs) that influence the normal range of cognitive ability have yet to be identified. One route to finding these genes is a better appreciation of individual differences in the anatomy and function of the main organ of information processing, the brain.

Table 9.1. Twin Correlations for Full Scale IQ in Dutch Twins (and Their Siblings) for Ages 5 to 50 Years

Age	MZF	DZF	MZM	DZM	DOS	MZ	DZ	Test
5 years	.77	.72	.75	.56	.61	.77	.62	RAKIT
N	46	37	42	43	39	88	119	
7 years	.74	.59	.61	.40	.53	.68	.50	RAKIT
N	41	34	37	41	38	78	113	
10 years	.83	.47	.75	.66	.47	.82	.50	RAKIT
N	43	37	38	41	37	81	115	
12 years	.85	.66	.85	.58	.31	.85	.54	WISC
N	43	37	36	39	35	79	111	
16 years	.50	.35	.77	.24	.42	.66	.39	RAVEN
N	46	36	37	31	44	83	111	
18 years	.84	.44	.86	.19	.24	.85	.30	WAIS
N	46	36	37	31	44	83	111	
26 years	.87	.36	.88	.64	.45	.88	.45	WAIS–3R
N	29	97	25	76	110	54	283	
50 years	.85	.52	.85	.27	.42	.85	.42	WAIS–3R
N	26	85	22	62	95	48	242	

Note. For the adult twins, the ages 26 (*SD* = 4.19) and 50 (*SD* = 7.51) are average ages (as opposed to the other age groups in which all subjects are of the same age). Additional siblings are modeled as similar to DZ twins (i.e., sharing on average 50% of genes and 100% shared environment) at ages 26 and 50 years. MZF = monozygotic female; DZF = dizygotic female; MZM = monozygotic male; DZM = dizygotic male; DOS = dizygotic opposite sex; RAKIT = Revisie Amsterdamse Kinder Intelligentie Test; WISC = Wechsler Intelligence Scale for Children; RAVEN = Raven's Advanced Progressive Matrices; WAIS = Wechsler Adult Intelligence Scale; WAIS–3R = WAIS 3rd edition, Revised.

Anatomical Measures of Cognitive Ability

An obvious source of individual differences in cognitive abilities is the size of the brain. Since the second half of the 19th century, positive relations between head size and intelligence have been observed. Correlations generally range around 0.20 (Jensen, 1994; Posthuma, Neale, et al., 2001) but can be as high as 0.44 (van Valen, 1974). Head size is usually measured with a measuring tape as circumference of the head. A more accurate measure of the size of the brain can be obtained through magnetic resonance imaging (MRI).

Willerman, Schultz, Rutledge, and Bigler (1991) correlated brain size as measured through MRI with IQ (measured with the revised Wechsler Adult Intelligence Scale, WAIS–R; Wechsler, 1981) in a sample of 40 unrelated participants. They found a correlation of 0.51, which was higher in men (0.65) than in women (0.35). In a follow-up study, Willerman,

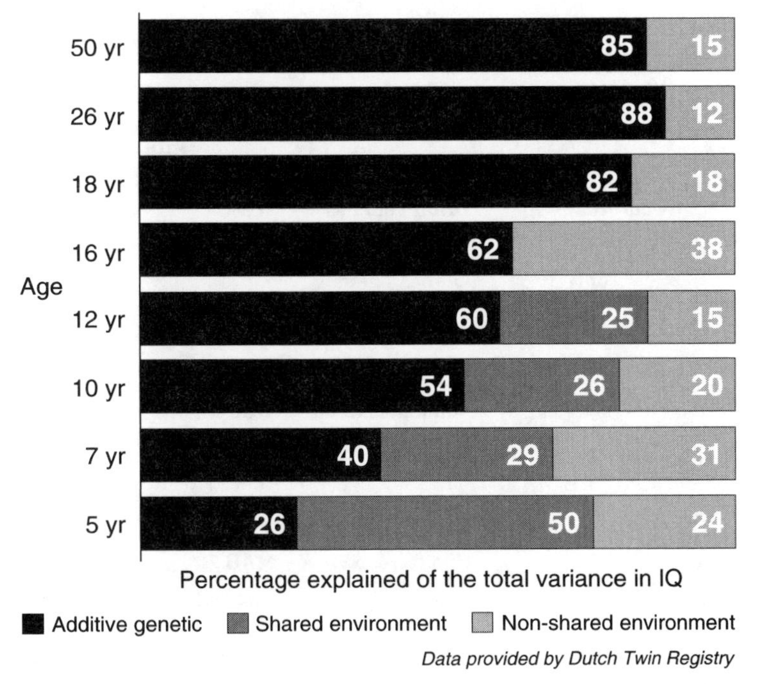

Figure 9.1. Decomposition of the variance in Full Scale IQ into additive genetic variance, shared environmental variance, and nonshared environmental variance, at different ages. Studies are specified in Table 9.1. Significance of additive genetic influences is .03 at age 5; of shared environmental influences it is .06 at age 7, .09 at age 10, and .08 at age 12.

Schultz, Rutledge and Bigler (1992) suggested that, in men, a relatively larger left hemisphere better predicted verbal IQ than it predicted performance IQ, whereas in women the opposite was true. Since then, several studies have provided confirmative evidence that brain volume and IQ correlate around 0.40 (e.g., Andreasen et al., 1993; Egan et al., 1994; Raz et al., 1993; Storfer, 1999; Wickett, Vernon, & Lee, 2000).

In a large MRI study including 111 twin pairs and 34 additional siblings, the heritability of volumes of several brain structures was investigated (Baaré et al., 2001; Posthuma et al., 2000). Heritability estimates for intracranial volume, total brain volume, gray-matter volume, white-matter volume, and cerebellar volume were all between 80% and 90%. Genetic intercorrelations between these measures were all very high, indicating that a largely overlapping set of genes is responsible for individual differences in each of these measures.

To our knowledge, only two multivariate genetic studies have been conducted to investigate whether the relation between IQ and brain volumes is mediated through a common genetic pathway or through a common environmental pathway (see Table 9.2). The first study, often cited although only published as an abstract so far (Wickett, Vernon, & Lee, 1997), was based on MRI and IQ data from 68 adult men from 34 sibships and compared within-family correlations with between-family correlations

Table 9.2. Overview of Multivariate Genetic Studies Relating Neurophysiological

Study	Subjects	(Neuro-) physiological measure	Cognitive measure	Design
Brain size/volume				
Pennington et al. (2000)	Concordant or discordant for reading disorder/control 25/9 MZ 23/9 DZ	Brain volumes measured with magnetic resonance imaging	WISC WAIS–3R	Correlational study
Wickett, Vernon, and Lee (1997)	68 adult males (34 sibships)	Total brain volume measured with magnetic resonance imaging	MAB Trail Making Test Factor Referenced Tests	Correlation within families compared with correlation between families
Reaction times/speeded responses				
Luciano, Wright, et al. (2001)	166 MZ 190 DZ	Choice reaction time task	MAB	Multivariate genetic model fitting, including unreliability correction
Neubauer et al. (2000)	169 MZ 131 DZ	Sternberg's memory scanning; Posner's letter matching	RAVEN Leistungs Pruf System	Multivariate genetic model fitting
Finkel and Pedersen (2000)	45 MZA 67 MZT 94 DZA 86 DZT	Oral version of Digit Symbol and Figure Identification subtest (perceptual speed)	Cognitive factor constructed from 11 cognitive measures (including WAIS, Thurstone's Picture Memory, Card Rotations)	Multivariate genetic model fitting
Rijsdijk, Vernon, and Boomsma (1998)	82 MZ 109 DZ age 16	Donder's simple/two choice reaction time; Sternberg's Memory Scanning; Posner's Letter Matching	RAVEN	Multivariate genetic model fitting
Rijsdijk et al. (1998)	74 MZ 100 DZ age 18	Donder's simple/two choice reaction time; Sternberg's Memory Scanning; Posner's Letter Matching	WAIS	Multivariate genetic model fitting

Indices of Brain Structure/Function to Measures of Cognitive Abilities

Heritability of (neuro-) physiological measure	Heritability of cognitive measure	Phenotypic correlation	Contribution of common genetic factors
MZ correlations (.78–.98) > DZ correlations (.32–.65) in both samples	Not reported	Reading disorder/ healthy sample 0.42/0.31	Genetic correlation 0.48 (both samples combined)
Not reported	Not reported	Correlation IQ-brain volume within families 0.24; between families 0.50	Suggestive of genetic mediation
79%–90%	89%	−0.31 to −0.56	Genotypic correlations −0.45 to −0.70
11%–61%	39%–81%	−0.08 to −0.50 (More difficult tasks → higher correlation)	65% of phenotypic correlation due to common genes
56%	Cognitive factor 61% (not age corrected) 85% (age corrected)	0.66	70% of genetic variance in cognitive factor was due to genes shared with perceptual; speed; 61% of phenotypic correlation due to genetic mediation
40%–58%	58%	Around −0.20	Genetic correlations around −0.40
22%–57%	33%–74%	−0.14 to −0.28	Genetic correlations −0.46 and −0.42

Table 9.2 continues on next page

Table 9.2. *Continued*

Study	Subjects	(Neuro-) physiological measure	Cognitive measure	Design
Baker, Vernon, and Ho (1991)	50 MZ 32 SS DZ ages 15–57	Battery of eight different reaction time tasks	MAB	Multivariate genetic model fitting
Ho, Baker, and Decker (1988)	30 MZ 30 SS DZ ages 8–18	Rapid Automatic Naming tests; Colorado Perceptual Speed test	WISC–R WAIS–R	Multivariate genetic model fitting, using IQ score and composite scores of reaction time tests
Inspection time				
Luciano, Smith, et al. (2001)	184 MZ 206 DZ age 16 years	Inspection time (pi-paradigm)	MAB	Multivariate genetic model fitting
Posthuma, de Geus, and Boomsma (2001)	Cross-sectional, mean age 26/50: 47/44 MZ 253/192 DZ and sib pairs	Inspection time (pi-paradigm)	WAIS–3R	Multivariate genetic model fitting
Peripheral nerve conduction velocity				
Rijsdijk and Boomsma (1997)	83 MZ 111 DZ age 18	Peripheral nerve conduction velocity	WAIS	Multivariate genetic model fitting
Rijsdijk, Boomsma, and Vernon (1995)	89 MZ 122 DZ age 16	Peripheral nerve conduction velocity	RAVEN	Multivariate genetic model fitting
EEG alpha peak frequency				
Posthuma, Neale, Boomsma, and de Geus (2001)	Cross-sectional, mean age 26/50: 47/44 MZ 253/192 DZ and sib pairs	Individual alpha peak during resting condition	WAIS–3R dimensions (verbal comprehension, working memory, processing speed, perceptual organization)	Multivariate genetic model fitting
EEG coherence				
van Baal, Boomsma, and de Geus (2001)	Longitudinal measurement (age 5/age 7) 71/77 MZ 96/105 DZ	Theta coherence (frontal, long, posterior)	RAKIT-VIQ/PIQ	Multivariate genetic model fitting

Note. MZ = monozygotic twins; DZ = dizygotic twins; MZA = MZ raised apart; MZT = MZ raised together; DZA = DZ raised apart; DZT = DZ raised together; SS = same sex; WISC = Wechsler Intelligence Scale for Children; WAIS = Wechsler Adult Intelligence Scale;

Heritability of (neuro-) physiological measure	Heritability of cognitive measure	Phenotypic correlation	Contribution of common genetic factors
45%	68%–85%	−0.59	Genetic correlations −0.92–−1.00
49%–52%	78%	0.37–0.42	70%–100% of the phenotypic correlation is due to common genetic factors
36%	VIQ 81% PIQ 73%	−0.26 with VIQ −0.35 with PIQ	Genetic correlations −0.47 with VIQ −0.65 with PIQ
46%	VIQ 85% PIQ 69%	−0.19 with VIQ −0.27 with PIQ	Genetic correlations −0.31 with VIQ −0.47 with PIQ
66%	81%	0.15	Genetic correlation 0.20
77%	65%	No correlation	
71% age 26 83% age 50	66% to 83%	No correlation	
Age 5/7 frontal 55%/42% long 70%/81% posterior 57%/71%	Age 5/7 VIQ 56%/14% PIQ 63%/64%	Frontal and VIQ: −0.13/−0.13 (age 5/7) Frontal and PIQ: −0.10 (age 7) Long and PIQ: 0.07 (age 7)	Genetic correlations (accounted for 100% of the phenotypic correlation); Frontal and VIQ age 5 −0.16 age 7 −0.26

WAIS–3R = WAIS 3rd Edition, Revised; MAB = Multidimensional Aptitude Battery; RAVEN = Raven's Advanced Progressive Matrices; RAKIT = Revisie Amsterdamse Kinder Intelligentie Test; VIQ = verbal IQ; PIQ = performance IQ.

of brain volume and IQ. A within-family correlation of 0.24 and a between-family correlation of 0.50 were reported, suggesting that some, but not all, of the phenotypic correlation between brain volume and IQ is due to a common underlying set of genes.

The second study (Pennington et al., 2000) specifically addressed the relation of reading disorder with brain volume but also included measures of IQ (Wechsler Intelligence Scale for Children [WISC; Wechsler, 1974] and the 3rd edition of the WAIS–R [WAIS–3R] scores [Wechsler, 1981]). In this study, both a reading disorder sample (25 monozygotic [MZ], 23 dizygotic [DZ]) and a non-reading disorder sample (9 MZ, 9 DZ) were included. The MZ and DZ correlations in the reading disorder sample and the non-reading disorder sample were comparable and suggested high heritability of brain volume (90%), which is in line with the larger study of Baaré et al. (2001). Phenotypic correlations between cerebral brain volume and IQ were 0.31 in the non-reading disorder sample and 0.42 in the reading disorder sample. The genetic correlation, as calculated from the cross-twin correlations, was 0.48 in the combined sample. This indicates that about half of the genetic influences on either cerebral brain volume or IQ is due to genetic factors influencing both. Put differently, 80% of the phenotypic correlation is explained by genetic mediation.

Functional Measures of Cognitive Ability

Structural brain volumes seem to provide a first important point of entry to the genetic sources of individual differences in cognitive abilities. Most differences in cognitive ability, however, will arise from functional aspects of the brain. Several behavioral and neurophysiological indices of brain function have repeatedly been shown to be influenced by the cognitive demands of the tasks in which they are elicited. For example, increasing the information-processing load of a task results in prolonged reaction times within the same subject (Hick, 1952). The same holds for evoked brain potentials elicited in a working-memory task; increasing working-memory load results in a longer latency of the evoked brain potentials (McEvoy, Smith, & Gevins, 1998). These and many other clear links between cognition and behavioral or neurophysiological measures in cognitive neuroscience largely are based on within-subject observations. Whatever is causing *within*-person variability does not necessarily also cause variability *between* people. To detect genetic sources of individual variation in cognitive abilities, behavioral and neurophysiological indices of information processing are needed that are associated with differences in cognitive abilities between people. Below we give an overview of our current armament of such "endophenotypes" of cognitive ability, where we restrict ourselves to discussing only those measures for which the bivariate genetic architecture with measures of psychometric intelligence has already been investigated.

Behavioral Measures of Speed of Information Processing

Galton (1883) was the first to propose that reaction time is correlated with general intelligence and may be used as a measure of it. His observations and the results of empirical studies afterward led to the general belief in the speed-of-processing theory of intelligence: The faster the accomplishment of basic cognitive operations, the more intelligent a person will be (Eysenck, 1986; Vernon, 1987). Since then, reaction times consistently have been negatively related to intelligence (e.g., Deary, Der, & Ford, 2001; Vernon, 1987); that is, a shorter reaction time corresponds to a higher IQ. Correlations with IQ generally range between −0.20 and −0.40 but can be as high as −0.60 (Fry & Hale, 1996). Higher correlations between reaction times and IQ usually are found when more complex reaction time tasks are used, although this effect is not unequivocally confirmed in empirical studies (Mackintosh, 1986).

Results from twin studies suggest heritabilities for reaction time of the same magnitude as those for IQ. McGue and Bouchard (1989) observed heritabilities of 54% and 58% for basic and spatial speed factors in a sample of MZ (N = 49) and DZ (N = 25) twins reared apart. For a general speed factor based on eight complex reaction time tests, Vernon (1989) found a heritability of 49% in 50 MZ and 52 DZ twins. In the same study, it was also found that reaction time tests requiring more complex mental operations show higher heritabilities. A bivariate analysis of these data with IQ in 50 MZ and 32 same-sex DZ pairs (15 to 57 years) was reported by Baker, Vernon, and Ho (1991). Phenotypic correlations of verbal and performance IQ with general speed were both −0.59 and were entirely mediated by genetic factors. Genetic correlations were estimated at −0.92 and −1.00. This is in line with results from an earlier study in which phenotypic correlations between reaction time (measured as the total number of correct responses on a timed task; see Table 9.2) and IQ ranged between 0.37 and 0.42, from which 70%–100% was attributed to genetic factors influencing both reaction time and IQ (Ho, Baker, & Decker, 1988).

More recently, Rijsdijk, Vernon, and Boomsma (1998) conducted a multivariate genetic analysis on reaction time data and IQ data, using 213 twin pairs measured at ages 16 and 18. Heritabilities were reported for age 16 of 58%, 57%, and 58% for simple reaction time, choice reaction time, and IQ (the Raven Advanced Progressive Matrices Test; Raven, 1958), respectively. Phenotypic correlations of simple reaction time and choice reaction time with IQ were −0.21 and −0.22, respectively, and were completely mediated by common genetic factors. Virtually the same picture was shown at age 18, where the same reaction time battery was correlated with IQ as measured with the WAIS (see Table 9.2).

Finkel and Pedersen (2000) investigated the underlying covariance structure of measures of speed and measures of cognition in a sample of 292 reared-together and reared-apart MZ and DZ twins (ages 40–84 years). Speed was measured by oral versions of the Digit Symbol and Picture Identification subtests of the WAIS. A cognitive factor was constructed based on several standard IQ tests. The phenotypic correlation between the speed factor and the cognitive factor was 0.66, of which 61% was due to correlated genetic factors between the two. Also, Finkel and Pedersen reported that 70% of the genetic variance in the cognitive factor was shared with the speed factor.

Neubauer et al. (2000) reported heritability estimates of reaction time data and IQ (Raven) ranging from 11% to 61% and 39% to 81%, respectively. Phenotypic correlations between reaction time data and IQ (Raven, 1958) data were between −0.08 and −0.50, in which higher correlations with IQ were found for more complex reaction time tasks. Common genetic influences on reaction time and IQ accounted for 65% of the observed phenotypic correlation.

Evidence for a genetic mediation between reaction time and IQ also emerged from a recent large twin study by Luciano, Wright, et al. (2001). Using reaction time data and IQ data from 166 MZ pairs and 190 DZ pairs, Luciano, Wright, et al. (2001) reported high heritabilities for both reaction time (79%–90%) and IQ (89%), with phenotypic correlations between −0.31 and −0.56 and genotypic correlations between −0.45 and −0.70. In other words, common genetic influences explained at least 70% of the observed phenotypic correlation between reaction times and IQ.

In summary, nongenetic studies have shown a consistent and stable relation between reaction time and IQ; interindividual variance in reaction time explains about 10%–30% of IQ test variance. This has been confirmed by results from genetic studies, which additionally showed that between 65% and 100% of this covariance is explained by a common underlying genetic mechanism.

Inspection time is a measure of central nervous system (CNS) processing and is defined as the minimum display time a subject needs for making an accurate perceptual discrimination on an obvious stimulus. It is distinct from reaction time because there is no need to make the discrimination quickly; all that is required is an accurate response. Visual inspection time can easily be measured in a computerized version of the Π-paradigm in which participants are asked to decide which leg of the Π-figure is longest. Visual inspection time is generally thought to reflect speed of apprehension or perceptual speed. A meta-analysis conducted by Kranzler and Jensen (1989) on almost all studies until 1989 investigating the relation between inspection time and intelligence indicated that inspection time and IQ correlate around −0.50: The less time a person needs to make an accurate decision on an obvious stimulus, the higher his or her IQ. The overall consensus on the relation between inspection time and IQ is given by Deary and Stough (1996): Inspection time accounts for approximately 20% of intelligence test variance.

Recently, two large twin studies have investigated whether the relation between inspection time and IQ is mediated by shared genetic factors or by shared environmental factors (Luciano, Smith, et al., 2001; Posthuma, de Geus, et al., 2001). These two studies were also the first to report on the heritability of inspection time per se. Using 184 MZ pairs and 206 DZ pairs age 16, Luciano, Smith, et al. (2001) reported a heritability estimate of inspection time of 36% and of IQ measures between 73% and

81%. Posthuma, de Geus, et al. (2001) reported a slightly higher heritability estimate of inspection time (46%) and similar heritability estimates of IQ measures (WAIS–3R) ranging from 69% to 85%. The latter sample consisted of 102 MZ pairs and 525 DZ/sibpairs belonging to two age cohorts (mean ages = 26 and 50 years; *SD* = 4.2 and 7.5, respectively).

Luciano, Smith, et al. (2001) reported a correlation between inspection time and performance IQ of −0.35 and between inspection time and verbal IQ of −0.26. Posthuma, de Geus, et al. (2001) reported slightly lower correlations: −0.27 and −0.19, respectively. Both studies unanimously found that the phenotypic correlations between inspection time and performance IQ/verbal IQ were completely mediated by common genetic factors. This meant that in the study by Luciano et al. the genetic correlation between inspection time and performance IQ was −0.65 and between inspection time and verbal IQ was −0.47. In the study by Posthuma et al., the genetic correlations were −0.47 and −0.31, respectively. Thus, the genes shared with inspection time are, across studies, estimated to explain between 10% and 42% of the total genetic variance in IQ.

Peripheral nerve conduction velocity (PNCV) is a measure of the speed with which action potentials travel the length of the axons in the peripheral nervous system; a high PNCV corresponds to faster conduction. The relation of PNCV with IQ does not go undebated; in some studies no relation was found (Barret, Daum, & Eysenck, 1990; Reed & Jensen, 1992; Rijsdijk et al., 1995) or a positive relation was found only in men (Tan, 1996; Wickett & Vernon, 1994). The general agreement, however, is that PNCV relates positively to measures of intelligence (Vernon & Mori, 1992), thereby providing evidence for the speed-of-information processing theory of intelligence (Eysenck, 1986; Vernon, 1987).

The first and only study reporting on genetic influences on PNCV was a large longitudinal study conducted by Rijsdijk and colleagues (Rijsdijk & Boomsma, 1997; Rijsdijk et al., 1995). Rijsdijk and colleagues determined PNCV for the wrist–elbow segment of the median nerve of the right arm for 213 sixteen-year-old twin pairs. By using structural equation, they estimated modeling heritability at 76% for PNCV at age 16 and 66% at age 18. At age 16, no correlation between PCNV and IQ scores as measured with the Raven test was found. However, at age 18, IQ (as measured with the WAIS) correlated .15 with PCNV, and this correlation was entirely mediated by common genetic factors (Rijsdijk & Boomsma, 1997).

Electrophysiological Measures of Speed of Information Processing

Electrophysiological indices of speed of information processing are considered to be a more direct measure of the speed of brain functioning than behavioral measures. Several noninvasive techniques, such as functional MRI, positron emission tomography, and electroencephalography (EEG), have repeatedly shown that quantitative differences in brain functioning are related to differences in cognitive functioning (Fabiani, Gratton, &

Coles, 2000; Reiman, Lane, Petten, & van Bandettini, 2000). Despite the vast amount of brain imaging studies investigating the brain during the execution of distinct cognitive operations (Gazzaniga, 2000), very little have done so using a genetic design. The only indices of which the relation with cognitive functioning have been investigated in a genetic design are two measures derived from the EEG: coherence and alpha peak frequency.

EEG recording is a noninvasive technique to measure electrical activity of the brain. EEG activity can be analyzed according to the frequency spectrum that is obtained when a Fourier transformation is performed on an EEG time series. Generally, five frequencies are distinguished in the EEG power spectrum: delta (0.5–4 cycles per second), theta (4–8 cycles per second), alpha (8–13 cycles per second), beta (13–30 cycles per second), and gamma (>30 cycles per second). Since the 1990s, the underlying biological mechanisms of the different frequencies, especially the alpha and beta rhythms, are well understood and have been described in the literature (Lopes da Silva, 1991; Steriade, Gloor, Llinás, Lopes da Silva, & Mesulam, 1990).

The dominant frequency in an adult human EEG spectrum lies in the alpha range, around 10 cycles per second. The alpha peak frequency has been related to cognitive abilities in general and to (working) memory in particular. Lebedev (1990, 1994) proposed a functional role for the human alpha peak frequency in stating that "cyclical oscillations in an alpha rhythm determine the capacity and speed of working memory. The higher the frequency the greater the capacity and the speed of memory" (p. 254; 1994). In addition, Klimesch (1997) argued that thalamo-cortical feedback loops oscillating within the alpha frequency range allow searching and identification of encoded information. He speculated that faster oscillating feedback loops would correspond to faster access to encoded information. These theories are supported by the results of some recent studies. Klimesch found that the alpha peak frequency of good working-memory performers lies about 1 Hz higher than that of bad working-memory performers. Anokhin and Vogel (1996) reported a correlation of 0.35 between alpha peak frequency and verbal abilities. In addition, Klimesch (1999) found that, within the same subject, alpha peak frequencies increase with increasing cognitive load of the task in which they are measured.

Results from a few small twin studies have suggested that alpha peak frequency is influenced by genetic factors (Christian et al., 1996), and it has also been speculated that its relation with IQ is due to a genetic basis (e.g., Vogel, 2000, p. 117). In only one multivariate genetic study, however, the nature of the relation between alpha peak frequency and cognitive abilities is formally investigated. Including 102 MZ pairs and 525 DZ/sib pairs from two age cohorts (M = 26 and 50 years; SD = 4.2 and 7.5, respectively), Posthuma, Neale, et al. (2001) found that alpha peak frequency is highly heritable. In young adults (M = 26 years; SD = 4.2 years), heritability was estimated at 71%; in older adults (M = 50 years; SD = 7.5 years), heritability was somewhat higher at 83%. Heritabilities for the WAIS–3R dimensions ranged from 66% to 83%. Surprisingly, no correlation was found between alpha peak frequency and IQ (WAIS–3R), thereby dismissing alpha peak frequency as a valuable endophenotype for specific cognitive abilities as measured with the WAIS–3R. However, alpha peak frequency may still be of importance for cognitive functioning, especially for memory functioning. Although working memory is an important aspect of IQ (Engle, Tuholski, Laughlin, & Conway, 1999; Kyllonen & Christal, 1990; Necka, 1992), it also is suggested that it comprises more than psychometric IQ as measured with the WAIS dimensions. A genetic relation between alpha peak frequency and specific aspects of (working) memory has not been investigated yet.

A second measure that can be extracted from the Fourier-transformed EEG is the degree of connectivity between certain brain areas. It is sometimes speculated that efficient interconnectivity of the brain relates positively to IQ. Studies relating coherence (a measure of connectivity of the brain) to IQ have indeed reported a relation between efficient connectivity (i.e., low coherence) and measures of intelligence (e.g., Anokhin, Lutzenberger, & Birbaumer, 1999; Jausovec & Jausovec, 2000; van Baal, Boomsma, & de Geus, 2001).

The amount of EEG coherence in the different frequency bands has previously been found to be under the influence of genetic factors (van Beijsterveldt & Boomsma, 1994). Van Beijsterveldt, Molenaar, de Geus, and Boomsma (1998) reported a mean heritability of 60% for coherence in 213 twin pairs age 16 years. A longitudinal study by van Baal et al. (2001) and van Baal, de Geus, and Boomsma (1998) included 70 MZ twin pairs and 97 DZ twin pairs who were measured at ages 5 and 7 years. In children, the dominant frequency is in the theta range. For frontal, long, and posterior theta interhemispheric coherence, heritabilities of 55%, 70%, and 57%, respectively, were reported at age 5. At age 7, these were 42%, 81%, and 71%, respectively. At age 5, frontal theta coherence and verbal IQ correlated −0.13, which was completely explained by common underlying genetic factors influencing both traits (van Baal et al., 2001). At age 7, significant correlations were found between frontal theta coherence and verbal IQ (−0.13), frontal theta coherence and performance IQ (−0.10), and long distance theta coherence and performance IQ (0.07). Again, these correlations were entirely ascribed to correlated genetic influences (van Baal et al., 2001).

Candidate Genes Derive From Anatomical and Functional Measures of Cognitive Ability

Summarizing the above discussion, we highlight the following:

- Brain size and IQ correlate around 0.40. About 80% of this correlation is due to common underlying genetic factors.
- Reaction times and IQ correlate in the range −0.20 to −0.40, depending on task complexity. Of this correlation, 70% to 100% is due to common underlying genetic factors.

- Inspection time and IQ correlate in the range −0.20 to −0.40. Of this correlation, 100% is due to common underlying genetic factors.
- Frontal theta coherence in children is correlated −0.13 with IQ. Of this correlation, 100% is due to common underlying genetic factors.
- Alpha peak frequency does not correlate with IQ as measured with the WAIS–3R.

This suggests that genes important for brain size, reaction times, inspection time, and theta coherence may also be important for intelligence, which fits very well in the *myelination hypothesis* as formulated by Miller (1994). According to this hypothesis, generally, the relation between speed and intelligence can be explained if part of the interindividual variance in intelligence can be ascribed to interindividual variance in the degree of myelination of cortico–cortical connections. If true, this could explain why more intelligent brains show faster nerve conduction, faster reaction times, and faster inspection times. And, all other things equal, thicker myelin sheaths will result in larger brain volume, thus explaining the positive relation between brain size and IQ (Miller, 1994). Although it is unlikely that the myelination hypothesis explains all observed anatomical, behavioral, and physiological relations with cognitive functioning, it provides theoretical guidance in the choice of candidate genes for cognition: Genes important for myelination also may be important for cognition. Several such genes have been implicated from animal models, some of which are known to cause dysmyelination in humans as well. The *Plp* gene (Xq22.3), for example, codes for two membrane proteins important for myelination. Disruption of expression of the *Plp* gene in mice causes a disruption in the assembly of the myelin sheath, which leads to a profound reduction in conduction velocity of CNS axons (Boison & Stoffel, 1994; Griffiths, Montague, & Dickinson, 1995, Ikenaka & Kagawa, 1995; Lemke, 1993). Mutations in the same gene in humans are known to result in Pelizaeus–Merzbacher disease (PMD; e.g., Anderson et al., 1999; Griffiths et al., 1995; Woodward & Malcolm, 1999). PMD is a hypomyelination disease which, in its mildest form, may lead to optic atrophy and dementia. Other genes implicated to be important for myelination in knock-out mouse studies are the *cgt* gene (Stoffel & Bosio, 1997), the *MAG* gene (for reviews, see Bartsch, 1996; Fujita et al., 1998, Sheikh et al., 1999), and the *tn-r* gene (Weber et al., 1999).

Thus, genetic correlations between IQ scores and behavioral and neurophysiological indices of brain structure and function may provide a theoretical framework from which candidate genes from cognitive functioning may be proposed. A second advantage of finding good "endophenotypes" for cognitive functioning is that they are more "upstream," as Kosslyn and Plomin (2001) put it, and thus are likely to be influenced by a smaller number of genes. When such behavioral and electrophysiological endophenotypes are included in genome screens aimed at detecting QTLs important for cognition, they can boost the statistical power to detect these QTLs (Boomsma, Anokhin, & de Geus, 1997; de Geus & Boomsma, 2002; Leboyer et al., 1998).

A few examples of the usefulness of this approach are already available from the field of alcohol research. The P3 event-related potential (ERP) is generally believed to provide a sensitive electrophysiological index of the attentional and working-memory demands of a task (Donchin & Coles, 1988; Picton, 1992). Because disturbed information processing is a feature in alcoholism, a genetic association between the DRD2 gene, of which the A1 allele is more frequent in alcoholics as compared with non-alcoholics (Kono et al., 1997; Noble et al., 1998), and P3 latency was tested in a study that included 32 sons of active alcoholic fathers, 36 sons of recovering alcoholic fathers, and 30 sons of social drinker fathers (mean age of children = 12.5 years, *SD* = 0.1 years; Noble, Berman, Ozkaragoz, & Ritchie, 1994). The frequency of the A1 allele of the DRD2 gene was highest in the group of sons of active alcoholic fathers, followed by the sons of recovering alcoholic fathers and sons of social drinker fathers, respectively. It is interesting to note that the A1 allele was also associated with prolonged P3 latency (42 ms). This result confirms the expected involvement of the dopaminergic system in cognitive functioning.

Williams et al. (1999) reported linkage on the chromosome 4 region near the class I alcohol dehydrogenase locus ADH3 for alcoholism and P3 amplitude. Several other loci influencing P3 amplitude have also been reported (Almasy et al., 2001; Begleiter et al., 1998). Another linkage result for EEG measures has been reported for a rare condition known as *low voltage EEG,* the near absence of an alpha rhythm (Anokhin et al., 1992; Steinlein, Anokhin, Yping, Schalt, & Vogel, 1992).

Endophenotypes Galore

It should be kept in mind that this chapter only discusses behavioral and neurophysiological indices of which the *genetic* relation with cognitive abilities had already been investigated. There are many other behavioral or neurophysiological measures that correlate with IQ, and because some of these measures have been shown to be (highly) heritable, they may correlate with IQ through a common genetic pathway. Examples of such measures include glucose metabolism in the brain (Haier et al., 1988; Haier, Siegel, Tang, Abel, & Buchsbaum, 1992), ERPs (Gevins & Smith, 2000; Hansell et al., 2001; Wright et al., 2001), and brain wave complexity (Blinkhorn & Hendrickson, 1982). Especially the field of ERPs has provided a vast number of studies relating latency and amplitude of evoked brain potentials to cognitive ability (for reviews, see Deary & Caryl, 1997; Fabiani et al., 2000). Heritability estimates are generally moderate to high (van Beijsterveldt & Boomsma, 1994) and vary for specific ERP components. For example, the slow wave is a late, long-duration ERP component that has been shown to reflect working-memory processes (Ruchkin, Canoune, Johnson, & Ritter, 1995). Hansell et al. (2001) investigated the heritability of the slow wave elicited in a working-memory task for 391 adolescent twin pairs. Thirty percent of the total variance and around 50% of the reliable variance in the slow wave was due to genetic influences. In a sample of 335 adolescent twin pairs and 48 siblings, Wright et al. (2001) estimated the heritabilities of the P3 amplitude and latency in the ranges 48%–61% and 44%–50%, respectively.

These neurophysiological indices of information processing that have already been shown to be related to cognitive functioning and also show moderate to high heritability present promising endophenotypes for cognition. Before measuring them in large samples and including them in actual DNA-based gene hunting, however, the bi- (or multi)variate genetic architecture of these indices with measures of higher order cognitive abilities needs to be established.

References

Almasy, L., Porjesz, B., Blangero, J., Goate, A., Edenberg, H., Chorlian, D., et al. (2001). Genetics of event-related brain potentials in response to a semantic priming paradigm in families with a history of alcoholism. *American Journal of Human Genetics, 68*, 128–135.

Anderson, T. J., Klugmann, M., Thomson, C. E., Schneider, A., Readhead, C., Nave, K., et al. (1999). Distinct phenotypes associated with increasing dosage of the PLP gene: Implications for CMT1A due to PMP22 gene duplication. *Annals of the New York Academy of Sciences, 14*(883), 234–246.

Andreasen, N. C., Flaum, M., Swayze, V., II, O'Leary, D. S., Alliger, R., Cohen, G., et al. (1993). Intelligence and brain structure in normal individuals. *American Journal of Psychiatry, 150*, 130–134.

Anokhin, A. P., Lutzenberger, W., & Birbaumer, N. (1999). Spatiotemporal organization of brain dynamics and intelligence: An EEG study in adolescents. *International Journal of Psychophysiology, 33*, 259–273.

Anokhin, A. P., Steinlein, O., Fischer, C., Mao, Y., Vogt, P., Schalt, E., et al. (1992). A genetic study of the human low-voltage electroencephalogram. *Human Genetics, 90*(1–2), 99–112.

Anokhin, A. P., & Vogel, F. (1996). EEG alpha rhythm frequency and intelligence in normal adults. *Intelligence, 23*, 1–14.

Baaré, W. F. C, Hulshoff Pol, H. E., Boomsma, D. I., Posthuma, D., de Geus, E. J. C., Schnack, H. G., et al. (2001). Genetic and environmental individual differences in human brain morphology. *Cerebral Cortex, 11*, 816–824.

Baker, L. A., Vernon, P. A., & Ho, H. Z. (1991). The genetic correlation between intelligence and speed of information processing. *Behavior Genetics, 21*, 351–367.

Barret, P. T., Daum, I., & Eysenck, H. J. (1990). Sensory nerve conduction and intelligence: A methodological study. *Journal of Psychophysiology, 4*, 1–13.

Bartels, M., Rietveld, M. J. H., van Baal, G. C. M., & Boomsma, D. I. (in press). Genetic and environmental influences on the development of intelligence. *Behavior Genetics*.

Bartsch, U. (1996). Myelination and axonal regeneration in the central nervous system of mice deficient in the myelin-associated glycoprotein. *Journal of Neurocytology, 25*, 303–313.

Begleiter, H., Porjesz, B., Reich, T., Edenberg, H. J., Goate, A., Blangero, J., et al. (1998). Quantitative trait loci analysis of human event-related brain potentials: P3 voltage. *Evoked Potential–Electroencephalography and Clinical Neurophysiology, 108*, 244–250.

Blinkhorn, S. F., & Hendrickson, D. E. (1982). Average evoked responses and psychometric intelligence. *Nature, 298*, 596–597.

Boison, D., & Stoffel, W. (1994). Disruption of the compacted myelin sheath of axons of the central nervous system in proteolipid protein-deficient mice. *Proceedings of the National Academy of Sciences, 91*(24), 11709–11713.

Boomsma, D. I., Anokhin, A., & de Geus, E. J. C. (1997). Genetics of electrophysiology:

Linking genes, brain, and behavior. *Current Directions in Psychological Science, 6*, 106–110.

Boomsma, D. I., & van Baal, G. C. M. (1998). Genetic influences on childhood IQ in 5- and 7-year old Dutch twins. *Developmental Neuropsychology, 14*, 115–126.

Bouchard, T. J., Jr., & McGue, M. (1981). Familial studies of intelligence: A review. *Science, 212*, 1055–1059.

Cherny, S., & Cardon, L. (1994). General cognitive ability. In J. DeFries, R. Plomin, & D. Fulker (Eds.), *Nature and nurture during middle childhood* (pp. 46–56). Oxford, England: Blackwell.

Christian, J. C., Morzorati, S., Norton, J. A., Jr., Williams, C. J., O'Connor, S., & Li, T. K. (1996). Genetic analysis of the resting electroencephalographic power spectrum in human twins. *Psychophysiology, 33*, 584–591.

Deary, I. J., & Caryl, P. G. (1997). Neurosciences and human intelligence. *Trends in Neurosciences, 20*, 365–371.

Deary, I. J., Der, G., & Ford, G. (2001). Reaction times and intelligence differences. A population-based cohort study. *Intelligence, 29*, 389–399.

Deary, I. J., & Stough, C. (1996). Intelligence and inspection time: Achievements, prospects, and problems. *American Psychologist, 51*, 599–608.

de Geus, E. J. C., & Boomsma, D. I. (2002). A genetic neuroscience approach to human cognition. *European Psychologist, 6*, 241–253.

Donchin, E., & Coles, M. (1988). Is the P300 component a manifestation of context updating? *Behavioral Brain Sciences, 11*, 357–427.

Egan, V., Chiswick, A., Santosh, C., Naidu, K., Rimmington, J. E., & Best J. J. K. (1994). Size isn't everything: A study of brain volume, intelligence and auditory-evoked potentials. *Personality and Individual Differences, 17*, 357–367.

Engle, R. W., Tuholski, S. W., Laughlin, J. E., & Conway, A. R. (1999). Working memory, short-term memory, and general fluid intelligence: A latent-variable approach. *Journal of Experimental Psychology: General, 128*, 309–331.

Eysenck, H. J. (1986). Toward a new model of intelligence. *Personality and Individual Differences, 7*, 731–736.

Fabiani, M., Gratton, G., & Coles, M. G. H. (2000). Event-related brain potentials. In J. T. Cacioppo, L. G. Tassinary, & G. G. Berntson (Eds.), *Handbook of psychophysiology* (2nd ed., pp. 53–84). Cambridge, England: Cambridge University Press.

Finkel, D., & Pedersen, N. L. (2000). Contribution of age, genes, and environment to the relationship between perceptual speed and cognitive ability. *Psychology and Aging, 5*, 56–64.

Flint, J. (1999). The genetic basis of cognition. *Brain, 122*, 2015–2031.

Fry, A. F., & Hale, S. (1996). Processing speed, working memory, and fluid intelligence: Evidence for a developmental cascade. *Psychological Science, 7*, 237–241.

Fujita, N., Kemper, A., Dupree, J., Nakayasu, H., Bartsch, U., Schachner, M., et al. (1998). The cytoplasmic domain of the large myelin-associated glycoprotein isoform is needed for proper CNS but not peripheral nervous system myelination. *Journal of Neuroscience, 15*, 1970–1978.

Galton, F. (1883). *Inquiries into human faculty and its development*. London: Macmillan, Everyman's Library.

Gazzaniga, M. S. (Ed.). (2000). *The new cognitive neurosciences*. Cambridge, MA: MIT Press.

Gevins, A., & Smith, M. E. (2000). Neurophysiological measures of working memory and individual differences in cognitive ability and cognitive style. *Cerebral Cortex, 10*, 829–839.

Griffiths, I. R., Montague, P., & Dickinson, P. (1995). The proteolipid protein gene. *Neuropathology and Applied Neurobiology, 21*, 85–96.

Haier, R. J., Siegel, B. V., Jr., Nuechterlein, K. H., Hazlett, E., Wu, J. C., Paek, J., et al. (1988). Cortical glucose metabolic-rate correlates of abstract reasoning and attention studied with positron emission tomography. *Intelligence, 12*, 199–217.

Haier, R. J., Siegel, B., Tang, C., Abel, L., & Buchsbaum, M. S. (1992). Intelligence and changes in regional cerebral glucose metabolic-rate following learning. *Intelligence, 16*(3–4), 415–426.

Hansell, N. K., Wright, M. J., Geffen, G. M., Geffen, L. B., Smith, G. A., & Martin, N. G.

(2001). Genetic influence on ERP slow wave measures of working memory. *Behavior Genetics,* 31, 603–614.
Hick, W. E. (1952). On the rate of gain of information. *Quarterly Journal of Experimental Psychology, 4,* 11–26.
Ho, H. Z., Baker, L. A., & Decker, S. N. (1988). Covariation between intelligence and speed of cognitive processing: Genetic and environmental influences. *Behavior Genetics, 8,* 247–261.
Ikenaka, K., & Kagawa, T. (1995). Transgenic systems in studying myelin gene expression. *Developmental Neuroscience, 17,* 127–136.
Jausovec, N., & Jausovec, K. (2000). Differences in resting EEG related to ability. *Brain Topography, 12,* 229–240.
Jensen, A. R. (1994). Psychometric g related to differences in head size. *Personality and Individual Differences, 17,* 597–606.
Klimesch, W. (1997). EEG-alpha rhythms and memory processes. *International Journal of Psychophysiology, 26,* 319–340.
Klimesch, W. (1999). EEG alpha and theta oscillations reflect cognitive and memory performance: A review and analysis. *Brain Research Review, 29*(2–3), 169–195.
Kono, Y., Yoneda, H., Sakai, T., Nonomura, Y., Inayama, Y., Koh, J., et al. (1997). Association between early-onset alcoholism and the dopamine D2 receptor gene. *American Journal of Medical Genetics, 74,* 179–182.
Kosslyn, S. M., & Plomin, R. (2001). Towards a neurocognitive genetics: Goals and issues. In D. Dougherty, S. L. Rauch, & J. F. Rosenbaum (Eds.), *Psychiatric neuroimaging strategies: Research and clinical applications* (pp. 491–515). Washington, DC: American Psychiatric Press.
Kranzler, J. K., & Jensen, A. R. (1989). Inspection time and intelligence: A meta-analysis. *Intelligence, 13,* 329–347.
Kyllonen, P. C., & Christal, R. E. (1990). Reasoning ability is (little more than) working-memory capacity? *Intelligence, 14,* 389–433.
Lebedev, A. N. (1990). Cyclical neural codes of human memory and some quantitative regularities in experimental psychology. In H. G. Geissler (Ed.), *Psychophysical explorations of mental structures* (pp. 303–310). Göttingen, Germany: Hogrefe & Huber.
Lebedev, A. N. (1994). The neurophysiological parameters of human memory. *Neuroscience and Behavioral Physiology, 24,* 254–259.
Leboyer, M., Bellivier, F., Nosten-Bertrand, M., Jouvent, R., Pauls, D., & Mallet, J. (1998). Psychiatric genetics: Search for phenotypes. *Trends in Neuroscience, 21,* 102–105.
Lemke, G. (1993). The molecular genetics of myelination: An update. *Glia, 7,* 263–271.
Lopes da Silva, F. H. (1991). Neural mechanisms underlying brain waves: From neural membranes to networks. *Electroencephalography and Clinical Neurophysiology, 79,* 81–93.
Luciano, M., Smith, G. A., Wright, M. J., Geffen, G. M., Geffen, L. B., & Martin, N. M. (2001). On the heritability of inspection time and its covariance with IQ: A twin study. *Intelligence*, 29, 443–457.
Luciano, M., Wright, M. J., Smith, G. A., Geffen, G. M., Geffen, L. B., & Martin, N. G. (2001). Genetic covariance amongst measures of information processing speed, working memory and IQ. *Behavior Genetics, 31,* 581–592.
Mackintosh, N. J. (1986). The biology of intelligence? *British Journal of Psychology, 77,* 1–18.
McEvoy, L. K., Smith, M. E., & Gevins, A. (1998). Dynamic cortical networks of verbal and spatial working memory: Effects of memory load and task practice. *Cerebral Cortex, 8,* 563–574.
McGue, M., & Bouchard, T. J. (1989). Genetic and environmental determinants of information processing and special mental abilities: A twin analysis. In R. J. Sternberg (Ed.), *Advances in the psychology of human intelligence* (Vol. 5, pp. 7–45). Hillsdale, NJ: Erlbaum.
McKusick, V. A. (1998). *Mendelian inheritance in man: Catalogs of human genes and genetic disorders.* Baltimore: Johns Hopkins University Press.
Miller, E. M. (1994). Intelligence and brain myelination: A hypothesis. *Personality and Individual Differences, 17,* 803–833.

Necka, E. (1992). Cognitive analysis of intelligence: The significance of working memory processes. *Personality and Individual Differences, 13*, 1031–1046.
Neubauer, A. C., Spinavh, F. M., Rieman, R., Angleitner, A., & Borkenau, P. (2000). Genetic and environmental influences on two measures of speed of information processing and their relation to psychometric intelligence: Evidence from the German Observational Study of Adult Twins. *Intelligence, 28,* 267–289.
Noble, E. P., Berman, S. M., Ozkaragoz, T. Z., & Ritchie, T. (1994). Prolonged P300 latency in children with the D2 dopamine receptor A1 allele. *American Journal of Human Genetics, 54*, 658–668.
Noble, E. P., Zhang, X., Ritchie, T., Lawford, B. R., Grosser, S. C., Young, R. M., et al. (1998). D2 dopamine receptor and GABA(A) receptor beta3 subunit genes and alcoholism. *Psychiatry Research, 81*, 133–147.
Pennington, B. F., Filipek, P. A., Lefly, D., Chhabildas, N., Kennedy, D. N., Simon, J. H., et al. (2000). A twin MRI study of size variations in human brain. *Journal of Cognitive Neuroscience, 12*, 223–232.
Picton, T. W. (1992). The P300 wave of the human event-related potential. *Journal of Clinical Neurophysiology, 9*, 456–479.
Plomin, R., Chipuer, H. M., & Neiderhiser, J. M. (1994). Behavioral genetic evidence for the importance of nonshared environment. In E. M. Hetherington & D. Reiss (Eds.), *Separate social worlds of siblings: The impact of nonshared environment on development* (pp. 1–31). Hillsdale, NJ: Erlbaum.
Plomin, R., DeFries, J. C., & McClearn, G. E. (1990). *Behavioral genetics: A primer*. New York: Freeman.
Plomin, R., Owen, M. J., & McGuffin, P. (1994). The genetic basis of complex human behaviors. *Science, 264,* 1733–1739.
Plomin, R., Pedersen, N. L., Lichtenstein, P., & McClearn, G. E. (1994). Variability and stability in cognitve abilities are largely genetic later in life. *Behavior Genetics, 24*, 207–215.
Posthuma, D., de Geus, E. J. C., & Boomsma, D. I. (2001). Perceptual speed and IQ are associated through common genetic factors. *Behavior Genetics,* 81, 593–602.
Posthuma, D., de Geus, E. J., Neale, M. C., Hulshoff Pol, H. E., Baaré, W. F. C., Kahn, R. S., et al. (2000). Multivariate genetic analysis of brain structure in an extended twin design. *Behavior Genetics, 30*, 311–319.
Posthuma, D., Neale, M. C., Boomsma, D. I., & de Geus, E. J. C. (2001). Are smarter brains running faster? Heritability of alpha peak frequency and IQ and their interrelation. *Behavior Genetics, 31,* 567–579.
Raven, J. C. (1958). *Standard progressive matrices*. London: M. K. Lewis & Co.
Raz, N., Torres, I. J., Spencer, W. D., Millman, D., Baertschi, J. C., & Sarpel, G. (1993). Neuroanatomical correlates of age-sensitive and age-invariant cognitive-abilities: An invivo MRI investigation. *Intelligence, 17*, 407–422.
Reed, T. E., & Jensen, A. R. (1992). Conduction velocity in a brain nerve pathway of normal adults correlates with intelligence level. *Intelligence, 16*, 259–272.
Reiman, E. M., Lane, R. D., Petten, C., & van Bandettini, P. A. (2000). Positron emission tomography and functional magnetic resonance imaging. In J. T. Cacioppo, L. G. Tassinary, & G. G. Berntson (Eds.), *Handbook of psychophysiology* (2nd ed., pp. 85–118). Cambridge, England: Cambridge University Press.
Rietveld, M. J. H., Dolan, C. V., van Baal, G. C. M., & Boomsma, D. I. (in press). A twin study of differentiation of cognitive abilities in childhood. *Child Development*. Manuscript submitted for publication.
Rijsdijk, F. V., & Boomsma, D. I. (1997). Genetic mediation of the correlation between peripheral nerve conduction velocity and IQ. *Behavior Genetics, 27*, 87–98.
Rijsdijk, F. V., Boomsma, D. I., & Vernon, P. A. (1995). Genetic analysis of peripheral nerve conduction velocity in twins. *Behavior Genetics, 25*, 341–348.
Rijsdijk, F. V., Vernon, P. A., & Boomsma, D. I. (1998). The genetic basis of the relation between speed-of-information-processing and IQ. *Behavioral Brain Research, 95*, 77–84.
Ruchkin, D. S., Canoune, H. L., Johnson, R., Jr., & Ritter, W. (1995). Working memory and

preparation elicit different patterns of slow wave event-related brain potentials. *Psychophysiology, 32*, 399–410.

Sheikh, K. A., Sun, J., Liu, Y., Kawai, H., Crawford, T. O., Proia, R. L., et al. (1999). Mice lacking complex gangliosides develop Wallerian degeneration and myelination defects. *Proceedings of the National Academy of Sciences, 96*, 7532–7537.

Steinlein, O., Anokhin, A., Yping, M., Schalt, E., & Vogel, F. (1992). Localization of a gene for the human low-voltage EEG on 20q and genetic heterogeneity. *Genomics, 12*, 69–73.

Steriade, M., Gloor, P., Llinás, R. R., Lopes da Silva, F. H., & Mesulam, M. M. (1990). Basic mechanisms of cerebral rhythmic activities. *Electroencephalography and Clinical Neurophysiology, 76*, 481–508.

Stoffel, W., & Bosio, A. (1997). Myelin glycolipids and their functions. *Current Opinion in Neurobiology, 7*, 654–661.

Storfer, M. (1999). Myopia, intelligence, and the expanding human neocortex: Behavioral influences and evolutionary implications. *International Journal of Neuroscience, 98*, 153–276.

Tan, U. (1996). Correlations between nonverbal intelligence and peripheral nerve conduction velocity in right-handed subjects: Sex-related differences. *International Journal of Psychophysiology, 22*, 123–128.

van Baal, G. C. M., Boomsma, D. I., & de Geus, E. J. C. (2001). Genetics of EEG coherence and IQ in young twins [Abstract]. *Behavior Genetics, 31*, 471.

van Baal, G. C. M., de Geus, E. J. C., & Boomsma, D. I. (1998). Genetic influences on EEG coherence in 5-year-old twins. *Behavior Genetics, 28*, 9–19.

van Beijsterveldt, C. E., & Boomsma, D. I. (1994). Genetics of the human electroencephalogram (EEG) and event-related brain potentials (ERPs): A review. *Human Genetics, 94*, 319–330.

van Beijsterveldt, C. E., Molenaar, P. C. M., de Geus, E. J. C., & Boomsma, D. I. (1998). Genetic and environmental influences on EEG coherence. *Behavior Genetics, 28*, 443–453.

van Valen, L. (1974). Brain size and intelligence in man. *American Journal of Physical Anthropology, 40*, 417–423.

Vernon, P. A. (1987). *Speed of information-processing and intelligence*. Norwood, NJ: Ablex.

Vernon, P. A. (1989). The heritability of measures of speed of information processing. *Personality and Individual Differences, 10*, 573–576.

Vernon, P. A., & Mori, M. (1992). Intelligence, reaction times, and peripheral nerve conduction velocity. *Intelligence, 8*, 273–288.

Vogel, F. (2000). *Genetics and the electroencephalogram*. Berlin: Springer-Verlag.

Weber, P., Bartsch, U., Rasband, M. N., Czaniera, R., Lang, Y., Bluethmann, H., et al. (1999). Mice deficient for tenascin-R display alterations of the extracellular matrix and decreased axonal conduction velocities in the CNS. *Journal of Neuroscience, 19*, 4245–4262.

Wechsler, D. (1974). *WISC: Manual for the Wechsler Intelligence Scale for Children*. New York: Psychological Corporation.

Wechsler, D. (1981). *WAIS–R manual for the Wechsler Adult Intelligence Scale*. New York: Psychological Corporation.

Wickett, J. C., & Vernon, P. A. (1994). Peripheral nerve conduction velocity, reaction time, and intelligence: An attempt to replicate Vernon and Mori. *Intelligence, 18*, 127–132.

Wickett, J. C., Vernon, P. A., & Lee, D. H. (1997). Within family correlations between general intelligence and MRI-measured brain volume in head size in male adult siblings [Abstract]. *Behavior Genetics, 27*, 611.

Wickett, J. C., Vernon, P. A., & Lee, D. H. (2000). Relationships between factors of intelligence and brain volume. *Personality and Individual Differences, 29*, 1095–1122.

Willerman, L., Schultz, R., Rutledge, J. N., & Bigler, E. D. (1991). In vivo brain size and intelligence. *Intelligence, 15*, 223–228.

Willerman, L., Schultz, R., Rutledge, J. N., & Bigler, E. O. (1992). Hemisphere size asymmetry predicts relative verbal and nonverbal intelligence differently in the sexes: An MRI study of structure–function relations. *Intelligence, 16*(3–4), 315–328.

Williams, J. T., Begleiter, H., Porjesz, B., Edenberg, H. J., Foroud, T., Reich, T., et al. (1999). Joint multipoint linkage analysis of multivariate qualitative and quantitative traits: II. Alcoholism and event-related potentials. *American Journal of Human Genetics, 65*, 1148–1160.

Woodward, K., & Malcolm, S. (1999). Proteolipid protein gene: Pelizaeus-Merzbacher disease in humans and neurodegeneration in mice. *Trends in Genetics, 15*, 125–128.

Wright, M. J., Hansell, N. K., Geffen, G. M., Geffen, L. B., Smith, G. A., & Martin, N. G. (2001). Genetic influence on the variance in P3 amplitude and latency. *Behavior Genetics,* 16, 315–328.

10

Genetic Covariance Between Processing Speed and IQ

Michelle Luciano, Margaret J. Wright, Glen A. Smith, Gina M. Geffen, Laurie B. Geffen, and Nicholas G. Martin

Behavioral genetic studies of cognition have turned increasingly to lower level cognitive processes, as measured by elementary cognitive tasks, to help understand the genetic structure of human mental ability (Baker, Vernon, & Ho, 1991; Ho, Baker, & Decker, 1988; Rijsdijk, Vernon, & Boomsma, 1998; Wright, Smith, Geffen, Geffen, & Martin, 2000). Elementary cognitive tasks often require a speeded response, and as such they have been largely considered to reflect mental speed, or speed of information processing (Jensen, 1998). In general, reaction time (RT) measures (including variability of responding) demonstrate correlations of approximately −0.30 with IQ, whereas measures without an RT component, such as inspection time, reach correlations of −0.50 with IQ (Deary & Stough, 1996; Jensen, 1993b). More complex elementary cognitive tasks correlate more highly with IQ; for example, choice RT confers a greater correlation with IQ than does simple RT (Larson, Merritt, & Williams, 1988).

In this chapter, we present the multivariate analyses of two processing speed measures (inspection time and choice RT) with IQ. First, we explore the complexity effect by including choice RT from three choice conditions in an analysis with full-scale IQ. Next, we explore the genetic covariance between inspection time and IQ, including a comparison of verbal and performance scales. Finally, we examine whether inspection time and choice RT relate to IQ through the same genetic factor. As the genes for processing speed overlap with genes for IQ, processing speed measures may provide a tractable framework through which to search for quantitative trait loci (QTLs) influencing intelligence. Herein lies the true value of the behavioral genetic study of basic information processing in the post-genomic era.

The earliest direct genetic study of basic processing speed and IQ was

This project is supported by grants from the Australian Research Council (Grants A79600334, A79906588, and A79801419) and the Human Frontier Science Program (Grant RG0154/1998-B). We express our gratitude to Ann Eldridge and Marlene Grace for data collection and to the twins in the study for their cooperation.

conducted by Ho et al. (1988) in a sample of 60 twins ages 8 to 18 years. The processing speed factor was defined by tests of rapid automatic naming (colors, numbers, letters, and pictures) and symbol processing (Colorado Perceptual Speed Test; Decker, & DeFries, 1981). Ho et al. found that the relationship between processing speed and IQ was almost completely mediated by genetic factors. Symbol processing was more strongly genetically correlated with IQ (0.67) than was rapid automatic naming (0.46). Nonshared environmental effects showed some influence, whereas shared environment was not important. Baker et al. (1991) replicated this finding of genetic overlap in processing speed and IQ by using a battery of RT tasks (as described by Vernon, 1989). A latent factor determined by additive genes influenced the RT component, verbal IQ, and performance IQ. Specific genes additionally contributed to the variance in performance IQ, whereas a specific common environment factor influenced verbal IQ. RT was somewhat determined by common environment (17%) but more so by unique environment (37%). Its genetic correlation with verbal and performance IQ was 1.00 and 0.92, respectively, indicating that nearly all the genetic variance of RT overlapped with IQ.

In other studies, elementary cognitive tasks that have demonstrated a substantial genetic relationship with IQ (genetic covariance ranging from 81% to 100%) include Sternberg's Memory Scanning, Posner's Letter Matching, and choice RT (Neubauer, Spinath, Riemann, Angleitner, & Borkenau, 2000; Petrill, Luo, Thompson, & Detterman, 1996; Rijsdijk et al., 1998). On the basis of their extensive use in intelligence research, choice RT and inspection time were investigated in our ongoing research on twins. Choice RT is a measure of the speed of response to the appearance of a single stimulus from an array of others. We sampled RT in three choice conditions (two-, four-, and eight-choice). The inspection time task differs fundamentally from choice RT in that it does not require a speeded response. Inspection time measures the minimum amount of time required by the participant to discriminate accurately between two lines of noticeably different length; thus, the stimulus duration is the time that is measured.

The genetic relationship between two-choice RT and IQ in adolescents has been established. Rijsdijk et al. (1998) used simple and two-choice RT in their longitudinal study of processing speed and IQ in 213 pairs of twins. Heritabilities for choice RT were lower on the second than the first occasion (0.49 vs. 0.62), perhaps owing to a change in task parameters and administration that may have produced the faster RTs and increased error rate observed on second test. The correlation of −0.22 between choice RT and Raven IQ was completely genetically mediated, demonstrating a genetic correlation of −0.36. In their analysis of choice RT and Wechsler Adult Intelligence Scale (WAIS) IQ subtest scores, phenotypic correlations ranged between −0.05 and −0.23, whereas genetic correlations ranged from −0.18 to −0.40 (higher for verbal than performance subtests). Simple and choice RT loaded on a general genetic factor (17% simple RT; 11% choice RT), but they loaded more highly on their own genetic factor (26% simple RT; 20% choice RT), indicating that although simple and choice RT overlap genetically with IQ, there is considerable unique genetic variance. Nonetheless, the relationship between choice RT and IQ in adolescents appears to be entirely mediated by genes.

We sought to replicate this genetic relationship using RT and variability indices in the largest sample of twins studied to date for RT measures. On the basis of findings that genetic effects increase proportionately with a test's *g* loading, it was expected further that this genetic relationship would be greatest for the more complex eight-choice condition and smallest for the two-choice condition, with the four-choice condition intermediate.

Ours was the first study of the genetic relationship between inspection time and full-scale IQ. We found that the phenotypic association between inspection time and IQ also was mediated entirely by genes (genetic correlation of −0.63), and furthermore, that the genetic covariance with inspection time was greater for performance IQ than verbal IQ (Luciano, Wright, et al., 2001). An additional genetic model (common pathway) of the covariance among inspection time, performance IQ, and verbal IQ is presented later in this chapter. Posthuma, de Geus, and Boomsma (2001) replicated our findings on inspection time and IQ in an extended twin design that included 688 family members (twins and siblings) from 271 families. They showed that the common genetic factor explained more variance in performance (22%) than verbal (10%) IQ. This result mirrors phenotypic findings of a higher association between inspection time with performance IQ than with verbal IQ (Deary, 1993; Kranzler & Jensen, 1989).

As diverse elementary cognitive tasks are moderately intercorrelated (Barrett, Alexander, Doverspike, Cellar, & Thomas, 1982; Saccuzzo, Johnson, & Guertin, 1994; Vernon, 1983), it has been suggested that the relationship between different elementary tasks and IQ is due to the same factor, for instance, neural transmission speed (e.g., Jensen, 1993a). This notion of a unitary speed factor has been supported by multiple regression analyses, which show that inspection time and choice RT do not make independent contributions to the prediction of IQ (Larson & Saccuzzo, 1989; Vernon, 1983). However, evidence to the contrary also has been reported showing that inspection time and reaction time factors from a battery of elementary cognitive tasks do make independent contributions to the prediction of IQ (Kranzler & Jensen, 1991). Behavioral genetic studies (e.g., Vernon, 1989) have shown differing strengths and patterns of genetic and environmental contributions to variance across diverse elementary cognitive tasks, indicating that there may not be a single processing mechanism influencing intelligence but rather different component processes (e.g., speed of perceptual apprehension, speed of short-term memory retrieval). The genes influencing these component processes may constitute the genes for intelligence (see Plomin, chapter 11, this volume).

It is possible then that the genes affecting choice RT are different from those affecting inspection time, despite the fact that both tasks index processing speed of some kind. Different genes may correspond to the different cognitive processes that are tapped by each task rather than the predominance of a single genetic factor controlling some basic biological mechanism. By using choice RT and inspection time as individual task measures rather than as a less informative unitary processing speed factor, a multivariate genetic analysis with IQ will establish whether the relationship is mediated by the same or separate genetic factors.

Method

Sample and Measures

Data were collected in the context of the ongoing Brisbane Memory, Attention, and Problem-Solving (MAPS) twin study (see Wright et al., 2001). Here we report data from the first 390 twin pairs: 97 monozygotic (MZ) females, 87 MZ males, 52 dizygotic (DZ) females, 48 DZ males, and 106 DZ opposite-sex pairs. Zygosity was determined by ABO, MN, and Rh blood groups and by nine independent polymorphic DNA markers. Twin pairs were excluded if either one had a history of significant head injury, neurological or psychiatric illness, or substance dependence or if they were currently taking long-term medications with central nervous system effects. Participants had normal or corrected-to-normal vision (better than 6/12 Snellen equivalent). The twins were mostly in their penultimate year of secondary school and were between ages 15 and 18 years (M = 16.17 years, SD = 0.34).

The choice RT task, inspection time task, and IQ test were part of a psychometric battery about 1.5 hours in length and were either preceded or followed by a testing session of similar duration that involved the measurement of event-related potentials during a delayed response task. The choice RT task was presented to the participants in the pseudo-game form of dripping taps, in which the participant had to quickly press the appropriate computer key to stop a tap from dripping. Participants aligned and rested their fingers on a keyboard rather than a home key (as used by Jensen, 1987). Different colored taps corresponded to the same fingers on both hands to aid tap and finger alliance; for example, the taps matching the index fingers were both red. The amount of water saved was indicated on the bottom left of the screen. Ninety-six, 48, and 96 trials were presented in the respective two-, four-, and eight-choice conditions. Individual trials were excluded if RT was less than 150 ms or greater than 2,000 ms. Output measures for each of the choice conditions included the mean RT and standard deviation (*SD*) of correct responses. For a more detailed description of this task, see Luciano, Smith, et al. (2001).

Inspection time was tested by a line discrimination task, which was presented as a pseudo-computer game of choosing the longer of two worms to go fishing. The two lines of comparison (see Figure 10.1) were described as worms that would quickly burrow into the ground (i.e., appearance of masking stimulus). The participant's task was to identify the longer worm in an effort to catch the most fish by pressing the corresponding left or

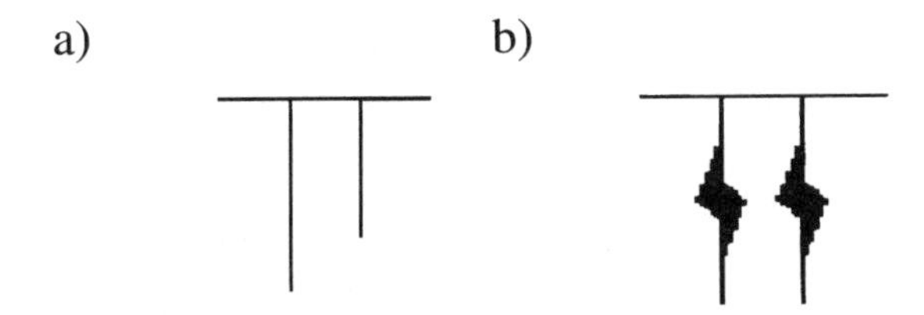

Figure 10.1. The pi figure stimulus (a) is presented briefly (duration ranges from 14.2 ms to 2,000 ms) in the center of the screen, and then (b) is hidden by a flash mask, consisting of two vertical lines shaped as lightning bolts, presented for a period of 300 ms.

right arrow key on the keyboard. Feedback in the form of a fish appeared at the lower left-hand side of the screen following every five correct judgments. The importance of accuracy and not reaction time was stressed verbally by the experimenter before beginning the task.

Inspection time was estimated post hoc by fitting a cumulative normal curve ($M = 0$) to accuracy as a function of stimulus onset asynchrony (SOA). Inspection time is commonly estimated through curve extrapolation so that performance at any desired accuracy level can be attained (see Nettelbeck, 1987). The statistic of interest is the SD of the curve, which is the SOA at which 84% accuracy is achieved. Participants whose data provided a poor fit to the cumulative normal curve ($R^2 < 0.95$) were excluded (i.e., 21 participants, or 2.7% of the sample).

A shortened version of the Multidimensional Aptitude Battery (Jackson, 1984, 1998) was used to measure IQ and included three verbal subtests (Information, Arithmetic, and Vocabulary) and two performance subtests (Spatial and Object Assembly). All subtests had a multiple-choice format and were limited to 7 minutes each. Administration and scoring were computerized. The Multidimensional Aptitude Battery was patterned after the Wechsler Adult Intelligence Scale–Revised (WAIS–R; Wechsler, 1981) and possesses good psychometric properties (Jackson, 1984, 1998).

Statistical Procedure

As speed (i.e., mean RT) and variability (i.e., SD) indices were both derived from the choice RT task, it was necessary to establish whether there would be any gain in information by including both indices in the multivariate analyses. Preliminary analyses involved deriving phenotypic correlations across RT and SD variables within each choice condition and inspecting them for excessive collinearity, an indication that the cognitive processes tapped by these measures were highly similar. If RT and SD were highly correlated, RT indices would be preferred over SD indices because of their higher test–retest reliability. Collinearity also was assessed across choice conditions to evaluate the similarity of these measures.

Multivariate genetic modeling progressed from the approach of a Cholesky (see Neale & Cardon, 1992) decomposition of the additive genetic

(A), common environmental (C), and unique environmental (E) variance contributing to RT in all choice conditions and full-scale IQ. More simplified models (independent pathways and common pathway) were compared with the best-fitting Cholesky model.

A trivariate analysis of inspection time, performance IQ, and verbal IQ (following from the phenotypic findings of a stronger relationship between inspection time with performance IQ than verbal IQ) was performed. In this analysis, a common pathway model was compared with the best-fitting Cholesky model.

To address the question of whether a single factor explained the relationship among choice RT, inspection time, and IQ, we performed a multivariate analysis of these variables. As inspection time has been shown to have different strengths of association with verbal and performance IQ, the subscales were investigated rather than the full-scale score. To arrive at a simpler model solution with equally weighted contributions from inspection time and choice RT, we chose only one choice RT measure to include with inspection time, performance IQ, and verbal IQ.

Results

Choice Reaction Time and IQ

Using the criterion of exceeding ±3.5 SDs from the mean, we excluded eight outliers from the two-choice condition, seven outliers from the four-choice condition, and five outliers from the eight-choice condition. In a contrasts analysis of mean and variances across birth order and zygosity, a multivariate outlier (twin pair) and three individual outliers in the eight-choice condition were removed. Following a $\log_{10}$ transformation, the number of outliers for SD in the two-choice, four-choice, and eight-choice conditions were eight, one, and two, respectively. All analyses using choice RT included a regression coefficient for accuracy in the means model to account for significant speed–accuracy trade-off effects. Computer or experimenter error resulted in the loss of five unrelated participants' verbal IQ scores (0.64%). IQ data were normally distributed with no outliers. As twins were tested as close as possible to their 16th birthdays, months of schooling differed across individual twin pairs, and so mean IQ was adjusted for months of schooling completed since the beginning of Grade 10.

The correlations between RT and SD within each choice condition were very high (see Table 10.1). The four- and eight-choice conditions showed excessive collinearity between RT and SD, which suggests that they are largely measuring the same process. This was further indicated by the comparison of the MZ and DZ cross-variable cotwin correlations in which genetic identity between RT and SD was apparent. Hence, SD variables from the four- and eight-choice conditions were not used in further analyses as RT measures were more reliable (data not shown.)

In the two-choice condition, the correlation between RT and SD was

Table 10.1. Phenotypic Correlations Between RT and SD Within Each Choice Condition and Cotwin Cross Variable (RT and SD) Correlations Within Each Choice Condition (95% Confidence Intervals in Parentheses)

Condition	Phenotypic r (N = 780 individuals)	MZ r (N = 184 twin pairs)	DZ r[a] (N = 206 twin pairs)
2 choice	0.73 (0.69–0.76)	0.34 (0.23–0.43)	0.19 (0.08–0.30)
4 choice	0.80 (0.77–0.83)	0.50 (0.41–0.58)	0.37 (0.26–0.46)
8 choice	0.87 (0.85–0.89)	0.60 (0.53–0.67)	0.39 (0.27–0.49)

Note. RT = reaction time; SD = standard deviation; MZ = monozygotic twins; DZ = dizygotic twins.
[a]Includes opposite-sex twins.

lower than those correlations from the four- and eight-choice conditions. A bivariate genetic analysis therefore was performed on two-choice RT and SD to assess whether the measures were tapping the same or separate genetic factors. There was insufficient power to differentiate between additive genetic and unique environmental (AE) and common environmental and unique environmental (CE) models in this analysis. For a gain of 3 degrees of freedom, increases in the −2 log-likelihood (LL) ratio from the additive genetic, common environmental, and unique environmental (ACE) model were 7.55 and 4.07 for respective AE and CE models. Importantly, the ACE model showed that a single genetic factor accounting for 26% of variance in SD and 29% of variance in RT explained the total genetic variation in the measures. Because RT and SD were influenced by the same genetic factor and to the same degree, we decided to only use RT, the more stable measure of the two, in subsequent multivariate analyses.

Phenotypic correlations between the RTs in the differing choice conditions were expected to be high as a result of the linear relationship between the variables (known as Hick's Law; Hick, 1952). The phenotypic and cotwin correlations across variables are displayed in Table 10.2. The correlation between two- and four-choice RT was slightly lower than either two- or four-choice correlations with eight-choice RT. Across all variable pairings, MZ cotwin correlations were higher than DZ cotwin correlations.

Table 10.2. Maximum Likelihood Estimates of Correlations Between RTs Obtained in the Different Choice Conditions and Their Respective Cross-Variable Cotwin Correlations

Relationship	Phenotypic r (N = 780 individuals)	MZ r (N = 184 twin pairs)	DZ r[a] (N = 206 twin pairs)
2 choice–4 choice	0.55 (0.50–0.60)	0.45 (0.36–0.52)	0.33 (0.23–0.42)
2 choice–8 choice	0.66 (0.61–0.70)	0.46 (0.38–0.54)	0.34 (0.24–0.44)
4 choice–8 choice	0.66 (0.62–0.71)	0.57 (0.49–0.63)	0.39 (0.29–0.48)

Note. RT = reaction time; MZ = monozygotic twins; DZ = dizygotic twins.
[a]Includes opposite-sex twins.

The phenotypic correlations between the RT measures and full-scale IQ are presented in Table 10.3, separately for female and male participants (although these estimates did not differ as indicated by the overlapping confidence intervals). The highest correlations with IQ were in the four-choice condition, followed by the eight-choice, then two-choice conditions.

The cotwin correlations across RT and IQ pairings were higher for MZs (−0.30, −0.54, −0.45) than DZs (−0.16, −0.26, −0.23) for the respective two-, four-, and eight-choice conditions. A multivariate analysis including the differing choice RT conditions and full-scale IQ was performed to determine whether a single genetic RT (or speed) factor could explain the shared variance. Three multivariate outliers (twin pairs) were removed from the analysis, as these observations were not consistent with the base model (ACE Cholesky). Goodness-of-fit statistics for the various models are reported in Table 10.4. The model of best fit was an AE Cholesky model in which nonsignificant parameters were dropped (Figure 10.2). In this reduced AE model, only two parameters could be dropped from the model: a genetic path coefficient from the third factor to full-scale IQ and a unique environment path coefficient from the third factor to full-scale IQ. The first genetic factor explained between 14% and 53% of variance in each of the measures. The second genetic factor explained 23%, 17%, and 30% of variance in four-choice RT, eight-choice RT, and full-scale IQ, respectively. Individual genetic factors accounted for 15% of the variance in eight-choice RT and 38% of variance in full-scale IQ.

Genetic factors almost completely accounted for the phenotypic correlations between each of the RT measures and IQ. Genetic and unique environment correlations among the choice RT variables and full-scale IQ are displayed in Table 10.5. These are derived from the full AE model.

Table 10.3. Maximum Likelihood Estimates (95% Confidence Intervals) of Phenotypic Correlations Among Choice Reaction Time (CRT) Variables and Full Scale IQ (FIQ) for Female (Below Diagonal) and Male (Above Diagonal) Participants

Variable	1	2	3	4
1. 2 CRT	—	0.54 (0.46–0.61)	0.66 (0.59–0.71)	−0.32 (−0.22–−0.41)
2. 4 CRT	0.51 (0.43–0.58)	—	0.68 (0.62–0.73)	−0.56 (−0.49–−0.63)
3. 8 CRT	0.62 (0.56–0.68)	0.65 (0.58–0.70)	—	−0.50 (−0.42–−0.58)
4. FIQ	−0.31 (−0.22–−0.40)	−0.53 (−0.45–−0.60)	−0.46 (−0.38–−0.53)	—

Note. Female participants, N = 396–399; male participants, N = 366–370.

Table 10.4. Goodness-of-Fit Statistics for the Multivariate Model of Two-Choice RT, Four-Choice RT, Eight-Choice RT, and Full Scale IQ: Fit (−2 LL) of the Cholesky Model and Change in Fit (χ^2) of the Independent Pathways and Common Pathway Models

Model	vs.	−2 LL	*df*	$\Delta\chi^2$	Δdf	*p*
i. ACE Cholesky decomposition		30,315.67	3025			
ii. AE Cholesky decomposition	i	30,325.36	3035	9.69	10	.47
iii. Reduced AE Cholesky decomposition	**ii**	**30,326.79**	**3037**	**1.43**	**2**	**.49**
iv. CE Cholesky decomposition	i	30,393.16	3035	77.49	10	<.01
v. AE independent pathways (32 parameters)	ii	30,376.97	3039	61.3	4	<.01
vi. AE common pathway (30 parameters)	ii	30,704.07	3041	388.4	6	<.01

Note. LL = log likelihood; A = additive genetic; C = common environmental; E = unique environmental. Boldface indicates the most parsimonious model.

Inspection Time and IQ

This next analysis was directed to the genetic association among inspection time, performance IQ, and verbal IQ. SD of the inspection time curve was positively skewed so a logarithmic transformation was applied to the data. Eleven outliers were removed. The phenotypic correlation between inspection time and performance IQ (−0.35) was of similar magnitude to that of verbal IQ (−0.26). Results of model fitting, in which a common pathway model was compared with the best-fitting Cholesky model, are displayed in Table 10.6. Because the common pathway model explained

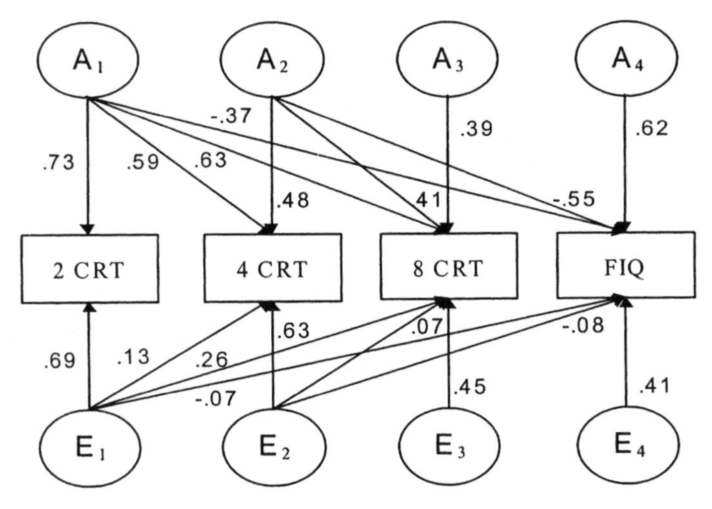

Figure 10.2. Path diagram depicting standardized path coefficients for the reduced additive genetic (A) and unique environment (E) Cholesky factorization of two-choice reaction time (2 CRT), four-choice reaction time (4 CRT), eight-choice reaction time (8 CRT), and full scale IQ (FIQ).

Table 10.5. Genetic (Below Diagonal) and Unique Environment (Above Diagonal) Correlations Between Choice Reaction Time (CRT) Variables and Full Scale IQ (FIQ) Estimated From the Full Additive Genetic and Unique Environment Model

Variable	1	2	3	4
1. 2 CRT	—	0.19	0.50	−0.16
2. 4 CRT	0.79	—	0.22	−0.21
3. 8 CRT	0.74	0.90	—	−0.16
4. FIQ	−0.41	−0.71	−0.58	—

the data most parsimoniously, it is presented as a path diagram in Figure 10.3. The relationships among inspection time, performance IQ, and verbal IQ were mediated by a latent factor that was primarily influenced by genes (92%). This latent factor explained the most variance in performance IQ (64%), then verbal IQ (37%), then inspection time (18%). The specific genetic variance contributing to each measure also was substantial, accounting for 14%, 46%, and 24% of variance in performance IQ, verbal IQ, and inspection time, respectively. To test whether the latent factor explained more variance in performance IQ than verbal IQ, we equated the proportions of variance explained by the latent factor on performance IQ and verbal IQ. This led to a significant change in the −2 LL ratio of 252.07 for one degree of freedom, confirming that the latent factor influenced performance IQ more than verbal IQ.

The genetic correlation between inspection time and performance IQ was −0.59, whereas for inspection time and verbal IQ it was −0.42; performance IQ and verbal IQ showed a genetic correlation of 0.61. The genetic correlations between inspection time and the IQ measures were fairly strong, substantiating our hypothesis that variation in genes that produce faster inspection times are strongly related to the variation in genes that promote higher IQs.

Table 10.6. Goodness-of-Fit Statistics for the Multivariate Model of Inspection Time, Performance IQ, and Verbal IQ: Fit (−2 LL) of the Cholesky Models and Change in Fit (χ^2) of the Common Pathway Model

Model	vs.	−2 LL	*df*	$\Delta\chi^2$	Δdf	*p*
i. ACE Cholesky decomposition		11,570.13	2262			
ii. AE Cholesky decomposition	i	11,581.78	2268	11.65	6	.07
iii. CE Cholesky decomposition	i	11,639.48	2268	69.35	6	<.01
iv. AE Common pathway model	**ii**	**11,585.40**	**2269**	**3.62**	**1**	**.06**

Note. LL = log likelihood; A = additive genetic; C = common environmental; E = unique environmental. Boldface indicates the most parsimonious model.

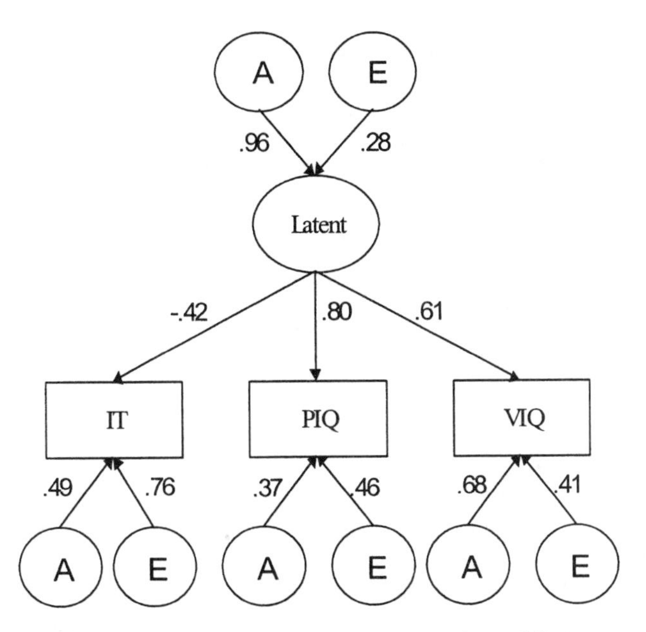

Figure 10.3. Common pathway model depicting the additive genetic (A) and unique environmental (E) relationship among inspection time (IT), performance IQ (PIQ), and verbal IQ (VIQ).

Choice Reaction Time, Inspection Time, and IQ

The maximum likelihood estimates of the phenotypic correlations among inspection time, choice RT measures, and verbal and performance IQ are displayed in Table 10.7, separately for female and male participants, although confidence intervals overlapped indicating that estimates did not differ across sex.

Two-choice RT was selected for use in the multivariate analysis with inspection time and IQ for several reasons. First, its correlation with IQ was more aligned with previous findings, suggesting that it was most likely tapping the same process as that measured in other choice RT studies (especially home-key paradigms). Second, it was the least complex of the choice conditions and as such could be argued to be a purer measure of processing speed (tapped fewer information processes).

Table 10.8 displays the goodness-of-fit statistics for the multivariate analysis of inspection time, choice RT, performance IQ, and verbal IQ. Common environment could be dropped from the Cholesky model with no significant change in fit of the model. The model of best fit was an AE common pathway model, and it is depicted in Figure 10.4. Additive genes accounted for 92% of the variance in the latent factor. The latent factor explained the most variance in performance IQ (53%), then verbal IQ (44%), then inspection time (21%), and then choice RT (15%). Specific additive genetic effects were of the same magnitude for verbal IQ and choice RT (40% of the variance) and were roughly the same for inspection time (21%) and performance IQ (25%). The genetic correlation between inspec-

Table 10.7. Maximum Likelihood Estimates (and 95% Confidence Intervals) of Phenotypic Correlations Among Inspection Time (IT), Choice Reaction Time (CRT) Variables, Verbal IQ (VIQ), and Performance IQ (PIQ) for Female (Below Diagonal) and Male (Above Diagonal) Participants, Assuming Independence of Cotwins

	1	2	3	4	5	6
1. IT	—	0.20 (0.10–0.30)	0.24 (0.14–0.34)	0.27 (0.17–0.36)	−0.29 (−0.19–−0.38)	−0.35 (−0.26–−0.44)
2. 2 CRT	0.20 (0.10–0.30)	—	0.54 (0.46–0.61)	0.66 (0.59–0.71)	−0.31 (−0.21–−0.40)	−0.27 (−0.17–−0.36)
3. 4 CRT	0.24 (0.14–0.34)	0.48 (0.39–0.55)	—	0.68 (0.62–0.73)	−0.45 (−0.36–−0.53)	−0.53 (−0.45–−0.60)
4. 8 CRT	0.23 (0.13–0.32)	0.61 (0.55–0.67)	0.58 (0.51–0.64)	—	−0.42 (−0.33–−0.50)	−0.47 (−0.38–−0.54)
5. VIQ	−0.20 (−0.10–−0.30)	−0.26 (−0.17–−0.35)	−0.45 (−0.37–−0.52)	−0.28 (−0.19–−0.37)	—	0.49 (0.41–0.56)
6. PIQ	−0.34 (−0.24–−0.42)	−0.24 (−0.14–−0.33)	−0.52 (−0.45–−0.59)	−0.35 (−0.25–−0.43)	0.55 (0.48–0.61)	—

Note. Female participants, N = 375–401; male participants, N = 354–370.

tion time and choice RT was 0.35, whereas the environmental correlation was effectively zero.

Discussion

Choice RT and IQ

The significant correlation between choice RT and psychometric intelligence is well established (Jensen & Munro, 1979; Neubauer, Riemann, Mayer, & Angleitner, 1997; Saccuzzo et al., 1994) and replicated in the

Table 10.8. Goodness-of-Fit Statistics for the Multivariate Model of Inspection Time, Choice Reaction Time, Performance IQ, and Verbal IQ: Fit (−2 LL) of the Cholesky Models and Change in Fit (χ^2) of the Common Pathway Model

Model	vs.	−2 LL	*df*	$\Delta\chi^2$	Δdf	*p*
i. ACE Cholesky decomposition		18,583.36	2986			
ii. AE Cholesky decomposition	i	18,599.51	2996	16.16	10	.09
iii. CE Cholesky decomposition	i	18,657.68	2996	74.32	10	<.01
iv. AE Common pathway model	**ii**	**18,611.63**	**3002**	**12.11**	**6**	**.06**

Note. LL = log likelihood; A = additive genetic; C = common environmental; E = unique environmental. Boldface indicates the most parsimonious model.

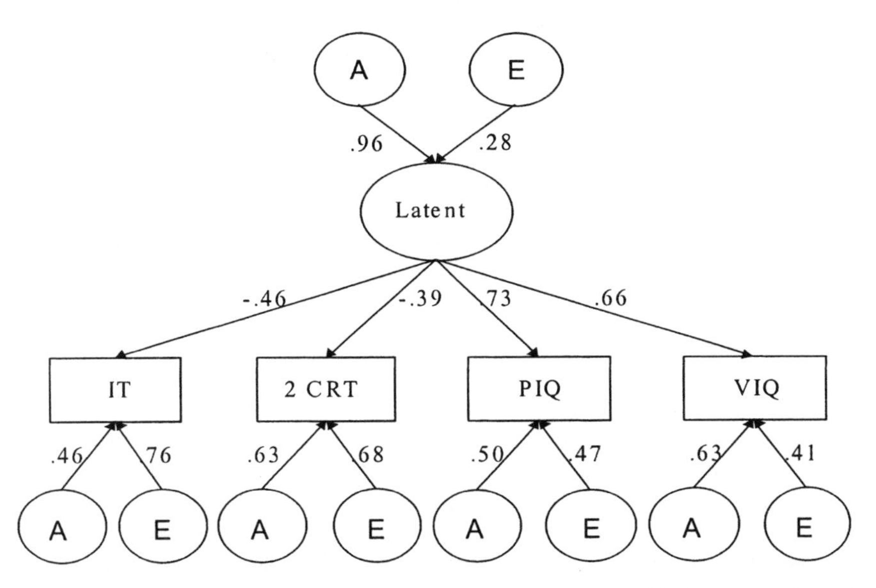

Figure 10.4. A common pathway model depicting the additive genetic (A) and unique environmental (E) relationship among inspection time (IT), choice reaction time (CRT), performance IQ (PIQ), and verbal IQ (VIQ).

present study. The correlation between two-choice RT and IQ was consistent with previous research. However, correlations between four-choice RT and IQ and eight-choice RT and IQ were at the top end of the range reported (Jensen, 1987). It may be that the correlation between four-choice RT and IQ was confounded by a learning effect as the four-choice condition was presented first without any previous practice. As for the high correlation between eight-choice RT and IQ, it is possible that the requirement of coordinating the different fingers with the varying stimulus positions actually increased the complexity of the task, invoking cognitive processes not normally required in the home-key paradigm. The high correlations between RT and SD measures within the same condition suggested that the processes measured by each were the same rather than different (see Carroll, 1993; Stankov, Roberts, & Spilsbury, 1994), and this was further implied by the genetic analysis of the two-choice condition, which found that a single factor explained the entire genetic variance in each of these measures.

This study confirmed that the relationship between choice RT and IQ was influenced by a common genetic factor as first reported by Rijsdijk et al. (1998), in contrast to an earlier study in children by Petrill et al. (1996), who found a more important role for common environment. Multivariate model fitting indicated that the covariance between choice RT variables and full-scale IQ was mediated entirely by genes, although a single genetic factor was not sufficient to explain the covariance between all choice RT measures and IQ. The first genetic factor, which had the largest loading on two-choice RT, explained more of the variance in four- and eight-choice RTs but less of the variance in IQ than did the second genetic factor. This first genetic factor may reflect a basic processing speed factor as the largest loading was on two-choice RT, the simplest of the choice conditions. The second genetic factor, which loaded most highly on IQ, might represent a set of information (component) processes needed to execute more complex cognitive operations. Alternatively, it may reflect strategy use, which enables more efficient performance in conditions of higher choice, and those with higher IQs are better able to form these strategies. There was no additional genetic factor influencing the relationship between eight-choice RT and IQ. This indicated that the component processes invoked by the eight-choice condition and associated with IQ were the same as those required in the four-choice condition or that the strategies used in each condition were highly similar.

The proportions of overlapping genetic variance between the RT measures and IQ were similar in magnitude to those reported by Risjdijk et al. (1998) for simple RT and two-choice RT. The genetic correlation between two-choice RT and IQ (−0.41) obtained in the present study also was comparable with theirs (−0.36) in the two-choice condition. There was no indication of a common environment factor influencing choice RT nor the relationship between choice RT and IQ as reported by Petrill et al. (1996) in a younger sample of twins (ages 6–13 years). In fact, the point estimates indicated the presence of small shared environmental (or assortative mating) effects for both IQ and RT, although the analysis did not have power to detect them simultaneously with additive genetic effects. Further research within a longitudinal context is needed to clarify the change in influence of common environment from childhood to adolescence.

The results showed that a further genetic factor (explaining 38% of variance) was needed to explain the genetic effects on full-scale IQ, suggesting that choice RT processes are related to specific components of intelligence and independent of others. The remaining variance was composed of unique environment, which was correlated as well as specific to each measure. The specific components of unique environment were far more influential than the common factors, especially in their effects on IQ. For choice RT variables, unique environment was perhaps mostly in the form of measurement error as test–retest reliability is lower than for IQ.

Inspection Time and IQ

Confirming previous studies (Deary & Stough, 1996; Kranzler & Jensen, 1989; Nettelbeck, 1987), significant correlations between inspection time with verbal and performance IQs were found, although they were toward the lower range of estimates reported. This study is perhaps the largest study conducted on inspection time, and the unselected nature of the sample suggests that the correlation estimates are unbiased and stable.

Like others who have investigated elementary cognitive tasks (Baker et al., 1991; Rijsdijk et al., 1998; Wright et al., 2000), we have demonstrated in the present study that inspection time shares a substantial genetic relationship with IQ. The genetic relationship was mediated by a latent factor that showed a larger influence on performance than verbal IQ. A psychometric factor model also adequately described the relationship among RT factor, verbal IQ, and performance IQ, but a differential genetic relationship between performance and verbal IQ was not found (Baker et al., 1991). Our finding can be accommodated within a *g* factor framework, in which the latent factor influencing inspection time, performance IQ, and verbal IQ is akin to general intelligence. Gustafsson (1984) suggested that fluid intelligence (tapped by performance subtests) is actually the *g* factor, and it may be for this reason that the latent factor influences performance IQ to a greater extent than verbal IQ. Fluid intelligence may depend on the overall proficiency of different information processes, of which inspection time is one, or may relate to an underlying biological speed mechanism which inspection time captures better than do RT variables. Alternatively, a top-down explanation, whereby higher fluid intelligence enables better inspection time performance, can be proffered.

As demonstrated, the unique environment effects on inspection time were large, but these were only weakly correlated with IQ. Strategy use (involving apparent motion cues) was invoked to explain the large uncorrelated unique environment effects in inspection time. Research has confirmed that the correlation between inspection time and IQ increases when participants using apparent motion strategies are excluded (Mackenzie & Bingham, 1985; Mackenzie & Cumming, 1986), and this may be consistent with the present genetic findings (which show that unique environment contributes slightly to the relationship between inspection time and IQ) if apparent motion cue use is actually influenced by unique environment rather than genes.

In Posthuma et al.'s (2001) study of the genetic relationship between inspection time and IQ, the performance index consisted of different tests (picture completion, block design, matrix reasoning) from those used in the present study, but a similar amount of genetic covariance between inspection time and performance IQ was explained. This confirms the generality of the genetic association across diverse IQ subtests and further suggests that the strength of the genetic association with inspection time may be dependent on a generalized fluid ability rather than specific performance group factors. Posthuma et al. also found that the variance in inspection time was largely determined by an independent unique environmental factor.

Choice RT, Inspection Time, and IQ

In the present study, the correlations between inspection time and the RT variables from the choice RT task ranged between 0.20 and 0.27, agreeing with past findings of significant, albeit modest, correlations (0.11 to 0.37) between the two (Larson, 1989; Vernon, 1983). The relationship among inspection time, choice RT, performance IQ, and verbal IQ was dependent on the same latent factor, which was mostly influenced by genes. This analysis may have capitalized on the substantial covariance between the

IQ composite scores, which forced a strong underlying latent factor. This latent factor represented the *g* factor of intelligence, which presides over the performance and verbal group factors (or in this case scores), and its influence appeared greater on inspection time than choice RT.

In the analysis of inspection time, performance IQ, and verbal IQ, a latent factor also had best described the data. The genetic variance shared by inspection time and verbal IQ in this previous analysis was only slightly lower (0.24 vs. 0.28) than that obtained when choice RT also was included, whereas the genetic covariance between inspection time and performance IQ remained unchanged (0.31). This implied that the latent factor influencing inspection time and IQ was essentially the same as the latent factor influencing inspection time, choice RT, and IQ.

Although this study showed that a unitary factor was able to explain the relationship among inspection time, choice RT, and IQ, it may be that an analysis of IQ subtest data demonstrates different genetic associations between choice RT and inspection time with various subtests. For instance, inspection time has been claimed to predict a perceptual speed factor (e.g., tapped by the digit–symbol substitution) rather than the general factor (Mackintosh, 1998). Hence, a study including subtest scores rather than composite scores will provide clearer evidence for or against the presence of a single genetic factor (or latent factor influenced by genes) mediating the relationship between diverse elementary cognitive tasks and intelligence.

That genetic factors mediated the relationship among choice RT, inspection time, and IQ supports the claim for a biological basis affecting both processing speed and higher order cognition. Jensen (1998) advocated a neural efficiency model in which factors such as oscillation speed of neuronal excitatory potentials and myelination of neurons determine the speed of information processes. A faster oscillation frequency causes the action potential to be nearer to the threshold of excitation, resulting in a faster response. Greater myelination of neurons also might promote a faster speed and efficiency of information processing because myelinated fibers are responsible for transmitting neural information to differing regions of the brain, and across the corpus callosum. Although these hypotheses have not been directly investigated, the results from the present study indicate that biological avenues of inquiry are indeed useful.

Conclusion and Future Prospects

This study demonstrated that the significant phenotypic relationship among choice RT, inspection time, and IQ was primarily genetically mediated. A model with a single latent factor (mostly determined by additive genes) influencing choice RT, inspection time, and verbal and performance IQs could account for the covariation of these measures of elementary and higher order cognitive processes, although other models also are consistent with the data. A future analysis using IQ subtest scores will enable a clearer understanding of whether choice RT and inspection time relate to a general genetic factor or whether they have different genetic associations with specific IQ subtests. The findings from multivariate genetic analyses of processing speed indices and IQ are informative to theorists wishing to elaborate information-processing models of intelligence.

However, greater value of the study of processing speed variables may be their key role in the molecular detection of genes influencing intelligence. The variance in IQ is most likely due to the combined action of many genes of small effect, some of which will overlap with genes for processing speed. QTL studies of cognition have thus far focused on a single measure, IQ, but much is to be gained by including multiple measures of intelligence in a QTL analysis. For instance, linkage analysis has low power to detect QTLs unless they are of major effect (20%–30% of variance), but by including multiple measures of a phenotype, the power to detect a QTL may be increased, and the prospective candidate region can be narrowed (Boomsma & Dolan, 2000; Williams et al., 1999). The increased power to detect linkage in a multivariate analysis derives from the underlying correlational structure of the phenotypes, provided this covariation is partly due to the QTL. Thus, by making use of the covariation between processing speed measures and IQ, the power to detect linkage with QTLs influencing both these facets of cognition should be increased.

As processing speed measures are theorized to index the neural speed of the brain, molecular approaches focusing on genes coding for basic structural aspects of neural wiring such as connectivity, myelin sheathing, number of ion channels, and efficiency of synaptic transmission may prove worthwhile. Explicit modeling of the multivariate genetic covariance of the processing speed and IQ measures will allow a direct test of whether candidate genes show pleiotropic effects (same gene influences variation in several measures) in multiple systems or whether gene effects on more elementary processes affect "downstream" cognition.

References

Baker, L. A., Vernon, P. A., & Ho, H. (1991). The genetic correlation between intelligence and speed of information processing. *Behavior Genetics, 21*, 351–367.

Barrett, G. V., Alexander, R. A., Doverspike, D., Cellar, D., & Thomas, J. C. (1982). The development and application of a computerized information-processing test battery. *Applied Psychology Measurement, 6*, 13–29.

Boomsma, D. I., & Dolan, C. V. (2000). Multivariate QTL analysis using structural equation modelling: A look at power under simple conditions. In T. D. Spector, H. Snieder, & A. J. MacGregor (Eds.), *Advances in twin and sib-pair analysis* (pp. 203–218). London: Greenwich Medical Media.

Carroll, J. B. (1993). *Human cognitive abilities: A survey of factor-analytic studies*. Cambridge, England: Cambridge University Press.

Deary, I. J. (1993). Inspection time and WAIS–R IQ subtypes: A confirmatory factor analysis study. *Intelligence, 17*, 223–236.

Deary, I. J., & Stough, C. (1996). Intelligence and inspection time. *American Psychologist, 51*, 599–608.

Decker, S. N., & DeFries, J. C. (1981). Cognitive ability profiles in families of reading-disabled children. *Developmental Medicine and Child Neurology, 23*, 217–227.

Gustafsson, J.-E. (1984). A unifying model for the structure of intellectual abilities. *Intelligence, 8*, 179–203.

Hick, W. E. (1952). On the rate of gain of information. *Quarterly Journal of Experimental Psychology, 4*, 11–26.

Ho, H., Baker, L. A., & Decker, S. N. (1988). Covariation between intelligence and speed of cognitive processing: Genetic and environmental influences. *Behavior Genetics, 18*, 247–261.

Jackson, D. N. (1984). *Manual for the Multidimensional Aptitude Battery*. Port Huron, MI: Research Psychologists Press.

Jackson, D. N. (1998). *Multidimensional Aptitude Battery II*. Port Huron, MI: Sigma Assessment Systems.

Jensen, A. R. (1987). Individual differences in the Hick paradigm. In P. A. Vernon (Ed.), *Speed of information-processing and intelligence* (pp. 101–175). Norwood, NJ: Ablex.

Jensen, A. R. (1993a). Spearman's *g*: Links between psychometrics and biology. In F. M. Crinella & J. Yu (Eds.), *Brain mechanisms: Papers in memory of Robert Thompson* (Vol. 702, pp. 103–130). New York: New York Academy of Sciences.

Jensen, A. R. (1993b). Why is reaction time correlated with psychometric *g*. *Current Directions in Psychological Science, 2*(2), 53–56.

Jensen, A. R. (1998). *The g factor*. Westport, CT: Praeger.

Jensen, A. R., & Munro, E. (1979). Reaction time, movement time, and intelligence. *Intelligence, 3*, 121–126.

Kranzler, J. H., & Jensen, A. R. (1989). Inspection time and intelligence: A meta-analysis. *Intelligence, 13*, 329–347.

Kranzler, J. H., & Jensen, A. R. (1991). The nature of psychometric *g*: Unitary process or a number of independent processes? *Intelligence, 15*, 397–422.

Larson, G. E. (1989). A brief note on coincidence timing. *Intelligence, 13*, 361–367.

Larson, G. E., Merritt, C. R., & Williams, S. E. (1988). Information processing and intelligence: Some implications of task complexity. *Intelligence, 12*, 133–147.

Larson, G. E., & Saccuzzo, D. P. (1989). Cognitive correlates of general intelligence: Toward a process theory of *g*. *Intelligence, 13*, 5–31.

Luciano, M., Smith, G. A., Wright, M. J., Geffen, G. M., Geffen, L. B., & Martin, N. G. (2001). On the heritability of inspection time and its covariance with IQ: A twin study. *Intelligence, 29*, 443–457.

Luciano, M., Wright, M. J., Smith, G. A., Geffen, G. M., Geffen, L. B., & Martin, N. G. (2001). Genetic covariance amongst measures of information processing, working memory, and IQ. *Behavior Genetics, 31*, 581–592.

Mackenzie, B., & Bingham, E. (1985). IQ, inspection time, and response strategies in a university population. *Australian Journal of Psychology, 37*, 257–268.

Mackenzie, B., & Cumming, S. (1986). How fragile is the relationship between inspection time and intelligence: The effects of apparent-motion cues and previous experience. *Personality and Individual Differences, 7*, 721–729.

Mackintosh, N. J. (1998). *IQ and human intelligence*. New York: Oxford University Press.

Neale, M. C., & Cardon, L. R. (1992). *Methodology for genetic studies of twins and families.* Dordrecht, NL: Kluwer Academic Publishers.

Nettelbeck, T. (1987). Inspection time and intelligence. In P. A. Vernon (Ed.), *Speed of information-processing and intelligence* (pp. 295–346). Norwood, NJ: Ablex.

Neubauer, A. C., Riemann, R., Mayer, R., & Angleitner, A. (1997). Intelligence and reaction time in the Hick, Sternberg, and Posner paradigms. *Personality and Individual Differences, 22*, 885–894.

Neubauer, A. C., Spinath, F. M., Riemann, R., Angleitner, A., & Borkenau, P. (2000). Genetic and environmental influences on two measures of speed of information processing and their relation to psychometric intelligence: Evidence from the German Observational Study of Adult Twins. *Intelligence, 28*, 267–289.

Petrill, S. A., Luo, D., Thompson, L. A., & Detterman, D. K. (1996). The independent prediction of general intelligence by elementary cognitive tasks: Genetic and environmental influences. *Behavior Genetics, 26*, 135–147.

Posthuma, D., de Geus, E. J. C., & Boomsma, D. I. (2001). Perceptual speed and IQ are associated through common genetic factors. *Behavior Genetics, 31*, 593–602.

Rijsdijk, F. V., Vernon, P. A., & Boomsma, D. I. (1998). The genetic basis of the relation between speed-of-information-processing and IQ. *Behavioural Brain Research, 95*, 77–84.

Saccuzzo, D. P., Johnson, N. E., & Guertin, T. L. (1994). Information processing in gifted versus nongifted African American, Latino, Filipino, and White children: Speeded versus nonspeeded paradigms. *Intelligence, 19*, 219–243.

Stankov, L., Roberts, R., & Spilsbury, G. (1994). Attentional variables and speed of test-taking in intelligence and aging. *Personality and Individual Differences, 16*, 423–434.

Vernon, P. A. (1983). Speed of information processing and general intelligence. *Intelligence, 7*, 53–70.

Vernon, P. A. (1989). The heritability of measures of speed of information-processing. *Personality and Individual Differences, 10*, 573–576.

Wechsler, D. (1981). *Wechsler Adult Intelligence Scale–Revised Manual.* New York: Psychological Corporation.

Williams, J. T., Begleiter, H., Porjesz, B., Edenberg, H. J., Foroud, T., Reich, T., et al. (1999). Joint multipoint linkage analysis of multivariate quantitative and qualitative traits: II. Alcoholism and event-related potentials. *American Journal of Human Genetics, 65*, 1148–1160.

Wright, M. J., Boomsma, D., De Geus, E., Posthuma, D., Van Baal, C., Luciano, M., et al. (2001). Genetics of cognition: Outline of collaborative twin study. *Twin Research, 4*(1), 48–56.

Wright, M. J., Smith, G. A., Geffen, G. M., Geffen, L. B., & Martin, N. G. (2000). Genetic influence on the variance in coincidence timing and its covariance with IQ: A twin study. *Intelligence, 28,* 239–250.

11

General Cognitive Ability

Robert Plomin

Complex quantitative traits will be the major beneficiary of the postgenomic era in which the 3 billion DNA bases of the human genome sequence are known as well as all variations in the DNA sequence, which is the stuff of hereditary transmission of individual differences. Behavior is the most complex of complex traits, and general cognitive ability (*g*), often called *intelligence,* is the most complex of all. Why study the genetics of such a complex and controversial trait? Would it not be the last trait to which one would want to apply molecular genetic techniques that attempt to identify specific genes and then try to understand how those genes function? This chapter attempts to address these questions by discussing what *g* is, what is known from quantitative genetic research, and the results from the first molecular genetic study of *g*.

General Cognitive Ability

g is a quantitative trait that varies from a low end of mild mental retardation to a high end of gifted individuals (Plomin, 1999b). This chapter is about the normal range of variation in *g*—there is surprisingly little known about genetic and environmental etiologies of mental retardation (Plomin, 1999a) or high ability (Plomin & Price, in press). Although intelligence means different things to different people, *g* has a more precise definition: *g* is what diverse cognitive abilities have in common. One of the most consistent findings from individual differences research on human cognitive abilities and disabilities during the 20th century is that diverse cognitive processes intercorrelate, including tests as different as reasoning, spatial ability, verbal ability, and memory. Despite the diversity of such tests, individuals who perform well on one test tend to do well on other tests. In a meta-analysis of 322 studies that included hundreds of cognitive tests, the average correlation among the tests was about 0.30 (Carroll, 1993). Factor analysis, a technique in which a composite score is created that represents what is shared in common among the measures, indicates that *g* accounts for about 40% of the total variance of cognitive tests (Jensen, 1998). However, *g* is not just a statistical abstraction—one can simply look at a matrix of correlations among such measures and see that there is a positive manifold among all tests and that some measures (e.g., spatial and verbal ability) intercorrelate more highly on average than do other measures (e.g., nonverbal memory tests). Because all of these measures intercorrelate to some extent, *g* also is indexed reasonably well by a simple total score on a diverse set of cognitive measures, as is done in IQ tests. This overlap emerges not only for traditional measures of cognitive abilities of the sort seen on IQ tests but also for information-processing tasks that rely on reaction time and other cognitive tasks used to assess, for example, working memory (M. Anderson, 1992; Baddeley & Gathercole, 1999; Deary, 2000; Stauffer, Ree, & Carretta, 1996) and theory of mind (Hughes & Cutting, 1999).

General cognitive ability was recognized nearly a century ago by Charles Spearman (1904, 1927), who used *g* as a neutral signifier that avoided the many connotations of the word *intelligence*: *g* is one of the most reliable and valid traits in the behavioral domain (Jensen, 1998), its long-term stability after childhood is greater than for any other behavioral trait (Deary, Whalley, Lemmon, Crawford, & Starr, 2000), it predicts important social outcomes such as educational and occupational levels far better than any other trait (Gottfredson, 1997), and it is a key factor in cognitive aging (Salthouse & Czaja, 2000). As discussed later, although *g* has not yet entered the lexicon of neuroscience, *g* needs to be incorporated in research on learning and memory.

Although the concept of *g* is widely accepted (Carroll, 1997; Neisser et al., 1996; Snyderman & Rothman, 1987), acceptance is not universal (Jensen, 1998). The arguments against *g* include ideological issues such as political concerns and the notion that *g* merely reflects knowledge and skills that happen to be valued by the dominant culture (Gould, 1996). Objections of a more scientific nature include theories that focus on specific abilities (Gardner, 1983; Sternberg, 1985). However, when these theories are examined empirically, *g* shines through. For example, Sternberg and Gardner (1983), advocates of a "componential" view of cognitive processing, conceded, "We interpret the preponderance of evidence as overwhelmingly supporting the existence of some kind of general factor in human intelligence. Indeed, we are unable to find any convincing evidence at all that militates against this view" (p. 249). There are of course many other important noncognitive abilities such as athletic ability, but nothing seems to be gained by lumping all such abilities together as is done with the popular notion of "multiple intelligences" (Gardner, 1983). Also, *g* by no means guarantees success either in school or in the workplace; achievement also requires personality, motivation, and social skills as emphasized in the fashionable concept of "emotional intelligence" (Goleman, 1995). Thus, *g* is not the whole story, but trying to tell the story of cognitive abilities without *g* loses the plot entirely.

Quantitative Genetic Research on *g*: Going Beyond Heritability

A year before the publication of Gregor Mendel's seminal paper on the laws of heredity, Francis Galton (1865), the father of behavioral genetics, published the first family study of *g*-like ability and other abilities. The first twin and adoption studies in the 1920s also investigated *g* (Burks, 1928; Freeman, Holzinger, & Mitchell, 1928; Theis, 1924). Subsequently, more research has addressed the genetics of *g* than any other human characteristic other than personality assessed by self-report questionnaires. Dozens of studies including more than 8,000 parent–offspring pairs, 25,000 pairs of siblings, 10,000 twin pairs, and hundreds of adoptive families all converge on the conclusion that genetic factors contribute substantially to *g* (Plomin, DeFries, McClearn, & McGuffin, 2001). Estimates of the effect size, called *heritability*, vary from 40% to 80%, but estimates based on the entire body of data are about 50%, indicating that genes account for about half of the variance in *g*.

The conclusion that genetic contributions to individual differences in IQ test scores are significant and substantial is one of the most important facts that has been uncovered in research on intelligence. Even when this conclusion is fully accepted, however, we are much closer to the beginning than to the end of the story of heredity and intelligence. The rest of the story requires going beyond anonymous genetic and environmental components of variance. Rather than reviewing the research literature on the heritability of *g*, it is more in keeping with the theme of the present volume to consider ways in which quantitative genetic research has been used to go beyond demonstrating the heritability of *g*. In this section, three examples of research that goes beyond heritability are mentioned: developmental, multivariate, and environmental analyses. These examples also point to the increasing importance of quantitative genetics in the postgenomic era, showing how quantitative genetics can chart the course for molecular genetic research.

Developmental Genetic Analysis

The predecessor of the present volume, *Nature, Nurture, and Psychology* (Plomin & McClearn, 1993), included a section on cognitive abilities and disabilities with chapters that focused on two developmental issues. The first issue is that heritability can differ throughout the lifespan. One of the most striking recent findings about *g* is that when the results are sorted by age, heritability increases from about .20 in infancy to about .40 in childhood to .60 or higher later in life (McGue, Bouchard, Iacono, & Lykken, 1993), even for individuals more than 80 years old (McClearn et al., 1997). This increase in heritability throughout the lifespan is interesting because it is counterintuitive in relation to Shakespeare's "slings and arrows of outrageous fortune" accumulating over time. The causes of this increase in heritability are not known, but a likely culprit is genotype–environment correlation (discussed later): As they live their lives, people actively select, modify, and even create environments conducive to the development of their genetic proclivities (Bouchard, Lykken, McGue, & Tellegen, 1997; Scarr, 1992). For this reason, it may be more appropriate to think about *g* as an appetite rather than an aptitude.

The second developmental issue concerns genetic influence on age-to-age change and continuity in longitudinal analyses. Change in genetic effects from age to age does not necessarily mean that genes are turned on and off during development, although this does happen. Genetic change simply means that genetic effects at one age differ from genetic effects at another age. For example, the same genes could have different effects in the brains of 8-year-olds as compared with 4-year-olds. In the previous volume, a longitudinal genetic model was applied to twin and adoptive sibling data on *g* from infancy to middle childhood (Fulker, Cherny, & Cardon, 1993). Although most genetic effects on *g* contribute to continuity from one age to the next, evidence for genetic change was found at two important developmental transitions. The first is the transition from infancy to early childhood, an age when cognitive ability rapidly changes as language develops. The second is the transition from early to middle childhood, at age 7 years. It is no coincidence that children begin formal schooling at this age—all theories of cognitive development recognize this as a major transition. Other longitudinal genetic analyses suggest that, despite such interesting examples of change, a surprising amount of genetic influence on *g* in childhood overlaps with genetic influence even into adulthood (Plomin, 1986).

Multivariate Genetic Analysis

One of the most important findings about *g* comes from multivariate genetic analysis, which is used to examine covariance among specific cognitive abilities rather than the variance of each trait considered separately. It yields a statistic called the *genetic correlation*, which is an estimate of the extent to which genetic effects on one trait correlate with genetic effects on another trait independent of the heritability of the two traits. That is, although all cognitive abilities are moderately heritable, the genetic correlations between them could be anywhere from 0.0, indicating complete independence, to 1.0, indicating that the same genes influence different cognitive abilities. Multivariate genetic analyses have consistently found that genetic correlations among cognitive abilities are very high, close to 1.0 (Petrill, 1997). In other words, if a gene were found that is associated with a particular cognitive ability, the same gene would be expected to be associated with all other cognitive abilities as well. As noted earlier, *g* accounts for about 40% of the total variance of cognitive tests. In contrast, multivariate genetic research indicates that *g* accounts for nearly *all* of the genetic variance of cognitive tests. That is, what is in common among cognitive abilities is almost completely genetic in origin. This finding of genetic *g* has the interesting converse implication that what is specific to each cognitive test is largely environmental—what makes us good at all tests is largely genetic but what makes us better at some tests than others is largely environmental. The predecessor of the present volume (Plomin & McClearn, 1993) included a chapter that combined longitudinal genetic analysis described in the previous section with multivariate genetic analysis of specific cognitive abilities (Cardon & Fulker, 1993). The results suggested that genetic *g* increases in explanatory power from infancy to early childhood to middle childhood, a finding confirmed in subsequent research (Dale, Dionne, Eley, & Plomin, 2000; Price, Eley, Dale, Stevenson, & Plomin, 2000; Rietveld, van Baal, Dolan, & Boomsma, 2000).

The multivariate genetic finding of genetic *g* is not limited to standard psychometric tests. Cognitive psychologists have assumed that *g* emerges from psychometric tests because such complex tests enlist many independent elementary cognitive processes such as information processing and working memory (Detterman, 2000). However, these elementary cognitive processes are not independent and yield a *g* factor (Deary, 2000). Moreover, genetic research indicates that elementary cognitive processes are heritable (Neubauer, Spinath, Riemann, Borkenau, & Angleitner, 2000; Petrill, Thompson, & Detterman, 1995; Vernon, 1989) and are correlated with *g* largely for genetic reasons (Baker, Vernon, & Ho, 1991; Petrill, Luo, Thompson, & Detterman., 1996; Rijsdijk, Vernon, & Boomsma, 1998). The same genetic *g* appears to underlie elementary cognitive tasks and more complex psychometric tests (Luciano et al., 2001).

Another *g*-related example of multivariate genetic analysis involves tests of school achievement, which also show substantial genetic influence. Multivariate genetic analyses indicate that genetic effects on *g* account completely for genetic effects on tests of school achievement (Thompson, Detterman, & Plomin, 1991; Wadsworth, 1994). Conversely, differences between *g* scores and tests of school achievement are exclusively environmental in origin. This finding suggests two important implications. First, school achievement scores independent of intelligence test scores appear to be largely devoid of genetic influence and would thus provide a better measure of the effectiveness of the child's environment in fostering achievement. Second, the finding suggests that underachievement, defined as a discrepancy between ability and achievement, is largely due to environmental factors.

The importance of genetic *g* as shown in multivariate genetic research also has important implications for understanding how the brain works from an individual differences perspective. Spearman, who first described *g* in 1904, noted that ultimate understanding of *g* "must needs come from the most profound and detailed study of the human brain in its purely physical and chemical aspects" (Spearman, 1927, p. 403). The simplest brain model of genetic *g* is that there is a single fundamental brain process that permeates all other brain processing, such as neural speed (e.g., myelination), power (e.g., number of neurons), or fidelity (e.g., density of dendritic spines). The opposite model is that there are many brain processes that are uncorrelated phenotypically and genetically but lead to a genetic correlation in performance on cognitive tasks because all of these brain processes are enlisted by the cognitive tasks. Although cognitive neuroscience seldom considers individual differences, current normative theories assume a model of this latter sort in which the brain consists of independent modules (Fodor, 1983; Pinker, 1994). A middle position is that multiple brain processes underlie *g* in cognitive tasks, but these processes are correlated phenotypically and genetically. That is, genetic *g* might exist in the brain as well as the mind (Plomin & Spinath, 2002). To test these different models about brain mechanisms responsible for *g*, it is necessary to identify reliable individual differences in brain processes and investigate the phenotypic and genetic relationships among these processes. What is known so far is surprising: The strongest correlate of *g* is brain volume. This association appears to be due to brain volume throughout the brain rather than to particular regions of the brain. Individual differences in brain volume are highly heritable, and genetic factors appear to mediate substantially the phenotypic correlation with *g* (Pennington et al., 2000; Posthuma et al., 2002). The only other brain mechanisms known to correlate with *g* are electroencephalography measures, event-related potentials (especially P-300 latency), and cerebral glucose metabolism rate as assessed by positron emissions tomography (Vernon, Wickett, Banzana, & Stelmack, 2000). Multivariate genetic research is needed to investigate the extent to which genetic factors mediate such phenotypic associations between brain mechanisms and *g*, as exemplified in research on brain volume and *g* (Pennington et al., 2000; Posthuma et al., 2002). Multivariate genetic research also is needed to explore the genetic links among brain mechanisms to assess the extent to which *g* exists in the brain as well as in the mind.

Human research on *g* can make progress toward understanding brain mechanisms using neuroimaging techniques (Kosslyn & Plomin, 2001). However, mouse models of *g* would facilitate the precise analysis of basic brain mechanisms using techniques such as single-cell recordings, microstimulation, targeted gene mutations, antisense DNA that disrupts gene transcription, and DNA expression studies. Clearly, there are major differences in brain and mind between the human species and other animals, most notably in the use of language and the highly developed prefrontal cortex in the human species. However, *g* in humans does not depend on the use of language—a strong *g* factor emerges from a battery of completely nonverbal tests (Jensen, 1998). Moreover, low-level tasks such as information-processing tasks assessed by reaction time contribute to *g* (Deary, 2000). Indeed, *g* can be used as a criterion to identify animal models of individual differences in cognitive processes. If *g* represents the way in which genetically driven components of the brain work together to solve problems, it would not be unreasonable to hypothesize that *g* exists in all animals (B. Anderson, 2000). Although much less well documented than *g* in humans, increasing evidence exists for a *g* factor in mice across diverse tasks of learning, memory, and problem solving (Plomin, 2001).

Environmental Genetic Analysis

A major emphasis of the previous volume (Plomin & McClearn, 1993) was an attempt to seek a rapprochement between behavioral geneticists and environmental researchers. Behavioral geneticists discussed the interface between nature and nurture, and prominent environmental theorists proposed environmental theories that attempted to encompass genetic influences. Although reading these chapters nearly a decade after they were written suggests an image of ships passing in the night, there was general agreement about the need to incorporate specific measures of the environment in behavioral genetic research and the need to study environmental processes using genetically sensitive designs, an exhortation followed in a recent large-scale study of family environment (Reiss, Neiderhiser, Hetherington, & Plomin, 2000). As indicated in chapter 1, quantitative genetic research is at least as informative about nurture as it is about nature and has led to two of the most important discoveries about environmental influences. The first finding concerns the importance of nonshared environmental influences that make children growing up in the same family as different as children growing up in different families (Plomin, Asbury, & Dunn, 2001). *g* was thought to be an exception to this rule. Model-fitting estimates of the role of shared environment for *g* suggest that it accounts for about 20% of the variance for parents and offspring analyses, about 25% for siblings, and as much as 40% for twins (Chipuer, Rovine, & Plomin, 1990). For example, adoptive siblings correlate about 0.30 (Loehlin, 1989). Because adoptive siblings are genetically unrelated children reared in the same adoptive family, their resemblance can only be due to shared environmental influence in the absence of selective placement. In these studies, little attention was paid to the fact that the adoptive siblings were children until the first study of older adoptive siblings found a correlation of −0.03 (Scarr & Weinberg, 1978). Other studies of older adoptive siblings also have found similarly low correlations for *g*. The most impressive evidence comes from a 10-year longitudinal follow-up study of more than 200 pairs of adoptive siblings. At the average age of 8 years, the *g* correlation was 0.26. Ten years later, their correlation was near zero (Loehlin, Horn, & Willerman, 1989). Thus, although shared environment is an important influence on *g* during childhood when children are living at home, its importance fades after adolescence as influences outside the family become more salient. Next to nothing is known about these mysterious nonshared environmental factors that make siblings growing up in the same family so different after adolescence.

The second finding is that most measures of the environment show genetic influence (reviewed by Plomin, 1994). For example, the most widely used measure of the home environment relevant to cognitive development is the Home Observation for Measurement of the Environment (HOME; Caldwell & Bradley, 1978). In an adoption study of the HOME, correlations for nonadoptive and adoptive siblings were compared when each child was 1 year old and again when each child was 2 years old (Braungart, Fulker, & Plomin, 1992). HOME scores are more similar for nonadoptive siblings than for adoptive siblings at both years (.58 vs. .35 at 1 year and .57 vs. .40 at 2 years), results suggesting genetic influence on the HOME. Dozens of studies using diverse measures of the environment in addition to family environment such as life events and social support—and even television-viewing, accidents, and divorce—find consistent evidence for genetic influence.

Given that environmental as well as behavioral measures show genetic influence, it is reasonable to ask whether associations between environmental measures and behavioral measures are mediated genetically. Multivariate genetic analysis can be used to analyze genetic and environmental contributions to the correlation between environmental measures and behavioral measures. For example, the adoption study just mentioned found that the HOME correlated substantially with children's *g*, which the HOME was designed to do (Braungart et al., 1992). Socialization researchers had reasonably assumed that such HOME measures as parental responsivity, encouraging developmental advance, and provision of toys were environmental causes of children's cognitive development. However, multivariate genetic analysis indicated that about half of the phenotypic correlation between the HOME and children's *g* is mediated genetically. One possible explanation of this result is that parents respond to genetically influenced *g* in their children. A general rule in research of this sort is that genetic factors are responsible for about half of the phenotypic correlation between measures of the environment and measures of behavior. No longer can environmental measures be assumed to be entirely environmental just because they are called environmental. Taking this argument to the extreme, two recent books have argued that socialization research has been fundamentally flawed because it has not considered the role of genetics (Harris, 1998; Rowe, 1994). A far-reaching implication of this research is that it supports a shift from passive models of how the environment affects individuals toward models that recognize the active role individuals play in selecting, modifying, constructing, and reconstructing in memory their experiences (Plomin, 1994).

Molecular Genetic Research on *g*

There is a lot of life left in the old workhorse of quantitative genetics. Quantitative genetics research that goes beyond heritability to address developmental, multivariate, and environmental issues as described in the preceding section will be increasingly important in charting the course for molecular genetic analyses that attempt to identify specific genes responsible for the ubiquitous heritability of complex traits such as *g*. For example, multivariate genetic analysis suggests that despite its apparent complexity, *g* is a good target for molecular genetic analysis because genetic effects on cognitive processes are general. Another example is that quantitative genetic research has been important in showing that genetic variation contributes importantly to differences among individuals in the normal range of variability as well as abnormal behavior. Moreover, as discussed in chapter 1, it is increasingly accepted that many complex disorders, especially common ones, may represent the quantitative extreme of the same genetic and environmental factors responsible for variation in the normal range. That is, genetic influence on common disorders such as mild mental retardation may not be due to genes specific to the disorder but rather to genes that contribute to the normal range of individual differences in *g*. This quantitative trait loci (QTL) perspective implies that some common disorders may not be disorders at all but rather the extremes of normal distributions; this has far-reaching implications for molecular genetics and for neuroscience. If many genes of small effect are involved, it will be much more difficult to find them. It also will be much more difficult to explore the brain mechanisms that mediate genetic effects on *g*.

As compared with research on psychiatric disorders (see Kalidindi & McGuffin, chapter 24, and Owen & O'Donovan, chapter 23, this volume), molecular genetic studies in the cognitive domain were begun relatively recently and have used QTL approaches from the start, which may contribute to the quicker successes in this domain. An example of a behavioral QTL, as mentioned in chapter 1, is the association between apolipoprotein-E and late-onset dementia (Corder et al., 1993), an association that has been replicated in scores of studies and remains the only known predictor of this common disorder in later life (see Williams, chapter 25, this volume). Although a particular allele in this gene leads to a fivefold increased risk for dementia, it is a QTL in the sense that many people with this allele do not have dementia, and most people with dementia do not have this genetic risk factor. Another cognitive trait, reading disability, is the best example of a QTL linkage with a complex trait (see Willcutt et al., chapter 13, this volume).

The advantage of linkage approaches, including QTL linkage approaches, is that they can systematically scan the genome for linkages using just a few hundred DNA markers (see Sham, chapter 3, this volume). QTL linkage designs use many small families (usually siblings) rather than a few large families as in traditional linkage designs. The disadvantage is that they can only detect QTLs that account for a substantial amount (at best about 10%) of the genetic variance given reasonable sample sizes. In contrast, allelic association can detect QTLs that account for 1% of the variance, but thousands of DNA markers are needed to screen the genome systematically for association because association can only be detected if a DNA marker is very close to a QTL. In other words, linkage is systematic but not powerful, and allelic association is powerful but not systematic (Risch & Merikangas, 1996). For this reason, allelic association approaches have largely been limited to studies of "candidate" genes. If, as is usually the case, a DNA marker in or near a candidate gene is not itself functional (i.e., it does not produce a coding difference), the marker may be close enough to a functional QTL to be in linkage disequilibrium with it and thus yield an indirect association with the disorder. One problem with a candidate gene approach is that, for behavioral disorders, any of the thousands of genes expressed in the brain could be viewed as candidate genes. The way out of this conundrum is to conduct a systematic scan of the genome for allelic association using many thousands of DNA markers, although tens of thousands or even hundreds of thousands of DNA markers would be needed to identify or exclude all QTL associations

(Kruglyak, 1999). Such systematic large-scale genome scans or scans of all known candidate gene polymorphisms are becoming possible with new technologies that can quickly genotype thousands of DNA markers (as described by Craig & McClay, chapter 2, this volume).

Progress in identifying genes associated with complex traits has been slower than expected in part because research to date has been underpowered for finding QTLs of small effect size, especially in linkage studies. Very large studies are needed to identify QTLs of small effect size (Cardon & Bell, 2001). A daunting target for molecular genetic research on complex traits such as *g* is to design research powerful enough to detect QTLs that account for 1% of the variance while providing protection against false-positive results in genome scans of thousands of genes. To break the 1% QTL barrier, samples of many thousands of individuals are needed for research on disorders (comparing cases and controls) and on dimensions (assessing individual differences in an unselected and, it is hoped, representative sample). Another factor in the slow progress to date is that only a few candidate gene markers have been examined rather than systematic scans of gene systems or of the entire genome. The Human Genome Project will accelerate progress toward identifying all functional DNA variants expressed in the brain, especially those for entire neurotransmitter pathways (see Craig & McClay, chapter 2, this volume).

The IQ QTL Project

Although systematic QTL studies of mild mental retardation have not been reported, a QTL perspective suggests that QTL studies of normal IQ or even high IQ could identify QTLs that also are associated with mild mental retardation. This is part of the rationale for an allelic association study comparing high-IQ and control individuals called the IQ QTL Project (Plomin, 2000). The first phase of the project used an allelic association strategy with DNA markers in or near candidate genes likely to be relevant to neurological functioning, such as genes for neuroreceptors. Allelic association results were reported for 100 DNA markers for such candidate genes (Plomin et al., 1995). Although several significant associations were found in an original sample, only one association was replicated in an independent sample. This finding was a chance result expected when 100 markers are investigated, and follow-up studies failed to replicate the result (Petrill, Ball, Eley, Hill, & Plomin, 1998). The 100 markers included markers for genes for which QTL associations with cognitive ability have recently been reported, such as apolipoprotein E as mentioned above and catechol-O-methyltransferase (Egan et al., 2001). For both of these genes, results from the IQ QTL Project were in the same direction but not significant (Plomin et al., 1995; Turic, Fisher, Plomin, & Owen, 2001), which may indicate problems of statistical power to detect QTLs of small effect size.

Attempts to find QTL associations with complex traits have begun to go beyond candidate genes to conduct systematic genome scans using dense maps of DNA markers. As part of the IQ QTL Project, a preliminary attempt to use a genome scan strategy to identify QTLs associated with IQ focused on the long arm of chromosome 6 and found replicated associations for a DNA marker that happened to be in the gene for insulinlike growth factor-2 receptor (IGF2R; Chorney et al., 1998), which has been shown to be especially active in brain regions most involved in learning and memory (Wickelgren, 1998). More replication is needed before we can conclude that IGF2R is an IQ QTL, but other attempts to replicate this association have not as yet been reported.

The problem with using a dense map of markers for a genome scan is the amount of genotyping required. To scan the entire genome at 1 million DNA base-pair intervals (1 Mb), one needs to genotype about 3,500 DNA markers. This would require 700,000 genotypings in a study of 100 individuals with high *g* and 100 control individuals. With markers at 1 Mb intervals, no QTL would be farther than 500,000 base pairs from a marker. Moreover, it is generally accepted that 10 to even 100 times as many markers would be needed to detect all QTLs (Abecasis et al., 2001; Kruglyak, 1999; Reich et al., 2001). Despite the daunting amount of genotyping required for a systematic genome scan, this approach has been fueled by the promise of "SNPs on chips," which can quickly genotype thousands of DNA markers of the single nucleotide polymorphism (SNP) variety.

DNA pooling, developed for use in the IQ QTL Project, provides a low-cost and flexible alternative to SNPs on chips for screening the genome for QTL associations (Daniels, Holmans, Plomin, McGuffin, & Owen, 1998). DNA pooling greatly reduces the need for genotyping by pooling DNA from all individuals in each group and comparing the pooled groups so that only 14,000 genotypings are required to scan the genome in the previous example involving 3,500 DNA markers. The IQ QTL Project has recently reported a genome scan using DNA pooling to screen 1,842 simple-sequence repeat (SSR) DNA markers that were approximately evenly spaced at 2 cM throughout the genome (Plomin et al., 2001). In an attempt to boost power, samples were selected for cases with high *g* and control cases with average *g*, and a multiple-stage design consisting of two case control studies and a within-family transmission disequilibrium design of parent–offspring trios permitted relaxed alpha levels, which accepted false-positive results in the early stages that were removed in later stages. The three samples were genotyped in five stages:

1. Case control DNA pooling (101 cases with mean IQ of 136 and 101 control individuals with mean IQ of 100).
2. Case control DNA pooling (96 cases with IQ > 160 and 100 control individuals with mean IQ of 100).
3. Individual genotyping of Stage 1 sample.
4. Individual genotyping of Stage 2 sample.
5. Transmission disequilibrium test (TDT; Sham, chapter 3, this volume); 196 parent–child trios for offspring with IQ > 160).

The overall Type I error rate is 0.000125, which robustly protects against false-positive results. The numbers of markers surviving each stage using a conservative allele-specific directional test were 10, 8, 6, 4, 2, and 0, respectively, for the five stages. A genomic control test using DNA pooling suggested that the failure to replicate the positive case control results in the TDT analysis was not due to ethnic stratification. Several markers that were close to significance at all stages are being investigated further.

The criteria for replication used in this multiple-stage study were conservative if not extreme. No other study has demanded clean replication in three samples using two different designs (case control and parent–offspring trios). However, a conservative approach seems warranted given problems in the literature with failure to replicate QTL association results (Cardon & Bell, 2001). Although the multiple-stage design with three extreme selected samples attempts to balance false positives and false negatives in an effort to detect QTLs of small effect size, the sample sizes and number of markers genotyped tip the balance very much in favor of avoiding false positives rather than false negatives. The design protects against any false-positive results in that the overall Type I error rate is 0.000125 using a nominal p value of .05 at each stage. However, as discussed below, the present study has at best skimmed the surface for possible QTLs of small effect size. Thus, the present results should not be taken as an indictment of the QTL approach but rather as a warning of the exorbitant demands of power to detect QTLs of small effect size in a genome scan based on indirect association.

What is responsible for the failure to find replicable QTLs that survive our stringent criteria? An important part of the answer is that many more markers are needed for a genome scan for allelic association relying on linkage disequilibrium. Compared with the 300 markers needed for a genome scan for linkage, 1,842 markers seem like a lot of markers; however, 100,000 markers may be needed to exclude QTL association in an indirect association design. The problem for allelic association analysis is that power drops off precipitously when a marker is not very close to the QTL (Plomin et al., 2001). When D′ (an index of linkage disequilibrium) is 1.0 (indicating complete linkage disequilibrium), the power to detect a QTL with 5%, 2.5%, and 1% heritability is, respectively, 100%, 93%, and 56% for the original case control sample; 100%, 100%, and 98% for the replication case control sample; and 100%, 100%, and 99% for the TDT sample. However, when D′ is 0.50 rather than 1.0, the power estimates decline to 73%, 42%, and 19% for the original case control sample; 100%, 92%, and 54% for the replication case control sample; and 100%, 96%, and 64% for the TDT sample. Data simulations, which agree with empirical linkage disequilibrium data from the human genome sequence, indicate that with a 2-cM SSR map, D′ falls to 0.50 at 700 kb for new mutations (100 generations old) and 70 kb for ancient mutations (1,000 generations old). These simulations suggest that a 2-cM screen may provide as much as 70% coverage with D′ of 0.50 for new mutations but less than 10% coverage for old mutations.

If the 50% heritability of g is due to 50 QTLs with the average effect size of 1%, and if we have only 50% coverage with D′ = 0.5, then the IQ

QTL Project would have been lucky to detect a single QTL. Thus, although the design attempts to balance false positives and false negatives in the quest for QTLs of small effect size, it does a much better job of protecting against false positives than false negatives. The study is novel in that it is the first attempt to begin to screen the genome systematically for allelic association even though it is merely a first step in that direction, it uses DNA pooling as an efficient method to screen many markers, and it uses extreme selected groups to boost power to detect QTLs of small effect. However, the study was begun in 1998, before the attenuation of power of indirect association based on linkage disequilibrium was understood.

With the recent mining of SNPs, it will soon be possible to conduct a genome scan using 100,000 evenly spaced common SNPs. However, the pattern of linkage disequilibrium depends on several factors such as marker type, allele frequency, mutation, and recombination, and recent data suggest that it is highly variable between different regions of the genome (Zavattari et al., 2000) and in different ethnic groups (Reich et al., 2001). In addition, it is clear from many studies that whereas the average strength of linkage disequilibrium declines with increasing physical distance, many pairs of closely adjacent markers show little or no linkage disequilibrium (e.g., Abecasis et al., 2001). For this reason, we believe that the most appropriate strategy at the present time is to focus on potentially functional polymorphisms. However, rather than focusing on a few candidate genes or gene systems, we can look forward to a systematic search using all functional polymorphisms in coding sequences (cSNPs) and in regulatory regions. These tens of thousands of SNPs can be genotyped using high-throughput techniques such as DNA pooling, which has been extended to SNPs (Hoogendoorn et al., 2000). In the meantime, it is possible to use nonsynonymous cSNPs (SNPs in coding regions) as well as functional SNPs in regulatory regions as they become available. For example, focusing on 225 genes on chromosome 21, a total of 337 cSNPs were suggested using bioinformatic approaches of which 78% have been confirmed (Deutsch, Iseli, Bucher, Antonarakis, & Scott, 2001), although caution is warranted because only about 15% of putative SNPs in some databases have been shown to be polymorphic in any population (Marth et al., 2001). Although the problem remains that a QTL association with a functional SNP might in fact be due to another nearby SNP, it is a reasonable assumption that a functional SNP *is* the QTL (i.e., $D' = 1.0$), which greatly increases the power of QTL association. The IQ QTL Project is now using such markers with its multiple-stage design to identify QTL associations that meet strict criteria for replicated significance.

A gloomier prospect is that QTLs for *g* account for less than 1% of the variance. Although the distribution of effect sizes is not known for *g* or for any other complex trait, if QTL heritabilities are less than 1% or if QTLs interact epistatically, it will be difficult to detect them reliably. Nonetheless, the convergence of evidence for the strong heritability of *g* from family, twin, and adoption studies indicates that *g*-relevant DNA polymorphisms exist. The solution of course is that power will need to be increased to detect the QTLs responsible for the heritability of *g*, even if the QTL heritabilities are less than 1%. DNA pooling will be useful in this context because it costs no more to genotype 1,000 individuals than 100 individuals.

Behavioral Genomics

Despite the slow progress to date in finding genes associated with general cognitive ability as well as other complex traits, I am confident that the pace of QTL discovery will pick up as the Human Genome Project continues to shower the field with new information and new technologies and as studies are designed to break the 1% QTL barrier. The greatest impact for the behavioral sciences will come after genes have been identified as we begin to use DNA in our research and eventually in our clinics. In research, we can investigate the developmental, multivariate, and environmental issues discussed earlier in this chapter with much greater precision if we can use specific genes. The apolipoprotein E story will be revisited in many areas of the behavioral sciences in which it will become *de rigueur* to genotype subjects for genes associated with a dimension or disorder. In clinics, genotyping will become routine if a genetic risk factor is found to predict differential response to interventions or treatments.

As discussed in chapter 1 of this volume, the future of genetic research lies in moving from finding genes (genomics) to finding out how genes work (functional genomics). As an antidote to the tendency to define functional genomics at the cellular level of analysis, the phrase *behavioral genomics* connotes the importance of another level of analysis: the behavioral level of analysis of the whole organism as it develops and interacts and correlates with its environment (Plomin & Crabbe, 2000). In addition to producing indisputable evidence of genetic influence, the identification of specific genes associated with learning and language delays and disorders will revolutionize genetic research by providing measured genotypes for investigating genetic links between behaviors, for tracking the developmental course of genetic effects, and for identifying interactions and correlations between genotype and environment (Plomin & Rutter, 1998). Identification of specific genes also might provide greater precision in diagnoses and individually tailored educational programs and therapies. For example, gene-based classification of disorders may bear little resemblance to the current symptom-based diagnostic systems. Indeed, from a QTL perspective, common disorders may be the quantitative extreme of normal genetic variation rather than qualitatively different, as mentioned earlier. Most importantly, gene-based prediction of early-onset disorders offers the hope for preventive intervention rather than waiting to intervene when language and learning problems begin to cast a long and wide shadow over children's development.

Conclusion

The evidence for a strong genetic contribution to *g* is clearer than for any other area of the behavioral sciences. Few scientists now seriously dispute the conclusion that *g* shows significant genetic influence. The magnitude of genetic influence is still not universally appreciated, however. Taken together, this extensive body of research suggests that about half of the total variance of *g* can be accounted for by genetic factors.

The case for heritability of *g* is so strong that quantitative genetic research has moved beyond estimating heritability to addressing developmental, multivariate, and environmental issues related to *g*. The heritability of *g* increases during the lifespan, reaching levels in adulthood comparable with the heritability of height. Longitudinal genetic analyses of *g* suggest that genetic factors primarily contribute to continuity, although some evidence for genetic change has been found, for example, in the transition from early to middle childhood. Multivariate genetic analysis shows that genetic effects on diverse cognitive abilities are surprisingly general—genetic *g* accounts for nearly all of the genetic variance on cognitive abilities. Research at the nature–nurture interface reveals that the influence of shared environment diminishes sharply after adolescence and that *g*-related family environments are substantially mediated genetically, which suggests a more active model of the environment in which genetic propensities affect people's experiences.

An exciting new direction for behavioral genetic research on *g* is to identify some of the QTLs responsible for the heritability of *g*. Although progress to date has been slow, advances from the Human Genome Project, new techniques, and appreciation of the need for sufficient power to break the 1% QTL barrier makes me optimistic that replicable QTLs will be identified and the field will move on to behavioral genomics.

References

Abecasis, G. R., Noguchi, E., Heinzmann, A., Traherne, J. A., Bhattacharyya, S., Leaves, N. I., et al. (2001). Extent and distribution of linkage disequilibrium in three genomic regions. *American Journal of Human Genetics, 68,* 191–197.

Anderson, B. (2000). The *g* factor in non-human animals. In G. R. Bock, J. A. Goode, & K. Webb (Eds.), *The nature of intelligence* (Novartis Foundation Symposium 233, pp. 79–95). Chichester, England: Wiley.

Anderson, M. (1992). *The development of intelligence: Studies in developmental psychology.* Hove, England: Psychology Press.

Baddeley, A., & Gathercole, S. (1999). Individual differences in learning and memory: Psychometrics and the single case. In P. L. Ackerman, P. C. Kyllonen, & R. D. Roberts (Eds.), *Learning and individual differences: Process, trait, and content determinants* (pp. 31–55). Washington, DC: American Psychological Association.

Baker, L., Vernon, P. A., & Ho, H. (1991). The genetic correlation between intelligence and speed of information processing. *Behavior Genetics, 21,* 351–368.

Bouchard, T. J., Jr., Lykken, D. T., McGue, M., & Tellegen, A. (1997). Genes, drives, environment, and experience: EPD theory revised. In C. P. Benbow & D. Lubinski (Eds.), *Intellectual talent: Psychometric and social issues* (pp. 5–43). Baltimore: John Hopkins University Press.

Braungart, J. M., Fulker, D. W., & Plomin, R. (1992). Genetic mediation of the home environment during infancy: A sibling adoption study of the HOME. *Developmental Psychology, 28,* 1048–1055.

Burks, B. (1928). The relative influence of nature and nurture upon mental development:

A comparative study on foster parent–foster child resemblance. *Yearbook of the National Society for the Study of Education (Part 1), 27,* 219–316.

Caldwell, B. M., & Bradley, R. H. (1978). *Home observation for measurement of the environment.* Little Rock: University of Arkansas Press.

Cardon, L. R., & Bell, J. (2001). Association study designs for complex diseases. *Nature Genetics, 2,* 91–99.

Cardon, L. R., & Fulker, D. W. (1993). Genetics of specific cognitive abilities. In R. Plomin & G. E. McClearn (Eds.), *Nature, nurture, and psychology* (pp. 99–120). Washington, DC: American Psychological Association.

Carroll, J. B. (1993). *Human cognitive abilities.* New York: Cambridge University Press.

Carroll, J. B. (1997). Psychometrics, intelligence, and public policy. *Intelligence, 24,* 25–52.

Chipuer, H. M., Rovine, M. J., & Plomin, R. (1990). LISREL modeling: Genetic and environmental influences on IQ revisited. *Intelligence, 14,* 11–29.

Chorney, M. J., Chorney, K., Seese, N., Owen, M. J., Daniels, J., McGuffin, P., et al. (1998). A quantitative trait locus (QTL) associated with cognitive ability in children. *Psychological Science, 9,* 1–8.

Corder, E. H., Saunders, A. M., Strittmatter, W. J., Schmechel, D. E., Gaskell, P. C., Small, G. W., et al. (1993). Gene dose of apolipoprotein E type 4 allele and the risk of Alzheimer's disease in late onset families. *Science, 261,* 921–923.

Dale, P. S., Dionne, G., Eley, T. C., & Plomin, R. (2000). Lexical and grammatical development: A behavioral genetic perspective. *Journal of Child Language, 27,* 619–642.

Daniels, J., Holmans, P., Plomin, R., McGuffin, P., & Owen, M. J. (1998). A simple method for analyzing microsatellite allele image patterns generated from DNA pools and its application to allelic association studies. *American Journal of Human Genetics, 62,* 1189–1197.

Deary, I. (2000). *Looking down on human intelligence: From psychometrics to the brain.* London: Oxford University Press.

Deary, I. J., Whalley, L. J., Lemmon, H., Crawford, J. R., & Starr, J. M. (2000). The stability of individual differences in mental ability from childhood to old age: Follow-up of the 1932 Scottish Mental Survey. *Intelligence, 28,* 49–55.

Detterman, D. K. (2000). General intelligence and the definition of phenotypes. In G. R. Bock et al. (Eds.), *The nature of intelligence* (Novartis Foundation Symposium 233; pp. 136–148). London: John Wiley & Sons.

Deutsch, S., Iseli, C., Bucher, P., Antonarakis, S. E., & Scott, H. S. A. (2001). A cSNP map and database for human chromosome 21. *Genome Research, 11,* 300–307.

Egan, M. F., Goldberg, T. E., Kolachana, B. S., Callicott, J. H., Mazzanti, C. M., Straub, R. E., et al. (2001). Effect of COMT $Val^{108/158}$ Met genotype on frontal lobe function and risk for schizophrenia. *Proceedings of the National Academy of Sciences, 98,* 6917–6922.

Fodor, J. A. (1983). *The modularity of mind.* Cambridge, MA: MIT Press.

Freeman, F. N., Holzinger, K. J., & Mitchell, B. (1928). The influence of environment on the intelligence, school achievement, and conduct of foster children. *Yearbook of the National Society for the Study of Education, 27,* 103–217.

Fulker, D. W., Cherny, S. S., & Cardon, L. R. (1993). Continuity and change in cognitive development. In R. Plomin & G. E. McClearn (Eds.), *Nature, nurture, and psychology* (pp. 77–97). Washington, DC: American Psychological Association.

Galton, F. (1865). Heredity talent and character. *Macmillan's Magazine, 12,* 157–166, 318–327.

Gardner, H. (1983). *Frames of mind: The theory of multiple intelligences.* New York: Basic Books.

Goleman, D. (1995). *Emotional intelligence.* New York: Bantam Books.

Gottfredson, L. S. (1997). Why *g* matters: The complexity of everyday life. *Intelligence, 24,* 79–132.

Gould, S. J. (1996). *The mismeasure of man* (2nd ed.). New York: Norton.

Harris, J. R. (1998). *The nurture assumption: Why children turn out the way they do.* New York: Free Press.

Hoogendoorn, B., Norton, N., Kirov, G., Williams, N., Hamshere, M. L., Spurlock, G., et al. (2000). Cheap, accurate and rapid allele frequency estimation of single nucleotide polymorphisms by primer extension and DHPLC in DNA pools. *Human Genetics, 107,* 488–493.

Hughes, C., & Cutting, A. L. (1999). Nature, nurture and individual differences in early understanding of mind. *Psychological Science, 10,* 429–432.

Jensen, A. R. (1998). *The g factor: The science of mental ability.* New York: Praeger.

Kosslyn, S., & Plomin, R. (2001). Towards a neuro-cognitive genetics: Goals and issues. In D. Dougherty, S. L. Rauch, & J. F. Rosenbaum (Eds.), *Psychiatric neuroimaging research: Contemporary strategies* (pp. 491–515). Washington, DC: American Psychiatric Press.

Kruglyak, L. (1999). Prospects for whole-genome linkage disequilibrium mapping of common disease genes. *Nature Genetics, 22,* 139–144.

Loehlin, J. C. (1989). Partitioning environmental and genetic contributions to behavioral development. *American Psychologist, 44,* 1285–1292.

Loehlin, J. C., Horn, J. M., & Willerman, L. (1989). Modeling IQ change: Evidence from the Texas Adoption Project. *Child Development, 60,* 993–1004.

Luciano, M., Smith, G. A., Wright, M. J., Geffen, G. M., Geffen, L. B., & Martin, N. G. (2001). Genetic influence on the relationship between choice reaction time and IQ: A twin adoption study. *Moscow Journal of Psychology, 14,* 49–59.

Marth, G., Yeh, R., Minton, M., Donaldson, R., Li, Q., Duan, S., et al. (2001). Single-nucleotide polymorphisms in the public domain: How useful are they? *Nature Genetics, 27,* 371–372.

McClearn, G. E., Johansson, B., Berg, S., Pedersen, N. L., Ahern, F., Petrill, S. A., et al. (1997). Substantial genetic influence on cognitive abilities in twins 80+ years old. *Science, 276,* 1560–1563.

McGue, M., Bouchard, T. J., Jr., Iacono, W. G., & Lykken, D. T. (1993). Behavioral genetics of cognitive ability: A life-span perspective. In R. Plomin & G. E. McClearn (Eds.), *Nature, nurture, and psychology* (pp. 59–76). Washington, DC: American Psychological Association.

Neisser, U., Boodoo, G., Bouchard, T. J., Jr., Boykin, A. W., Brody, N., Ceci, S. J., et al. (1996). Intelligence: Knowns and unknowns. *American Psychologist, 51,* 77–101.

Neubauer, A. C., Spinath, F. M., Riemann, R., Borkenau, P., & Angleitner, A. (2000). Genetic (and environmental) influence on two measures of speed of information processing and their relation to psychometric intelligence: Evidence from the German Observational Study of Adult Twins. *Intelligence, 28,* 267–289.

Pennington, B. C., Filipek, P. A., Lefly, D., Chhabildas, N., Kennedy, D. N., Simon, J. H., et al. (2000). A twin MRI study of size variations in the human brain. *Journal of Cognitive Neuroscience, 12,* 223–232.

Petrill, S. A. (1997). Molarity versus modularity of cognitive functioning? A behavioral genetic perspective. *Current Directions in Psychological Science, 6,* 96–99.

Petrill, S. A., Ball, D. M., Eley, T. C., Hill, L., & Plomin, R. (1998). Failure to replicate a QTL association between a DNA marker identified by EST00083 and IQ. *Intelligence, 25,* 179–184.

Petrill, S. A., Luo, D., Thompson, L. A., & Detterman, D. K. (1996). The independent prediction of general intelligence by elementary cognitive tasks: Genetic and environmental influences. *Behavior Genetics, 26,* 135–147.

Petrill, S. A., Thompson, L. A., & Detterman, D. K. (1995). The genetic and environmental variance underlying elementary cognitive tasks. *Behavior Genetics, 25,* 199–209.

Pinker, S. (1994). *The language instinct.* New York: William Morrow.

Plomin, R. (1986). *Development, genetics, and psychology.* Hillsdale, NJ: Erlbaum.

Plomin, R. (1994). *Genetics and experience: The interplay between nature and nurture.* Thousand Oaks, CA: Sage.

Plomin, R. (1999a). Genetic research on general cognitive ability as a model for mild mental retardation. *International Review of Psychiatry, 11,* 34–36.

Plomin, R. (1999b). Genetics and general cognitive ability. *Nature, 402,* C25–C29.

Plomin, R. (2000). Quantitative trait loci and general cognitive ability. In J. Benjamin, R. P. Ebstein, & R. H. Belmaker (Eds.), *Molecular genetics and the human personality* (pp. 211–230). Washington, DC: American Psychiatric Press.

Plomin, R. (2001). The genetics of *g* in human and mouse. *Nature Reviews Neuroscience, 2,* 136–141.

Plomin, R., Asbury, K., & Dunn, J. (2001). Why are children in the same family so different? Nonshared environment a decade later. *Canadian Journal of Psychiatry, 46,* 225–233.

Plomin, R., & Crabbe, J. C. (2000). DNA. *Psychological Bulletin, 126,* 806–828.

Plomin, R., DeFries, J. C., McClearn, G. E., & McGuffin, P. (2001). *Behavioral genetics* (4th ed.). New York: Worth.

Plomin, R., Hill, L., Craig, I., McGuffin, P., Purcell, S., Sham, P., et al. (2001). A genome-wide scan of 1842 DNA markers for allelic associations with general cognitive ability: A five-stage design using DNA pooling. *Behavior Genetics, 31,* 497–509.

Plomin, R., & McClearn, G. E. (1993). *Nature, nurture, and psychology.* Washington, DC: American Psychological Association.

Plomin, R., McClearn, G. E., Smith, D. L., Skuder, P., Vignetti, S., Chorney, M. J., et al. (1995). Allelic associations between 100 DNA markers and high versus low IQ. *Intelligence, 21,* 31–48.

Plomin, R., & Price, T. (in press). Genetics and intelligence. In N. Colangelo & G. A. Davis (Eds.), *Handbook of gifted education* (3rd ed.). Boston: Allyn & Bacon.

Plomin, R., & Rutter, M. (1998). Child development, molecular genetics, and what to do with genes once they are found. *Child Development, 69,* 1223–1242.

Plomin, R., & Spinath, F. M. (2002). Genetics and general cognitive ability (g). *Trends in Cognitive Science, 6,* 169–176.

Posthuma, D., de Geus, E. J. C., Baare, W. F. C., Pol, H. E. H., Kahn, R. S., & Boomsma, D. I. (2002). The association between brain volume and intelligence is of genetic origin. *Nature Neuroscience, 5,* 83–84.

Price, T. S., Eley, T. C., Dale, P. S., Stevenson, J., & Plomin, R. (2000). Genetic and environmental covariation between verbal and non-verbal cognitive development in infancy. *Child Development, 71,* 948–959.

Reich, D. E., Cargill, M., Bolk, S., Ireland, J., Sabeti, P. C., Richter, D. J., et al. (2001). Linkage disequilibrium in the human genome. *Nature, 411,* 199–204.

Reiss, D., Neiderhiser, J. M., Hetherington, E. M., & Plomin, R. (2000). *The relationship code: Deciphering genetic and social patterns in adolescent development.* Cambridge, MA: Harvard University Press.

Rietveld, M. J. H., van Baal, G. C. M., Dolan, C. V., & Boomsma, D. I. (2000). Genetic factor analyses of specific cognitive abilities in 5-year-old Dutch children. *Behavior Genetics, 30,* 29–40.

Rijsdijk, F. V., Vernon, P. A., & Boomsma, P. A. (1998). The genetic basis of the relation between speed-of-information-processing and IQ. *Behavioural Brain Research, 95,* 77–84.

Risch, N., & Merikangas, K. R. (1996). The future of genetic studies of complex human diseases. *Science, 273,* 1516–1517.

Rowe, D. C. (1994). *The limits of family influence: Genes, experience, and behaviour* (Vol. 5). New York: Guilford Press.

Salthouse, T. A., & Czaja, S. J. (2000). Structural constraints on process explanations in cognitive aging. *Psychology and Aging, 15,* 44–55.

Scarr, S. (1992). Developmental theories for the 1990s: Development and individual differences. *Child Development, 63,* 1–19.

Scarr, S., & Weinberg, R. A. (1978). The influence of "family background" on intellectual attainment. *American Sociological Review, 43,* 674–692.

Snyderman, M., & Rothman, S. (1987). Survey of expert opinion on intelligence and aptitude testing. *American Psychologist, 42,* 137–144.

Spearman, C. (1927). *The abilities of man: Their nature and measurement.* New York: Macmillan.

Spearman, C. (1904). General intelligence, objectively determined and measured. *American Journal of Psychology, 15,* 201–293.

Stauffer, J. M., Ree, M. J., & Carretta, T. R. (1996). Cognitive-components tests are not much more than *g*: An extension of Kyllonen's analyses. *Journal of General Psychology* 193–205.

Sternberg, R. J. (1985). *Beyond IQ: A triarchic theory of human intelligence.* Cambridge, England: Cambridge University Press.

Sternberg, R. J., & Gardner, M. K. (1983). A componential interpretation of the general factor in human intelligence. In H. J. Eysenck (Ed.), *A model for intelligence* (pp. 231–254). Berlin: Springer Verlag.

Theis, S. V. S. (1924). *How foster children turn out* (Publication No. 165). New York: State Charities Aid Association.

Thompson, L. A., Detterman, D. K., & Plomin, R. (1991). Associations between cognitive abilities and scholastic achievement: Genetic overlap but environmental differences. *Psychological Science, 2,* 158–165.

Turic, D., Fisher, P. J., Plomin, R., & Owen, M. J. (2001). No association between apolipoprotein E polymorphisms and general cognitive ability in children. *Neuroscience Letters, 299,* 97–100.

Vernon, P. A. (1989). The heritability of measures of speed of information-processing. *Personality and Individual Differences, 10,* 575–576.

Vernon, P. A., Wickett, J. C., Banzana, P. G. & Stelmack, R. M. (2000). The neuropsychology and psychophysiology of human intelligence. In R. J. Sternberg (Ed.), *Handbook of intelligence* (pp. 245–264). Cambridge, England: Cambridge University Press.

Wadsworth, S. J. (1994). School achievement. In R. Plomin, J. C. DeFries, & D. W. Fulker (Eds.), *Nature and nurture during middle childhood* (pp. 86–101). Oxford, England: Blackwell.

Wickelgren, I. (1998). Tracking insulin to the mind. *Science, 280,* 517–519.

Zavattari, P., Deidda, E., Whalen, M., Lampis, R., Mulargia, A., Loddo, M., et al. (2000). Major factors influencing linkage disequilibrium by analysis of different chromosome regions in distinct populations: Demography, chromosome recombination frequency and selection. *Human Molecular Genetics, 9,* 2947–2957.

Part V

Cognitive Disabilities

12

Isolation of the Genetic Factors Underlying Speech and Language Disorders

Simon E. Fisher

Speech and language are uniquely human traits that develop rapidly in the majority of children, without conscious effort and in the absence of formal training. The average 4-year-old child has access to a vocabulary of several thousand words and can assemble these into an unlimited number of complex structured sentences. The ease with which most children achieve this remarkable feat suggests that aspects of humans' ability to acquire speech and language may be innately encoded (see Pinker, 1994). However, a small but significant proportion of children have unexplained difficulty developing language skills, even though they are of normal intelligence, have received adequate environmental stimulation, and show no obvious sensory or neurological deficits. A substantial discrepancy between language ability and that in other cognitive domains is often referred to as *specific language impairment* (SLI) or *developmental dysphasia* (Bishop, 1987). Although this definition might imply that SLI is a single clinical entity, in practice the diagnosis can encompass a range of differing phenotypic profiles. Some children affected with SLI have problems with language production while their comprehension of language remains relatively intact; others may have significant deficits in both expressive and receptive domains. In addition, expressive impairments may be accompanied by problems with *articulation* (production of speech sounds). A further complication for the investigation of children with SLI is that the cognitive profile of any individual is not fixed but changes over time, such that the language difficulties of an affected child may resolve as he or she gets older. Despite extensive studies, it remains a matter of debate whether the variability in manifestation of speech and language disorders reflects distinct subtypes with differing underlying causes (Bishop, North, & Donlan, 1995).

Early speculation regarding the biological basis of SLI focused on environmental risk factors, such as perinatal damage to higher cortical cen-

I am grateful to Dianne Newbury and Anthony Monaco for their helpful comments on this chapter.

ters of the brain or poor language stimulation from parents. However, perinatal brain injury rarely leads to disproportionate impairment of language abilities (Bishop, 1987), and there is little evidence to implicate impoverished language input from parents as a causal factor in SLI; rather it appears that parents adjust their communication to appropriately meet the needs of their child (Conti-Ramsden & Friel-Patti, 1984). Furthermore, although speech and language disorders do cluster in families (as discussed below), there are often affected and unaffected children in the same sibship, despite receiving apparently similar levels of parental stimulation (e.g., Tomblin, 1989). Therefore, in recent years there has been a shift toward alternative explanations of SLI that are centered on the possible role of genetic factors. A large body of evidence has now accumulated from a variety of study designs indicating that genes play an important part in the etiology of speech and language disorders (reviewed in Bishop, 2001). However, as for most behavioral and cognitive traits, complexity at both the phenotypic and genotypic levels represents a major challenge for those attempting to isolate such genes.

This chapter highlights the extraordinary advances that are being made in this area of research, which exploit newly developed genotyping technology, novel statistical methodology, and the wealth of sequence data generated by the Human Genome Project. I begin by giving an overview of the compelling results from family, twin, and adoption studies supporting genetic involvement and then go on to outline progress in a number of genetic mapping efforts that have been recently completed or are currently under way. Remarkably, although it is only a few years since the report of the first linkage of a chromosomal region to a speech and language disorder (Fisher, Vargha-Khadem, Watkins, Monaco, & Pembrey, 1998), it has already been possible for geneticists to pinpoint the specific mutation responsible (Lai, Fisher, Hurst, Vargha-Khadem, & Monaco, 2001). This provides a pertinent example of how the availability of human genomic sequence data can greatly accelerate the pace of disease gene discovery. Finally, I discuss future prospects on how molecular genetics may offer new insight into the etiology underlying speech and language disorders, leading, one hopes, to improvements in diagnosis and treatment. Perhaps the most fascinating aspect of this work is its potential for dissecting the neurological mechanisms that mediate speech and language acquisition.

Genetic Factors Increase the Risk of Developing a Speech and Language Disorder

A series of studies have documented increased incidence of speech and language problems in relatives of children with SLI, as compared with that found in relatives of matched control individuals (Lahey & Edwards, 1995; Lewis, Ekelman, & Aram, 1989; Neils & Aram, 1986; Tallal, Ross, & Curtiss, 1989; Tomblin, 1989). This familial aggregation is consistent with genetic involvement but could equally well be explained by environmental factors that are shared between family members. Using twin-based methods, it is possible to separately assess the relative contributions made by shared family environment and genetic influences to the trait of interest. Typically, such studies derive estimates of the genetic contribution on the basis of phenotypic comparisons between monozygotic (MZ) twins (who have a virtually identical genetic makeup) and dizygotic (DZ) twins (who share approximately half their genes). Twin studies of SLI, using dichotomous classification of affection status, have demonstrated that concordance for the disorder is much higher in MZ twins than DZ twins, indicating a significant genetic influence (Bishop et al., 1995; Lewis & Thompson, 1992; Tomblin & Buckwalter, 1998). For example, Bishop et al. (1995) investigated a sample of 90 twin pairs in which at least one twin met strictly defined criteria for developmental speech or language disorder and observed 70% concordance in MZ pairs versus 46% in DZ pairs for this strict diagnosis. Broadening the affection status to include children with less substantial deficits in language ability plus resolved cases of children who had a history of speech and language problems led to an estimate of MZ concordance that increased to nearly 100%, compared with a DZ concordance of around 50% (Bishop et al., 1995). It is important to realize that the twin approach assumes environmental similarity is not elevated in MZ pairs versus DZ pairs. Nevertheless, data suggest that this assumption is likely to be sound, at least for studies of SLI (Bishop, 2001).

Investigation of phenotypic outcomes in adopted children provides another powerful method for exploring genetic and environmental influences. Felsenfeld and Plomin (1997) assessed 156 adopted and nonadopted children who were at varying risk of developing speech problems, on the basis of their biological and adoptive parent histories. They found that positive biological parental background was a significant risk factor in determining the speech outcome of a child, whereas positive adoptive parental history was not. Note, however, that estimates of genetic effect from this type of approach may be confounded by environmental factors that act at early stages of development.

Measures of Extreme Deficit in Language Ability Are Highly Heritable

There can be a number of drawbacks associated with using categorical definitions of SLI in genetic studies. Commonly used diagnostic criteria (such as *ICD–10* [the 10th edition of the *International Classification of Diseases,* World Health Organization, 1993] or *DSM–IV* [the 4th edition of the *Diagnostic and Statistical Manual of Mental Disorders*; American Psychiatric Association, 1994]) classify children as affected if they score below a specific cutoff on standardized tests of language ability. In addition, there has to be a significant level of discrepancy between a child's language skills and his or her nonverbal IQ. The recommended cutoffs and discrepancy levels are arbitrarily defined relative to distributions in normal populations (e.g., *ICD–10* suggests a 1 standard deviation [*SD*] difference between language and IQ). As a consequence, there will be some children who have substantial problems with speech and language but are classified as unaffected because they do not meet diagnostic criteria. Furthermore, the validity of IQ discrepancy-based definitions for specific learning disorders has been questioned on a number of grounds (Cole, Dale, & Mills, 1992; Siegel & Himel, 1998). As noted above, another problem with classifying SLI is that there may be considerable phenotypic diversity among those individuals who are considered affected, but attempts to define discrete subtypes have been largely unsuccessful (Bishop, 2001).

These issues of phenotypic complexity and diagnostic limitation are commonly encountered problems for genetic studies of behavioral and cognitive traits. Various strategies have been developed over recent years to handle uncertainty in affection status. Some approaches dispense with qualitative definitions completely, opting instead for direct investigation of some quantitative index of trait severity. These techniques have been particularly successful in exploring the genetic etiology of developmental dyslexia (specific reading disability), another childhood learning disorder that is closely related to, and often comorbid with, SLI (reviewed in Fisher & Smith, 2001). One such quantitative-based method, known as *DeFries–Fulker (DF) regression,* allows the estimation of heritability of extreme deficits in a measure (also referred to as *group heritability*) using twin data (DeFries & Fulker, 1985; see also Willcutt et al., chapter 13, this volume). Essentially, this involves selecting subjects with extreme deficits and then assessing whether the scores of their cotwins regress toward the mean of the normal population as a function of relatedness (MZ vs. DZ). Using DF analysis in their twin sample, Bishop et al. (1995) obtained significant group heritabilities (close to 100%) for several measures of absolute language deficit. Interestingly, they found no significant genetic influence for measures of discrepancy between language and nonverbal ability.

In a follow-up to this work, the twin study of Bishop et al. (1999) estimated the heritabilities of two language-related measures that are central to differing theoretical accounts of the biological processes underlying SLI. The first measure was a test of auditory temporal processing, which evaluated a subject's success at discriminating sequences of tones presented at varying rates. This test was chosen because it has been proposed that SLI is the result of a difficulty with discriminating brief or rapid auditory stimuli and is indicative of a fundamental deficit in temporal resolution of the nervous system (Tallal & Piercy, 1973). The second measure, nonword repetition, assessed an individual's ability to repeat back meaningless nonwords of varying complexity. Children with SLI perform particularly poorly on this task, which has been taken as evidence that the disorder is caused by deficits in phonological working memory, a system specialized for temporary storage of speech sound sequences (Gathercole, Willis, Baddeley, & Emslie, 1994).

There has been a great deal of investigation into these temporal processing and phonological working-memory accounts of SLI, which are not reviewed here. Although the theories may appear to be conflicting, some have suggested that deficits on auditory and nonword repetition tasks could both stem from a single underlying mechanism with a heritable basis; for example, difficulty with nonword repetition might be a secondary consequence of impairment in temporal processing of speech sounds. However, this was not borne out by the quantitative twin analyses of Bishop et al. (1999). The two tests were each able to differentiate language-impaired children from unaffected control children, as predicted by theory, but although deficits in the nonword repetition task were very highly heritable, there was no significant genetic influence on the auditory-processing measure (Bishop et al., 1999). For the latter, it appeared that shared environmental factors played the most important role. The data from this study suggest that there is a complex interplay between deficits in phonological working memory, which are largely influenced by genes, and those in auditory temporal processing, which are mainly influenced by environmental factors, together contributing to increased risk of developing SLI. (Note, however, that the analysis of group deficits in this study involved small sample sizes; larger scale studies will be necessary to confirm differential heritability and provide further insight into the relationship between auditory and nonword repetition tasks.) This work illustrates the importance of quantitative genetic approaches for dissecting biological pathways underlying a trait. Moreover, it demonstrates that just because a measure (e.g., auditory processing) is a good predictor of a highly heritable disorder like SLI, it does not follow per se that the measure itself constitutes a genetically determined core deficit.

Similar quantitative twin-based strategies have been used to address an unresolved issue that has received much attention in recent years. Is SLI really a qualitatively distinct clinical syndrome, or does it represent the extreme lower end of the continuous distribution of language ability in the normal population? This is not merely a question of semantics. If speech and language impairment is qualitatively indistinct from variability in the entire population, this implies that any specific risk factors identified by studies of disorder will also be relevant for normal language development. Dale et al. (1998) studied a measure of productive vocabulary in 3,039 twin pairs of 2-year-olds, taken from a sample of all twins born in England and Wales in 1994. Using model-fitting analyses, Dale et al. found that genetic factors accounted for about 25% of the variance in vocabulary across the entire distribution. There was a much more substantial influence from shared environment, which accounted for about 69% of the total variance. DF analysis of the extreme end of the distribution, corresponding to the bottom 5% of the population, revealed quite different results, yielding a group heritability estimate of about 73%, with only about 18% of the variance attributed to shared environment. Dale et al. interpreted these results as support for the view of SLI as a qualitatively distinct disorder and suggested that the majority of genes that contribute to disrupted language development are unlikely to be associated with individual variation in the normal range of language ability. However, it is not clear how these findings may relate to clinically diagnosed cases of SLI, because poor vocabulary level at age 2 years is not the best predictor of future speech and language difficulties (Bishop, 2001). Furthermore, it should be noted that the differential magnitudes of overall genetic contribution observed by Dale et al. do not necessarily imply that the specific genes involved in disorder are different from those that are important for normal variation.

An alternative explanation of their data is that the same *genes* influence language throughout the entire normal range but that the specific *alleles* operating at the extremes have a greater impact on phenotype. For example, nonsense mutations that result in loss of function of a particular gene might have dramatic effects on language acquisition, giving substantially increased risk of developing SLI. Sequence variants in the promoter of this same gene could alter expression levels and lead to more subtle effects on language ability that would be relevant to individual variability in the normal range. Such a gene could be regarded as important for both SLI and normal variability. These kinds of allele-specific effects might contribute to the elevated group heritability reported by Dale et al. (1998). Perhaps the only way to fully address this issue is to isolate specific genes predisposing to speech and language impairment and examine the effects of sequence variants in these genes among normal individuals. Regardless, one key message from Dale et al.'s study is that one can maximize power to find genes involved in speech and language pathways by focusing on deficits.

Localizing Genes Involved in Speech and Language Disorders

In the absence of any knowledge of the biological mechanisms underlying speech and language impairment, it is difficult to make predictions about the nature of the genetic risk factors involved. Therefore, for this particular phenotype, a pure candidate gene approach, in which a gene is targeted for investigation solely on the basis of its known function, is somewhat akin to a lottery and is unlikely to be successful. A more lucrative strategy in this case is to perform a genomewide search using highly polymorphic markers in families segregating SLI, to identify chromosomal regions that are linked to the inheritance of speech and language deficits. The advantage of such linkage-based methods is that they can detect a risk locus that maps several million bases away from any particular marker, thus enabling the entire genome to be investigated with only a few hundred markers. The intervals implicated by linkage analysis, which may typically contain around a hundred or so genes, can then be targeted for further study. As outlined elsewhere in this chapter, once a chromosomal region has been implicated, there are a number of ways of pinpointing the particular gene that acts as a risk factor for the disorder, a process that is now made easier by the availability of human genomic sequence. Note that linkage analysis lacks power for detection of loci with small effect sizes (see Cardon, chapter 4, this volume). As an alternative to linkage-based approaches, it will soon be feasible to apply association-based methods at high density across the entire genome. However, these have not yet been used for the study of SLI and thus are not discussed further here.

As described earlier, the familial clustering of speech and language difficulties is well documented, and much of this is due to genetic factors. However, in the vast majority of families, transmission patterns are of a complex nature, due to factors such as *phenocopy* (when an individual with a low-risk genotype still develops the disorder) and *incomplete penetrance* (when an individual with a high-risk genotype does not develop the disorder). There also may be *heterogeneity,* with different genes being implicated in different families, and *oligogenic inheritance,* with multiple genes interacting in a single family to increase risk of impairment. This breakdown in genotype–phenotype correspondence can be problematic for traditional model-based linkage mapping, which assumes monogenic inheritance and relies on accurate estimates of parameters like penetrance levels and phenocopy rates (Lander & Schork, 1994). In light of these issues, two contrasting, but complementary, approaches are being adopted to successfully map genes influencing speech and language disorders.

Monogenically Inherited Speech and Language Disorder: Localizing SPCH1

Although in most cases of speech and language disorder the underlying genetic basis of impairment is likely to be complex, there is one rare exception that has been the subject of extensive study. This is a three-generation pedigree, known as KE, in which about half of the members are affected with a severe form of speech and language disorder (Figure 12.1). As first noted in the initial report describing this family (Hurst, Baraitser, Auger, Graham, & Norell, 1990), the inheritance pattern suggests that the disorder is caused by a fully penetrant single locus with an autosomal dominant mode of transmission. Given this extraordinary finding, which implies that mutation of a single gene could have a specific detrimental effect on speech and language abilities, it is not surprising that the phenotype of affected members of the KE family has been the subject of intense scrutiny by a number of research groups. This has led to considerable debate regarding the precise nature of the family's disorder, which has become central for discussions regarding the role of innate factors in language acquisition.

In the original description of the family, Hurst et al. (1990) characterized the disorder as a *developmental verbal dyspraxia*, focusing on the severe articulation impairment of affected individuals that can make their speech difficult to understand. In a contrasting account, Gopnik and Crago (1991) emphasized deficits in linguistic ability, proposing that the affected members have a specific problem with the generation of word endings in accordance with grammatical rules. This latter viewpoint has received wide publicity and has often been cited as supporting evidence for the existence of genes that selectively influence grammatical abilities (e.g., Pinker, 1994). However, a study by Vargha-Khadem, Watkins, Alcock,

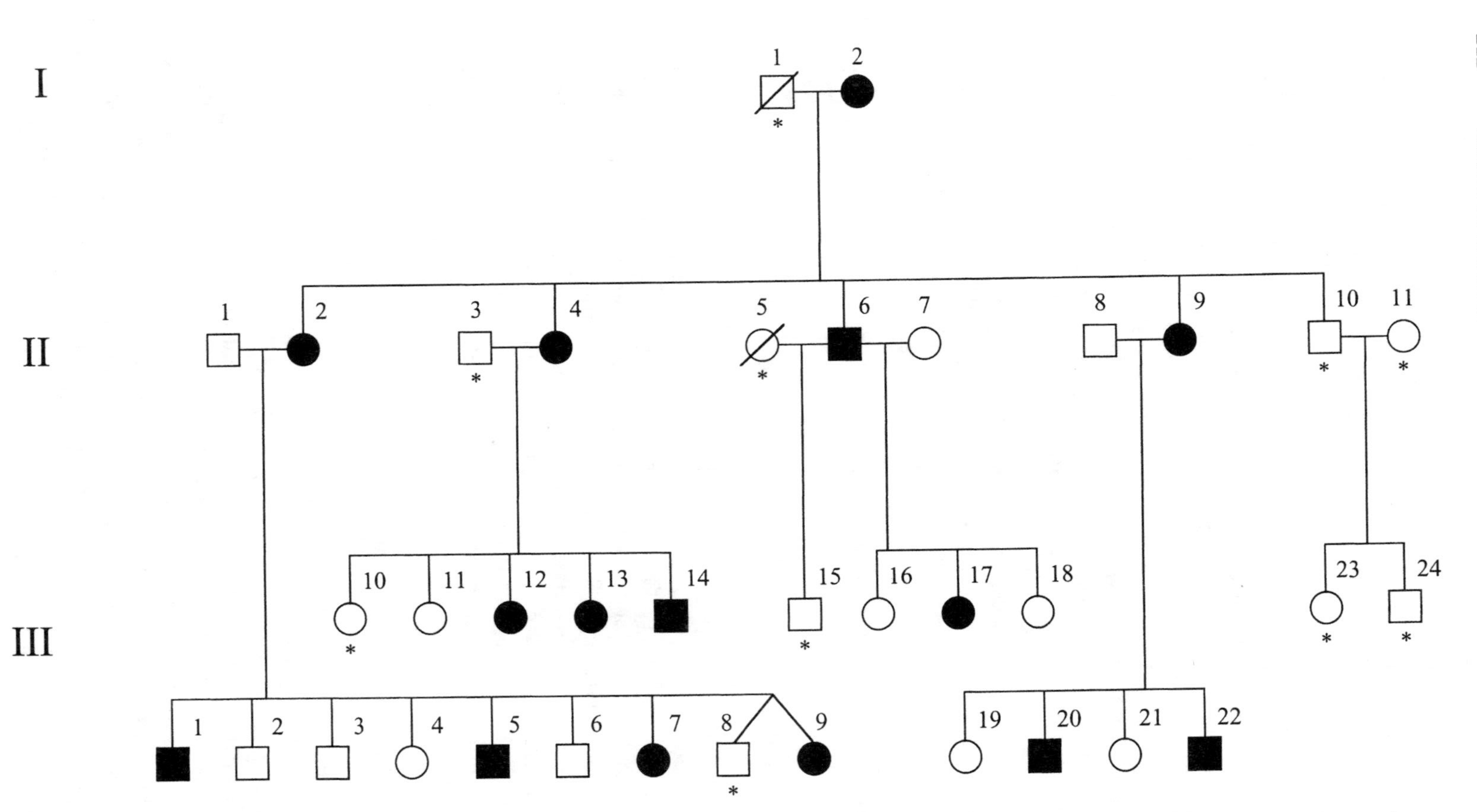

Figure 12.1. Pedigree of the speech and language disorder family KE. Affected individuals are shaded; asterisks indicate those individuals who were unavailable for linkage analysis. From "Localisation of a Gene Implicated in a Severe Speech and Language Disorder," by S. E. Fisher, F. Vargha-Khadem, K. E. Watkins, A. P. Monaco, and M. E. Pembrey, 1998, *Nature Genetics, 18,* p. 168.

Fletcher, and Passingham (1995) demonstrated that the affected individuals of the KE family manifest deficits in multiple facets of language processing and grammatical skill. In addition, Vargha-Khadem et al. further investigated the speech difficulties associated with the disorder and found that there was an underlying impairment in orofacial praxis, which also affected nonspeech movements. Later work by Vargha-Khadem et al. (1998) revealed that tests of word/nonword repetition and orofacial motor coordination provided the best index of the disorder. This is interesting considering the high heritability of nonword repetition deficits reported by Bishop et al. (1999).

Another question that has been raised in studies of the KE phenotype is whether cognitive impairment is confined to the verbal domain. Vargha-Khadem et al. (1995) reported that the mean nonverbal IQ of the affected members of the family was lower than that of the unaffected members. They noted that several affected individuals, and one unaffected individual, had nonverbal IQ scores below 85, which some researchers view as an exclusionary criterion for SLI diagnosis. However, it is important to realize that there were affected individuals in the family who had nonverbal ability that was actually above the population average, despite the presence of severe speech and language difficulties, so nonverbal deficits cannot be considered as characteristic of the disorder. Perhaps these data reveal more about the limitations of current diagnostic schemes than the nature of the core deficit in the KE family. It may be somewhat naïve to assume that a substantial impairment in speech and language (even if this is itself the result of selectively acting genetic factors) will make no impact at all on a child's developmental course in other domains. Indeed, it has been shown that children with developmental speech and language disorders have an increased risk of educational, behavioral, and social difficulties (Rutter & Mawhood, 1991).

In sum, although there is conflict over what constitutes the exact core deficit segregating in the KE family, all of the studies have agreed that the most prominent feature of the phenotype is impairment in speech and/or language abilities. Given the apparent monogenic inheritance pattern, this family offered a unique opportunity to apply traditional model-based linkage methods for localization of the gene responsible. Fisher et al. (1998) therefore initiated a genomewide scan for linkage in the family, assuming autosomal dominant inheritance with full penetrance, and thereby identified a 5.6-cM region on chromosome 7q31 that cosegregated perfectly with the speech and language disorder (Figure 12.2). The maximum logarithm of the odds ratio (LOD) score yielded by 7q31 markers was 6.62 (corresponding to a p value $< 2 \times 10^{-8}$), which dramatically exceeds the significance threshold recommended for parametric linkage analysis (LOD = 3; p value $< 1 \times 10^{-4}$). This investigation thus provided formal proof that disruption of a single genetic locus (referred to as SPCH1) can result in a speech and language disorder and represented the first step toward identifying the gene.

As yet, there have been no other large pedigrees like the KE family reported in the literature, so this remains the only clear case of monogenic

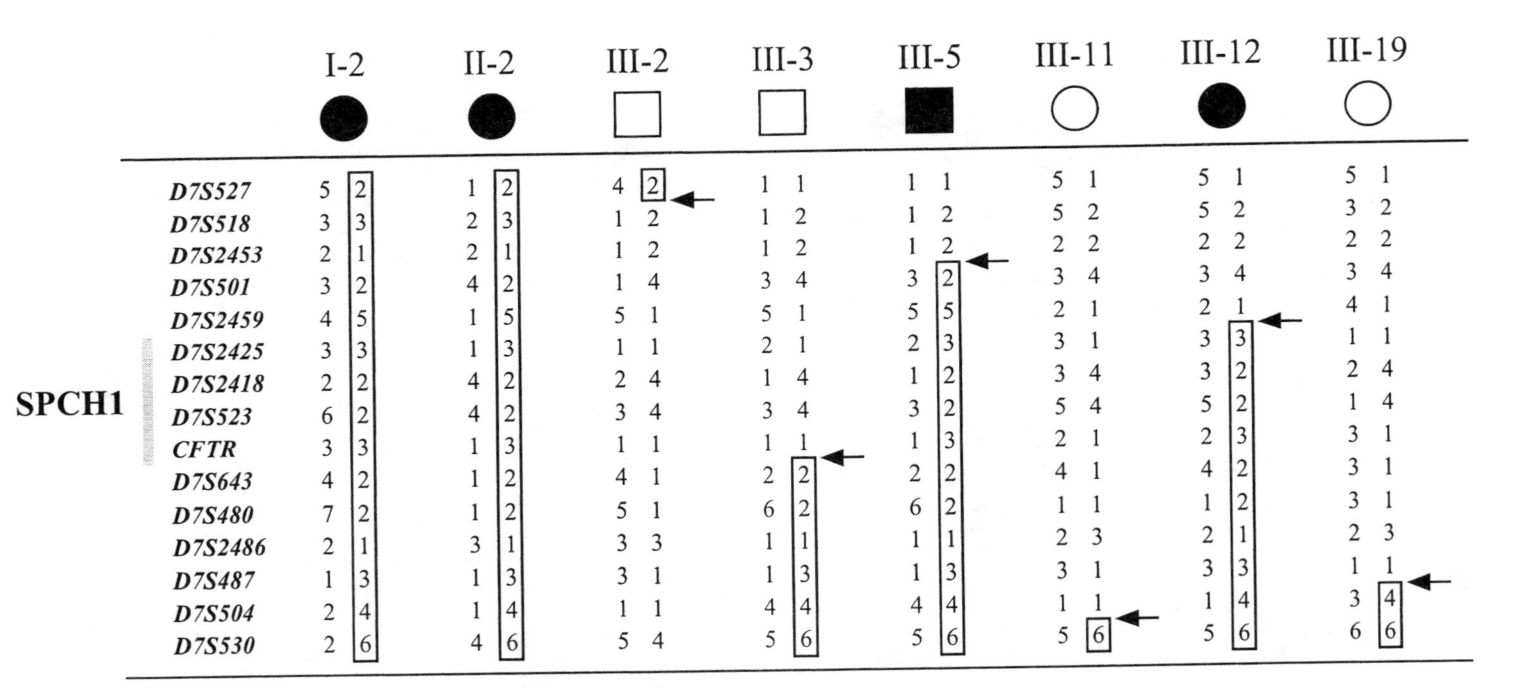

Figure 12.2. Haplotypes of markers from the D7S527–D7S530 region of chromosome 7 for key members of the KE pedigree. Numbers used to identify individuals correspond to those in Figure 12.1. Boxed areas represent the haplotype that cosegregates with the disorder. Arrows show the positions of recombination events involving the disease chromosome. The region implicated by this analysis is indicated by the shaded line to the left. All affected members that are not shown in this figure have inherited the nonrecombinant disorder-associated haplotype for this region. From "Localisation of a Gene Implicated in a Severe Speech and Language Disorder," by S. E. Fisher, F. Vargha-Khadem, K. E. Watkins, A. P. Monaco, and M. E. Pembrey, 1998, *Nature Genetics, 18,* p. 168. Copyright 1998 by Nature America Inc. Reprinted with permission.

inheritance for speech and language impairment. Furthermore, although the study of this kind of family may be a powerful route toward identification of an interesting gene involved in neurological development, a key question is whether that gene will be relevant to general SLI or only restricted to very rare cases of severe speech and language disorder. Preliminary findings, presented at an international conference in 1998, suggested that markers in the SPCH1 interval might be associated with spoken language abilities in a large cohort of children with SLI (Tomblin, Nishimura, Zhang, & Murray, 1998), but these results have not yet been reported in detail. In contrast, a recently completed genomewide search for genes influencing SLI in large numbers of small nuclear families did not detect linkage to the SPCH1 interval (SLI Consortium, 2002). It also is interesting to note that several linkage studies of autistic disorder, a condition that is often associated with language delay, have implicated regions on chromosome 7 that overlap with, or are close to, the SPCH1 region (see Lai et al., 2000). However, the nature of the speech and language problems of the KE family are rather different from those found in individuals with autism, and none of the affected members show any autistic features. Therefore, the significance of SPCH1 for the broader population of children with SLI or for individuals with autistic disorder remains open to question. This will be easier to address in the coming years now that the SPCH1 gene itself has been identified (see below).

Dealing With Complexity: Genome Scans for SLI in Large Samples of Families

In recent years there has been much interest in the development of study designs and statistical methods that are optimized for handling genetic complexity (Lander & Schork, 1994). Analytical techniques exist that do not require knowledge of parameters such as penetrance levels and phenocopy rates, do not rely on assumptions of monogenic inheritance, and cope well with a certain degree of heterogeneity (these methods are sometimes referred to as *nonparametric* or *model-free*). However, to provide sufficient power for genetic mapping, such techniques are dependent on the availability of large sample sizes. For example, a typical study based on allele sharing of sibling pairs (see below) might involve the collection and analysis of several hundred small nuclear families. Until a few years ago, genomewide searches in samples of this size were highly labor intensive and time consuming, and well beyond the capabilities of most genetics laboratories. The advent of high-throughput semiautomated genotyping technology now allows the rapid generation of genetic data from hundreds of samples, so that large-scale genome scanning for complex cognitive–behavioral traits is a feasible option (Fisher, Stein, & Monaco, 1999).

I digress for a moment to provide a little background on some of the statistical techniques that are currently being used for genetic mapping of SLI. The simplest nonparametric linkage methods evaluate whether affected individuals inherit identical copies of a genetic marker from a common ancestor more often than predicted by chance. At any marker, sibling pairs will, on average, share about 50% of their alleles identical by descent (IBD) as a result of random Mendelian segregation. If a pair of siblings who are both diagnosed with the trait of interest demonstrate elevated IBD at a specific marker, this implies linkage between the trait and the marker. Qualitative nonparametric approaches may address problems of genetic complexity to a limited extent, but their major shortcoming is that they still treat a complicated phenotype as an all-or-nothing phenomenon. (It is true that the traditional linkage analysis of the KE family also used qualitative definition of affection status, but this was within a powerful model-based framework in which all assumptions were met and in this case had little consequence for genetic mapping of the trait.)

As highlighted in previous sections of this chapter, it can sometimes be more fruitful to exploit quantitative measures of language ability in genetic studies of traits such as SLI. Quantitative trait locus (QTL) linkage mapping across the entire genome has already proved to be a powerful strategy for dissecting the genetic etiology of complex cognitive traits, at least for genes of moderate effect size (Fisher et al., 2002; see also Willcutt et al., chapter 13, this volume). (Detection of genes of small effect size will most likely require the use of quantitative association-based methods that have recently been developed, which is discussed in Cardon, chapter 4, this volume.) As with qualitative nonparametric methods, QTL linkage analysis depends on estimating the proportion of IBD allele sharing between relatives, usually siblings. With sibling pairs, a QTL test for linkage assesses whether there is a correlation between phenotypic similarity (indexed by comparison of quantitative scores of each sibling) and genetic similarity (as determined by IBD allele sharing) at a particular marker. A straightforward way of doing this is to regress the square of the difference in phenotypic scores against the level of IBD; a standard *t* test of the regression coefficient then provides an estimate for the significance of linkage (Haseman & Elston, 1972). Similar tests, based on the DF regression method, are useful for analyses of extreme selected samples (Cardon et al., 1994). Although easy to apply, these sibpair trait difference techniques do not make full use of the available phenotypic information and do not accommodate multiple sibships in the most appropriate manner. A more powerful approach based on variance components (VCs) has now been developed; this exploits almost all of the trait information and simultaneously incorporates all relationships in a pedigree (Amos, 1994). For VC analyses, maximum-likelihood techniques are used to partition the full variability of a phenotypic measure into components that are due to major gene, unlinked polygenic, and residual environmental factors. At each marker, a statistical test compares the likelihood of the data when allowing for linkage (i.e., an unconstrained major gene component) with that under the null hypothesis of no linkage (i.e., no major gene effect). VC-based linkage approaches were first applied to human psychometric test data in a study that explored linkage of reading-related measures to markers on the short arm of chromosome 6 (Fisher, Marlow, et al., 1999). A key limitation of VC analysis is that it assumes multivariate normality, which is unlikely to be valid for selected samples. Violation of this assumption may lead to false-positive evidence for linkage, but this can be overcome with the use of simulation-based methods that provide empirical estimates of significance (Fisher et al., 2002).

By using the approaches described above, the first genomewide scan for loci influencing quantitative measures of SLI was recently completed. This collaborative effort has so far investigated nearly 100 nuclear families from two U.K.-based sources: an epidemiological study by the Cambridge Language and Speech Project and a clinical sample collected at Guy's Hospital, London (SLI Consortium, 2002). Each family included at least one child with standard language scores 1.5 *SD* below the mean for their chronological age, but quantitative psychometric test data were collected from all siblings regardless of affection status. QTL analyses were performed for composite scores of expressive and receptive language ability, as well as for the nonword repetition task that had been previously shown to be highly heritable by Bishop et al. (1999). Complementary methods based on sibpair trait differences and variance components were used, accompanied by simulations for assessing significance of linkage. The genome scan clearly implicated two distinct loci on chromosomes 16 and 19 but found no evidence for linkage to the SPCH1 region on chromosome 7 (SLI Consortium, 2002). Evidence supporting the chromosome 16 and 19 loci was remarkably consistent for the epidemiological and clinical family samples, suggesting that these loci are relevant for a broad population of children affected with SLI. The study is being expanded to include additional collections of families in which the loci implicated by the scan can be further investigated.

Another genome scan study that is currently under way is using both qualitative and quantitative methodologies to investigate a sample of 14 families (93 participants) segregating SLI (Bartlett et al., 2000). Regions showing weak evidence for linkage are being followed up in a second set of 5 large families (93 participants). This study has thus far reported on only chromosomes 1, 6, and 15, chromosomes that have been suggested to harbor risk factors for developmental dyslexia. It is not known if there is evidence supporting involvement of SPCH1 on 7q31 in their sample. An interesting result from the initial findings of this scan is that a quantitative measure of word identification may be linked to a region on chromosome 15 (Bartlett et al., 2000). This region apparently overlaps with one that has previously been implicated in susceptibility to developmental dyslexia (see Fisher & Smith, 2001). It remains to be determined whether this convergence reflects a common genetic factor on chromosome 15 influencing both SLI and dyslexia.

This area of research is still very much in its infancy. Over the coming years, there are likely to be many intriguing findings reported from these large-scale mapping studies of SLI. Early indications suggest that these investigations will be successful in identifying the approximate locations of a number of genetic risk factors involved in SLI. This leads to a key question. Given this information, how does one move toward isolating the specific gene variants that are responsible for this trait?

Exploiting Genomic Sequence Data: The Search for SPCH1

The recent isolation of the mutated SPCH1 gene on chromosome 7 is an apt illustration of how genetic risk factors may be pinpointed in the post-genomic era. Initially, linkage analysis of the KE family enabled SPCH1 to be localized within 5.6cM on 7q31 (Fisher et al., 1998) in a region that could contain a hundred or so genes. The next step was to construct a comprehensive transcript map of the interval to identify candidate genes that could then be tested for mutations. Fortunately, the 7q31 band was already well characterized at the physical level and extensively covered by a series of contiguous large genomic clones. Several known genes and expressed sequence tags (ESTs) had already been mapped to the region. Most importantly, at the time, this band was in the process of being sequenced by Washington University as part of the large-scale sequencing of chromosome 7. In this transitional period, although a large (and rapidly growing) volume of 7q31 sequence information was available in public databases, much of it was fragmented and not fully cross-referenced. In addition, there were only limited annotations regarding positions of anonymous markers, genes, and repetitive elements relative to these reference sequences. Using a computer-based approach that evaluated similarity between separate sequence entries, Lai et al. (2000) were able to assemble all of the available fragments from the SPCH1 interval into a detailed map containing nearly 8 million bases of completed sequence (see Figure 12.3). Further bioinformatic analyses, involving similarity searches and gene prediction programs, allowed them to precisely localize 20 known genes (see Table 12.1) and 50 anonymous transcripts relative to the sequence map. It should be noted that the draft sequence data that are now available from the Human Genome Project have been assembled and annotated to a much greater degree than they were at the time of the Lai et al. (2000) study. An overview of any chromosomal region of interest can now be quickly and easily obtained using interfaces accessed through the World Wide Web (e.g., http://genome.ucsc.edu/ or http://www.ensembl.org/).

Lai et al.'s (2000) study highlighted the value of a detailed sequence map in a number of ways. First, a sequence map can be used to reliably establish the order of and distances between polymorphic genetic markers, which may be crucial for the success of fine-mapping linkage and association studies. Second, transcripts can be accurately placed in relation to these markers, allowing one to prioritize candidate genes that map closest to peaks of linkage or association and exclude those that map outside of a critical interval. Third, sequence data can be used to generate new polymorphic markers from the interval (e.g., by searching for novel microsatellite repeats) to aid fine mapping. Adopting this approach, Lai et al. identified several novel polymorphic markers and analyzed them in the KE family, thereby narrowing the SPCH1 linkage interval by several million bases. Fourth, comparison between the mRNA sequence of a candidate gene and the corresponding genomic DNA sequence enables the rapid in silico (i.e., at the computer) determination of its exon/intron structure. (Many human genes are divided into regions containing protein coding

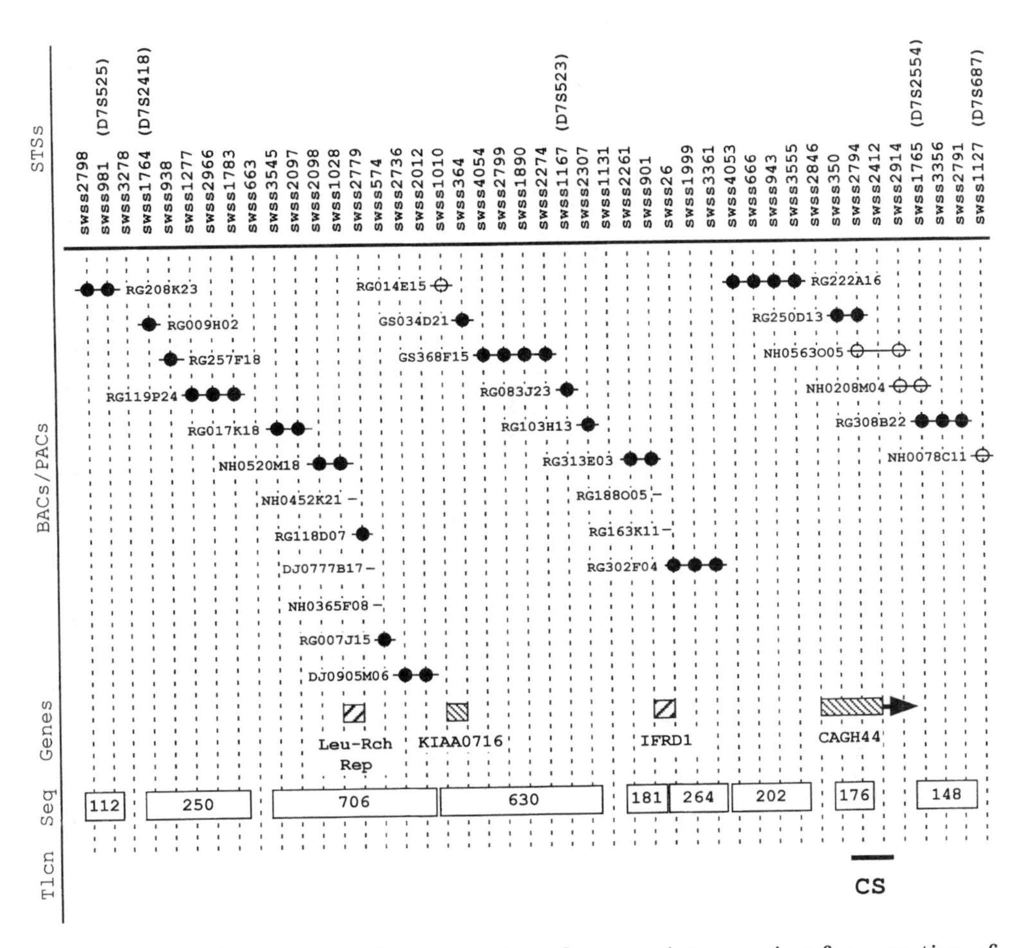

Figure 12.3. Illustration of sequence-based transcript mapping for a section of the SPCH1 interval. Sequence tagged sites (STSs), aligned along the top, provided the framework for map construction. Some STSs correspond to polymorphic genetic markers (such as D7S525). Sequence contigs, assembled using data from large genomic clones (represented by circles), are aligned beneath the ordered STSs. White circles indicate that only partial sequence was available at the time of analysis. Sequences could be assembled into large contiguous blocks, as indicated by white rectangles beneath the contigs (with sizes shown in kilobases). Positions of known genes are marked with shaded rectangles. CAGH44 had only been partially characterized at the time of map construction. The position of patient CS's translocation breakpoint is given at the bottom of the map. From "The SPCH1 Region on Human 7q31: Genomic Characterization of the Critical Interval and Localization of Translocations Associated With Speech and Language Disorder," by C. S. L. Lai et al., 2000, *American Journal of Human Genetics, 67,* p. 360. Copyright 2000 by the American Society of Human Genetics. Adapted with permission.

Table 12.1. Known Genes Residing Between D7S2459 and D7S643

Gene	Protein	Function	Accession
PDS	Pendrin	Anion transporter; mutated in Pendred syndrome	AF030880
DRA	Down-regulated in adenoma	Anion transporter; mutated in congenital chloride diarrhea	L02785
DLD	Dihydrolipoamide dehydrogenase	Mitochondrial matrix protein; mutated in infantile lactic acidosis	NM_000108
LAMB1	Laminin beta-1 chain precursor	Component of extracellular matrix	M61916
LAMB1R	Novel laminin beta-1 related protein	Highly similar to LAMB1	AF172277
Bravo-NrCAM	Neuronal cell adhesion molecule	Implicated in axonal pathfinding in developing nervous system	NM_005010
MDG1	Microvascular endothelial differentiation gene 1	Expression of rat homologue increased during tube-forming process of microvascular endothelial cells	AB026908
HF12	Zinc finger protein	Involved in cell differentiation/proliferation	X07290
Leu-Rch Rep	Leucine-rich repeat protein	Possible role in neural development via protein–protein interactions	AC004142
KIAA0716	Large protein expressed in brain	Unknown	AB018259
IFRD1	Interferon-related developmental regulator	Growth factor-sensitive positive regulator of cell differentiation	Y10313

CAGH44	Polyglutamine repeat protein from brain	Long polyglutamine tract containing 40 consecutive glutamines	U80741
CAV2	Caveolin 2	Component of caveolae membranes; upregulated in response to neuronal cell injury	AF035752
CAV1	Caveolin 1	Component of caveolae membranes; interacts directly with G-protein alpha subunits	Z18951
MET	Hepatocyte growth factor receptor	Tyrosine protein kinase; mutated in papillary renal carcinoma	J02958
CAPZA2	F-actin capping protein	Binds growing ends of actin filaments; blocks exchange of subunits	U03269
WNT2	Wingless-type MMTV integration site 2	May be signaling protein affecting development of discrete regions of tissues	X07876
CFTR	Cystic fibrosis transmembrane conductance regulator	ATP-binding chloride transporter; mutated in cystic fibrosis	NM_000492
TSA806	Testis-specific ankyrin motif protein	Possible role in cell cycle control or cell fate determination	D78334
KCND2	Brain-specific potassium channel KV4.2	Involved in dendritic spike propagation	AF121104

Note. From "The SPCH1 Region on Human 7q31: Genomic Characterization of the Critical Interval and Localization of Translocations Associated With Speech and Language Disorder," by C. S. L. Lai et al., 2000, *American Journal of Human Genetics, 67,* p. 363.

information, known as *exons,* and intervening sequences of unknown function, referred to as *introns.*) This information is important for any DNA-based mutation searches.

Another strategy that complements those already discussed in this chapter is the investigation of chromosomal rearrangements in patients with speech and language impairment. When such rearrangements involve a region that has already been implicated by a linkage study, mapping of chromosomal breakpoints can provide important clues to the genetic etiology. Again, this is an area in which a sequence-based transcript map proves highly valuable. Lai et al. (2000) identified an unrelated patient CS, who has a de novo balanced translocation involving 7q31, [t(5;7)(q22;q31.2)], associated with a speech and language disorder that is strikingly similar to the KE phenotype. Using fluorescence in-situ hybridization with genomic clones identified from the sequence map, Lai et al. demonstrated that the 7q31 breakpoint of CS was very close to a partially characterized candidate gene, known as CAGH44. This gene is expressed in brain and encodes a protein with a long stretch of consecutive polyglutamine residues, an interesting feature given that several neurodegenerative diseases (e.g., Huntington's chorea) are caused by expansion of polyglutamine tracts (Cummings & Zoghbi, 2000). Demonstrating that a translocation breakpoint coincides with a candidate gene is encouraging, but it does not provide proof that disruption of that particular gene causes the disorder. Chromosomal rearrangements can sometimes alter the expression of genes that map up to 1 million bases away from the breakpoint, via a *position effect* (Kleinjan & van Heyningen, 1998). Because some chromosomal abnormalities may have no phenotypic consequences at all, it is even possible that the breakpoint is completely unrelated to the observation of disorder in the patient. (There is an ascertainment bias for chromosomal abnormalities in children with developmental disorders, because in such cases cytogenetic testing is routinely carried out.) Therefore, converging evidence from a number of sources (e.g., multiple translocation cases, point mutations in other affected individuals) is needed to confirm the role of a gene implicated by studies of chromosomal rearrangements.

Given that the CAGH44 gene appeared to be disrupted in the CS translocation case, Lai et al. (2000) searched for mutations in the known gene sequence in affected members of the KE family but found no sequence variant that cosegregated with the speech and language disorder. However, at that time only a few hundred bases of the 5′ part of the CAGH44 mRNA had been characterized, and there was a gap in the genomic sequence exactly where the more 3′ regions should lie. A few months after Lai et al.'s (2000) study was published, the remaining genomic sequence from this gap was finished. Lai et al. (2001) were then able to follow up their previous report, using this genomic sequence to assemble a computer-based prediction of the complete candidate gene, which they verified with laboratory methods. They discovered that the 3′ end of the gene encoded an 84 amino-acid segment showing high similarity to the characteristic DNA-binding domain of the forkhead/winged-helix (FOX) family of transcription factors (see Kaufmann & Knochel, 1996). Therefore, the complete gene was officially renamed FOXP2, in accordance with standardized nomenclature guidelines proposed by Kaestner, Knochel, and Martinez (2000). Many FOX proteins have been shown to be key regulators of embryogenesis (Kaufmann & Knochel, 1996). Furthermore, mutations and chromosomal rearrangements involving several of these forkhead-domain transcription factors are known to cause various human disorders, including glaucoma (FOXC1; Nishimura et al., 1998) and ovarian failure (FOXL2; Crisponi et al., 2001).

Sequencing of the entire FOXP2 gene in the KE family revealed a point mutation in the region encoding the third alpha-helix of the forkhead domain, and this was found to cosegregate perfectly with affection status (Lai et al., 2001). This mutation alters an arginine residue that is normally an invariant feature of this type of DNA-binding domain and is likely to be critical for protein function. Lai et al.'s results suggest that FOXP2, which is strongly expressed in human fetal brain, corresponds to the SPCH1 gene. In both the KE and CS cases, speech and language disorder probably results from a lack of sufficient functional protein at a key stage of fetal brain development.

The Future

This is a particularly exciting time for genetic studies of language impairment. The isolation of the SPCH1 gene provides the first molecular target for investigating the pathways underlying speech and language disorders. Functional studies of FOXP2 will explore its influence on the expression of other genes during fetal brain development and could eventually shed light on some of the neural networks that are important for normal speech and language acquisition. It will also be fascinating to compare FOXP2 sequence in different species (particularly other great apes) and find out if it has come under any novel selective pressures since the divergence of human and chimpanzee lineages. Of course, this gene is only one of many factors that are likely to be implicated in these complex neurological mechanisms. It is not even known yet whether FOXP2 is important for a broad population of children affected with SLI, let alone for speech and language variation in the normal range. Even if this is only significant for a rare severe form of disorder, analysis of the gene will nevertheless yield novel insights into brain development.

Over the coming years, large-scale genetic mapping studies will identify approximate chromosomal locations of a number of additional genetic risk factors influencing SLI. To pinpoint the specific gene variants that give increased risk, investigations of chromosomal rearrangements and association-based mapping methods will be increasingly important. In the postgenomic era, the human genomic sequence represents a powerful resource to aid in these efforts, as demonstrated by the isolation of FOXP2. However, it is possible that many of the genetic changes that are involved in SLI will not have effects that are predictable by studies of sequence alone. Knowledge of the consequences of specific sequence changes for function of a candidate gene may become critical for success. Furthermore, it is clear that we are not ready to close the book on phenotype definition for speech and language disorders; considerable work remains to be done in this complex area. Ultimately, it is hoped that information gathered from molecular genetics will increase the understanding of SLI, allow early identification of children at risk, and help us to find ways of aiding their speech and language acquisition.

References

American Psychiatric Association. (1994). *Diagnostic and statistical manual of mental disorders* (4th ed.). Washington, DC: Author.

Amos, C. I. (1994). Robust variance-components approach for assessing genetic linkage in pedigrees. *American Journal of Human Genetics, 54,* 535–543.

Bartlett, C. W., Flax, J., Yabut, O., Hirsch, L., Li, W., Tallal, P., et al. (2000). A genome-scan for linkage of specific language impairment: Report on chromosomes 1, 6 and 15. *American Journal of Human Genetics, 67*(Suppl.), 309.

Bishop, D. V. M. (1987). The causes of specific developmental language disorder ("development dysphasia"). *Journal of Child Psychology and Psychiatry, 28,* 1–8.

Bishop, D. V. M. (2001). Genetic and environmental risks for specific language impairment in children. *Philosophical Transactions: Biological Sciences, 356,* 369–380.

Bishop, D. V. M., Bishop, S. J., Bright, P., James, C., Delaney, T., & Tallal, P. (1999). Different origin of auditory and phonological processing problems in children with language impairment: Evidence from a twin study. *Journal of Speech, Language, and Hearing Research, 42,* 155–168.

Bishop, D. V. M., North, T., & Donlan, C. (1995). Genetic basis for specific language impairment: Evidence from a twin study. *Developmental Medicine and Child Neurology, 37,* 56–71.

Cardon, L. R., Smith, S. D., Fulker, D. W., Kimberling, W. J., Pennington, B. F., & DeFries, J. C. (1994). Quantitative trait locus for reading disability on chromosome 6. *Science 266,* 276–279.

Cole, K. N., Dale, P. S., & Mills, P. E. (1992). Stability of the intelligence quotient–language quotient relation: Is discrepancy modeling based on a myth? *American Journal on Mental Retardation, 97,* 131–143.

Conti-Ramsden, G., & Friel-Patti, S. (1984). Mother–child dialogues: A comparison of normal and language impaired children. *Journal of Communication Disorders, 17,* 19–35.

Crisponi, L., Deiana, M., Loi, A., Chiappe, F., Uda, M., Amati, P., et al. (2001). The putative forkhead transcription factor FOXL2 is mutated in blepharophimosis/ptosis/epicanthus inversus syndrome. *Nature Genetics, 27,* 159–166.

Cummings, C. J., & Zoghbi, H. Y. (2000). Fourteen and counting: Unraveling trinucleotide repeat diseases. *Human Molecular Genetics, 9,* 909–916.

Dale, P. S., Simonoff, E., Bishop, D. V. M., Eley, T. C., Oliver, B., Price, T. S., et al. (1998). Genetic influence on language delay in two-year-old children. *Nature Neuroscience, 1,* 324–328.

DeFries, J. C., & Fulker, D. W. (1985). Multiple regression analysis of twin data. *Behavior Genetics, 15,* 467–473.

Felsenfeld, S., & Plomin, R. (1997). Epidemiological and offspring analyses of developmental speech disorders using data from the Colorado Adoption Project. *Journal of Speech, Language, and Hearing Research, 40,* 778–791.

Fisher, S. E., Francks, C., Marlow, A. J., MacPhie, I. L., Newbury, D. F., Cardon, L. R., et al. (2002). Independent genome-wide scans identify a chromosome 18 quantitative-trait locus influencing dyslexia. *Nature Genetics, 30,* 86–91.

Fisher, S. E., Marlow, A. J., Lamb, J., Maestrini, E., Williams, D. F., Richardson, A. J., et al. (1999). A quantitative trait locus on chromosome 6p influences different aspects of developmental dyslexia. *American Journal of Human Genetics, 64,* 146–156.

Fisher, S. E., & Smith, S. D. (2001). Progress towards the identification of genes influencing developmental dyslexia. In A. J. Fawcett (Ed.), *Dyslexia: Theory and good practice* (pp. 39–64). London: Whurr.

Fisher, S. E., Stein, J. F., & Monaco, A. P. (1999). A genome-wide search strategy for identifying quantitative trait loci involved in reading and spelling disability (developmental dyslexia). *European Child and Adolescent Psychiatry, 8*(S3), 47–51.

Fisher, S. E., Vargha-Khadem, F., Watkins, K. E., Monaco, A. P., & Pembrey, M. E. (1998). Localisation of a gene implicated in a severe speech and language disorder. *Nature Genetics, 18,* 168–170.

Gathercole, S. E., Willis, C. S., Baddeley, A. D., & Emslie, H. (1994). The Children's Test of Nonword Repetition: A test of phonological working memory. *Memory, 2,* 103–127.

Gopnik, M., & Crago, M. B. (1991). Familial aggregation of a developmental language disorder. *Cognition, 39,* 1–50.

Haseman, J. K., & Elston, R. C. (1972). The investigation of linkage between a quantitative trait and a marker locus. *Behavior Genetics, 2,* 3–19.

Hurst, J. A., Baraitser, M., Auger, E., Graham, F., & Norell, S. (1990). An extended family with a dominantly inherited speech disorder. *Developmental Medicine and Child Neurology, 32,* 347–355.

Kaestner, K. H., Knochel, W., & Martinez, D. E. (2000). Unified nomenclature for the winged helix/forkhead transcription factors. *Genes and Development, 14,* 142–146.

Kaufmann, E., & Knochel, W. (1996). Five years on the wings of fork head. *Mechanisms of Development, 57,* 3–20.

Kleinjan, D. J., & van Heyningen, V. (1998). Position effect in human genetic disease. *Human Molecular Genetics, 7,* 1611–1618.

Lahey, M., & Edwards, J. (1995). Specific language impairment: Preliminary investigation of factors associated with family history and with patterns of language performance. *Journal of Speech and Hearing Research, 38,* 643–657.

Lai, C. S. L., Fisher, S. E., Hurst, J. A., Levy, E. R., Hodgson, S., Fox, M., et al. (2000). The SPCH1 region on human 7q31: Genomic characterization of the critical interval and localization of translocations associated with speech and language disorder. *American Journal of Human Genetics, 67,* 357–368.

Lai, C. S. L., Fisher, S. E., Hurst, J. A., Vargha-Khadem, F., & Monaco, A. P. (2001). A forkhead-domain gene is mutated in a severe speech and language disorder. *Nature, 413,* 519–523.

Lander, E. S., & Schork, N. J. (1994). Genetic dissection of complex traits. *Science, 265,* 2037–2048.

Lewis, B. A., Ekelman, B. L., & Aram, D. M. (1989). A familial study of severe phonological disorders. *Journal of Speech and Hearing Research, 32,* 713–724.

Lewis, B. A., & Thompson, L. A. (1992). A study of developmental speech and language disorders in twins. *Journal of Speech and Hearing Research, 35,* 1086–1094.

Neils, J., & Aram, D. M. (1986). Family history of children with developmental language disorders. *Perceptual and Motor Skills, 63,* 655–658.

Nishimura, D. Y., Swiderski, R. E., Alward, W. L., Searby, C. C., Patil, S. R., Bennet, S. R., et al. (1998). The forkhead transcription factor gene FKHL7 is responsible for glaucoma phenotypes which map to 6p25. *Nature Genetics, 19,* 140–147.

Pinker, S. (1994). *The language instinct.* London: Allen Lane.

Rutter, M., & Mawhood, L. (1991). The long-term psychosocial sequelae of specific developmental disorders of speech and language. In M. Rutter & P. Casaer (Eds.), *Biological risk factors for psychosocial disorders* (pp. 233–259). Cambridge, England: Cambridge University Press.

Siegel, L. S., & Himel, N. (1998). Socioeconomic status, age and the classification of dyslexics and poor readers: The dangers of using IQ scores in the definition of reading disability. *Dyslexia, 4,* 90–103.

SLI Consortium. (2002). A genomewide scan identifies two novel loci involved in specific language impairment. *American Journal of Human Genetics, 70,* 384–398.

Tallal, P., & Piercy, M. (1973). Defects of non-verbal auditory perception in children with developmental aphasia. *Nature, 241,* 468–469.

Tallal, P., Ross, R., & Curtiss, S. (1989). Familial aggregation in specific language impairment. *Journal of Speech and Hearing Disorders, 54,* 167–173.

Tomblin, J. B. (1989). Familial concentration of developmental language impairment. *Journal of Speech and Hearing Disorders, 54,* 287–295.

Tomblin, J. B., & Buckwalter, P. R. (1998). Heritability of poor language achievement among twins. *Journal of Speech, Language, and Hearing Research, 41,* 188–199.

Tomblin, J. B., Nishimura, C., Zhang, X., & Murray, J. C. (1998). Association of developmental language impairment with loci at 7q3. *American Journal of Human Genetics, 63*(Suppl.), A312.

Vargha-Khadem, F., Watkins, K., Alcock, K., Fletcher, P., & Passingham, R. (1995). Praxic and nonverbal cognitive deficits in a large family with a genetically transmitted speech and language disorder. *Proceedings of the National Academy of Sciences USA, 92,* 930–933.

Vargha-Khadem, F., Watkins, K. E., Price, C. J., Ashburner, J., Alcock, K. J., Connelly, A., et al. (1998). Neural basis of an inherited speech and language disorder. *Proceedings of the National Academy of Sciences USA, 95,* 12695–12700.

World Health Organization. (1993). *The ICD–10 classification for mental and behavioural disorders: Diagnostic criteria for research.* Geneva, Switzerland: Author.

13

Genetic Etiology of Comorbid Reading Difficulties and ADHD

Erik G. Willcutt, John C. DeFries, Bruce F. Pennington, Shelley D. Smith, Lon R. Cardon, and Richard K. Olson

Plomin and McClearn's (1993) precursor to the present volume introduced psychologists to recent advances in behavioral genetics and suggested directions for future research. In the concluding chapter, Plomin (1993) highlighted several new research tools that were "fueling the current momentum of the field" (p. 467). These methodological advances included the use of multiple regression to understand the etiology of extreme phenotypic scores, the application of multivariate genetic analyses to test for pleiotropic effects of genes on two or more traits, and a paradigm shift toward the use of molecular genetic techniques to identify specific genes that influence behavior.

In a chapter in that previous volume, DeFries and Gillis (1993) summarized existing knowledge regarding the etiology of reading disability (RD). At that time, reading deficits had been shown to be familial, but studies that could disentangle genetic and environmental influences in families were still in their infancy. Results of several small twin studies were somewhat inconsistent but suggested tentatively that reading deficits were due at least in part to genetic influences. In a brief section at the conclusion of the chapter, DeFries and Gillis described preliminary evidence indicating that reading disability might be linked to quantitative trait loci (QTL) on chromosomes 6 and 15.

Subsequently, twin studies of substantially larger samples have provided conclusive evidence that reading disability is heritable, and linkage studies have localized possible genes for reading disability to at least four

This work was supported in part by program project and center grants from the National Institute of Child Health and Human Development (NICHD; HD-11681 and HD-27802) to John C. DeFries, grants from the National Institute of Mental Health (NIMH; MH-38820) and NICHD (HD-04024) to Bruce F. Pennington, and NIMH Individual National Research Service Award (MH-12100) to Erik G. Willcutt. We thank the school personnel and families who participated in the study, and Helen E. Datta, Terry Grupp, and Deborah Porter for database management.

chromosomal regions. The rapid accumulation of new knowledge about the genetic etiologies of reading disability and other related disorders illustrates the impact of both behavioral and molecular genetic methods on the development of knowledge about factors that influence both normal and abnormal development. However, results that have emerged in the past decade also reveal the complexity of the etiological pathways to reading difficulties and other multifactorial behaviors and underscore how much is yet to be learned.

We have four primary objectives in this chapter: (a) to review briefly previous family and twin studies of reading disability and to summarize the most recent results from the Colorado Learning Disabilities Research Center (CLDRC) twin study; (b) to summarize the results of genetic linkage studies of reading disability from our laboratory and others; (c) to describe results of studies that have applied behavioral and molecular genetic methods to understand the etiology of comorbidity between reading disability and attention deficit hyperactivity disorder (ADHD); and (d) to discuss future research directions for both behavioral and molecular genetic studies in this area.

Behavioral Genetic Studies of Reading Disability

Although the specific etiological mechanisms that lead to reading disability are still unknown, significant advances have been made in understanding the extent to which reading deficits are attributable to genetic or environmental influences. Previous studies demonstrate clearly that reading difficulties are familial (e.g., DeFries & Decker, 1982; DeFries, Vogler, & LaBuda, 1986; Finucci & Childs, 1983; Pennington et al., 1991). Specifically, the relative risk for reading disability is four to eight times higher in first-degree relatives of probands with a reading disability than in relatives of individuals without one (e.g., Gilger, Pennington, & DeFries, 1991; Pennington & Lefly, 2001). However, because members of intact nuclear families share both genetic and environmental influences, familial resemblance alone does not provide conclusive evidence for heritable variation.

By comparing the similarity of monozygotic (MZ) twins, who share all of their genes, with dizygotic (DZ) twins, who share half of their segregating genes on average, twin studies estimate the extent to which familiality is due to familial environmental versus genetic influences. Several early twin studies compared the probandwise concordance rates for reading disability in relatively small samples of MZ and DZ twin pairs (Bakwin, 1973; Stevenson, Graham, Fredman, & McLoughlin, 1984, 1987; Zerbin-Rüdin, 1967). Although results varied somewhat across these studies, all found that MZ twins were more likely than DZ twins to be concordant for reading disability, providing tentative evidence that reading deficits are due at least in part to genetic influences.

CLDRC Twin Study

Because of the paucity of well-designed twin studies of reading disability, a twin study was initiated in 1982 as part of the Colorado Reading Project (Decker & Vandenberg, 1985; DeFries, 1985). In 1991, this ongoing twin study was incorporated as one component of the CLDRC, a six-site research center devoted to the study of the etiology, assessment, and treatment of reading and other learning disabilities.

To minimize the possibility of referral bias, the researchers contact parents of all twins between ages 8 and 18 in 27 local school districts and obtain permission from parents to review the school records of both members of each pair of twins for evidence of reading problems. If either member of a twin pair manifests a positive history of reading problems (e.g., low reading achievement test scores, referral to a reading therapist because of poor reading performance, or reports by classroom teachers or school psychologists), both members of the pair are invited to complete an extensive battery of tests in our laboratories at the University of Colorado and the University of Denver.

The test battery includes the Wechsler Intelligence Scale for Children–Revised (Wechsler, 1974) or Wechsler Adult Intelligence Scale–Revised (Wechsler, 1981) and the Peabody Individual Achievement Test (PIAT; Dunn & Markwardt, 1970). A reading achievement composite score is calculated on the basis of a discriminant function analysis of the PIAT Reading Recognition, Reading Comprehension, and Spelling subtests in a separate sample of nontwin individuals with and without a history of significant reading problems (DeFries, 1985). In addition, each participant completes several measures of language processes that are related to reading ability, and parents complete a battery of interviews regarding each twin's emotional and behavioral functioning, including measures of ADHD.

The comparison of concordance rates in MZ and DZ twin pairs provides an initial test for genetic etiology of a discrete disorder. As of May 2001, 245 pairs of MZ twins and 195 pairs of same-sex DZ twins had been selected because at least one twin in the pair met criteria for reading disability based on a positive school history of reading difficulties and a score at least 1.5 standard deviations (*SD*) below the estimated population mean on the reading composite score. The probandwise concordance rate for reading disability is significantly higher in MZ twin pairs (65%) than in DZ twin pairs (35%), providing additional evidence that reading disability is at least partially attributable to genetic influences.

Multiple Regression Analysis of Twin Data

Although the simplicity of a comparison of concordance rates is appealing, increasing evidence suggests that reading disability, ADHD, and most other disorders are defined on the basis of arbitrary diagnostic cutoff points on a quantitative measure (e.g., Barkley, 1998; Stevenson et al.,

1987; Willcutt, Pennington, & DeFries, 2000a). Transformation of a continuous measure such as reading performance into a categorical variable (e.g., reading disabled vs. unaffected) results in the loss of important information pertaining to the continuum of variation in reading performance. In contrast, multiple regression analysis of twin data provides a versatile and powerful test of the etiology of extreme scores on a continuous trait (DeFries & Fulker, 1985, 1988).

The Basic Regression Model

The DeFries–Fulker (DF) model is based on the regression of MZ and DZ cotwin scores toward the population mean when probands are selected because of extreme scores on a phenotype of interest (e.g., reading deficits). Although scores of both MZ and DZ cotwins would be expected to regress toward the mean of the unselected population (μ), scores of DZ cotwins should regress further than scores of MZ co-twins to the extent that extreme scores are influenced by genes (Figure 13.1). The magnitude of this differential regression by zygosity provides an estimate of the heritability of the proband group deficit in reading (h^2_g).

To illustrate the DF method, we fitted a multiple regression model to reading composite data from the current sample of probands and co-twins in the CLDRC twin study. All twins who scored more than 1.5 *SD* below the mean of the population on the reading discriminant function score were selected as probands. For pairs that were concordant for reading deficits, each twin was entered once as the proband and once as the cotwin. This "double-entry" procedure facilitates the most valid estimates of h^2_g when a sample has been ascertained using truncate selection (e.g., DeFries & Alarcón, 1996; McGue, 1992). Because the double entry of con-

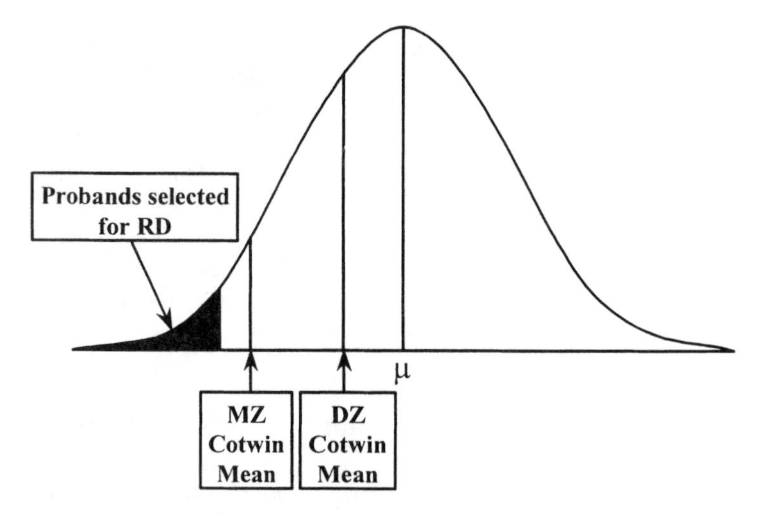

Figure 13.1. The distribution of reading achievement in the population: differential regression of monozygotic (MZ) and dizygotic (DZ) cotwin means toward the population mean (μ). RD = reading disability.

cordant pairs artificially inflates the sample size, the standard errors of the regression coefficients and resulting t values were corrected for double entry to obtain unbiased tests of significance (e.g., Stevenson, Pennington, Gilger, DeFries, & Gillis, 1993).

Before the multiple regression analysis was conducted, standardized scores were created based on the mean and standard deviation of the non–reading disability sample to facilitate the estimation of h_g^2. The standardized scores for the selected sample of MZ probands and cotwins were then divided by the MZ proband mean, and the DZ proband and cotwin scores were divided by the DZ proband mean. This procedure ensures that the MZ and DZ probands are equally divergent from the mean of the population prior to the regression analysis.

The basic regression model for the univariate case is as follows:

$$C = B_1P + B_2R + K, \tag{1}$$

where C is the expected cotwin score, P is the proband score, R is the coefficient of relationship (1 for MZ pairs, 0.5 for DZ pairs), and K is the regression constant. The B_1 coefficient represents the partial regression of the cotwin's score on the proband's score and provides a measure of twin resemblance irrespective of zygosity. The B_2 parameter represents the partial regression of the cotwin's score on the coefficient of relationship, and after appropriate transformation of the data provides a direct estimate of h_g^2. After adjustment of the standard errors of the regression coefficients to correct for the double entry of concordant pairs, the significance of the B_2 parameter provides a statistical test of the extent to which extreme scores are attributable to genetic influences.

The means of MZ and DZ probands on the reading composite score were almost identical, and both means were over 2.5 *SD* below the mean of the comparison sample of twins without reading disability (see Table 13.1). The mean reading score of the MZ co-twins regressed less toward the population mean than the reading score of the DZ co-twins, consistent with what would be expected if genetic influences contribute to deficits in reading. Indeed, when Equation 1 was fitted to the reading data, the resulting estimate of extreme group heritability was highly significant: $h_g^2 = 0.55$ (.08), $t = 6.88$, $p = 2.3 \times 10^{-11}$.

Table 13.1. Mean Reading Discriminant Scores of 245 Pairs of Monozygotic Twins and 195 Pairs of Dizygotic Twins

	Proband		Cotwin	
Twin pair	*M*[a]	*SD*	*M*[a]	*SD*
Monozygotic	−2.61	0.85	−2.39	1.03
Dizygotic	−2.60	0.84	−1.67	1.28

[a]Expressed as standardized deviations from the mean of 996 control twins.

Extensions of the Basic DF Model

The DF model is versatile and can be elaborated to test for the influence of other independent variables on the etiology of the trait under consideration. For example, the model to test whether h^2_g for reading difficulties differs as a function of IQ is as follows:

$$C_{RD} = B_1P_{RD} + B_2R + B_3P_{IQ} + B_4P_{IQ} \times P_{RD} + B_5P_{IQ} \times R + K, \quad (2)$$

where C_{RD} is the expected cotwin reading score, P_{RD} is the proband reading score, R is the coefficient of relationship, P_{IQ} is the IQ of the proband, $P_{IQ} \times P_{RD}$ is the product of the proband's IQ score and the proband's reading score, and $P_{IQ} \times R$ is the product of the proband's IQ and the coefficient of relationship. The B_3 term represents a test for the main effect of IQ on reading across zygosity, the B_4 coefficient tests for differential twin resemblance as a function of IQ, and the B_5 coefficient tests for differential h^2_g of reading as a function of IQ.

The CLDRC dataset has recently been used to test whether the heritability of reading ability varies as a function of gender or IQ (Wadsworth, Knopik, & DeFries, 2000; Wadsworth, Olson, Pennington, & DeFries, 2000). Wadsworth, Knopik, and DeFries (2000) obtained almost identical estimates of the heritability of reading deficits in girls (h^2_g = .59) and boys (h^2_g = .58), providing little support for the hypothesis that genetic influences play a larger role in reading deficits in girls than in boys (Geschwind, 1981).

In contrast to the results for gender, Wadsworth, Olson, et al. (2000) found that the heritability of the group deficit in reading varied significantly as a function of IQ. Twin pairs were divided into pairs with an average IQ greater than 100 and pairs with an average IQ below 100, and Equation 2 was then fitted to data from both groups simultaneously. Results revealed that the estimate of h^2_g was significantly lower in the group with IQ scores below 100 (h^2_g = .43) than the group with IQ scores of 100 or higher (h^2_g = .72). Similarly, the heritability of the reading deficit also increased significantly as a linear function of IQ ($p < .01$), suggesting that environmental influences may be a more salient cause of reading difficulties in children with lower general cognitive ability. These results demonstrate the potential utility of the DF method for identifying and understanding etiologically meaningful subtypes within a disorder.

Genetic Linkage Analysis

If a trait is found to be significantly heritable, genetic linkage analyses can be conducted to localize the genes that are associated with the trait. Linkage analysis is based on the extent to which alleles at two genetic loci are inherited together (for additional reading on linkage analysis, see Faraone, Tsuang, & Tsuang, 1999; Plomin, DeFries, McClearn, & McGuffin, 2001). If two genes are close together on the same chromosome, alleles at the two loci are likely to be transmitted together from parent to child. In contrast, if the two genes are on different chromosomes or are far apart on the same chromosome, alleles at the two loci will be transmitted independently. Classical linkage analysis (CLA) tests whether a marker allele recombines with a putative gene for a categorical phenotype significantly less frequently than expected by chance across several generations of a family.

Although CLA is a powerful analytic technique in some situations, this method has several disadvantages for complex, heterogeneous disorders such as reading disability and ADHD. First, CLA is optimized for use with categorical definitions of disease and is best able to identify a "major gene" that accounts for a large amount of variance in the phenotype. As noted previously, a growing consensus suggests that disorders such as reading disability and ADHD may be conceptualized more appropriately as extreme scores on continuous symptom dimensions. Second, parametric linkage analyses require that the mode of inheritance and the penetrance of the locus be specified in the model, which may be difficult for a complex trait such as reading disability. Finally, CLA requires a reliable and valid measure of the phenotype in two or more consecutive generations of individuals in a large familial pedigree. If the nature of the phenotype changes significantly across development, this may further complicate the analysis.

Multiple Regression Linkage Analysis

As an alternative to CLA, sibling–pair linkage methods have been developed to localize QTL. QTLs are genes that influence susceptibility to a disorder but are not typically a necessary or sufficient cause of the disorder (see Plomin & Crabbe, 2000).

Haseman and Elston (1972) first advocated that regression analysis with pairs of siblings could be used to test for the effect of a QTL on individual differences in a dimensional trait. Fulker et al. (1991) then showed that when a sample is selected because at least one sibling exhibits an extreme score on a trait, an adaptation of the multiple regression model described by DeFries and Fulker (1985, 1988) provides a versatile and powerful test for linkage. If a QTL close to a chromosomal marker influences the selected trait, the scores of cosibs of selected probands should regress differentially toward the population mean as a function of the number of marker alleles shared identical by descent (IBD; i.e., the same allele inherited from the same parent) with the proband. To adapt the basic DF multiple regression model for linkage analysis, the coefficient of relationship in Equation 1 is replaced by the estimated proportion of alleles shared IBD ($\hat{\pi}$) by the pair of siblings.

The regression model for univariate QTL analysis is

$$C = B_1P + B_2\hat{\pi} + K, \tag{3}$$

where C is the expected cosib score, P is the proband score, and $\hat{\pi}$ is the estimated proportion of alleles shared IBD at the marker. After appropriate adjustment of the standard errors of the regression coefficients to correct for the double entry of concordant sibling pairs (Gayán et al., 1999), the significance of the B_2 parameter provides a statistical test of the extent to which extreme scores are influenced by a QTL linked to the marker. Cardon and Fulker (1994) then developed this approach further to map the interval between markers. By analyzing multiple genetic markers simultaneously, this multipoint interval mapping procedure estimates the proportion of alleles shared IBD between siblings at points on the chromosome between available DNA markers. On the basis of these estimates, a continuous line is obtained indicating the significance of linkage at all points along a chromosomal region of interest.

Linkage Studies of Reading Disability

Linkage studies have identified several chromosomal regions that may contain genes that increase susceptibility to reading disability. There have been single reports of linkage to genes on chromosome 1 (Grigorenko et al., 1998), 2 (Fagerheim et al., 1999), 6q (Petryshen, Kaplan, & Field, 1999), and 18 (Fisher et al., 2002); QTLs on chromosome 6p21.3 and 15q21 have now been found in several independent samples.

An initial study of nine nuclear families suggested that reading disability may be associated with a gene near the centromere of chromosome 15 (Smith, Kimberling, Pennington, & Lubs, 1983), although this result was weakened when the sample was expanded to 21 families (Smith, Kimberling, & Pennington, 1991). Subsequently, several studies have reported significant linkage to markers in a different region of the chromosome (15q21) for reading disability (Fulker et al., 1991; Grigorenko et al., 1997) and spelling disability (Nöthen et al., 1999). Moreover, using a family-based association approach, Morris et al. (1999) reported a significant association between reading disability and a three-marker haplotype within the linkage region identified in the previous studies.

Smith et al. (1991) first reported evidence suggesting that a gene on chromosome 6 might influence reading deficits. Subsequently, Cardon et al. (1994, 1995) used the interval mapping technique described previously to localize a QTL for reading disability to chromosome 6p21.3. Although one subsequent study did not find evidence of a QTL in this region (Petryshen, Kaplan, Liu, & Field, 2000), this localization has now been confirmed in three independent samples (Fisher et al., 1999; Gayán et al., 1999; Grigorenko et al., 1997). Moreover, these subsequent studies revealed significant linkage for both overall reading ability and several specific reading and language skills that influence reading ability. Thus, this QTL represents one of the most consistently replicated linkages in genetic studies of complex traits with respect to both phenotype definition and chromosomal refinement (e.g., Flint, 1999). The fact that the evidence for this QTL has replicated more consistently than linkages for other complex traits such as schizophrenia (see review by Riley & McGuffin, 2000) or bipolar disorder (Craddock & Jones, 2001) may be attributable to its relatively large effect (Knopik et al., in press) or to the power gained through analyses of continuous phenotypic scores versus the categorical phenotype used in linkage studies of psychiatric illnesses.

In summary, substantial progress has been made in understanding the etiology of reading disability since the publication of the previous volume (Plomin & McClearn, 1993). Twin studies of larger samples have provided conclusive evidence that reading disability is heritable, and linkage studies have localized putative QTLs for reading disability to regions on chromosomes 1, 2, 6, 15, and 18. In the next section, we turn to studies that have applied these methods to understand the relation between reading disability and other disorders such as ADHD.

Etiology of Comorbidity Between Reading Disability and ADHD

The genetics of ADHD is reviewed in more detail elsewhere in this volume (see Thapar, chapter 22). Here we focus on the finding that reading disability and ADHD co-occur more frequently than expected by chance; 25%–40% of children with either reading disability or ADHD also meet criteria for the other disorder (e.g., August & Garfinkel, 1990; Semrud-Clikeman et al., 1992; Willcutt, Chhabildas, & Pennington, 2001; Willcutt & Pennington, 2000). However, the etiology of this association is not well understood. Reading disability and ADHD are significantly comorbid in both clinical and community samples, indicating that it is not a selection artifact. Similarly, because reading disability is assessed by cognitive tests, whereas ADHD is usually assessed only by behavioral ratings, the relation between reading disability and ADHD does not appear to be due to shared method variance (e.g., Willcutt, Pennington, Boada, et al., 2001). In the following two sections, we describe a series of analyses that we have conducted over the past several years to test if comorbidity of reading disability and ADHD is attributable to common genetic influences.

Applying the DF Model to Bivariate Data

A simple generalization allows the univariate multiple regression models described in Equations 1 and 2 to be applied to bivariate twin data. Rather than comparing the relative similarity of MZ and DZ twins for the same trait, bivariate analyses compare the relation between the proband's score on one trait and the cotwin's score on a second trait across zygosity. Therefore, if common genetic influences contribute to the association between reading disability and ADHD, the ADHD score of the cotwins of MZ probands with reading disability would be expected to regress less toward the population mean than the ADHD score of DZ cotwins.

The regression equation to apply the DF method to bivariate data is expressed as follows:

$$C_{ADHD} = B_1P_{RD} + B_2R + K, \quad (4)$$

where C_{ADHD} is the expected cotwin score on the nonselected measure (ADHD), P_{RD} is the proband score on the selected measure (reading disability), R is the coefficient of the relationship, and K is the regression constant. The B_2 coefficient provides a direct estimate of the bivariate heritability of reading disability and ADHD ($h^2_{g[RD/ADHD]}$), the extent to which the proband reading deficit is attributable to genetic influences that are also associated with elevations of ADHD.

Our most recent article used a sample of 313 same-sex twin pairs (183 MZ, 130 DZ) from the CLDRC twin study in which at least one twin met criteria for reading disability (Willcutt, Pennington, & DeFries, 2000b). The *DSM–III* (*Diagnostic and Statistical Manual of Mental Disorders,* 3rd ed.; American Psychiatric Association, 1980) version of the Diagnostic Interview for Children and Adolescents was used to assess symptoms of ADHD because it had been completed by the largest proportion of participants, and inattention and hyperactivity–impulsivity factor scores were computed on the basis of results of a factor analysis conducted previously (Willcutt & Pennington, 2000). The overall ADHD composite and the inattention and hyperactivity–impulsivity factor scores were all highly heritable in this sample (h^2_g = .78–.94; Willcutt, Pennington, & DeFries, 2000a).

Phenotypic analyses indicated that individuals with reading disability exhibited significantly more symptoms of both inattention and hyperactivity–impulsivity than individuals without reading disability but that this effect was significantly larger for inattention scores. When Equation 4 was fitted to bivariate twin data, results revealed significant bivariate heritability for reading disability and the overall ADHD symptom count ($h^2_{g(RD/ADHD)}$ = .23, $p < .05$). However, additional analyses indicated that the etiology of the relation between reading disability and ADHD varied as a function of the ADHD symptom dimension. Specifically, the bivariate heritability of reading disability and inattention symptoms was significant ($h^2_{g(RD/Inatt)}$ = 0.39, $p < .01$), whereas reading disability and hyperactivity–impulsivity symptoms were not significantly coheritable ($h^2_{g(RD/Hyp)}$ = 0.05).

Interpretation of our study and other previous studies of the relation between reading disability and ADHD is complicated by recent changes in the diagnostic criteria for ADHD. Because our initial results were based on a measure of *DSM–III* attention deficit disorder (American Psychiatric Association, 1980), additional analyses were conducted for this chapter to test if these results replicated when a measure of *DSM–IV* ADHD (American Psychiatric Association, 1994) was used (see Table 13.2). Results revealed significant bivariate heritability of reading disability and parent ratings of total *DSM–IV* ADHD symptoms ($h^2_{g(RD/ADHD)}$ = .42). The estimate of bivariate heritability was similar for teacher ratings of ADHD ($h^2_{g(RD/ADHD)}$ = .37), but this result was not significant because teacher ratings were only available for a smaller subset of the sample.

When symptoms of inattention and hyperactivity–impulsivity were analyzed separately, estimates of bivariate heritability were again higher

Table 13.2. Bivariate Heritability of RD and *DSM–IV* ADHD in Twin Pairs Selected for RD

	MZ cotwins		DZ cotwins		Bivariate		
Cotwin measure	*M*	*SD*	*M*	*SD*	h^2_g	*SE*	t^c
Parent ratings[a]							
Total ADHD	1.02	1.63	0.49	1.45	0.42	.20	2.12*
Hyperactivity–impulsivity	0.59	1.45	0.25	1.15	0.25	.17	1.47†
Inattention	1.19	1.54	0.57	1.42	0.49	.21	2.32*
Teacher ratings[b]							
Total ADHD	1.58	1.96	1.10	1.70	0.37	.34	1.10
Hyperactivity–impulsivity	0.96	2.16	0.69	1.88	0.20	.36	0.55
Inattention	2.03	1.60	1.49	1.70	0.42	.30	1.41†

Note. RD = reading disability; *DSM–IV* = *Diagnostic and Statistical Manual of Mental Disorders* (4th ed.; American Psychiatric Association, 1994); ADHD = attention deficit hyperactivity disorder.

[a]Expressed as standardized deviations from the mean of 367 control twins. *N* = 66 MZ pairs, 69 DZ pairs. [b]Expressed as standardized deviations from the mean of 202 control twins. *N* = 34 MZ pairs, 36 DZ pairs. [c]Degrees of freedom are 133 for parent ratings and 67 for teacher ratings.

$*p < .05$. $\dagger p < .10$.

for reading disability and inattention ($h^2_{g(RD/Inatt)}$ = .42 and .49) than reading disability and hyperactivity–impulsivity ($h^2_{g(RD/Hyp)}$ = .20 and .25), although those differences were not significant. Moreover, a similar pattern of results was obtained when probands were selected for *DSM–IV* ADHD ($h^2_{g(Inatt/RD)}$ = .36, $h^2_{g(Hyp/RD)}$ = .14), although the sample of probands with ADHD is still relatively small.

In summary, our results suggest that common genetic influences contribute to comorbidity of reading disability and ADHD and that these common genes may influence symptoms of inattention more strongly than symptoms of hyperactivity–impulsivity. In the next section, we describe results of linkage analyses that we conducted recently to attempt to localize a QTL that influences both disorders.

Bivariate Linkage Analyses

The influence of genes on behavior is likely to be pleiotropic, such that the same genes affect more than one phenotype (e.g., Falconer & MacKay, 1996). For example, the gene for albinism in mice is also associated with significantly higher levels of emotionality (e.g., DeFries, Hegmann, & Weir, 1966; Turri, Datta, DeFries, Henderson, & Flint, 2001). However, genes that influence both reading disability and ADHD have not been previously localized. On the basis of the finding that common genetic influences contribute to comorbidity of reading disability and ADHD, we recently tested whether the well-replicated QTL for reading disability on chromosome 6p is also associated with increased susceptibility to ADHD (Willcutt et al., 2002).

The univariate DF model for linkage analysis can be applied to bivariate data to test whether a QTL has pleiotropic effects on two traits. Rather than comparing the relative similarity of siblings for the same trait, bivariate analyses test if the relation between the proband score on the selected trait and the cosib score on a second, unselected trait varies as a function of $\hat{\pi}$. Therefore, if the QTL for reading disability on chromosome 6p is also a susceptibility locus for ADHD, the number of symptoms of ADHD exhibited by cosibs of reading disability probands would be expected to increase with allele sharing at DNA markers close to the susceptibility locus.

The regression model for the bivariate case is

$$C_{ADHD} = B_1 P_{RD} + B_2 \hat{\pi} + K, \quad (5)$$

where C_{ADHD} is the predicted cosib score on the nonselected measure (ADHD) and B_2 tests for pleiotropic effects of the QTL on the two phenotypes.

Eight informative DNA markers spanning a region of 14.7 centimorgans (cM) on the short arm of chromosome 6 were used for these analyses (see Gayán et al., 1999). These markers span the region reported in previous 6p linkage studies of reading disability (Cardon et al., 1994, 1995; Fisher et al., 1999; Gayán et al., 1999; Grigorenko et al., 1997; Petryshen et al., 2000). Siblings and both parents from each family were genotyped for these markers from blood/cheek samples using published methods (Idury & Cardon, 1997). MAPMAKER/SIBS, a statistical package for genetic linkage analysis (Kruglyak & Lander, 1995), was used to estimate $\hat{\pi}$ or single-marker analyses. For multipoint analyses, MAPMAKER/SIBS uses available information at all markers to estimate the proportion of alleles-shared IBD at positions between markers ($\hat{\pi}_q$). Multipoint tests for linkage were conducted at 0.5 cM intervals across the region of interest.

Initial univariate analyses were conducted to test if the QTL on chromosome 6p21.3 is also a susceptibility locus for ADHD. Multipoint analyses revealed significant linkage for the overall ADHD composite and for both inattention and hyperactivity symptoms (see Figure 13.2). The area of peak linkage fell between markers D6S276 and D6S105 for all phenotypes. The maximum log of the odds (LOD) score for the ADHD measures (LOD = 1.8) was somewhat lower than the LOD scores obtained for the reading phenotypes (LOD = 1.9–3.2; Gayán et al., 1999), suggesting that this QTL may have a stronger effect on reading than symptoms of ADHD.

To test for pleiotropic effects of the QTL on reading disability and ADHD, we conducted bivariate linkage analyses for each reading measure based on the regression model in Equation 5. For each analysis, probands were selected for a score below the 1.5 *SD* cutoff on the relevant reading measure, and the cosib's ADHD score was regressed onto the proband's reading score and the proportion of alleles-shared IBD. Multipoint analyses revealed significant bivariate linkage for ADHD and each of the three reading phenotypes (Figures 13.3A–C), with the strongest results for the orthographic choice measure. Moreover, in contrast to the results of the twin analyses, the evidence for bivariate linkage with reading was some-

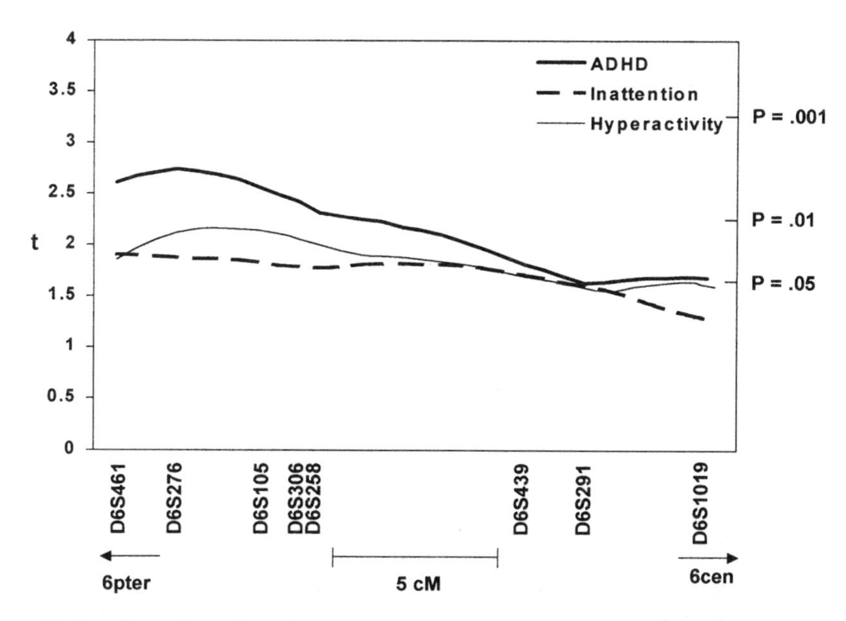

Figure 13.2. Results of multipoint linkage analyses of overall attention deficit hyperactivity disorder (ADHD) composite, inattention symptoms, and hyperactivity symptoms to eight DNA markers on chromosome 6p ($N = 46–50$ pairs). P values indicate one-tailed significance levels.

what higher for symptoms of hyperactivity, although the general pattern of results was similar for all three ADHD phenotypes. Because of power considerations (Fulker et al., 1991), however, much larger samples would be required to test for differential bivariate QTL heritability.

Future Directions

Our results indicate that both reading disability and ADHD are significantly heritable and that the frequent comorbidity of reading disability and ADHD is attributable substantially to common genetic influences. Linkage analyses indicate that the QTL for reading disability on chromosome 6p is also a susceptibility locus for ADHD and suggest that comorbidity between reading disability and ADHD may be due at least in part to pleiotropic effects of this QTL. In the remainder of this chapter, we discuss future directions for research on the etiology of reading disability and ADHD and their comorbidity.

Localization of QTLs

Linkage analysis is an essential first step toward the localization of susceptibility loci for a complex trait. However, the region of significant linkage for a QTL is frequently large, often including hundreds of genes (e.g., Morris et al., 1999). In contrast to family-based linkage analysis, associ-

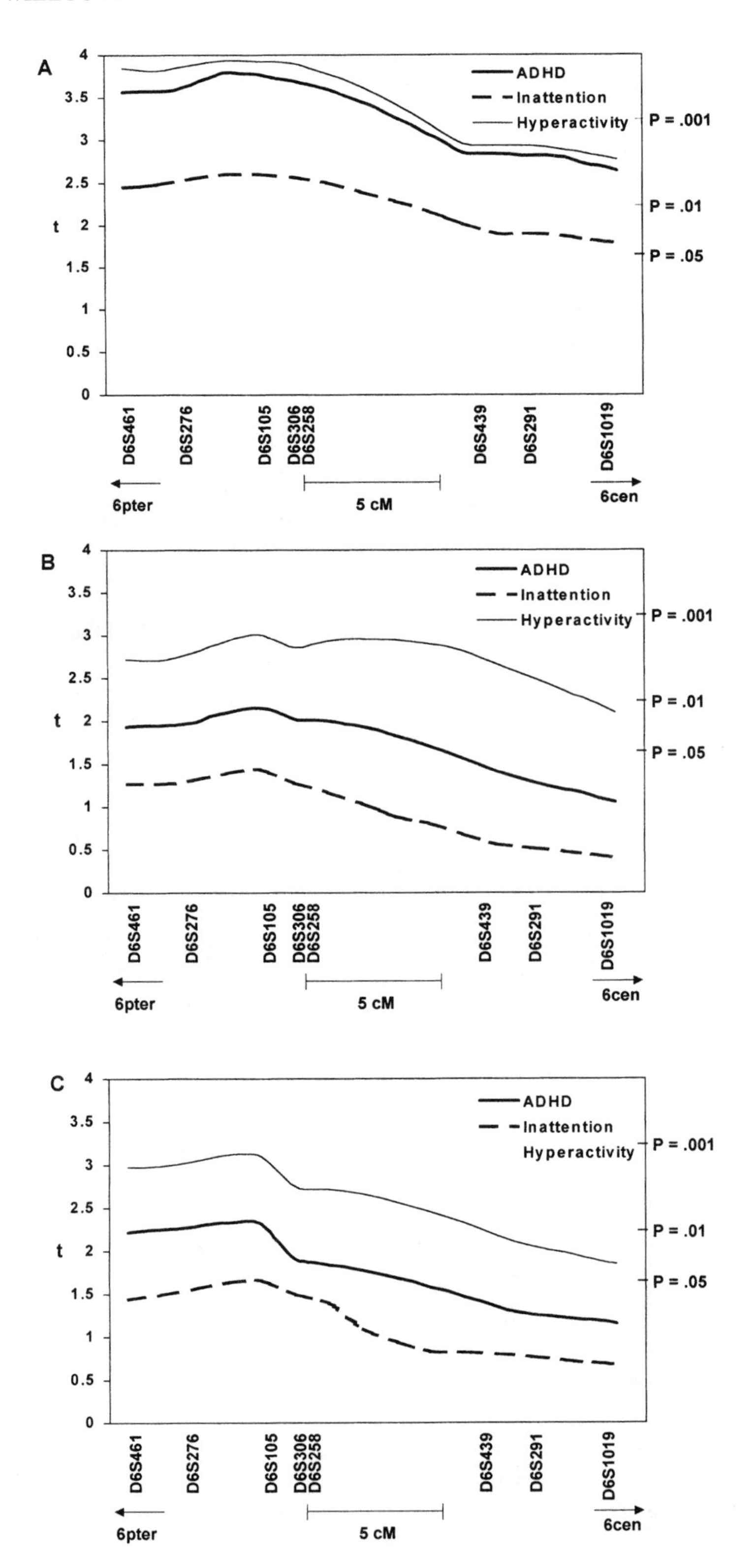
A
ADHD
Inattention
Hyperactivity
P = .001
P = .01
P = .05
t
D6S461
D6S276
D6S105
D6S306
D6S258
D6S439
D6S291
D6S1019
6pter
5 cM
6cen
B
ADHD
Inattention
Hyperactivity
P = .001
P = .01
P = .05
t
D6S461
D6S276
D6S105
D6S306
D6S258
D6S439
D6S291
D6S1019
6pter
5 cM
6cen
C
ADHD
Inattention
Hyperactivity
P = .001
P = .01
P = .05
t
D6S461
D6S276
D6S105
D6S306
D6S258
D6S439
D6S291
D6S1019
6pter
5 cM
6cen

ation and linkage disequilibrium occur when alleles at a marker locus cosegregate with alleles at a susceptibility locus more frequently than expected by chance in an entire population (e.g., Cardon & Bell, 2001). Because linkage disequilibrium extends a relatively short distance in most regions of the genome, a DNA marker must be close to a QTL that influences a trait to detect significant association. Therefore, after linkage analysis has been used to identify chromosomal regions that are likely to contain a QTL for a trait, association methods can be used to refine the estimated chromosomal location of the QTL.

Evidence for linkage disequilibrium can be further enhanced by testing if a disorder is associated with a haplotype, a specific combination of alleles at a set of closely spaced markers. As noted previously, Morris et al. (1999) tested for an association between reading disability and eight markers on chromosome 15q, the site of a putative QTL for reading disability (e.g., Grigorenko et al., 1997). Their results revealed a significant association between reading disability and a three-marker haplotype in the region of significant linkage identified in previous studies, providing additional evidence that a QTL in this region increases susceptibility to reading difficulties. Moreover, because linkage disequilibrium rarely extends more than 0.5 cM from a marker (e.g., Abecasis et al., 2001), this result identifies a chromosomal region that is narrower than the 5–8 cM region identified in linkage studies (Grigorenko et al., 1997; Schulte-Korne et al., 1998).

As increasing numbers of DNA markers become available and genotyping techniques continue to become more efficient and less expensive, association methods will provide a powerful method to identify the location of QTLs for complex traits. We currently are genotyping additional markers in our sample across the region of significant linkage on chromosome 6p and flanking regions of the chromosome. Moreover, although no obvious candidate genes for reading disability or ADHD have been identified in this region to date, future molecular genetic studies will determine the function of additional genes in this region. On the basis of these findings, it is likely that plausible candidate genes will be identified that are expressed in the brain or influence other developmental processes related to reading disability or ADHD.

Multivariate QTL Analyses

In addition to the QTL on chromosome 6, possible QTLs for reading disability have been localized to regions on chromosomes 1 (e.g., Rabin et al.,

---

Figure 13.3. Results of bivariate multipoint linkage analyses of reading deficits and the overall attention deficit hyperactivity disorder (ADHD) composite, inattention symptoms, and hyperactivity symptoms. A: Probands selected for deficits in orthographic choice ($N = 50$ pairs); B: Probands selected for deficits in phonological decoding ($N = 58$ pairs); C: Probands selected for deficits on the reading composite score ($N = 69$ pairs). P values indicate one-tailed significance levels.

1993), 2 (Fagerheim et al., 1999), and 15 (e.g., Fulker et al., 1991; Grigorenko et al., 1997; Smith et al., 1983), and additional QTLs are likely to be identified in genomewide scans that are currently under way. Moreover, it is plausible that epistatic interactions among subsets of these genes may contribute to susceptibility to reading difficulties (e.g., Grigorenko, chapter 14, this volume). Similarly, candidate gene studies have demonstrated a significant association between ADHD and polymorphisms in at least five genes in the dopamine system (e.g., Barr et al., 2000; Cook et al., 1995; Daly, Hawi, Fitzgerald, & Gill, 1999; Eisenberg et al., 1999; Faraone et al., 1999), although these results have not been replicated in all samples (e.g., Asherson et al., 1998; Castellanos et al., 1998). These results underscore the complexity of the genetic etiology of reading disability and ADHD and suggest that additional genes are likely to contribute to comorbidity of these disorders. In future studies, it will be useful to obtain genotypes for markers in these chromosomal regions and for each of these candidate genes. These genotypes may then be included in multivariate analyses to test which genes contribute independently to reading disability, ADHD, or their comorbidity and whether epistatic interactions among any of the loci play a significant role in the etiology of these disorders or their overlap.

Conclusion

Our findings regarding reading disability and ADHD suggest that the boundaries between putatively distinct diagnoses may prove to be blurry, with the same genetic influences conferring risk for more than one disorder. In some cases, such findings may indicate that two disorders may be better conceptualized as alternate forms of the same disorder, whereas in other cases common risk factors may contribute to two or more distinct disorders. In either case, etiologically informative methods such as those described in this chapter provide an important tool that can be used in future studies to improve the validity of the diagnostic nosology of psychiatric disorders by revealing the etiology of comorbidity between complex syndromes.

References

Abecasis, G. R., Noguchi, E., Heinzmann, A., Traherne, J. A., Bhattacharyya, S., Leaves, N. I., et al. (2001). Extent and distribution of linkage disequilibrium in three genomic regions. *American Journal of Human Genetics, 68,* 191–197.

American Psychiatric Association. (1980). *Diagnostic and statistical manual of mental disorders* (3rd ed.). Washington, DC: Author.

American Psychiatric Association. (1994). *Diagnostic and statistical manual of mental disorders* (4th ed.). Washington, DC: Author.

Asherson, P., Virdee, V., Curran, S., Ebersole, S., Freeman, B., Craig, I., et al. (1998). Association of *DSM–IV* attention deficit hyperactivity disorder and monoamine pathway genes. *American Journal of Medical Genetics, 81,* 549.

August, G. J., & Garfinkel, B. D. (1990). Comorbidity of ADHD and reading disability among clinic referred children. *Journal of Abnormal Child Psychology, 18,* 29–45.

Bakwin, H. (1973). Reading disability in twins. *Developmental Medicine and Child Neurology, 15,* 184–187.

Barkley, R. A. (1998). *Attention-deficit hyperactivity disorder: A handbook for diagnosis and treatment* (2nd ed.). New York: Guilford Press.

Barr, C. L., Wigg, K. G., Bloom, S., Schachar, R., Tannock, R., Roberts, W., et al. (2000). Further evidence from haplotype analysis for linkage of the dopamine D4 receptor gene and attention-deficit hyperactivity disorder. *American Journal of Medical Genetics, 96,* 262–267.

Cardon, L. R., & Bell, J. I. (2001). Association study designs for complex diseases. *Nature Reviews Genetics, 2,* 91–99.

Cardon, L., & Fulker, D. (1994). The power of interval mapping of quantitative trait loci, using selected sib pairs. *American Journal of Human Genetics, 55,* 825–833.

Cardon, L. R., Smith, S. D., Fulker, D. W., Kimberling, W. J., Pennington, B. F., & DeFries, J. C. (1994). Quantitative trait locus for reading disability on chromosome 6. *Science, 226,* 276–279.

Cardon, L. R., Smith, S. D., Fulker, D. W., Kimberling, W. J., Pennington, B. F., & DeFries, J. C. (1995). Quantitative trait locus for reading disability: A correction. *Science, 268,* 1553.

Castellanos, F. X., Lau, E., Tayebi, N., Lee, P., Long, R. E., Giedd, J. N., et al. (1998). Lack of an association between a dopamine-4 receptor polymorphism and attention-deficit/hyperactivity disorder: Genetic and brain morphometric analyses. *Molecular Psychiatry, 3,* 431–434.

Cook, E., Stein, M., Krasowski, M., Cox, N., Olkon, D., Kieffer, J., et al. (1995). Association of attention deficit disorder and the dopamine transporter gene. *American Journal of Human Genetics, 56,* 993–998.

Craddock, N., & Jones, I. (2001). Molecular genetics of bipolar disorder. *British Journal of Psychiatry, 178,* S128-S133.

Daly, G., Hawi, Z., Fitzgerald, M., & Gill, M. (1999). Mapping susceptibility loci in attention deficit hyperactivity disorder: Preferential transmission of parental alleles at DAT1, DBH, and DRD5 to affected children. *Molecular Psychiatry, 4,* 192–196.

Decker, S. N., & Vandenberg, S. G. (1985). Colorado twin study of reading disability. In D. B. Gray & J. F. Kavanagh (Eds.), *Biobehavioral measures of dyslexia* (pp. 123–135). Parkton, MD: York Press.

DeFries, J. C. (1985). Colorado reading project. In D. B. Gray & J. F. Kavanagh (Eds.), *Biobehavioral measures of dyslexia* (pp. 107–122). Parkton, MD: York Press.

DeFries, J. C., & Alarcón, M. (1996). Genetics of specific reading disability. *Mental Retardation and Developmental Disabilities Research Reviews, 2,* 39–47.

DeFries, J. C., & Decker, S. N. (1982). Genetics aspects of reading disability: A family study. In R. N. Malatesha & P. G. Aaron (Eds.), *Reading disorders: Varieties and treatments* (pp. 255–279). New York: Academic Press.

DeFries, J. C., & Fulker, D. F. (1985). Multiple regression analysis of twin data. *Behavior Genetics, 15,* 467–473.

DeFries, J. C., & Fulker, D. F. (1988). Multiple regression analysis of twin data: Etiology of deviant scores versus individual differences. *Acta Geneticae Medicae et Gemellologiae: Twin Research, 37,* 205–216.

DeFries, J. C., & Gillis, J. J. (1993). Genetics of reading disability. In R. Plomin & G. E. McClearn (Eds.), *Nature, nurture, and psychology* (pp. 121–145). Washington, DC: American Psychological Association.

DeFries, J. C., Hegmann, J. P., & Weir, M. W. (1966). Open-field behavior in mice: Evidence for a major gene effect mediated by the visual system. *Science, 154,* 1577–1579.

DeFries, J. C., Vogler, G. P., & LaBuda, M. C. (1986). Colorado family reading study: An overview. In J. L. Fuller & E. C. Simmel (Eds.), *Behavior genetics: Principles and applications II* (pp. 25–56). Hillsdale, NJ: Erlbaum.

Dunn, L. M., & Markwardt, F. C. (1970). *Examiner's manual: Peabody Individual Achievement Test.* Circle Pines, MN: American Guidance Service.

Eisenberg, J., Mei-Tal, G., Steinberg, A., Tartakovsky, E., Zohar, A., Gritsenko, I., et al. (1999). Haplotype relative risk study of catachol-O-methyltransferase and attention deficit hyperactivity disorder (ADHD): Association of the high enzyme activity Val allele with ADHD impulsive-hyperactive type. *American Journal of Medical Genetics (Neuropsychiatric Genetics), 88,* 497–502.

Fagerheim, T., Raeymaekers, P., Tonnessen, F. E., Pedersen, M., Tranebjaerg, L., & Lubs, H. A. (1999). A new gene (DYX3) for dyslexia is located on chromosome 2. *Journal of Medical Genetics, 36,* 664–669.

Falconer, D., & MacKay, T. (1996). *Introduction to quantitative genetics* (4th ed.). London: Longman Scientific and Technical.

Faraone, S. V., Biederman, J., Weiffenbach, B., Keith, T., Chu, M. P., Weaver, A., et al. (1999). Dopamine D4 gene 7-repeat allele and attention-deficit hyperactivity disorder. *American Journal of Psychiatry, 156,* 768–770.

Faraone, S. V., Tsuang, M. T., & Tsuang, D. W. (1999). *Genetics of mental disorders.* New York: Guilford Press.

Finucci, J. M., & Childs, B. (1983). Dyslexia: Family studies. In C. L. Ludlow & J. A. Cooper (Eds.), *Genetic aspects of speech and language disorders* (pp. 157–167). New York: Academic Press.

Fisher, S. E., Francks, C., Marlow, A. J., MacPhie, I. L., Newbury, D. F., Cardon, L. R., et al. (2002). Independent genome-wide scans identify a chromosome 18 quantitative-trait locus influencing dyslexia. *Nature Genetics, 30,* 86–91.

Fisher, S. E., Marlow, A. J., Lamb, J., Maestrini, E., Williams, D. F., Richardson, A. J., et al. (1999). A quantitative-trait locus on chromosome 6p influences different aspects of developmental dyslexia. *American Journal of Human Genetics, 64,* 146–154.

Flint, J. (1999). The genetic basis of cognition. *Brain, 122,* 2015–2031.

Fulker, D., Cardon, L., DeFries, J., Kimberling, W., Pennington, B., & Smith, S. (1991). Multiple regression analysis of sib-pair data on reading to detect quantitative trait loci. *Reading and Writing: An Interdisciplinary Journal, 3,* 299–313.

Gayán, J., Smith, S. D., Cherny, S. S., Cardon, L. R., Fulker, D. W., Bower, A. M., et al. (1999). Quantitative trait locus for specific language and reading deficits on chromosome 6p. *American Journal of Human Genetics, 64,* 157–164.

Geschwind, N. (1981). A reaction to the conference on sex differences and dyslexia. In A. Ansara., N. Geschwind, A. Galaburda, M. Albert, & N. Gartrell (Eds.), *Sex differences in dyslexia* (pp. xiii–xviii). Towson, MD: Orton Dyslexia Society.

Gilger, J., Pennington, B. F., & DeFries, J. C. (1991). Risk for reading disability as a function of parental history in three family studies. *Reading and Writing, 3,* 205–217.

Grigorenko, E., Wood, F., Meyer, M., Hart, L., Speed, W., Shuster, A., et al. (1997). Susceptibility loci for distinct components of developmental dyslexia on chromosomes 6 and 15. *American Journal of Human Genetics, 60,* 27–39.

Grigorenko, E. L., Wood, F. B., Meyer, M. S., Pauls, J. E. D., Hart, L. A., & Pauls, D. L. (1998). Linkage studies suggest a possible locus for dyslexia near the Rh region on chromosome 1 [Abstract]. *Behavior Genetics, 28,* 470.

Haseman, J., & Elston, C. (1972). The investigation of linkage between a quantitative trait and a marker locus. *Behavior Genetics, 2,* 3–19.

Idury, R. M., & Cardon, L. R. (1997). A simple method for automated allele binning in microsatellite markers. *Genome Research, 7,* 1104–1109.

Knopik, V. S., Smith, S. D., Cardon, L. R., Pennington, B. F., Gayán, J., Olson, R. K., et al. (in press). Differential genetic etiology of reading component processes as a function of IQ. *Behavior Genetics.*

Kruglyak, L., & Lander, E. S. (1995). Complete multipoint sib pair analysis of qualitative and quantitative traits. *American Journal of Human Genetics, 57,* 439–454.

McGue, M. K. (1992). When assessing twin concordance, use the probandwise and not the pairwise rate. *Schizophrenia Bulletin, 18,* 171–176.

Morris, D. W., Nöthen, M. M., Schulte-Körne, G., Grimm, T., Cichon, S., Vogt, I. R., Müller-Myshok, B., et al. (1999). Genetic linkage analysis with dyslexia: Evidence for linkage of spelling disability to chromosome 15. *European Child and Adolescent Psychiatry, 8*(Suppl.), 56–59.

Pennington, B. F., Gilger, J. W., Pauls, D., Smith, S. A., Smith, S. D., & DeFries, J. C. (1991). Evidence for major gene transmission of developmental dyslexia. *Journal of the American Medical Association, 266,* 1527–1534.

Pennington, B. F., & Lefly, D. L. (2001). Early reading development in children at family risk for dyslexia. *Child Development, 72,* 816–833.

Petryshen, T. L., Kaplan, B. J., & Field, L. (1999). Evidence for a susceptibility locus for phonological coding dyslexia on chromosome 6q13-q16.2. *American Journal of Human Genetics, 65,* 32.

Petryshen, T. L., Kaplan, B. J., Liu, M. F., & Field, L. L. (2000). Absence of significant linkage between phonological coding dyslexia and chromosome 6p23-21.3, as determined by use of quantitative-trait methods: Confirmation of qualitative analyses. *American Journal of Human Genetics, 66,* 708–714.

Plomin, R. (1993). Nature and nurture: Perspective and prospective. In R. Plomin & G. E. McClearn (Eds.), *Nature, nurture, and psychology* (pp. 459–485). Washington, DC: American Psychological Association.

Plomin, R., & Crabbe, J. (2000). DNA. *Psychological Bulletin, 126,* 806–828.

Plomin, R., DeFries, J. C., McClearn, G. E., & McGuffin, P. (2001). *Behavioral genetics* (4th ed.). New York: Freeman.

Plomin, R., & McClearn, G. E. (1993). *Nature, nurture, and psychology.* Washington, DC: American Psychological Association.

Rabin, M., Wen, X. L., Hepburn, M., Lubs, H. A., Feldman, E., & Duara, R. (1993). Suggestive linkage of developmental dyslexia to chromosome 1p34-p36. *Lancet, 342,* 178.

Riley, B. P., & McGuffin, P. (2000). Linkage and association studies of schizophrenia. *American Journal of Medical Genetics, 97,* 23–44.

Schulte-Korne, G., Grimm, T., Nöthen, N. M., Muller-Myshok, B., Cichon, S., Vogt, I. R., et al. (1998). Evidence for linkage of spelling disability to chromosome 15. *American Journal of Human Genetics, 63,* 279–282.

Semrud-Clikeman, M., Biederman, J., Sprich-Buckminster, S., Lehman, B. K., Faraone, S. V., & Norman, D. (1992). Comorbidity between ADDH and LD: A review and report in a clinically referred sample. *Journal of the American Academy of Child and Adolescent Psychiatry, 31,* 439–448.

Smith, S. D., Kimberling, W. J., & Pennington, B. F. (1991). Screening for multiple genes influencing dyslexia. *Reading and Writing, 3,* 285–298.

Smith, S. D., Kimberling, W. J., Pennington, B. F., & Lubs, H. A. (1983). Specific reading disability: Identification of an inherited form through linkage analysis. *Science, 219,* 1345–1347.

Stevenson, J., Graham, P., Fredman, G., & McLoughlin, V. (1984). The genetics of reading disability. In C. J. Turner & H. B. Miles (Eds.), *The biology of human intelligence* (pp. 85–97). Nafferton, England: Nafferton Books.

Stevenson, J., Graham, P., Fredman, G., & McLoughlin, V. (1987). A twin study of genetic influences on reading and spelling ability and disability. *Journal of Child Psychology and Psychiatry, 28,* 229–247.

Stevenson, J., Pennington, B. F., Gilger, J. W., DeFries, J. C., & Gillis, J. J. (1993). Hyperactivity and spelling disability: Testing for shared genetic aetiology. *Journal of Child Psychology and Psychiatry, 14,* 1137–1152.

Turri, M. G., Datta, S., DeFries, J. C., Henderson, N. D., & Flint, J. (2001). QTL analysis identifies multiple behavioral dimensions in ethological tests of anxiety in laboratory mice. *Current Biology, 11,* 725–734.

Wadsworth, S. J., Knopik, V. S., & DeFries, J. C. (2000). Reading disability in boys and girls: No evidence for differential genetic etiology. *Reading and Writing: An Interdisciplinary Journal, 13,* 133–145.

Wadsworth, S. J., Olson, R. K., Pennington, B. F., & DeFries, J. C. (2000). Differential genetic etiology of reading disability as a function of IQ. *Journal of Learning Disabilities, 33,* 192–199.

Wechsler, D. (1974). *Examiner's manual: Wechsler Intelligence Scale for Children–Revised.* New York: Psychological Corporation.

Wechsler, D. (1981). *Manual for the Wechsler Adult Intelligence Scale–Revised.* San Antonio, TX: Psychological Corporation.

Willcutt, E. G., Chhabildas, N., & Pennington, B. F. (2001). Validity of the *DSM–IV* subtypes of ADHD. *The ADHD Report, 9,* 2–5.

Willcutt, E. G., & Pennington, B. F. (2000). Comorbidity of reading disability and attention-deficit/hyperactivity disorder: Differences by gender and subtype. *Journal of Learning Disabilities, 33,* 179–191.

Willcutt, E. G., Pennington, B. F., Boada, R., Tunick, R. A., Ogline, J., Chhabildas, N. A., & Olson, R. K. (2001). A comparison of the cognitive deficits in reading disability and attention-deficit/hyperactivity disorder. *Journal of Abnormal Psychology, 110,* 157–172.

Willcutt, E. G., Pennington, B. F., & DeFries, J. C. (2000a). Etiology of inattention and hyperactivity/impulsivity in a community sample of twins with learning difficulties. *Journal of Abnormal Child Psychology, 28,* 149–159.

Willcutt, E. G., Pennington, B. F., & DeFries, J. C. (2000b). A twin study of the etiology of comorbidity between reading disability and attention-deficit/hyperactivity disorder. *American Journal of Medical Genetics (Neuropsychiatric Genetics), 96,* 293–301.

Willcutt, E. G., Pennington, B. F., Smith, S. D., Cardon, L. R., Gayán, J., Knopik, V. S., et al. (2002). Quantitative trait locus for reading disability on chromosome 6p is pleiotropic for attention deficit hyperactivity disorder. *American Journal of Medical Genetics (Neuropsychiatric Genetics), 114,* 260–268.

Zerbin-Rüdin, E. (1967). Kongenitale Wortblindheit oder spezifische Dyslexie [Congenital word-blindness]. *Bulletin of the Orton Society, 17,* 47–56.

14

Epistasis and the Genetics of Complex Traits

Elena L. Grigorenko

It is well known that classical quantitative genetic theory is largely a theory of additive effects. This basic assumption does not mean that *all* genetic effects are additive but that the additive components of genotype–phenotype covariance are of particular importance. Until recently, the role of interactive effects between genes *(epistasis)* has been viewed as secondary or as a noise term relative to the additive effects.

Historically, there have been two differential theoretical emphases on the relevant importance of the additive and interactive effects of different loci that, in turn, differentially defined the role of epistasis—one taken by Fisher (1918) and the other by Wright (1965; see Provine, 1971, 1986). Specifically, Fisher assumed that the interactions of alleles at different loci are relatively unimportant and that the effect of an allele and its evolutionary future can be adequately summarized by its effect averaged across genetic backgrounds. Wright, in contrast, assumed that allele interactions are very important in genetic architecture and that they define both the effect of an allele and the evolutionary fate of this allele. These two differential emphases are not mutually exclusive—in fact they are themselves additive. The more one learns about specific realizations of genetic mechanisms in particular complex conditions, the more one understands the importance of various types of genetic influences. There are possibilities for numerous types of combinations. For example, it is possible that epistatic interactions might be widespread but account for a small portion of genetic variance (Falconer & Mackay, 1996). However, it is also possible that, while the additive contribution of a particular mutation in a given gene might be relatively low, the contribution of interactive effects involving this gene might be of substantial magnitude (MacCluer, Blangero, Dyer, & Speer, 1997).

Nevertheless, this differential accentuation introduced by Fisher and Wright has had a major impact on both modern evolutionary theory and quantitative genetics. The Fisherian view and, therefore, the emphasis on the role of additive effects, have dominated the field of quantitative genetics for over 50 years. This, of course, does not mean that epistatic effects have not been studied; they have, both theoretically (e.g., Mather & Jinks, 1971) and empirically (Game & Cox, 1972). Yet, the ratio of studies of additive and nonallelic interactive genetic effects has been hundreds to few. Among others, one important reason for this lack of balance is related to power considerations (i.e., statistical modeling of epistatic effects requires large samples).

Although the field, since the 1930s, has primarily focused on the description and quantification of additive genetic effects, many data have accumulated stressing the importance of gene interaction for many evolutionary processes (e.g., the evolution of sex, the evolution of recombination, mutation, inbreeding depression, and the evolution of reproductive isolation among species; for a review, see Wolf, Brodie, & Wade, 2000). Yet, the role of epistatic interactions in the etiology of complex traits has been barely studied.

The more one learns about gene interactions, the more one appreciates the complexity of interactive processes. For example, certain combinations of alleles at different loci interact to cause phenotypes that differ substantially from those anticipated by considering the effects of various alleles in isolation. The main ramification here is that little is known about the nature of a particular interaction from the mean interaction effect. Thus, it appears that, as a field, in using the "additive" approach to interaction effects, we might have been underestimating the importance of gene interactions.

Epistatic effects and epistasis-related variance have recently received theoretical (e.g., Otto & Feldman, 1997) and empirical (e.g., Elena & Lenski, 1997) attention. Several articles have stressed the evolutionary importance of epistasis, explaining such phenomena as mutational load, evolution of recombination, drift of and interactions among deleterious mutations, and speciation. Peck and Waxman (2000), for example, referred to epistasis as the key to the puzzle of the present state of nature. Many authors (Frank, 1999; Long et al., 1995; Omholt, Plahte, Oyehaug, & Xiang, 2000) have referred to epistasis as one of the essential features of complex genetic networks.

In this chapter, I attempt to conceptualize epistasis in a way that encompasses the complex interplay among genes that must exist in real systems underlying the genetics of complex neuropsychiatric conditions. This conceptualization is not novel in any way—I simply review the literature available, primarily in the fields of evolutionary genetics and clinical genetics and relate it to the current status of understanding the genetic basis of human abilities and disabilities. Specifically, I present many epistasis-related concepts, ways of quantifying epistasis, and data (so far rather limited) on the role of gene interaction in the genetics of complex human traits. I conclude the chapter with a brief discussion of the potential role of epistasis in human abilities and disabilities.

Definitions

A complex genetic trait is a trait whose phenotypic variation is clearly influenced by genetic variation but that does not obey a simple Mendelian inheritance pattern. It has been long recognized in human genetics, molecular psychiatry, neuropsychiatric genetics, and behavior genetics that the vast majority of traits of interest for these fields of study are complex in this sense (e.g., Czeizel, 1989). These deviations from Mendelian patterns can be accounted for by the interactions between genes or between genes and environmental factors involved in the manifestation of a given trait.

The interactions between genes are referred to as genetic *epistasis*. Thus, by definition, epistasis arises only within a system with complex genetic architecture (i.e., that involves enough genes for interactions to take place). At this point, we do know that most complex traits involve complex genetic architecture; what we do not know (or are only beginning to understand) is how often these genetic systems develop interactions and what role these interactions play in the mapping from genotype to phenotype.

According to Phillips (1998), the term *epistasis* was introduced by William Bateson at the beginning of the 20th century to refer to the masking expression of one locus by the alleles at another locus. Subsequently, Fisher (1918) expanded this definition by including quantitative differences among genotypes—he referred to any deviation from the additive combination of single-locus genotype as *epistacy*. Since the beginning of the century, the field has developed many definitions of epistasis (e.g., Fenster, Galloway, & Chao, 1997; Wagner, Laubichler, & Bagheri-Chaichian, 1998), including but not limited to the following: (a) The effect of the allele in one locus depends on which allele is present in the other locus (or other loci); (b) the effect of one locus depends on the content of the other locus (or other loci); (c) the influence of the genotype at one locus on the effects of a mutation at the other locus (or other loci); and (d) the deviation of two-locus genotypic values from the sum of the contributing single-locus genotypic values. Trying to reduce the concept of epistasis to a more precise meaning, Phillips (1998) recommended not using the word *epistasis* as a generic term to describe gene interactions. Currently, there are multiple classifications of epistatic effects (e.g., Wolf et al., 2000), which lead to a realization that there is no generic type of epistasis and that consequences of each type of interaction can be quite different and therefore should be considered separately (Phillips, 1998). For the purpose of the discussion in this chapter, I mention only one distinction often seen in the evolutionary literature: Epistasis is said to be *synergistic* or *reinforcing* if the effect of a mutation is stronger in the presence of other mutations than in their absence and is said to be *antagonistic* or of *diminishing returns* when the effect of mutations is weaker in the presence of other mutations than in their absence (J. F. Crow & Kimura, 1970). The purpose of my mentioning this distinction here is to make it obvious (without bringing in any mathematical derivations) that epistatic effects can have different signs and, although independently may be statistically significant and substantial, when averaged, can effectively cancel each other out.

Measures of Epistasis

Given the variety of its definitions, one can easily deduce that there are many ways to quantify epistasis. Again, in appreciation of the variety of these measures, I limit my discussion here to only three levels: (a) at the level of genotypic values, where epistatic effects emerge from the molecular and physiological context in which genes are expressed; (b) at the population level, where epistatic variance is estimated as a fraction of the total amount of genotypic variation found in the population; and (c) at the level of specific samples analyses, where, effectively, some features of approaches in the first and second levels are combined.

With regard to the first estimate, the measure of epistasis can be obtained within an individual organism and describes variance between different sets of loci in their interactive effects. Thus, no averaging is technically correct—this measure of epistasis dwells at the functional level of these interactions. Developmental genetics has both observed and predicted the variability with which different loci interact (e.g., Avery & Wasserman, 1992). The studies that have reported variance in epistatic effects used the technology of isolating genetic changes against a fixed background—specifically, using targeted mutations to pinpoint the effects of specific loci (Clark & Wang, 1997; de Visser, Hoekstra, & van den Ende, 1997; Elena & Lenski, 1997). These methods result in precise estimates of epistatic effects for a specific allelic combination but fail to make a link between the observed epistatic effects and the overall population context in which these interactions arise and exist.

With regard to the second estimate of epistasis (i.e., epistatic variance), researchers have raised several issues. First, these estimates are dependent on gene frequency in a given population (Phillips, Otto, & Whitlock, 2000). Second, the estimates are usually confounded by different sources of genetic effects: The tremendous number of potential genetic backgrounds overwhelms the detection of the interaction between any specific locus pair, with various interactions, on average, canceling each other (e.g., there could be substantial amounts of genetic interaction, but the value of epistatic variance due to the averaging over the entire genome might be zero or near zero; Whitlock, Phillips, Moore, & Tonsor, 1995). As mentioned earlier, the Fisherian measures that dominate classical quantitative genetics have been defined in such a way as to make epistasis primarily contribute to the "additive" effects (Cheverud & Routman, 1995). Correspondingly, the statistical method of detecting epistasis might produce results according to which there is no epistatic effect when in fact there is. Third, the detection of the epistatic component of the variance calls for large samples and specialized designs that might not be practical (Wade, 1992). Fourth, these estimates of the epistatic components of the variance are usually so removed from the physiological meaning of genetic interaction that they are often viewed as useless in establishing the link between this statistic and the corresponding biological mechanism.

As mentioned earlier, the third approach combined elements of the first approach (by "working" with measured genotypes) and elements of the second approach (by making certain assumptions about the population/sample context in which epistasis unfolds). With regard to this methodology, two subtypes of this approach should be mentioned here: the linkage marker approach and the candidate locus approach.

The linkage marker approach strategy measures genotypes only as markers of regions of chromosomes and makes no claim that the markers themselves directly contribute to the phenotypic variation of interest. During the 1990s, significant advances in genomics and statistical genetics have permitted the unbiased estimation of genotypic values for loci with small to moderate effects on a phenotype (quantitative trait loci [QTLs]; Cordell, Todd, & Lathrop, 1998; Elston, Buxbaum, Jacobs, & Olson, 2000; Ghosh & Majumber, 2000; Haley & Knott, 1992; Lander & Botstein, 1989; Lynch & Walsh, 1998; Zeng, Kao, & Basten, 1999). These advances, in turn, led to the development of methodologies of testing for epistatic interactions using linear combinations of genotypic value error variances (Routman & Cheverud, 1997). Thus, one can localize QTLs into chromosomal regions and test epistatic interactions between these regions. Two points have to be made with regard to this strategy. First, within this approach, for epistatic interactions even to be estimated, the contributing regions should independently show a statistically significant (or at least borderline-significant) linkage. Note, however, that statistically it is possible to have a significant interaction in absence of the main effects of the interacting variables. Templeton, Sing, and Brokaw (1976) demonstrated that much epistasis exists between chromosomal regions that show no significant effects; consequently, studies that attempt to detect epistasis only after detecting main effects are strongly biased against detecting epistasis (Templeton, 2000). Second, because of mathematical, statistical, and computational limits, the overwhelming majority of the studies test only for pairwise interactions (e.g., Goldstein & Andrieu, 1999; Yang, Khoury, Sun, & Flanders, 1999). Very few studies have attempted to measure three-way interactions (e.g., Templeton et al., 1976); these studies revealed the presence of three-way interactions that were not inferable from the corresponding two-way interactions.

The second strategy within this methodology (the candidate locus approach) relies on prior knowledge of the phenotype of interest and plausible candidate loci. Genetic variation at such candidate loci can then be tested for association with phenotypic variation. Similar to the linkage strategy described above, this strategy does not automatically detect epistasis—one has to follow up Fisherian additive tests with tests targeted to do so. This approach shares limitations with linkage analyses but has a couple of distinct advantages. First, specific loci are identified, not on the basis of statistical indicators but on prior information that assumes the association between the trait of interest and some specific biological mechanism (i.e., providing insight into the specific genetic architecture of the trait). Second, theoretically, various genotypes at a candidate locus can be fixed as "states" so that potential epistatic interactions with other candidate loci can be investigated (irrespective of whether these others have significant statistical effects themselves).

Having defined epistasis and briefly reviewed the major methods used to detect it, I now review the empirical evidence supporting the importance of epistasis in research on complex human traits.

Illustrations of Epistasis

The illustrations presented below constitute a selective collection of examples from both animal and human literature. Note that the idea of screening for epistasis is rather new (just review the dates of the references), so there are not that many studies doing these screens, and the majority of these studies have been conducted in clinical genetics and animal genetics. Despite the diverse contents of these illustrations, there is one common thread running through all of them: Nothing about any of these examples indicated the involvement of epistasis a priori. In other words, all of these discoveries of the presence of epistasis are empirically, not theoretically, driven.

With regard to animal studies, let me describe just two studies. I refer readers for more examples to a wonderful book edited by Wolf et al. (2000).

Long et al. (1995) found a large number of significant epistatic interactions among the observed QTLs in a study of bristle-number variation in *Drosophila melanogaster*. An interesting outcome of this study is the amount of variability within these interactions: There were interactions of different signs even within the same chromosome regions (this finding indicates why it might be so difficult to detect genetic interactions statistically).

Routman and Cheverud (1997) studied body weight QTLs in mice and discovered extensive interactions within QTL regions; these interactions also demonstrated variability from locus to locus. Specifically, 76 polymorphic markers were placed at equal intervals on the 19 mouse autosomes. The authors carried out all pairwise comparisons and revealed a total of 216 epistasis tests significant at the .05 level. These instances of epistasis were 2.3 times as common as would be expected due to chance alone (93.5; Cheverud, 2000). However, only 3 of these 216 significant values exceed the appropriate Bonferroni corrections (the conservative nature of these corrections makes these results unlikely to be false positives).

Thus, these two examples from the animal work suggest that epistatic effects are common and that their detection is often missed with the application of traditional statistical methodologies. So, what is known about the role of epistasis in complex human traits?

The role of epistasis in various human phenotypes becomes immediately apparent in the consideration of such a classical Mendelian trait as sickle-cell anemia. The textbook version of the disease presents it as a single nucleotide trait: The phenotype arises when an individual is a carrier of two copies of the mutant *S* allele, which is characterized by a single nucleotide change in the sixth codon of the β-chain of hemoglobin (coded by β-locus). What is remarkable, however, is the wide variability in clinical severity of the disease among people who are homozygous for the *S* allele (i.e., affected by the conventional definition of sickle-cell anemia). It appears that the main source of this variability is epistasis (Berry, Grosveld, & Dillon, 1992; Patrinos, Loutradianagnostou, & Papadakis, 1996). Specifically, the *S* allele is observed at high frequency in many Mediterranean populations, but individuals homozygous for this allele often manifest mild phenotypes of the disease. Studies of the physiology of the disease (e.g., Berry et al., 1992) have demonstrated that persistence of fetal hemoglobin (coded by the γ-locus closely linked to the β-locus) into the adult can amend the severity of sickle-cell anemia. The normal physiology of hemoglobin production assumes that the γ-genes are turned off after birth and the β-gene is activated; this mechanism ensures the transition from fetal to adult hemoglobin. However, researchers have found several mutations in or near γ-genes that can cause the persistence of fetal hemoglobin into adulthood. Thus, the effect of two copies of the *S* allele can be ameliorated by the mutant copies of the γ-genes. Such mutations have been found in Greek (Patrinos et al., 1996) and Arabian (el-Hazmi, Warsey, Addar, & Babae, 1994) populations. Thus, this apparently simple Mendelian system is in reality a complex system involving epistatic interactions.

If epistasis is important for “simple Mendelian traits,” its importance for complex traits must be overwhelming. But, given that the phenotype is influenced by two or more loci, do interactions among the component loci play a significant role in the mapping from genotype to phenotype? Indeed, there is more and more evidence that the role of interactions between genes is equally or, in some cases, more important than the main effects of these genes. Specifically, within the last decade, there has been a string of research pointing out the importance of gene–gene interaction in the development and manifestation of complex diseases such as myocardial infarction (Fomicheva, Gukova, Larionova-Vasina, Kovalev, & Schwartz, 2000), Alzheimer's disease (Bagli et al., 2000), asthma and atopy (Barnes, 1999), early coronary artery disease (CAD; Alvarez et al., 1998), colorectal cancer (Bell et al., 1995), childhood acute lymphoblastic leukemia (Krajinovic, Richer, Sinnett, Labuda, & Sinnett, 2000), lung cancer (Hou et al., 2001), familial melanoma (Goldstein, Goldin, Dracopoli, Clark, & Tucker, 1996), inflammatory bowel disease (Cho et al., 1998), and polycystic ovary syndrome (Diamanti-Kandarakis, Bartzis, Bergiele, Tsianateli, & Kouli, 2000). As a detailed illustration of the role of epistasis in a non-Mendelian trait, let us consider the example of CAD (as summarized in Templeton, 2000).

CAD is of great public health interest because it is the most common cause of death in the developed world. According to many phenotype decomposition and segregation studies, CAD is inherited as a polygenic trait with a strong additive component (e.g., Sing & Moll, 1990). Thus, under Fisherian variance-components assumptions, the role of epistasis in the genetic variation of CAD is insignificant.

Over recent years, many candidate loci have been identified as principal risk factors for CAD and associated phenotypes. One of these loci is apolipoprotein E (ApoE) (Cummings & Robertson, 1984; Sing & Davignon,

1985; Utermann, Kindermann, Kaffarnik, & Steinmetz, 1984). There are three common alleles in this locus: ε_2 (generally associated with lowered total serum and low-density lipoprotein cholesterol levels), ε_3, and ε_4 (generally associated with elevated total serum and lower density lipoprotein cholesterol levels). It has been shown, by means of cohort studies, that the ε_4 allele is a risk factor, because CAD is a disproportionately frequently observed cause of death for individuals who bear the ε_4 allele (Stengard et al., 1995). According to Fisherian variance-components applications, the additive variance of ApoE is responsible for 6.9% of the total phenotypic variation in cholesterol levels and about 14% of the total polygenic variance (Sing & Davignon, 1985). Another locus of interest is the low-density lipoprotein receptor (LDLR) locus. This locus has two common alleles, A_1 and A_2, and has been shown to have a small effect on the manifestation of CAD. However, Pedersen and Berg (1989, 1990) and Berg (1990) conducted direct examinations of interactions between these two loci and showed that the ε_4 allele of ApoE is associated with significantly elevated cholesterol levels *only* in individuals who are homozygous (A_2/A_2) at the LDLR locus. This epistasis is not surprising because these proteins have been known to interact directly (de Knijff, Vandenmaagdenberg, Frants, & Havekes, 1994; Zannis & Breslow, 1985).

It is important to mention that ApoE is also one of the risk factors for Alzheimer's disease (Bedlack, Strittmatter, & Morgenlander, 2000). Specifically, the ε_4 allele has been named as one of the main alleles contributing additively to the phenotype of Alzheimer's disease (Saunders, 2000; see also Plomin, DeFries, Craig, & McGuffin, chapter 1, this volume). This phenotype, in turn, has been shown to be related to complex epistatic interactions of other candidate loci (Bagli et al., 2000; Kamboh et al., 1997; Kamboh, Sanghera, Ferrell, & Dekosky, 1995; Lehmann, Johnston, & Smith, 1997; Morgan et al., 1997; Okuizumi et al., 1995). Thus, the phenotypes of Alzheimer's disease and CAD are both associated with the ε_4 allele as a common focal allele for epistasis, and the evidence available so far indicates that this allele is involved in different networks of interacting alleles for these disorders. In summary, the ApoE locus is one of the additive contributors to the phenotypes of CAD and Alzheimer's disease, but further investigation has revealed the involvement of this locus in a complex and tangled web of pleiotropic effects and epistasis.

Obviously, ApoE is not the only example of a gene involvement in such complex webs. Ukkola and colleagues have studied epistatic effects among a set of adrenergic receptor genes with regard to obesity phenotypes and have reported that these effects contribute to the phenotypic variability in abdominal obesity and plasma lipid and lipoprotein (Ukkola et al., 2000). The same group has also investigated associations between epistatic effects among adrenergic genes, the glucocorticoid receptor, and lipoprotein lipase and phenotypes of plasma insulin and lipid levels (Ukkola et al., 2001). The researchers have revealed the presence of significant contributions of first-order and higher order epistatic interactions among the studied genes in the absence of main effects of these genes.

Botto and Mastroiacovo (1998) explored the role of epistatic interaction in the etiology of neural tube defect. Because of the protective role of folic acid in the development of neural tube, two genes encoding enzymes involved in folate-related metabolism have been investigated in this study. The results suggest that coexisting mutations at two loci might influence the risk of neural tube defect and that the combined effects of these mutations might be larger than expected on the basis of main effects of each locus. Specifically, homozygosity for the harmful mutation in the first gene of interest (5,10-methylene-tetrahydrofolate reductase) was associated with a twofold increased risk for the defect, whereas homozygosity for the mutation in the second gene of interest (cystathionine-β synthase) alone was not a risk factor. However, homozygous individuals for the mutations at both loci had a fivefold greater risk for the neural tube defect than those with a reference genotype.

Cox et al. (1999) investigated epistatic interactions between a major susceptibility gene for noninsulin-dependent diabetes mellitus 1 on chromosome 2 (NIDDM1) and other autosomal regions containing genes for the associated phenotype, maturity onset diabetes of the young (specifically, the gene CYP19 on chromosome 15). The indicators of linkage for CYP19 were substantially stronger when conditioned by evidence for linkage to NIDDM1; thus, the conclusion was drawn that in the studied sample of Mexican American families there is a substantial epistatic interaction of NIDDM1 on chromosome 2 and a locus near CYP19 on chromosome 15.

Nelson, Kardia, Ferrell, and Sing (1998, as cited in Templeton, 2000) conducted a study of genetic variation in a human population using 18 markers in six candidate regions among which were ApoE and LDLR (see above for description). The phenotypes of interest in this research were various lipid-related traits. The goal was to detect all pairwise combinations of markers that explained a significant proportion (with $p < .01$) of the observed phenotypic variance. (Note that this study has been biased against the detection of epistasis to begin with, because Nelson et al. screened only for first-order interactions.) To control for false positives, Nelson et al. subjected the results of the initial search to cross-validation by bootstrapping (using statistical procedures to establish the replicability of the findings), and only then were main effects and interaction terms estimated. They found that the validated pairs of loci that explained the greatest amount of phenotypic variability showed strong evidence for epistasis for most traits examined. Moreover, they pointed out that the two-stage methodology (initial identification of genes contributing additively, followed by a secondary search for interactions among only these genes) routinely fails to identify the combinations of loci associated with the greatest amount of quantitative trait variability. For example, the percentage of variability in serum triglyceride levels explained by main effects of ApoB and ApoE in females is 1.6% and 1.8%, respectively. However, the epistatic effects between alleles at ApoB and ApoE explain 8.2% of the variance. In addition, Nelson et al. encountered many situations in which there were no statistically significant effects at one or both loci in the presence of substantial epistatic effects originating from these loci.

To my knowledge, there are only two published examples of the role of epistasis in human nonclinical traits. One of these examples points out the presence of epistasis in the genetic etiology of the personality traits Harm Avoidance, Reward Dependence, Novelty Seeking, and Persistence (Cloninger et al., 1998). In this study, the hypothesis of the presence of epistatic interactions between various genes involved in the manifestation of personality traits was generated a priori, based on the observed discrepancies in heritability estimates for personality traits obtained in twin samples (ranging from 40% to 60%) and those estimates obtained in samples of biological parents and their siblings adopted away (ranging from 20% to 30%). This discrepancy can be explained by many factors: within-family environment, dominant genetic effects, or epistatic genetic effects. In this study, based on some previous suggestions in the literature (e.g., Finkel & McGue, 1997; Plomin, Corley, Caspi, Fulker, & DeFries, 1998) and due to the availability of a vast amount of genetic information originated from a whole-genome screen performed with 291 markers, Cloninger et al. concentrated their attention on the effects of epistasis. The results provided strong evidence that a genetic locus on 8p21-23 contributed to the phenotypic variance in Harm Avoidance additively and explains 20%–30% of this variance. However, when epistatic interactions with other loci with main effects below traditionally accepted levels of significance were considered (those on chromosome 21q21-22.1, 18p, and 20p), the amount of the explained variance reached 54%–66%.

The other example of the search for epistasis in an attempt to unveil the genetic etiology of complex traits comes from studies in developmental dyslexia. In this study, a number of dyslexia-related cognitive processes have been investigated for linkage with a set of markers on chromosome 1 (Grigorenko et al., 2001). These dyslexia-related processes are assumed to form a complex hierarchical structure such that a particular outcome of developmental dyslexia is not solely predicted by any specific process independently but rather a result of a complex combination of weaknesses in many processes. It has been long noticed that although these processes are related to each other, the correlations are of surprisingly modest magnitude (Hulme & Joshi, 1998). Moreover, there is evidence in the literature that there are different subtypes of dyslexia that assume more expressed weaknesses in some processes while other processes appear to be relatively intact (Duane, 1999). Furthermore, all reading-related processes appear to be heritable, although heritability estimates vary for specific processes (Gayán & Olson, 1999). On the basis of this information Grigorenko et al. suggested that the observed complex patterns of phenotypic correlations between reading-related processes can be understood, at least partially, as a complex net of pleiotropic and epistatic effects produced by the genes involved in the manifestation of these phenotypes. This study was a first attempt at disentangling this net of interactions. The analyses revealed many main effects, suggesting that some reading-related processes of the dyslexia spectrum (specifically, Phonemic Awareness and Rapid Naming) were linked to the region on chromosome 1. However, these and other processes have been previously linked to the 6p21.3 region (see Willcutt et al., chapter 13, this volume). Both regions contain genes with similar biological functions, and therefore the hypothesis that these genes might interact appeared to be biologically plausible. Grigorenko et al. tested this hypothesis by carrying out many epistatic interactions analyses. When epistatic interactions were considered, the results differentiated the composition of genetic contributions to different reading-related phenotypes. Specifically, taking into account the interaction between the loci on chromosome 6 and 1 resulted in an almost fivefold increase in the linkage scores for one of the dyslexia-related cognitive processes (Rapid Naming). Thus, it is plausible that there is an epistatic interaction between the chromosome 6 and 1 genes contributing to the manifestation of deficient Rapid Naming. Quite on the contrary, the modeling of epistatic interaction for the phenotype of Phonemic Awareness did not result in the change in linkage scores—the main effects of interest continued to be on both chromosomes 6 and 1, but there was no evidence of interaction. The fact that the same loci appear to be contributing to different phenotypes suggests the presence of pleiotropic effects in the manifestation of reading-related deficiencies.

Thus, epistatic effects appear to be rather common, despite the numerous biases that exist against their detection. It appears that the literature on epistasis is rather scarce, not because there are no such effects on complex traits but because these effects are not sought for routinely. Frankel and Schork (1996) suggested that the primary reason why many complex traits are not reported to be associated with or to be products of epistatic networks is simply that many researchers use designs or analytical methods that exclude or that are unable to detect epistasis. For example, the Fisherian quantitative-genetic measures operate in such a way that the fit is estimated first by a test for an additive component, then by a test for additive and dominant components, and only then are models tested that are designed to attribute the leftover variance to epistatic effects. This "doggy bag" (let's-go-for-leftovers) approach results in the impression that epistatic effects are difficult to detect and that they are weak in comparison with additive effects (Cheverud & Routman, 1995; Goodnight, 1995; Wade, 1992).

DeRisi, Iyer, and Brown (1997) studied a single mutation in yeast that resulted in altering the expression of 355 other genes. In total, yeast has about 6,000 genes, so, through epistasis, this single mutation affects 5% of the yeast genome. With the availability of such dramatic evidence, there is a definite justification for paying careful attention to epistatic effects in our attempts to understand the genetic etiology of complex human traits.

Considering Future Steps: Epistatic Interactions and Cognitive Abilities and Disabilities

Given the above, what relevant issues can be brought into the discussion of the future research on the etiology of cognitive abilities and disabilities?

The first relevant observation has to do with a necessity for modifying the single-candidate gene approach, which is currently the dominant approach in the genetics of cognitive abilities and disabilities, in favor of a system-based approach. For example, there have been genetic association studies of both the holistic phenotype of attention deficit hyperactivity disorder (ADHD) and its components (inattention and hyperactivity) with many genes involved in the systemic operation of one of the brain neurotransmitters, the dopaminergic system—dopamine receptor genes, the dopamine transporter gene, and selected genes involved in the metabolic pathways of dopamine. Specifically, many researchers have investigated the association between ADHD and the allele 7 in the dopamine receptor gene 4 (DRD4; e.g., La Hoste et al., 1996; Puumala et al., 1996; Tahir et al., 2000). The results of these studies have been mixed—some researchers have been able to establish the association whereas others have failed to support it. To comprehend these findings, meta-analytic data summarizing approaches have been recruited. Specifically, Faraone and colleagues (Faraone, Doyle, Mick, & Biederman, 2001) have carried out such meta-analysis for ADHD and DRD4 allele 7 and have concluded that, although the association between ADHD and DRD4 is small, it is real.

Also, there have been genetic association studies of the genetic variability in other dopamine-related genes. Specifically, researchers have established an association between the dopamine receptor 5 gene and ADHD (Daly, Hawi, Fitzgerald, & Gill, 1999; Tahir et al., 2000). Moreover, there have been many association studies between ADHD and the dopamine transporter (DAT). However, the outcomes of these studies are variable. Specifically, many authors (Cook et al., 1995; Gill, Daly, Heron, Hawi, & Fitzgerald, 1997) have found an association between DAT and ADHD, whereas others have failed to find evidence for this association (Bailey, 1997; Daly et al., 1999). Researchers have also investigated associations between COMT (catechol-O-methyl transferase, one of the enzymes that is important to the initial steps of metabolic transformation of dopamine) and ADHD; similarly, the results of these investigations have been controversial (Eisenberg et al., 1999; Hawi, Millar, Daly, Fitzgerald, & Gill, 2000). The contradictory nature of these findings might possibly be relevant to the discussion above: It is possible that the instability of the results is due to the fact that researchers have looked only for main effects of these genes, not venturing into the world of interactions between the genes. However, the fact that all these genes are involved in dopaminergic pathway and are extremely interactive at the physiological levels is inviting, considering the importance of epistatic effects in the manifestation of the ADHD phenotype.

The second relevant observation has to do with the evidence accumulating within the experimental neuroscience, especially in the work with animals, and has to do with the interactive relationships between systems of neurotransmitters. Recent advances in neuroscience have led to the identification of a large number of neurotransmitter receptors (e.g., there are 24 known receptor subunits in the glutamatergic system) and have enriched our appreciation of how molecular diversity leads to the functional diversity of synaptic transmission in the brain. The more we know about the signaling in the brain, the more we appreciate how distinct the interactive nature of this signaling is. It is not an overstatement to say that all neurotransmitter metabolic pathways are interactive in that they involve many proteins the production of which is controlled by different genes. Moreover, various neurotransmitter systems interact with each other to form the biological basis of the brain circuitry. However, the presence of interactions between different proteins does not assume its interactive etiology at the gene level.

Yet, because neurotransmitters are intimately involved in the mechanisms of learning and memory, the genetic systems of neurotransmitters seem to be good candidates for studying epistatic interactions and their involvement in cognitive abilities and disabilities. Although there are still some unmapped genes involved with various neurotransmitter systems, the majority of these genes have been mapped and cloned. Researchers have long been involved in capturing and describing the effects of various mutations in these genes, thereby looking primarily for the main effects of these mutations. However, there is virtually no research investigating epistasis between different genes involved in given neurotransmitting systems. The evidence presented below does not prove the existence of epistasis but suggests that testing for epistasis might help us understand the genetic etiology of complex human traits.

Over the past couple of decades, the search for the molecular-genetic mechanisms of learning and memory has been focused on experimental models of long-term potentiation (LTP). For example, extensive investigations have repeatedly shown that the induction of LTP in the hippocampus requires the activation of the NMDA (*N*-methyl-D-asparate) receptor (one of the forms of the glutamate receptor; for a review, see Tsien, 2000). In addition, other investigations have shown that experimentally induced synaptic changes involve a whole set of protein kinases and immediate-early genes (Malenka & Nicoll, 1999; Sweatt, 1999).

Tsien et al. (1996) deleted the NMDA receptor subunit 1 (NR1) gene, encoding a core subunit for the NMDA receptor complex in a specific area of the hippocampus. This knock out resulted in a complete loss of NMDA currents and many other losses, including the loss of LTP. This elegant cell-type-specific knock out has stressed the importance of NMDA in LTP. However, as it appears now, the NMDA story has only partially unveiled the genetic mechanism of LTP.

A few observations, made in numerous knock-out studies, are of relevance here. The common feature of these observations is that the modification of the LTP can be accomplished by genetic experiments with other than NMDA-related genes. For example, recent study of the GluR1 gene (a gene, encoding another type of glutamate receptor, AMPA [α-amino-3-hydroxy-5-methyl-4-isoxazole propionic acid]) in knock-out mice reported that these mice have impaired LTP. To add complexity to the picture, researchers have investigated mutations resulting in enhanced LTP. For example, null mutations of another AMPA gene GluR2 results in a markedly enhanced LTP in the presence of multiple behavioral anomalies in experimental mice (Gerlai, Henderson, Roder, & Jia, 1998; Jia et al., 1996).

Consequently, other genetic interventions of signaling enzymes have been reported to lead to enhanced LTP. In transgenic mice, for example, overexpressing a transgene encoding constitutively active mutant Fyn kinase in the forebrain (Lu et al., 1999) elicited significant LTP in the hippocampus.

These experimental models have allowed researchers to appreciate the complexity and interactive nature of the signaling mechanism underlying LTP. This evidence stresses the salience of interactive mechanisms at the level of neurotransmitters for the realization of brain function. Future explorations will be necessary to determine whether this complexity is supported by epistatic networks of relevant genes.

The hypothesis, however, appears to be plausible. Given the amount of interaction between different neurotransmitters at the physiological level, it appears to be paramount to test for interactions between genes of different neurotransmitter systems. Thus, this point is an extension of the first point made above: We need to look for interactions between genes constituting parts of certain neurotransmitter systems (e.g., interactions between various dopamine receptors); but we also need to consider screening for interactions between different genes involved in different subtypes of a neurotransmitter system (e.g., interactions between NMDA and AMPA receptors) or even different neurotransmitter systems.

Although the examples above are intended to be only illustrations of possibilities, they do support the assertion that, most likely, the genetic variability related to variability in cognitive abilities and disabilities will be generated by both additive and nonadditive (interactive) factors descriptive of the diversity at the genetic level. This assertion brings us back to where we started: Whatever genetic factors underlie human traits, especially cognitive traits, they are complex, exhibiting systemic interactions and are open to environmental impact (namely, this openness ensures the plasticity of cognitive functioning).

In summary, the immediate "lesson learned" from the experience with epistasis in other fields of genetics, especially evolutionary and clinical genetics, suggests the importance of systematic screening for epistatic effects—and not only for the genes that are located close to each other on a chromosome but also between genes at different "corners" of the genome whose interactions might be physiologically meaningful.

In Lieu of a Conclusion

Therefore, based on the evidence presented in this chapter, the most obvious conclusion is that "where complex genetic traits loom, epistasis is not far behind" (Frankel & Schork, 1996, p. 371). An appreciation of the importance of epistatic effects for staging a breakthrough in understanding complex behavioral phenotypes is also important for a reconciliation with the discovery-nonreplication phenomenon so relevant to the field of neuropsychiatric psychiatry. One of the implications of epistasis lies in creating a "genetic context" for an expressed allele. In other words, epistasis means that the expression of a given allele differs depending on the genetic surroundings of this allele. This context dependency of gene expression may be one of the main factors in explaining the phenomenon of "disappearing genes" (T. J. Crow, 1997)—genes that have been shown to be associated with phenotypes in some studies but not others. The point made by Crow is that these genes appear far too frequently to be explained by multiple false positives. Templeton (2000) argued that this phenomenon can be easily explained by epistasis, when different populations have different allele frequencies at epistatic loci. To supplement his argument, Templeton referred to the following example.

In families with four or more cases of breast cancer, certain mutations at the BRCA1 locus are predictive of 52%–76% of the cases (Ford et al., 1998). However, such families are not typical of the general population. Specifically, mutations at the BRCA1 locus are predictive of 7.2% of the cases in families with one to two affected individuals before the age of 45 (Malone et al., 1998), and of only 1.4% of sporadic cases, representing the bulk of the disease in the general population (Newman et al., 1998). This variability in the predictive values of mutations at the BRCA1 locus can be explained by the presence of a variety of interactive agents at the population level somehow altering the expression of mutant alleles. These interactive agents can operate both at the level of main effects and at the level of interactions (increase or decrease), depending on the overall genetic architecture in a given sample (Fallin et al., 1997; Kamboh et al., 1995, 1997; Thomas, Green, Lumlum, & Humphries, 1995). The stunning conclusion here is that, most likely, for many complex human conditions, we will not be able to detect the "defective gene"; it is the defective *network* of genes we should be looking for (Templeton, 2000).

The accumulating evidence of the importance of epistatic effects for understanding the etiology of complex human behavioral traits imposes an important challenge to the current dominant research paradigm of screening for main effects first and considering epistasis later (if ever). The supremacy of this research paradigm is explainable by mathematical and computation constraints, although the more we learn about the biological and physiological realities of complex human traits, the more we understand the very interactive nature of these effects. Our anecdotal and intuitive understanding that interaction is the norm rather than an exception has not been translated into the mathematical models and statistical techniques available to a large circle of researchers in the fields of molecular psychiatry and neurobehavioral genetics.

References

Alvarez, R., Reguero, J. R., Batalla, A., Iglesias-Cubero, G., Cortina, A., Alvarez, V., et al. (1998). Angiotensin-converting enzyme and angiotensin II receptor 1 polymorphisms: Association with early coronary disease. *Cardiovascular Research, 40,* 375–379.

Avery, L., & Wasserman, S. (1992). Ordering gene function: The interpretation of epistasis in regulatory hierarchies. *Trends in Genetics, 8,* 312–316.

Bagli, M., Papassotiropoulos, A., Jessen, F., Rao, M. L., Maier, W., & Heun, R. (2000). Gene–

gene interaction between interleukin-6 and alpha2-macroglobulin influences the risk for Alzheimer's disease. *Annals of Neurology, 47*, 138–139.
Bailey, J. N. (1997). DRD4 gene susceptibility to attention deficit hyperactivity disorder: Differences in familial and sporadic cases. *American Journal of Medical Genetics, 74,* 623.
Barnes, K. C. (1999). Gene–environment and gene–gene interaction studies in the molecular genetic analysis of asthma and atopy. *Clinical and Experimental Allergy, 29,* 47–51.
Bedlack, R. S., Strittmatter, W. J., & Morgenlander, J. C. (2000). Apolipoprotein E and neuromuscular disease: A critical review of the literature. *Archives of Neurology, 57,* 1561–1565.
Bell, D. A., Stephens, E. A., Castranio, T., Umbach, D. M., Watson, M., Deakin, M., et al. (1995). Polyadenylation polymorphism in the acetyltransferase 1 gene (NAT1) increases risk of colorectal cancer. *Cancer Research, 55*, 3537–3542.
Berg, K. (1990). Risk factor variability and coronary heart disease. *Acta Geneticae Medicae er Gemellologiae, 39,* 15–24.
Berry, M., Grosveld, F., & Dillon, N. (1992). A single point mutation is the cause of the Greek form of hereditary persistence of fetal haemoglobin. *Nature 358,* 499–502.
Botto, L. D., & Mastroiacovo, P. (1998). Exploring gene–gene interactions in the etiology of neural tube defects. *Clinical Genetics, 53*, 456–459.
Cheverud, J. M. (2000). Detecting epistasis among quantitative trait loci. In J. B. Wolf, E. D. Brodie III, & M J. Wade (Eds.), *Epistasis and the evolutionary process* (pp. 58–81). New York: Oxford University Press.
Cheverud, J. M., & Routman, E. J. (1995). Epistasis and its contribution to genetic variance components. *Genetics, 139,* 1455–1461.
Cho, J. H., Nicolae, D. L., Gold, L. H., Fields, C. T., LaBuda, M. L., Rohal, P. M., et al. (1998). Identification of novel susceptibility loci for inflammatory bowel disease on chromosomes 1p, 3q, and 4q: Evidence for epistasis between 1p and IBD1. *Proceedings of the National Academy of Sciences USA, 95,* 7502–7507.
Clark, A. G., & Wang, L., (1997). Epistasis in measured genotypes: *Drosophila* P-element insertions. *Genetics, 147,* 157–163.
Cloninger, C. R., Van Eerdewegh, P., Goate, A., Edenberg, H. J., Blangero, J., Hesselbrock, V., et al. (1998). Anxiety proneness linked to epistatic loci in genome scan of human personality traits. *American Journal of Medical Genetics, 81,* 313–317.
Cook, E. H., Jr., Stein, M. A., Krasowski, M. D., Cox, N. J., Olkon, D. M., Kieffer, J. E., et al. (1995). Association of attention-deficit disorder and the dopamine transporter gene. *American Journal of Human Genetics, 56*, 993–998.
Cordell, H. J., Todd, J. A., & Lathrop, G. M. (1998). Mapping multiple linked quantitative trait loci in non-obese diabetic mice using a stepwise regression strategy. *Genetics Research, 71,* 51–64.
Cox, N. J., Frigge, M., Nicolae, D. L., Concannon, P., Hanis, C. L., Bell, G. I., et al. (1999). Loci on chromosomes 2 (NIDDM1) and 15 interact to increase susceptibility to diabetes in Mexican Americans. *Nature Genetics, 21*, 213–215.
Crow, J. F., & Kimura, M. (1970). *An introduction to population genetics theory*. Minneapolis, MN: Burgess.
Crow, T. J. (1997). Current status of linkage for schizophrenia: Polygenes of vanishingly small effect or multiple false positives. *American Journal of Medical Genetics, 74,* 99–103.
Cummings, A. M., & Robertson, F. W. (1984). Polymorphism at the apolipoprotein-E locus in relation to risk of coronary disease. *Clinical Genetics, 25,* 310–313.
Czeizel, A. (1989). Application of DNA analysis in diagnosis and control of human diseases. *Biologiches Zentralblatt, 108,* 295–301.
Daly, G., Hawi, Z., Fitzgerald, M., & Gill, M. (1999). Mapping susceptibility loci in attention deficit hyperactivity disorder: Preferential transmission of parental alleles at DAT1, DBH, and DRD5 to affected children. *Molecular Psychiatry, 4,* 192–196.
de Knijff, P., Vandenmaagdenberg, A., Frants, R. R., & Havekes, L. M. (1994). Genetic heterogeneity of apolipoprotein E and its influence on plasma lipid and lipoprotein levels. *Human Mutation, 4,* 178–194.

DeRisi, J. L., Iyer, V. R., & Brown, P. O. (1997). Exploring the metabolic and genetic control of gene expression on a genomic scale. *Science, 278,* 680–686.

de Visser, J. A. G. M., Hoekstra, R. F., & van den Ende, H. (1997). An experimental test for synergistic epistasis and its application in *Chlamydomonas. Genetics, 145,* 815–819.

Diamanti-Kandarakis, E., Bartzis, M. I., Bergiele, A. T., Tsianateli, T. C., & Kouli, C. R. (2000). Microsatellite polymorphism (TTTA)n at −528 base pairs of gene CYP11A influences hyperandrogenemia in patients with polycystic ovary syndrome. *Fertility and Sterility, 73,* 735–741.

Duane, D. D. (Ed.). (1999). *Reading and attention disorders.* Baltimore, MD: York Press.

Eisenberg, J., Mei-Tal, G., Steinberg, A., Tartakovsky, E., Zohar, A., Gritsenko, I., et al. (1999). A haplotype relative risk study of catechol-O-methyltransferase (COMT) and attention deficit disorder (ADHD): Association of the high-enzyme activity val allele with ADHD impulsive-hyperactive phenotype. *American Journal of Medical Genetics (Neuropsychiatric Genetics),* 497–502.

Elena, S. F., & Lenski, R. E., (1997). Test of synergistic interactions among deleterious mutations in bacteria. *Nature, 390,* 395–398.

el-Hazmi, M. A., Warsey, A. S., Addar, M. H., & Babae, Z. (1994). Fetal haemoglobin level: Effect of gender, age and haemoglobin disorders. *Molecular and Cellular Biochemistry, 135,* 181–186.

Elston, R. C., Buxbaum, S., Jacobs, K. B., & Olson, J. M. (2000). Hasemand and Elston revised. *Genetic Epidemiology, 19,* 1–17.

Falconer, D. S., & Mackay, T. F. C. (1996). *Quantitative genetics.* Essex, England: Addison Wesley Longman.

Fallin, D., Reading, S., Schinka, J., Hoyne, J., Scibelli, P., Gold. M., et al. (1997). No interaction between the ApoE and alpha-1-antichymotrypsin genes on risk for Alzheimer's disease. *American Journal of Medical Genetics, 74,* 192–194.

Faraone, S. V., Doyle, A. E., Mick, E., & Biederman, J. (2001). Meta-analysis of the association between DRD4 7-repeat allele and ADHD. *American Journal of Psychiatry, 158,* 1052–1057.

Fenster, C. B., Galloway, L. F., & Chao, L. (1997). Epistasis and its consequences for the evolution of natural populations. *Trends in Ecology and Evolution, 12,* 282–286.

Finkel, D., & McGue, M. (1997). Sex differences and nonadditivity in heritability of Multidimensional Personality Questionnaire scales. *Journal of Personality and Social Psychology, 72,* 929–938.

Fisher, R. A. (1918). The correlations between relatives on the supposition of Mendelian inheritance. *Transactions of the Royal Society of Edinburgh, 52,* 399–433.

Fomicheva, E. V., Gukova, S. P., Larionova-Vasina, V. I., Kovalev, Y. R., & Schwartz, E. I. (2000). Gene–gene interaction in the RAS system in the predisposition to myocardial infarction in elder population of St. Petersburg (Russia). *Molecular Genetics and Metabolism, 69,* 76–80.

Ford, D., Easton, D. F., Stratton, M., Narod, S., Goldgar, D., Devilee, P., et al. (1998). Genetic heterogeneity and penetrance analysis of the BRCA 1 and BRCA2 genes in breast cancer families. *American Journal of Human Genetics, 62,* 676–689.

Frank, S. A. (1999). Population and quantitative genetics of regulatory networks. *Journal of Theoretical Biology, 197,* 281–294.

Frankel, W. N., & Schork, N. J. (1996). Who's afraid of epistasis. *Nature Genetics 14,* 371–373.

Game, J. C., & Cox, B. S. (1972). Epistatic interactions between four rad loci in yeast. *Mutation Research, 16,* 353–362.

Gayán, J., & Olson, R. K. (1999). Reading disability: Evidence for a genetic etiology. *European Child and Adolescent Psychiatry, 8,* 52–55.

Gerlai, R., Henderson, J. T., Roder, J. C., & Jia, Z. (1998). Multiple behavioral anomalies in GluR2 mutant mice exhibiting enhanced LTP. *Behavioral Brain Research, 95,* 37–45.

Ghosh, S., & Majumber, P. P. (2000). A two-stage variable-stringency semiparametric method for mapping quantitative-trait loci with the use of genomewide-scan data on sib pairs. *American Journal of Human Genetics, 66,* 1046–1061.

Gill, M., Daly, G., Heron, S., Hawi, Z., & Fitzgerald, M. (1997). Confirmation of association between attention deficit hyperactivity disorder and a dopamine transporter polymorphism. *Molecular Psychiatry, 2*, 311–313.

Goldstein, A. M., & Andrieu, N. (1999). Detection of interaction involving identified genes: Available study designs. *Journal of the National Cancer Institute: Monographs, 26*, 49–54.

Goldstein, A. M., Goldin, L. R., Dracopoli, N. C., Clark, W. H., Jr., & Tucker, M. A. (1996). Two-locus linkage analysis of cutaneous malignant melanoma/Dysplastic Nevi. *American Journal of Human Genetics, 58,* 1050–1056.

Goodnight, C. J. (1995). Epistasis and the increase in additive genetic variance: Implications for Phase 1 of Wright's shifting-balance process. *Evolution, 49,* 502–511.

Grigorenko, E. L., Wood, F. B., Meyer, M. S., Pauls, J. E. D., Hart, L. A., & Pauls, D. L. (2001). Linkage studies suggest a possible locus for developmental dyslexia on chromosome 1p. *American Journal of Medical Genetics (Neuropsychiatric Genetics), 105,* 120–129.

Haley, C. S., & Knott, S. A. (1992). A simple regression method for mapping quantitative trait loci in line crosses using flanking markers. *Heredity, 69,* 315–324.

Hawi, Z., Millar, N., Daly, G., Fitzgerald, M., & Gill, M. (2000). No association between cathecol-O-methyltransferase (COMT) gene polymorphism and attention deficit hyperactivity disorder (ADHD) in an Irish sample. *American Journal of Medical Genetics, 96,* 282–284.

Hou, S. M., Falt, S., Yang, K., Nyberg, F., Pershagen, G., Hemminki, K., et al. (2001). Differential interactions between GSTM1 and NAT2 genotypes on aromatic DNA adduct level and HPRT mutant frequency in lung cancer patients and population controls. *Cancer Epidemiology, Biomarkers and Prevention, 10*, 133–140.

Hulme, C., & Joshi, R. M. (Eds.). (1998). *Reading and spelling.* Mahwah, NJ: Erlbaum.

Jia, Z., Agopyan, N., Miu, P., Xiong, Z., Henderson, J. T., Gerlai, R., et al. (1996). Enhanced LTP in mice deficient in the AMPA receptor, GLUR2. *Neuron, 17,* 945–956.

Kamboh, M. I., Sanghera, D. K., Aston, C. E., Bunker, C. H., Hamman, R. F., Ferrell, R. E., et al. (1997). Gender specific nonrandom association between the alpha-1-antichymotrypsin and apolipoprotein E polymorphisms in the general population and its implication for the risk of Alzheimer's disease. *Genetic Epidemiology, 14,* 160–180.

Kamboh, M. I., Sanghera, D. K., Ferrell, R. E., & Dekosky, S. T., (1995). APOE*-4 associated Alzheimer's disease risk is modified by alpha-1-antichymotrypsin polymorphisms. *Nature Genetics, 10,* 486–488.

Krajinovic, M., Richer, C., Sinnett, H., Labuda, D., & Sinnett, D. (2000). Genetic polymorphisms of N-acetyltransferases 1 and 2 and gene–gene interaction in the susceptibility to childhood acute lymphoblastic leukemia. *Cancer Epidemiology, Biomarkers and Prevention, 9*, 557–562.

La Hoste, G. J., Swanson, J. M., Wigal, S., Glabe, C., Wigal, T., King N., et al. (1996). Dopamine D4 receptor gene polymorphism is associated with attention-deficit hyperactivity disorder. *Molecular Psychiatry, 1,* 121–124.

Lander, E. S., & Botstein, D. (1989). Mapping Mendelian factors underlying quantitative traits using RFLP linkage maps. *Genetics, 121,* 185–199.

Lehmann, D. J., Johnston, C., & Smith, A. D. (1997). Synergy between the genes for butrylcholinestrerase K variant and apolipoprotein e4 in late-onset confirmed Alzheimer's disease. *Human Molecular Genetics, 6,* 1933–1936.

Long, A. D., Mullaney, S. L., Reid, L. A., Fry, J. D., Langley, C. H., & Mackay, T. F. C. (1995). High resolution mapping of genetic factors affecting abdominal bristle number in *Drosophila melanogaster. Genetics, 43,* 1273–1291.

Lu, Y. F., Kojima, N., Tomizawa, K., Moriwaki, A., Matsushita, M., Obata, K., et al. (1999). Enhanced synaptic transmission and reduced threshold for LTP induction in fyn-transgenic mice. *European Journal of Neuroscience, 11,* 75–82.

Lynch, M., & Walsh, B. (1998). *Genetics and analysis of quantitative traits*. Sunderland, MA: Sinauer Associates.

MacCluer, J. W., Blangero, J., Dyer, T. D., & Speer, M. C. (1997). GAW10: Simulated family data for a common oligogenic disease with quantitative risk factors. *Genetic Epidemiology, 14*, 737–742.

Malenka, R. C., & Nicoll, R. A. (1999). Long-term potentiation: A decade of progress? *Science, 285,* 1870–1874.

Malone, K. E., Daling, J. R., Thompson, J. D., Obrien, C. A., Francisco, L. V., & Ostrander, E. A. (1998). BRCA1 mutations and breast cancer in the general population: Analyses in women before age 35 years and in women before age 45 years with first-degree family history. *Journal of the American Medical Association, 279,* 922–929.

Mather, K., & Jinks, J. L. (1971). *Biometrical genetics.* London: Chapman & Hall.

Morgan, K., Morgan, L., Carpenter, K., Lowe, J., Lam, L., Cave, S., et al. (1997). Microsatellite polymorphisms of the alpha(1)-antichymotrypsin gene locus associated with sporadic Alzheimer's disease. *Human Genetics, 99,* 27–31.

Nelson, M. R., Kardia, S. L. R., Ferrell, R. E., & Sing, C. F. (1998, October). *A combinatorial optimization approach to identify multilocus genotypic classes associated with quantitative trait variability.* Paper presented at the meeting of the American Society of Human Genetics, Denver, CO.

Newman, B. H., Mu, L., Butler, M., Millikan, R. C., Moorman, P. G., & King, M. C. (1998). Frequency of breast cancer attributable to BRAC1 in a population-based series of American women. *Journal of the American Medical Association, 279,* 915–921.

Okuizumi, K., Onodera, O., Namba, Y., Ikeda, K., Yamamoto, T., Seiki, K., et al. (1995). Genetic association of the very low density lipoprotein (VLDL) receptor gene with sporadic Alzheimer's disease. *Nature Genetics, 11,* 207–209.

Omholt, S., Plahte, E., Oyehaug, L., & Xiang, K. (2000). Gene regulatory networks generating the phenomena of additivity, dominance, and epistasis. *Genetics, 155,* 969–980.

Otto, S. P., & Feldman, M. W. (1997). Deleterious mutations, variable epistatic interactions, and the evolution of recombination. *Theoretical Population Biology, 51,* 529–545.

Patrinos, G. P., Loutradianagnostou, A., & Papadakis, M. N. (1996). A new base susbstitution in the 5′ regulatory region of the human (a) gamma globin gene is linked with the beta (S) gene. *Human Genetics, 97,* 357–358.

Peck, J. R., & Waxman D. (2000). Mutation and sex in a competitive world. *Nature, 406,* 399–404.

Pedersen, J. C., & Berg, K. (1989). Interaction between low-density lipoprotein receptor (LDLR) and apolipoprotein-E (ApoE) alleles contributes to normal variation in lipid level. *Clinical Genetics, 35,* 331–337.

Pedersen, J. C., & Berg, K. (1990). Gene–gene interaction between the low density lipoprotein receptor and apolipoprotein E loci affects lipid levels. *Clinical Genetics, 38,* 287–294.

Phillips, P. C. (1998). The language of gene interaction. *Genetics, 149,* 1167–1171.

Phillips, P. C., Otto, S. P., & Whitlock, M. C. (2000). Beyond the average: The evolutionary importance of gene interactions and variability of epistatic effects. In J. B. Wolf, E. D. Brodie III, & M. J. Wade (Eds.), *Epistasis and the evolutionary process* (pp. 20–38). New York: Oxford University Press.

Plomin, R., Corley, R., Caspi, A., Fulker, D. W., & DeFries, J. (1998). Adoption results for self-reported personality: Evidence for nonadditive genetic effects? *Journal of Personality and Social Psychology, 75,* 211–218.

Provine, W. B. (1971). *The origins of theoretical population genetics.* Chicago: University of Chicago Press.

Provine, W. P. (1986). *Sewall Wright and evolutionary biology.* Chicago: University of Chicago Press.

Puumala, T., Ruotsalainen, S., Jakala, P., Koivisto, E., Riekkinen, P., Jr., & Sirvio, J. (1996). Behavioral and pharmacological studies on the validation of a new animal model for attention deficit hyperactivity disorder. *Neurobiology of Learning and Memory, 66,* 198–211.

Routman, E. J., & Cheverud, J. M. (1997). Gene effects on a quantitative trait: Two-locus epistatic effects measured at microsatellite markers and at estimated QTL. *Evolution, 51,* 1654–1662.

Saunders, A. M. (2000). Apolipoprotein E and Alzheimer disease: An update on genetic and functional analyses. *Journal of Neuropathology and Experimental Neurology, 59,* 751–758.

Sing, C. F., & Davignon, J. (1985). Role of the apoliprotein E polymorphism in determining

normal plasma lipid and lipoprotein variation. *American Journal of Human Genetics, 37,* 268–285.
Sing, C. F., & Moll, P. P. (1990). Genetics of antherosclerosis. *Annual Review of Genetics, 24,* 171–188.
Stengard, J. H., Zebra, K. E., Pekkanen, J., Ehnholm, C., Nissen, A., & Sing, C. F. (1995). Apolipoprotein E polymorphism predicts death from coronary heart disease in a longitudinal study of elderly Finnish men. *Circulation, 91,* 265–269.
Sweatt, J. D. (1999). Toward a molecular explanation for long-term potentiation. *Learning and Memory, 6,* 399–416.
Tahir, E., Yazgan, Y., Cirakogu, B., Ozbay, F., Waldman, I., & Asherson, P. J. (2000). Association and linkage of DRD4 and DRD5 with attention deficit hyperactivity disorder (ADHD) in a sample of Turkish children. *Molecular Psychiatry, 5,* 396–404.
Templeton, A. R. (2000). Epistasis and complex traits. In J. B. Wolf, E. D. Brodie III, & M. J. Wade (Eds.), *Epistasis and the evolutionary process* (pp. 41–57). New York: Oxford University Press.
Templeton, A. R., Sing, C. F., & Brokaw, B. (1976). The unit of selection in *Drosophila mercatorum*: I. The interaction of selection and meiosis in parthenogenetic strains. *Genetics, 82,* 349–376.
Thomas, A. E., Green, F. R., Lumlum, H., & Humphries, S. E. (1995). The association of combined alpha and beta fibrinogen genotype on plasma fibrinogen levels in smokers and non-smokers. *Journal of Medical Genetics, 32,* 585–589.
Tsien, J. Z. (2000). Linking Hebb's coincidence-detection to memory formation. *Current Opinion in Neurobiology, 10,* 266–273.
Tsien, J. Z., Chen, D. F., Gerber, D., Tom, C., Mercer, E. H., Anderson, D. J., et al. (1996). Subregion- and cell type-restricted gene knockout in mouse brain. *Cell, 87,* 1317–1326.
Ukkola, O., Perusse, L., Weisnagel, S. J., Bergeron, J., Despres, J. P., Rao, D. C., et al. (2001). Interactions among the glucocorticoid receptor, lipoprotein lipase, and adrenergic receptor genes and plasma insulin and lipid levels in the Quebec Family Study. *Metabolism: Clinical and Experimental, 50*, 246–252.
Ukkola, O., Rankinen, T., Weisnagel, S. J., Sun, G., Perusse, L., Chagnon, Y. C., et al. (2000). Interactions among the alpha2-, beta2-, and beta3-adrenergic receptor genes and obesity-related phenotypes in the Quebec Family Study. *Metabolism: Clinical and Experimental, 49,* 1063–1070.
Utermann, G., Kindermann, I., Kaffarnik, H., & Steinmetz, A. (1984). Apoliprotein E phenotypes and hyperlipidemia. *Human Genetics, 65,* 232–236.
Wade, M. J. (1992). Sewall Wright: Gene interaction and the shifting balance theory. In D. Futuyama & J. Antonovics (Eds.), *Oxford surveys in evolutionary biology* (pp. 35–62). New York: Oxford University Press.
Wagner, G. P., Laubichler, M. D., & Bagheri-Chaichian, H. (1998). Genetic measurement theory of epistatic effects. *Genetica, 102/103,* 569–580.
Whitlock, M. C., Phillips, P. C., Moore, F. B.-G., & Tonsor, S. J. (1995). Multiple fitness peaks and epistasis. *Annual Review in Ecology and Systematics, 26,* 601–629.
Wolf, J. B., Brodie, E. D. III, & Wade, M. J. (Eds.). (2000). *Epistasis and the evolutionary process*. New York: Oxford University Press.
Wright, S. (1965). Factor interaction and linkage in evolution. *Proceedings of the Royal Society of London, 162,* 80–104.
Yang, Q., Khoury, M. J., Sun, F., & Flanders, W. D. (1999). Case-only design to measure gene–gene interaction. *Epidemiology, 10,* 167–170.
Zannis, V. I., & Breslow, J. L. (1985). Genetic mutations affecting human lipoprotein metabolis. *Advances in Human Genetics, 14,* 125–215.
Zeng, Z.-B., Kao, C.-H., & Basten, C. J. (1999). Estimating the genetic architecture of quantitative traits. *Genetics Research, 74,* 279–289.

15

The Genetics of Autistic Disorder

Margaret A. Pericak-Vance

Nearly 60 years have passed since Leo Kanner first described autism as a distinct clinical entity (Kanner, 1943). *Autism* is a colloquial term, which usually refers to a spectrum of disorders known as the pervasive developmental disorders (PDDs; American Psychiatric Association, 1994). Autistic disorder is one of the better understood of the five PDDs and is the primary focus of this chapter. This chapter reviews the approaches to identifying genes in complex disorders such as autistic disorder and discusses some of the recent advances in research in the genetics of autistic disorder.

Autistic disorder is a neurodevelopmental disorder characterized by significant disturbances in social, communicative, and behavioral functioning. Onset of autistic disorder is in early childhood (usually less than 3 years of age), with symptoms continuing throughout life (Folstein & Piven, 1991; Lotspeich & Ciranello, 1993). The cause of autistic disorder is unknown. Prevalence estimates for autistic disorder range from 2 to 10 in 10,000, making it almost as frequent as Down syndrome (1 in 1000; Bailey et al., 1995). Autistic disorder is characterized by an increased male-female ratio of 3 or 4 to 1 (McLennan, Lord, & Schopler, 1993; Volkmar, Szatmari, & Sparrow, 1993). However, X-linked transmission appears likely to account for only a small proportion of the genetic effect in autistic disorder (Cuccaro et al., 2000). The severe impairment and associated behavioral problems in autistic disorder confer a profound burden on patients, their parents, and families. Thus, research to dissect autistic disorder etiology, which could eventually result in successful therapeutic intervention, is of critical importance.

Clinical Studies

Autistic disorder is a clinical diagnosis that is made on the basis of developmental history and behavioral observations. Presently, there are no confirmed biological markers that aid in diagnosis. Autistic disorder is categorized as one of five PDDs, which encompass a spectrum of significant impairments in social communication. The PDDs include autistic disorder, Asperger's disorder, pervasive developmental disorder not otherwise specified, Rett's disorder, and childhood disintegrative disorder (American Psychiatric Association, 1994). Distinctions among the PDDs are made on the basis of several factors, including the number of symptoms present, differences in language development, and the course of the disorder.

Although genes for many inherited disorders have been identified, the search for the genes associated with autistic disorder was stymied for several confounding reasons. The most notable reason was the lack of diagnostic consensus for autistic disorder until the early 1990s. A consensus diagnostic classification for autism has been refined during the past few years. This diagnostic consensus enables researchers to reliably diagnose this disorder; accurate diagnosis is the cornerstone of genetic research. Specifically, the diagnostic criteria for autistic disorder are now explicit and are the same for both the *Diagnostic and Statistical Manual of Mental Disorders* (4th ed., *DSM–IV*; American Psychiatric Association, 1994) and the *International Classification of Diseases (ICD–10)* criteria, thereby reducing diagnostic misclassification. Furthermore, the development of the Autistic Diagnostic Interview–Revised (Lord, Rutter, & LeCouteur, 1994) has helped define the particular aspects of the phenotype, allowing adequate discrimination between individuals with autistic disorder and without autistic disorder (Volkmar et al., 1994).

Gene discovery in autistic disorder also has been delayed by the fact that autistic disorder is a complex genetic disorder (Haines & Pericak-Vance, 1998; Lander & Schork, 1994). There are two general classes of genetic disease: diseases of simple genetic architecture (Mendelian disorders) and diseases of complex genetic architecture. In diseases of simple genetic architecture, how the trait is inherited in a family follows a recognizable pattern. In addition, there is only one genetic mutation per family causing the disease; this mutated gene is referred to as a *causative gene*. A gene is designated as causative when mutations in the gene lead directly to the disease phenotype. In complex diseases, there is no clear pattern of inheritance, there is moderate to high evidence that the disorder is inherited, and there is evidence that multiple genes and the environment may interact to cause disease. Complex diseases are common in the general population; the genes involved are usually referred to as *susceptibility genes*. A susceptibility gene confers an increased risk of disease but by itself is not sufficient to result in the disease. The susceptibility gene could even be a common variant allele in the population. Such is the case with the apolipoprotein E allele, a susceptibility gene for Alzheimer's disease (Corder et al., 1993). Some examples of complex genetic disorders include asthma, cardiovascular disease, diabetes mellitus, and autistic disorder. Figure 15.1 schematically depicts the relative contributions of genes and environment in several Mendelian versus complex (including infectious) diseases.

Genetic Epidemiological Studies

It was not until many years after Leo Kanner described autism as a clinical entity in 1943 that evidence was first published suggesting some forms of autism might have a genetic basis (Folstein & Rutter, 1977). Since the

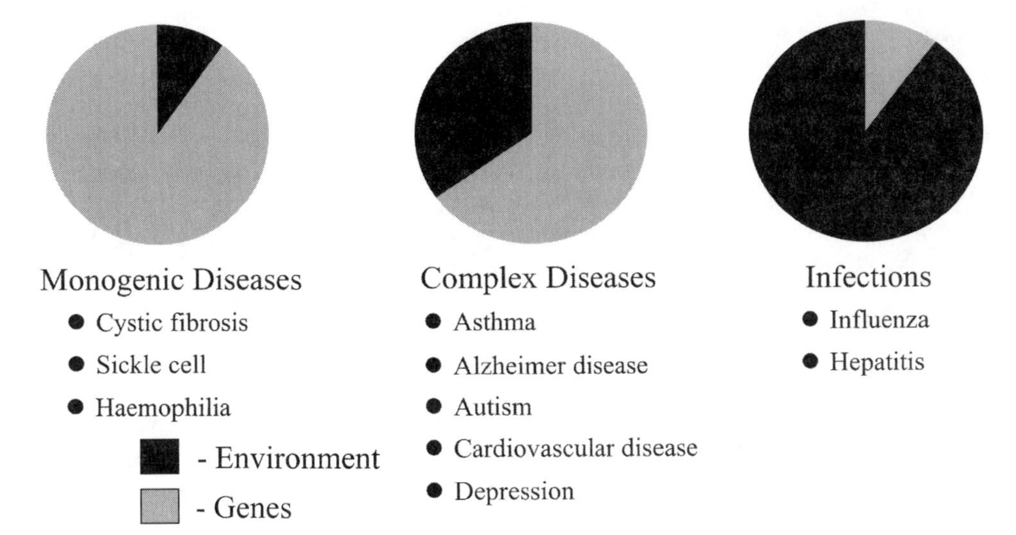

Figure 15.1. Relative contribution of genes and environment to disease.

1970s, data from multiple studies have accumulated to show that autistic disorder has a genetic basis. These first reports included observational studies such as case reports and studies of familial clustering of autism. As genetic research became more sophisticated, twin studies and chromosomal and statistical evidence also emerged, further supporting the genetic contribution to the development of autistic disorder. Today, several genomic regions, potentially containing autistic disorder susceptibility genes, have already been identified. I next review the genetic basis for autistic disorder and the recent genetic research breakthroughs for this common-occurring disorder.

Compelling evidence exists that autistic disorder is a genetic disorder. Epidemiological data from several studies consistently have implicated genetic factors in the etiology of autistic disorder (Folstein & Piven, 1991; Lotspeich & Ciaranello, 1993; Rutter, Bolton, et al., 1990; Rutter, Macdonald, et al., 1990; Smalley, Asarnow, & Spence, 1988). Pooled estimates of concordance rates for monozygotic (MZ) versus dizygotic (DZ) twins are 64% and about 3%, respectively (Folstein & Rutter, 1977; Ritvo, Freeman, Mason-Brothers, Mo, & Ritvo, 1985; Rutter, Bolton, et al., 1990; Rutter, Macdonald, et al., 1990; Smalley et al., 1988; Steffenburg et al., 1989). Heritability estimates for autistic disorder range from over 90% (Bailey et al., 1995; Smalley et al., 1988) to 100% (LaBuda, Gottesman, & Pauls, 1993). Sibling recurrence risk ratio (λ_s) estimates range from 100 to 200. This estimate is at least 10-fold higher than that seen in other complex genetic disorders. The consensus of recent studies suggests that there are anywhere from 2 to 10 loci underlying the genetic form of autistic disorder (Jorde et al., 1991). Given the high λ_s, it may seem surprising that more progress on finding the genes for autistic disorder has not been made. Indeed, with four genes of equal autistic disorder additive effect, the λ_s for each locus would be 37.5, and these genes should be identifiable even in relatively small datasets. However, the rapid decrease in λ_s with increasing degree of relationship in autistic disorder suggests an epistatic interaction of these multiple loci (Jorde et al., 1991). Four genes of equal multiplicative effect would have a λ_s of only 3.5 each. Although such genes would be harder to detect, simulations suggest that there is high power to detect genes under either of these simplistic autistic disorder models, or even more complex models with gene-specific effects of $\lambda_s \sim 2$ (Hauser, Boehnke, Guo, & Risch, 1996; Risch, 1990b).

Gene Identification

Almost since the first suggestions of a genetic component in autistic disorder, studies have been conducted to identify the genes involved in autistic disorder. The two basic approaches that have been used are candidate gene (association) studies and linkage analysis studies. With the advent of restriction fragment length polymorphisms (RFLPs; Botstein, White, Skolnick, & Davis, 1980) and microsatellite markers (Weber & May, 1989), the entire genome was opened for examination through linkage analysis. In addition, through the auspices of the Human Genome Project, huge numbers of genes have been identified and at least partially described (Deloukas et al., 1998), thus allowing many more candidate genes to be tested.

Methods for Testing Candidate Genes

Genes can become candidates for autistic disorder in two ways: as functional candidates or locational candidates. Functional candidate genes are defined by what is known of the biology of the gene and of autistic disorder. Thus, any gene that affects (or, more commonly, could be hypothesized to affect) that biology becomes a candidate. The strength and weakness of this approach arises from the confidence placed on the role these genes play. If the evidence is strong that a direct role is played, only one or a few such genes may need to be tested to find a true susceptibility gene. If the evidence is more circumstantial, then many genes may have an equal chance of being involved and not much has been gained over a random genomic screen.

Locational candidate genes are defined simply by residing in a chromosomal region that is interesting for other reasons, not because of any known function the gene might have. Chromosomal regions may be of interest because an initial genomic screen has defined several "interesting" regions, or a chromosomal abnormality (e.g., a duplication, deletion, or translocation such as has been seen on chromosomes 7 and 15 in autistic disorder) has been observed in one or more patients.

At the population level, there are several ways to test candidate genes for an effect in autistic disorder. The first, and most common, is the case control design. This has the advantage of being able to use many, if not most, of the patients coming through the clinic. In the simplest formulation, the allele frequencies for a polymorphism within the gene are compared between the case and control groups. However, the case control design is very sensitive to sample size, stringency of diagnosis, and, particularly, appropriate matching of the control group. In addition, the case control design cannot distinguish between a true genetic effect (either linkage disequilibrium [LD] or biological interaction) and a marker of population admixture that is biologically irrelevant to autistic disorder.

The second approach for testing candidate genes is to use genetic linkage analysis for polymorphisms within or very near the gene in question. If significant evidence of linkage is found (using any one of several different linkage methods), then a role for that gene is supported, although not concluded, since the linkage could represent the effect of nearby genes. This approach has the advantage of not requiring a control group. However, it requires families with multiple affected individuals, and these can be difficult to find.

The third approach for testing candidate genes is to apply the new family-based association study designs. This combines the advantages of the case control design (only one affected per family) with the family approach, by using genetically well-matched controls (either the parents or siblings of the patient). To alleviate some of these problems and to control for any possible ethnic or population structure variations, researchers have developed family-based association methods. The most commonly applied is the transmission disequilibrium test (TDT; Spielman, McGinnis, & Ewens, 1993), which examines the frequencies of alleles transmitted and not transmitted to children. This allows the use of single affected individuals but also requires sampling their parents. The power of this simple test arises from its reliance on LD, which usually exists over only very small intervals (usually less than 1 cM). Thus it is complementary to standard linkage analysis, where the effect is broad (up to 30 cM) but its discriminative power over small regions is low. The TDT is best applied to the candidate genes that will fall within the critical regions. An issue that arises when testing for association with the TDT is that only a single affected child may be selected from any family, which means that information from other affected relatives (which is often the case with autistic disorder) is discarded. To ease this restriction, Martin and colleagues (2000) developed the pedigree disequilibrium test (PDT). This test allows one to use a combination of family types, including multiple affected siblings. The PDT compares genotypes at a single marker of affected individuals with their unaffected siblings, cousin pairs, and avuncular pairs. If parental data are available, the PDT also will make use of transmissions from parents to affected offspring. Transmit (TMT; Clayton & Jones, 1999) is a likelihood-based, family-based association approach that also uses extended family data.

Association analysis has its greatest applicability in the fine localization and identification of loci, as it examines directly the effect of a candidate locus, not just a diffused effect spread across a large genomic region (Fisher, Vargha-Khadem, Watkins, Monaco, & Pembrey, 1998; Risch

& Zhang, 1996). The power and simplicity of association tests depend on the availability of numerous polymorphic markers in any candidate gene or region, a situation only recently realized. Evaluation of candidate genes in association studies holds great potential with development of single nucleotide polymorphisms (SNPs). It is estimated that SNPs occur on average every 1,000 base pairs and have a low mutation rate, both of which may be advantages in association studies. Several million SNPs have been submitted to the National Center for Biotechnology Information SNP database, and efforts are expected to continue until a complete SNP map is available (Collins, Brooks, & Chakravarti, 1998; Marshall, 1999).

Methods for Genomic Screening

The alternative to trying to understand the biology of autistic disorder before knowing the genes is to try to know the genes before understanding the biology. This can be accomplished by testing families with multiple cases of autistic disorder for linkage to polymorphic markers spread throughout the genome. The type of family to be tested varies from study to study, but the most commonly tested family type is the affected sibpair and parents. This is the most common constellation of affected individuals in a family.

The linkage analysis approach has its greatest applicability in initial localization of candidate loci. Parametric log of the odds (LOD) score analysis is the most powerful method of linkage analysis if the genetic model is known. However, so little is known about complex diseases such as autistic disorder that accurate model specification is impossible. Although LOD score methods may be robust to errors underlying assumptions in specific instances, this is not always the case. Nonparametric methods are an attractive alternative. As the interest in mapping complex diseases has blossomed, nonparametric methods have experienced improvements and innovations. Efficient multipoint sibpair analysis using the likelihood approach (MLS; Risch, 1990a, 1990b, 1990c) is now available in several different computer programs. In addition, methods that incorporate data on other affected relative pairs, including the nonparametric linkage (NPL) method implemented in the GENEHUNTER package (Kruglyak, Daly, Reeve-Daly, & Lander, 1996) and extended by Kong and Cox (1997; Cox, Frigge, et al., 1999) in GENEHUNTER+ and ALLEGRO (Gudbjartsson, Jonasson, Frigge, & Kong, 2000), are now commonly used.

A good random genomic screen will attempt to cover the entire human genome using markers evenly spaced across a given genetic map. The most common spacing of markers is 10 cM, resulting in screens using between 300 and 400 markers. It does not require any knowledge of the function of any genes or even of the biology of autistic disorder. The goal in an initial genomic screen is to quickly leaf through the genome and identify those regions most likely to harbor autistic disorder genes. This approach is purposely liberal in that many regions will be false-positive results, but this allows high power, in that regions that truly do harbor autistic disorder genes are far less likely to be missed. To eliminate the false-positive results, researchers usually test a second dataset only for those regions that were identified in the first dataset. Once confirmed, the region is further dissected to identify candidate genes for specific study.

Genomic Screens in Autistic Disorder

There have been at least seven genome screens in autism to date, by the International Molecular Genetic Study of Autism Consortium (IMGSAC; 1998, 2001), the Collaborative Linkage Study of Autism (CLSA; 1999), the Paris Autism Research International Sib-Pair Study (Philippe et al., 1999), Stanford University (Risch et al., 1999), the studies of my colleagues and I—the Collaborative Autism Team (CAT; Shao, Wolpert, et al., 2002), Mt. Sinai Medical Center (Buxbaum et al., 2001), and the Autism Genetic Resource Exchange (AGRE; http://www.agre.org/; Liu et al., 2001). Several regions of interest, some of which overlap, have been identified.

IMGSAC (1998) reported the first genome screen with 316 microsatellite markers on 99 White families from Europe and the United States. Regions on six chromosomes (4, 7, 10, 16, 19, and 22) were identified, with a critical level of multipoint MLS > 1.00. Chromosome 7q gave the highest score, with an MLS of 2.53 near markers D7S530 and D7S684. An area on chromosome 16p near the telomere was the next most significant region, with an MLS of 1.99. Subsequent analyses from IMGSAC included evidence for an MLS $= 2.20$ on chromosome 2 and an MLS > 2.00 on chromosome 17. Philippe et al. (1999) performed a genomic screen in 51 multiplex autistic disorder families using affected sibpair analysis. Eleven regions were identified in this screen, with critical level of nominal p values $\leq .05$: 2q, 4q, 5p, 6q, 7q, 10q, 15q, 16p, 18q, 19p, and Xp. The most significant peak was close to marker D6S283 (MLS $= 2.23$). In another genomic screen, Risch et al. (1999) identified nine locations using 90 multiplex families with MLS ≥ 1.00. The CLSA (1999) reported a genomic screen for autism in 75 multiplex families, with their strongest evidence of linkage to chromosomes 13 and 7. CLSA peak markers were D13S800, with an MLOD (maximium LOD score) of 3.00 and D7S1813 with an MLOD of 2.20. Recently, Buxbaum et al. (2001) reported their strongest evidence for linkage from their two-stage genomic screen of 95 autistic disorder families to be on chromosome 2q, with a heterogeneity LOD score of 1.96.

Several critical regions from the CAT genomic screen (CAT) were consistent with the regions reported by other autistic disorder genomic screen studies. The most interesting results (MLOD or MLS > 1) for the CAT dataset are on chromosomes 2, 3, 7, 15, 19, and X. Our region on chromosome 2 potentially overlapped with those reported by the IMGSAC (1998), Phillippe et al. (1999), and Buxbaum et al. (2001). Their peak on chromosome 2 was localized to the same region as we observed. The 7q region identified by our CAT study appeared to overlap with results from the IMGSAC study, the CLSA, and the Paris group. The IMGSAC study gave their peak score as an MLS of 2.53 on chromosome 7q32-35. The increased LOD scores from their subsequent investigation with a dense set of markers further supported 7q as an autistic disorder susceptibility region (IMGSAC, 1998). The peak on chromosome 7 from the Paris group was also on this region. On chromosome 15, Philippe et al. (1999) identified a potential susceptibility region, with positive results for three adjacent markers, which may overlap with our identified region 15q11-q13. Regions on 19 and X potentially overlap with those reported by IMGSAC, the Paris group, and Liu's group (Liu et al., 2001). In addition, our CAT study revealed a unique region potentially involved in autistic disorder on chromosome 3. The X chromosome findings are of interest because of the increased male-female ratio of 3 or 4 to 1 seen in autistic disorder, which has been described in both simplex and multiplex autistic disorder families (McLennan et al., 1993; Volkmar et al., 1993). Despite this abnormal sex ratio, previous studies of X-linkage in autistic disorder have concluded that if X-linked transmission exists, it is likely to account for only a subset of autistic disorder families (Cuccaro et al., 2000; Hallmayer et al., 1996). The genomic screen data suggest the possibility that a subset of families may be X-linked.

In summary, the convergence of linkage findings is promising and emphasizes the need for further analysis of these areas. Some of these additional analyses are described in chromosome-specific detail below.

Chromosome 15 Studies

In addition to the linkage findings, chromosome 15 is the site of frequent chromosomal anomalies associated with autistic disorder. Chromosome 15 q11-q13 duplication and inversion have been found in 20+ independent autistic disorder patients, including 3 new individuals with autistic disorder with duplications in the 15q11-13 region identified in my own group's dataset (Wolpert et al., 2000). In many cases, the new rearrangements are thought to be maternal in origin (Lindgren et al., 1996). Chromosome 15q also contains genes that are associated with Prader-Willi and Angelman syndromes, two disorders in which patients display developmental delays and, in some cases, autisticlike behavior (Butler, Meaney, & Palmer, 1986; Demb & Papola, 1995; Magenis, Brown, Lacy, Budden, & LaFranchi, 1987; Matsuura et al., 1997; Ozcelik et al., 1992; Penner, Johnston, Faircloth, Irish, & Williams, 1993; Steffenburg, Gillberg, Steffenburg, & Kyllerman, 1996; Summers, Allison, Lynch, & Sandler, 1995).

The neurotransmitter gamma-aminobutyric acid (GABA) mediates synaptic inhibition in the adult brain through its interaction with the postsynaptic $GABA_A$ receptors. These receptors are usually pentameric structures comprising several different homologous subunits that together form a ligand-gated ion channel (a transmembrane ion channel whose permeability is increased by the binding of a specific ligand, typically a neurotransmitter at a chemical synapse). The functional properties of a particular $GABA_A$ receptor are dependent on the subunit composition (Smith &

Olsen, 1995; Tyndale, Olsen, & Tobin, 1994). Genes for three of the receptor subunits, GABRB3, GABRA5, and GABRG3, are located on chromosome 15q11-q13 (Greger et al., 1995; Knoll et al., 1993; Wagstaff et al., 1991). Family-based association analysis with microsatellite markers in the GABRB3 gene has provided evidence in support of LD of autistic disorder to this gene (Cook et al., 1998; Martin, Menold, et al., 2000). Cook et al. (1998) found LD with a marker in intron 3 of GABRB3, and Martin, Menold, et al. (2000) found suggestive LD with an independent microsatellite marker located just beyond the 3′ end of the gene. Recently, SNPs were identified in all three of the $GABA_A$ receptor subunit genes on chromosome 15 (www.ncbi.nlm.nih.gov/SNP/). To further define the association between these genes and autistic disorder, CAT analyzed 16 of the SNPs in the CAT dataset (see Figure 15.2) using either oligonucleotide ligation assay (Samiotaki, Kwiatkowski, & Landegren, 1996) or single-strand conformation polymorphism followed by DNA sequencing. We found significant evidence of LD in the γ subunit of GABA (Menold et al., in press; see Table 15.1). The most interesting results were obtained from the SNPs in the GABRG3 gene. Analysis of SNPs in GABRG3 revealed an area of the gene that appeared to be in LD with an autistic disorder susceptibility locus. SNPs in both exon 5 (exon 5_539T/C) and intron 5 (intron 5_687T/C) gave the most significant evidence for association. In addition, examination of the data showed a trend in two other SNPs that are located upstream in the GABRG3 gene (intron 1_403C/G and intron 4_421G/A). Subsequent studies of exon 5 and intron 5 SNPs in an additional 90 families from AGRE family resource resulted in increased significance (Menold et al., in press). Collectively, these data suggest that the $GABA_A$ receptor subunit gene, GABRG3, located on proximal 15q, may be involved in the genetic risk of autistic disorder. Considering the complicated nature of this region (Christian, Fantes, Mewborn, Huang, & Ledbetter, 1999; Maddox et al., 1999; Sutcliffe et al., 1994) and the differing patterns of expression of the subunits during development (Barker, Behar, Ma, Maric, & Maric,

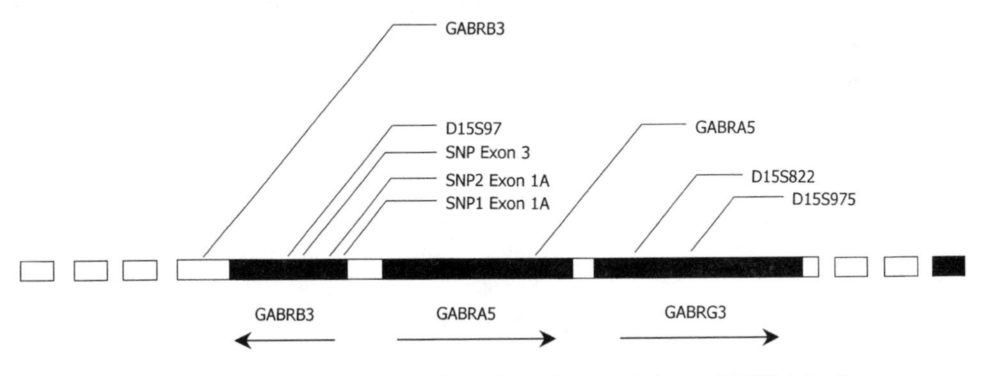

Figure 15.2. Location of single nucleotide polymorphisms (SNPs) in the gamma-aminobutyric acid A (GABA) receptor subunit genes on chromosome 15 analyzed for association.

Table 15.1. Examination of Linkage Disequilibrium Between Autistic Disorder and the GABA Receptor Gene Complex

Gene	SNP location	p-value
GABRB3	5′UTR_30C/G	0.78
	Exon 1A_170C/T	0.73
	Intron 3_1427C/T	0.06
GABRA5	Intron 7_68G/A	0.83
	Exon 8_213T/C	0.96
	Intron 9_44T/A	0.87
	Exon 10_168T/C	0.88
GABRG3	Intron 1_320T/A	0.22
	Intron 1_403C/G	0.13
	Intron 4_421G/A	0.10
	Intron 4_558A/C	0.47
	Exon 5_539T/C	**0.02**
	Intron 5_687T/C	**0.03**
	Intron 5_414A/T/C	0.44
	Intron 6_115A/C	0.32
	Exon 8_685T/C	0.77

Note. Boldface values indicate significant p-values (≤.05).
GABA = gamma-aminobutyric acid; SNP = single nucleotide polymorphism.

2000), it is very likely that the regulation of the GABRG3 gene is a multifactorial process. Molecular studies are needed to elucidate the elaborate nature of these mechanisms and to precisely define their role in the genetic risk and development of autistic disorder.

Chromosome 7 Studies

CAT has followed up on the initial region of linkage interest identified by IMGSAC, demonstrating additional support for linkage to chromosome 7 (Ashley-Koch et al., 1999). In this same report, we also show evidence for a paternal effect for autistic disorder and chromosome 7 (see Table 15.2), as well as evidence for increased recombination in autistic disorder families. Cytogenetic evidence supports the presence of a locus on chromosome 7 in autistic disorder as well (Ashley-Koch et al., 1999; Vincent, Paterson, Strong, Petronis, & Kennedy, 2000). CAT has identified a family with three children carrying a paracentric inversion on the long arm of chromosome 7, Inv (7) (q22q31.2) (see Figure 15.3), which overlaps the linkage results (Ashley-Koch et al., 1999). Two of the three children (males) have autistic disorder, and the third and youngest child (female) suffers from significant developmental delays and expressive language disorder. Interestingly, Fisher et al. (1998) found a locus (SPCH1) on 7q31 that segregated with a severe speech and language disorder in a large three-generation pedigree. Recently, this group identified the causative gene for this phenotype, the FOX2P gene (Lai, Fisher, Hurst, Vargha-Khadem, & Monaco, 2001).

Table 15.2. Identity by Descent (IBD) Sharing Among Affected Sibs Supporting Excess Sharing of Paternal Alleles

Marker	Paternal p	Maternal p	Overall p
D7S1799	1.00	.30	.45
D7S1817	.17	.82	.29
D7S2847	.41	.41	.27
D7S2527	.03*	.33	.03*
D7S530	.25	.73	.29
D7S640	.007**	.75	.03*
D7S495	.02*	.32	.02*
D7S684	.09	.19	.03*
D7S1824	.35	.18	.11

*$p < .05$. **$p < .01$.

Preliminary results indicate that this FOX2P gene is not associated with autistic disorder (Newbury et al., 2002). Santangelo et al. (2000) also reported an association with a speech phenotype and chromosome 7. They showed that stratification by presence or absence of speech problems improved the evidence for linkage to this region and that the majority of the linkage signal originated from the subset of families that presented speech problems. The subset of CLSA families in which both probands had severe language delay gave higher LOD scores than that for the overall dataset, further supporting the involvement of chromosome 7 in autistic disorder (Nurmi et al., 2001).

As a result of the linkage findings on chromosome 7, candidate gene studies also have emerged. Recent reports suggest two genes, WNT2 at 7q31 (Wassink et al., 2001) and RELN at 7q22 (Persico et al., 2001), influ-

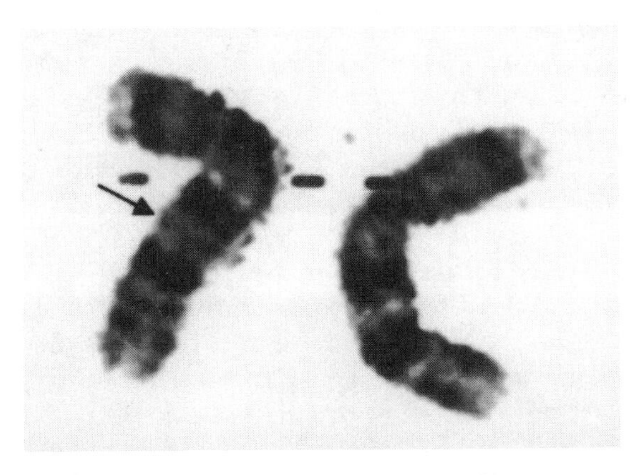

Figure 15.3. This depicts the paracentric inversion (inv(7)(q22.1-q31.2)) in Collaborative Autism Team family 7543. This metaphase spread shows the inverted chromosome on the left with an arrow marking the location of the inversion. The patient's normal chromosome 7 is located on the right.

ence genetic risk in autistic disorder. Both genes are also good functional candidates. The WNT2 gene is one of more than a dozen WNT genes that are active during human development. WNT genes encode several signaling proteins with regulatory roles in embryonic patterning, cell proliferation, and cell determination (Brown & Moon, 1998; Wodarz & Nusse, 1998). The WNT signaling pathway also included several receptors that are expressed in both developing and mature nervous systems (Salinas, 1999). Because autistic disorder is a developmental disorder that affects the nervous system, one could hypothesize that a change in a WNT protein might contribute to the development of autistic disorder. Wassink et al. (2001) reported finding that variations of the WNT2 gene were associated with autistic disorder ($p = .013$). They found the association was stronger when they examined the genes of a subset of people with autistic disorder and severe language delays consistent with other studies on chromosome 7 that suggest that language delay could play a role in autistic disorder. They also found two nonconservative missense mutations that segregate with autistic disorder in two families.

RELN regulates neuronal migration during brain development (D'Arcangelo et al., 1995). In mice, RELN also has been suggested to be involved in schizophrenia (Impagnatiello et al., 1998). Several research groups are currently studying the role of the human RELN in brain development and human disease (Persico et al., 2001). Persico et al. found that certain variations in the size of the trinucleotide repeat (>10 repeats) occur more often in individuals *with* autistic disorder than in individuals *without* autistic disorder. This finding suggests that this gene, specifically larger repeat length polymorphism, may be associated with some cases of autistic disorder.

Using two different but complementary methods of family-based association analysis, PDT and TMT, CAT tested for genetic association of two WNT2 variants in the 5′UTR and 3′UTR regions of WNT2. No significant association was found between autistic disorder and the WNT2 genotypes. My colleagues and I also screened the coding region of WNT2 for two reported autism mutations (T→G mutation in the signal domain of Exon 1 of WNT2 produced Leu5Arg and a C→T mutation in Exon 5 produced Arg299Trp) and failed to identify these WNT2 mutations in our dataset. Our data suggest that the WNT2 gene is not a major contributor to genetic risk in autistic disorder (McCoy et al., 2002).

In addition, CAT examined the association of a 5′UTR polymorphism in RELN ($(GCC)_n$) in a dataset that included both the CAT (multiplex and singleton) as well as 90 AGRE multiplex families. PDT and TMT analysis of the combined data found no evidence for association ($p = .35$ and $p = .07$, respectively). However, examining the AGRE families independently resulted in support of association to RELN ($p = .005$ for PDT; $p = .009$ for TMT). Analysis of SNP in Exon 62 of RELN also gave suggestive evidence for association in the confined dataset (PDT = .03, TMT = .07). These data suggest that additional analysis of the RELN gene is necessary to elucidate its potential involvement as an autistic disorder risk gene (McCoy et al., 2002).

Table 15.3. Linkage Results Subsetting by Delayed Onset of Phrase Speech (PSD) as Reported by Buxbaum et al. (2001)

D2S364 (186cM)	Overall (N = 95 families)	PSD subset (N = 49 families)
HLOD (dominant model)	1.96	2.99
NPL	2.39	2.32

Note. PSD = delayed onset of phrase speech >36 months; HLOD = LOD score allowing for heterogeneity; NPL = nonparametric linkage score. Data from Buxbaum, Silverman, Smith, Kilifarsky, Reichert, Hollander, et al. (2001).

Chromosome 2 Studies

In addition to linkage findings from multiple genomic screens, there is evidence for allele-specific association on chromosome 2 in my laboratory's preliminary dataset for marker D2S116 (Bass et al., 2000), a marker associated with Celiac disease (Holopainen et al., 1999). This is of interest because gastrointestinal complaints have been reported in autistic disorder patients (Horvath, Papadimitriou, Rabsztyn, Drachenberg, & Tildon, 1999; Kawashima et al., 2000). Recently, Buxbaum et al. (2001) published their data in support of linkage to chromosome 2. They reported a maximum heterogeneity LOD (HLOD) score of 1.96 and an NPL of 2.39. When they restricted their analysis to the subset of families with delayed onset (>36 months) of phrased speech (PSD), their scores increased to HLOD = 2.99 and NPL = 3.32 (see Table 15.3). Using this same stratification strategy on our dataset resulted in an increase of the NPL from 0.71 to 3.12 at D2S116 and from 0.71 to 2.65 at D2S23098. The HLOD score (dominant model) increased from 0.53 to 2.12 at D2S116 and from 0.74 to 2.17 at D2S2309 when we restricted analysis to the subset of families with PSD (see Table 15.4). Permutation analysis conducted to determine the significance of the increase in the NPL and HLOD resulted in p values of .028 and .007, respectively. Permutation analysis supported the strength of our conclusions that the findings on chromosome 2 with respect to PSD are significant (Shao, Raiford, et al., 2002). The convergence of linkage findings by PSD stratification in two independent datasets is critical.

Table 15.4. Linkage Results Subsetting by Delayed Onset of Phrase Speech (PSD) in the CAT Sample

Marker	Statistical measurement	All sibs (N = 82 families)	PSD sibs (N = 45 families)
D2S116 (198cM)	HLOD (dominant model)	0.53	2.12
	NPL	1.18	3.12
D2S2309 (198cM)	HLOD (dominant model)	0.74	2.17
	NPL	0.71	2.65

Note. HLOD = LOD score allowing for heterogeneity; NPL = nonparametric linkage score; CAT = Collaborative Autism Team.

These data both support evidence for a gene on chromosome 2 and suggest that examining the PSD subgroup in more detail may be critical in identifying the chromosome 2 autistic disorder risk gene.

Chromosome 3 Studies

Genomic screen analysis of our 99 multiplex families (Shao, Wolpert, et al., 2002) revealed evidence for linkage on chromosome 3 with D3S3680 (36.1 cM) resulting in a peak two-point MLS = 2.02. Subsequently, we genotyped an additional 90 multiplex autistic disorder families from AGRE for markers in this region. With the new data, the two-point MLS score for D3S3680 increased to 3.16. Several other markers in the region also gave support for linkage in the expanded dataset, including D3S4545 (26.3 cM), D3S3589 (32.3 cM), and D3S1259 (36.7 cM) with MLS = 2.32, 1.95, and 1.16, respectively (see Table 15.5). The peak multipoint MLS was at D3S1259 (MLS = 2.66). The oxytocin gene is involved in the facilitation of affiliative and social behaviors (Michelini, Urbanek, Dean, & Goldman, 1995). The oxytocin receptor gene (OXTR) is located on chromosome 3p25-p26 near our peak linkage region. Animal models have shown that OXTR is expressed in brain as well as uterus and mammary gland and has an effect on neural development. Mutations in OXTR could disturb the function of oxytocin. To evaluate the OXTR as a potential autistic disorder candidate gene, we genotyped several SNPs in OXTR and tested for association using both PDT and TMT in the combined (CAT and AGRE) autistic disorder dataset. We found no evidence for association with either method in any of the OXTR SNPs (see Table 15.6). These data suggest that the OXTR gene is not a major contributor to risk in autistic disorder and is not the chromosome 3 autistic disorder gene.

Table 15.5. Chromosome 3 and Autistic Disorder Linkage Results

		CAT		AGRE		Combined	
Marker	cM	Max HLOD	MLS	Max HLOD	MLS	Max HLOD	MLS
D3S4545	26.25	0.14	0.02	2.18	2.63	1.99	2.32
D3S3589	32.3	1.25	0.70	1.59	1.54	2.32	1.95
D3S1263	36.1	0.34	0.18	1.64	1.80	1.45	1.50
D3S3680	36.1	0.90	2.02	0.80	0.86	1.46	*3.16*
D3S1259	36.7	1.76	0.56	0.99	0.75	*2.38*	1.16
D3S2385	38.5	0.37	0.48	0.42	0.43	0.66	0.80
D3S3613	41.6	0.32	0.42	0.78	0.65	0.93	0.99

Note. CAT = Collaborative Autism Team; AGRE = Autism Genetic Resource Exchange; HLOD = LOD score allowing for heterogeneity; MLS = multipoint sibpair analysis using the likelihood approach.

Table 15.6. Association Results for Autistic Disorder and Chromosome 3p25 Oxytocin Receptor Gene (OXTR)

	CAT		AGRE		Combined	
OXTR gene	PDT	Transmit	PDT	Transmit	PDT	Transmit
oxtr_snp5	0.83	0.61	0.09	0.17	0.18	0.19
oxtr_snp6	0.48	0.73	0.84	0.83	0.51	0.68
oxtr_snp7	0.74	0.90	0.07	0.08	0.38	0.28
OXTR haplotype						
oxtr_snp5-snp6						0.31
oxtr_snp6-snp7						0.73

Note. CAT = Collaborative Autism Team; AGRE = Autism Genetic Resource Exchange; PDT = pedigree disequilbrium test.

Other Genes of Interest in Autistic Disorder

The logical progression of genetic research for the PDDs is to investigate the genetic basis for the remaining four PDDs: Asperger disorder, Rett disorder (RTT), pervasive developmental disorder not otherwise specified, and childhood disintegrative disorder. Rett disorder (RTT; MIM 312750) is a PDD and a Mendelian disorder in which the causative gene has been identified (Amir et al., 1999). Rett disorder is the only PDD for which a genetic link has been found, and the only PDD other than autism on which significant genetics research has been done. Genetics research into PDD-NOS, Asperger disorder, and childhood disintegrative disorder is currently in its infancy.

RTT is an X-linked disorder characterized by normal neurological development until 6 to 18 months of age, followed by a progressive developmental regression. RTT occurs almost exclusively in females. Incidence rates range from 0.25 to 1 per 10,000 female births (Kerr, 1995; Terai et al., 1995). Skjeldal and colleagues (1997) reported a rate of 2.17 per 10,000 female births when the full spectrum of RTT syndrome variants was included. Developmental regression is followed by a deceleration of head growth, and loss of purposeful hand movements and followed by the appearance of midline, stereotypic hand movements (American Psychiatric Association, 1994). Additional features observed in RTT patients include seizures, ataxia, hyperventilation, and social communication problems. These social engagement problems are similar to those observed in autistic disorder and therefore can result in an initial misdiagnosis of autistic disorder in some RTT patients (Gillberg, 1989). Often the appearance of the additional findings seen in RTT clarifies the diagnosis in these patients.

In 1999, Amir and colleagues demonstrated that mutations in the X chromosome gene methyl-CpG-binding protein 2 (MeCP2) cause the majority of cases of RTT (Amir et al., 1999). The MeCP2 protein binds methylated cytosine residues, eventually leading to transcriptional repression (Buyse et al., 2000). Subsequent studies have found over 60 unique mutations in MeCP2, with mutations noted in 80%–100% of RTT patients

(Auranen et al., 2001; Cheadle et al., 2000; Dragich, Houwink-Manville, & Schanen, 2000; Huppke, Laccone, Kramer, Engel, & Hanefeld, 2000). As previously mentioned, until recently, RTT had only been observed in females and was believed to be lethal in males. However, since the identification of MeCP2 as the gene defect, two males with RTT have now been reported (Villard et al., 2000). In addition, increasing evidence points to a wider range of severity and heterogeneity of phenotype (Huppke et al., 2000; Lam et al., 2000).

Data now suggest that there is a clinical overlap between RTT and autistic disorder, particularly in early childhood. In a recent study by CAT, mutations in the MeCP2 gene were identified in two females (ages 10 and 16) who meet criteria for the diagnosis of autistic disorder (Carney et al., 2001). Neither of the patients evaluated exhibit the classic RTT features (Rett Syndrome Diagnostic Criteria Work Group, 1988). In another study, Lam et al. (2000) screened 21 patients with "autism and mental retardation" and found 1 patient with an MeCP2 mutation. These data suggest that the phenotype associated with MeCP2 mutations is highly variable and that individuals with MeCP2 mutations can present with a milder clinical phenotype such as autistic disorder. Thus, in some cases, MeCP2 testing may be warranted for females presenting with autistic disorder, especially those presenting with any associated symptoms of RTT.

Future Endeavors

To date, there have been seven genome screens in autism, indicating several consensus regions of interest, including chromosomes 2, 3, 7, 15, and 19. There also have been multiple candidate gene studies examining both functional and combined functional/locational candidate loci. It also is possible that epistatic effects among multiple genes play an important role in determining risk of autistic disorder, and for some genes the interactive effects may be stronger than the independent single-gene effects. One of the goals of genetic studies is to determine the role of epistasis, or gene–gene interaction resulting from nonadditivity of main effects, on the risk of autistic disorder. Finally, fully understanding autistic disorder as a complex multifactorial disorder ultimately will require moving beyond genetic studies to include environmental factors as well. The power to identify such risk factors will be increased by controlling for as many known genetic susceptibility loci as possible.

The etiology of autistic disorder is now ripe for dissection using newly developed statistical and molecular genetic tools. The long-range goal of these studies is to identify all major, moderate, and epistatic genetic effects in autistic disorder. It is hoped that the isolation of the genes for autistic disorder and related disorders, and the dissection of how these genes interact, will result in earlier diagnosis and successful therapeutic interventions.

References

American Psychiatric Association. (1994). *Diagnostic and statistical manual of mental disorders* (4th ed.). Washington, DC: Author.

Amir, R. E., Van den Veyver, I. B., Wan, M., Tran, C. Q., Francke, U., & Zoghbi, H. Y. (1999). Rett syndrome is caused by mutations in X-linked MECP2, encoding methyl-CpG-binding protein 2. *Nature Genetics, 23*, 185–188.

Ashley-Koch, A., Wolpert, C. M., Menold, M. M., Zaeem, L., Basu, S., Donnelly, S. L., et al. (1999). Genetic studies of autistic disorder and chromosome 7. *Genomics, 61*, 227–236.

Auranen, M., Vanhala, R., Vosman, M., Levander, M., Varilo, T., Hietala, M., et al. (2001). MECP2 gene analysis in classical Rett syndrome and in patients with Rett-like features. *Neurology, 56*, 611–617.

Bailey, A., Le Couteur, A., Gottesman, I., Bolton, P., Simonoff, E., Yuzda, E., et al. (1995). Autism as a strongly genetic disorder: Evidence from a British twin study. *Psychological Medicine, 25*, 63–77.

Barker, J. L., Behar, T. N., Ma, W., Maric, D., & Maric, I. (2000). GABA emerges as a developmental signal during neurogenesis of the rat central nervous system. In D. L. Martin & R. W. Olsen (Eds.), *GABA in the nervous system: The view at fifty years* (pp. 245–263). Philadelphia: Lippincott Williams & Wilkins.

Bass, M. P., Menold, M. M., Joyner, K. L., Wolpert, C. M., Donnelly, S. L., Ravan, S. A., et al. (2000). Association analysis of GI candidate genes in autistic disorder [Abstract]. *American Journal of Human Genetics, 67*, 349:1946A.

Botstein, D., White, R. L., Skolnick, M., & Davis, R. W. (1980). Construction of a genetic linkage map in man using restriction fragment length polymorphisms. *American Journal of Human Genetics, 32*, 314–331.

Brown, J. D., & Moon, R. T. (1998). Wnt signaling: Why is everything so negative? *Current Opinion in Cell Biology, 10*, 182–187.

Butler, M. G., Meaney, F. J., & Palmer, C. G. (1986). Clinical and cytogenetic survey of 39 individuals with Prader-Labhart-Willi syndrome. *American Journal of Medical Genetics, 23*, 793–809.

Buxbaum, J., Silverman, J. M., Smith, C. J., Kilifarsky, M., Reichert, J., Hollander, E., et al. (2001). Evidence for a susceptibility gene for autism on chromosome 2 and for genetic heterogeneity. *American Journal of Human Genetics, 68*, 1514–1520.

Buyse, I. M., Fang, P., Hoon, K. T., Amir, R. E., Zoghbi, H. Y., & Roa, B. B. (2000). Diagnostic testing for Rett syndrome by DHPLC and direct sequencing analysis of the MECP2 gene: Identification of several novel mutations and polymorphisms. *American Journal of Human Genetics, 67*, 1428–1436.

Carney, R. J., Vance, J. M., Dancel, R. D., Wolpert, C. M., DeLong, G. R., McClain, C., et al. (2001, May). Screening for MECP2 mutations in females with autistic disorder [Abstract]. *10th International Congress of Human Genetics*.

Cheadle, J. P., Gill, H., Fleming, N., Maynard, J., Kerr, A., Leonard, H., et al. (2000). Long-read sequence analysis of the MECP2 gene in Rett syndrome patients: Correlation of disease severity with mutation type and location. *Human Molecular Genetics, 9*, 1119–1129.

Christian, S. L., Fantes, J. A., Mewborn, S. K., Huang, B., & Ledbetter, D. H. (1999). Large genomics duplicons map to sites of instability in the Prader-Willi/Angelman syndrome chromosome region (15q11-q13). *Human Molecular Genetics, 8*, 1025–1037.

Clayton, D., & Jones, H. (1999). Transmission/disequilibrium tests for extended marker haplotypes. *American Journal of Human Genetics, 65*, 1161–1169.

Collaborative Linkage Study of Autism. (1999). An autosomal screen for autism. *American Journal of Medical Genetics, 88*, 609–615.

Collins, F. S., Brooks, L. D., & Chakravarti, A. (1998). A DNA polymorphism discovery resource for research on human genetic variation. *Genome Research, 8*, 1229–1231.

Cook, E. H., Courchesne, R. Y., Cox, N. J., Lord, C., Gonen, D., Guter, S. J., et al. (1998). Linkage-disequilibrium mapping of autistic disorder, with 15q11-13 markers. *American Journal of Human Genetics, 62*, 1077–1083.

Corder, E. H., Saunders, A. M., Strittmatter, W. J., Schmechel, D. E., Gaskell, P. C., Small,

G. W., et al. (1993). Gene dose of apolipoprotein E type 4 allele and the risk of Alzheimer's disease in late onset families. *Science, 261*, 921–923.

Cox, N. J., Frigge, M., Nicolae, D. L., Concannon, P., Hanis, C. L., Bell, G. I., & Kong, A. (1999). Loci on chromosome 2 (NIDDM1) and 15 interact to increase susceptibility to diabetes in Mexican Americans. *Nature Genetics, 21*, 213–215.

Cuccaro, M., Bass, M. P., Abramson, R. K., Ravan, S. A., Wright, H. H., Wolpert, C. M., et al. (2000). Factor-analysis of behavioral characteristics in autistic disorder (AD) [Abstract]. *American Journal of Human Genetics, 67,* 302:1665A.

D'Arcangelo, G., Miao, G. G., Chen, S. C., Soares, H. D., Morgan, J. I., & Curran, T. (1995). A protein related to extracellular matrix proteins deleted in the mouse mutant reeler. *Nature, 374*, 719–723.

Deloukas, P., Schuler, G. D., Gyapay, G., Beasley, E. M., Soderlund, C., Rodriguez-Tomé, P., et al. (1998). A physical map of 30,000 human genes. *Science, 282*, 744–746.

Demb, H. B., & Papola, P. (1995). PDD and Prader-Willi syndrome. *Journal of the American Academy of Child and Adolescent Psychiatry, 34*, 539–540.

Dragich, J., Houwink-Manville, I., & Schanen, C. (2000). Rett syndrome: A surprising result of mutation in MECP2. *Human Molecular Genetics, 9*, 2365–2375.

Fisher, S. E., Vargha-Khadem, F., Watkins, K. E., Monaco, A. P., & Pembrey, M. E. (1998). Localisation of a gene implicated in a severe speech and language disorder. *Nature Genetics, 18*, 168–170.

Folstein, S. E., & Piven, J. (1991). Etiology of autism: Genetic influences. *Pediatrics, 87*(5, Part 2), 767–773.

Folstein, S., & Rutter, M. (1977). Infantile autism: A genetic study of 21 twin pairs. *Journal of Child Psychology and Psychiatry and Allied Disciplines, 18*, 297–321.

Gillberg, C. (1989). The borderland of autism and Rett syndrome: Five case histories to highlight diagnostic difficulties. *Journal of Autism and Developmental Disorders, 19,* 545–559.

Greger, V., Knoll, J. H. M., Woolf, E., Glatt, K., Tyndale, R. F., DeLorey, T. M., et al. (1995). The γ-aminobutyric acid receptor γ3 subunit gene (GABRG3) is tightly linked to the α5 subunit gene (GABRA5) on human chromosome 15q11-q13 and is transcribed in the same orientation. *Genomics, 26*, 258–264.

Gudbjartsson, D. F., Jonasson, K., Frigge, M. L., & Kong, A. (2000). Allegro, a new computer program for multipoint linkage analysis. *Nature Genetics, 25*, 12–13.

Haines, J. L., & Pericak-Vance, M. A. (1998). Overview of mapping common and genetically complex human disease traits. In J. L. Haines & M. A. Pericak-Vance (Eds.), *Approaches to gene mapping in complex human diseases* (pp. 1–12). New York: Wiley.

Hallmayer, J., Spiker, D., Lotspeich, L., McMahon, W. M., Petersen, P. B., Nicholas, P., et al. (1996). Affected-sib-pair interval mapping and exclusion for complex genetic traits: Sampling considerations. *Genetic Epidemiology, 13*, 117–137.

Holopainen, P., Arvas, M., Sistonen, P., Mustalahti, K., Collin, P., Maki, M., & Partanen, J. (1999). CD28/CTLA4 gene region on chromosome 2q33 confers genetic susceptibility to celiac disease: A linkage and family-based association study. *Tissue Antigens, 53*, 470–475.

Horvath, K., Papadimitriou, J. C., Rabsztyn, A., Drachenberg, C., & Tildon, J. T. (1999). Gastrointestinal abnormalities in children with autistic disorder. *Journal of Pediatrics, 135,* 559–563.

Huppke, P., Laccone, F., Kramer, N., Engel, W., & Hanefeld, F. (2000). Rett syndrome: Analysis of MECP2 and clinical characterization of 31 patients. *Human Molecular Genetics, 9*, 1369–1375.

Impagnatiello, F., Guidotti, A. R., Pesold, C., Dwivedi, Y., Caruncho, H., Pisu, M. G., et al. (1998). A decrease of reelin expression as a putative vulnerability factor in schizophrenia. *Proceedings of the National Academy of Sciences USA, 95*(26), 15718–15723.

International Molecular Genetic Study of Autism Consortium. (1998). A full genome screen for autism with evidence for linkage to a region on chromosome 7q. *Human Molecular Genetics, 7*, 571–578.

International Molecular Genetic Study of Autism Consortium. (2001). A genomewide screen for autism: Strong evidence for linkage to chromosomes 2q, 7q, and 16p. *American Journal of Human Genetics, 69*, 570–581.

Jorde, L. B., Hasstedt, S. J., Ritvo, E. R., Mason-Brothers, A., Freeman, B. J., Pingree, C., et al. (1991). Complex segregation analysis of autism. *American Journal of Human Genetics, 49*, 932–938.

Kanner, L. (1943). Review of *Psychiatric Progress 1942: Child Psychiatry*. *American Journal of Psychiatry, 99,* 608–610.

Kawashima, H., Mori, T., Kashiwagi, Y., Takekuma, K., Hoshika, A., & Wakefield, A. (2000). Detection and sequencing of measles virus from peripheral mononuclear cells from patients with inflammatory bowel disease and autism. *Digestive Diseases and Sciences, 45,* 723–729.

Kerr, A. M. (1995). Early clinical signs in the Rett disorder. *Neuropediatrics, 26*, 67–71.

Knoll, J. H. M., Sinnett, D., Wagstaff, J., Glatt, K., Schantz-Wilcox, A., Whiting, P. M., et al. (1993). FISH ordering of reference markers and of the gene for the α5 subunit of the γ-aminobutyric acid receptor (GABRA5) within the Angelman and Prader-Willi syndrome chromosomal regions. *Human Molecular Genetics, 2*, 183–189.

Kong, A., & Cox, N. J. (1997). Allele-sharing models: LOD scores and accurate linkage tests. *American Journal of Human Genetics, 61*, 1179–1188.

Kruglyak, L., Daly, M. J., Reeve-Daly, M. P., & Lander, E. S. (1996). Parametric and nonparametric linkage analysis: A unified multipoint approach. *American Journal of Human Genetics, 58*, 1347–1363.

LaBuda, M. C., Gottesman, I. T., & Pauls, D. L. (1993). Usefulness of twin studies for exploring the etiology of childhood and adolescent psychiatric disorders. *American Journal of Medical Genetics, 48*, 47–59.

Lai, C. S., Fisher, S. E., Hurst, J. A., Vargha-Khadem, F., & Monaco, A. P. (2001). A forkhead-domain gene is mutated in a severe speech and language disorder. *Nature, 413*, 519–523.

Lam, C. W., Yeung, W. L., Ko, C. H., Poon, P. M., Tong, S. F., Chan, K. Y., et al. (2000). Spectrum of mutations in the MECP2 gene in patients with infantile autism and Rett syndrome. *Journal of Medical Genetics, 37*(12), E41.

Lander, E. S., & Schork, N. J. (1994). Genetic dissection of complex traits. *Science, 265*, 2037–2048.

Lindgren, V., Cook, E. H., Jr., Leventhal, B. L., Courchesne, R., Lincoln, A., Shulman, C., et al. (1996). Maternal origin of proximal 15q duplication in autism. *American Journal of Human Genetics, 39*(Suppl.), 688.

Liu, J., Nyholt, D. R., Magnussen, P., Parano, E., Pavone, P., Geschwind, D., et al. (2001). A genomewide screen for autism susceptibility loci. *American Journal of Human Genetics, 67,* 327–340.

Lord, C., Rutter, M., & LeCouteur, A. (1994). Autism diagnostic interview-revised: A revised version of a diagnostic interview for caregivers of individuals with possible pervasive developmental disorders. *Journal of Autism and Developmental Disorders, 24*, 659–685.

Lotspeich, L. J., & Ciaranello, R. D. (1993). The neurobiology and genetics of infantile autism [Review]. *International Review of Neurobiology, 35*, 87–129.

Maddox, L. O., Menold, M. M., Bass, M. P., Rogala, A. R., Pericak-Vance, M. A., Vance, J. M., et al. (1999). Autistic disorder and chromosome 15q11-q13: Construction and analysis of a BAC/PAC contig. *Genomics, 62*, 325–331.

Magenis, R. E., Brown, M. G., Lacy, D. A., Budden, S., & LaFranchi, S. (1987). Is Angelman syndrome an alternate result of del(15)(q11q13)? *American Journal of Medical Genetics, 28*, 829–838.

Marshall, E. (1999). Genomics: Drug firms to create public database of genetic mutations. *Science, 284*, 406–407.

Martin, E. R., Menold, M. M., Wolpert, C. M., Bass, M. P., Donnelly, S. L., Ravan, S. A., et al. (2000). Analysis of linkage disequilibrium in gamma-aminobutyric acid receptor subunit genes in autistic disorder. *American Journal of Medical Genetics, 96*, 43–48.

Martin, E. R., Monks, S. A., Warren, L. L., & Kaplan, N. L. (2000). A test for linkage and association in general pedigrees: The pedigree disequilibrium test. *American Journal of Human Genetics, 69,* 327–340.

Matsuura, T., Sutcliffe, J. S., Fang, P., Galjaard, R. J., Jiang, Y. H., Benton, C. S., et al. (1997). De novo truncating mutations in E6-AP ubiquitin-protein ligase gene (UBE3A) in Angelman syndrome. *Nature Genetics, 15,* 74–77.

McCoy, P. A., Shao, Y. J., Wolpert, C. M., Donnelly, S. L., Ashley-Koch, A. E., Abel, H. L., et al. (2002). No association between the WNT2 gene and autistic disorder. *American Journal of Human Genetics, 114,* 106–109.

McLennan, J. D., Lord, C., & Schopler, E. (1993). Sex differences in higher functioning people with autism. *Journal of Autism and Developmental Disorders, 23,* 217–227.

Menold, M. M., Shao, Y. J., Wolpert, C. M., Donnelly, S. L., Raiford, K. L., Martin, E. R., et al. (in press). Association analysis of chromosome 15 GABA-A receptor subunit genes in autistic disorder. *Journal of Neurogenetics.*

Michelini, S., Urbanek, M., Dean, M., & Goldman, D. (1995). Polymorphism and genetic mapping of the human oxytocin receptor gene on chromosome 3. *American Journal of Medical Genetics (Neuropsychiatric Genetics), 60,* 183–187.

Newbury, D. F., Bonora, E., Lamb, J. A., Fisher, S. E., Lai, C. S., Baird, G., et al. (2002). FOXP2 is not a major susceptibility gene for autism or specific language impairment. *American Journal of Human Genetics, 70,* 1318–1327.

Nurmi, E. L., Bradford, Y., Chen, Y., Hall, J., Arnone, B., Gardiner, M. B., et al. (2001). Linkage disequilibrium at the Angelman syndrome gene UBE3A in autism families. *Genomics, 77*(1/2), 105–113.

Ozcelik, T., Leff, S., Robinson, W., Donlon, T., Lalande, M., Sanjines, E., et al. (1992). Small nuclear ribonucleoprotein polypeptide N (SNRPN), an expressed gene in the Prader-Willi syndrome critical region. *Nature Genetics, 2,* 265–269.

Penner, K. A., Johnston, J., Faircloth, B. H., Irish, P., & Williams, C. A. (1993). Communication, cognition, and social interaction in the Angelman syndrome. *American Journal of Medical Genetics, 46,* 34–39.

Persico, A. M., D'Agruma, L., Maiorano, N., Totaro, A., Militerni, R., Bravaccio, C., et al. (2001). Reelin gene alleles and haplotypes as a factor predisposing to autistic disorder. *Molecular Psychiatry, 6,* 150–159.

Philippe, A., Martinez, M., Guilloud-Bataille, M., Gillbert, C., Rastam, M., Sponheim, E., et al. (1999). Genome-wide scan for autism susceptibility genes: Paris Autism Research International Sibpair Study. *Human Molecular Genetics, 8,* 805–812.

Rett Syndrome Diagnostic Criteria Work Group. (1988). Diagnostic criteria for Rett syndrome. *Annals of Neurology, 23,* 425–428.

Risch, N. (1990a). Linkage strategies for genetically complex traits: I. Multilocus models. *American Journal of Human Genetics, 46,* 222–228.

Risch, N. (1990b). Linkage strategies for genetically complex traits: II. The power of affected relative pairs. *American Journal of Human Genetics, 46,* 229–241.

Risch, N. (1990c). Linkage strategies for genetically complex traits: III. The effect of marker polymorphism on analysis of affected pairs. *American Journal of Human Genetics, 46,* 242–253.

Risch, N., Spiker, D., Lotspeich, L., Nouri, N., Hinds, D., Hallmayer, J., et al. (1999). A genomic screen of autism: Evidence for a multilocus etiology. *American Journal of Human Genetics, 65,* 493–507.

Risch, N., & Zhang, H. (1996). Mapping quantitative trait loci with extreme discordant sib pairs: Sample size considerations. *American Journal of Human Genetics, 58,* 836–843.

Ritvo, E. R., Freeman, B. J., Mason-Brothers, A., Mo, A., & Ritvo, A. M. (1985). Concordance for the syndrome of autism in 40 pairs of afflicted twins. *American Journal of Psychiatry, 142,* 74–77.

Rutter, M., Bolton, P., Harrington, R., Le Couteur, A., Macdonald, H., & Simonoff, E. (1990). Genetic factors in child psychiatric disorders: I. A review of research strategies [Review]. *Journal of Child Psychology and Psychiatry and Allied Disciplines, 31,* 3–37.

Rutter, M., Macdonald, H., Le Couteur, A., Harrington, R., Bolton, P., & Bailey, A. (1990). Genetic factors in child psychiatric disorders: II. Empirical findings. *Journal of Child Psychology and Psychiatry, 31,* 39–83.

Salinas, P. C. (1999). Wnt factors in axonal remodeling and synaptogenesis. *Biochemical Society Symposia, 65*, 101–109.

Samiotaki, M., Kwiatkowski, M., & Landegren, U. (1996). OLA: Dual-color oligonucleotide ligation assay. In U. Landegren (Ed.), *Laboratory protocols for mutation detection* (pp. 96–100). New York: Oxford University Press.

Santangelo, S. L., Collaborative Linkage Study of Autism, Buxbaum, J., Silverman, J., Pericak-Vance, M. A., & Ashley-Koch, A. (2000). Confirmatory evidence of linkage for autism to 7q based on combined analysis of three independent data sets [Abstract]. *American Journal of Human Genetics, 67.*

Shao, Y. J., Raiford, K. L., Wolpert, C. M., Cope, H. A., Ravan, S. A., Ashley-Koch, et al. (2002). Phenotypic homogeneity provides increased support for linkage on chromosome 2 in autistic disorder. *Journal of Human Genetics, 70,* 1058–1061.

Shao, Y. J., Wolpert, C. M., Raiford, K. L., Menold, M. M., Donnelly, S. L., Ravan, S. A., et al. (2002). Genomic screen and follow-up analysis for autistic disorder. *American Journal of Human Genetics, 114,* 99–105.

Skjeldal, O. H., von Tetzchner, S., Asplelund, F., Herder, G. A., & Lofterld, B. (1997). Rett syndrome: Geographic variation in prevalence in Norway. *Brain and Development, 19,* 258–261.

Smalley, S. L., Asarnow, R. F., & Spence, M. A. (1988). Autism and genetics: A decade of research. *Archives of General Psychiatry, 45*, 953–961.

Smith, G. B., & Olsen, R. W. (1995). Functional domains of $GABA_A$ receptors. *Trends in Pharmacological Sciences, 16*, 162–168.

Spielman, R. S., McGinnis, R. E., & Ewens, W. J. (1993). Transmission test for linkage disequilibrium: The insulin gene region and insulin-dependent diabetes mellitus (IDDM). *American Journal of Human Genetics, 52*, 506–516.

Steffenburg, S., Gillberg, C., Hellgren, L., Andersson, L., Gillberg, I. C., Jakobsson, G., et al. (1989). A twin study of autism in Denmark, Finland, Iceland, Norway, and Sweden. *Journal of Child Psychology and Psychiatry and Allied Disciplines, 30*, 405–416.

Steffenburg, S., Gillberg, C. L., Steffenburg, U., & Kyllerman, M. (1996). Autism in Angelman syndrome: A population-based study. *Pediatric Neurology, 14*, 131–136.

Summers, J. A., Allison, D. B., Lynch, P. S., & Sandler, L. (1995). Behavior problems in Angelman syndrome. *Journal of Intellectual Disability Research, 39*, 97–106.

Sutcliffe, J. S., Nakao, M., Christian, S., Orstavik, K. H., Tommerup, N., Ledbetter, D. H., et al. (1994). Deletions of a differentially methylated CpG island at the SNRPN gene define a putative imprinting control region. *Nature Genetics, 8*, 52–58.

Terai, K., Munesue, T., Hiratani, M., Jiang, Z. Y., Jibiki, I., & Yamaguchi, N. (1995). The prevalence of Rett syndrome in Fukui prefecture. *Brain and Development, 17*, 153–154.

Tyndale, R. F., Olsen, R. W., & Tobin, A. J. (1994). $GABA_A$ receptors. In S. J. Peroutka (Ed.), *Handbook of receptors and channels* (pp. 265–290). London: CRC Press.

Villard, L., Kpebe, A., Cardoso, C., Chelly, P. J., Tardieu, P. M., & Fontes, M. (2000). Two affected boys in a Rett syndrome family: Clinical and molecular findings [Abstract]. *Neurology, 55,* 1188–1193.

Vincent, J. B., Paterson, A. D., Strong, E., Petronis, A., & Kennedy, J. L. (2000). The unstable trinucleotide repeat story of major psychosis [Review]. *American Journal of Medical Genetics, 97*, 77–97.

Volkmar, F. R., Klin, A., Siegel, B., Szatmari, P., Lord, C., Campbell, M., et al. (1994). Field trial for autistic disorder in *DSM–IV. American Journal of Psychiatry, 151*, 1361–1367.

Volkmar, F. R., Szatmari, P., & Sparrow, S. S. (1993). Sex differences in pervasive developmental disorders. *Journal of Autism and Developmental Disorders, 23*, 579–591.

Wagstaff, J., Knoll, J. H. M., Fleming, J., Kirkness, E. F., Martin-Gallardo, A., Greenberg, F., et al. (1991). Localization of the gene encoding the $GABA_A$ receptor $\beta 3$ subunit to the Angelman/Prader-Willi region of human chromosome 15. *American Journal of Human Genetics, 49*, 330–337.

Wassink, T. H., Piven, J., Vieland, V. J., Huang, J., Swiderski, R. E., Pietila, J., et al. (2001). Evidence supporting WNT2 as an autism susceptibility gene. *American Journal of Medical Genetics, 105,* 406–413.

Weber, J. L., & May, P. E. (1989). Abundant class of human DNA polymorphisms which can be typed using the polymerase chain reaction. *American Journal of Human Genetics, 44*, 388–396.

Wodarz, A., & Nusse, R. (1998). Mechanisms of Wnt signaling in development. *Annual Review of Cell Development Biology, 14*, 59–88.

Wolpert, C. M., Menold, M. M., Bass, M. P., Qumsiyeh, M. B., Donnelly, S. L., Ravan, S. A., et al. (2000). Three probands with autistic disorder and isodicentric chromosome 15. *American Journal of Medical Genetics, 96*, 365–372.

Part VI

Psychopharmacology

16

Finding Genes for Complex Behaviors: Progress in Mouse Models of the Addictions

John C. Crabbe

A predisposition to alcoholism, smoking, or abuse of other drugs can be inherited. Definitive twin and adoption studies suggest that the heritability of substance abuse risk is on the order of 0.4–0.6 (Prescott & Kendler, 1999; Rose, 1995; Tsuang et al., 1996). However, until relatively recently, risk could be assigned only statistically.

One does not inherit risk: One inherits specific risk-promoting or risk-protective genes. There are three ways in which to begin to identify the specific genes that underlie individual differences in risk. One approach searches for differences in gene sequence arising from polymorphisms; the second seeks gene expression differences; and the third targets a specific gene, manipulates it, and examines the physiological and behavioral consequences. The introduction to this volume points out that complex traits (which vary quantitatively in populations) are affected by multiple genes, each of which exerts a relatively small effect on the trait (see Plomin, DeFries, Craig, & McGuffin, chapter 1, this volume). New molecular methods have made it possible to move beyond assigning statistical risk to the pursuit of specific risk-promoting and risk-ameliorating genes. For traits essentially controlled by a single gene, mapping methods have successfully identified many such genes, starting with the gene controlling production of huntingtin protein, the cause of Huntington's disease. For complex traits, the intermediary step of identifying the rough genomic locations of quantitative trait loci (QTLs) is a necessary first step. Because of the long history of the application of mouse and rat genetic animal models to the study of the addictions, application of these new QTL mapping methods to complex traits was applied to animal models of addictive behaviors, and the neuroadaptations occurring with chronic drug treatments, at an early stage (Plomin, McClearn, Gora-Maslak, & Neiderhiser, 1991).

Other classical animal genetic methods relying on gene sequence dif-

---

This research is supported by the Department of Veterans Affairs, the National Institute on Alcohol Abuse and Alcoholism, and the National Institute on Drug Abuse. I thank Pamela Metten for Figure 16.1.

ferences, such as exploring inbred strains and selectively breeding on the basis of extreme phenotypic responders, are still widely used and continue to provide their customary advantages and disadvantages for understanding both the genetics and neurobiology of complex traits (Crabbe, 1999). The goal of this chapter, however, is to focus on recent advances in identifying specific genes and their function. Thus, I concentrate on QTL mapping and the alternative strategy of randomly induced mutagenesis in this part of the chapter.

The second approach recognizes that individuals may differ because of differences in the expression of genes as well rather than in the structure of their coded protein. Gene expression arrays (gene chips) are just beginning to be used as a gene discovery tool, and the limited data available are discussed. Finally, candidate gene studies using transgenic technologies to over- or underexpress specific genes in mice are mentioned. Because many relevant genes have been targeted for a variety of drug responses, a systematic review of this literature is beyond the scope of this chapter, but such reviews are indicated.

Throughout the chapter, I plead the case for a behavioral genomics approach to understanding gene function. *Behavioral genomics* may be defined as the attempt to study the effects of specific genes on the whole, behaving organism, in contrast to the currently dominant functional genomics or proteomics approaches that seek the explanations of a gene's effects at the cellular or protein level (Plomin & Crabbe, 2000). A behavioral genomics perspective depends on the careful analysis of phenotypic variation in its environmental context and on the rigorous mapping of specific behaviors to the psychological constructs they are intended to represent.

Animal Models of the Addictions

Drug responses have proved to be a useful paradigm for complex traits. In part because drugs abused by humans are not readily available in nature, natural selection has not favored sensitive or insensitive genotypes, leaving a great deal of heritable genetic influence on behavioral responses to drugs. In addition, with the exception of alcohol, drugs generally have rather specific neurobiological targets, starting with the receptor protein to which they bind and the pathways in which the targeted neurons are found. This has supported the simultaneous development of genetic strategies and neurobiological, mechanism-oriented research, to the benefit of both.

The study of drug responses generally is framed by a pharmacological perspective. Studies of the addictive drugs have therefore used models that account for the pharmacological as well as the psychological aspects of drug dependence and abuse. To understand an individual's genetic susceptibility to an abused drug, one needs to at least model the following: sensitivity to the initial drug challenge, the development of tolerance or its converse, sensitization; physical dependence, inferred from withdrawal when chronic drug is removed; the reinforcing effects of the drug, be they positive or negative; and individual differences in how efficiently the drug is metabolized.

Inbred Strains

Behavioral genetics has benefited from the ready availability of numerous inbred strains of mice (and, to a lesser extent, rats). More than 100 mouse strains are maintained by The Jackson Laboratory (Bar Harbor, Maine), and a handful have been in wide use for many years. Because each inbred strain comprises genetically identical individuals and remains quite genetically stable over time, the pleiotropic influence of genes can readily be assessed by comparing strains for multiple drug responses in a controlled environment. This has allowed investigators to compare results with inbred strains across laboratories, and over many years. In 1959, McClearn and Rodgers reported that the differences among several inbred strains of mice in their propensity to drink ethanol solutions far exceeded the within-strain differences. The specific finding that C57BL/6 strain mice were high preferrers, whereas DBA/2 strain mice were nearly complete abstainers, has been replicated many times since (Fuller, 1964; Phillips, Belknap, Buck, & Cunningham, 1998; Tarantino, McClearn, Rodriguez, & Plomin, 1998).

Many other studies have compared multiple strains for sensitivity to alcohol and other drug effects (for reviews, see Crabbe & Harris, 1991; Crawley et al., 1997; Mogil et al., 1996; Mohammed, 2000). These correlational analyses have taught us much about codetermination of genetic influence (Crabbe, Belknap, & Buck, 1994). One recent example reported a very high, negative correlation between the tendency of 15 mouse strains to drink alcohol and the severity of acute alcohol withdrawal (Metten et al., 1998). A similar comparison required mice to inhibit a rewarded nosepoke response when an auditory signal was presented. Efficiency of response inhibition in the signaled nosepoke task was highly predictive of low ethanol consumption across inbred strains (Logue, Swartz, & Wehner, 1998). Specifically, these results are consistent with the hypotheses that a genetic predisposition to express severe withdrawal and to be able to inhibit responding are two behavioral sequelae that lead some strains to elect not to self-administer alcohol when it is offered. As with any correlation-based hypothesis, testing this proposed causal relationship would require additional experiments. It also should be remembered that the specific genes responsible for influencing multiple drug effects are unknown.

The Mouse Phenome Project is a new initiative undertaken by the Jackson Laboratory to support systematic collection of behavioral and physiological data in a large number (currently 40) of inbred strains. As such data are gleaned from the literature and collected in new studies, they are being assembled in a relational database that will make it simple to access such genetic relationships (Paigen & Eppig, 2000). Although there are many rat inbred strains, they have been studied much less systematically with respect to their psychopharmacological responses.

Selected Lines

Artificial selection, or selective breeding, is the oldest technique in behavioral genetics. Starting with a genetically segregating population with many polymorphic genes, extreme-scoring individuals are mated to produce lines of mice or rats that differ genetically in sensitivity, tolerance, dependence, or preference for alcohol and several other drugs of abuse. Such selected lines are found to differ on many other behavioral and physiological traits. In an ideal selection experiment, this is because of the pleiotropic effects of the genes important for the selected trait. However, many caveats are appropriate before subscribing to this interpretation, mostly related to the small population sizes involved (Crabbe, 1999). As is the case in studies with inbred strains, the most frequently selected trait has been the tendency to drink ethanol solutions, and several rat lines have been developed. Studies with these selected lines have been reviewed (Browman, Crabbe, & Li, 2000; Eriksson, 1972; Lumeng, Murphy, McBride, & Li, 1995). All selections essentially have offered animals a choice between 10% ethanol in tap water and plain water over a period of weeks. Because several sets of rat lines (and now new sets of mouse lines) have all been selected for virtually the same trait, it has proved useful to compare results across selection experiments. It is interesting to note that when operant paradigms are used to assess the avidity of these rat lines for alcohol under different scheduling and procedural conditions, some rather striking differences have emerged across selection experiments (Files, Samson, Denning, & Marvin, 1998; Samson, Files, Denning, & Marvin, 1998). Nonetheless, some similarities also have been seen, such as the tendency for most of the genetically high-drinking lines and strains to display lowered brain serotonin levels or function, consistent with evidence in humans and animal models.

Studies with other lines selected to be alcohol Withdrawal Seizure-Prone (WSP) or -Resistant (WSR), sensitive (Long-Sleep, LS) or resistant (Short-Sleep, SS) to ethanol sedation, and differentially responsive to many effects of several drugs have been reviewed elsewhere (Browman et al., 2000; Deitrich & Erwin, 1996; Marley, Arros, Henricks, Marley, & Miner, 1998; Mogil et al., 1995; Mohammed, 2000).

Use of the technique of artificial selection continues. In a later section, I show that it can serve as an aid to gene mapping. As more aspects of drug sensitivity phenotypes are modeled through selection, it will become clear which parts of the complex drug dependence phenotypes are genetically coregulated. One approach that has not yet been systematically applied is to breed selectively for behavioral traits thought to be relevant for drug sensitivity. For example, antisocial behavior, depression, and impulsivity are highly comorbid with alcoholism and drug dependence. If lines of impulsive and reserved mice were bred, they could be compared for their drug sensitivity or tendency to self-administer drugs. Examples of the application of this approach to anxiety can be found in the chapter by Flint (chapter 21, this volume).

QTL Mapping

The basic methods used for finding genes that affect a behavior have been described elsewhere (Crabbe, Phillips, Buck, Cunningham, & Belknap, 1999; Moore & Nagle, 2000; Plomin et al., 1991; Whatley, Johnson, & Erwin, 1999). All depend on linkage between genetic markers and behavioral phenotypes. Spontaneously occurring mutations were used as markers to map traits from the early days of genetics. Examples of the application of QTL methods to the study of learning and memory and anxiety can be found in other chapters in this volume by Flint (chapter 21) and by Wehner and Balogh (chapter 7). In the addictions, nearly all research has been conducted with mice, and much of that has used various populations derived from the genetic cross between two inbred strains: C57BL/6J (B6) and DBA/2J (D2). Much research has shown that these strains differ in sensitivity to nearly all effects of nearly all drugs (Crabbe & Harris, 1991). Beginning immediately after the strategy was proposed (Plomin et al., 1991), several groups began to map QTLs by phenotyping the set of 25 recombinant inbred (RI) strains derived from the F2 cross of B6 and D2. Because these RI strains had been genotyped for more than 1,500 markers whose genomic positions are known, only phenotyping was necessary, which offered a great advantage to laboratories expert in behavioral analysis but initially without the molecular biological expertise necessary to genotype. Because there are few genotypes in the BXD RI panel (each strain represents a single genotype), it is insufficiently powerful to detect QTLs for all but those genes with the strongest effects on the phenotype. Thus, additional populations are used to follow up promising associations discovered in the BXD RIs. Such populations include individual B6D2F2 animals, animals from the backcross to either progenitor inbred strain (B6 or D2) or lines derived from the F2 that have been selected as phenotypic extremes (see Figure 16.1).

In an early review of progress toward QTLs relevant for drugs of abuse in the BXD RI panel, our laboratory identified 187 associations with 38 behavioral responses to six drugs, ranging from acute sensitivity to morphine-induced analgesia to the severity of alcohol withdrawal (Crabbe et al., 1994). After several more years of work using follow-up populations and some other genetic crosses as well, 24 QTLs could be placed with certainty on the mouse genome. These included three alcohol withdrawal loci, six alcohol preference loci (one in rats), five loci for ethanol-induced loss of righting reflex, and others for cocaine, morphine, and pentobarbital responses (Crabbe, Phillips, et al., 1999). Many new loci have been added since.

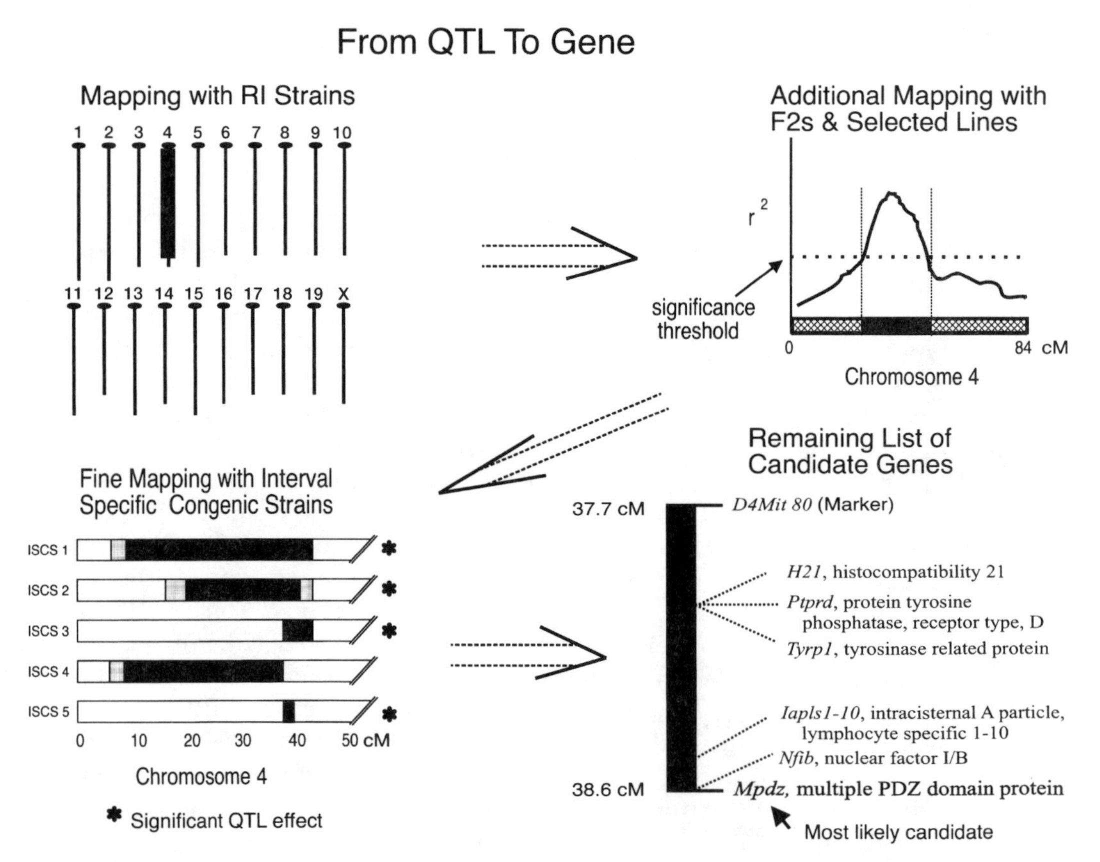

Figure 16.1. From QTL to Gene. Steps depicting the stages in the process of identifying the location of a gene are shown. The mapped trait is the severity of the withdrawal reaction following acute treatment with the drug, pentobarbital. The first stage was to test 25 BXD RI strains of mice and their C57BL/6J and DBA/2J progenitors (first panel). Comparison with mapped markers showed the presence of an influential gene somewhere on chromosome 4. Additional mapping with B6D2F2 mice and short-term selectively bred lines confirmed the significance of association with markers on chromosome 4 (r^2) and delimited the region of association to that shown in the black bar in the second panel. A panel of interval-specific congenic strains was then tested, each bearing a short section of chromosome 4 (shown in panel 3 in black) introgressed from the C57BL/6J onto the DBA/2J background. Four of the five congenic strains (indicated with black stars) showed significantly lower withdrawal than the DBA/2J stain, despite the incorporation of less than 2% C57BL/6J DNA. The region of less than 1 cM introgressed onto ISCS 5 contained about 20 genes. Some possible candidates within this region are depicted in the final panel. For reasons discussed in the text, *Mpdz* is a particularly promising candidate gene. Data from Buck et al., 1999 and Fehr et al., 2002.

From QTL to Gene

This substantial amount of progress, as yet, has failed to identify unequivocally a single gene definitively accepted to be the source of a QTL association. This is because the number of genotypes studied for any one trait is still relatively modest, and the margin of error (i.e., the 90% confidence interval for the location of the QTL) is relatively large. A typical confidence interval during the early stages of a QTL mapping project contains more than 100 genes, any one of which could be responsible for the QTL association. Progress in the area of QTL mapping has now shifted toward the use of congenic strains (Darvasi, 1997). These are created by backcrossing one inbred strain onto another for more than 10 generations, each time keeping track of the genotype surrounding the QTL confidence interval. Gradually, the DNA from the donor strain is diluted (by approximately half each generation) until only the 1%–2% of the genome surrounding the QTL has been transferred (or introgressed) to the recipient, or background, strain (see Browman & Crabbe, 1999; Crabbe, Phillips, et al., 1999). When the congenic strain is tested for the phenotype versus the background strain and found to differ significantly, the capture of the gene responsible for the QTL within the small introgressed region is proved, and the job of gene identification becomes more approachable (see Figure 16.1). The congenic strain can be backcrossed to the background strain, and when recombinations occur, they can be selected so that a panel of interval-specific congenic strains can be created. Systematic analysis of which strains possess the phenotypic effect leads to an ever-diminishing confidence interval (Darvasi, 1997).

One of the most advanced cases of this method is seen in the efforts of Kari Buck to map QTLs for acute alcohol and pentobarbital withdrawal. Using the enhanced handling-induced convulsion seen several hours after acute injection of a high dose of drug to index dependence severity, our group first showed that several genomic regions were associated with a genetic predisposition to alcohol and nitrous oxide withdrawal in the BXD RI strains (Belknap et al., 1993). Using three mapping populations (RIs, B6D2F2s, and short-term selected lines), Buck's group was able to provide clear-cut statistical evidence for three QTLs and suggestive evidence for two others (Buck, Metten, Belknap, & Crabbe, 1997). Subsequent studies have found substantial overlap in the regions similarly mapped for acute pentobarbital withdrawal (Buck, Metten, Belknap, & Crabbe, 1999) and provisionally for chronic ethanol withdrawal (Crabbe, 1998). Buck and Belknap have recently used a set of congenic strains to demonstrate that the region of chromosome 4 harboring one of the pentobarbital and alcohol withdrawal QTLs can be limited to less than 1 cM. This region contains fewer than 20 genes (Fehr, Shirley, Belknap, Crabbe, & Buck, 2002; see Figure 16.1).

One gene, a zinc finger protein called *Mpdz*, appears to be an attractive candidate among the eight known genes and various other DNA sequences remaining in the confidence interval. *Mpdz* was further implicated when an examination of single-nucleotide polymorphisms (SNPs; see

Wehner & Balogh, chapter 7, and other references in this volume) within the *Mpdz* gene in 15 inbred strains found 18 SNPs, arrayed in six patterns (haplotypes). Three different protein variants of *Mpdz* were identified. The haplotypes and protein variants correlated significantly with the ethanol withdrawal severity previously determined in the strains, suggesting that *Mpdz* is a strong candidate for being the gene underlying the QTL association (Fehr et al., 2002).

Another drug response project that has made significant strides toward mapping QTLs has been the search for QTLs underlying sensitivity to ethanol-induced loss of righting reflex. Several of the QTLs have been verified in congenic strains, and the confidence intervals surrounding their location narrowed. A review of all known existing congenic strains in development for alcohol- and drug-related responses also contains an excellent descriptive discussion of congenic development and use (Bennett, 2000). Multiple groups have mapped QTLs related to alcohol preference drinking. Progress in these mapping efforts has been discussed in detail elsewhere (Foroud et al., 2000; Phillips et al., 1998).

New Developments in QTL Mapping

Congenic strains (and the even-smaller interval-specific and small donor segment congenics) are not the only route available for fine genetic mapping. Flint (chapter 21, this volume) discusses the use of an alternative strategy to map fear conditioning or anxiety loci based on use of a genetically heterogeneous stock of mice, and heterogeneous stocks also have been used to map the locomotor stimulant response to alcohol (Demarest, Koyner, McCaughran, Cipp, & Hitzemann, 2001). A thoughtful review (Flint & Mott, 2001) argues rather pessimistically that QTL mapping efforts based on inbred strains, crosses, and congenics are likely to fail to resolve the effects of single genes with sufficient precision to allow gene capture. A major concern they raise is that as the confidence intervals for individual QTLs gradually are narrowed, each QTL may represent the actions of multiple genes, each of even smaller effect. This was seen for the alcohol locomotor stimulant QTLs, but it did not preclude their finer mapping (Demarest et al., 2001). Similar arguments devaluing QTL approaches in favor of others such as random mutagenesis (see next section) have been raised elsewhere (Nadeau & Frankel, 2000). The arguments for and against the various methods of gene discovery are rather technical. Although QTL mapping efforts have yet to capture a gene, my colleagues and I are more sanguine about the potential for success using the current methods (Belknap et al., 2001; Demarest et al., 2001). The alternative methods proposed also have not yet succeeded in capturing a gene, so it would seem prudent to wait for the dust to settle before declaring which methods work best.

Two new resources were reported that could shorten the time from trait measurement to initial rough QTL mapping. Two sets, each with more than 60 congenic strains, have been generated, each comprising strains bearing C57BL/6J genotype except for a region introgressed from either DBA/2J or CAST/Ei mice (Iakoubova et al., 2001). The 2–4 congenics for each chromosome have overlapping segments and span the entire genome. Thus, an individual could test the 60 congenic strains versus the background C57BL/6J for a trait of interest and identify, in a straightforward fashion, the rough chromosomal location of QTLs. A similar logic was applied to develop chromosome substitution strains, in which an entire chromosome was introgressed for each strain (Nadeau, Singer, Matin, & Lander, 2000).

Finally, one group has recently reported a novel approach to mapping QTLs (Grupe et al., 2001). Grupe et al. first analyzed 15 standard inbred mouse strains for SNPs and identified 500 scattered across the genome. To these data they added published allele information about nearly 3,000 additional SNPs. They sought associations between SNPs and published strain differences in behavioral and other responses and found several "hits" in regions mapped by other methods. The method has yet to prove its specificity or reliability, and as with most gene mapping methods, the sensitivity of this approach currently is limited by the relatively few genotypes (strains) in the database. However, if the Mouse Phenome Project leads to a proliferation of data on multiple strains, such computational methods could replace some of the more tedious early steps of screening the genome for regions harboring genes. A third newer method based on gene trapping (Mitchell et al., 2001) is more technical than can be described in this chapter.

Random Mutagenesis

An alternative to QTL analyses is the use of a mutagenic chemical such as N-ethyl-N-nitrosourea (ENU) to induce mutations at random throughout the genome. Offspring of mutated males are then screened to see whether there are phenotypic effects, that is, whether the mouse is phenotypically deviant. That mouse is then bred, its offspring tested, and if they are also deviant, the mutation that is the source of altered behavior can then be mapped and identified using methods such as QTL mapping described earlier. Several very large projects currently are mutating thousands of mice (Hrabe de Angelis et al., 2000; Nolan et al., 2000). Because of the costs, mice in such a project typically are screened in a number of very simple, high-throughput assays. Interesting mutants will then be tested for related behaviors that are more complex and relevant. Several of the ongoing screens include behavioral assays relevant to drug or alcohol sensitivity. The ENU screens are very new and have strong supporters, some of whom have argued that they are a much better gene discovery alternative than QTL mapping (Nadeau et al., 2000). As neither approach has yet to put a gene on the table, and each has distinct strengths and weaknesses (Belknap et al., 2001), it seems premature to decide on theoretical grounds which will ultimately prove to be more useful.

Gene Expression Analyses

Many neurobiological studies over the years have used painstaking biochemical and molecular biological methods such as Northern analysis, Western analysis, Southern blotting, and in situ hybridization to ascertain whether the abundance of specific DNAs, messenger RNAs (mRNA), or proteins are affected by drug treatment or drug susceptibility. All such differences can ultimately be traced to greater or lesser expression of the protein's gene. These studies are reviewed frequently for a range of psychopharmacological traits. For example, an excellent review of the role of gene expression vis-à-vis alcohol considers the substantial evidence that gamma-aminobutyric acid (GABA) and G-protein coupled receptor systems are markedly influenced by alcohol (Reilly, Fehr, & Buck, 2001). Such studies are slow because each gene's product must be targeted with specific biochemical ligands. These traditional approaches depend on the nomination of a candidate gene of interest.

Expression Array Profiling

A modern alternative is expression array profiling, or the analysis of gene chips. This allows the investigator to study thousands of genes simultaneously, without prior targeting other than deciding which genes are represented on the chip (Watson & Akil, 1999). Currently, at least 6,000 mouse brain DNA probes are represented on one or another type of gene chip, and other chips contain ligands for several thousand human genes. Nestler (2000) discussed how changes in gene expression are expected to be of importance to addictive processes. One study has reported a specific pattern of gene expression changes (e.g., in protein kinases) in nucleus accumbens tissue from primates exposed to cocaine for over a year (Freeman et al., 2001). Two studies have used the technology to screen widely for gene expression changes relevant for alcohol responses. In a comparison of frontal cortical tissue from alcoholics versus control patients (Lewohl et al., 2000), several genes that play a role in the cell cycle were found to be expressed differentially. A similar study used adrenal tissue from mice acutely withdrawing from ethanol, or cultured cells, and found that expression of dopamine beta hydroxylase was highly activated during ethanol withdrawal (Thibault et al., 2000). In an initial attempt to screen more broadly for differentially expressed genes, Xu, Ehringer, Yang, and Sikela (2001) compared whole-brain mRNA samples from naive inbred LS and SS mice using two different high-density arrays containing more than 18,000 and nearly 9,000 probes. Forty-one genes (or expressed sequence tags, ESTs) were found to be expressed differentially, including several with interesting functions (Xu et al., 2001).

A behavioral genomics perspective will have much to offer to the gene expression approach as it develops exponentially over the next few years. Each expression array study typically identifies dozens of genes whose expression is increased or decreased as a function of the diagnostic or treatment group compared with the control group. Although there are exceptions, expression typically is changed about 100%–200%. Sorting the wheat from the chaff is daunting, and many of the challenges are similar to those facing QTL analysis. To name a few, there are numbers of false-positive findings, some due to the thousands of multiple comparisons performed; there is difficulty detecting changes in genes with intrinsically low expression levels; and determining that a given gene is indeed regulated requires additional experiments, a problem when there are dozens of genes differentially expressed.

The principal hurdle facing expression profiling, though, is truly the behavioral genomics question: What are all these genes doing? The new discipline charged with sorting all this out typically is described as *bioinformatics*, and the current state of the art for synthesizing an array output comprises classifying the differentially expressed genes categorically (e.g., *cellular metabolism* or *cellular signaling*). A behavioral genomics approach to such a classification presents a dilemma. Would it be reasonable to state that cocaine dysregulated expression of genes involved in "reinforcement"? Conceivably, one could perform an experiment in which animals were trained to self-administer cocaine and compare brain tissue from those animals self-administering with tissue from those to whom cocaine administrations had been given in equal amounts, but noncontingently (i.e., so-called yoked controls). If one found that chronic cocaine affected expression of genes coding for the enzymes of monoamine synthesis and metabolism and of monoamine receptor genes, and specifically in the brain areas thought to subserve reward circuitry, and differentially in those animals actively self-administering cocaine, one might be comfortable in making such an assertion. For the moment, however, the intrinsic orientation of scientists using these techniques tends to be toward a reductionist, proteomics perspective. That is, the genes affected in fact code for proteins, and for the time being we will see the explanations offered at the level of the immediate environment in which each protein is synthesized and acts (i.e., a cellular–molecular level).

Targeted Mutagenesis

The new, "classical" method for analyzing single gene function in neuroscience is no longer the analysis of naturally occurring mutants but the generation of new mutants. As described by Wehner and Balogh (chapter 7, this volume), targeting strategies have been applied to genes thought to be highly relevant to the trait being mapped based on other neurobiological evidence. Many such overexpression, underexpression, or null mutants have been studied for drug-related responses (for review, see Buck, Crabbe, & Belknap, 2000; Mogil, Yu, & Basbaum, 2000). It is beyond the scope of this chapter to review progress in such studies, but as the second and subsequent generations (e.g., temporally conditional and tissue-specific gene knock outs) of this technology mature, the information gained will be increasingly valuable. As another example of future developments, sometimes a QTL confidence interval contains a gene whose function appears to be important, and a knock-out model can then be used to test that possibility. Discussions of the limitations of interpretation of such gene targeting studies have appeared elsewhere (Wehner & Balogh, chapter 7, this volume; see also Gerlai, 1996).

Complexities Facing Behavioral Genomics

Identifying the genes responsible for Mendelian traits has become a straightforward task. However, this chapter now turns to considerations of the genetic complexity of behavior. Individuals, and animals possessing different genotypes such as inbred strains or mutants, respond differently to the same environment. Directly from this fact follows the quantitative genetic truism that a heritability estimate is a description of a trait in a population, as well as the recognition that careful definition of the trait is also crucial. Rowe (chapter 5, this volume) discusses this complexity in more detail and deals with the separate issue of gene–environment correlation as well. I cite three examples of complexity (in addition to pleiotropy) relevant to addiction genetics.

Environmental Dependence of Genetic Effects

Even though both genotype and environment can be manipulated by the experimenter in animal model research, the specific environmental variables responsible for causing genotypes to respond differently may be difficult to specify. Given that inbred strain differences in behavior are well documented, my colleagues and I recently performed an experiment asking whether the same strain differences would be evident in three laboratories. We tried to equate as many environmental and genetic variables as possible (Crabbe, Wahlsten, & Dudek, 1999). We obtained mice from several inbred strains and null mutants for the serotonin receptor subtype 5-HT$_{1B}$ at exactly the same age, fed them the same chow, and treated them as identically as we could. At exactly the same age, mice were tested in each laboratory on a battery of six behaviors, using identical apparatus (e.g., identical elevated plus mazes were used to assess "anxiety") and test protocols, which specified how animals were handled, time of day they were tested, and so on. There were, of course, some variables we could not standardize, such as the local water and air in the animal facility and the different individuals in the three locations performing the experiments.

Strains differed markedly and significantly in all behaviors, an expected outcome. For some tests (e.g., open-arm exploration of the elevated plus maze, an index of anxiety) there were significant Strain × Laboratory interactions. Taking the example of the 5-HT$_{1B}$ null mutants, one lab would have concluded that they were more anxious than wild types, a second lab less anxious, and the third, not different. However, for alcohol preference drinking, the strain pattern was very similar across laboratories (Crabbe, Wahlsten, & Dudek, 1999). Because the reliability of an internal replication was high at each site, we interpret these results to mean that it is difficult to discern what may be important variations in environmental conditions; therefore, before drawing conclusions about gene–behavior relationships, examination in multiple laboratories and under varying test conditions is prudent.

Phenotypic Specificity of Genetic Influence

The dependence of genetic differences on the environmental milieu immediately suggests an additional complexity, one that turns out to be very important as well. The most salient feature of the "environment" is the test apparatus and protocol. Because we cannot interrogate our animals about their subjective experiences, we devise apparatus and tests that appear to us to tap the physiological and behavioral domains of interest. There is a natural tendency among behavioral scientists to reify their assays of behavior, which should be strenuously resisted. Particularly in gene-finding experiments, we must not conclude that a gene associated with increased open-arm entries in an elevated plus maze is a gene "for anxiety" (Lederhendler & Schulkin, 2000). Even to approach such a generalization would require convergent evidence from a battery of tests.

In another study, my colleagues and I reported that null mutants lacking 5-HT_{1B} receptors were less susceptible to acute alcohol intoxication using the grid test of motor incoordination. When we gathered additional data with this mutant, we found that knock outs were less sensitive to alcohol on some, but not all of several tasks, all of which putatively assess alcohol's incoordinating effects (e.g., the rotarod, a balance beam; Boehm II, Schafer, Phillips, Browman, & Crabbe, 2000). Therefore, not all laboratory assays described by the experimenters as if they assessed a single, monolithic domain are measuring the same thing. We are currently exploring the particular attributes of a single task, the accelerating rotarod, in a panel of inbred strains. By varying such parameters as testing protocol, rod diameter, and acceleration rate, we will be able to determine whether strain sensitivities are correlated or whether these variants of a single task in fact represent different behaviors at the level of the genes that influence them.

Epistasis

The last complexity to be considered here is just beginning to be explored. Multiple genes of relatively small effect influence complex traits. All the current methods essentially assume that each gene's effect can be isolated and studied. However, genes interact, a condition termed *epistasis* (see Grigorenko, chapter 14, this volume). I have discussed a recent example likely to be representative of epistasis elsewhere (Crabbe, 2001). Three groups studying the stress axis recently independently produced null mutants for the corticotropin-releasing hormone receptor-2 (*Crhr2*) gene. All groups compared knock outs and wild-types on multiple tests of anxiety, and three different conclusions were drawn. In each case, the laboratory's results were consistent internally. For example, the first group (Coste et al., 2000) found no differences on three separate tasks related to anxiety, whereas the second (Bale et al., 2000) found significant differences in all tests, and the third found sex-specific differences (Kishimoto et al., 2000). The likely source of the different results is due to epistatic interactions between genes in the background strains harboring the mutation (because genetically distinct substrains of the 129 inbred strains were used in the three laboratories) or other genes in the transgenic construct introduced, although other possibilities cannot be ruled out (Crabbe, 2001).

A more straightforward demonstration of an epistatic effect is a QTL analysis for acute pentobarbital withdrawal severity. Mice homozygous for DBA/2J strain alleles in a region of chromosome 11 and those homozygous for C57BL/6J alleles differed significantly in withdrawal, indicating the presence of QTL in this region (Buck et al., 1999). However, when mice were genotyped in another region on the distal end of chromosome 1, only mice with DBA/2J genes showed the difference in withdrawal. If mice had C57BL/6J alleles on distal chromosome 1, there was no difference in withdrawal between mice with C57BL/6J versus DBA/2J genomes on chromosome 11 (Hood, Crabbe, Belknap, & Buck, 2001). As expected, there are different opinions about the implications of epistasis for QTL analysis. Some feel that epistasis makes QTL analysis likely to fail (Flint & Mott, 2001; Nadeau et al., 2000), whereas others see a better understanding of epistasis as a step toward the better articulated, behavioral genomics description of gene effects on integrated, whole-organism responses (Belknap et al., 2001).

Conclusion

Studies with genetic animal models have made sustained contributions to our understanding of the neurobiology of addictions. Much of this occurred during a period when genetic studies were neither fashionable nor widespread. Genetics is now both, and genetic studies of complex traits is the current frontier. Although many difficulties have been raised in this chapter, the message to the reader should not be one of pessimism, rather one of caution. Interpretation of genetic effects on behavior is neither more nor less difficult now than before the genomics revolution. What has changed is that its practice has expanded, and it remains too tempting to many scientists to overinterpret their data—hence, the caution recommended.

One goal currently driving the engine of gene discovery is the discovery of novel therapeutic agents (Hefti, 2001). Using specific knowledge of an individual's genotype, scientists could tailor drugs to maximize an individual's response. This of course raises ethical questions, including genetic privacy issues. A recent commentary discusses many of these issues in thoughtful detail (Freeman et al., 2001). The genetic tools available early in the 21st century are dazzling, but they cannot substitute for careful biology and behavioral analysis. The next generation of addiction scientists should be able to solve at least some of the riddles governing individual differences in susceptibility to abuse drugs.

References

Bale, T. L., Contarino, A., Smith, G. W., Chan, R., Gold, L. H., Sawchenko, P. E., et al. (2000). Mice deficient for corticotropin-releasing hormone receptor-2 display anxiety-like behaviour and are hypersensitive to stress. *Nature Genetics, 24*, 410–414.

Belknap, J. K., Hitzemann, R., Crabbe, J. C., Phillips, T. J., Buck, K. J., & Williams, R. W. (2001). QTL analysis and genome-wide mutagenesis in mice: Complementary genetic approaches to the dissection of complex traits. *Behavior Genetics, 31*, 5–15.

Belknap, J. K., Metten, P., Helms, M. L., O'Toole, L. A., Angeli-Gade, S., Crabbe, J. C., et al. (1993). Quantitative trait loci (QTL) applications to substances of abuse: Physical dependence studies with nitrous oxide and ethanol in BXD mice. *Behavior Genetics, 23*, 213–222.

Bennett, B. (2000). Congenic strains developed for alcohol- and drug-related phenotypes. *Pharmacology, Biochemistry and Behavior, 67*, 671–681.

Boehm II, S. L., Schafer, G. L., Phillips, T. J., Browman, K. E., & Crabbe, J. C. (2000). Sensitivity to ethanol-induced motor incoordination in 5-HT(1B) receptor null mutant mice is task-dependent: Implications for behavioral assessment of genetically altered mice. *Behavioral Neuroscience, 114*, 401–409.

Browman, K. E., & Crabbe, J. C. (1999). Alcohol and genetics: New animal models. *Molecular Medicine Today, 5*, 310–318.

Browman, K. E., Crabbe, J. C., & Li, T.-K. (2000). Genetic strategies in preclinical substance abuse research. In F. E. Bloom & D. J. Kupfer (Eds.), *Psychopharmacology: A fourth generation of progress* [CD-ROM Version 3]. New York: Lippincott, Williams & Wilkins.

Buck, K. J., Crabbe, J. C., & Belknap, J. K. (2000). Alcohol and other abused drugs. In D. W. Pfaff, W. H. Berrettini, T. H. Joh, & S. C. Maxson (Eds.), *Genetic influences on neural and behavioral functions* (pp. 159–183). Boca Raton, FL: CRC Press.

Buck, K. J., Metten, P., Belknap, J. K., & Crabbe, J. C. (1997). Quantitative trait loci involved in genetic predisposition to acute alcohol withdrawal in mice. *Journal of Neuroscience, 17*, 3946–3955.

Buck, K., Metten, P., Belknap, J., & Crabbe, J. (1999). Quantitative trait loci affecting risk for pentobarbital withdrawal map near alcohol withdrawal loci on mouse chromosomes 1, 4, and 11. *Mammalian Genome, 10*, 431–437.

Coste, S. C., Kesterson, R. A., Heldwein, K. A., Stevens, S. L., Heard, A. D., Hollis, J. H., et al. (2000). Abnormal adaptations to stress and impaired cardiovascular function in mice lacking corticotropin-releasing hormone receptor-2. *Nature Genetics, 24*, 403–409.

Crabbe, J. C. (1998). Provisional mapping of quantitative trait loci for chronic ethanol withdrawal severity in BXD recombinant inbred mice. *Journal of Pharmacology Experimental Therapeutics, 286*, 263–271.

Crabbe, J. C. (1999). Animal models in neurobehavioral genetics: Methods for estimating genetic correlation. In P. Mormède & B. C. Jones (Eds.), *Neurobehavioral genetics: Methods and applications* (pp. 121–138). Boca Raton, FL: CRC Press.

Crabbe, J. C. (2001). Use of genetic analyses to refine phenotypes related to alcohol tolerance and dependence. *Alcoholism: Clinical and Experimental Research, 25*, 288–292.

Crabbe, J. C., Belknap, J. K., & Buck, K. J. (1994). Genetic animal models of alcohol and drug abuse. *Science, 264*, 1715–1723.

Crabbe, J. C., & Harris, R. A. (1991). *The genetic basis of alcohol and drug actions*. New York: Plenum Press.

Crabbe, J. C., Phillips, T. J., Buck, K. J., Cunningham, C. L., & Belknap, J. K. (1999). Identifying genes for alcohol and drug sensitivity: Recent progress and future directions. *Trends in Neuroscience, 22*, 173–179.

Crabbe, J. C., Wahlsten, D., & Dudek, B. C. (1999). Genetics of mouse behavior: Interactions with laboratory environment. *Science, 284*, 1670–1672.

Crawley, J. N., Belknap, J. K., Collins, A., Crabbe, J. C., Frankel, W., Henderson, N., et al. (1997). Behavioral phenotypes of inbred mouse strains: Implications and recommendations for molecular studies. *Psychopharmacology, 132*, 107–124.

Darvasi, A. (1997). Interval-specific congenic strains (ISCS): An experimental design for mapping a QTL into a 1-centimorgan interval. *Mammalian Genome, 8*, 163–167.

Deitrich, R. A., & Erwin, V. G. (1996). *Pharmacological effects of ethanol on the nervous system*. Boca Raton, FL: CRC Press.

Demarest, K., Koyner, J., McCaughran, J. J., Cipp, L., & Hitzemann, R. (2001). Further characterization and high-resolution mapping of quantitative trait loci for ethanol-induced locomotor activity. *Behavior Genetics, 31*, 79–91.

Eriksson, K. (1972). Behavioral and physiological differences among rat strains specially selected for their alcohol consumption. *Annals of the New York Academy of Sciences, 197*, 32–41.

Fehr, C., Shirley, R. L., Belknap, J. K., Crabbe, J. C., & Buck, K. J. (2002). Congenic mapping of alcohol and pentobarbital withdrawal liability loci to a <1cM interval of murine chromosome 4: Identification of *Mpdz* as a candidate gene. *Journal of Neuroscience, 22*, 3730–3738.

Files, F. J., Samson, H. H., Denning, C. E., & Marvin, S. (1998). Comparison of alcohol-preferring and nonpreferring selectively bred rat lines: II. Operant self-administration in a continuous-access situation. *Alcoholism: Clinical and Experimental Research, 22*, 2147–2158.

Flint, J., & Mott, R. (2001). Finding the molecular basis of quantitative traits: Successes and pitfalls. *Nature Reviews Genetics, 2*, 437–445.

Foroud, T., Bice, P., Castelluccio, P., Bo, R., Miller, L., Ritchotte, A., et al. (2000). Identification of quantitative trait loci influencing alcohol consumption in the high alcohol drinking and low alcohol drinking rat lines. *Behavior Genetics, 30*, 131–140.

Freeman, W. M., Nader, M. A., Nader, S. H., Robertson, D. J., Gioia, L., Mitchell, S. M., et al. (2001). Chronic cocaine-mediated changes in non-human primate nucleus accumbens gene expression. *Journal of Neurochemistry, 77*, 542–549.

Fuller, J. L. (1964). Measurement of alcohol preference in genetic experiments. *Journal of Comparative Psychology, 57*, 85–88.

Gerlai, R. (1996). Gene-targeting studies of mammalian behavior: Is it the mutation or the background genotype? *Trends in Neuroscience, 19*, 177–181.

Grupe, A., Germer, S., Usuka, J., Aud, D., Belknap, J. K., Klein, R. F., et al. (2001). *In silico* mapping of complex disease-related traits in mice. *Science, 292*, 1915–1918.

Hefti, F. (2001). From genes to effective drugs for neurological and psychiatric disease. *Trends in Pharmacological Science, 22*, 159–160.

Hood, H. M., Crabbe, J. C., Belknap, J. K., & Buck, K. J. (2001). Epistatic interaction between quantitative trait loci affects the genetic predisposition for physical dependence on pentobarbital in mice. *Behavior Genetics, 31*, 93–100.

Hrabe de Angelis, M. H., Flaswinkel, H., Fuchs, H., Rathkolb, B., Soewarto, D., Marschall, S., et al. (2000). Genome-wide, large-scale production of mutant mice by ENU mutagenesis. *Nature Genetics, 25*, 444–447.

Iakoubova,O. A., Olsson, C. L., Dains, K. M., Ross, D. A., Andalibi, A., Lau, K., et al. (2001). Genome-tagged mice (gtm): Two sets of genome-wide congenic strains. *Genomics, 74*, 89–104.

Kishimoto, T., Radulovic, J., Radulovic, M., Lin, C. R., Schrick, C., Hooshmand, F., et al. (2000). Deletion of *crhr2* reveals an anxiolytic role for corticotropin-releasing hormone receptor-2. *Nature Genetics, 24*, 415–419.

Lederhendler, I., & Schulkin, J. (2000). Behavioral neuroscience: Challenges for the era of molecular biology. *Trends in Neuroscience, 23*, 451–454.

Lewohl, J. M., Wang, L., Miles, M. F., Zhang, L., Dodd, P. R., & Harris, R. A. (2000). Gene expression in human alcoholism: Microarray analysis of frontal cortex. *Alcoholism: Clinical and Experimental Research, 24*, 1873–1882.

Logue, S. F., Swartz, R. J., & Wehner, J. M. (1998). Genetic correlation between performance on an appetitive-signaled nosepoke task and voluntary ethanol consumption. *Alcoholism: Clinical and Experimental Research, 22*, 1912–1920.

Lumeng, L., Murphy, J. M., McBride, W. J., & Li, T.-K. (1995). Genetic influences on alcohol preference in animals. In H. Begleiter & B. Kissin (Eds.), *The genetics of alcoholism* (pp. 165–201). New York: Oxford University Press.

Marley, R. J., Arros, D. M., Henricks, K. K., Marley, M. E., & Miner, L. L. (1998). Sensitivity to cocaine and amphetamine among mice selectively bred for differential cocaine sensitivity. *Psychopharmacology, 140*, 42–51.

McClearn, G. E., & Rodgers, D. A. (1959). Differences in alcohol preference among inbred strains of mice. *Quarterly Journal of Studies on Alcohol, 20*, 691–695.

Metten, P., Phillips, T. J., Crabbe, J. C., Tarantino, L. M., McClearn, G. E., Plomin, R., et al. (1998). High genetic susceptibility to ethanol withdrawal predicts low ethanol consumption. *Mammalian Genome, 9*, 983–990.

Mitchell, K. J., Pinson, K. I., Kelly, O. G., Brennan, J., Zupicich, J., Scherz, P., et al. (2001). Functional analysis of secreted and transmembrane proteins critical to mouse development. *Nature Genetics, 28*, 241–249.

Mogil, J. S., Flodman, P., Spence, M. A., Sternberg, W. F., Kest, B., Sadowski, B., et al. (1995). Oligogenic determination of morphine analgesic magnitude: A genetic analysis of selectively bred mouse lines. *Behavior Genetics, 25*, 397–406.

Mogil, J. S., Sternberg, W. F., Marek, P., Sadowski, B., Belknap, J. K., & Liebeskind, J. C. (1996). The genetics of pain and pain inhibition. *Proceedings of the National Academy of Sciences USA, 93*, 3048–3055.

Mogil, J. S., Yu, L., & Basbaum, A. I. (2000). Pain genes? Natural variation and transgenic mutants. *Annual Review of Neuroscience, 23*, 777–811.

Mohammed, A. H. (2000). Genetic dissection of nicotine-related behaviour: A review of animal studies. *Behavioral Brain Research, 113*(1–2), 35–41.

Moore, K. J., & Nagle, D. L. (2000). Complex trait analysis in the mouse: The strengths, the limitations and the promise yet to come. *Annual Review of Genetics, 34*, 653–686.

Nadeau, J. H., & Frankel, W. N. (2000). The roads from phenotypic variation to gene discovery: Mutagenesis versus QTLs. *Nature Genetics, 25*, 381–384.

Nadeau, J. H., Singer, J. B., Matin, A., & Lander, E. S. (2000). Analysing complex genetic traits with chromosome substitution strains. *Nature Genetics, 24*, 221–225.

Nestler, E. J. (2000). Genes and addiction. *Nature Genetics, 26*, 277–281.

Nolan, P. M., Peters, J., Strivens, M., Rogers, D., Hagan, J., Spurr, N., et al. (2000). A systematic, genome-wide, phenotype-driven mutagenesis programme for gene function studies in the mouse. *Nature Genetics, 25*, 440–443.

Paigen, K., & Eppig, J. T. (2000). A mouse phenome project. *Mammalian Genome, 11*, 715–717.

Phillips, T. J., Belknap, J. K., Buck, K. J., & Cunningham, C. L. (1998). Genes on mouse chromosomes 2 and 9 determine variation in ethanol consumption. *Mammalian Genome, 9*(12), 936–941.

Plomin, R., & Crabbe, J. (2000). DNA. *Psychological Bulletin, 126*, 806–828.

Plomin, R., McClearn, G. E., Gora-Maslak, G., & Neiderhiser, J. M. (1991). Use of recombinant inbred strains to detect quantitative trait loci associated with behavior. *Behavior Genetics, 21*, 99–116.

Prescott, C. A., & Kendler, K. S. (1999). Genetic and environmental contributions to alcohol abuse and dependence in a population-based sample of male twins. *American Journal of Psychiatry, 156*, 34–40.

Reilly, M. T., Fehr, C., & Buck, K. J. (2001). Alcohol and gene expression in the central nervous system. In N. Moussa-Moustaid & C. D. Berdanier (Eds.), *Nutrient–gene interactions in health and disease* (pp. 131–162). Boca Raton, FL: CRC Press.

Rose, R. J. (1995). Genes and human behavior. *Annual Review of Psychology, 46*, 625–654.

Samson, H. H., Files, F. J., Denning, C., & Marvin, S. (1998). Comparison of alcohol-preferring and nonpreferring selectively bred rat lines: I. Ethanol initiation and limited access operant self-administration. *Alcoholism: Clinical and Experimental Research, 22*, 2133–2146.

Tarantino, L. M., McClearn, G. E., Rodriguez, L. A., & Plomin, R. (1998). Confirmation of quantitative trait loci for alcohol preference in mice. *Alcoholism: Clinical and Experimental Research, 22*, 1099–1105.

Thibault, C., Lai, C., Wilke, N., Duong, B., Olive, M. F., Rahman, S., et al. (2000). Expression profiling of neural cells reveals specific patterns of ethanol-responsive gene expression. *Molecular Pharmacology, 58*, 1593–1600.

Tsuang, M. T., Lyons, M. J., Eisen, S. A., Goldberg, J., True, W., Lin, N., et al. (1996). Genetic influences on *DSM–III–R* drug abuse and dependence: A study of 3,372 twin pairs. *American Journal of Medical Genetics, 67*, 473–477.

Watson, S. J., & Akil, H. (1999). Gene chips and arrays revealed: A primer on their power and their uses. *Biological Psychiatry, 45*, 533–543.

Whatley, V. J., Johnson, T. E., & Erwin, V. G. (1999). Identification and confirmation of quantitative trait loci regulating alcohol consumption in congenic strains of mice. *Alcoholism: Clinical and Experimental Research, 23*, 1262–1271.

Xu, Y., Ehringer, M., Yang, F., & Sikela, J. M. (2001). Comparison of global brain gene expression profiles between inbred long-sleep and inbred short-sleep mice by high-density gene array hybridization. *Alcoholism: Clinical and Experimental Research, 25*, 810–818.

17

Genetic and Environmental Risks of Dependence on Alcohol, Tobacco, and Other Drugs

Andrew C. Heath, Pamela A. F. Madden, Kathleen K. Bucholz, Elliot C. Nelson, Alexandre Todorov, Rumi Kato Price, John B. Whitfield, and Nicholas G. Martin

The field of genetic research on addiction vulnerability, defined here as research designed to understand individual differences in risk of dependence on, or persistent harmful use of, alcohol, tobacco, or illicit drugs, offers great opportunities but also considerable challenges. Because genes have been identified that influence the metabolism of alcohol and have effects on alcohol-dependence risk (ALDH2, ADH2) or that influence the metabolism of nicotine (CYP2A6), the addiction field already provides illustration of the complexities that are to be anticipated as we identify other individual gene effects on human behavior. As reviewed elsewhere (see Crabbe, chapter 16, this volume), the ability to study drug effects in mice, rats, and other experimental organisms makes possible controlled breeding experiments that ultimately may identify the involvement of previously unsuspected genes in the regulation of responses to individual licit or illicit drugs, and also makes possible the dissection of the central nervous system effects of these drugs in ways that would not be possible in humans. Controlled dosing of humans, in drug challenge studies conducted in the human experimental laboratory, offers the possibility of identifying aspects of drug response (e.g., postalcohol "ataxia," or body sway; subjective, cardiovascular, and other physiological reactions to alcohol, nicotine, or other drugs) that are under simpler genetic control than more complex constructs such as dependence. At the same time, the challenges of understanding the complex interplay of genetic and environmental influences on behavior are particularly great for human drug use and dependence, where environmental influences may begin prenatally (through fetal alcohol effects; Streissguth et al., 1994; Yates, Cadoret, Troughton,

This work was supported by National Institutes of Health Grants AA07728, AA11224, AA11998, AA13321, CA75581, and DA/CA12854.

Stewart, & Giunta, 1998) and continue throughout the life span through parent, peer, and partner/spouse effects; where drug availability and use may vary markedly across different cultures and between different birth cohorts; and where the pertinent environmental risk factors may vary as a function of stage of progression of substance use or problems. The addictions field is thus fertile for those seeking to understand the interactions of individual genetic and environmental risk factors. In the present chapter, we highlight some of these issues that must be confronted as we attempt to understand individual gene effects on human addictive behaviors.

What Is Drug Dependence?

Drug dependence is defined in contemporary psychiatric practice (*Diagnostic and Statistical Manual of Mental Disorders* [*DSM–IV*]; American Psychiatric Association, 1994) by use of a drug class (e.g., alcohol, cannabis, or stimulants, which we refer to as *drug*) leading to the experiencing, within the same 12-month period, of at least three symptoms from a list of seven: (a) acquisition of tolerance to the drug; (b) withdrawal symptoms when drug use is reduced or stopped; (c) inability to quit or a persistent desire to reduce or stop using the drug; (d) loss of control over drug use as indicated by using much more than originally intended; (e) spending a great deal of time using the drug, getting the drug, or recovering from its effects; (f) giving up important occupational, social, or recreational activities because of use of the drug; and (g) continued use despite knowledge of the adverse physical or emotional consequences that the drug is having. Although very high interrater reliability can be achieved using standardized diagnostic interview assessments of drug dependence (e.g., Bucholz, Cadoret, et al., 1996; Hesselbrock, Eaton, Bucholz, Schuckit, & Hesselbrock, 1999), there remain unresolved questions about how best to operationalize individual diagnostic criteria for particular drug classes, or indeed about whether individual criteria are appropriate for some drug classes. Thus the existence of a cannabis withdrawal syndrome, denied in *DSM–IV*, remains controversial (e.g., Wiesbeck et al., 1996). The discrimination between tolerance to alcohol associated with central nervous system changes occurring after chronic heavy alcohol use and behavioral adaptations to the effects of alcohol in adolescents who are learning to drink, and sometimes drinking very heavily, remains beyond the resolving power of many diagnostic interviews. The extent to which specific genetic (or environmental) risk factors may differentially predispose to particular symptoms (e.g., withdrawal) in humans remains unknown, although findings in other species support this possibility (e.g., Buck, Metten, Belknap, & Crabbe, 1997). Tackling such questions, however, is likely to require a move away from reliance on current approaches to the diagnosis and assessment of substance dependence in genetic research.

Defined by contemporary diagnostic criteria, dependence on alcohol (alcoholism), tobacco, or illicit drugs (substance dependence) is widespread, with lifetime rates of dependence (i.e., the proportion of individuals reporting a history of dependence, though not necessarily currently affected) increasing in younger birth cohorts (e.g., Grant, 1997; Kessler et al., 1994; Kandel, Chen, Warner, Kessler, & Grant, 1997). Reported lifetime prevalence of *DSM–III–R* (American Psychiatric Association, 1987) alcohol dependence in the U.S. National Comorbidity Survey, for example, was 20.1% for males and 8.2% for females, with corresponding figures of 9.2% and 5.9% for illicit drug dependence (Kessler et al., 1994); and 20% lifetime prevalence of *DSM–III–R* nicotine dependence was reported for a sample of young adult men and women (Breslau, Kilbey, & Andreski, 1991). This high prevalence of dependence itself poses a challenge to genetic researchers, because the increase in risk to the full siblings of a substance-dependent proband, compared with risk in the general population, though highly significant, for *DSM–IV* dependence criteria, is of the order of only 1.5- to 2-fold, a magnitude of effect that predicts relatively poor power for use of traditional linkage methods (Risch, 1990). Substance dependence is typically of early onset, particularly in recent birth cohorts, with median age of individual alcoholism symptoms, for example, reported to be age 20 in the recent U.S. National Comorbidity Survey (Nelson, Little, Heath, & Kessler, 1996).

There has been a long history of attempts to subclassify drug dependence cases, most notably in the alcohol dependence field (Babor, 1996). Cloninger (1987), for example, proposed a framework that distinguished between an early-onset highly heritable subtype, seen predominantly in men and associated with antisocial features such as fighting, and a later onset type that was more moderately heritable, seen in women as well as men and associated with features such as guilt about drinking. Successful subtyping would facilitate the task of identifying individual genes associated with dependence vulnerability if it identified subtypes that were more strongly heritable or had more homogeneous inheritance patterns. Although apparent support has been found for such subtypes in Scandinavian data (e.g., Cloninger, Bohman, & Sigvardsson, 1981; Kendler, Karkowski, Prescott, & Pedersen, 1998), and variant classifications have been proposed (e.g., Babor et al., 1992), support for the use of such classification schemes in genetic research using diagnostic interview-based data has been scant: The Cloninger and Kendler studies used official records for analysis: Swedish Temperance Board registrations for alcohol-related offenses. The empirical basis for distinguishing high heritability and intermediate heritability subtypes in the original Stockholm adoption study also does not appear strong (Heath, Slutske, & Madden, 1997).

Latent class analysis (e.g., McCutcheon, 1987), which may be viewed as a categorical variant of factor analysis and which hypothesizes the existence of categorical rather than continuous latent variables, has been used in an attempt to define rigorous classification algorithms to subtype substance-dependent cases (e.g., Bucholz, Heath, et al., 1996; Heath et al., 1994; Kendler et al., 1998; Madden et al., 1997). Factor analysis implies the hypothesis that the observed correlations between a set of symptoms or other variables can be explained by the influence of a much smaller number of continuously distributed latent factors; latent class analysis implies that symptom correlations can be explained by the existence of a small number of discrete, mutually exclusive classes, having class-specific symptom endorsement probabilities. Parameters of the latent class model are class membership probabilities (i.e., prevalence estimates) and class-specific symptom endorsement probabilities; and from these can be derived conditional probabilities of membership in each of the estimated latent classes, given any observed profile of symptoms. Advantages of the latent class approach are that it thus allows probabilistic classification, taking into account the fact that some cases can be assigned to a given class with probability close to unity, whereas for others membership in two or more different classes may be equally probable; and that it can, in principle, be extended to incorporate a genetic modeling approach (Eaves et al., 1993).

Figure 17.1 compares two hypothetical latent class solutions. The first (Figure 17.1a) corresponds to the case in which there are genuine subtypes, with one subtype having especially high endorsement of one subset of symptoms and the second subtype having high endorsement of a second subset of symptoms. This type of pattern is seen for symptoms of attention deficit hyperactivity disorder (ADHD; e.g., Hudziak et al., 1998; Neuman et al., 1999) and broadly supports the subclassification of ADHD into in-

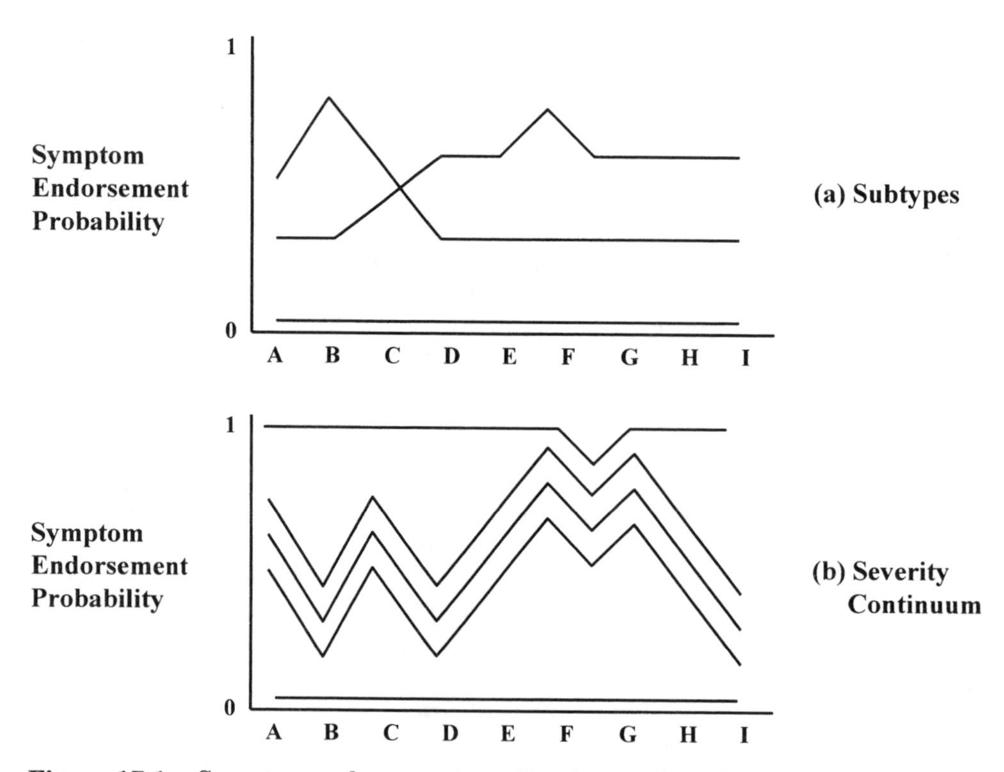

Figure 17.1. Symptom endorsement profiles for two hypothetical latent class solutions: (a) genuine subtypes and (b) cases in which classes differ in severity. Upper case letters denote different symptoms. Lines are used to show symptom endorsement probabilities for different classes.

attentive, hyperactive-impulsive, and combined types. The second hypothetical solution (Figure 17.1b) corresponds to the case in which classes differ in their severity, with symptom endorsement probabilities varying for different symptoms and also increasing from least severe to most severely affected classes, but in this idealized case with no crossing over of the symptom endorsement probability plots for the different classes. Latent class analysis has been applied most frequently in the alcohol field, although findings appear to be similar when applied to tobacco or other drug classes. Very consistently, whether applied to alcoholism symptoms reported by general community samples (e.g., female and male twins from an Australian twin study; Heath et al., 1994; Heath, Slutske, Bucholz, Madden, & Martin, 1997) or reported by severe alcoholic probands and their relatives from a family study (Bucholz, Heath, et al., 1996), it is the second pattern that is observed. Some symptoms (e.g., alcohol withdrawal, unsuccessful efforts to cut down on drinking) are endorsed with high probability by only the most severe classes. Others (e.g., alcohol tolerance, use in greater amounts than intended, persistent desire to quit or cut down) are endorsed even by relatively mild problem classes that endorse few or no other symptoms. Thus there is no evidence to support subtyping using dependence symptoms, at least in the sense implied by Figure 17.1(a).

Many traditional diagnostic distinctions are also not supported by results of these same latent class studies. Thus *DSM–IV* makes a distinction between substance *abuse* and *dependence*, with the former to be diagnosed in those not meeting criteria for dependence and sometimes considered a milder form of problems ("maladaptive pattern of substance use"; American Psychiatric Association, 1994). In the case of alcohol-related problems, with the exception of recurrent hazardous substance use (e.g., drunk driving), which in some cultures is widely reported by men even with no other problem use, other abuse symptoms are endorsed with high probability only by the more severe alcoholism classes (e.g., Bucholz et al., 1996). Likewise, the subclassification of dependence into subtypes with and without physiological dependence (American Psychiatric Association, 1994), on the basis of symptoms of either tolerance or withdrawal, is not supported by the finding that symptoms of tolerance are commonly reported by those with otherwise mild problems, while withdrawal is typically reported only by the most severely affected individuals. Finally, the use for the same diagnostic criterion of symptoms that have very different endorsement patterns—for example, a history of unsuccessful attempts to quit, and persistent desire to quit, which as we have noted above are endorsed by severe versus even moderate problem classes—also introduces considerable heterogeneity into the range of individuals who are being classified as dependent. All of these issues raise concerns about unthinking acceptance of traditional diagnostic criteria for genetic research.

The results obtained by latent class analyses of alcoholism symptoms, taking the form of Figure 17.1(b) rather than Figure 17.1(a), are consistent with two quite distinct (but not necessarily exclusive) etiologic models, which we may summarize as the *continuum* and *stage* models. Under the stage model, it is hypothesized that dependence progresses through a series of stages, from initial mild dependence progressing to severe dependence with withdrawal, with risk factors (genetic or environmental) for progression to the mildest stages not necessarily the same as risk factors for progression to the most severe stages (e.g., Bucholz, Heath, & Madden, 2000; Nelson, Heath, & Kessler, 1998). For the purposes of genetic research, the stage model would suggest that much narrower definitions of drug dependence (e.g., requiring the presence of severe symptoms such as alcohol withdrawal) will be required to discover genes that play a critical role in transition to the most severe stages of dependence. Under the continuum model, it is hypothesized that there is a continuum of severity of substance-related problems, and that it is quite possible that the same risk factors are operating throughout the range of dependence severity, implying that use of quantitative symptom count measures may be more useful than an arbitrary dichotomization into dependent and unaffected cases. More progress in clarifying these questions can be made using traditional quantitative genetic or genetic epidemiologic methods (e.g., Neale & Cardon, 1992), although the clearest insights will come from studying individual gene effects on progression of symptoms. Moving beyond the arbitrary dichotomization into dependent and unaffected cases is essential if progress is to be made in addressing such questions. For the purposes of genetic research, we still do not have convincing answers to the question: What is drug dependence?

Lessons to Be Learned From Existing Studies

Gene Effects Associated With Metabolism

The best example of how a polymorphism at a single genetic locus may lead to important differences in addiction risk is provided by the example of the ALDH2 (aldehyde dehydrogenase) locus effects on alcoholism risk, documented in a series of studies conducted by Higuchi and colleagues in Japan. It is worthwhile to consider this research in some detail, for the broader implications that it carries for studying single gene effects on human behavior, as well as its particular implications for single gene effects on human addictive behaviors.

Ethanol is converted by the enzyme alcohol dehydrogenase to the toxic metabolite acetaldehyde, which is in turn converted by the enzyme acetaldehyde dehydrogenase to acetic acid. A single point mutation in the gene for ALDH2, which is located on chromosome 12 (Hsu, Yoshida, & Mohandas, 1986), leads to an inactive enzyme. Those who are homozygous for the mutant allele (ALDH2*2/*2) or who are heterozygotes (ALDH2*1/*2) have substantially elevated blood acetaldehyde concentrations after ingestion of alcohol (Wall et al., 1997) and experience a characteristic flushing response. Compared with frequencies of 58%, 35%, and 7% for the *1/*1, *1/*2, and *2/*2 genotypes in a community control sample ($N = 461$), Higuchi et al. (1994) reported genotype frequencies in a large series of Japanese male alcoholics ($N = 655$) of 88%, 12%, and 0%, respectively, demonstrating significant reductions in risk associated with the *1/*2 and *2/*2 genotypes. Similar results have been obtained in Chinese and Taiwanese alcoholic series (Chen et al., 1999; Shen et al., 1997). Although it is commonly presumed that single gene effects on behavior are always probabilistic, the risk of alcohol dependence in ALDH2*2/*2 homozygotes is close to zero, with very few cases of individuals with this genotype who have become alcohol dependent known to alcohol researchers.

The Japanese data, as well as supporting data from other Asian populations, illustrate several important principles: (a) the multigenic determination of dependence risk and relatively modest contribution of any single gene to risk of dependence, (b) the increased power that may be obtained by considering gene effects on quantitative measures compared with dichotomous disease outcomes, (c) the importance of Genotype × Environment interaction effects, (d) the importance of identifying mediating variables that intervene between genes and the final outcome of substance dependence, (e) the fact that individual risk factors (whether genetic or environmental) may have effects specific to a particular stage in the onset and progression of substance use and dependence, and (f) the possible confounding role of population stratification effects that may complicate interpretation of an association between genotype and outcome measures. Let us consider these issues in turn.

Multigenic determination. In addition to the effects of the ALDH2 locus, significant though smaller effects of a second locus, ADH2, on alcoholism risk have also been demonstrated. Alcohol dehydrogenase (ADH), responsible for most of the conversion of alcohol to acetaldehyde, is formed by a random combination of three different subunits (alpha, beta, and gamma) that are encoded by three closely linked loci on chromosome 4 (ADH1, ADH2, and AHD3 in traditional rotation), of which only the two latter are known to be polymorphic. The low-risk ADH2*2 allele encodes the Beta2 subunit associated with faster metabolism of alcohol (i.e., leading to a faster increase in toxic acetaldehyde levels; Thomasson et al., 1991) than the Beta1 subunit associated with the ADH2*1 allele. Because ALDH2 effects on risk are much greater than those of ADH2, Higuchi et al. (1994) compared genotype frequencies in alcoholics and community controls who were ALDH2*1/*1 homozygotes, finding that the ADH2*1/*1 homozygotes were significantly more common in the alcoholic than control series (30.4% vs. 7.3%) and the ADH2*2/*2 homozygotes correspondingly less common (35.8% vs. 58.1%). This association has also been confirmed in Han Chinese (Muramatsu et al., 1995; Shen et al., 1997; Thomasson et al., 1991) and Atayal Taiwanese (Thomasson et al., 1994). Although Higuchi et al. also noted an association between ADH3 genotype and alcohol dependence risk, interpretation of this latter association is complicated by the strong linkage disequilibrium that exists between ADH2 and ADH3 loci (high positive correlation between occurrence of the ADH2*1 and ADH3*2 alleles), with more recent research failing to find an association with ADH3 once this linkage disequilibrium is allowed for (Osier et al., 1999).

The relatively modest contributions of the ALDH2 and ADH2 loci to risk of alcoholism are documented by additional data collected by Higuchi and colleagues from a small community sample. Using a questionnaire measure of problems with alcohol, Higuchi, Matsushita, Muramatsu, Murayama, and Hayashida (1996) found that only 16% of men with the high-risk ALDH2*1/*1 genotype were alcoholic by the questionnaire measure, a small but significant increased risk compared with the 10% overall prevalence of alcoholism in the sample. Among individuals with the ALDH2* 1/*1 genotype, risk was as high as 29% among those with the ADH2*1/*1 genotype or as low as 7% in those with the ADH2*2/*2 genotype. In other words, knowledge of the ALDH2 and ADH2 genotypes of Japanese males living in Japan is giving important information about differences in alcoholism risk, but not much more so than, say, knowledge of the history of smoking (a risk factor strongly correlated with alcoholism risk; Madden, Bucholz, Martin, & Heath, 2000) in males of European ancestry.

Quantitative effects. In contrast to the modest effect on risk of alcoholism, defined as a dichotomous measure, seen for the ALDH2 locus in the Higuchi et al. general community sample, a much more potent effect on simple quantitative indices of alcohol consumption is observed (Higuchi et al., 1996), with a 10-fold difference in average monthly alcohol consumption between ALDH2*1/*1 homozygotes and ALDH2*2/*2 homozygotes, and a 2.7-fold increase in consumption of the former compared with ALDH2*1/*2 heterozygotes. From the distribution of frequency of alcohol consumption reported by these authors, we may compute that, even without adjustment for unreliability of measurement, approximately one third of the total variance in alcohol consumption levels in this sample is explained by this single genetic locus. In psychological research, it is extremely rare to find individual risk factors with effects on quantitative indices of behavior of this magnitude!

Genotype × environment interaction effects. Higuchi et al. (1994) also provided what is arguably the most striking example of a Genotype × Environment interaction effect on human behavior. In comparisons of genotype frequencies in Japanese alcoholics from patient samples acquired at three times (1979, 1986, and 1992), they documented a highly significant increase in the proportion of patients who were ALDH2*1/*2 heterozygotes, from 2.5% to 8.0% and finally 13.0% (in other words, a diminishing protective effect of having a single *2 allele). Higuchi et al. suggested that this change is associated with increased social pressures for drinking with colleagues after work experienced by Japanese men. The dangers of ignoring sociocultural influences on drinking patterns, and their possible interaction with genotype, are also demonstrated by the gender differences observed by Higuchi et al. (1996) in their analysis of average monthly alcohol consumption as a function of ALDH2 genotype. They reported for their female respondents similar marked increase in levels of consumption, compared with ALDH2*2/*2 homozygotes, in *1/*1 homozygotes (a 33-fold increase) or *1/*2 heterozygotes (a 2-fold increase).

However, Japanese women with the highest risk ALDH2*1/*1 genotype were reporting alcohol consumption levels similar in magnitude to those of Japanese men with the lowest risk ALDH2*2/*2 genotype. Overall, average monthly alcohol consumption levels by Japanese women in Japan were 5–10 times lower than those of Japanese men. In contrast, among Americans of Japanese ancestry, living in a culture in which there is near universal acceptance of drinking by women, a much smaller gender difference in consumption levels is seen.

Mediating variables. Taken together, the ALDH2 genotype frequency data from the alcoholic case control series of Higuchi and colleagues, the association between the ALDH2*2 allele and the flushing response, and the associations between ALDH2 genotype and average monthly alcohol consumption suggest a very simple *mediating variable* model for the association between genotype and alcoholism risk. Impaired metabolism of alcohol associated with possession of a single ALDH2*2 allele may lead to unpleasant reactions after ingestion of alcohol, hence decreased average levels of alcohol consumption, and therefore decreased risk of alcohol dependence. While this is likely to be an oversimplification—for example, there is some suggestion that ALDH2*1/*2 heterozygotes find low doses of alcohol more pleasant, rather than less pleasant, than *1/*1 homozygotes (Wall, Thomasson, Schuckit, & Ehlers, 1992; Luczak, Elvrie-Kreis, Shea, Carr, & Wall, 2002)—it generates a model that can be tested prospectively (e.g., beginning in adolescence) to examine the role of genetic effects on consumption in accounting for genetic effects on alcoholism risk. In contrast to the association between ALDH2 and consumption levels, Higuchi et al. (1996) failed to find a significant association between ADH2 genotype and consumption levels, leading them to the conclusion that ADH2 effects must be mediated through other effects, perhaps only seen at much higher levels of alcohol consumption than would be seen in a general community sample. Establishing that a particular gene has a significant effect on risk of substance dependence still leaves unanswered many questions, including behavioral questions, about how that effect arises.

Stage-specific risk factor effects. Other studies on Japanese and other Asian populations show that while possession of a single ALDH2*2 allele on average has a protective effect on level of alcohol consumption and on alcoholism risk in those who progress to heavy drinking, presumably because of the increased exposure to acetaldehyde associated with impaired metabolism, possession of an ALDH2*2 allele becomes associated with increased risk of adverse medical consequences, including increased frequency of alcohol-related cancers (e.g., Hori, Kawano, Endo, & Yuasa, 1997; Yokoyama et al., 1996). This extreme example, in which a protective factor at one stage (the progression to heavy drinking) becomes a risk factor for a later stage (the experiencing of alcohol-related medical complications), illustrates the fundamental principle that risk factors, genetic or environmental, may have effects that are limited to particular stages in the progression of addictive behaviors.

Population stratification effects. Compared with most Western societies, which comprise individuals of very diverse ancestral origins, Japan is considered a relatively genetically homogeneous society. This is an important consideration in the interpretation of case control data, such as the Higuchi series of alcoholics and community controls, because if there are sociocultural differences between individuals of different ancestral origins (e.g., Scandinavian Lutherans vs. Irish Roman Catholics), as well as gene frequency differences, then these sociocultural differences may be important confounding factors that can create false-positive associations, or alternatively mask associations, with individual genetic loci. In an extreme case, we may even find a genetic association for a nongenetic trait (e.g., a "Lutheran" religious affiliation gene).

The practical importance of such confounding effects (termed *population stratification effects* by geneticists) has sometimes been questioned (e.g., Morton & Collins, 1998; see also Cardon, chapter 4, this volume). However, sociocultural confounding factors are likely to be particularly important for behaviors related to alcohol, tobacco, or other drug use that show pronounced prevalence differences between different societies, or even between different age cohorts within societies. For this reason, despite the greater statistical power of traditional case control tests for single gene effects (e.g., Risch, 2000), in most studies on Western populations it is usual to include at least a validation stage that uses strategies to control for such potential confounding effects. This may be achieved by using multiple unlinked genetic markers as a "genomic control" (Devlin & Roeder, 1999; Pritchard & Rosenberg, 1999)—the principle here being that with a sufficient number of unlinked genetic markers, it should be possible to unmix a genetically heterogeneous population into more homogeneous genetic groups, in the same way that latent class analysis or other latent structure methods may be used to identify subtypes on the basis of symptom profiles, and thus determine whether genetic background is a potential confounding factor in an observed genetic association. Alternatively, it may be achieved by using within-family comparisons, such as the transmission disequilibrium test (which obtains DNA from affected cases and both biological parents and tests for significantly more frequent transmission from the heterozygous parent to affected offspring of a candidate allele—hypothetically the ALDH2*1 allele, for example—than other alleles at that locus), or its sibship extensions (Spielman & Ewens, 1998; Spielman, McGinnis, & Ewens, 1993). Such comparisons achieve perfect matching for ancestry, because full siblings, who share the same two biological parents, necessarily share the same ancestral origins, and the biological child of those parents must likewise be of the same ancestral background. All of the Japanese studies that we have cited have used standard case control methods, assuming that confounding effects are likely to be minimal in this more genetically homogeneous society. However, even within Japan, it now is apparent that there are regional differences in ALDH2 gene frequency that correlate with regional differences in alcohol consumption levels.

More limited data are available concerning the metabolism of nicotine and its involvement in genetically determined differences in smoking behavior. Preliminary reports of an association between a CYP2A6 polymorphism and smoking behavior in a sample of European ancestry (Pianezza, Sellers, & Tyndale, 1998) were found to be marred by a genotyping error (Oscarson, 2001). As in the case of loci with effects on alcohol metabolism, it does now appear that CYP2A6 is more highly polymorphic in Asian than in European ancestry populations (Oscarson, 2001), so that it is likely that findings on associations between CYP2A6 genotype, nicotine metabolism, and smoking behavior will emerge initially from Asian populations (cf. Kwon et al., 2001; Nakajima et al., 2001).

Genetic Effects in Europeans: Implications of Twin and Adoption Studies

Results from the first large-sample alcohol challenge twin study conducted using young adult Australian twins in the late 1970s (Martin, Oakeshott, et al., 1985; Martin, Perl, et al., 1985) suggest that there are also important genetic effects on alcohol metabolism in those of European ancestry. (Quantitative genetic studies of nicotine metabolism in humans, using adequately large sample sizes to detect significant genetic effects, are only just beginning, and to our knowledge similar studies of other drugs have not yet been undertaken.) On the basis of long-term follow-up of the Australian alcohol challenge sample, it appears that these differences in alcohol metabolism may be associated with differences in risk of alcohol dependence (Whitfield et al., 1998). However, the ALDH2*2 allele has not been reliably observed in individuals of solely European ancestry and thus cannot contribute to genetic risk of alcohol dependence in this group. The ADH2*2 allele is seen only at relatively low frequency (approximately 5% gene frequency), so that there have been few studies of associations with alcohol dependence in European populations. One study of Australians of European ancestry did observe the predicted decreased frequency of the ADH2*1/*2 genotype (there were no ADH2*2/*2 homozygotes observed) in alcohol-dependent men compared with controls (2.7% vs. 15.1%), but there was a nonsignificant trend in the opposite direction in women (11.1% vs. 4.6%; Whitfield et al., 1998). Furthermore, interpretation of this traditional case control comparison in men is complicated by evidence for significant population stratification effects, with the low-risk ADH2*2 allele associated particularly with English ancestry (e.g., significantly associated with a religious affiliation of Church of England, and with endorsement of Socialist social attitudes; Heath et al., 2001). Within-sibship comparisons identified five male like-sex sibpairs discordant for ADH2 genotype: In all five cases, quantitative measures of alcohol use and problems were elevated in the dizygotic (DZ) twin with the (high-risk) ADH2*1/*1 genotype compared with the heterozygous cotwin (Whitfield et al., 1998). Thus the evidence for an ADH2 genotype effect on alcoholism in those of European ancestry is at best marginal (but see Neale et al., 1999, for stronger evidence for ADH2 effects on alcohol consumption levels). As noted earlier, the CYP2A6 locus, when properly genotyped, shows only a very low frequency variant in those of European ancestry, which greatly limits the power to detect associations with smoking behavior in these populations.

More indirect evidence that there are important genetic effects on risk of alcohol, tobacco, or other drug dependence in those of European ancestry (there are still remarkably little data on minority groups such as those of African American ancestry) is provided by a series of large-sample twin and adoption studies. Although the earliest twin studies of alcohol and tobacco use, in particular, began in the 1950s and 1960s, the most compelling evidence for genetic effects on risk of substance dependence has been from more recent research. Extensive review of the twin and adoption data on genetic contributions to substance dependence is beyond the scope of this chapter (for reviews, see, e.g., Heath, 1995a, 1995b; Heath & Madden, 1997). Here we limit ourselves to consideration of three key issues and their implications for other genetic research: (a) shared environment effects and Genotype × Environment interaction effects on dependence risk; (b) mediators of genetic influence; and (c) genetic effects on natural history or treatment or intervention response.

Shared Environment, Genotype × Environment Interaction Effects

Adoption studies have consistently found increased risk of alcoholism or other drug dependence in the offspring of alcoholic parents compared with the offspring of control pairs (for a review of individual studies, see, e.g., Heath, Slutske, & Madden, 1997). The pioneering Danish adoption study (Goodwin et al., 1974; Goodwin, Schulsinger, Knop, Mednick, & Guze, 1977) was the first to clearly demonstrate increased risk of alcoholism to the adopted-away male offspring of alcoholic biologic parents (18%) but not to the adopted-away offspring of control parents (5%). Uniquely, the study was also able to obtain interviews with some biologic offspring who had been reared by alcoholic parents who had given other children up for adoption, finding no increased risk (17%) in this admittedly small group compared with the high-risk adoptees. Thus the importance of genetic factors in the intergenerational transmission of alcoholism (at least in males) was established. Inferences about the importance of environmental effects are limited, however, by the more limited range of environmental adversity experienced in adoptive families. In most Western societies, adoptive parents are typically older and screened for the absence of major psychopathology, and are therefore unlikely to be experiencing severe alcohol or drug dependence at the time they are rearing adoptive offspring. This limitation of the adoption design is apparent from the second major Scandinavian adoption study, the Stockholm Adoption Study (e.g., Cloninger et al., 1981; Cloninger, Bohman, Sigvardsson, & Von Knorring, 1985), which noted a very low rate of alcohol-related temperance board registrations in adoptive parents (fewer than 4% of adoptive families had one or more parents with at least one temperance board registration, in contrast to approximately 14% of all Swedish males and 32% of the biological fathers of the adopted-away offspring). For the purpose of understanding intergenerational environmental transmission of risk of alcoholism or other drug dependence, the traditional adoption design may be considered far from ideal.

In the case of the twin study, at the simplest level, inference of significant genetic effects relies on the observation, in the case of quantitative traits, of higher twin pair correlations in monozygotic (MZ) pairs, who are genetically identical, than in fraternal pairs who on average share only one half of their genes in common; or, in the case of binary traits such as history of substance dependence, significantly higher risk to the MZ cotwins of affected individuals than to the DZ cotwins of affected individuals, after controlling for any overall zygosity differences in prevalence (this latter requirement, unfortunately, is often overlooked in studies that rely exclusively on twin samples ascertained through clinically affected individuals or probands). Provided that there is no evidence that MZ pairs have had more similar early trait-relevant environmental experiences than DZ pairs (except insofar as these are elicited by the behaviors of the twins themselves)—an issue that has been studied many times without finding major effects (see, e.g., Kendler, Neale, Kessler, Heath, & Eaves, 1993)—genetic influences may be inferred. Very consistently, in studies of quantitative measures of alcohol consumption (Heath, 1995b), categorical measures of alcoholism variously defined (e.g., from treatment or other official records, from interview data, or from questionnaire data; Heath, 1995a; Heath et al., 1997), categorical measures of smoking and nicotine dependence (Heath & Madden, 1997; Kendler et al., 1998; True et al., 1999), and measures of illicit drug use (e.g., Tsuang et al., 1999) or abuse or dependence (Kendler, Bulik, et al., 2000; Tsuang et al., 1996), evidence for significant genetic effects is found.

This simplicity disappears, however, when we wish to make inferences from twin data about shared environmental influences (such as parent-offspring environmental influences, neighborhood influences, school inferences, shared peer influences, or other environmental influences shared by two same-age siblings growing up together), because of the complicating factors of genetic nonadditivity and Genotype × Environment interaction effects. In an ideal twin study, data would be collected on MZ and DZ twin pairs reared together and MZ and DZ twin pairs reared apart. In practice, given the very low numbers of separated twin pairs (particularly in the case of contemporary birth cohorts) and the large numbers of pairs required for testing genetic hypotheses using binary phenotypes of intermediate heritability, most twin studies of substance dependence risk have been limited to twin pairs reared together. Table 17.1 summarizes the contributions of additive and nonadditive (i.e., dominance or epistatic) genetic effects, shared and nonshared environmental effects, as well as Genotype × Environment interaction effects, to the variances and covariances of MZ and DZ twin pairs reared together and reared apart. Inter-

Table 17.1. Contributions of Additive and Nonadditive Genetic Affects, Shared and Nonshared Environment Effects, and Genotype × Environment Interaction Effects to Twin Pair Covariances and Variance

Variance/covariance	Additive genetic variance	Nonadditive genetic variance	Shared environmental variance	Additive genetic × shared environment interaction variance	Nonshared environmental variance	Additive genetic × nonshared environment interaction variance
MZ pairs reared together	1	1	1	1	0	0
DZ pairs reared together	1/2	1/4	1	1/2	0	0
MZ pairs reared apart	1	1	0	0	0	0
DZ pairs reared apart	1/2	1/4	0	0	0	0
Phenotypic variance	1	1	1	1	1	1

Note. MZ = monozygotic; DZ = dizygotic.

actions between nonadditive genetic effects and environmental effects are ignored in the table, because these effects are likely to be slight.

In a seminal paper, Jinks and Fulker (1970) showed that data from MZ and DZ twin pairs reared together and at least one twin group (ideally two groups) reared apart were sufficient to permit estimation of additive and nonadditive genetic and shared and nonshared environmental main effects (a fact that can be clearly seen from the coefficients in Table 17.1). In the absence of separated twin pairs, however, shared environmental effects and nonadditive genetic effects will be confounded, with genetic nonadditivity decreasing the DZ covariance below one half the MZ covariance, and shared environmental effects increasing the DZ covariance above one half the MZ covariance. If nonadditive genetic effects are assumed small in magnitude, then ignoring genetic nonadditivity and using only data from twin pairs reared together should yield a reasonable (albeit only approximate) estimate of the importance of genetic effects. This strategy has been used successfully by behavioral geneticists for a wide range of quantitative traits, yielding results from twin data that are in good agreement with results from adoption or other designs. It is important to recognize, however, the approximation that is being used. Thus a recent meta-analysis of twin studies of major depression (which occurs at greater than chance rates in alcoholics, i.e., is commonly comorbid with alcoholism) concluded that the shared environmental contribution to risk of depression was consistently estimated at zero, with a very narrow 95% confidence interval (0%–5%; Sullivan, Neale, & Kendler, 2000). Yet if twin pair resemblance were entirely explained by additive genetic effects, we would not expect to observe that the shared environmental variance would be consistently estimated at its lower bound of zero across a number of different studies: Because of sampling variation, we would expect to observe at least some studies with a small positive estimate for the shared environmental variance. A more plausible summary of this pattern of findings would thus be that there are both additive and nonadditive genetic influences on risk of major depression and that the possibility of shared environmental effects on risk of depression that are being masked by genetic nonadditivity cannot be excluded.

Consideration of the possibility of Genotype × Environment interaction effects, and in particular the possibility that the effects of environmental risk factors on risk of substance dependence depend on the individual's genotype, introduces an additional complication (e.g., Eaves, Last, Martin, & Jinks, 1977; Jinks & Fulker, 1970). As shown in Table 17.1, the coefficients of the interaction components are simply given by the product of the corresponding main effects. The principal implication of this, though pointed out previously, deserves emphasis. To the extent that the environmental impact of, say, parental alcoholism or drug dependence depends on offspring genotype, this Genotype × Shared Environment interaction effect will be confounded with the additive genetic main effect in data limited to MZ and DZ twin pairs reared together (they can in principle be resolved with additional data from both separated MZ and DZ twin pairs; e.g., see Heath et al., 2002). Thus, although we may safely conclude that there are important genetic effects on risk of substance dependence, the relative balance of additive genetic main effects and Genotype × Shared Environment interaction effects remained undetermined. Given the well-established association between parental alcoholism and early trauma such as early childhood sexual abuse, which in turn predict increased risk of substance use disorders and psychopathology (e.g., Dinwiddie et al., 2000; Kendler, Bulik, et al., 2000; Nelson et al., 2002), there is reason for concern that the importance of Genotype × Environment interaction effects in the etiology of substance use disorders has received insufficient attention.

Consideration of the limitations of the traditional adoption design for resolving environmental consequences of parental substance use disorders, as well as the confounding of additive genetic and Genotype Family Environmental interaction effects, has led to a new generation of behavioral genetic studies using an offspring-of-twins design (Jacob et al., 2001; for discussion of the offspring-of-twins designs, see, e.g., Cloninger, Rice, & Reich, 1979; Heath, Kendler, Eaves, & Markell, 1985; Nance & Corey, 1976). As summarized in Table 17.2, by studying offspring of twin pairs who are concordant or discordant for alcoholism or other drug dependence and of control pairs, one can distinguish between individuals at high genetic and high environmental risk (raised by a drug-dependent biological parent), individuals at much lower environmental risk but high genetic risk (raised by nondrug-dependent parents, one of whom is the MZ cotwin of a drug-dependent twin) or intermediate genetic risk (a nondependent parent is the DZ cotwin or full sibling of a drug-dependent individual), or individuals at low genetic and low environmental risk (from control families). Most importantly, the confounding of genetic effects and Genotype × Family Environment effects is thus avoided, because offspring at high genetic risk are observed under both high-risk and low-risk environmental conditions. Thus as progress is made in the coming years in the identification of genes that contribute to risk of substance dependence, we may also anticipate increased progress in identifying environmental risk factors that may have genotype-dependent effects.

Table 17.2. Genetic and Environmental Risk to Offspring in the Offspring-of-Twins Design, as a Function of Drug Dependence History of Parent and Parent's MZ or DZ Cotwin (see Jacob et al., 2001, for further details)

		Offspring risk	
Parental status	Cotwin status	Genetic	Environmental
Drug dependent	Any	High	High
Unaffected	Drug dependent, MZ cotwin	High	Low
Unaffected	Drug dependent, DZ cotwin	Intermediate	Low
Unaffected	Unaffected	Low	Low

Note. MZ = monozygotic; DZ = dizygotic.

Mediators of Genetic Influence

The question of whether there are important genetic effects on risk of substance use disorders has been answered convincingly in the case of alcohol and tobacco dependence; and, in the case of illicit drug use and dependence, data are accumulating to suggest that important genetic effects will also be a consistent finding. To date, behavioral geneticists have been less successful in addressing the mediating-variable questions about how genetic effects on addiction vulnerability arise. Improved insights into how genetic effects lead ultimately to differences in risk of substance use disorders should help accelerate the identification of individual genes that contribute to differences in risk; and, at the same time, successful identification of genetic risk or protective factors may be expected to lead to new insights into how gene effects on risk can arise. We know, for example, that history of alcohol dependence commonly co-occurs with affective and anxiety disorders and with antisocial personality disorder (Kessler et al., 1997). To what extent can these other disorders explain the inheritance of alcohol or other drug dependence? From the work of Schuckit and Smith (1996), we know that differences in reactions to a controlled dose of alcohol, observed in young adult sons of alcoholics and controls, are predictive of later differences in risk of alcohol dependence. Do genetically determined differences in reactions to alcohol (or other drugs) play a more important role than personality or other heritable behavioral differences in predicting risk of dependence? How does the predictive power of known genetic risk or protective factors compare with that of other known risk factors? To begin to address these questions, we need to draw together the research traditions of genetics, epidemiology, and studies in the human experimental laboratory.

We have begun this integrative process using data from the Australian twin panel 1981 cohort, who were first assessed by mailed questionnaire in 1980–1982, and an overlapping sample who participated in the alcohol challenge study described previously. Diagnostic interview follow-up data were obtained from this panel over the period 1992–1994 (Heath, Madden, et al., 1999), and those who also participated in the challenge study were genotyped at the ADH2 locus (Whitfield et al., 1998). Unexpectedly, we found that male participants in the alcohol challenge study differed only in minor respects from same-age males from the entire twin panel, although female participants in the challenge study were more likely to be heavy drinkers (Heath et al., 1999). Using a simple count of the number of *DSM–III–R* alcohol dependence symptoms as the dependent variable, we fitted a series of regression models to identify significant predictors of risk. A dummy variable was used to identify individuals who had not participated in the original challenge study, and these individuals were coded as "0" on dummy variables for the challenge study and genotype measures. In this way, it was possible to combine data from challenge study participants and other individuals from the original twin panel in the same analysis (Heath et al., 2001).

Table 17.3 summarizes the major predictors that we have identified,

Table 17.3. Predictors of Log-Transformed Alcohol Dependence Symptom Count in the Australian Twin Study (see Heath et al., 2001, for further details)

	Men		Women	
Predictor	β	SE	β	SE
High alcohol sensitivity	−0.42	0.11	−0.13[a]	0.09
ADH2*2 allele	−0.39	0.15	0.14[a]	0.16
Childhood conduct disorder	0.33	0.04	0.47	0.06
History of major depression	0.23	0.04	0.16	0.02
History of regular smoking	0.23	0.03	0.27	0.02
"Other Protestant" religious affiliation	−0.19	0.04	−0.11	0.02
Not participant in alcohol challenge study	−0.13[a]	0.06	−0.14	0.05

Note. Regression coefficients estimated under a multiple logistic regression model are shown. [a]Not significant.

showing regression coefficients from a multiple linear regression analysis that included smoking history in addition to other risk and protective factors. As noted previously based on analyses of history of *DSM–III–R* alcohol dependence (Heath et al., 1999), high alcohol sensitivity in men (scoring in the top 25% of the distribution of a composite score derived from alcohol challenge study measures of subjective intoxication and post-alcohol increase in body-sway after first alcohol dose—a phenotype that has shown moderately high heritability) was associated with significantly reduced alcohol problems at long-term follow-up. Possession of a single ADH2*2 allele had a similarly strong protective effect in men. Neither of these protective factors was a significant predictor in women. History of childhood conduct disorder was the next most potent predictor in men and the most potent predictor in women, with histories of major depression and regular smoking having more modest risk-increasing effects, and having a religious affiliation of "Other Protestant" (which would include some groups with prohibitions against alcohol use) having a modest protective effect. With the exception of religious affiliation, all of these phenotypes have previously been found to show significant heritability in this sample (Bierut et al., 1999; Heath & Madden, 1997; Slutske, Heath, Dinwiddie, Madden, & Bucholz, 1998) and thus may account for a small piece of the "genetic puzzle" of alcoholism. As noted previously, however (Heath, Bucholz, et al., 1997; Heath et al., 2001), substantial residual genetic variance in alcohol dependence risk remains when these predictors are controlled for.

Genetic Effects on Natural History, Treatment/Intervention Response

Most twin and adoption studies of substance use disorders have been cross-sectional in design and have considered only lifetime occurrence of a disorder. It is clear that there are important genetic influences on who becomes dependent. To date, quite basic questions remain unanswered about (a) whether there are important genetic influences on risk of drug dependence vulnerability, once genetic effects on level of exposure (e.g., amount of alcohol consumed) are controlled for; (b) whether there are important genetic influences on risk of progression to more severe dependence in those who become dependent, or whether chance or environmental risk factors play a more important role in determining this outcome; (c) whether there are important genetic influences on probability of remission of problems without treatment or intervention; and (d) whether there are important genetic influences on probability of a favorable response to treatment.

Conclusions: Where Next?

There is robust evidence for important genetic influences on risk of alcohol or tobacco dependence, at least in those of European or Asian ancestry, and growing evidence that genes are important predictors of risk of illicit drug dependence. As the transition is made from questions about whether genetic influences are important to discovery of which genes are important and how their effects are behaviorally mediated (e.g., through effects on substance use patterns) or moderated by specific environmental risk factors (e.g., experiencing early trauma), the field of behavioral genetics faces a number of major challenges. From this brief overview of genetic effects (and potential Genotype × Environment interaction effects) on addiction vulnerability, we may draw several conclusions about priorities for future research, which we discuss below.

Quantitative Trait Approaches to Genetic Linkage Studies

Active debate continues in the human genetics research community about the relative merits of different approaches to gene mapping, with various investigators advocating genomewide linkage versus genomewide association approaches (e.g., Risch, 2000; Risch & Merikangas, 1996) versus candidate gene approaches. For a drug such as ethanol, which has many and diverse central nervous system effects (e.g., U.S. Department of Health and Human Services, 2000), a comprehensive candidate gene approach is likely to involve studying vast numbers of candidate genes and yet still miss genes with as yet unknown effects on dependence vulnerability. Although a genetic linkage strategy is unlikely to discover genes of small effect, there is a considerable advantage to being able to identify chromosomal regions likely to contain genes of relatively major effect. Some positive linkage signals are already emerging from traditional linkage studies using clinically ascertained alcohol-dependent probands and their relatives (e.g., Long et al., 1998; Reich et al., 1998), which have identified some priority chromosomal regions for follow-up genetic association studies. There is now growing recognition of the considerable potential power of QTL or variance-components linkage approaches when a quantitative risk factor or index of risk can be defined, particularly where large sibships (e.g., Sham, Cherny, Purcell, & Hewitt, 2000) or sibships selected through an extreme-discordant sibpair (e.g., Risch & Zhang, 1995, 1996) can be identified. The feasibility of using quantitative symptom count measures or quantitative consumption indices for gene mapping studies of addiction vulnerability deserves greater consideration than it has as yet received.

Genetic Association Studies Revisited

The traditional case control study, although potentially susceptible to false-positive findings because of population stratification effects, remains the most powerful design for detecting genuine associations (e.g., Risch, 2000). Because of the specific need to uncover gene effects on treatment response, which may ultimately lead to improved tailoring of pharmacotherapies, as well as more general effects on dependence vulnerability, it is likely that treatment trials will become a major setting for genetic studies. While space limitations have not allowed us to review the somewhat chaotic literature on candidate gene associations with substance use disorders, we may note that most studies have been smaller by at least an order of magnitude than would be desirable for a complex multigenic disorder. A drawback of exclusive reliance on dependence cases ascertained through treatment settings is that a high proportion of substance-dependent individuals do not receive treatment, with the co-occurrence of other mental disorders predicting increased likelihood of treatment contacts. Thus in the U.S. National Comorbidity Survey (Kessler et al., 1996), of those reporting a history of alcohol dependence or drug dependence in the preceding 12 months, with no co-occurring mental disorder, only 19% and 26%, respectively, reported any treatment (in contrast to 29% and 47% of those who also reported mental disorder). There is therefore also great need for genotyping of substantial series of cases and controls identified through large-scale community surveys.

Toward a Molecular Epidemiology of Drug Addiction and Other Psychopathology

Uncovering effects of environmental risk factors, understanding the mediating role of psychopathology or normal behavioral variation (e.g., personality differences), and dissecting the interplay between specific genes and specific environmental risk factors can best be achieved prospectively. Case control comparisons of treatment samples and community controls, though a useful first step, have important limitations, including the following: (a) uncertainty about generalizability of findings to the general population, because a high proportion of individuals with drug dependence do not receive treatment (Kessler et al., 1996); (b) possible false-positive associations with apparent environmental risk factors or environmental moderators of genetic influence due to recall bias, with affected individuals more likely to acknowledge or perceive adverse early experiences; and (c) difficulty in reconstructing, on the basis of retrospective report, the sequence of events that includes onset of drug dependence (e.g., did onset of depression predate or follow onset of nicotine dependence?). Retrospective studies of general community samples, as well as retrospective case control studies, will always have an important role to play in genetic epidemiology. There are some risk factors that are inherently difficult to assess prospectively: An alcoholic father who is repeatedly raping his adolescent daughter or son, or abusing his partner, is unlikely to volunteer the family to participate in research. Nonetheless, for many purposes, prospective research remains the ideal for understanding risk factor and outcome associations. A particularly strong case can be made for establishing prospective studies in those of Asian ancestry, because three major genetic protective factors against alcoholism or tobacco dependence have already been identified (ALDH2, AHD2, and CYP2A6), but many basic questions about the effects of these genetic factors in the context of differing predisposing risk factors (e.g., early onset depression, childhood conduct disorder) and differing environmental exposures remain unanswered.

In contrast to many areas of medicine, prospective addiction research, and psychiatric research more broadly, must necessarily begin as pediatric research. Because onset of alcohol, tobacco, and other drug use commonly occurs by early adolescence, the conventional strategy of beginning with adult samples will inevitably lead to research that is largely retrospective in nature. Given the large sample sizes required for robust detection of genetic association, particularly when allowance is made for poststratification by ethnic origin and gender, required sample sizes are likely to be of the magnitude of 100–200,000 and higher (depending on the rarity of disorders that are to be included in the investigation). Although the logistical challenges of obtaining parental consents for participation of their children in a research project (and particularly a prospective study that will include collection of DNA for genotyping) must be faced, pediatric research has important advantages, including the feasibility of school-based assessment of most children at the earliest age groups, the feasibility of obtaining sibling and peer assessments, and the feasibility of obtaining, for at least subsamples of high-risk and random control families, teacher reports and parental assessments. The use of self-administered computer-based diagnostic interviews, while also desirable for the purpose of minimizing underreporting by adolescents, allows for economies of scale that would not be possible using conventional in-person interviewing. Studies of adolescent substance use and other psychopathology have to date tended to be either large in scale but limited to predominantly non-diagnostic assessments and only rarely designed to allow genotyping, or, if diagnostic, and particularly if involving genotyping, to be much smaller in scale. It is, however, realistic to anticipate a new generation of prospective genetic epidemiologic studies, beginning at least as young as early adolescence, that will help better define the interplay between genetic and environmental risk factors in the etiology of addictive and other psychiatric disorders as well as early onset chronic physical disorders and learning disorders.

References

American Psychiatric Association. (1987). *Diagnostic and statistical manual of mental disorders* (3rd ed., Rev.). Washington, DC: Author.

American Psychiatric Association. (1994). *Diagnostic and statistical manual of mental disorders* (4th ed.). Washington, DC: Author.

Babor, T. F. (1996). The classification of alcoholics: Typology theories from the 19th century to the present. *Alcohol Health & Research World, 20*(1), 6–17.

Babor, T. F., Hoffman, M., DelBoca, F. K., Hesselbrock, V., Meyer, R. E., Dolinsky, Z. S., & Rounsaville, B. (1992). Types of alcoholics: I. Evidence for an empirically derived typology based on indicators of vulnerability and severity. *Archives of General Psychiatry, 49,* 599–608.

Bierut, L. B., Heath, A. C., Bucholz, K. K., Dinwiddie, S. H., Madden, P. A. F., Statham, D. J., et al. (1999). Major depressive disorder in a community-based twin sample: Are there different genetic and environmental contributions for men and women? *Archives of General Psychiatry, 56,* 557–563.

Breslau, N., Kilbey, M., & Andreski, P. (1991). Nicotine dependence, major depression and anxiety in young adults. *Archives of General Psychiatry, 48,* 1069–1074.

Bucholz, K. K., Cadoret, R., Cloninger, C. R., Dinwiddie, S. H., Hesselbrock, V. M., Nurnberger, J. L., et al. (1996). A new semi-structured psychiatric interview for use in genetic linkage studies: A report on the reliability of the SSAGA. *Journal of Studies on Alcohol, 55,* 149–158.

Bucholz, K. K., Heath, A. C., & Madden, P. A. F. (2000). Transitions in drinking in adolescent females: Evidence from the Missouri Female Adolescent Twin Study. *Alcoholism: Clinical and Experimental Research, 24,* 914–923.

Bucholz, K. K., Heath, A. C., Reich, T., Hesselbrock, V. M., Kramer, J. R., Nurnberger, J. I., Jr., & Schuckit, M. A. (1996). Can we subtype alcoholism? A latent class analysis of data from relatives of alcoholics in a multi-center family study of alcoholism. *Alcoholism: Clinical and Experimental Research, 20,* 1462–1471.

Buck, K. J., Metten, P., Belknap, J. K., & Crabbe, J. C. (1997). Quantitative trait loci involved in genetic predisposition to acute alcohol withdrawal in mice. *Journal of Neuroscience, 17,* 3946–3955.

Chen, C. C., Lu, R. B., Chen, Y. C., Wang, M. F., Chang, Y. C., Li, T. K., et al. (1999). Interaction between the functional polymorphisms of the alcohol metabolism genes in protection against alcoholism. *American Journal of Human Genetics, 65,* 795–807.

Cloninger, C. R. (1987). Neurogenetic adaptive mechanisms in alcoholism. *Science, 236,* 410–416.

Cloninger, C. R., Bohman, M., & Sigvardsson, S. (1981). Inheritance of alcohol abuse: Cross-fostering analysis of adopted men. *Archives of General Psychiatry, 38,* 861–868.

Cloninger, C. R., Bohman, M., Sigvardsson, S., & Von Knorring, A. L. (1985). Psychopathology in adopted-out children of alcoholics: The Stockholm Adoption Study. In M. Galanter (Ed.), *Recent developments in alcoholism* (Vol. 3, pp. 37–51). New York: Raven Press.

Cloninger, C. R., Rice, J. P., & Reich, T. (1979). Multifactorial inheritance with cultural transmission and assortative mating: II. A general model of combined polygenic and cultural inheritance. *American Journal of Human Genetics, 31,* 176–198.

Devlin, B., & Roeder, K. (1999). Genomic control for association studies. *Biometrics, 55,* 997–1004.

Dinwiddie, S. H., Heath, A. C., Dunne, M. P., Bucholz, K. K., Madden, P. A. F., Slutske, W., et al. (2000). Early sexual abuse and lifetime psychopathology: A cotwin-control study. *Psychological Medicine, 30,* 41–52.

Eaves, L. J., Last, K. A., Martin, N. G., & Jinks, J. L. (1977). A progressive approach to non-additivity and genotype–environment covariance in the analysis of human differences. *British Journal of Mathematical and Statistical Psychology, 30,* 1–42.

Eaves, L. J., Silberg, J. L., Hewitt, J. K., Rutter, M., Meyer, J. M., Neale, M. C., & Pickles, A. (1993). Analyzing twin resemblance in multisymptom data: Genetic applications of a latent class model for symptoms of conduct disorder in juvenile boys. *Behavior Genetics, 23,* 5–20.

Goodwin, D. W., Schulsinger, F., Knop, J., Mednick, S., & Guze, S. B. (1977). Psychopathology in adopted and non-adopted daughters of alcoholics. *Archives of General Psychiatry, 31,* 164–169.

Goodwin, D. W., Schulsinger, F., Moller, N., Hermansen, L., Winokur, G., & Guze, S. B. (1974). Drinking problems in adopted and nonadopted sons of alcoholics. *Archives of General Psychiatry, 31,* 164–169.

Grant, B. F. (1997). Prevalence and correlates of alcohol use and *DSM–IV* alcohol dependence in the United States: Results of the National Longitudinal Alcohol Epidemiologic Survey. *Journal of Studies on Alcohol, 58,* 464–473.

Heath, A. C. (1995a). Genetic influences on alcoholism risk: A review of adoption and twin studies. *Alcohol Health and Research World, 19,* 166–171.

Heath, A. C. (1995b). Genetic influences on drinking behavior in humans. In H. Begleiter & B. Kissin (Eds.), *Alcohol and alcoholism: Vol. 1. The genetics of alcoholism* (pp. 82–121). New York: Oxford University Press.

Heath, A. C., Bucholz, K. K., Madden, P. A. F., Dinwiddie, S. H., Slutske, W. S., Statham, D. J., et al. (1997). Genetic and environmental contributions to alcohol dependence risk in a national twin sample: Consistency of findings in men and women. *Psychological Medicine, 27,* 1381–1396.

Heath, A. C., Bucholz, K. K., Slutske, W. S., Madden, P. A. F., Dinwiddie, S. H., Dunne, M. P., et al. (1994). The assessment of alcoholism in surveys of the general community: What are we measuring? Some insights from the Australian twin panel survey. *International Review of Psychiatry, 6,* 295–307.

Heath, A. C., Bucholz, K. K., Whitfield, J. B., Madden, P. A. F., Dinwiddie, S. H., Slutske, W. S., et al. (2001). Towards a molecular epidemiology of alcohol dependence: Analyzing the interplay of genetic and environmental risk-factors. *British Journal of Psychiatry, 178*(Suppl. 40), 533–540.

Heath, A. C., Kendler, K. S., Eaves, L. J., & Markell, D. (1985). The resolution of cultural and biological inheritance: Informativeness of different relationships. *Behavior Genetics, 15,* 439–465.

Heath, A. C., & Madden, P. A. F. (1997). Genetic influences on smoking behavior. In J. R. Turner, L. R. Cardon, & J. K. Hewitt (Eds.), *Behavior genetic approaches in behavioral medicine* (pp. 45–65). New York: Plenum Press.

Heath, A. C., Madden, P. A. F., Bucholz, K. K., Dinwiddie, S. H., Slutske, W. S., Bierut, L. J., et al. (1999). Genetic differences in alcohol sensitivity and the inheritance of alcoholism risk. *Psychological Medicine, 29,* 1069–1081.

Heath, A. C., Slutske, W. S., Bucholz, K. K., Madden, P. A. F., & Martin, N. G. (1997). Behavioral genetic methods in prevention research: An overview. In K. Bryant, M. Windle, & S. G. West (Eds.), *The science of prevention: Methodological advances from alcohol and substance abuse research* (pp. 123–163). Washington, DC: American Psychological Association.

Heath, A. C., Slutske, W. S., & Madden, P. A. F. (1997). Gender differences in the genetic contribution to alcoholism risk and to alcohol consumption patterns. In R. W. Wilsnack & S. C. Wilsnack (Eds.), *Gender and alcohol* (pp. 114–149). Rutgers, NJ: Rutgers University Press.

Heath, A. C., Todorov, A. A., Nelson, E. C., Madden, P. A. F., Bucholz, K. K., & Martin, N. G. (2002). Gene-environment interaction effects on behavioral variation and risk of complex disorders: The example of alcholism and other psychiatric disorders. *Twin Research, 5,* 30–37.

Hesselbrock, V. M., Eaton, C., Bucholz, K. K., Schuckit, M., & Hesselbrock, V. (1999). A validity study of the SSAGA: A comparison with the SCAN. *Addiction, 94,* 1361–1370.

Higuchi, S., Matsushita, S., Imazeki, H., Kinoshita, T., Takagi, S., & Kono, H. (1994). Aldehyde dehydrogenase genotypes in Japanese alcoholics. *The Lancet, 343,* 741–742.

Higuchi, S., Matsushita, S., Muramatsu, T., Murayama, M., & Hayashida, M. (1996). Alcohol and aldehyde dehydrogenase genotypes and drinking behavior in Japanese. *Alcoholism: Clinical and Experimental Research, 20,* 493–497.

Hori, H., Kawano, T., Endo, M., & Yuasa, Y. (1997). Genetic polymorphisms of tobacco- and alcohol-related metabolizing enzymes and human esophageal squamous cell carcinoma susceptibility. *Journal of Clinical Gastroenterology, 25,* 568–575.

Hsu, L. C., Yoshida, A., & Mohandas, T. (1986). Chromosomal assignment of the genes for human aldehyde dehydrogenase-1 and aldehyde dehydrogenase-2. *American Journal of Human Genetics, 38,* 641–648.

Hudziak, J., Heath, A. C., Madden, P. A. F., Reich, W., Bucholz, K. K., Slutske, W. S., et al. (1998). The latent class and factor analysis of *DSM–IV* ADHD: A twin study of female adolescents. *Journal of the American Academy of Child and Adolescent Psychiatry, 37,* 848–857.

Jacob, T., Sher, K. J., Bucholz, K. K., True, W. T., Sirevaag, E. J., Rohrbaugh, J., et al. (2001). An integrative approach for studying the etiology of alcoholism and other addictions. *Twin Research, 4,* 103–118.

Jinks, J. L., & Fulker, D. W. (1970). Comparison of the biometrical genetical, MAVA and classical approaches to the analysis of human behavior. *Psychological Bulletin, 73,* 311–349.

Kandel, D., Chen, K., Warner, L. A., Kessler, R. C., & Grant, B. (1997). Prevalence and demographic correlates of symptoms of last year dependence on alcohol, nicotine, marijuana, and cocaine in the U.S. population. *Drug and Alcohol Dependence, 10,* 11–24.

Kendler, K. S., Bulik, C. M., Silberg, J., Hettema, J. M., Myers, J., & Prescott, C. A. (2000). Childhood sexual abuse and adult psychiatric and substance use disorders in women: An epidemiological and cotwin control analysis. *Archives of General Psychiatry, 57,* 953–959.

Kendler, K. S., Karkowski, L. M., Prescott, C. A., & Pedersen, N. L. (1998). Latent class analysis of temperance board registrations in Swedish male–male twins pairs born 1902 to 1949: Searching for subtypes of alcoholism. *Psychological Medicine, 28,* 803–813.

Kendler, K. S., Neale, M. C., Kessler, R. C., Heath, A. C., & Eaves, L. J. (1993). A test of the equal-environment assumption in twin studies of psychiatric illness. *Behavior Genetics, 23,* 21–28.

Kessler, R. C., Crum, R. M., Warner, L. A., Nelson, C. B., Schulenberg, J., & Anthony, J. C. (1997). Lifetime co-occurrence of *DSM–III–R* alcohol abuse and dependence with other psychiatric disorders in the National Comorbidity Survey. *Archives of General Psychiatry, 54,* 313–321.

Kessler, R. C., McGonagle, K. A., Zhao, S., Nelson, C. B., Hughes, M., Eshelman, S., et al. (1994). Lifetime and 12-month prevalence of *DSM–III–R* psychiatric disorders in the United States: Results from the National Comorbidity Study. *Archives of General Psychiatry, 51,* 8–19.

Kessler, R. C., Nelson, C. B., McGonagle, K. A., Edlund, M. J., Frank, R. G., & Leaf, P. J. (1996). The epidemiology of co-occurring addictive and mental disorders: Implications for prevention and service utilization. *American Journal of Orthopsychiatry, 66,* 17–31.

Kwon, J. T., Nakajima, M., Chai, S., Yom, Y. K., Kim, H. K., Yamazaki, H., et al. (2001). Nicotine metabolism and CYP2A6 allele frequencies in Koreans. *Pharmacogenetics, 11,* 317–323.

Long, J. C., Knowler, W. C., Hanson, R. L., Robin, R. W., Urbanek, M., Moore, E., et al. (1998). Evidence for genetic linkage to alcohol dependence of chromosomes 4 and 11 from an autosome-wide scan in an American Indian population. *American Journal of Medical Genetics, 81,* 216–221.

Luczak, S. E., Elvine-Kreis, B., Shea, S. H., Carr, L. G., & Wall, T. L. (2002). Genetic risk for alcoholism relates to level of response to alcohol in Asian-American men and women. *Journal for the Study of Alcoholism, 63*(1), 74–82.

Madden, P. A. F., Bucholz, K. K., Dinwiddie, S. H., Slutske, W. S., Bierut, L. J., Statham, D. J., et al. (1997). Nicotine withdrawal in women. *Addiction, 92,* 889–902.

Madden, P. A. F., Bucholz, K. K., Martin, N. G., & Heath, A. C. (2000). Smoking and the genetic contribution to alcohol dependence risk. *Alcohol Research and Health, 24,* 209–214.

Martin, N. G., Oakeshott, J. G., Gibson, J. B., Starmer, G. A., Perl, J., & Wilks, A. V. (1985). A twin study of psychomotor and physiological responses to an acute dose of alcohol. *Behavior Genetics, 15,* 305–347.

Martin, N. G., Perl, J., Oakeshott, J. G., Gibson, J. B., Starmer, G. A., & Wilks, A. V. (1985). A twin study of ethanol metabolism. *Behavior Genetics, 15,* 93–109.

McCutcheon, A. L. (1987). *Latent class analysis.* Beverly Hills, CA: Sage.

Morton, N. E., & Collins, A. (1998). Tests and estimates of allelic association in complex inheritance. *Proceedings of the National Academy of Sciences USA, 95,* 11389–11393.
Muramatsu, T., Zu-cheng, W., Yi-Ru, F., Kou-Bao, H., Hequin, Y., Yamada, K., et al. (1995). Alcohol and aldehyde dehydrogenase genotypes and drinking behavior of Chinese living in Shanghai. *Human Genetics, 96,* 151–154.
Nakajima, M., Kwon, J. T., Tanaka, N., Zenta, T., Yamomoto, Y., Yamomoto, H., et al. (2001). Relationship between interindividual differences in nicotine metabolism and CYP2A6 genetic polymorphisms in humans. *Clinical Pharmacology and Therapeutics, 69,* 72–78.
Nance, W. E., & Corey, L. A. (1976). Genetic models for the analysis of data from the families of identical twins. *Genetics, 83,* 811–826.
Neale, M. C., & Cardon, L. R. (1992). *Methodology for genetic studies of twins and families.* Dordrecht, The Netherlands: Kluwer Academic.
Neale, M. C., Cherny, S. S., Sham, P., Whitfield, J. B., Heath, A. C., Birley, A. J., & Martin, N. G. (1999). Distinguishing population stratification from genuine allelic effects linkage with Mx: Association of ADH2 with alcohol consumption. *Behavior Genetics, 29,* 233–243.
Nelson, C. B., Heath, A. C., & Kessler, R. C. (1998). Temporal progression of alcohol dependence symptoms in the U.S. household population: Results from the National Comorbidity Survey. *Journal of Consulting and Clinical Psychology, 13,* 6–17.
Nelson, E. C., Heath, A. C., Madden, P. A. F., Cooper, L. C., Dinwiddie, S. H., Glowinski, A., et al. (2002). The correlates of childhood sexual abuse: A retrospective examination using the twin study design. *Archives of General Psychiatry, 59,* 139–145.
Nelson, C. B., Little, R. J. A., Heath, A. C., & Kessler, R. C. (1996). Patterns of *DSM–III–R* alcohol dependence symptom progression in a general population survey. *Psychological Medicine, 26,* 449–460.
Neuman, R. J., Todd, R. D., Heath, A. C., Reich, W., Hudziak, J. J., Bucholz, K. K., et al. (1999). Evaluation of ADHD typology in three contrasting samples: A latent class approach. *Journal of the American Academy of Child and Adolescent Psychiatry, 38,* 25–33.
Oscarson, M. (2001). Genetic polymorphisms in the cytochrome P450 2A6 (CYP2A6) gene: Implications for interindividual differences in nicotine metabolism. *Drug Metabolism and Disposition, 29,* 91–95.
Osier, M., Pakstis, A. J., Kidd, J. R., Lee, J. F., Yin, S. J., Ko, H. C., et al. (1999). Linkage disequilibrium at the ADH2 and ADH3 loci and risk of alcoholism. *American Journal of Human Genetics, 64,* 1147–1157.
Pianezza, M. L., Sellers, E. M., & Tyndale, R. F. (1998). Nicotine metabolism defect reduces smoking. *Nature, 393,* 750.
Pritchard, J. K., & Rosenberg, N. A. (1999). Use of unlinked markers to detect population stratification in association studies. *American Journal of Human Genetics, 65,* 220–228.
Reich, T., Edenberg, H., Goate, A., Williams, T. J., Rice, J., Van Eerdwegh, P., et al. (1998). A genome-wide search for genes affecting the risk for alcohol dependence. *American Journal of Medical Genetics, 81,* 207–215.
Risch, N. J. (1990). Linkage strategies for genetically complex traits: II. The power of affected relative pairs. *American Journal of Human Genetics, 46,* 229–241.
Risch, N. J. (2000). Searching for genetic determinants in the new millennium. *Nature, 405,* 847–856.
Risch, N. J., & Merikangas, K. (1996). The future of genetic studies of complex human diseases. *Science, 273,* 1516–1517.
Risch, N. J., & Zhang, H. (1995). Extreme discordant sib pairs for mapping quantitative trait loci in humans. *Science, 268,* 1584–1589.
Risch, N. J., & Zhang, H. (1996). Mapping quantitative trait loci with extreme discordant sib pairs: Sampling considerations. *American Journal of Human Genetics, 58,* 836–843.
Schuckit, M. A., & Smith, T. (1996). An 8-year follow-up of 450 sons of alcoholic and control subjects. *Archives of General Psychiatry, 53,* 202–210.
Sham, P. C., Cherny, S. S., Purcell, S., & Hewitt, J. K. (2000). Power of linkage versus association analysis of quantitative traits, by use of variance-components models. *American Journal of Human Genetics, 66,* 1616–1630.
Shen, Y. C., Fan, J. H., Edenberg, H. J., Li, T.-K., Cui, Y. H., Wang, Y. F., et al. (1997).

Polymorphism of ADH and ALDH genes among 4 ethnic groups in China and effects upon the risk for alcoholism. *Alcoholism: Clinical and Experimental Research, 21,* 1272–1277.

Slutske, W. S., Heath, A. C., Dinwiddie, S. H., Madden, P. A. F., & Bucholz, K. K. (1998). Common genetic risk factors for conduct disorder and alcohol dependence. *Journal of Abnormal Psychology, 107,* 363–374.

Spielman, R. S., & Ewens, W. J. (1998). A sibship test for linkage in the presence of association: The sib transmission/disequilibrium test. *American Journal of Human Genetics, 62,* 450–458.

Spielman, R. S., McGinnis, R. E., & Ewens, W. J. (1993). Transmission test for linkage disequilibrium: The insulin gene region and insulin-dependent diabetes mellitus (IDDM). *American Journal of Human Genetics, 52,* 506–516.

Streissguth, A. P., Sampson, P. D., Olson, H. C., Bookstein, F. L., Barr, H. M., Scott, M., et al. (1994). Maternal drinking during pregnancy: Attention and short-term memory in 14-year-old offspring: A longitudinal prospective study. *Alcoholism: Clinical and Experimental Research, 18,* 202–218.

Sullivan, P. F., Neale, M. C., & Kendler, K. S. (2000). Genetic epidemiology of major depression: Review and meta-analysis. *American Journal of Psychiatry, 157,* 1552–1562.

Thomasson, H. R., Crabb, D. W., Edenberg, H. J., Li, T. K., Hwu, H. G., Chen, C. C., et al. (1994). Low frequency of the ADH2*2 allele among Atayal natives of Taiwan with alcohol use disorders. *Alcoholism: Clinical and Experimental Research, 18,* 640–643.

Thomasson, H., Edenberg, H. J., Crabb, D. W., Mai, X. L., Jerome, R. E., Li, T.-K., et al. (1991). Alcohol and aldehyde dehydrogenase genotypes and alcoholism in Chinese men. *American Journal of Human Genetics, 48,* 677–681.

True, W. R., Xian, H., Scherrer, J. F., Madden, P. A. F., Bucholz, K. K., Heath, A. C., et al. (1999). Common genetic vulnerability for nicotine and alcohol dependence. *Archives of General Psychiatry, 56,* 655–661.

Tsuang, M. T., Lyons, M. J., Eisen, S. A., Goldberg, J., True, W., Lin, N., et al. (1996). Genetic influences on *DSM–III–R* drug abuse and dependence: A study of 3,372 twin pairs. *American Journal of Medical Genetics, 67,* 473–477.

Tsuang, M. T., Lyons, M. J., Harley, R. M., Xian, H., Eisen, S., Goldberg, J., et al. (1999). Genetic and environmental influences on transitions in drug use. *Behavior Genetics, 29,* 473–479.

U.S. Department of Health and Human Services. (2000). *10th Special Report to the U.S. Congress on Alcohol and Health.* Washington, DC: National Institute on Alcohol Abuse and Alcoholism.

Wall, T. L., Peterson, C. M., Peterson, K. P., Johnson, M. L., Thomasson, H., Cole, M., & Ehlers, C. L. (1997). Alcohol metabolism in Asian-American men with genetic polymorphisms of aldehyde dehydrogenase. *Annals of Internal Medicine, 127,* 401–402.

Wall, T. L., Thomasson, H. R., Schuckit, M. A., & Ehlers, C. L. (1992). Subjective feelings of alcohol intoxication in Asians with genetic variations of ALDH2 alleles. *Alcoholism: Clinical and Experimental Research, 16,* 991–995.

Whitfield, J. B., Nightingale, B. N., Bucholz, K. K., Madden, P. A. F., Heath, A. C., & Martin, N. G. (1998). ADH genotypes and alcohol use and dependence in Europeans. *Alcoholism: Clinical and Experimental Research, 22,* 1463–1469.

Whitfield, J. B., Zhu, G., Duffy, D., Birley, A. J., Madden, P. A. F., Heath, A. C., et al. (2001). Variation in alcohol metabolism as a risk factor for alcohol dependence. *Alcoholism: Clinical and Experimental Research, 25,* 1257–1263.

Wiesbeck, G. A., Schuckit, M. A., Kalmijn, J. A., Tipp, T. E., Bucholz, K. K., & Smith, T. L. (1996). An evaluation of the history of a marijuana withdrawal syndrome in a large population. *Addiction, 91,* 1469–1478.

Yates, W. R., Cadoret, R. J., Troughton, E. P., Stewart, M., & Giunta, T. S. (1998). Effect of fetal alcohol exposure on adult symptoms of nicotine, alcohol and drug dependence. *Alcoholism: Clinical and Experimental Research, 22,* 914–920.

Yokoyama, A., Ohmeri, T., Muramatsu, T., Higuchi, S., Yokoyama, T., Matsushita, S., et al. (1996). Cancer screening of upper aerodigestive tract in Japanese alcoholics with reference to drinking and smoking habits and aldehyde dehydrogenase-2 genotype. *International Journal of Cancer, 68,* 313–316.

18

Pharmacogenetics in the Postgenomic Era

Katherine J. Aitchison and Michael Gill

The response of an individual organism (human or animal) to a given psychotropic agent is determined by a complex interplay of multiple factors. This is demonstrated most simply by the existence of the placebo effect (Keck, Welge, Strakowski, Arnold, & McElroy, 2000; Spallone & Wilkie, 2000). Factors that are genetic in origin are studied in the field known as *pharmacogenetics*. Vogel (1959) defined pharmacogenetics as the study of heritable differences in the metabolism and activity of exogenous agents such as drugs or environmental toxins. More recently, this has been understood as comprising the study of variations in the coding or regulatory regions of genes that lead to interindividual variability in drug effect (efficacy) or in adverse effect profile (toxicity). Pharmacogenetics therefore includes the study of both pharmacodynamic genetic factors (affecting the drug response at the level of the target organ) and pharmacokinetic genetic factors (affecting drug absorption, metabolism, and excretion; Aitchison, Jordan, & Sharma, 2000; Masellis et al., 2000). There are various indicators of drug response at the level of the target organ, or endophenotypes, such as changes in severity of clinical symptoms as measured by standardized rating scales (e.g., the Hamilton Depression Rating Scale; Hamilton, 1967), changes in physiological gating mechanisms (e.g., prepulse inhibition), or changes in standardized behavioral paradigms in the case of animal studies.

With the advent of the "new genetics," the term *pharmacogenomics* has arisen, which has been variously defined (Housman & Ledley, 1998; Masellis et al., 2000; Persidis, 1998; Regalado, 1999; Rioux, 2000) but encompasses the application of genomics to the study of pharmacogenetics, or pharmacogenetics in the postgenomic era. It may be considered a branch of functional genomics. The Human Genome Project and recent advances in molecular genetics now present an unprecedented opportunity to study all genes in the human genome, including genes encoding drug metabolizing enzymes, drug targets, regulatory mechanisms, and postreceptor messenger machinery in relation to drug safety and efficacy (Ozdemir, Shear, & Kalow, 2001). Pharmacogenomics includes the utilization of techniques such as high-throughput genotyping of genomewide markers in order to identify genomic "hotspots" corresponding to susceptibility loci contributing to response or adverse effect profiles. It also includes bioinformatics studies analyzing the wealth of sequence data now available, to identify novel sequence variants in or near candidate genes, followed by experimental confirmation of such variants, with functional characterization (e.g., of variants in promoter sequences, 5′ or 3′ untranslated regions, or intronic sequence and their impact on levels of gene expression/messenger RNA [mRNA] splicing). More recently, it has been shown that gene expression microarrays can be used to correlate patterns of gene expression (upregulation or downregulation) with drug effect or toxicity; this has been termed *transcriptome profiling* and has been employed with some promising results (Alen, 2001).

The short-term goal of pharmacogenomics is the same as pharmacogenetics, that is, to elucidate how individual genetic makeup influences the response to or side effects of a particular drug (Masellis et al., 2000). A methodological difference between the two fields is that pharmacogenetics is hypothesis driven, requiring a priori knowledge of drug pharmacology, whereas (e.g., in the search for genomic "hotspots") pharmacogenomics may not be. The long-term vision of pharmacogenomics is that the discovery of genetic influences on drug action will lead to the development of new diagnostic procedures and therapeutic products that enable drugs to be prescribed selectively to patients for whom they will be effective and safe (Housman & Ledley, 1998). In the United States, in 1994, adverse drug reactions (ADRs) were between the fourth and sixth leading cause of death (ahead of pneumonia and diabetes mellitus), causing 106,000 deaths among hospitalized patients (Lazarou, Pomeranz, & Corey, 1998). The expectation of pharmacogenomics is that subgroups of patient populations might be identifiable—those who are likely to respond better to a given drug than the rest of the population, and those who are less likely to have an ADR—and that drugs will then be able to be selectively marketed to those patient groups (population segmentation or stratification by genetic profile). This could lead to significant cost savings, both in the cost of drug development, and costs to the individual and to the health care provider. Clinical trial design might become more efficient, and drugs might reach the market faster and stay there longer (Spallone & Wilkie, 1999). If a patient could be identified as being unlikely to respond well to a given drug, then the cost of months of ineffective treatment—costs both personal and economic—could be avoided. Similarly, those at high risk of an ADR for a given drug could be saved resulting morbidity (i.e., negative health consequences) or even mortality (i.e., a fatal adverse drug effect). Pharmaceutical companies might even resurrect drugs that have been taken off the market, if subgroups of patients who were at risk of the particular offending ADR could be identified and excluded from the prescribing license. The antipsychotic clozapine, for example, has a small but significant risk of the potentially fatal complication of agranulocytosis (0.7% in the first 12 weeks of treatment; Atkin et al., 1996), and may only be prescribed in the United States and United Kingdom under a restricted license, with strict hematological monitoring. If the subgroup of patients likely to be at high risk of agranulocytosis could be identified, then this patient group could be excluded from the prescribing license, with the result that hematological monitoring might not be required for the rest of the population prescribed clozapine or be required at a lower monitoring frequency. This, of course, assumes a test for susceptibility to agranulocytosis with a high level of sensitivity and specificity. Furthermore, if this test is to be genetically based, this presupposes that there is a genetic contribution to susceptibility to agranulocytosis. There is some evidence in support of this (Corzo et al., 1995; Lieberman et al., 1990). Assuming such tests could be developed, the prospect raises some ethical considerations.

Ethical Considerations

Ethical considerations have been highlighted by Spallone and Wilkie (1999, 2000) and include the following. Firstly, what are the costs to the individual of population segmentation based on pharmacogenomic profile? For example, what might the psychological consequence be for someone to realize that he or she was excluded from a given treatment, for example, clozapine, which is arguably the most effective antipsychotic to date? An individual might prefer to take the risk of agranulocytosis and have hematological monitoring, depending on the magnitude of the risk, whereas a pharmaceutical company might not be prepared to take that risk: The risks acceptable to an individual and to a company might differ significantly. Similarly, to continue with this example, an effective treatment for agranulocytosis might be developed at some point in the future, and, if that were to happen, would the pharmaceutical company remove the exclusion from the particular population subgroup deemed to be at high risk of agranulocytosis, or would they consider that the possible associated risks or costs were still too high for them?

A related question is the likely size of the population subgroups resulting from population segmentation. The size of the groups resulting from a profile designed to predict drug efficacy is likely to be significantly larger than the size designed to predict drug toxicity. Pharmaceutical companies might well be keen to identify relatively large subgroups likely to respond well to a given drug, but will it be cost-effective for a drug company to market a diagnostic test for a small percentage of individuals at risk of drug toxicity? What will be the threshold percentage, threshold cost of the test, and threshold severity of the potential ADR? With the economics of drug development already constrained to the point that many approved drugs never recover their development costs, it is unlikely that any pharmacogenomic strategies that require any significant increase in the scope or cost of drug development will be adopted by the industry. It has therefore been said that the challenge, then, for pharmacogenomics is to invent and implement the novel technologies that can meet drug development's needs (Ledley, 1999).

Secondly, there are ethical considerations regarding informatics. There is the question of informed consent, when the current knowledge base regarding pharmacogenomics is relatively poor. What exactly does one tell the patient or volunteer—that one is consenting for a study or clinical trial? To consent for a pharmacogenomics study requires a much broader consent provision than a pharmacogenetics study, the latter tending to involve specific candidate genes, the function of which one could make an attempt at explaining; the former requires consent for analysis of markers whose functional effect may not be known, and the possible consequences to the individual of having a particular known genetic marker profile in the future may be very difficult to predict. Some drug companies are making efforts to address these issues in written information provided to both investigators and patients in pharmacogenetics studies that are part of clinical trials (Spallone & Wilkie, 1999).

Then there is the question of protecting confidentiality for those about whom medical data is collected, relevant to any genetic study, and not unique to pharmacogenetics or pharmacogenomics, but perhaps worth outlining here as collaborations with pharmaceutical companies or biotechnology companies may render the issue particularly pertinent. There are three methods of handling data in order to enhance or ensure confidentiality (Spallone & Wilkie, 2000): anonymity, encoding, and encryption. Data is anonymous if all information capable of identifying the individual to whom the data relates has been removed and destroyed. Further information pertaining to that individual may then never be added to the appropriate record in the database. Data is encoded if a serial number or other code is attached to the data and a key to this is held elsewhere. Encoded data might be effectively anonymous to the research team working on it (e.g., laboratory research workers), if they did not have access to the key, but the data would not be truly anonymous, as someone would be able to link the two. Encoding would allow updating of an individual's record, for example, in the course of a longitudinal study. Encryption means turning the data into meaningless strings of numbers or letters, which are only decipherable by someone with the key. This latter method may be useful not so much for reasons of confidentiality toward the subject as for reasons of commercial security, to prevent unauthorized access by rivals to commercially significant collections of data. For pharmacogenomics studies, in which it may be necessary to link detailed clinical information with genetic data, encoding would seem the most appropriate method of preserving subject confidentiality. This allows further clinical data to be incorporated in the database in time should research indicate that analysis with respect to other clinical variables might be informative. In the case of truly anonymized data, the sample can only be used once, then discarded, which may, of course, represent the loss of a potentially highly valuable genetic resource.

Finally, in terms of ethical considerations for pharmacogenomics, commercialization, issue of intellectual property, and patent regimes for biotechnology may affect the doctor–patient relationship, which is vulnerable at a time when new partnerships developing between academia and industry may blur the boundaries between public and private interest. The involvement of industry in pharmacogenomics may constrain academic freedom, restrict the use of any diagnostic tests resulting from the pharmacogenomics research, or restrict access to important resources, including genetic sequencing databases and DNA banks. The last has been a concern regarding, for example, the deCODE-sponsored databank in Iceland (Spallone & Wilkie, 2000).

Evidence for Heritability of Drug Effect

As early as 1967, it was observed that in a small group of patients on a standard dose of the tricyclic antidepressant (TCA) desipramine, a very large (36-fold) variation in the steady-state plasma concentrations existed (Sjöqvist et al., 1967). The importance of genetic factors in determining the steady-state levels of the TCA nortriptyline was shown in twin and family studies (Alexanderson, 1973; Alexanderson, Evans, & Sjöqvist, 1969; Åsberg, Evans, & Sjöqvist, 1971). Family-based studies have also indicated the heritability of response to other classes of antidepressants. O'Reilly, Bogue, and Singh (1994) reported a family multiply affected with major depressive disorder, four members of whom did not respond to either TCAs or various new generation antidepressants, but did respond subsequently to the monoamine oxidase inhibitor, tranylcypromine. Similarly, Franchini, Serretti, Gasperini, and Smeraldi (1998) described familial concordance of fluvoxamine response. A study of treatment-resistant depression in which data on various possible factors influencing response were collected noted that the only covariate that reached significance in terms of influencing whether an individual was responsive or resistant to antidepressant therapy was family history of affective disorder (Smith, personal communication, October 2, 2001).

With respect to antipsychotic response, typical antipsychotic sensitivity has been reported as cosegregating with CYP2D6 phenotype (Aitchison et al., 1995), and Vojvoda, Grimmel, Sernyak, and Mazure (1996) reported a set of monozygotic twins, each with a *DSM–IV* diagnosis of schizophrenia (4th edition of the *Diagnostic and Statistical Manual of Mental Disorders;* American Psychiatric Association, 1994), who were concordant for treatment refractoriness to various typical antipsychotic agents, but who had a positive response to clozapine. Furthermore, a heritable component to antipsychotic response has been demonstrated by studies with mouse strains that differ in antipsychotic responsiveness (Dains, Hitzemann, & Hitzemann, 1996) and indicated by a human family study (Sautter, McDermott, & Garver, 1993) and studies indicating differential response to antipsychotics in different ethnicity groups (Aitchison, Jordan, & Sharma, 2000; Frackiewicz, Sramek, Herrera, Kurtz, & Cutler, 1997).

There is, however, a relative paucity of studies investigating the heritability of drug response, and there are several reasons for this. Firstly, the phenotype to be measured is a transient, temporary trait, the response or adverse effects being elicited only in the presence of the drug. This means that it may well be difficult to find, for instance, other family members or cotwins who are also taking the drug in question at the same time as the proband being investigated. If a family member has taken the same drug at some time in the past, it may be possible to perform a retrospective clinical assessment of their response or side effects, but this may be subject to inaccuracies in their recall and so on. Secondly, the range of psychotropic medication available on the market has changed radically over the past 50 years and continues to develop rapidly, so that what was available to a clinician treating a family member 30 years ago may be very different from the range of choices now available. Of note, the drug in the study of O'Reilly et al. (1994) was tranylcypromine, which due to its stimulant action is now regarded as the most hazardous of the monoamine oxidase inhibitors, and is rarely prescribed in the United Kingdom. Similarly, even if there are two or more family members simultaneously affected by a particular psychiatric disorder, owing to the large choice of possible pharmacological agents now available for the treatment of psychiatric disorders such as depression, the affected members may be prescribed different drugs, especially if they are not in the same geographical area. Thirdly, in a study of families multiply affected with psychiatric illness, the precise disorder may vary among different family members, due to the phenomenon of incomplete penetrance (Farmer, McGuffin, & Gottesman, 1987), which may again lead to variation in the psychotropic agents being used. Fourthly, polypharmacy, that is, the use of more than one psychotropic agent simultaneously (e.g., a mood stabilizer and an antidepressant), is common these days, and this may lead to further difficulties in finding family members who are being treated only with a particular drug.

As there is a relative paucity of data regarding heritability, and good prospective data are hard to collect for the reasons outlined above, a case could be made for a retrospective collection of data regarding treatment history (including response and adverse effect profile to all current and previous agents used) for new genetic collections involving particular psychiatric disorders such as sibpair or twin collections, with care being taken regarding the method of statistical analysis for the data. Nonetheless, a number of investigators have endeavored to provide informative pharmacogenetic data, largely using the candidate gene association approach, and the results of some illustrative examples will now be outlined.

Examples of Pharmacogenetic Studies to Date

Pharmacogenetic studies to date that have generated at least partially informative results include studies of thiopurine methyltransferase (TPMT) and toxic response to agents used in the treatment of leukemia, studies of the cytochrome P450 enzyme CYP2C9 and warfarin therapy, studies of the cytochrome P450s CYP2D6 and CYP2C19 and tricyclic antidepressant response, studies of CYP1A2 and clozapine response, the 5-HT_{2C} receptor and antipsychotic response, the serotonin transporter and response to proserotonergic antidepressants, and studies that have investigated an interactive effect at more than one genetic locus.

TPMT and Response to Thiopurines

TPMT catalyzes the methylation of cytotoxic thiopurine drugs, such as thioguanine, mercaptopurine, and azathioprine (Weinshilboum & Sladek, 1980). Three major TPMT phenotypes have been identified: high, intermediate, and deficient methylators. As a result, individuals differ greatly in detoxification of the thiopurine drugs to 6-mercaptopurine, which is correlated with adverse effects and therapeutic efficacy. Children with leukemia treated with these agents who are deficient methylators exhibit severe hematopoietic toxicity, whereas those who are high methylators require high dose to achieve any clinical benefit. The TPMT gene (including promoter and 3′ flanking region) has been screened for mutations (Spire-Vayron de la Moureyre et al., 1998), and a VNTR (variable number of tandem repeats region, like the DRD3 exon 3 polymorphism) was found in the promoter region, with a negative correlation being found between the number of repeats and TPMT activity. In addition, exonic mutations were found, and the TPMT phenotype was correctly predicted by genotyping in 87% of individuals. More recent publications demonstrate the accuracy of the genotyping method for predicting individuals at highest risk of thiopurine toxicity (McLeod et al., 1999; Relling et al., 1999). This may be preferable to relying on phenotyping given the problems of interpreting phenotyping data when patients have received a red blood cell transfusion (Rioux, 2000).

CYP2C9 and Warfarin Dose Requirement and Toxicity

An association between *CYP2C9* polymorphisms and warfarin dose requirement and risk of bleeding complications has been described (Aithal, Day, Kesteven, & Daly, 1999). CYP2C9 catalyzes the conversion of S-warfarin to inactive 6-hydroxy and 7-hydroxy metabolites, and there are two allelic variants, *CYP2C9*2* and *CYP2C9*3*, which show less than 5% and approximately 12% of wild-type CYP2C9 activity, respectively (Crespi & Miller, 1997; Haining, Hunter, Veronese, Trager, & Rettie, 1996; Rettie, Wienkers, Gonzalez, Trager, & Korzekwa, 1994). Genotyping for the *CYP2C9*2* and *CYP2C9*3* alleles in a group of patients with a low warfarin dose requirement (1.5 mg or less), a group of randomly selected warfarin clinic control patients, and 100 healthy population control individuals was conducted. This showed an odds ratio of 6.21 (95% confidence interval, or CI = 2.48–15.6) for individuals with a low warfarin dose requirement having one or more *CYP2C9* low activity alleles compared with the normal population. Patients in the low-dose group were more likely to have difficulties at the time of induction of warfarin therapy, including prolongation of length of inpatient admission when compared with the clinic control group (odds ratio = 5.97, 95% CI = 2.26–15.82), and had an increased risk of major bleeding complications when compared with this group (rate ratio = 3.68, 95% CI = 1.43–9.50). Preprescribing genotyping might therefore identify individuals more likely to have problems on induction, and lead to altered dose escalation procedures for these patients. In addition, the likely increased risk of major bleeding complications (in general later in the course of therapy) might lead either to increased monitoring of these clients or a careful assessment of the risk–benefit ratio. The *CYP2C9*3* allele has also been shown to be associated with reduced clearance of phenytoin (an anticonvulsant) and tolbutamide (a hypoglycemic agent; Miners & Birkett, 1998).

CYP2D6 and CYP2C19 and Tricyclic Antidepressant Response

CYP2D6 and CYP2C19 are hepatic cytochrome P450 enzymes, involved in the metabolism of TCAs (CYP2D6 catalyzing hydroxylation, and CYP2C19 being involved in demethylation), and are functionally polymorphic, that is, show substantial interindividual variation in level of activity (reviewed in Kirchheiner et al., 2001). In early studies, the steady-state concentrations of desipramine and nortriptyline were shown to be related to CYP2D6 activity (Bertilsson & Åberg-Wistedt, 1983; Nordin, Siwers, Benitz, & Bertilsson, 1985), and the hydroxylation of the TCAs amitriptyline and clomipramine was also shown to be associated with polymorphic CYP2D6 activity (Balant-Gorgia et al., 1982, 1986). A recent study (Morita et al., 2000) has shown that the number of mutant *CYP2D6* alleles accounts for 41% of the variance in logarithm_{10} (nortriptyline level corrected for dose and body weight). Several studies support the existence of a concentration–effect relationship for different TCAs and/or their active metabolites (Aitchison et al., 2001; Gram et al., 1984; Perry, Pfohl, & Holstad, 1987; Preskorn & Jerkovich, 1990; Sjöqvist, Bertilsson, & Åsberg, 1980). Concentration-dependent side effects have also been described in individuals deficient in CYP2D6 treated with TCAs (Balant-Gorgia, Balant, & Garrone, 1989; Bertilsson, Mellström, Sjöqvist, Mårtensson, & Åsberg, 1981; Sjöqvist & Bertilsson, 1984). A recent study (Aitchison et al., 2001) has shown an association between the *N*-demethylation index (ratio of concentration of demethylated metabolite to parent TCA) and number of functional *CYP2C19* genes and response as measured by percentage change in Hamilton Depression Score rating. A review of 32 antidepressants marketed in the United States, Canada, and Europe in which the pharmacokinetic parameters of the antidepressants were compared among extensive, intermediate, and poor metabolizers of either CYP2D6 or CYP2C19 led to distinct dose recommendations for the three metabolic phenotype groups at these two loci (Kirchheiner et al., 2001).

CYP1A2 and Clozapine Response

There is marked interindividual variability in the pharmacokinetics of clozapine (Cheng, Lundberg, Bondesson, Lindstrom, & Gabrielsson, 1988; Choc et al., 1987, 1990; Olesen et al., 1995; Perry, Miller, Arndt, & Cadoret, 1991; Potkin et al., 1994). Clozapine is metabolized mainly in the liver by cytochrome P450s CYP1A2 and CYP3A4, with both in vitro and in vivo work indicating the importance of CYP1A2 in contributing to the variance in the distribution of clozapine (Bertilsson et al., 1994; Carrillo, Herraiz, Ramos, & Benitez, 1998; Eiermann, Engel, Johansson, Zanger, & Bertilsson, 1997; Ozdemir, Kalow, et al., 2001; Tugnait, Hawes, McKay, Eichelbaum, & Midya, 1999). A study of the pharmacodynamics and pharmacokinetics of clozapine in CYP1A2-null mice (mice lacking a functional *CYP1A2* gene) as compared with wild-type mice showed that the CYP1A2-null mice had significantly higher clozapine concentrations and lower desmethylclozapine concentrations, with 61% of the clozapine clearance in wild-type mice being mediated by CYP1A2 (Aitchison, Jann, et al., 2000; see Figures 18.1 and 18.2). Furthermore, the CYP1A2-null mice were significantly more drowsy and showed more motor impairment and myoclonus than the wild-type mice ($p = .0145$; see Figures 18.3 and 18.4 and Exhibit 18.1). CYP1A2 shows wide interindividual variation in activity in humans (reviewed in Aitchison, Jann, et al., 2000; Aitchison, Jordan, & Sharma, 2000), which is likely to be due to both genetic variation and environmental effects such as induction by cigarette smoking. The CYP1A2-null mouse study suggests that individuals with relatively low CYP1A2 activity might be more susceptible than those with high activity to dose-related adverse effects of clozapine, such as sedation, myoclonus, and seizures.

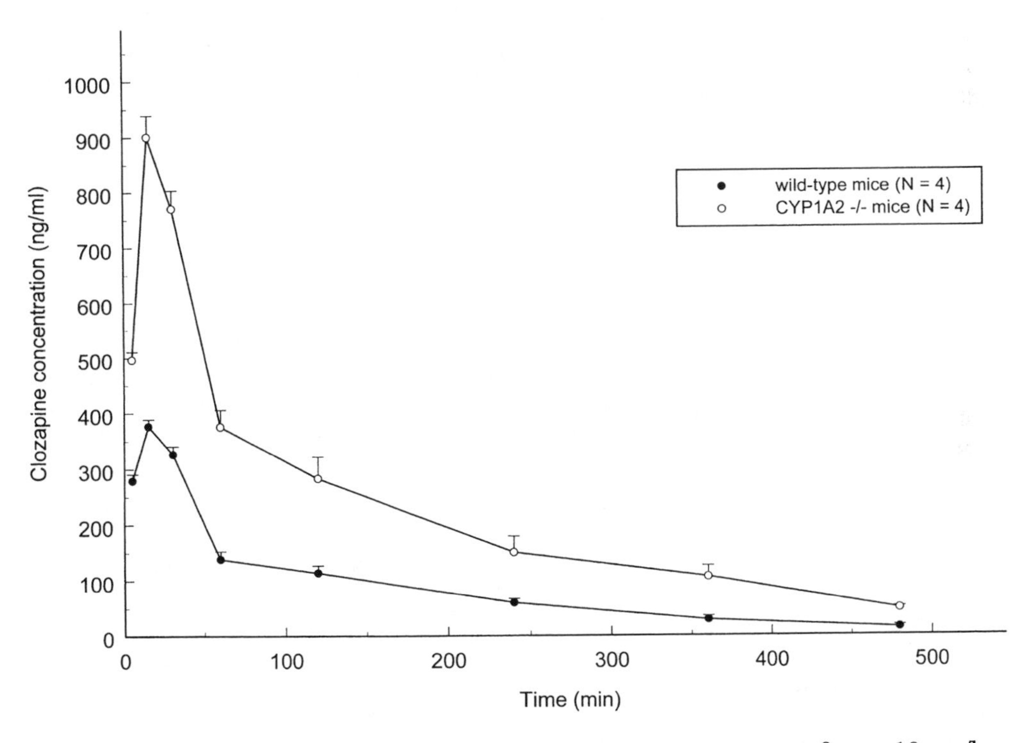

Figure 18.1. Whole blood clozapine concentration-time curves after a 10-mg/kg intraperitoneal dose of clozapine to 4 male wild-type and 4 CYP1A2 −/− mice, administered at Time 0. Mean values + *SD* are given.

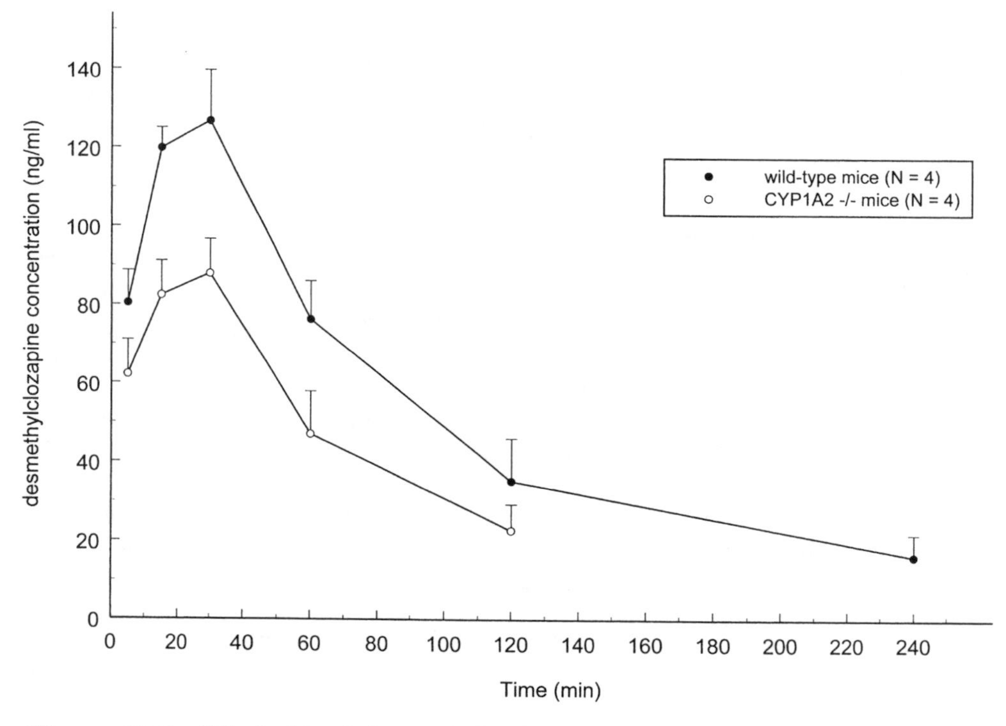

Figure 18.2. Whole blood desmethylclozapine concentration-time curves after the 10-mg/kg intraperitoneal dose of clozapine. For the wild-type mice at 240 min, the mean of two values is reported (the other two values were below the limit of detection, i.e., less than 5.0 ng/ml), otherwise the data are mean of four values + *SD*s.

Serotonin 5-HT$_{2C}$ Receptor and Antipsychotic Response

Clozapine has a relatively high affinity for the 5-HT$_{2C}$ receptor, a postsynaptic neuronal receptor enriched in the prefrontal cortex of individuals with schizophrenia (Sodhi & Murray, 1997; Weinberger, 1987). A preliminary meta-analysis (Weele, Anderson, & Cook, 2000) of the published association studies of the *HTR2C* Cys23Ser allele and clozapine response indicates an association between the 23Ser allele and good response to clozapine (p = .031), despite the fact that in one of the studies, a trend was found in the opposite direction (Malhotra, Goldman, Ozaki, Rooney, et al., 1996). There is a significant difference in allele frequencies between Whites and African Americans (Masellis et al., 1998), and Malhotra, Goldman, Ozaki, Rooney, et al. (1996) did not specify the ethnicity of their sample.

Investigation of association between 5-HT$_{2C}$R variants and adverse effects of clozapine might, however, be more fruitful. Meltzer (1999) pointed out the relationship between the affinity of an atypical antipsychotic for the 5-HT$_{2C}$ receptor and the potential to induce weight gain. Indeed, the *HTR2C* knock-out mouse (i.e., *HTR2C*-null mouse) is overweight, with a diminished response to the satiating effect of the 5-HT

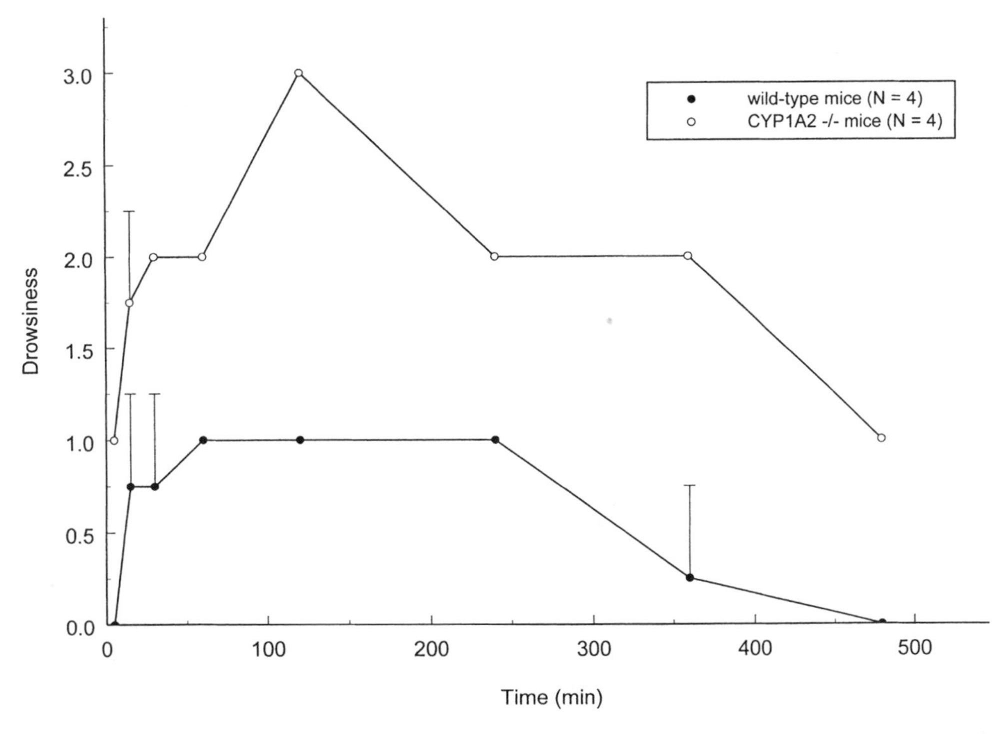

Figure 18.3. Degree of drowsiness versus time after the 10-mg/kg intraperitoneal dose of clozapine. Mice were rated for drowsiness using the criteria given in Exhibit 18.1. Mean scores + *SD* are given.

releaser D-fenfluramine (Tecott et al., 1995; Vickers, Clifton, Dourish, & Tecott, 1999). The *HTR2C* knock-out mouse also shows an increased susceptibility to both spontaneous and chemically induced seizures (Tecott et al., 1995). A recent report describes an association between the -759C/T polymorphism in *HTR2C* promotor and antipsychotic-induced weight gain (Reynolds, Zhang, & Zhang, 2002).

In addition to allelic variation, 5-HT_{2C}R heterogeneity also arises from alternative RNA splicing (Stam et al., 1994; Wang et al., 2000) and RNA editing (Burns et al., 1997). RNA editing is a process whereby adenosine bases in the mRNA are deaminated to inosine, which base pairs with cytosine, and is read as guanosine during translation (Scott, 1995). Hybridization of adjacent complementary sequences in pre-mRNA to form a double-stranded RNA loop is a prerequisite for this process (Burns et al., 1997), which therefore does not occur with all transcripts. In the case of the 5-HT_{2C}R, RNA editing has been demonstrated to occur in the human brain, at five adenosine sites near the exon 5/intron 5 boundary (Burns et al., 1997; Fitzgerald et al., 1999), which leads to pharmacologically distinct variants (reviewed in Sodhi, Burnet, Makoff, Kerwin, & Harrison, 2001). Reduced 5-HT_{2C}R editing, with increased expression of the unedited isoform (5-$HT_{2C\text{-}INI}$), has been noted in schizophrenia (Sodhi et al., 2001). Given that there seem to be far fewer genes in the human genome than

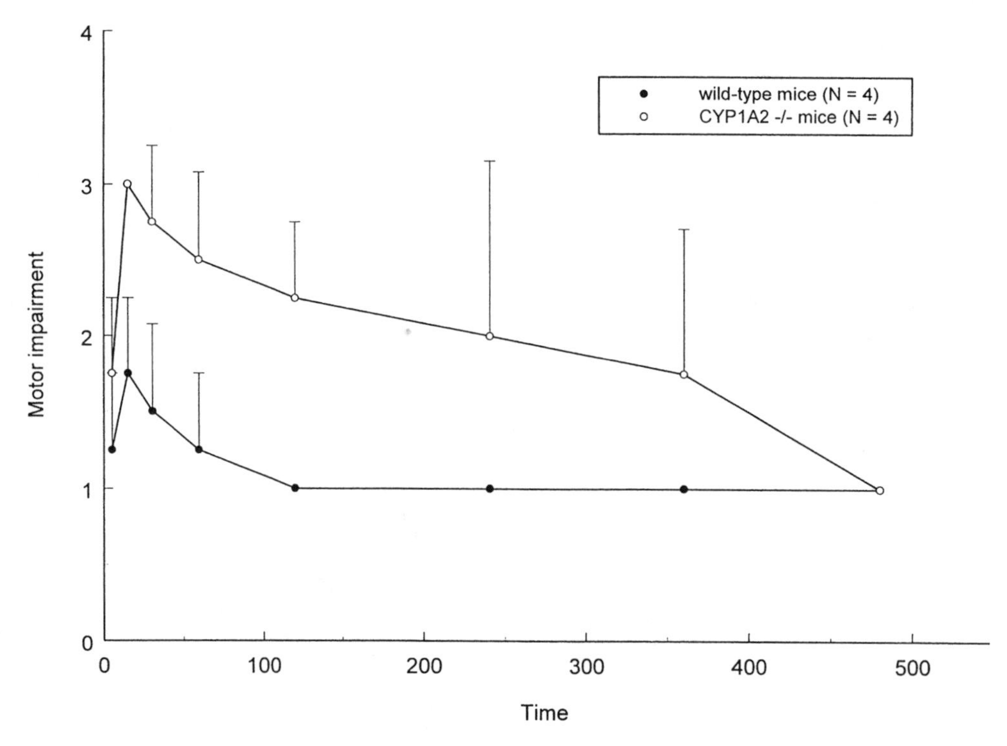

Figure 18.4. Degree of motor impairment versus time after the 10-mg/kg intraperitoneal dose of clozapine. Motor impairment in the mice was rated using the criteria given in Exhibit 18.1. Mean scores + *SD* are given.

was expected (see Plomin, DeFries, Craig, & McGuffin, chapter 1, this volume), researchers in pharmacogenomics might be well advised to investigate the effects of posttranscriptional modifications such as alternative splicing and RNA editing on drug response and adverse effects.

Serotonin Transporter and Response to Proserotonergic Antidepressants

Two functional polymorphic regions in the serotonin transporter gene (*5-HTT*, or SCL6A4) have been identified, one in the promoter region at −1376 to −1027 (Heils et al., 1996; Lesch et al., 1996; Little et al., 1998) and one in intron 2 (Battersby et al., 1996; MacKenzie & Quinn, 1999; Fiskerstrand, Lovejoy, & Quinn, 1999). The *5-HTT* promoter polymorphism (5-HTT gene linked polymorphic region, or 5-HTTLPR) consists of a GC-rich 20–23 base pair repeat element, the common allelic variants comprising 14 (known as *s*) or 16 (known as *l*) repeats, with the *s* variant being associated with a lower expression of 5-HTT sites and a reduced efficiency of 5-HTT reuptake (Hanna et al., 1998; Lesch et al., 1996; Little et al., 1998). An association between *s*/*s* homozygosity and diminished response to fluvoxamine (Smeraldi et al., 1998) or paroxetine (Zanardi, Benedetti, Di Bella, Catalano, & Smeraldi, 2000) in Italians has been re-

Exhibit 18.1. Scale for Rating Behavioral Effects of Clozapine in Mice

Clozapine behavioral effects ratings in mice
Drowsiness
0 Fully alert
1 Eyes half closed at rest
2 Eyes closed at rest, but easily rousable
3 Eyes closed at rest, difficult to rouse
Motor impairment
0 Normal posture at rest, move normally in cage on handling
1 Mild splaying of legs at rest, movements slowed and jerky, mild reduction in struggle on handling
2 Moderate splaying of legs at rest, little movement in cage, moderate reduction in struggle on handling
3 Prominent splaying of legs at rest (sprawled), no movement in cage, minimal struggle on handling

ported, consistent with a report that *s/s* homozygosity is associated with a lower prolactin response to clomipramine (Whale, Questad, Laver, Harrison, & Cowen, 2000), and a poorer response to brain-derived neurotrophic factor (BDNF, Mössner et al., 2000). However, another study, in Korean subjects, found that *s/s* homozygosity was *positively* associated with response to fluoxetine or paroxetine (Kim et al., 2000). Furthermore, a study of the 5-HTTLPR in antidepressant-induced mania in bipolar disorder found an excess of the short allele in individuals with at least one manic or hypomanic episode induced by treatment with proserotonergic antidepressants (Mundo, Walker, Cate, Macciardi, & Kennedy, 2001), which would be consistent with a *positive* association of the *s* allele with proserotonergic antidepressant response.

There are several possible reasons for the discrepancy in the above findings. Firstly, the allele frequencies of the 5-HTTLPR variants vary considerably between different ethnic groups (Gelernter, Cubells, Kidd, Pakstis, & Kidd, 1999), with the distribution in Koreans differing significantly from that in Caucasians. Although the study design of the sample of Mundo et al. (2001) included matching for ethnicity between those with and without antidepressant-induced mania, it is possible that the ethnic composition of the total group led to results that were similar to those of Kim et al. (2000). Indeed, approximately 40% of the sample were Asian (Mundo, 2002). Secondly, the disparity in findings could be due to other heterogeneous factors between the studies, including diagnostic criteria, definition of treatment response, and antidepressant drug dose (see below). Indeed, monoamine transporter levels are regulated by psychoactive drugs including antidepressants (reviewed in Ordway, 2000), and therefore factors such as differences in antidepressant dose may be confounders. Finally, it is possible that the significant association of the polymorphism, in opposite directions in different studies, could mean that the 5-HTT is indeed associated with response to proserotonergic antidepressants but that a different polymorphism, which has different degrees of linkage disequilibrium with the promoter polymorphism in different ethnic groups, is the polymorphism of main effect. Flattem and Blakely (2000) reported a 381-base pair insert between the 5-HTTLPR and the transcription start site, a sequence that appears to be related to the differential transcriptional activity of the promoter region alleles.

Studies Investigating Interactive Effects

The latter example (i.e., the example of discrepancy in findings regarding the 5-HTT) may indicate the importance of controlling for factors (e.g., dose) that might seem to be relevant only to pharmacokinetic genetic factors, but that are in fact also relevant to pharmacodynamic genetic factors. Additionally, the analysis of both pharmacokinetic and pharmacodynamic genetic factors in the same study may be particularly informative. A preliminary study has indicated an interactive effect between a cytochrome P450 17hydroxylase (*CYP17*) genotype and the dopamine *DRD3* 9Gly allele and severity of tardive dyskinesia, an adverse effect of typical antipsychotics (Segman et al., 2002). CYP17 is responsible for the conversion of pregnanolone to dehydroepiandrosterone (Brown, Cascio, & Papadopoulos, 2000; Zwain & Yen, 1999), and Segman et al. (2002) suggest that the effect of the *CYP17* polymorphism on susceptibility to tardive dyskinesia may be mediated by altered neuroprotective capacity (Maurice, Phan, Sandillon, & Urani, 2000) or modulation of central dopamine release by pregnanolone (Zwain & Yen, 1999). A similar interactive effect between CYP1A2 and DRD3 in the genesis of tardive dyskinesia has been proposed (Ozdemir, Basile, Masellis, & Kennedy, 2001).

Methodological Issues in Pharmacogenetics Research

The examples of pharmacogenetics research outlined above represent, however, a small percentage of the worldwide research effort in pharmacogenetics, with much of the data generated so far being disappointing. There are a number of methodological issues that may have contributed to this, as has been discussed by various authors (Elston, 1998; Hodge, 1994; Malhotra & Goldman, 1999; Paterson, Sunohara, & Kennedy, 1999; Weele et al., 2000), and include heterogeneity in clinical, genetic, and statistical methodology employed by different research groups, and the complexity of the phenotype of drug response. All of these issues are well illustrated by studies on the pharmacogenetics of response to clozapine (Collier et al., in press; Masellis et al., 2000), and we discuss each of these below.

Heterogeneity in Clinical Methodology

The various studies on response to clozapine included patients with different diagnoses, some including schizoaffective disorder and others restricting the diagnosis to schizophrenia. Not all of the studies used the same international diagnostic criteria (e.g., *DSM–IV*; American Psychiatric Association, 1994) or established diagnosis by means of a structured interview (e.g., the Structured Clinical Interview for *DSM–III–R* [SCID] diagnoses; Spitzer, William, Gibbon, & First, 1990; or the Schedules for Clinical Assessment in Neuropsychiatry [SCAN Version 2.1]; World Health Organization, 1999). Furthermore, some groups assessed for clinical response to clozapine retrospectively, others prospectively, some employing an antipsychotic washout period, others not. Some studies attempted to control for dosing or plasma levels of clozapine given that a threshold plasma concentration exists that predicts response (Bell, McLaren, Galanos, & Copolov, 1998), and the length of treatment with clozapine before response was assessed varied between 28 days and 6 months. Finally, the clinical rating scales used to measure response differed, and response was variously defined as a dichotomous trait or a continuous variable.

Heterogeneity in Genetic Methodology

Different $5HT_{2A}$-receptor polymorphisms have been investigated in different studies in relation to clozapine response. An initial report of an association between a $5HT_{2A}$-receptor polymorphism (102 T/C) and clozapine response (Arranz et al., 1995) was not replicated in several independent studies (Malhotra, Goldman, Ozaki, Breier, et al., 1996; Masellis et al., 1995, 1998; Nöthen et al., 1995). The 102 T/C substitution is a silent or synonymous substitution that does not alter the amino acid sequence and would not be expected to alter the level of gene transcription. It has, however, been demonstrated to be in near-perfect linkage disequilibrium with a −1438 G/A single nucleotide polymorphism (SNP) in the $5HT_{2A}$ promoter region (Arranz et al., 2000; Masellis et al., 1998), which some work suggests may be functional (Turecki et al., 1999) but other work suggests is likely not to be (Nakamura et al., 1999; Spurlock et al., 1998).

Association studies utilizing variants with no known functional effect have a significantly lower prior probability of detecting valid associations than studies using functional variants (Malhotra & Goldman, 1999). Positive findings with nonfunctional markers are more likely to be secondary to chance or, in the case of studies using a basic case control genetic association design, to inadvertent ethnic stratification of cases and control (see below). Arranz, Munro, Owen, et al. (1998) and Masellis et al. (1998) did, however, find an association between a functional $5HT_{2A}$ variant (His452Tyr) and clozapine response, and a meta-analysis of both the 102 T/C and the His452Tyr polymorphisms showed a trend for an association between these polymorphisms and response (Arranz, Munro, Sham, et al., 1998). Nonetheless, the significance of the result for the 102C allele disappeared when the original study (Arranz et al., 1995) was excluded. Moreover, this meta-analysis has been criticized because it did not properly address the issue of heterogeneity among studies, which is important in meta-analyses (Colditz, Burdick, & Mosteller, 1995; DerSimonian &

Laird, 1986). In addition, it has been suggested that the apparent association of the T102C polymorphism with clozapine response may in fact represent a nonspecific association with poor response to antipsychotics in general, with poor long-term outcome in schizophrenia (Joober et al., 1999). The predicted population frequency of the 452Tyr/Tyr genotype is approximately 1%, and the negative predictive value (a measure of an inverse association with response) of this genotype is 0.85, with 95% confidence intervals of 0.54–0.97 (Weele et al., 2000), which would mean that screening for this allele alone would not be justifiable.

Problems With Statistical Methodology

In regard to problems of statistical methodology in these studies of the pharmacogenetics of response to clozapine, the following points may be made. Firstly, there are power considerations. Many of the studies did not have power to detect a significant gene effect (with the small effect size suggested), due to inadequate sample sizes. Secondly, in studies that have investigated more than one polymorphism in the same gene, or polymorphic sites in different genes, it is important to perform a statistical correction for multiple testing (e.g., Bonferroni correction). One study, which analyzed 19 genetic polymorphisms in nine putative pharmacological targets of clozapine (Arranz et al., 2000) gave a very low (and therefore apparently highly significant) p value ($p = .0001$) for the combination of the six polymorphisms showing the strongest association with response but failed to correct for multiple testing. Not surprisingly, a study in a German population of those six polymorphisms failed to replicate the findings of Arranz et al. (Schumacher, Schulze, Wienker, Rietschel, & Nöthen, 2000).

Additional statistical difficulties arise from the failure to adequately address important confounding variables, such as response to previous antipsychotic treatments and severity of symptoms at the end of the preceding treatment, precise symptom complex (e.g., positive symptoms predominating), duration of illness, ethnicity, gender, age at onset of illness, compliance, plasma clozapine level, and comedications (e.g., mood stabilizers; Rietschel, Kennedy, Macciardi, & Meltzer, 1999). Good study design thus requires detailed clinical data collection as well as adequate statistical methodology. The retrospective design of some studies has precluded the inclusion of many of the above factors.

One potential pitfall of all genetic association studies is population stratification (Elston, 1998; Hodge, 1994; Malhotra & Goldman, 1999; Paterson et al., 1999). An allele at a candidate gene may occur at different frequencies in different ethnic groups (as, has, for example been found for the T102C polymorphism; Nakamura et al., 1999). If cases and controls are not carefully matched, an apparent but spurious association with the phenotype of interest can occur. One method of controlling for population differences is to restrict analysis to ethnically homogeneous samples; however, samples that may appear homogeneous may in fact be deceptive (e.g., Finnish people have a dual Asiatic and European origin; Kittles et al.,

1999). Another method is to collect DNA from the parents of subjects and conduct analysis using the transmission disequilibrium test (TDT; Spielman & Ewens, 1996), as discussed in more detail below. There are also other methods for estimating the degree of population stratification in an apparently homogeneous population, for example, *genomic control* (GC), using 20–60 SNPs spread throughout the genome (Bacanu, Devlin, & Roeder, 2000), or unlinked microsatellite markers (Pritchard, Stephens, Rosenberg, & Donnelly, 2000). The significance level of a candidate gene polymorphism is then adjusted to reflect true association in the face of population substructure.

Finally, in terms of statistical methodology, there have been errors in the grouping of alleles of genotypes in order to reduce the number of degrees of freedom in the analysis (and hence increase the likelihood of a significant result). For example, the dopamine D4 receptor (DRD4) has a hypervariable region in the third cytoplasmic loop (exon 3) consisting of 2–10 imperfect 48 base pair repeats (Van Tol et al., 1992), with the most common alleles having 2, 4, or 7 repeats. However, there does not seem to be a simple linear relationship between the repeat length and functional activity, with the 10-repeat variant being 2- to 3-fold more potent in dopamine-mediated coupling to adenylate cyclase than the 2-repeat variant (Jovanovic, Guan, & Van Tol, 1999), and the 2- and 4-repeat variants being 2- to 3-fold more potent than the 7-repeat (Ashgari et al., 1995). Some clozapine pharmacogenetics studies have divided subjects into two groups: having "short" repeats (2–6) or "long" repeats (7–12 repeats; Benjamin, Patterson, Greenberg, Murphy, & Hamer, 1996), which would place the 10- and 7-repeat variants, which appear to have degrees of activity at opposite ends of the spectrum, in the same group. Furthermore, there are sequence variations even in the alleles of the same length, which are entirely overlooked by such pooling strategies (Wong, Buckle, & Van Tol, 2000). Grouping errors in statistical analyses have also occurred when biallelic systems have been tested, and heterozygotes have been grouped with one of the homozygote groups, under the assumption that one of the alleles acts dominantly, sometimes with little supporting data (Malhotra & Goldman, 1999).

Complexity of the Phenotype of Drug Response

Apart from the heterogeneity in the methodology among the studies, the inconsistent findings are also likely to be related to the complex nature of the phenotype of clozapine response (Masellis et al., 2000). One might speculate that variability in a particular candidate gene might contribute to the phenotype of response only within particular subgroups of patients, whereas different candidate genes might be relevant to other subgroups (i.e., genetic heterogeneity). Alternatively, polymorphisms across several candidate genes may interact, and thus each may contribute a small proportion to the total variance observed in the phenotype of clozapine response (i.e., polygenic inheritance), which could lead to Type I errors (i.e., false negative results). These concepts are clearly plausible given the known genetic heterogeneity of schizophrenia and the involvement of multiple loci in the disorder (McGuffin, Owen, & Farmer, 1995), in addition to the complex mechanisms of action of clozapine (Sodhi & Murray, 1997).

Future Directions

From the discussion in this chapter, a number of recommendations may be made regarding future pharmacogenetics and pharmacogenomics studies. Firstly, the design of the clinical dataset is crucial. Prospective studies that are specifically designed for pharmacogenetic or pharmacogenomic analysis and include the power and rigor of design necessary to test pharmacogenetic hypotheses are most likely to yield informative data (Masellis et al., 2000; Pickar & Rubinow, 2001). Ideally, standardized diagnostic interviews should be adopted; a clearly defined washout period should be used, with prior treatment response and baseline symptom level recordings, and prospective recordings of clinical response and adverse effects using standardized rating instruments should be made at predefined, theoretically justifiable time points. In addition, the drug dose and plasma drug levels (both serially, at different response times) should be determined, and details of standard demographic variables (age, sex, and ethnicity), other clinical variables (e.g., age at onset of diagnosis, comorbid diagnosis, family history, symptom complex), and any others pertinent to the psychotropic drug being evaluated should be determined. Exclusion criteria should be defined a priori, as should the criteria for clinical response (e.g., mean change in clinical score over time, controlling for baseline; Basile et al., 1999).

Second, some recommendations regarding statistical issues may be made. In the case of pharmaco*genetic* studies, calculations of sample size required for adequate power should also be conducted a priori, taking into consideration allele frequencies for the specific polymorphisms to be examined and the number of candidate genes to be tested (Masellis et al., 2000). This may not, however, be possible for pharmaco*genomics* studies, in which, for example, marker allele frequencies (particularly in disease populations) may not be known. Similarly, it may be appropriate to define a priori an appropriate significance level given the number of candidate loci and hypotheses to be tested. Finally, in the data analysis, the relative effect size contributed by each candidate gene polymorphism should be estimated (Masellis et al., 2000).

To control for population stratification, methods such as the TDT, GC control, or the method of Pritchard et al. (2000) may be used (see Cardon, chapter 4, this volume). The TDT has been extended for use with quantitative traits (and response to drug treatment measured on a linear scale may be viewed as a quantitative trait), is resistant to confounding (Allison, 1997; Rabinowitz, 1997), and has particularly high power if the most extreme 20% of the phenotypic distribution is selectively sampled. However, adequate sample sizes may be difficult to attain, given that it can be difficult or impossible to collect clinical information and DNA from parents of relatively old affected individuals, and also, with TDT, only heterozygous parents are informative. Whenever parents are homozygous, the genotyping data are effectively useless, with consequent loss of study power. An alternative TDT approach to the standard parents–child trio uses samples of sibling pairs to test for association (Spielman & Ewens, 1996). The analysis may also be improved by genotyping for a number of different polymorphic sites at loci of interest, so that haplotypes may be constructed (a haplotype being the combination of more than one allelic variant, e.g., a particular 5-HTTLPR variant plus an intron 2 variant). Some high-throughput genotyping strategies (especially those involving pooling) may not automatically yield haplotype data, but by using haplotype estimation maximization algorithms (Long, Williams, & Urbanek, 1995), it is possible to assign haplotypes objectively. Usually haplotype analysis provides the benefit of more information about the likelihood that a particular gene is associated with the phenotype of interest. The power and validity of haplotype-based analysis is increased by the grouping of haplotypes according to their most likely common origin (Templeton, Boerwinkle, & Sing, 1987).

Thirdly, there are recommendations to be made in regard to the candidate genes chosen for analysis. Both pharmacokinetic and pharmacodynamic genetic factors should be investigated. An extension of this is candidate pathway analysis (see Craig & McClay, chapter 2, this volume). For example, for a pharmacogenetic analysis of response to proserotonergic antidepressants, one might include the genes involved in serotonin synthesis and breakdown, in addition to the serotonin transporter and regulators of serotonin transporter response (e.g., the presynaptic $5HT_{1A}$-receptor; Smeraldi et al., 1998). This suggestion, though, may lead to the inclusion of a large number of candidate genes (e.g., BDNF [Stahl, 2000]; third messenger systems such as kinases; and other neurotransmitter systems functionally connected to the serotonergic system [Schafer, 1999]), and a balance may have to be struck between what is desirable to investigate and what is practical, given the sample size and power considerations. If possible, a hierarchy of which genes are presumed to be "better candidates" than others should be constructed, using the existing literature, which will not be easy unless pilot pharmacogenetic studies have been conducted.

Finally, some general comments regarding study design may be made. Pilot pharmacogenetic studies, including retrospective studies or those that do not conform to the "ideal" pharmacogenetic or pharmacogenomic design, may be useful. The "ideal" study design would be most easily used in the context of a clinical trial, predesigned as a pharmacogenetics or pharmacogenomics study. Unless pharmaceutical companies collaborate with academics to design such trials, trials of sufficient sample size may be difficult to construct and, furthermore, may take a considerable time to generate adequate data. It has been recommended that subject DNA collections should become a standard part of drug research protocols in Phase II and Phase III studies (Pickar & Rubinow, 2001); however, this alone will not suffice. Less than ideal study designs, for example, those with a combination of retrospective and prospective data collection (retrospective baseline values and prospective response ratings), may therefore have to be acceptable to funding bodies, especially for the purpose of generating hypotheses on which more rigorous studies may be built. Any initial findings should be replicated in independent samples, preferably in a group of comparable ethnicity and in those of different ethnicity. Animal studies may also be particularly useful, both for generating and testing hypotheses.

Ultimately, pharmacogenomics studies should be based on a priori hypotheses with selected populations stratified by genotype (Pickar & Rubinow, 2001). However, prior to this, pilot analyses and pharmacogenetics and pharmacogenomics studies in the context of clinical trials should be performed and replicated.

References

Aitchison, K. J., Checkley, S., Patel, M., Kinirons, M., Chapman, S., Collier, D. A., et al. (2001). Pharmacogenetic determinants of response to tricyclic antidepressants. *Journal of Psychopharmacology, 15*(Suppl.), A12.

Aitchison, K. J., Gonzalez, F. J., Quattrochi, L. C., Sapone, A., Zhao, J. H., Zaher, H., et al. (2000). Identification of novel polymorphisms in the 5′ flanking region of CYP1A2, characterisation of interethnic variability, and investigation of their functional significance. *Pharmacogenetics, 10,* 695–704.

Aitchison, K. J., Jann, M.W., Zhao, J. H., Sakai, T., Zaher, H., Wolff, K., et al. (2000). Clozapine pharmacokinetics and pharmacodynamics studied with CYP1A2-null mice. *Journal of Psychopharmacology, 14,* 353–359.

Aitchison, K. J., Jordan, B. D., & Sharma, T. (2000). The relevance of ethnic influences on pharmacogenetics to the treatment of psychosis. *Drug Metabolism and Drug Interactions, 16,* 15–38.

Aitchison, K. J., Patel, M., Taylor, M., Murray, R. M., Arranz, M. J., Collier, D. A., et al. (1995). Neuroleptic sensitivity and enzyme deficiency in two schizophrenic brothers: A case report. *Schizophrenia Research, 18,* 140.

Aithal, G. P., Day, C. P., Kesteven, P. J., & Daly, A. K. (1999). Association of polymorphisms in the cytochrome P450 CYP2C9 with warfarin dose requirement and risk of bleeding complications. *Lancet, 353,* 717–719.

American Psychiatric Association. (1994). *Diagnostic and statistical manual of mental disorders* (4th ed.). Washington, DC: Author.

Alen, P. (2001, May). *Toxicogenomics: Identification of biomarker genes to predict adverse drug reactions*. Paper presented at "Pharmacogenomics Europe," Munich, Germany.

Alexanderson, B. (1973). Prediction of steady-state plasma levels of nortriptyline from single oral dose kinetics: A study in twins. *European Journal of Clinical Pharmacology, 6,* 44–53.

Alexanderson, B., Evans, D. A. P., & Sjöqvist, F. (1969). Steady state plasma levels of nortriptyline in twins: Influence of genetic factors and drug therapy. *British Medical Journal, 4,* 174–180.

Allison, D. B. (1997). Transmission-Disequilibrium tests for quantitative traits. *American Journal of Human Genetics, 60,* 676–690.

Arranz, M. J., Collier, D., Sodhi, M., Ball, D., Roberts, G., Price, J., et al. (1995). Association between clozapine response and allelic variation in 5-HT2A receptor gene. *Lancet, 346,* 281–282.

Arranz, M. J., Munro, J., Birkett, J., Bolonna, A., Mancama, D., Sodhi, M., et al. (2000). Pharmacogenetic prediction of clozapine response. *Lancet, 355,* 1615–1616.

Arranz, M. J., Munro, J., Owen, M. J., Spurlock, G., Sham, P. C., Zhao, J., et al. (1998).

Evidence for association between polymorphisms in the promoter and coding regions of the 5-HT2A receptor gene and response to clozapine. *Molecular Psychiatry, 3,* 61–66.

Arranz, M. J., Munro, J., Sham, P., Kirov, G., Murray, R. M., Collier, D. A., et al. (1998). Meta-analysis of studies on variation in 5-HT2A receptors and clozapine response. *Schizophrenia Research, 32,* 93–99.

Åsberg, M., Evans, D. A. P., & Sjöqvist, F. (1971). Genetic control of nortriptyline kinetics in man: A study of relatives of propositi with high plasma concentrations. *Journal of Medical Genetics, 8,* 129–135.

Ashgari, V., Sanyal, S., Buchwaldt, S., Paterson, A., Jovanovic, V., & Van Tol, H. H. M. (1995). Modulation of intracellular cyclic AMP levels by different human dopamine D4 receptor variants. *Journal of Neurochemistry, 65,* 1157–1165.

Atkin, K., Kendall, F., Gould, D., Freeman, H., Lieberman, J., & O'Sullivan, D. (1996). *British Journal of Psychiatry, 169,* 483–488.

Bacanu, S.-A., Devlin, B., & Roeder, K. (2000). The power of genomic control. *American Journal of Human Genetics, 66,* 1933–1944.

Balant-Gorgia, A. E., Balant, L. P., & Garrone, G. (1989). High blood concentrations of imipramine or clomipramine and therapeutic failure: A case report study using drug monitoring data. *Therapeutic Drug Monitoring, 11,* 415–420.

Balant-Gorgia, A. E., Balant, L. P., Genet, C. H., Dayer, P., Aeschlimann, J. M., & Garraone, G. (1986). Importance of oxidative polymorphism and levomepromazine treatment on the steady-state blood concentrations of clomipramine and its major metabolites. *European Journal of Clinical Pharmacology, 31,* 415–420.

Balant-Gorgia, A. E., Schulz, P., Dayer, P., Balant, L., Kubli, A., Gertsch, C., et al. (1982). Role of oxidation polymorphism on blood and urine concentrations of amitriptyline and its metabolites in man. *Archiv für Psychiatrie und Nervenkrankheiten [Archives of Psychiatry and Neurological Sciences], 232,* 215–222.

Basile, V. S., Masellis, M., Badri, F., Paterson, A. D., Meltzer, H. Y., Lieberman, J. A., et al. (1999). Association of the MscI polymorphism of the dopamine D3 receptor gene with tardive dyskinesia in schizophrenia. *Neuropsychopharmacology, 21,* 17–27.

Battersby, S., Ogilvie, A. D., Smith, C. A. D., Blackwood, D. H. R., Muir, W. J., Quinn, J. P., et al. (1996). Structure of a variable number tandem repeat of the serotonin transporter gene and association with affective disorder. *Psychiatric Genetics, 6,* 177–181.

Bell, R., McLaren, A., Galanos, J., & Copolov, D. (1998). The clinical use of plasma clozapine levels. *Australian and New Zealand Journal of Psychiatry, 32,* 567–574.

Benjamin, J., Li, L., Patterson, C., Greenberg, B. D., Murphy, D. L., & Hamer, D. H. (1996). Population and familial association between the D4 dopamine receptor gene and measures of novelty seeking. *Nature Genetics, 12,* 81–84.

Bertilsson, L., & Åberg-Wistedt, A. (1983). The debrisoquine hydroxylation test predicts steady-state plasma levels of desipramine. *British Journal of Clinical Pharmacology, 15,* 388–390.

Bertilsson, L., Carrillo, J. A., Dahl, M. L., Llerena, A., Alm, C., Bondesson, U., et al. (1994). Clozapine disposition covaries with CYP1A2 activity determined by a caffeine test. *British Journal of Clinical Pharmacology, 38,* 471–473.

Bertilsson, L., Mellström, B., Sjöqvist, F., Mårtensson, B., & Åsberg, M. (1981). Slow hydroxylation of nortriptyline and concomitant poor debrisoquine hydroxylation: Clinical implications. *Lancet, i,* 560–561.

Brown, R. C., Cascio, C., & Papadopoulos, V. (2000). Pathways of neurosteroid biosynthesis in cell lines from human brain: Regulation of DHEA formation by oxidative stress and beta-amyloid peptide. *Journal of Neurochemistry, 74,* 847–859.

Burns, C. M., Chu, H., Rueter, S. M., Hutchinson, L. K., Canton, H., Sanders-Bush, E., et al. (1997). Regulation of serotonin-$_{2C}$ receptor G-protein coupling by RNA editing. *Nature, 387,* 303–308.

Carrillo, J. A., Herraiz, A. G., Ramos, S. I., & Benitez, J. (1998). Effects of caffeine withdrawal from the diet on the metabolism of clozapine in schizophrenic patients. *Journal of Clinical Psychopharmacology, 18,* 311–316.

Cheng, Y. F., Lundberg, T., Bondesson, U., Lindstrom, L., & Gabrielsson, J. (1988). Clinical pharmacokinetics of clozapine in chronic schizophrenic patients. *European Journal of Clinical Pharmacology, 34,* 445–449.

Choc, M. G., Hsuan, F., Honigfeld, G., Robinson, W. T., Ereshefsky, L., Crismon, M. L., et al. (1990). Single- vs multiple-dose pharmacokinetics of clozapine in psychiatric patients. *Pharmaceutical Research, 7,* 347–351.

Choc, M. G., Lehr, R. G., Hsuan, F., Honigfeld, G., Smith, H. T., Borison, R., et al. (1987). Multiple-dose pharmacokinetics of clozapine in patients. *Pharmaceutical Research, 4,* 402–405.

Colditz, G. A., Burdick, E., & Mosteller, F. (1995). Heterogeneity in meta-analysis of data from epidemiologic studies: A commentary. *American Journal of Epidemiology, 142,* 371–382.

Collier, D. A., Arranz, M. J., Osborne, S., Aitchison, K. J., Munro, J., Mancama, D., & Kerwin, R. W. (in press). The pharmacogenetics of response to clozapine: the influence of genetic variation in neurotransmitter receptor targets. In B. Lerer (Ed.), *Pharmacogenetics of psychotropic drugs*. Cambridge, England: Cambridge University Press.

Corzo, D., Yunis, J. J., Salazar, M., Lieberman, J. A., Howard, A., Awdeh, Z., et al. (1995). The major histocompatibility complex region marked by HSP70-1 and HSP70-2 variants is associated with clozapine-induced agranulocytosis in two different ethnic groups. *Blood, 86,* 3835–3840.

Crespi, C. L., & Miller, V. P. (1997). The R144C change in the *CYP2C9*2* allele alters interaction of the cytochrome P450 with NADPH:cytochrome P450 oxidoreductase. *Pharmacogenetics, 7,* 203–210.

Dains, K., Hitzemann, B., & Hitzemann, R. (1996). Genetics, neuroleptic response and the organization of cholinergic neurons in the mouse striatum. *Journal of Pharmacology and Experimental Therapeutics, 279,* 1430–1438.

DerSimonian, R., & Laird, N. (1986). Meta-analysis in clinical trials. *Control Clinical Trials, 7,* 177–188.

Eiermann, B., Engel, G., Johansson, I., Zanger, U. M., & Bertilsson, L. (1997). The involvement of CYP1A2 and CYP3A4 in the metabolism of clozapine. *British Journal of Clinical Pharmacology, 44,* 439–446.

Elston, R. C. (1998). Linkage and association. *Genetic Epidemiology, 15,* 565–576.

Farmer, A. E., McGuffin, P., & Gottesman, I. I. (1987). Twin concordance for *DSM–III* schizophrenia: Scrutinising the validity of the definition. *Archives of General Psychiatry, 44,* 634–641.

Fiskerstrand, C. E., Lovejoy, E. A., & Quinn, J. P. (1999). An intronic polymorphic domain often associated with susceptibility to affective disorders has allele dependent differential enhancer activity in embryonic stem cells. *FEBS Letters, 458,* 171–174.

Fitzgerald, L. W., Iyer, G., Conklin, D. S., Krause, C. M., Marshall, A., Patterson, J. P., et al. (1999). Messenger RNA editing of the human serotonin $5\text{-}HT_{2C}$ receptor. *Neuropsychopharmacology, 21,* 82S–90S.

Flattem, N. L., & Blakely, R. D. (2000). Modified structure of the human serotonin transporter promoter. *Molecular Psychiatry, 5,* 110–115.

Frackiewicz, E. U. J., Sramek, J. J., Herrera, J. M., Kurtz, N. M., & Cutler, N. R. (1997). Ethnicity and antipsychotic response. *Annals of Pharmacotherapy, 31,* 1360–1369.

Franchini, L., Serretti, A., Gasperini, M., & Smeraldi, E. (1998). Familial concordance of fluvoxamine response as a tool for differentiating mood disorder pedigrees. *Journal of Psychiatric Research, 32,* 255–259.

Gelernter, J., Cubells, J. F., Kidd, J. R., Pakstis, A. J., & Kidd, K. K. (1999). Population studies of polymorphisms of the serotonin transporter protein gene. *American Journal of Medical Genetics, 88,* 61–66.

Gram, L. F., Kragh-Sørensen, P., Kristensen, C. B., Møller, M., Pedersen, O. L., & Thayssen, P. (1984). Plasma level monitoring of antidepressants: Theoretical basis and clinical application. In E. Udin, M. Åsberg, L. Bertilsson, & F. Sjöqvist (Eds.), *Frontiers in biochemical and pharmacological research in depression* (pp. 399–411). New York: Raven Press.

Haining, R. L., Hunter, A. P., Veronese, M. E., Trager, W. F., & Rettie, A. E. (1996). Allelic variants of human cytochrome P450 2C9: Baculovirus-mediated expression purification, structural characterization, substrate stereoselectivity and prochiral selectivity of the wild-type and 1359L mutant forms. *Archives of Biochemistry and Biophysics, 333,* 447–458.

Hamilton, M. (1967). Development of a rating scale for primary depressive illness. *British Journal of the Society of Clinical Psychology, 6,* 278–296.

Hanna, G. L., Himle, H. A., Curtis, G. C., Koram, D. Q., Weele, J. V., Leventhal, B. L., & Cook, E. H., Jr. (1998). Serotonin transporter and seasonal variation in blood serotonin in families with obsessive–compulsive disorder. *Neuropsychopharmacology, 18,* 102–111.

Heils, A., Teufel, A., Petri, S., Stöber, G., Riederer, P., Bengel, D., & Lesch, K. P. (1996). Allelic variation of human serotonin transporter gene expression. *Journal of Neurochemistry, 66,* 2621–2624.

Hodge, S. E. (1994). What association analysis can and cannot tell us about the genetics of complex disease. *American Journal of Medical Genetics, 54,* 318–323.

Housman, D., & Ledley, F. D. (1998). Why pharmacogenomics? Why now? *Nature Biotechnology, 16,* 492–493.

Joober, R., Benkelfat, C., Brisebois, K., Toulouse, A., Turecki, G., Lal, S., et al. (1999). T102C polymorphism in the 5HT2A gene and schizophrenia: Relation to phenotype and drug response variability. *Journal of Psychiatry and Neuroscience, 24,* 141–146.

Jovanovic, V., Guan, H. C., & Van Tol, H. H. (1999). Comparative pharmacological and functional analysis of the human dopamine D4.2 and D4.10 receptor variants. *Pharmacogenetics, 9,* 561–568.

Keck, P. E., Welge, J. A., Strakowski, S. M., Arnold, L. M., & McElroy, S. L. (2000). Placebo effect in randomized, controlled maintenance studies of patients with bipolar disorder. *Biological Psychiatry, 47,* 756–761.

Kim, D. K., Lim, S.-W., Lee, S., Sohn, S. E., Kim, S., Hahn, C. G., & Carroll, B. J. (2000). Serotonin transporter gene polymorphism and antidepressant response. *NeuroReport, 11,* 215–219.

Kirchheiner, J., Brøsen, K., Dahl, M. L., Gram, L. F., Kasper, S., Roots, I., et al. (2001). CYP2D6 and CYP2C19 genotype-based dose recommendations for antidepressants: A first step towards subpopulation-specific dosages. *Acta Psychiatrica Scandinavica, 104,* 173–192.

Kittles, R. A., Bergen, A. W., Urbanek, M., Virkkunen, M., Linnoila, M., Goldman, D., & Long, J. C. (1999). Autosomal, mitochondrial, and Y chromosome DNA variation in Finland: Evidence for a male-specific bottleneck. *American Journal of Physical Anthropology, 108,* 381–399.

Lazarou, J., Pomeranz, B. H., & Corey, P. N. (1998). Incidence of adverse drug reactions in hospitalized patients: A meta-analysis of prospective studies. *Journal of the American Medical Association, 279,* 1200–1205.

Ledley, F. D. (1999). Can pharmacogenomics make a difference in drug development? *Nature Biotechnology, 17,* 731.

Lesch, K. P., Bengel, D., Heils, A., Sabol, S., Greenberg, B. D., Petri, S., et al. (1996). Association of anxiety-related traits with a polymorphism in the serotonin transporter gene regulatory region. *Science, 274,* 1527–1531.

Lieberman, J. A., Yunis, J., Egea, E., Canoso, R. T., Kane, J. M., & Yunis, E. J. (1990). HLA-B38, DR4, DQW3 and clozapine-induced agranulocytosis in Jewish patients with schizophrenia. *Archives of General Psychiatry, 47,* 945–948.

Little, K. Y., McLaughlin, D. P., Zhang, L., Livermore, C. S., Dalack, G. W., McFinton, P. R., et al. (1998). Cocaine, ethanol, and genotype effects on human midbrain serotonin transporter binding sites and mRNA levels. *American Journal of Psychiatry, 155,* 207–213.

Long, J. C., Williams, R. C., & Urbanek, M. (1995). An E-M algorithm and testing strategy for multiple-locus haplotypes. *American Journal of Human Genetics, 56,* 799–810.

MacKenzie, A., & Quinn, J. (1999). A serotonin transporter gene intron 2 polymorphic region, correlated with affective disorders, has allele-dependent differential enhancer-like properties in the mouse embryo. *Proceedings of the National Academy of Sciences, 96,* 15251–15255.

Malhotra, A. K., & Goldman, D. (1999). Benefits and pitfalls encountered in psychiatric genetic association studies. *Biological Psychiatry, 45,* 544–550.

Malhotra, A. K., Goldman, D., Ozaki, N., Breier, A., Buchanan, R., & Pickar, D. (1996). Lack of association between polymorphisms in the 5-HT2A receptor gene and the antipsychotic response to clozapine. *American Journal of Psychiatry, 153,* 1092–1094.

Malhotra, A. K., Goldman, D., Ozaki, N., Rooney, W., Clifton, A., Buchanan, R. W., et al. (1996). Clozapine response and the 5HT2CCys23Ser polymorphism. *NeuroReport, 7,* 2100–2102.

Masellis, M., Basile, V., Meltzer, H. Y., Lieberman, J. A., Sevy, S., Macciardi, F. M., et al. (1998). Serotonin subtype 2 receptor genes and clinical response to clozapine in schizophrenia patients. *Neuropsychopharmacology, 19,* 123–132.

Masellis, M., Basile, V. S., Özdemir, V., Meltzer, H. Y., Macciardi, F. M., & Kennedy, J. L. (2000). Pharmacogenetics of antipsychotic treatment: Lessons learned from clozapine. *Biological Psychiatry, 47,* 252–266.

Masellis, M., Paterson, A. D., Badri, F., Lieberman, J. A., Meltzer, H. Y., Cavazzoni, P., & Kennedy, J. L. (1995). Genetic variation of 5-HT2A receptor and response to clozapine. *Lancet, 346,* 1108.

Maurice, T., Phan, V. L., Sandillon, F., & Urani, A. (2000). Differential effect of DHEA and its steroid precursor pregnanolone against the behavioral deficits in CO-exposed mice. *European Journal of Pharmacology, 320,* 145–155.

McGuffin, P., Owen, M. J., & Farmer, A. E. (1995). Genetic basis of schizophrenia. *Lancet, 346,* 678–682.

McLeod, H. L., Pritchard, S. C., Githang'a, J., Indalo, A., Ameyaw, M. M., Powrie, R. H., et al. (1999). Ethnic difference in thiopurine methyltransferase pharmacogenetics: Evidence for allele specificity in Caucasians and Kenyan individuals. *Pharmacogenetics, 9,* 773–776.

Meltzer, H. Y. (1999). The role of serotonin in antipsychotic drug action. *Neuropsychopharmacology, 21,* 106S–115S.

Miners, J. O., & Birkett, D, J. (1998). Cytochrome P4502C9: An enzyme of major importance in human drug metabolism. *British Journal of Clinical Pharmacology, 45,* 525–538.

Morita, S., Shimoda, K., Someya, T., Yoshimura, Y., Kamijima, K., & Kato, N. (2000). Steady-state plasma levels of nortriptyline and its hydroxylated metabolites in Japanese patients: Impact of *CYP2D6* genotype on the hydroxylation of nortriptyline. *Journal of Clinical Psychopharmacology, 20,* 141–149.

Mössner, R., Daniel, S., Albert, D., Heils, A., Okladnova, O., Schmitt, A., & Lesch, K.-P. (2000). Serotonin transporter function is modulated by brain-derived neurotrophic factor (BDNF) but not nerve growth factor (NGF). *Neurochemistry International, 36,* 197–202.

Mundo, E., Walker, M., Cate, T., Macciardi, F., & Kennedy, J. L. (2001). The role of serotonin transporter protein gene in antidepressant-induced mania in bipolar disorder. *Archives of General Psychiatry, 58,* 539–544.

Mundo, E., Walker, M., Zai, G., Ni, X., Cate, T., Macciardi, F., et al. (2002). Specificity of the promoter polymorphism of the serotonin transporter gene in predicting antidepressant-induced mania in bipolar disorder. Oral presentation, April 12, 2002, First Annual Pharmacogenetics in Psychiatry Meeting, NY.

Nakamura, T., Matsushita, S., Nishiguchi, N., Kimura, M., Yoshino, A., & Higuchi, S. (1999). Association of a polymorphism of the 5HT2A receptor gene promoter region with alcohol dependence. *Molecular Psychiatry, 4,* 85–88.

Nordin, C., Siwers, B., Benitz, J., & Bertilsson, L. (1985). Plasma concentrations of nortriptyline and its 10-hydroxy metabolite in depressed patients: Relationship to the debrisoquine metabolic ratio. *British Journal of Clinical Pharmacology, 19,* 832–835.

Nöthen, M. M., Rietschel, M., Erdmann, J., Oberlander, H., Moller, H. J., Nober, D., & Propping, P. (1995). Genetic variation of the 5-HT2A receptor and response to clozapine. *Lancet, 346,* 908–909.

Olesen, O. V., Thomsen, K., Jensen, P. N., Wulff, C. H., Rasmussen, N. A., Refshammer, C., et al. (1995). Clozapine serum levels and side effects during steady state treatment of schizophrenic patients: A cross-sectional study. *Psychopharmacology, 113,* 371–378.

Ordway, G. A. (2000). Searching for the chicken's egg in transporter gene polymorphisms. *Archives of General Psychiatry, 57,* 739–740.

O'Reilly, R. L., Bogue, L., & Singh, S. M. (1994). Pharmacogenetic response to antidepressants in a multicase family with affective disorder. *Biological Psychiatry, 36,* 467–471.

Ozdemir, V., Basile, V. S., Masellis, M., & Kennedy, J. L. (2001). Pharmacogenetic assessment of antipsychotic-induced movement disorders: Contribution of the dopamine D3 receptor and cytochrome P450 1A2 genes. *Journal of Biochemistry and Biophysical Methods, 47,* 151–157.

Ozdemir, V., Kalow, W., Posner, P., Collins, E. J., Kennedy, J. L., Tang, B. K., et al. (2001). CYP1A2 activity as measured by a caffeine test predicts clozapine and active metabolite steady-state concentration in patients with schizophrenia. *Journal of Clinical Psychopharmacology, 21,* 398–407.

Ozdemir, V., Shear, N. H., & Kalow, W. (2001). What will be the role of pharmacogenetics in evaluating drug safety and minimising adverse effects? *Drug Safety, 24,* 75–85.

Paterson, A. D., Sunohara, G. A., & Kennedy, J. L. (1999). Dopamine D4 receptor gene: Novelty or nonsense? *Neuropsychopharmacology, 21,* 3–16.

Perry, P. J., Miller, D. D., Arndt, S. V., & Cadoret, R. J. (1991). Clozapine and norclozapine plasma concentrations and clinical response of treatment-refractory schizophrenic patients. *American Journal of Psychiatry, 148,* 231–235.

Perry, P. J., Pfohl, B. M., & Holstad, S. G. (1987). The relationship between antidepressant response and tricyclic antidepressant plasma concentrations: A retrospective analysis of the literature using logistic regression analysis. *Clinical Pharmacokinetics, 13,* 381–392.

Persidis, A. (1998). Pharmacogenomics and diagnostics. *Nature Biotechnology, 16,* 791–792.

Pickar, D., & Rubinow, K. (2001). Pharmacogenomics of psychiatric disorders. *Trends in Pharmacological Sciences, 22,* 75–83.

Potkin, S., Bera, R., Gulasekaram, B., Costa, J., Hayes, S., Jin, Y., et al. (1994). Plasma clozapine concentrations predict clinical response in treatment-resistant schizophrenia. *Journal of Clinical Psychiatry, 55,* 133–136.

Preskorn, S. H., & Jerkovich, G. S. (1990). Central nervous system toxicity of tricyclic antidepressants: Phenomenology, course, risk factors, and role of therapeutic drug monitoring. *Journal of Clinical Psychopharmacology, 10,* 88–95.

Pritchard, J., Stephens, M., Rosenberg, N., & Donnelly, P. (2000). Association mapping in structured populations. *American Journal of Human Genetics, 67,* 170–181.

Rabinowitz, D. (1997). A transmission disequilibrium test for quantitative trait loci. *Human Heredity, 47,* 342–350.

Regalado, A. (1999). Inventing the pharmacogenomics business. *American Journal of Health-Systems Pharmacists, 56,* 40–50.

Relling, M. V., Hancock, M. L., Rivera, G. K., Sandlund, J. T., Ribeiro, R. C., Krynetski, E. Y., et al. (1999). Mercaptopurine therapy intolerance and heterozygosity at the thiopurine *S*-methyltransferase gene locus. *Journal of the National Cancer Institute, 91,* 2001–2008.

Rettie, A. E., Wienkers, L. C., Gonzalez, F. J., Trager, W. F., & Korzekwa, K. R. (1994). Impaired *S*-warfarin metabolism catalyzed by R144C allelic variant of CYPC9. *Pharmacogenetics, 4,* 39–42.

Reynolds, G. P., Zhang, Z. J., & Zhang, X. B. (2002). Association of antipsychotic drug-induced weight gain with a polymorphism of the promoter region of the 5-HT2C receptor gene [Abstract]. *Schizophrenia Research, 53*(Suppl.), 158–159.

Rietschel, M., Kennedy, J. L., Macciardi, F., & Meltzer, H. Y. (1999). Application of pharmacogenetics to psychotic disorders: The first consensus conference. *Schizophrenia Research, 37,* 191–196.

Rioux, P. P. (2000). Clinical trials in pharmacogenetics and pharmacogenomics: Methods and applications. *American Journal of Health-Systems Pharmacists, 57,* 887–898.

Sautter, F., McDermott, G., & Garver, D. (1993). Familial differences between rapid neuroleptic response psychosis and delayed neuroleptic response psychosis. *Biological Psychiatry, 33,* 15–21.

Schafer, W. R. (1999). How do antidepressants work? Prospects for genetic analysis of drug mechanisms. *Cell, 98,* 551–554.

Schumacher, J., Schulze, T. G., Wienker, T. F., Rietschel, M., & Nöthen, M. M. (2000). Pharmacogenetics of clozapine response [Letter]. *Lancet, 356,* 506–514.

Scott, J. (1995). A place in the world for RNA editing. *Cell, 81,* 833–836.

Segman, R. H., Heresco-Levy, U., Yakir, A., Goltser, T., Strous, T., Greenberg, D., & Lerer, B. (2002). Interactive effect of cytochrome P450 17-hydroxylase and dopamine D3 receptor gene polymorphisms on abnormal involuntary movements in chronic schizophrenia. *Biological Psychiatry, 51,* 261–263.

Sjöqvist, F., & Bertilsson, L. (1984). Clinical pharmacology of antidepressant drugs: Pharmacogenetics. In E. Usdin, M. Åsberg, L. Bertilsson, & F. Sjöqvist (Eds.), *Frontiers in biochemical and pharmacological research in depression* (pp. 359–372). New York: Raven Press.

Sjöqvist, F., Bertilsson, L., & Åsberg, M. (1980). Monitoring tricyclic antidepressants. *Therapeutic Drug Monitoring, 2,* 85–93.

Sjöqvist, F., Hammer, Q., Ideström, C.-M., Lind, M., Tuck, D., & Åsberg, M. (1967). Plasma levels of monomethylated tricyclic antidepressants and side-effects in man. *Excerpta Medica International Congress, Series No. 45,* 246–257.

Smeraldi, E., Zanardi, R., Beneditti, F., Di Bella, D., Perez, J., & Catalano, M. (1998). Polymorphism within the promoter of the serotonin transporter gene and antidepressant efficacy of fluvoxamine. *Molecular Psychiatry, 3,* 508–511.

Sodhi, M. S., Burnet, P. W. J., Makoff, A. J., Kerwin, R. W., & Harrison, P. J. (2001). RNA editing of the 5-HT_{2C} receptor is reduced in schizophrenia. *Molecular Psychiatry, 6,* 373–379.

Sodhi, M. S., & Murray, R. M. (1997). Future therapies for schizophrenia. *Expert Opinion on Therapeutic Patents, 7,* 151–165.

Spallone, P., & Wilkie, T. (1999, November). *Social, ethical, and public policy implications of advances in the biomedical sciences: The Wellcome Trust's initiative on pharmacogenetics.* Paper presented at the European Workshop on Legal, Regulatory and Ethical Aspects in Pharmacogenetics, Berlin.

Spallone, P., & Wilkie, T. (2000). The research agenda in pharmacogenetics and biological sample collections: A view from the Wellcome Trust. *New Genetics and Society, 19,* 193–205.

Spielman, R. S., & Ewens, W. J. (1996). The TDT and other family-based tests for linkage disequilibrium and association. *American Journal of Human Genetics, 59,* 983–989.

Spire-Vayron de la Moureyre, C., Debuysere, H., Mastain, B., Vinner, E., Marez, D., Lo Guidice, J. M., et al. (1998). Genotypic and phenotypic analysis of the polymorphic thiopurine S-methyltransferase gene (TPMT) in a European population. *British Journal of Pharmacology, 125,* 879–887.

Spitzer, R. L., William, J. B. W., Gibbon, M., & First, M. G. (1990). *User's guide for the Structured Clinical Interview for DSM–III–R.* Washington, DC: American Psychiatric Press.

Spurlock, G., Heils, A., Holmans, P., Williams, J., D'Souza, U. M., Cardno, A., et al. (1998). A family based association study of T102C polymorphism in 5HT2A and schizophrenia plus identification of new polymorphisms in the promoter. *Molecular Psychiatry, 3,* 42–49.

Stahl, S. M. (2000). Blue genes and the monamine hypothesis of depression. *Journal of Clinical Psychiatry, 61,* 77–78.

Stam, N. J., Vanderheyden, P., van Aleback, C., Klomp, J., de Boer, T., van Delft, A., et al. (1994). Genomic organisation and functional expression of the gene encoding the human serotonin 5-HT_{2C} receptor. *European Journal of Pharmacology, 269,* 339–348.

Tecott, L., Sun, L., Akana, S., Strack, A., Lowenstein, D., Dallman, M., & Julius, D. (1995). Eating disorder and epilepsy in mice lacking 5-HT2C serotonin receptors. *Nature, 374,* 542–546.

Templeton, A. R., Boerwinkle, E., & Sing, C. F. (1987). A cladistic analysis of phenotypic associations with haplotypes inferred from restriction endonuclease mapping: I. Basic theory and an analysis of alcohol dehydrogenase activity in *Drosophila. Genetics, 117,* 343–351.

Tugnait, M., Hawes, E. M., McKay, G., Eichelbaum, M., & Midya, K. K. (1999). Characterization of the human hepatic cytochromes P450 involved in the in vitro oxidation of clozapine. *Chemico-Biological Interactions, 118,* 171–189.

Turecki, G., Briere, R., Dewar, K., Antonetti, T., Lesage, A. D., Seguin, M., et al. (1999). Prediction of level of serotonin 2A receptor binding by serotonin receptor 2A genetic variation in postmortem brain samples from subjects who did or did not commit suicide. *American Journal of Psychiatry, 156,* 1456–1458.

Van Tol, H. H., Wu, C. M., Guan, H. C., Ohara, K., Bunzow, J. R., Civelli, O., et al. (1992). Multiple dopamine D4 receptor variants in the human population. *Nature, 358,* 149–152.

Vickers, S. P., Clifton, P. G., Dourish, C. T., & Tecott, L. H. (1999). Reduced satiating effect of d-fenfluramine in serotonin 5-HT(2C) receptor mutant mice. *Psychopharmacology (Berlin), 143,* 309–314.

Vogel, F. (1959). Moderne Probleme der Humangenetik [Modern problems of human genetics]. *Ergebn Inn Med Kinderheilk, 12,* 52–125.

Vojvoda, D., Grimmel, K., Sernyak, M., & Mazure, C. M. (1996). Monozygotic twins concordant for response to clozapine. *Lancet, 347,* 61.

Wang, Q., O'Brien, P. J., Chen, C. X., Cho, D. S., Murray, J. M., & Nishikura, K. (2000). Altered G protein-coupling functions of RNA editing isoform and splicing variant serotonin$_{2C}$ receptors. *Journal of Neurochemistry, 74,* 1290–1300.

Weele, J. V.-V., Anderson, G. M., & Cook, E. H. (2000). Pharmacogenetics and the serotonin system: Initial studies and future directions. *European Journal of Pharmacology, 410,* 165–181.

Weinberger, D. R. (1987). Implications of normal brain development for the pathogenesis of schizophrenia. *Archives of General Psychiatry, 44,* 660–669.

Weinshilboum, R. M., & Sladek, S. L. (1980). Mercaptopurine pharmacogenetics: Monogenic inheritance of erythrocyte thiopurine methyltransferase activity. *American Journal of Human Genetics, 32,* 651–662.

Whale, R., Quested, D. J., Laver, D., Harrison, P. J., & Cowen, P. J. (2000). Serotonin transporter (5-HTT) promoter genotype may influence the prolactin response to clomipramine. *Psychopharmacology, 150,* 120–122.

Wong, A. H. C., Buckle, C. E., & Van Tol, H. H. M. (2000). Polymorphisms in dopamine receptors: What do they tell us? *European Journal of Pharmacology, 410,* 183–203.

World Health Organization. (1999). *Schedules for clinical assessment in neuropsychiatry (SCAN).* Geneva: Author.

Zanardi, R., Benedetti, F., Di Bella, D., Catalano, M., & Smeraldi, E. (2000). Efficacy of paroxetine in depression is influenced by a functional polymorphism within the promoter of the serotonin transporter gene [Letter]. *Journal of Clinical Psychopharmacology, 20,* 105–106.

Zwain, I. H., & Yen, S. S. (1999). Neurosteroidogenesis in astrocytes, oligodendrocytes, and neurons of cerebral cortex of rat brain. *Endocrinology, 140,* 3843–3852.

Part VII

Personality

19

Behavioral Genetics, Genomics, and Personality

Richard P. Ebstein, Jonathan Benjamin, and Robert H. Belmaker

Developments in behavioral psychology over the past several decades demonstrated the major contribution of heredity to many human behavioral traits including personality, but only recently has it become possible to assign specific genes to specific traits. In this chapter, we review the first fruits of the genomic revolution as applied to one aspect of human behavior: temperament. The determination of the DNA sequences of genes coding for brain-expressed proteins has led to the discovery of many common variants that may be candidates for contributing to personality. Although there are few criteria for choosing a particular polymorphism for studying with personality traits, functionality (e.g., Do the alleles differ in physiological action? Are the base pair changes in coding regions? Is the gene expressed in an appropriate brain region?) and a relevant role played by the coded gene in neurotransmission (as receptors, metabolic enzymes, or transporters) are good starting points toward selecting from the thousands of genes now known to be expressed in the brain.

Two genes in particular, the dopamine D4 receptor (DRD4) and the serotonin transporter promoter region (5-HTTLPR), satisfy criteria for likely candidates in behavioral genetic studies. Most of the studies we review in this chapter focus on these two polymorphisms and the evidence that they indeed contribute in part to human personality traits. Not unexpectedly, considering their distribution in the brain and the frequency of the allelic variants in the population, these two genes appear to have pleiotropic effects (i.e., having multiple phenotypic expressions) on behavior. Some of these behavioral effects may have a common basis that remains to be resolved: for example, the DRD4 association with adult novelty seeking and some childhood behaviors such as attachment disorganization or attention deficit/hyperactivity disorder (ADHD), and the association of 5-HTTLPR with both neuroticism and depression. Additionally, a few studies have simultaneously examined more than one gene for association with personality traits, and there is some evidence for

This research was supported in part by a grant from the Israel Science Foundation, founded by the Israel Academy of Sciences and Humanities.

epistatic interactions. In the future, multiple gene effects are likely to play an increasingly important role in understanding the genetic basis of temperament.

The completion of the first phase of the Human Genome Project and the availability of a dense genetic map based on single nucleotide polymorphisms (SNPs) coupled with increasingly sophisticated statistical tools to analyze complex behavioral traits including personality are harbingers of a productive future. Only one genomewide scan has been carried out in personality genetics, and additional scans undoubtedly will be undertaken in the coming years. The ability to survey broad chromosomal regions using a dense SNP map as well as the fine mapping of promising regions using haplotypes should enable us to identify many of the genes contributing to human temperament.

Self-Report Questionnaires Measure Personality Traits

Diverse definitions of personality exist, but most genetic studies are based on the following definition (Loehlin, 1992): *Personality* may be considered the typical manner or style of an individual's behavior as opposed to the goals toward which it is directed or the machinery of its execution (cognitive and motor skills). Many other definitions of personality exist, but most genetic studies are based on this definition. For example, Bonnie and Clyde were (in the movie) flamboyant bank robbers, and many college professors show a similar extraverted style in their lectures. However, the goals of the behavior (robbing a bank vs. teaching students) are very different with very different social consequences. Personality may be described as the vigor, temper, or persistence of behavior or the emotional expression that accompanies it, such as fearfulness, exuberance, aggressiveness, or self-restraint. The vast majority of genetic research on personality involves self-report questionnaires administered to adolescents and adults. People's responses to such questions are remarkably stable, even over several decades (Costa & McCrae, 1994). Additionally, personality trait structure appears to be universal and shows similarity across highly diverse cultures (McCrae & Costa, 1997).

A number of self-report questionnaires are in use, and they appear to measure similar personality traits. Descriptive taxonomists have agreed that a set of five dimensions can be identified consistently across methods. These "Big Five" dimensions of self-report and peer description have been labeled Neuroticism, Extraversion, Agreeableness, Openness, and Conscientiousness and are commonly assessed using the NEO Personality Inventory–Revised (NEO–PI–R; Costa & McCrae, 1994). Other self-report personality inventories include Eysenck's Personality Inventory (Eysenck, 1952), Zuckerman's Sensation Seeking Scale (Zuckerman, 1994), Cattell's 16 Personality Factor Questionnaire (16 PF; Cattell, 1946), Cloninger's Tridimensional Personality Questionnaire (TPQ; Cloninger, 1987), and Temperament Character Index (TCI; Cloninger, Svrakic, & Przybeck,

1993). Underlining the similarity between questionnaires are formulas that can be used to convert NEO–PI–R scores into TPQ Harm Avoidance and Novelty Seeking (Benjamin et al., 1996). To summarize, all the major self-report questionnaires appear to reliably assess similar personality factors, especially the higher categories of novelty seeking (extraversion, conscientiousness, sensation seeking, and psychoticism) and harm avoidance (neuroticism and anxiety-related traits).

One questionnaire that we have found particularly useful and has proven heuristic value in our genetic studies is Cloninger's TPQ. The TPQ is a 100-question instrument that requires yes or no answers. Most participants complete the form in about 30 minutes. Questions relating to Novelty Seeking include the following:

- I often try new things just for fun or thrills, even if most people think it is a waste of time. (Yes)
- I often do things based on how I feel at the moment without thinking about how they were done in the past. (Yes)
- I am much more controlled than most people. (No)

Questions relating to Harm Avoidance include the following:

- I usually am confident that everything will go well, even in situations that worry most people. (No)
- I often feel tense and worried in unfamiliar situations, even when others feel there is no danger at all. (Yes)
- I often have to stop what I'm doing because I start worrying about what might go wrong. (Yes)

The TPQ draws on human and animal work to suggest that behavior is mediated by certain neurotransmitters that underlie three basic and largely heritable dimensions: Novelty Seeking, Harm Avoidance, and Reward Dependence. In populations who filled out NEO–PI–R and TPQ questionnaires, TPQ Harm Avoidance was highly correlated with NEO–PI–R Neuroticism, and TPQ Novelty Seeking was positively correlated with NEO–PI–R Extraversion and negatively correlated with NEO–PI–R Conscientiousness. Novelty Seeking therefore taps aspects of impulsiveness, curiosity (or exploratory behavior in animals), and disorderliness. Reward Dependence is the TPQ term for attachment to others, sentimentality, and warmth; its opposites are pragmatism and tough-mindedness.

According to the biological theory behind the TPQ, harm avoidance is the inhibition of approach behaviors and the enhancement of escape behaviors, and these are principally served by the serotonin (5-HT) system in most animals. The trait has face-valid resemblance to human neuroticism and perhaps depression. Novelty seeking, the increased tendency to respond with approach to new and promising situations, is principally served by the meso-limbic dopamine system. Rewarding activities, including the use of illicit drugs, increase dopamine release or inhibit its reuptake in this system (Schultz, 1997). Reward dependence reflects the persistence of behavior that is likely to be intermittently, rather than immediately, rewarded, so that a key subscale of reward dependence in the TPQ is persistence.

Heritability of Personality Traits

Twin studies demonstrate that, first, personality traits measured by self-report questionnaires show moderate heritability and, second, although environmental influence is important, almost all the environmental variance is nonshared (Loehlin, 1992). Heritabilities in the 30%–50% range are typical. Extraversion is one of the most highly heritable categories (~50%), closely followed by neuroticism or emotional stability (~40%).

Association of Personality Traits With Specific Genes

Dopamine D4 Receptor (DRD4)

The first association study between a specific gene variant and a personality trait was carried out in 1996, and two independent studies suggested the association between the DRD4 and novelty seeking (Benjamin et al., 1996; Ebstein et al., 1996). These investigations catalyzed many other studies aimed at validating this first finding and identifying other genes that contribute to novelty seeking and other personality factors.

The DRD4 gene has a highly polymorphic 16 amino acid repeat region in the putative third cytoplasmic loop that varies between 2 and 10 repeats in most populations and changes the length of the receptor protein (Lichter et al., 1993; Van Tol et al., 1992). There is some evidence that the long and short forms of this protein have a moderate functional significance (Asghari, Sanyal, Buchwaldt, Paterson, Jovanovic, & Van Tol, 1995; Asghari et al., 1994), although a more recent study showed that the polymorphic region of the loop does not appear to affect the specificity or efficiency of G(i)alpha coupling (Kazmi, Snyder, Cypess, Graber, & Sakmar, 2000). G protein coupling is part of the chain of events between neurotransmitter binding to the cell surface and downstream actions eventually leading to changes in membrane potential. The Ebstein study was a case control design, whereas the National Institutes of Health (NIH) group's sample also included a number of sibpairs allowing for both a between- and a within-families design. Both studies suggested an association between novelty seeking and the long alleles of the DRD4. Although some subsequent studies confirmed an association between the DRD4 receptor and novelty seeking (Bau et al., 2001; Kuhn, Meyer, Nothen, Gansicke,

Papassotiropoulos, & Maier, 1999; Noble et al., 1998; Okuyama et al., 2000; Ono et al., 1997; Poston et al., 1998; Ronai et al., 2001; Strobel, Wehr, Michel, & Brocke, 1999; Swift, Larsen, Hawi, & Gill, 2000; Tomitaka et al., 1999), others did not (Bau, Roman, Almeida, & Hutz, 1999; Gebhardt et al., 2000; Gelernter et al., 1997; Herbst, Zonderman, McCrae, & Costa, 2000; Hill, Zezza, Wipprecht, Locke, & Neiswanger, 1999; Jonsson et al., 1997; Malhotra et al., 1996; Mitsuyasu et al., 2001; M. L. Persson et al., 2000; Pogue-Geile, Ferrell, Deka, Debski, & Manuck, 1998; Sander, Harms, Dufeu, et al., 1997; Sullivan et al., 1998; Vandenbergh, Zonderman, Wang, Uhl, & Costa, 1997).

Three recent studies appear to confirm the relevancy of the DRD4 gene to novelty seeking. A Japanese group has characterized several polymorphisms in the upstream promoter region of the DRD4 gene, and one of these was associated with higher novelty-seeking scores in a group of Japanese participants (Okuyama et al., 2000). This particular promoter region polymorphism is functionally significant, and the T variant of the −521C/T polymorphism reduces transcriptional efficiency. The association between the promoter region DRD4 polymorphism and novelty seeking also has been confirmed in a Hungarian population (Ronai et al., 2001). However, a second Japanese investigation did not observe an association between this polymorphism and novelty seeking (Mitsuyasu et al., 2001). A Finnish group has shown an association between the DRD4 exon III 2- and 5-repeat alleles and high novelty-seeking scores in a very large and homogeneous Finnish birth cohort of more than 4,700 participants (Ekelund, Lichtermann, Jarvelin, & Peltonen, 1999). Both the Finnish study and the Japanese promoter region results suggest that the original observation of an association between the DRD4 48 base pair repeat region and TPQ novelty seeking may have been due to linkage disequilibrium between the exon III region and a second polymorphism within the DRD4 gene (the promoter region) or in a neighboring gene. Finally, the first genomewide scan in personality genetics in a group of alcoholic families (Cloninger et al., 1998), which reported a linkage between anxiety-related traits and a region on chr 8p, also observed a linkage between novelty seeking and the region on chr 11 where the DRD4 gene is located (C. R. Cloninger, personal communication, 2001). This finding is being replicated in the second panel from this cohort.

To some extent, animal studies support a role for this receptor in novelty-seeking behavior. In addition to increased sensitivity to ethanol (Rubinstein et al., 1997), DRD4 knock-out mice also exhibit reduced behavioral response to novel stimuli (Dulawa, Grandy, Low, Paulus, & Geyer, 1999). The DRD4 receptor has also been analyzed in dog breeds, and the long D allele was dominant in the endogenous Japanese breed (Shiba) compared with golden retrievers (Niimi, Inoue-Murayama, Murayama, Ito, & Iwasaki, 1999). It is interesting that the Shiba is considered an aggressive species, and some kinds of aggression are associated with sensation/novelty-seeking personality traits in humans (Zuckerman, 1994).

Serotonin Transporter Promoter Region Polymorphism (5-HTTLPR)

Following our report of an association between DRD4 and novelty seeking, Lesch and his colleagues showed an association between a newly discovered 44 bp deletion in the promoter region of the serotonin transporter promoter region affecting transcription of the gene (Heils et al., 1996) and neuroticism or harm avoidance personality traits (Lesch et al., 1996). Some subsequent reports replicated the initial association (Du, Bakish, & Hrdina, 2000; Greenberg et al., 2000; Hu et al., 2000; Katsuragi et al., 1999; Lerman et al., 2000; Melke et al., 2001; Murakami et al., 1999; Osher, Hamer, & Benjamin, 2000; Ricketts et al., 1998), whereas two yielded ambiguous results (Gelernter, Kranzler, Coccaro, Siever, & New, 1998; Mazzanti et al., 1998). Many studies failed to observe any association between anxiety-related traits and the 5-HTTLPR polymorphism (Ball et al., 1997; Deary et al., 1999; Ebstein, Gritsenko, et al., 1997; Flory et al., 1999; Golimbet et al., 2001; Gustavsson et al., 1999; Hamilton et al., 1999; Jorm et al., 1998; Jorm et al., 2000; Kumakiri et al., 1999; Matsushita et al., 2001; Samochowiec et al., 2001).

Cloninger's psychobiological model of personality (Cloninger et al., 1993) proposed two broad domains: temperament, which is due largely to inherited variations in specific monoamine neurotransmitter systems; and character, which arises more from socioculturally learned differences in values, goals, and self-concepts and is the strongest predictor of personality disorders. Hamer, Greenberg, Sabol, and Murphy (1999) analyzed the association of temperament and character traits with the 5-HTTLPR polymorphism in 634 volunteer participants. Contrary to what might have been expected from Cloninger's hypothesis that character traits would show less genetic influence, the 5-HTTLPR was most associated with the character traits of cooperativeness and self-directedness. Associations with the temperament traits of reward dependence and harm avoidance were weaker and could be attributable largely to cross-correlations with the character traits and demographic variables. Psychometric analysis indicated that the serotonin transporter influences two broad areas of personality, negative affect and social disaffiliation, that are consistent across inventories but are more concisely described by the five-factor model of personality than by the biosocial model. Hamer et al. interpreted their results to indicate that there is no fundamental mechanistic distinction between character and temperament, at least in regard to the serotonin transporter gene, and that a single neurotransmitter can influence multiple personality traits.

This association between a "character" trait and the serotonin transporter is an example of a new approach to examining the heritability of complex behavior traits. Although historically the proof of the heritability of complex human traits was based on twin, family, and adoption studies, advances in molecular genetics may allow investigators to bypass this classical methodological troika, and by directly demonstrating association of a particular quantitative trait locus (QTL; as Hamer et al., 1999, did) with a specific gene, will establish the heritability of the trait.

The importance of these first studies of the DRD4 and 5-HTTLPR polymorphisms in contributing to personality traits is as much in the questions and problems they raise as in the provisional assignment of their specific contribution (which is small) to novelty seeking and anxiety-related traits, respectively. It is becoming apparent that the small effect size of QTLs contributing to personality traits creates considerable obstacles to replicating first findings both from a similar participant group and especially across ethnically diverse population samples that are characterized by considerable genetic diversity. For example, the Japanese and Chinese populations have very low frequencies of the DRD4 7-repeat compared with many other groups, signifying that the role of this specific repeat number in novelty seeking and related traits cannot be of general significance in East Asian participants.

Toward clarifying the role of these two polymorphisms in personality genetics, we suggest that a meta-analysis of the findings already reported in the literature would be worthwhile. The meta-analysis approach may highlight such issues as which variables (if any), such as questionnaire (TPQ vs. NEO–PI–R), ethnic groups, gender, and age, are significant in generating positive results. Additionally, investigators contemplating future personality genetics studies might consider designing their research so that the results are amenable to future meta-analyses.

As a means of identifying chromosomal regions that may be harboring personality genes, a genomewide survey of broad chromosomal regions, preferably in several ethnic groups, is indicated. Moreover, it is not unlikely that chromosomal regions already identified in linkage studies of the major psychoses (schizophrenia and bipolar disorder) may be prime targets for "personality genes" that also contribute to psychopathology. The information now available from genome scans in schizophrenia (Riley & McGuffin, 2000) and bipolar disorder (Gershon, 2000) is a resource for choosing candidate chromosomal regions for more intense survey.

An area of particular interest is the tip of chromosome 11 (11p15.5) where DRD4 is located. Microsatellites and SNPs in this region could be profitably inventoried in suitable populations for linkage to personality traits, especially novelty seeking. Such a design might sidestep the conundrum that the DRD4 7-repeat is possibly in linkage disequilibrium (LD) with another unknown element in this gene or with another nearby gene, thus explaining some of the conflicting results reported in the literature. It should be considered that only in some ethnic groups the 7-repeat is in LD with the "real" novelty-seeking gene. Moreover, recombination events could reverse the relationship between the 7-repeat and novelty, and it is not inconceivable that the short DRD4 repeat could in some populations be linked to this personality trait. Similar arguments could be made regarding the chromosomal region surrounding the serotonin transporter (17q11.1–12). It is interesting to note that more than 10 variants have been observed for the serotonin transporter in a Japanese study (Nakamura, Ueno, Sano, & Tanabe, 2000), and significant differences in allele frequency were observed between Japanese and White groups.

Another worthwhile strategy that has yet to be systematically undertaken in the emerging field of personality genetics is haplotype analysis. Rather than analyzing a single polymorphism for its association or linkage to a particular trait, it is more informative to examine haplotypes by combining allelic variants at different sites within the same gene or further upstream or downstream to the coding region. An example of such an approach is an article by Barr and her colleagues (2000) in Toronto, who analyzed several polymorphisms in the DRD4 gene as well as a nearby polymorphism in the tyrosine hydroxylase gene that further strengthened the evidence for a link between the DRD4 gene and ADHD. Similarly, another study used haplotype analysis to demonstrate an association between the Notch4 gene and schizophrenia (Wei & Hemmings, 2000).

Other Polymorphisms and Personality Traits

Another dopamine receptor that has been examined for a role in personality traits is the dopamine 3 receptor (DRD3). In a large sample of White people (Henderson et al., 2000), those in the first half of the split sample with at least one of the DRD3 Ser(9) alleles had significantly higher scores in neuroticism and behavioral inhibition. There was a trend, failing to meet the 1% significance criterion, for those with this genotype also to have higher depression and anxiety. However, when the second half of the split sample was analyzed, the results failed to obtain significance. Another group of investigators (Thome et al., 1999) found that alcohol-dependent patients with the genotype DRD3 ser/gly (heterozygotes) showed significantly higher novelty-seeking scores than patients with the genotype ser/ser (homozygotes). An association between the DRD3 receptor and novelty seeking was observed in a small group of bipolar patients (Staner et al., 1998). It has also been reported that in a group of bipolar patients, harm avoidance was associated with the serotonergic 5-HT_{2a} receptor polymorphism (Blairy et al., 2000). A single nucleotide polymorphism in exon 6 of the DRD2 was significantly associated with Cluster A personality disorders in a group of hypertensive male participants (Rosmond et al., 2001).

Gene Interactions

Some studies in personality genetics are now revealing the role of multiple gene effects in the determination of this complex phenotype. A word of caution is needed, however, regarding the design and interpretation of such studies. First, and most importantly, larger sample sizes are indicated because the availability of participants with particular genotype combinations, for example, DRD4 7-repeat (allele frequency ~30%) and 5-HTTLPR short allele (allele frequency ~50%), is the multiple of the allele frequencies of both alleles in the population ($0.5 \times 0.3 = 15\%$). Second, there is an increased chance of false positives because an increased number of statistical tests are carried out, for example, DRD4 + 5-HTTLPR + interaction, and statistical techniques need to be used that take into account these reservations.

Noble et al. (1998) observed an interaction between the DRD4 and dopamine receptor D2 (DRD2) polymorphisms on novelty seeking. Novelty seeking was significantly higher in boys having all three minor alleles of the DRD2 in common, compared with boys without any of these alleles. Boys with the DRD4 7-repeat allele also had a significantly higher novelty-seeking score than those without this allele. However, the greatest difference in novelty-seeking score was found when boys having all three minor DRD2 alleles and the DRD4 7-repeat allele were compared with those without any of these alleles. Neither the DRD2 nor the DRD4 polymorphisms differentiated total harm avoidance score.

Two groups (Ebstein, Segman, et al., 1997; Kuhn et al., 1999) observed an interaction between the serotonin 5-HT_{2c} receptor polymorphism and the DRD4 polymorphism on TPQ reward-dependent behavior. In our study (Ebstein, Segman, et al., 1997), when the combined group of male and female participants is examined by comparing partial correlation coefficients (controlling for age, sex, and ethnicity), the variance (r^2) explained by the 5-HT2C genotype (comparing cys/cys participants vs. cys/ser and ser/ser participants) is 3.2% ($p < .06$) for RD2 or persistence and 4.2% for RD134 ($p < .03$). However, when the same analysis is carried out in those participants having only the long repeat DRD4 allele, the variance explained by the 5-HT2C genotype is 29% for RD2 or persistence ($p < .001$) and 12.8% for RD134 ($p < .04$). We emphasize that these results should be interpreted cautiously because only a small number of individuals ($n = 6$) have both the less-common 5-HT2C polymorphism and the long repeat DRD4 allele, suggesting that a larger cohort must be used to substantiate the role played by multiple polymorphisms in the determination of behavioral traits.

Our group also examined the role of three functional polymorphisms: DRD4, 5-HTTLPR serotonin transporter, and catechol-O-methyltransferase polymorphisms, for association with TPQ personality factors in 455 participants (Benjamin, Osher, Kotler, et al., 2000). In the absence of the short 5-HTTLPR allele and in the presence of the high enzyme activity catechol-O-methyltransferase (COMT) val/val genotype, novelty-seeking scores are higher in the presence of the DRD4 7-repeat allele (Figure 19.1). The effect of these three polymorphisms on novelty seeking also was examined using a within-families design. Siblings who shared identical genotype groups for all three polymorphisms (COMT, DRD4, and 5-HTTLPR) had significantly correlated novelty-seeking scores (intraclass correlation coefficient = 0.39, $F = 2.26$, $p = .008$, $n = 49$), whereas sibs with dissimilar genotypes in at least one polymorphism showed no significant correlation for novelty-seeking scores (intraclass correlation coefficient = 0.177, $F = 1.43$, $p = .09$, $n = 110$). Similar interactions also were observed among these three polymorphisms and novelty seeking when the 150 in-

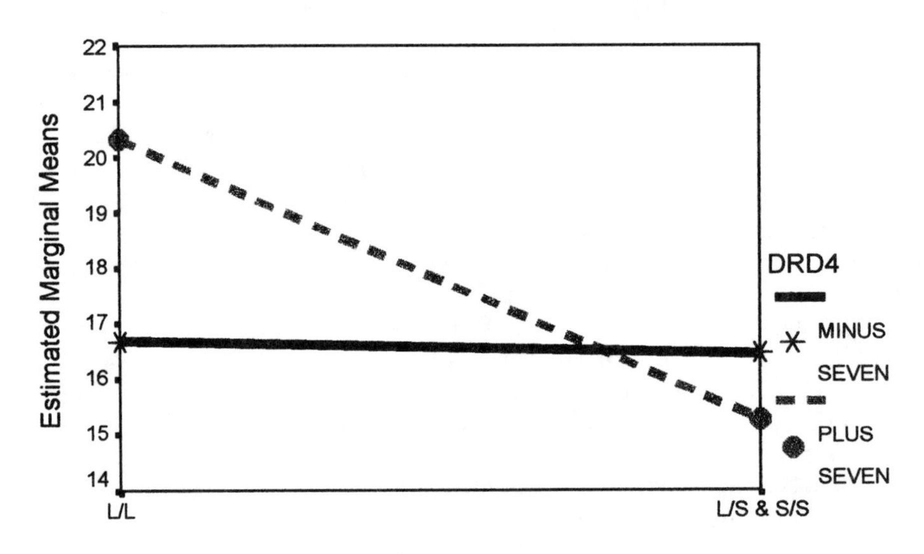

Figure 19.1. Tridimensional Personality Questionnaire (TPQ) novelty-seeking scores in the presence of three polymorphisms: DRD4, catechol-O-methyltransferase (COMT), and 5-HTTLPR (Benjamin, Osher, Kotler, et al., 2000). All participants represented were homozygous for the high-enzyme activity COMT val/val genotype. The *y*-axis represents TPQ novelty-seeking scores. Participants at the left *y*-axis are homozygous for the long/long 5-HTTLPR genotype, whereas participants at the right *y*-axis carry at least one short 5-HTTLPR allele (the so-called neuroticism allele). Note the relatively large difference in mean novelty-seeking scores between participants with the DRD4 7-repeat allele and participants without the 7-repeat in the absence of the short 5-HTTLPR allele.

dependently recruited and nonrelated participants were analyzed. We also observed an interaction between the COMT and 5-HTTLPR polymorphisms and TPQ persistence scores (Benjamin, Osher, Lichtenberg, et al., 2000). Participants with high persistence scores are characterized as industrious, hard-working, ambitious, and perfectionistic.

We note that only a handful of additional common polymorphisms have been examined for association with personality traits, including the transcription factor AP-2beta gene (Damberg et al., 2000), the α_{2A} adenoreceptor (Comings et al., 2000), and a tyrosine hydroxylase TCAT repeat (M. Persson et al., 2000). In a Brazilian study of male alcoholics, an association was observed between the TAQ A1 dopamine D2 receptor polymorphism and TPQ harm avoidance personality traits, stress-related traits, and facets of alcoholism (Bau, Almeida, & Hutz, 2000).

Considering that there are perhaps as many as 50,000 human genes and that a majority probably are expressed in the brain, the few candidate gene studies carried out so far in personality genetics barely have revealed the tip of the iceberg. Lots of work remains to be done before a picture of the molecular architecture of personality traits emerges.

Extension of Genetic Personality Studies to Infant Temperament

Individual differences in temperament, that is, individual differences in emotional, motoric, and attentional reactivity, are apparent from the first days of life. One of the most widely used methods to assess temperament in the first weeks of life is the Neonatal Behavioral Assessment Scale (NBAS; Brazelton & Nugent, 1995). The NBAS assesses the behavioral repertoire of infants from birth to 1 month. It attempts to capture individual differences in the complexity of behavioral responses to environmental stimuli as the newborn moves from sleep states to alert states to crying states.

Although the heritability of the NBAS-defined clusters, to our knowledge, has not been examined directly in twin studies, the role of genes in determining infant temperament at a somewhat later age has been studied using other testing instruments that measure related behavioral clusters similar to those assessed by the NBAS. For example, a Chinese twin study (Chen et al., 1990) demonstrated significant genetic variance for 6-month-old infant activity level, approach or withdrawal, intensity of reaction, quality of mood, and threshold of responsiveness with heritability >0.39. Similarly, in a second study, high estimates of heritability were found for eight of the nine temperament scales, and three scales showed significant effects of shared environmental influences (Cyphers, Phillips, Fulker, & Mrazek, 1990). The Carey instrument (Carey & McDevitt, 1978) used in Chen et al.'s study and Rothbart's questionnaire (Rothbart, 1981, 1986) contain an activity dimension reminiscent of motor functioning on the NBAS. Although such dimensions are not equivalent, they do show enough overlap to suggest the possibility of meaningful agreement. Convergence among some of these different instruments for assessing neonatal temperament was demonstrated by the observation of a number of significant correlations between instruments (including the NBAS). These results reinforce both the validity of measures of neonate behavior and the relevance of comparing results using different scales even at different developmental stages during infancy (Worobey, 1986, 1997).

Additional evidence for a genetic explanation of the variance in NBAS clusters is provided by the study of temperament differences between rhesus monkeys from genetically diverse backgrounds that significantly differed in NBAS cluster scores (Champoux, Suomi, & Schneider, 1994). Similarly, variance in NBAS scores also has been observed between human children from different cultural and racial backgrounds (Nugent, Lester, & Brazelton, 1989). Such differences may be determined partially by diversity in allele frequency of putative temperament genes among these populations. Overall, we believe that these studies suggest that neonatal temperament clusters measured by the NBAS may have some genetic basis. However, the only twin study of neonates found no evidence for genetic influences (Riese, 1994); the NBAS was not, however, used as the assessment instrument.

In a first study of its kind (Ebstein et al., 1998), we examined the role of two common polymorphisms on neonatal behavior, evaluated at 2 weeks of age, in 81 neonates (40 male and 41 female) using the NBAS. We also note that this initial investigation is the first in a longitudinal study of neonate, infant, and toddler behavior that we have initiated.

The effect of the DRD4 short (s) versus DRD4 long (l) and 5-HTTLPR short/short (s/s) versus 5-HTTLPR long/short (l/s) and long/long (l/l) on four of the NBAS temperament clusters was compared by multivariate analysis. A significant association of DRD4 across the NBAS clusters was observed. Univariate *F* tests showed significant effects of DRD4 on all four of the temperament clusters. Overall, the presence of the long forms of the DRD4 exon III repeat region is to raise the mean score for all four of the temperament clusters.

No direct effect of the 5-HTTLPR polymorphism was observed by either multivariate or univariate *F* tests on any of the temperament clusters. However, a significant multivariate interaction was observed between DRD4 and 5-HTTLPR, and univariate *F* tests showed a significant interaction between the two polymorphisms on orientation (Figure 19.2). The

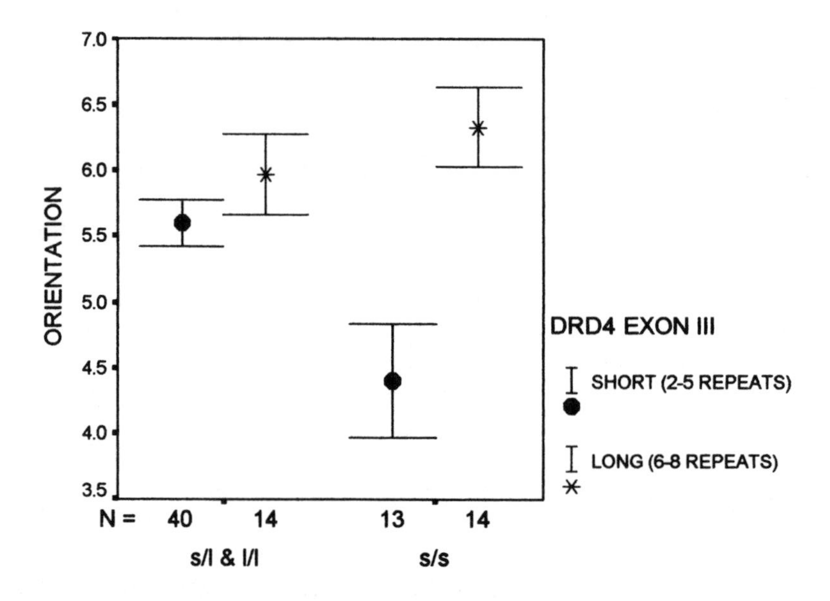

Figure 19.2. Neonatal Behavioral Assessment Scale (NBAS) orientation scores in 2-week-old infants sorted by the dopamine D4 receptor exon III repeat polymorphism (DRD4) and the serotonin transporter promoter region polymorphism (5-HTTLPR; Ebstein et al., 1998). Error bars represent mean values ± *SEM*. The effect size of DRD4 on orientation was much larger in those neonates homozygous for the short (s/s) 5-HTTLPR genotype compared with infants with either the s/l or l/l genotypes (partial eta-squared: 0.35 vs. 0.02).

effect size of DRD4 on orientation was much larger in those neonates homozygous for the s/s 5-HTTLPR genotype compared with infants with either the s/l or l/l genotypes. Moreover, no significant effect of l DRD4 on orientation was observed in neonates grouped by the l/l or l/s 5-HTTLPR, whereas the effect of l DRD4-repeat alleles was significant in neonates with the s/s 5-HTTLPR genotype. The effect of s/s 5-HTTLPR was to lower the orientation score for the group of neonates lacking the l DRD4.

In summary, the results indicate that infants with one or more long alleles of 5-HTTLPR, which is associated with less anxious traits in adults, show no further enhancing effect of the long DRD4 alleles, which also promote approach-type behaviors in adults. However, in infants with the inhibiting form (at least in adults) of 5-HTTLPR, l DRD4 alleles but not s DRD4 alleles oppose this inhibition.

We also examined the association of DRD4 and 5-HTTLPR and temperament in 76 two-month-old infants (39 boys and 37 girls) who had participated in the neonatal assessment described above (Auerbach et al., 1999). Mothers completed the Infant Behavior Questionnaire (IBQ; Rothbart, 1981). IBQ scores reflect the ease of elicitation of a given reaction and the intensity of that reaction. Both the stability and validity of the IBQ have been documented (Rothbart, 1986; Worobey & Blajda, 1989).

Temperament stability was examined by calculating bivariate correlation coefficients for the NBAS and the IBQ scales. There were significant negative correlations between several of the NBAS and the IBQ scales.

We next considered the effect of the DRD4 and 5-HTTLPR genotypes on the IBQ scales. Infants with long DRD4 alleles had significantly lower scores on negative emotionality and distress to limitations than infants with the short DRD4 alleles. In contrast, infants with the s/s 5-HTTLPR genotype had higher scores on negative emotionality and distress to limitations than infants with the l/s or l/l genotypes. A DRD4 main effect of distress to sudden or novel stimuli approached significance. Infants with the long DRD4 alleles were less distressed by sudden or novel stimuli than were infants with the short alleles. In those infants lacking the DRD4 long repeat alleles, the effect of the short homozygous form (s/s) of the 5-HTTLPR gene on negative emotionality and distress to limitations was significant, whereas no effect of this polymorphism was observed in infants possessing the long DRD4 alleles. In the absence of the long form of the DRD4 allele, the effect of the s/s form of 5-HTTLPR gene was to significantly raise the scores for negative emotionality and distress to limitations.

We have continued to examine this cohort and found that in a structured play situation and on an information-processing task, 1-year-old infants with the DRD4 7-repeat allele showed less sustained attention and novelty preference than do infants without the DRD4 7-repeat allele (Auerbach, Benjamin, Faroy, Geller, & Ebstein, 2001). There was also a significant interaction between DRD4 and the serotonin transporter promoter (5-HTTLPR) gene on a measure of sustained attention. Our results provide evidence for a possible developmental link between DRD4 and ADHD through early-sustained attention and information processing. It also points to the importance of considering the influence of more than one gene in studies of behavior as well as studying gene effects over developmental time.

A Hungarian group has published an interesting study of early infant behavior and the DRD4 polymorphism (Lakatos et al., 2000). About 15% of 1-year-old infants in nonclinical, low-risk and up to 80% in high-risk (e.g., maltreated) populations show extensive disorganized attachment (Solomon & George, 1999) behavior in the Strange Situation test (Ainsworth & Bell, 1970). It also has been reported that disorganization of early attachment is a major risk factor for the development of childhood behavior problems. The collapse of organized attachment strategy has been explained primarily by inappropriate caregiving, but recently, the contribution of child factors such as neurological impairments and neonatal behavioral organization also has been suggested. Attachment behavior of 90 infants was tested in the Strange Situation, and these infants were independently genotyped for the DRD4 exon III polymorphism. The 7-repeat allele was represented with a significantly higher frequency in infants classified as disorganized compared with nondisorganized infants: 12 of 17 (71%) versus 21 of 73 (29%) had at least one 7-repeat allele, $\chi^2(1) = 8.66$, $p < .005$. The estimated relative risk for disorganized attachment among children carrying the 7-repeat allele was 4.15. Lakatos et al. suggest that, in nonclinical, low social-risk populations, having a 7-repeat allele predisposes infants to attachment disorganization.

Pleiotropic Effects of Common Polymorphisms Affecting Human Behavior

Not surprisingly, three of the genes most intensely investigated in personality genetics—DRD4, 5-HTTLPR, and COMT—are known to affect a wide range of human behaviors. One of the most robust findings is the reported association between the 7-repeat allele and ADHD, a heterogeneous childhood disruptive behavior characterized by impulsive acts (see Thapar, chapter 22, this volume). Adults diagnosed with ADHD show higher TPQ novelty-seeking and harm avoidance scores (Downey, Stelson, Pomerleau, & Giordani, 1997). Many studies show an association between ADHD and the DRD4 7-repeat (see recent meta-analysis by Faraone, Doyle, Mick, & Biederman, 2001).

The DRD4 polymorphism also has been shown in some studies to be associated with addictive behavior. We note that substance abusers, including both alcoholics (Cloninger, Sigvardsson, Przybeck, & Svrakic, 1995) and heroin addicts (Mel et al., 1998), are distinguished by high TPQ novelty-seeking scores. We observed in a case control design an association between heroin addicts and the DRD4 7-repeat (Kotler et al., 1997; Mel et al., 1998), which has also been observed in a Chinese population (Li et al., 1997) and in an American population (Vandenbergh et al., 2000) but not in a middle European population (Franke et al., 2000). Goldman and his colleagues observed linkage between alcoholism in American Indians and a region on chromosome 11 very close to the DRD4 gene (Goldman et al., 1997). Also, a Japanese study (Muramatsu, Higuchi, Murayama, Matsushita, & Hayashida, 1996) observed an interaction between the DRD4 receptor and alcoholism in participants carrying the ALDH2(2) protective "flushing" allele, common in some Asian populations.

The 5-HTTLPR polymorphism has been associated in some studies with obsessive–compulsive disorder (Bengel et al., 1999; McDougle, Epperson, Price, & Gelernter, 1998), depression and anxiety in Parkinson's disease (Menza, Palermo, DiPaola, Sage, & Ricketts, 1999), alcoholism (Hallikainen et al., 1999; Ishiguro et al., 1999; Sander, Harms, Lesch, et al., 1997), compulsive buying (Devor, Magee, Dill-Devor, Gabel, & Black, 1999), autism (Cook et al., 1997; Klauck, Poustka, Benner, Lesch, & Poustka, 1997; Tordjman et al., 2001; Yirmiya et al., 2000; Yirmiya et al., 2001), and bipolar disorder (Furlong et al., 1998).

The COMT high/low enzyme activity val/met polymorphism also has been examined in a number of human behaviors. Two groups, including ours, found an association between the low enzyme activity met allele and violent behavior in schizophrenia (Kotler et al., 1999; Lachman, Nolan, Mohr, Saito, & Volavka, 1998; Strous, Bark, Parsia, Volavka, & Lachman, 1997). The association between the high enzyme val allele and heroin addiction first reported by Vandenbergh, Rodriguez, Miller, Uhl, and Lachman (1997) was confirmed by us in a family-based design (Horowitz et al., 2000). We (Alsobrook et al., 2002) also confirmed an association between this polymorphism and obsessive–compulsive disorder (Gogos et al., 1998; Karayiorgou et al., 1997, 1999). Finally, we note that we have observed a provisional association between the COMT polymorphism and hypnotizability (Lichtenberg, Bachner-Melman, Gritsenko, & Ebstein, 2000). Egan et al. (2001) elegantly demonstrated a role for COMT in cognitive ability.

A common theme underlying the pleiotropic effects of these three genes might be their role in modulating neurotransmitter systems underpinning similar facets of higher cortical and limbic functions subsumed under the heading of executive function. These three genes appear to regulate aspects of impulsivity and attention processes that are common to normal personality as well as disturbances reflected in such diverse disorders as autism, ADHD, schizophrenia, and addiction.

Future Directions

Finding genes contributing to human personality reflects the challenges encountered in genetic analysis of other complex human traits, including Type II diabetes (Alper, 2000), obesity (Arner, 2000), schizophrenia and bipolar disorder (Gershon, 2000), hypertension (Garbers & Dubois, 1999), and allergies (Ono, 2000), to name just a few areas of active research. Small effect size of individual genes and oligogenic interactions (i.e., when two or more genes work together to produce the phenotype) increase the difficulty to establish the role of common polymorphisms in contributing to these complex disorders.

As pointed out by Risch's (2000) review, the molecular methods successfully used to identify the genes underlying rare Mendelian syndromes are failing to find the numerous genes causing more common, familial, non-Mendelian diseases. However, there is room for optimism; especially with the completion of the human genome sequence, new opportunities are being presented for unraveling the complex genetic basis of non-Mendelian disorders based on large-scale genomewide studies.

In the past few years in the field of personality, common genetic polymorphisms have been identified that contribute to the determination of personality traits assessed by self-report questionnaires. Although the importance of genetic factors in determining human temperament has been recognized for two decades, recognition was lacking regarding the possibility that the revolution in molecular biology could also be applied to the mystery of human personality. Recent studies (Ebstein, Benjamin, & Belmaker, 2000) initially focused on a candidate gene approach to personality, and some common polymorphisms were associated with specific personality dimensions. This new field took an important turn with the first publication by Cloninger and his colleagues of a genomewide scan for chromosomal regions harboring genes contributing to the determination of human temperament (Cloninger et al., 1998). Not unexpectedly, it was apparent almost immediately that "personality" genes have a modest effect size and that several genes likely are contributing to any specific personality trait. A particularly exciting development is the possibility that the role of genes in human temperament could be recognized very early in development at a time when environmental influences are minimal (Auerbach et al., 2001; Auerbach et al., 1999; Ebstein & Auerbach, 2000; Ebstein et al., 1998; Lakatos et al., 2000). Finally, there is increasing awareness of the role played by personality factors in psychopathology and the possibility that personality may be a useful endophenotype in analyzing the genetics of mental illness.

References

Ainsworth, M. D., & Bell, S. M. (1970). Attachment, exploration, and separation: Illustrated by the behavior of one-year-olds in a Strange Situation. *Child Development, 41*, 49–67.

Alper, J. (2000). Biomedicine: New insights into Type 2 diabetes. *Science, 289*, 37–39.

Alsobrook, J. P., II., Zohar, A. H., Leboyer, M., Chabane, N., Ebstein, R. P., & Pauls, D. L. (2002). Association between the COMT locus and obsessive compulsive disorder in females but not males. *American Journal of Medical Genetics, 114,* 116–120.

Arner, P. (2000). Obesity: A genetic disease of adipose tissue? *British Journal of Nutrition, 83*(Suppl. 1), S9–S16.

Asghari, V., Sanyal, S., Buchwaldt, S., Paterson, A., Jovanovic, V., & Van Tol, H. H. (1995). Modulation of intracellular cyclic AMP levels by different human dopamine D4 receptor variants. *Journal of Neurochemistry, 65*, 1157–1165.

Asghari, V., Schoots, O., van Kats, S., Ohara, K., Jovanovic, V., Guan, H. C., et al. (1994). Dopamine D4 receptor repeat: Analysis of different native and mutant forms of the human and rat genes. *Molecular Pharmacology, 46*, 364–373.

Auerbach, J. G., Benjamin, J., Faroy, M., Geller, V., & Ebstein, R. (2001). DRD4 related to infant attention and information processing: A developmental link to ADHD? *Psychiatric Genetics, 11*, 31–35.

Auerbach, J., Geller, V., Lezer, S., Shinwell, E., Belmaker, R. H., Levine, J., et al. (1999). Dopamine D4 receptor (D4DR) and serotonin transporter promoter (5-HTTLPR) polymorphisms in the determination of temperament in 2-month-old infants. *Molecular Psychiatry, 4*, 369–373.

Ball, D., Hill, L., Freeman, B., Eley, T. C., Strelau, J., Riemann, R., et al. (1997). The serotonin transporter gene and peer-rated neuroticism. *Neuroreport, 8*, 1301–1304.

Barr, C. L., Wigg, K. G., Bloom, S., Schachar, R., Tannock, R., Roberts, W., et al. (2000). Further evidence from haplotype analysis for linkage of the dopamine D4 receptor gene and attention-deficit hyperactivity disorder. *American Journal of Medical Genetics, 96*, 262–267.

Bau, C. H., Almeida, S., Costa, F. T., Garcia, C. E., Elias, E. P., Ponso, A. C., et al. (2001). DRD4 and DAT1 as modifying genes in alcoholism: Interaction with novelty seeking on level of alcohol consumption. *Molecular Psychiatry, 6*, 7–9.

Bau, C. H., Almeida, S., & Hutz, M. H. (2000). The TaqI A1 allele of the dopamine D2 receptor gene and alcoholism in Brazil: Association and interaction with stress and harm avoidance on severity prediction. *American Journal of Medical Genetics, 96*, 302–306.

Bau, C. H., Roman, T., Almeida, S., & Hutz, M. H. (1999). Dopamine D4 receptor gene and personality dimensions in Brazilian male alcoholics. *Psychiatric Genetics, 9*, 139–143.

Bengel, D., Greenberg, B., Cora-Locatelli, G., Altemus, M., Heils, A., Li, Q., et al. (1999). Association of the serotonin transporter promoter regulatory region polymorphism and obsessive–compulsive disorder. *Molecular Psychiatry, 4*, 463–466.

Benjamin, J., Li, L., Patterson, C., Greenberg, B. D., Murphy, D. L., & Hamer, D. H. (1996). Population and familial association between the D4 dopamine receptor gene and measures of novelty seeking. *Nature Genetics, 12*, 81–84.

Benjamin, J., Osher, Y., Kotler, M., Gritsenko, I., Nemanov, L., Belmaker, R. H., et al. (2000). Association between tridimensional personality questionnaire (TPQ) traits and three functional polymorphisms: Dopamine receptor D4 (DRD4), serotonin transporter promoter region (5-HTTLPR) and catechol O-methyltransferase (COMT). *Molecular Psychiatry, 5*, 96–100.

Benjamin, J., Osher, Y., Lichtenberg, P., Bachner-Melman, R., Gritsenko, I., Kotler, M., et al. (2000). An interaction between the catechol O-methyltransferase and serotonin transporter promoter region polymorphisms contributes to Tridimensional Personality Questionnaire persistence scores in normal ssubjects. *Neuropsychobiology, 41*, 48–53.

Blairy, S., Massat, I., Staner, L., Le Bon, O., Van Gestel, S., Van Broeckhoven, C., et al. (2000). 5-HT2a receptor polymorphism gene in bipolar disorder and harm avoidance personality trait. *American Journal of Medical Genetics, 96*, 360–364.

Brazelton, T. B., & Nugent, J. K. (1995). *Neonatal Behavioral Assessment Scale*. Cambridge, England: Cambridge University Press.

Carey, W. B., & McDevitt, S. C. (1978). Stability and change in individual temperament diagnoses from infancy to early childhood. *Journal of American Academic Child Psychiatry, 17*, 331–337.

Cattell, R. B. (1946). *Description and measurement of personality*. Yonkers-on-Hudson, NY: World.

Champoux, M., Suomi, S. J., & Schneider, M. L. (1994). Temperament differences between captive Indian and Chinese-Indian hybrid rhesus macaque neonates. *Laboratory Animal Science, 44*, 351–357.

Chen, C. J., Yu, M. W., Wang, C. J., Tong, S. L., Tien, M., Lee, T. Y., et al. (1990). Genetic variance and heritability of temperament among Chinese twin infants. *Acta Geneticae, Medicae, et Gemellologiae (Roma), 39*, 485–490.

Cloninger, C. R. (1987). A systematic method for clinical description and classification of personality variants: A proposal. *Archives of General Psychiatry, 44*, 573–588.

Cloninger, C. R., Sigvardsson, S., Przybeck, T. R., & Svrakic, D. M. (1995). Personality antecedents of alcoholism in a national area probability sample. *European Archives of Psychiatry and Clinical Neuroscience, 245*, 239–244.

Cloninger, C. R., Svrakic, D. M., & Przybeck, T. R. (1993). A psychobiological model of temperament and character. *Archives of General Psychiatry, 50*, 975–990.

Cloninger, C. R., Van Eerdewegh, P., Goate, A., Edenberg, H. J., Blangero, J., Hesselbrock,

V., et al. (1998). Anxiety proneness linked to epistatic loci in genome scan of human personality traits. *American Journal of Medical Genetics, 81*, 313–317.

Comings, D. E., Johnson, J. P., Gonzalez, N. S., Huss, M., Saucier, G., McGue, M., et al. (2000). Association between the adrenergic alpha 2A receptor gene (ADRA2A) and measures of irritability, hostility, impulsivity and memory in normal subjects. *Psychiatric Genetics, 10*, 39–42.

Cook, E. H., Jr., Courchesne, R., Lord, C., Cox, N. J., Yan, S., Lincoln, A., et al. (1997). Evidence of linkage between the serotonin transporter and autistic disorder. *Molecular Psychiatry, 2*, 247–250.

Costa, P. T., & McCrae, R. R. (1994). Stability and change in personality from adolescence through adulthood. In C. F. Halverson, G. A. Kohnstamm, & R. P. Martin (Eds.), *The developing structure of temperament and personality from infancy to adulthood* (pp. 139–150). Hillsdale, NJ: Erlbaum.

Cyphers, L. H., Phillips, K., Fulker, D. W., & Mrazek, D. A. (1990). Twin temperament during the transition from infancy to early childhood. *Journal of American Academic Child and Adolescent Psychiatry, 29*, 392–397.

Damberg, M., Garpenstrand, H., Alfredsson, J., Ekblom, J., Forslund, K., Rylander, G., et al. (2000). A polymorphic region in the human transcription factor AP-2beta gene is associated with specific personality traits. *Molecular Psychiatry, 5*, 220–224.

Deary, I. J., Battersby, S., Whiteman, M. C., Connor, J. M., Fowkes, F. G., & Harmar, A. (1999). Neuroticism and polymorphisms in the serotonin transporter gene. *Psychological Medicine, 29*, 735–739.

Devor, E. J., Magee, H. J., Dill-Devor, R. M., Gabel, J., & Black, D. W. (1999). Serotonin transporter gene (5-HTT) polymorphisms and compulsive buying. *American Journal of Medical Genetics, 88*, 123–125.

Downey, K. K., Stelson, F. W., Pomerleau, O. F., & Giordani, B. (1997). Adult attention deficit hyperactivity disorder: Psychological test profiles in a clinical population. *Journal of Nervous and Mental Disease, 185*, 32–38.

Du, L., Bakish, D., & Hrdina, P. D. (2000). Gender differences in association between serotonin transporter gene polymorphism and personality traits. *Psychiatric Genetics, 10*, 159–164.

Dulawa, S. C., Grandy, D. K., Low, M. J., Paulus, M. P., & Geyer, M. A. (1999). Dopamine D4 receptor-knock-out mice exhibit reduced exploration of novel stimuli. *Journal of Neuroscience, 19*, 9550–9556.

Ebstein, R. P., & Auerbach, J. (2000). Dopamine D4 receptor (DRD4) and serotonin transporter promoter (5-HTTLPR) polymorphisms and temperament in early childhood. In J. Benjamin, R. P. Ebstein, & R. H. Belmaker (Eds.), *Molecular genetics and the human personality* (pp. 137–149). Washington, DC: American Psychiatric Press.

Ebstein, R. P., Benjamin, J., & Belmaker, R. H. (2000). Personality and polymorphisms of genes involved in aminergic neurotransmission. *European Journal of Pharmacology, 410*, 205–214.

Ebstein, R. P., Gritsenko, I., Nemanov, L., Frisch, A., Osher, Y., & Belmaker, R. H. (1997). No association between the serotonin transporter gene regulatory region polymorphism and the Tridimensional Personality Questionnaire (TPQ) temperament of harm avoidance. *Molecular Psychiatry, 2*, 224–226.

Ebstein, R. P., Levine, J., Geller, V., Auerbach, J., Gritsenko, I., & Belmaker, R. H. (1998). Dopamine D4 receptor and serotonin transporter promoter in the determination of neonatal temperament. *Molecular Psychiatry, 3*, 238–246.

Ebstein, R. P., Novick, O., Umansky, R., Priel, B., Osher, Y., Blaine, D., et al. (1996). Dopamine D4 receptor (D4DR) exon III polymorphism associated with the human personality trait of novelty seeking. *Nature Genetics, 12*, 78–80.

Ebstein, R. P., Segman, R., Benjamin, J., Osher, Y., Nemanov, L., & Belmaker, R. H. (1997). 5-HT2C (HTR2C) serotonin receptor gene polymorphism associated with the human personality trait of reward dependence: Interaction with dopamine D4 receptor (D4DR) and dopamine D3 receptor (D3DR) polymorphisms. *American Journal of Medical Genetics (Neuropsychiatric Genetics), 74*, 65–72.

Egan, M. F., Goldberg, T. E., Kolachana, B. S., Callicott, J. H., Mazzanti, C. M., Straub, R. E., et al. (2001). Effect of COMT Val108/158 Met genotype on frontal lobe function and risk for schizophrenia. *Proceedings of the National Academy of Sciences USA, 98*, 6917–6922.

Ekelund, J., Lichtermann, D., Jarvelin, M. R., & Peltonen, L. (1999). Association between novelty seeking and the Type 4 dopamine receptor gene in a large Finnish cohort sample. *American Journal of Psychiatry, 156*, 1453–1455.

Eysenck, H. J. (1952). *The scientific study of personality*. London: Routledge & Kegan Paul.

Faraone, S. V., Doyle, A. E., Mick, E., & Biederman, J. (2001). Meta-analysis of the association between the 7-repeat allele of the dopamine D(4) receptor gene and attention deficit hyperactivity disorder. *American Journal of Psychiatry, 158*, 1052–1057.

Flory, J. D., Manuck, S. B., Ferrell, R. E., Dent, K. M., Peters, D. G., & Muldoon, M. F. (1999). Neuroticism is not associated with the serotonin transporter (5-HTTLPR) polymorphism. *Molecular Psychiatry, 4*, 93–96.

Franke, P., Nothen, M. M., Wang, T., Knapp, M., Lichtermann, D., Neidt, H., et al. (2000). DRD4 exon III VNTR polymorphism-susceptibility factor for heroin dependence? Results of a case-control and a family-based association approach. *Molecular Psychiatry, 5*, 101–104.

Furlong, R. A., Ho, L., Walsh, C., Rubinsztein, J. S., Jain, S., Paykel, E. S., et al. (1998). Analysis and meta-analysis of two serotonin transporter gene polymorphisms in bipolar and unipolar affective disorders. *American Journal of Medical Genetics, 81*, 58–63.

Garbers, D. L., & Dubois, S. K. (1999). The molecular basis of hypertension. *Annual Review of Biochemistry, 68*, 127–155.

Gebhardt, C., Leisch, F., Schussler, P., Fuchs, K., Stompe, T., Sieghart, W., et al. (2000). Non-association of dopamine D4 and D2 receptor genes with personality in healthy individuals. *Psychiatric Genetics, 10*, 131–137.

Gelernter, J., Kranzler, H., Coccaro, E. F., Siever, L. J., & New, A. S. (1998). Serotonin transporter protein gene polymorphism and personality measures in African American and European American subjects. *American Journal of Psychiatry, 155*, 1332–1338.

Gelernter, J., Kranzler, H., Coccaro, E., Siever, L., New, A., & Mulgrew, C. L. (1997). D4 dopamine-receptor (DRD4) alleles and novelty seeking in substance-dependent, personality-disorder, and control subjects. *American Journal of Human Genetics, 61*, 1144–1152.

Gershon, E. S. (2000). Bipolar illness and schizophrenia as oligogenic diseases: Implications for the future. *Biological Psychiatry, 47*, 240–244.

Gogos, J. A., Morgan, M., Luine, V., Santha, M., Ogawa, S., Pfaff, D., et al. (1998). Catechol-O-methyltransferase-deficient mice exhibit sexually dimorphic changes in catecholamine levels and behavior. *Proceedings of the National Academy of Sciences USA, 95*, 9991–9996.

Goldman, D., Knowler, W. C., Hanson, R. L., Robin, R. W., Urbanek, M., Moore, E., et al. (1997). Autosome-wide scan for alcohol dependence in southwestern American Indians provides suggestive evidence for loci on chromosomes 4 and 11. *American Journal of Medical Genetics (Neuropsychiatric Genetics), 74*, 574–575.

Golimbet, V. E., Alfimova, V. M., Shcherbatykh, T. V., Abramova, L. I., Kaleda, V. G., & Rogaev, E. I. (2001). Insertion-deletion polymorphism of the serotonin carrier gene and evaluation of neurotism as a temperament trait in patients with affective disorders and mentally healthy people. *Molecular Biology (Mosk), 35*, 397–400.

Greenberg, B. D., Li, Q., Lucas, F. R., Hu, S., Sirota, L. A., Benjamin, J., et al. (2000). Association between the serotonin transporter promoter polymorphism and personality traits in a primarily female population sample. *American Journal of Medical Genetics, 96*, 202–216.

Gustavsson, J. P., Nothen, M. M., Jonsson, E. G., Neidt, H., Forslund, K., Rylander, G., et al. (1999). No association between serotonin transporter gene polymorphisms and personality traits. *Neuropsychiatric Genetics, 88*, 430–436.

Hallikainen, T., Saito, T., Lachman, H. M., Volavka, J., Pohjalainen, T., Ryynanen, O. P., et al. (1999). Association between low activity serotonin transporter promoter genotype and early onset alcoholism with habitual impulsive violent behavior. *Molecular Psychiatry, 4*, 385–388.

Hamer, D. H., Greenberg, B. D., Sabol, S. Z., & Murphy, D. L. (1999). Role of the serotonin transporter gene in temperament and character. *Journal of Personality Disorder, 13,* 312–327.

Hamilton, S. P., Heiman, G. A., Haghighi, F., Mick, S., Klein, D. F., Hodge, S. E., et al. (1999). Lack of genetic linkage or association between a functional serotonin transporter polymorphism and panic disorder. *Psychiatric Genetics, 9,* 1–6.

Heils, A., Teufel, A., Petri, S., Stober, G., Riederer, P., Bengel, D., et al. (1996). Allelic variation of human serotonin transporter gene expression. *Journal of Neurochemistry, 66,* 2621–2624.

Henderson, A. S., Korten, A. E., Jorm, A. F., Jacomb, P. A., Christensen, H., Rodgers, B., et al. (2000). COMT and DRD3 polymorphisms, environmental exposures, and personality traits related to common mental disorders. *American Journal of Medical Genetics, 96,* 102–107.

Herbst, J. H., Zonderman, A. B., McCrae, R. R., & Costa, P. T., Jr. (2000). Do the dimensions of the temperament and character inventory map a simple genetic architecture? Evidence from molecular genetics and factor analysis. *American Journal of Psychiatry, 157,* 1285–1290.

Hill, S. Y., Zezza, N., Wipprecht, G., Locke, J., & Neiswanger, K. (1999). Personality traits and dopamine receptors (D2 and D4): Lnkage studies in families of alcoholics. *American Journal of Medical Genetics, 88,* 634–641.

Horowitz, R., Kotler, M., Shufman, E., Aharoni, S., Kremer, I., Cohen, H., et al. (2000). Confirmation of an excess of the high enzyme activity COMT val allele in heroin addicts in a family-based haplotype relative risk study. *American Journal of Medical Genetics, 96,* 599–603.

Hu, S., Brody, C. L., Fisher, C., Gunzerath, L., Nelson, M. L., Sabol, S. Z., et al. (2000). Interaction between the serotonin transporter gene and neuroticism in cigarette smoking behavior. *Molecular Psychiatry, 5,* 181–188.

Ishiguro, H., Saito, T., Akazawa, S., Mitushio, H., Tada, K., Enomoto, M., et al. (1999). Association between drinking-related antisocial behavior and a polymorphism in the serotonin transporter gene in a Japanese population. *Alcoholism Clinical and Experimental Research, 23,* 1281–1284.

Jonsson, E. G., Nothen, M. M., Gustavsson, J. P., Neidt, H., Brene, S., Tylec, A., et al. (1997). Lack of evidence for allelic association between personality traits and the dopamine D4 receptor gene polymorphisms. *American Journal of Psychiatry, 154,* 697–699.

Jorm, A. F., Henderson, A. S., Jacomb, P. A., Christensen, H., Korten, A. E., Rodgers, B., et al. (1998). An association study of a functional polymorphism of the serotonin transporter gene with personality and psychiatric symptoms. *Molecular Psychiatry, 3,* 449–451.

Jorm, A. F., Prior, M., Sanson, A., Smart, D., Zhang, Y., & Easteal, S. (2000). Association of a functional polymorphism of the serotonin transporter gene with anxiety-related temperament and behavior problems in children: A longitudinal study from infancy to the mid-teens. *Molecular Psychiatry, 5,* 542–547.

Karayiorgou, M., Altemus, M., Galke, B. L., Goldman, D., Murphy, D. L., Ott, J., et al. (1997). Genotype determining low catechol-O-methyltransferase activity as a risk factor for obsessive–compulsive disorder. *Proceedings of the National Academy of Sciences USA, 94,* 4572–4575.

Karayiorgou, M., Sobin, C., Blundell, M. L., Galke, B. L., Malinova, L., Goldberg, P., et al. (1999). Family-based association studies support a sexually dimorphic effect of COMT and MAOA on genetic susceptibility to obsessive–compulsive disorder. *Biological Psychiatry, 45,* 1178–1189.

Katsuragi, S., Kunugi, H., Sano, A., Tsutsumi, T., Isogawa, K., Nanko, S., et al. (1999). Association between serotonin transporter gene polymorphism and anxiety-related traits. *Biological Psychiatry, 45,* 368–370.

Kazmi, M. A., Snyder, L. A., Cypess, A. M., Graber, S. G., & Sakmar, T. P. (2000). Selective reconstitution of human D4 dopamine receptor variants with Gi alpha subtypes. *Biochemistry, 39,* 3734–3744.

Klauck, S. M., Poustka, F., Benner, A., Lesch, K. P., & Poustka, A. (1997). Serotonin transporter (5-HTT) gene variants associated with autism? *Human Molecular Genetics, 6,* 2233–2238.

Kotler, M., Barak, P., Cohen, H., Averbuch, I. E., Grinshpoon, A., Gritsenko, I., et al. (1999). Homicidal behavior in schizophrenia associated with a genetic polymorphism determining low catechol O-methyltransferase (COMT) activity. *American Journal of Human Genetics, 88*, 628–633.

Kotler, M., Cohen, H., Segman, R., Gritsenko, I., Nemanov, L., Lerer, B., et al. (1997). Excess dopamine D4 receptor (D4DR) exon III seven repeat allele in opioid-dependent subjects. *Molecular Psychiatry, 2*, 251–254.

Kuhn, K. U., Meyer, K., Nothen, M. M., Gansicke, M., Papassotiropoulos, A., & Maier, W. (1999). Allelic variants of dopamine receptor D4 (DRD4) and serotonin receptor 5HT2c (HTR2c) and temperament factors: Replication tests. *American Journal of Medical Genetics, 88*, 168–172.

Kumakiri, C., Kodama, K., Shimizu, E., Yamanouchi, N., Okada, S., Noda, S., et al. (1999). Study of the association between the serotonin transporter gene regulatory region polymorphism and personality traits in a Japanese population. *Neuroscience Letters, 263*, 205–207.

Lachman, H. M., Nolan, K. A., Mohr, P., Saito, T., & Volavka, J. (1998). Association between catechol O-methyltransferase genotype and violence in schizophrenia and schizoaffective disorder. *American Journal of Psychiatry, 155*, 835–837.

Lakatos, K., Toth, I., Nemoda, Z., Ney, K., Sasvari-Szekely, M., & Gervai, J. (2000). Dopamine D4 receptor (DRD4) gene polymorphism is associated with attachment disorganization in infants. *Molecular Psychiatry, 5*, 633–637.

Lerman, C., Caporaso, N. E., Audrain, J., Main, D., Boyd, N. R., & Shields, P. G. (2000). Interacting effects of the serotonin transporter gene and neuroticism in smoking practices and nicotine dependence. *Molecular Psychiatry, 5*, 189–192.

Lesch, K. P., Bengel, D., Heils, A., Sabol, S. Z., Greenberg, B. D., Petri, S., et al. (1996). Association of anxiety-related traits with a polymorphism in the serotonin transporter gene regulatory region. *Science, 274*, 1527–1531.

Li, T., Xu, K., Deng, H., Cai, G., Liu, J., Liu, X., et al. (1997). Association analysis of the dopamine D4 gene exon III VNTR and heroin abuse in Chinese subjects. *Molecular Psychiatry, 2*, 413–416.

Lichtenberg, P., Bachner-Melman, R., Gritsenko, I., & Ebstein, R. P. (2000). Exploratory association study between catechol-O-methyltransferase (COMT) high/low enzyme activity polymorphism and hypnotizability. *American Journal of Human Genetics, 96*, 771–774.

Lichter, J. B., Barr, C. L., Kennedy, J. L., Van Tol, H. H., Kidd, K. K., & Livak, K. J. (1993). A hypervariable segment in the human dopamine receptor D4 (DRD4) gene. *Human Molecular Genetics, 2*, 767–773.

Loehlin, J. C. (1992). *Genes and environment in personality development*. Newbury Park, CA: Sage.

Malhotra, A. K., Virkkunen, M., Rooney, W., Eggert, M., Linnoila, M., & Goldman, D. (1996). The association between the dopamine D4 receptor (D4DR) 16 amino acid repeat polymorphism and novelty seeking. *Molecular Psychiatry, 1*, 388–391.

Matsushita, S., Yoshino, A., Murayama, M., Kimura, M., Muramatsu, T., & Higuchi, S. (2001). Association study of serotonin transporter gene regulatory region polymorphism and alcoholism. *American Journal of Medical Genetics, 105*, 446–450.

Mazzanti, C. M., Lappalainen, J., Long, J. C., Bengel, D., Naukkarinen, H., Eggert, M., et al. (1998). Role of the serotonin transporter promoter polymorphism in anxiety-related traits. *Archives of General Psychiatry, 55*, 936–940.

McCrae, R. R., & Costa, P. T., Jr. (1997). Personality trait structure as a human universal. *Australian Journal of Psychology, 52*, 509–516.

McDougle, C. J., Epperson, C. N., Price, L. H., & Gelernter, J. (1998). Evidence for linkage disequilibrium between serotonin transporter protein gene (SLC6A4) and obsessive compulsive disorder. *Molecular Psychiatry, 3*, 270–273.

Mel, H., Horowitz, R., Ohel, N., Kramer, I., Kotler, M., Cohen, H., et al. (1998). Additional evidence for an association between the dopamine D4 receptor (D4DR) exon III seven-repeat allele and substance abuse in opioid dependent subjects: Relationship of treatment retention to genotype and personality. *Addiction Biology, 3*, 473–481.

Melke, J., Landen, M., Baghei, F., Rosmond, R., Holm, G., Bjorntorp, P., et al. (2001). Serotonin transporter gene polymorphisms are associated with anxiety-related personality traits in women. *American Journal of Medical Genetics, 105*, 458–463.

Menza, M. A., Palermo, B., DiPaola, R., Sage, J. I., & Ricketts, M. H. (1999). Depression and anxiety in Parkinson's disease: Possible effect of genetic variation in the serotonin transporter. *Journal of Geriatric Psychiatry and Neurology, 12*(2), 49–52.

Mitsuyasu, H., Hirata, N., Sakai, Y., Shibata, H., Takeda, Y., Ninomiya, H., et al. (2001). Association analysis of polymorphisms in the upstream region of the human dopamine D4 receptor gene (DRD4) with schizophrenia and personality traits. *Journal of Human Genetics, 46*, 26–31.

Murakami, F., Shimomura, T., Kotani, K., Ikawa, S., Nanba, E., & Adachi, K. (1999). Anxiety traits associated with a polymorphism in the serotonin transporter gene regulatory region in the Japanese. *Journal of Human Genetics, 44*, 15–17.

Muramatsu, T., Higuchi, S., Murayama, M., Matsushita, S., & Hayashida, M. (1996). Association between alcoholism and the dopamine D4 receptor gene. *Journal of Medical Genetics, 33*, 113–115.

Nakamura, M., Ueno, S., Sano, A., & Tanabe, H. (2000). The human serotonin transporter gene linked polymorphism (5-HTTLPR) shows ten novel allelic variants. *Molecular Psychiatry, 5*, 32–38.

Niimi, Y., Inoue-Murayama, M., Murayama, Y., Ito, S., & Iwasaki, T. (1999). Allelic variation of the D4 dopamine receptor polymorphic region in two dog breeds, golden retriever and Shiba. *Journal of Veterinary Medical Science, 61*, 1281–1286.

Noble, E. P., Ozkaragoz, T. Z., Ritchie, T. L., Zhang, X., Belin, T. R., & Sparkes, R. S. (1998). D2 and D4 dopamine receptor polymorphisms and personality. *American Journal of Medical Genetics, 81*, 257–267.

Nugent, J. K., Lester, B. M., & Brazelton, T. B. (1989). *The cultural context of infancy.* Norwood, NJ: Ablex.

Okuyama, Y., Ishiguro, H., Nankai, M., Shibuya, H., Watanabe, A., & Arinami, T. (2000). Identification of a polymorphism in the promoter region of DRD4 associated with the human novelty seeking personality trait. *Molecular Psychiatry, 5*, 64–69.

Ono, S. J. (2000). Molecular genetics of allergic diseases. *Annual Review of Immunology, 18*, 347–366.

Ono, Y., Manki, H., Yoshimura, K., Muramatsu, T., Mizushima, H., Higuchi, S., et al. (1997). Association between dopamine D4 receptor (D4DR) exon III polymorphism and novelty seeking in Japanese subjects. *American Journal of Medical Genetics, 74*, 501–503.

Osher, Y., Hamer, D., & Benjamin, J. (2000). Association and linkage of anxiety-related traits with a functional polymorphism of the serotonin transporter gene regulatory region in Israeli sibling pairs. *Molecular Psychiatry, 5*, 216–219.

Persson, M., Wasserman, D., Jonsson, E. G., Bergman, H., Terenius, L., Gyllander, A., Neiman, J., et al. (2000). Search for the influence of the tyrosine hydroxylase (TCAT)(n) repeat polymorphism on personality traits. *Psychiatry Research, 95*, 1–8.

Persson, M. L., Wasserman, D., Geijer, T., Frisch, A., Rockah, R., Michaelovsky, E., et al. (2000). Dopamine D4 receptor gene polymorphism and personality traits in healthy volunteers. *European Archives of Psychiatry and Clinical Neuroscience, 250*, 203–206.

Pogue-Geile, M., Ferrell, R., Deka, R., Debski, T., & Manuck, S. (1998). Human novelty-seeking personality traits and dopamine D4 receptor polymorphisms: A twin and genetic association study. *American Journal of Medical Genetics, 81*, 44–48.

Poston, W. S., II., Ericsson, M., Linder, J., Haddock, C. K., Hanis, C. L., Nilsson, T., et al. (1998). D4 dopamine receptor gene exon III polymorphism and obesity risk. *Eating and Weight Disorders, 3*, 71–77.

Ricketts, M. H., Hamer, R. M., Sage, J. I., Manowitz, P., Feng, F., & Menza, M. A. (1998). Association of a serotonin transporter gene promoter polymorphism with harm avoidance behaviour in an elderly population. *Psychiatric Genetics, 8*(2), 41–44.

Riese, M. L. (1994). Neonatal temperament in full-term twin pairs discordant for birth weight. *Journal of Developmental and Behavorial Pediatrics, 15*, 342–347.

Riley, B. P., & McGuffin, P. (2000). Linkage and associated studies of schizophrenia. *American Journal of Medical Genetics, 97*, 23–44.

Risch, N. J. (2000). Searching for genetic determinants in the new millennium. *Nature, 405*, 847–856.

Ronai, Z., Szekely, A., Nemoda, Z., Lakatos, K., Gervai, J., Staub, M., et al. (2001). Association between novelty seeking and the −521 C/T polymorphism in the promoter region of the DRD4 gene. *Molecular Psychiatry, 6*, 35–38.

Rosmond, R., Rankinen, T., Chagnon, M., Perusse, L., Chagnon, Y. C., Bouchard, C., et al. (2001). Polymorphism in exon 6 of the dopamine D(2) receptor gene (DRD2) is associated with elevated blood pressure and personality disorders in men. *Journal of Human Hypertension, 15*, 553–558.

Rothbart, M. K. (1981). Measurement of temperament in infancy. *Child Development, 52*, 569–578.

Rothbart, M. K. (1986). Longitudinal observation of infant temperament. *Developmental Psychology, 22*, 356–365.

Rubinstein, M., Phillips, T. J., Bunzow, J. R., Falzone, T. L., Dziewczapolski, G., Zhang, G., et al. (1997). Mice lacking dopamine D4 receptors are supersensitive to ethanol, cocaine, and methamphetamine. *Cell, 90*, 991–1001.

Samochowiec, J., Rybakowski, F., Czerski, P., Zakrzewska, M., Stepien, G., Pelka-Wysiecka, J., et al. (2001). Polymorphisms in the dopamine, serotonin, and norepinephrine transporter genes and their relationship to temperamental dimensions measured by the Temperament and Character Inventory in healthy volunteers. *Neuropsychobiology, 43*, 248–253.

Sander, T., Harms, H., Dufeu, P., Kuhn, S., Rommelspacher, H., & Schmidt, L. G. (1997). Dopamine D4 receptor exon III alleles and variation of novelty seeking in alcoholics. *American Journal of Medical Genetics, 74*, 483–487.

Sander, T., Harms, H., Lesch, K. P., Dufeu, P., Kuhn, S., Hoehe, M., et al. (1997). Association analysis of a regulatory variation of the serotonin transporter gene with severe alcohol dependence. *Alcoholism Clinical and Experimental Research, 21*, 1356–1359.

Schultz, W. (1997). Dopamine neurons and their role in reward mechanisms. *Current Opinions in Neurobiology, 7*, 191–197.

Solomon, J., & George, C. (1999). *Attachment disorganization*. New York: Guilford Press.

Staner, L., Hilger, C., Hentges, F., Monreal, J., Hoffmann, A., Couturier, M., et al. (1998). Association between novelty-seeking and the dopamine D3 receptor gene in bipolar patients: A preliminary report. *American Journal of Medical Genetics, 81*, 192–194.

Strobel, A., Wehr, A., Michel, A., & Brocke, B. (1999). Association between the dopamine D4 receptor (DRD4) exon III polymorphism and measures of novelty seeking in a German population. *Molecular Psychiatry, 4*, 378–384.

Strous, R. D., Bark, N., Parsia, S. S., Volavka, J., & Lachman, H. M. (1997). Analysis of a functional catechol-O-methyltransferase gene polymorphism in schizophrenia: Evidence for association with aggressive and antisocial behavior. *Psychiatry Research, 69*(2–3), 71–77.

Sullivan, P. F., Fifield, W. J., Kennedy, M. A., Mulder, R. T., Sellman, J. D., & Joyce, P. R. (1998). No association between novelty seeking and the type 4 dopamine receptor gene (DRD4) in two New Zealand samples. *American Journal of Psychiatry, 155*, 98–101.

Swift, G., Larsen, B., Hawi, Z., & Gill, M. (2000). Novelty seeking traits and D4 dopamine receptors [Letter]. *American Journal of Medical Genetics, 96*, 222–223.

Thome, J., Weijers, H. G., Wiesbeck, G. A., Sian, J., Nara, K., Boning, J., et al. (1999). Dopamine D3 receptor gene polymorphism and alcohol dependence: Relation to personality rating. *Psychiatric Genetics, 9*, 17–21.

Tomitaka, M., Tomitaka, S., Otuka, Y., Kim, K., Matuki, H., Sakamoto, K., et al. (1999). Association between novelty seeking and dopamine receptor D4 (DRD4) exon III polymorphism in Japanese subjects. *American Journal of Medical Genetics, 88*, 469–471.

Tordjman, S., Gutknecht, L., Carlier, M., Spitz, E., Antoine, C., Slama, F., et al. (2001). Role of the serotonin transporter gene in the behavioral expression of autism. *Molecular Psychiatry, 6*, 434–439.

Van Tol, H. H., Wu, C. M., Guan, H. C., Ohara, K., Bunzow, J. R., Civelli, O., et al. (1992). Multiple dopamine D4 receptor variants in the human population. *Nature, 358*, 149–152.

Vandenbergh, D. J., Rodriguez, L. A., Hivert, E., Schiller, J. H., Villareal, G., Pugh, E. W., et al. (2000). Long forms of the dopamine receptor (DRD4) gene VNTR are more prevalent in substance abusers: No interaction with functional alleles of the catechol-o-methyltransferase (COMT) gene. *American Journal of Medical Genetics, 96*, 678–683.

Vandenbergh, D. J., Rodriguez, L. A., Miller, I. T., Uhl, G. R., & Lachman, H. M. (1997). High-activity catechol-O-methyltransferase allele is more prevalent in polysubstance abusers. *American Journal of Medical Genetics, 74*, 439–442.

Vandenbergh, D. J., Zonderman, A. B., Wang, J., Uhl, G. R., & Costa, P. T., Jr. (1997). No association between novelty seeking and dopamine D4 receptor (D4DR) exon III seven repeat alleles in Baltimore Longitudinal Study of Aging participants. *Molecular Psychiatry, 2*, 417–419.

Wei, J., & Hemmings, G. P. (2000). The NOTCH4 locus is associated with susceptibility to schizophrenia. *Nature Genetics, 25*, 376–377.

Worobey, J. (1986). Convergence among assessments of temperament in the first month. *Child Development, 57*, 47–55.

Worobey, J. (1997). Convergence between temperament ratings in early infancy. *Journal of Developmental and Behavioral Pediatrics, 18*, 260–263.

Worobey, J., & Blajda, V. M. (1989). Temperament ratings at 2 weeks, 2 months, and 1 year: Differential stability of activity and emotionality. *Developmental Psychology, 25*, 257–263.

Yirmiya, N., Pilowsky, T., Nemanov, L., Arbelle, S., Feinsilver, T., Fried, I., et al. (2000). Analysis of three coding region (dopamine D4 receptor, catechol-O-methyltransferase and serotonin transporter promoter region) polymorphisms in autism: Evidence for an association with the serotonin transporter. In E. Schopler, L. Marcus, C. Shulman, & N. Yirmiya (Eds.), *The research basis of autism intervention* (pp. 91–101). New York: Kluwer Academic/Plenum Press.

Yirmiya, N., Pilowsky, T., Nemanov, L., Arbelle, S., Feinsilver, T., Fried, I., et al. (2001). Evidence for an association with the serotonin transporter promoter region polymorphism and autism. *American Journal of Medical Genetics, 105*, 381–386.

Zuckerman, M. (1994). *Behavioral expressions and biosocial bases of sensation seeking*. New York: Cambridge University Press.

20

Neuroticism and Serotonin: A Developmental Genetic Perspective

K. Peter Lesch

The expression of anxiety, which constitutes many facets of human personality and behavior, is influenced by a complex interaction of biological, psychological, and social variables. Even though individual differences in anxiety and its behavioral consequences are substantially heritable, they ultimately result from an interplay between genetic variations and environmental factors (i.e., epigenesis). After all, it is no longer controversial whether nature or nurture shapes human development and personality; the issue is how complex genetic and environmental factors interact in the formation and expression of a behavioral phenotype. Although genetic research has typically focused either on normal personality characteristics or on behavioral disorders, with few investigations evaluating the genetic and environmental relationship between the two, it is of critical importance to resolve the disagreement whether a certain quantitative trait etiopathogenetically influences the disorder or whether the trait is a psychopathological dimension of a disorder. This controversy increasingly has encouraged the pursuit of dimensional approaches to developmental and behavioral genetics, in addition to the traditional strategy of studying individuals with categorically defined psychiatric conditions.

A complementary approach to genetic studies of anxiety and related disorders in humans involves investigation of expression of genes with emphasis on expression regulation (i.e., construction of transcriptome maps) and their protein products (i.e., application of proteomics) implicated in the brain neurocircuitry of anxiety in animal models. Based on findings that genetically driven variability of expression and function of proteins that regulate the function of brain neurotransmitter systems (e.g., receptors, ion channels, transporters, and enzymes) is associated with personality and other behavioral traits, research is also giving emphasis to the molecular psychobiological basis of anxiety-related behaviors in rodents and, increasingly, nonhuman primates. For example, conditioning of anxiety involves pathways transmitting information to and from the amygdala to various neural networks that control the expression of aggressive or defensive reactions, including behavioral, autonomic nervous system, and stress hormone responses. While pathways from the thalamus and cortex (sensory and prefrontal) project to the amygdala, inputs are

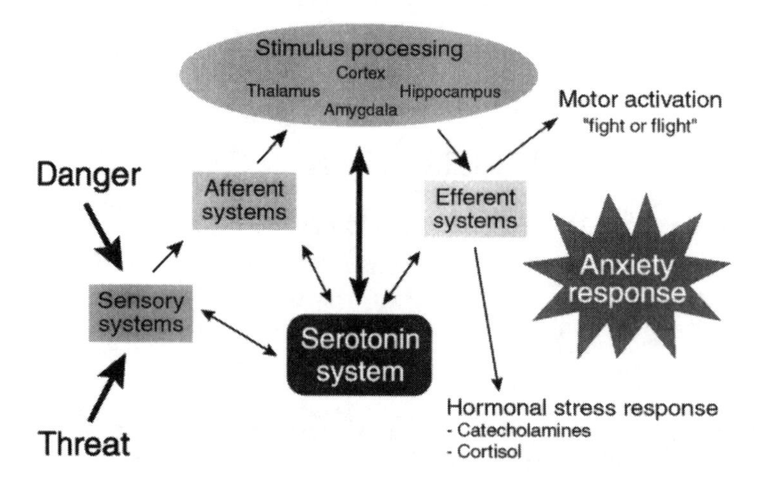

Figure 20.1. Functional neuroanatomy of anxiety-related responses.

processed within intra-amygdaloid circuitries and outputs are directed to the hippocampus, brain stem, hypothalamus, and other regions (Figure 20.1). Thus, the amygdala-associated neural network is critical to integration of the "fight or flight" response. Identification of molecular components of neural circuits involved in anxiety are currently leading to target genes of presumed pathophysiologic pathways, in addition to candidates that generally are derived from hypothesized pathogenetic mechanisms or from observations of therapeutic response.

Formation and integration of multiple neural networks depend on the actions of neurotransmitters, such as serotonin (5HT). The 5HT transporter (5HTT) fine-tunes serotonergic transmission as well as 5HT's morphogenetic actions in numerous projection fields throughout the brain, including regions crucial to emotional behavior, such as limbic and cortical areas involved in motor behavior, emotional experience, and memory. Consistent with this observation, brain 5HTT is the initial site of action of widely used 5HT reuptake inhibitor antidepressant and anxiolytic drugs. Based on converging lines of evidence that the 5HTT is a key player in the development of brain sensory systems, genetically driven variability of 5HTT function is likely to influence temperamental predispositions as well as phenotypic expression of behavioral traits, such as anxiety and aggressiveness.

This chapter reviews the genetics of anxiety-related traits with focus on the 5HTT, both in humans and nonhuman primates. Conceptual issues in the search for candidate genes for anxiety and in the generation of mouse models of human anxiety and related disorders are also discussed.

Anxiety-Related Traits and Serotonin

The analysis of genetic contributions to anxiety-related traits is difficult both conceptually and methodologically, so that consistent findings remain sparse. Because the contribution of a single locus to most personality traits is likely to be in the range 0.1%–5% (see Plomin, DeFries, Craig, & McGuffin, chapter 1, this volume) and because the power of linkage analysis to detect small gene effects is quite limited, at least with realistic samples, quantitative trait locus (QTL) research in humans has relied on association analysis using DNA variants in or near candidate genes that are functional. Gene variants with a significant impact on the functionality of components of brain monoamine neurotransmission, such as the 5HT system, are a rational beginning. The brainstem raphe 5HT system is the most widely distributed neurotransmitter system in the brain; serotonergic raphe neurons diffusely project to a variety of brain regions (e.g., cortex, hippocampus, and amygdala; Hensler, Ferry, Labow, Kovachich, & Frazer, 1994; Whitaker-Azmitia & Peroutka, 1990). In addition to its neurotransmitter role, 5HT is an important regulator of morphogenetic activities during early brain development as well as during adult neurogenesis and plasticity, including cell proliferation, migration, differentiation, and synaptogenesis (Azmitia & Whitaker-Azmitia, 1997; Gould, 1999; Lauder, 1990, 1993). In the brain of adult humans, nonhuman primates, and other mammals, 5HT neurotransmission is a major modulator of emotional behavior and integrates cognition, sensory processing, motor activity, and circadian rhythms including food intake, sleep, and reproductive activity. This diversity of physiological functions is due to the fact that 5HT orchestrates the activity and interaction of several other neurotransmitter systems. Although 5HT may be viewed as a master control neurotransmitter within this highly complex system of neural communication mediated by 14+ pre- and postsynaptic receptor subtypes with a multitude of isoforms and subunits, 5HT's action as a chemical messenger is primarily terminated by reuptake via the 5HTT.

The 5HTT plays a critical role in regulating serotonergic transmission in numerous projection fields throughout the brain, including regions such as limbic and cortical areas involved in motor behavior, emotional experience, and memory (Lesch, 1997). Consistent with this view, the brain 5HTT is the initial site of action of widely used 5HT reuptake inhibitor antidepressant and antianxiety drugs. Based on converging lines of evidence that 5HTT, as the prime modulator of central 5HT activity, is involved in myriad processes during brain development as well as synaptic plasticity in adulthood, genetically driven variability of 5HTT function is likely to influence behavioral traits (Lesch & Mössner, 1998). As a result, the contribution of a length variation of 44 base pairs in the 5′-flanking regulatory region of the gene encoding the human 5HTT (SLC6A4), the 5HTT gene-linked polymorphic region (5HTTLPR), to individual phenotypic differences in personality traits and the risk to develop behavioral disorders was explored in a remarkable number of population and family-based genetic studies.

5HTT Gene-Linked Polymorphic Region (5HTTLPR)

The human 5HTT is encoded by a single gene on chromosome 17q11 adjacent to the CPD and NF1 genes (Shen et al., 2000). It is composed of

14–15 exons spanning approximately 35 kb (Lesch et al., 1994). Transcriptional activity of the human 5HTT gene is modulated by a polymorphic repetitive element, the 5HTT gene-linked polymorphic region (5HTTLPR) located upstream of the transcription start site. Additional variations have been described in the 5′ untranslated region due to alternative splicing of exon 1B (Bradley & Blakely, 1997), in intron 2 (variable number of a 16/17-base pair tandem repeat, VNTR-17; Lesch et al., 1994), and in the 3′ untranslated region (Battersby et al., 1999). Comparison of different mammalian species confirmed the presence of the 5HTTLPR in platyrrhini and catarrhini (hominoids, cercopithecoids) but not in prosimian primates and other mammals (Lesch et al., 1997). Thus, the 5HTTLPR is unique to humans and simian primates. In humans, the majority of alleles are composed of either 14- or 16-repeat elements (short and long allele, respectively), although alleles with up to 22-repeat elements sometimes occur, as do variants with single-base insertions/deletions or substitutions within individual repeat elements. A predominantly White population displayed allele frequencies of 57% for the long (l) allele and 43% for the short (s) allele with a 5HTTLPR genotype distribution of 32% l/l, 49% l/s, and 19% s/s (Lesch et al., 1996); different allele and genotype distributions are found in other populations (Gelernter, Kranzler, & Cubells, 1997; Ishiguro et al., 1997; Kunugi et al., 1997).

The guanine- and cytosine-rich sequence of the 5HTTLPR is likely to give rise to the formation of DNA secondary structure that has the potential to regulate 5HTT gene transcription. The short and long 5HTTLPR variants differentially modulate transcriptional activity of the 5HTT gene promoter, 5HTT protein concentration, and 5HT uptake activity in human lymphoblastoid cell lines (Lesch et al., 1996). Membrane preparations from l/l lymphoblasts showed higher inhibitor binding than did s/s cells. Furthermore, the rate of specific 5HT uptake was more than twofold higher in cells homozygous for the l form of the 5HTTLPR than in cells carrying one or two copies of the s variant of the promoter. Further evidence from studies of 5HTT gene promoter activity in other cell lines (Mortensen, Thomassen, Larsen, Whittemore, & Wiborg, 1999), mRNA concentrations in the raphe complex of human postmortem brain (Little et al., 1998), platelet 5HT uptake and content (Greenberg et al., 1999; Hanna et al., 1998; Nobile et al., 1999), 5HT system responsivity elicited by pharmacologic challenge tests with clomipramine and fenfluramine (Reist, Mazzanti, Vu, Tran, & Goldman, 2001; Whale, Quested, Laver, Harrison, & Cowen, 2000), mood changes following tryptophan depletion (Neumeister et al., in press), and in vivo SPECT imaging of human brain 5HTT (Heinz et al., 1999) confirmed that the s variant is associated with lower 5HTT expression and function.

5HTTLPR and Genotype–Phenotype Correlations

The contribution of the 5HTTLPR to individual phenotypic differences in personality traits was explored in two independent population and family-based genetic studies. Anxiety-related and other personality traits were assessed by the NEO Personality Inventory–Revised (NEO–PI–R), a self-report inventory based on the five-factor model of personality ("Big Five"; Costa & McCrae, 1992). The five factors assessed are Neuroticism (emotional instability), Extraversion, Agreeableness (cooperation, reciprocal alliance formation), Openness (intellect, problem solving), and Conscientiousness (will to achieve).

In an initial study, population and within-family associations were discovered between the low-expressing s allele and Neuroticism, a trait related to anxiety, hostility, and depression, on the NEO–PI–R in a primarily male population (N = 505; Lesch et al., 1996). Individuals with either one or two copies of the short 5HTTLPR variant (Group S) had significantly greater levels of Neuroticism, defined as proneness to negative emotionality, including anxiety, hostility, and depression, than those homozygous for the long genotype (Group L) in the sample as a whole and also within sibships. Individuals with 5HTTLPR S genotypes also had significantly decreased NEO Agreeableness. In addition, the Group S individuals had increased scores on Anxiety on the separate 16-PF personality inventory, a trait related to NEO Neuroticism. In a more recent investigation, this association was also found in an independent sample (N = 397, 84% female, primarily sibpairs; Greenberg et al., 2000). The findings replicated the association between the 5HTTLPR and NEO Neuroticism. Combined data from the two studies (N = 902), which were corrected for ethnicity and age, gave a highly significant association between the s allele and anxiety-related traits both across individuals and within families, reflecting a genuine genetic influence rather than an artifact of ethnic admixture.

Another association encountered in the original study between the s allele and lower scores of NEO Agreeableness was also replicated and was even more robust in the primarily female replication sample. Gender-related differences in 5HTTLPR–personality trait associations are possible because several lines of evidence demonstrate gender-related differences in central 5HT system function in humans and in animals (Fink, Sumner, Rosie, Wilson, & McQueen, 1999; McQueen, Wilson, & Fink, 1997). These findings include effects of gonadal steroids on 5HTT expression in rodent brain and differences in anxiety-related behaviors in male and female 5HTT knock-out mice (Li, Wichems, Heils, Lesch, & Murphy, 2000).

Analysis of NEO subscales further defined the specific aspects of personality that were reproducibly associated with 5HTTLPR genotype. For the NEO Neuroticism subscales, the combined samples from Lesch et al. (1996) and Greenberg et al. (2000) demonstrated significant associations between 5HTTLPR S group genotypes and the facets of Depression and Angry Hostility, two of the three facets that showed the most significant associations in the initial sample (Lesch et al., 1996). In contrast, the NEO Neuroticism Anxiety facet, which was associated with genotype in the first study at a weaker significance level, was not significantly associated with 5HTTLPR genotype in the new cohort. With regard to the NEO Agreeableness subscales, the previously observed associations of 5HTTLPR S group genotypes with decreased Straightforwardness and Compliance were consistently replicated. It therefore appears more accurate to state that the 5HTTLPR is associated with traits of negative emotionality related to interpersonal hostility and depression. The relationship between these two aspects of negative emotionality is not unexpected in view of the previously observed negative correlation between Angry Hostility, a facet of Neuroticism, and Agreeableness (Costa & McCrae, 1992), indicating that both measures assess a behavioral predisposition toward uncooperative interpersonal behavior (Greenberg et al., 2000).

The effect sizes for the 5HTTLPR–personality associations, which were comparable in the two samples, indicate that this gene variation has only a modest influence on the behavioral predispositions of approximately 0.30 standard deviation units. This corresponds to 3%–4% of the total variance and 7%–9% of the genetic variance, based on estimates from twin studies using these and related measures that have consistently demonstrated that genetic factors contribute 40%–60% of the variance in neuroticism and other related personality traits. Although the genetic effect suggests that the 5HTTLPR represents an above-average sized QTL, the associations represent only a small portion of the genetic contribution to anxiety-related personality traits. If additional genes were hypothesized to contribute similar gene–dose effects to anxiety, at least 10–15 genes are predicted to be involved. Thus, the results are consistent with the view that the influence of a single, common polymorphism on continuously distributed traits is likely to be small in humans, and it possibly influences different quantitative characteristics in other species (Plomin, Owen, & McGuffin, 1994).

At first glance, association between the high-activity 5HTTLPR l allele with lower Neuroticism and related traits seems to be counterintuitive with regard to the known antidepressant and anxiolytic effects of 5HTT inhibitors (SRIs). Likewise, Knutson et al. (1998) reported that long-term inhibition of the 5HTT by the SRI paroxetine reduced indices of hostility through a more general decrease in negative affect, a personality dimension related to neuroticism. The same individuals also demonstrated an increase in directly measured social cooperation after paroxetine treatment, an interesting finding in view of the replicated association between 5HTTLPR genotype and agreeableness. That a drug which inhibits the 5HTT lessened negative emotionality and increased social cooperation appears to conflict with findings that the 5HTTLPR long allele, which confers higher 5HTT expression, is associated with lower NEO Neuroticism and higher NEO Agreeableness. However, this apparent contradiction may be due to the fact that both 5HT and 5HTT play critical roles in brain development that differ from their functions in regulating neurotransmission in the adult.

The conclusion that the 5HTT may affect personality traits via an influence on brain development and synaptic plasticity is strongly supported by recent findings in rodents and nonhuman primates. Studies in rats confirmed that the 5HTT gene is expressed in brain regions central to emotional behavior during fetal development but not later in life (Hansson,

Mezey, & Hoffman, 1998, 1999), hence enduring individual differences in personality could result from 5HTTLPR-driven differential 5HTT expression during pre- and perinatal life. In support of this notion, mice with a targeted disruption of the 5HTT display enhanced anxiety-related behaviors in models of avoidance and anxiety (also see below; Wichems et al., 2000).

In subsequent studies, only several attempts to replicate the original finding have been reported using large populations (Jorm et al., 1998; Mazzanti et al., 1998; see Table 20.1). Two of the three available large studies found evidence congruent with an influence of the 5HTTLPR on Neuroticism and related traits (Lesch et al., 1996; Mazzanti et al., 1998), whereas a large population study not using a within-family design (Jorm et al., 1998) did not. Smaller population-based studies have had variable but generally negative results. Interpretation of these studies is complicated by their use of relatively small or unusual samples and the lack of within-family designs that may minimize population stratification artifacts (for a review of the requirement of family-based approaches, see also Cardon, chapter 4, this volume). Other efforts to detect associations between the 5HTTLPR and personality traits have been complicated by the use of small sample sizes, heterogeneous subject populations, and differing methods of personality assessment (see Table 20.1). Furthermore, the subject selection in two studies (Ball et al., 1997; Deary et al., 1999) was unusual in that subjects at the high or low ends of the distribution for Neuroticism were selected on the assumption that the 5HTTLPR affects the trait uniformly across its distribution. However, reanalysis of the data from the initial study revealed that the contribution of the 5HTTLPR to NEO Neuroticism is greatest in the central range of the distribution and actually decreases at the extremes (Sirota, Greenberg, Murphy, & Hamer, 1999). This illustrates the need to obtain genotypes from individuals across the distribution of a trait and suggests caution in the use of extreme populations in attempting to establish genetic influences on traits that are continuously distributed in the population.

Finally, difficulties in interpretation of population-based association studies due to ethnic differences in 5HTTLPR allele frequencies, which are most prominent in the Japanese population with l/l genotype frequencies around 5%, have also been raised by more recent studies (Katsuragi et al., 1999; Kumakiri et al., 1999). From an evolutionary psychological perspective, anxiety is a pervasive and innately driven form of distress that arises in response to actual or threatened exclusion from social groups (Baumeister & Tice, 1990; Buss, 1991). Notably, Nakamura et al. (1997) discussed the higher prevalence of the anxiety and depression-related l/s and s/s genotypes in the context of extraordinary emotional restraint and interpersonal sensitivity in Japanese people as a possible population-typical adaptation to prevent social exclusion (Ono et al., 1996).

Mouse Models of Anxiety-Related Traits

Recent advances in transgenics and gene targeting (constitutive or conditional knock-out and knock-in techniques) are increasingly affecting our

Table 20.1. Overview of Published Data on the Association Between the 5HT Transporter Gene-Linked Polymorphic Region (5HTTLPR) and Anxiety-Related Traits in General Populations and Patient Samples

Study	*N*	Population and study design	Trait assessment/ inventory	Association of s allele, *p*	Family-based studies, *p*
Lesch et al. (1996)	505	General, American, males, two subsamples	NEO–PI–R, 16-PF	.02–.002	.03–.004
Ebstein et al. (1997)	120	General, Israeli	TPQ	.75, *ns*	—
Ball et al. (1997)	106	General, German, 5% of highest vs. lowest *N* scores	NEO–FFI	.89, *ns*	—
Nakamura et al. (1997)	203	General, Japanese, females	NEO–PI, TCI	s increased, l is rare	—
Mazzanti et al. (1998)	655	Controls, alcoholic violent offenders, Finnish	TPQ	*ns*	.45, but .003 for HA1 + HA2
Ebstein et al. (1998)	81	General, Israeli, neonates (2 weeks old)	NBAS	Interaction with DRD4[a]	—
Ricketts et al. (1998)	84	Controls, Parkinson's disease, American	TPQ	.04–.003	—
Gelernter et al. (1998)	185/322	Controls, substance dependence, personality disorder, American	NEO–PI TPQ	.47, *ns* .87, *ns*	—
Jorm et al. (1998)	759	General, European, Australian	EPQ–R	*ns*	—
Serretti et al. (1999)	132	Depression, bipolar disorder	HDRS, anxiety	.01	—
Murakami et al. (1999)	189	General, Japanese	SRQ–AD	.05	—

Flory et al. (1999)	225	General, American	NEO–PI–R	.97		—
Kumakiri et al. (1999)	144	General, Japanese	NEO–PI–R, TPQ	*ns*		—
Katsuragi et al. (1999)	101	General, Japanese	TPQ	.007		—
Auerbach et al. (1999)	76	General, Israeli, infants (2 months old), follow-up of Ebstein et al. (1998)	IBQ	.05–.005, interaction with DRD4[a]		—
Menza et al. (1999)	32	Parkinson's disease	HDRS	.05		—
Deary et al. (1999)	204	General, Scottish, 20% of highest vs. lowest *N* scores	NEO–FFI	*ns*		—
Gustavsson et al. (1999)	305	General, Swedish	KSP	*ns*		—
Osher et al. (1999)	148	General, Israeli	NEO–PI–R	—	.06	
			TPQ		.07	
Greenberg et al. (2000)	397	General, American, females	NEO–PI–R, 16-PF	.03–.01	.01	
Melke et al. (2001)	251	General, Swedish, females, same age	KSP	.002–.04		—
Du et al. (2000)	186	General, Canadian, males	NEO–PI–R	.018		—

Note. NEO–PI–R and NEO–FFI = Costa and McCrae's NEO Personality Inventory–Revised and Five-Factor Index; TPQ/TCI = Cloninger's Tridimensional Personality Questionnaire/Temperament and Character Inventory; 16-PF = Catell's 16 Personality Factor Inventory; EPQ–R = Eysenck Personality Questionnaire; KSP = Karolinska Scales of Personality; SRQ–AD = Self-Rating Questionnaire for anxiety and depression; HDRS = Hamilton Depression Rating Scale; IBQ = Rothbart's Infant Behavior Questionnaire; NBAS = Brazelton's Neonatal Assessment Scale. Dashes indicate values not determined.

[a]Dopamine receptor D4.

understanding of the neurobiologic basis of anxiety-related behavior in mice (Lesch & Mössner, 1999; Rudolph & Mohler, 1999). Mouse models make it possible to control and to manipulate the environment, which will facilitate elucidation of gene–gene and gene–environment interactions (Lesch, in press-a). In addition, mice provide practical models to study the impact of genetic mechanisms on development and plasticity of the brain, including regionalization and connectivity of the cerebral cortex as well as subcortical structures. Mouse models may therefore be essential for the dissection of the neurodevelopmental–behavioral genetic interface (Lesch, in press-b).

The design of a mouse model partially or completely lacking a gene of interest during all stages of development (constitutive knock out) is among the prime strategies directed at elucidating the role of genetic factors in anxiety. Within the serotonergic gene pathway, a possible role for the $5HT_{1A}$ receptor in the modulation of anxiety (and depression) as well as in the mode of action of anxiolytic and antidepressant drugs has been suspected for many years. $5HT_{1A}$ receptors operate both as somatodendritic autoreceptors and as postsynaptic receptors. Somatodendritic $5HT_{1A}$ autoreceptors are located predominantly on 5HT neurons and dendrites in the midbrain raphe complex, and their activation by 5HT or $5HT_{1A}$ agonists decreases the firing rate of serotonergic neurons and subsequently reduces the release of 5HT from nerve terminals. Postsynaptic $5HT_{1A}$ receptors are distributed widely in forebrain regions that receive serotonergic input, notably in the cortex, hippocampus, and hypothalamus. Their activation results in neuronal inhibition, the consequences of which are not well understood, and in physiological responses that depend on the function of the target cells.

A recent series of papers reported the generation of mice with a targeted inactivation of the $5HT_{1A}$ receptor (Heisler et al., 1998; Parks, Robinson, Sibille, Shenk, & Toth, 1998; Ramboz et al., 1998). $5HT_{1A}$ receptor knock-out mice display a spontaneous phenotype that is associated with a gender-modulated and gene/dose-dependent increase of anxiety-related behaviors. With the exception of an enhanced sensitivity of terminal $5HT_{1B}$ receptors and downregulation of gamma-aminobutyric acid (GABA-A) receptors, no major neuroadaptational changes were detected. Activation of presynaptic $5HT_{1A}$ receptors provides the brain with an autoinhibitory feedback system controlling 5HT neurotransmission. Thus, enhanced anxiety-related behavior most likely represents a consequence of increased terminal 5HT availability resulting from the lack or reduction in presynaptic somatodendritic $5HT_{1A}$ autoreceptor negative feedback function. Indirect evidence for increased presynaptic serotonergic activity is provided by the compensatory upregulation of terminal $5HT_{1B}$ receptors. This mechanism is also consistent with recent theoretical models of anxiety that are primarily based on pharmacologically derived data. Stimulation of postsynaptic $5HT_{1A}$ receptors has been proposed to elicit anxiogenic effects, whereas activation of $5HT_{1A}$ autoreceptors is thought to induce anxiolytic effects via suppression of serotonergic neuronal firing, resulting in attenuated 5HT release in limbic terminal fields.

Exaggerated serotonergic neurotransmission also may be implicated in increased anxiety-related behaviors recently found in 5HTT-deficient mice (Wichems et al., 2000). Mice with a disrupted 5HTT gene have been suggested as an alternative model to pharmacological studies of selective serotonin reuptake inhibitor (SSRI)-evoked antidepressant and anxiolytic mechanisms to assess the hypothesized association between 5HT uptake function and $5HT_{1A}$ receptor desensitization (Bengel et al., 1998). Excess serotonergic neurotransmission in mice lacking 5HTT results in desensitized and downregulated $5HT_{1A}$ receptors in the midbrain raphe complex but not in the hippocampus (Li et al., 2000) and is suspected to play a role in the increased anxiety-related behaviors in these mice using the light-dark and elevated-zero or plus-maze paradigms. These findings are consistent with the notion that increased 5HT availability may contribute to anxiety in rodents, and the studies reporting that anxiety-related traits in humans are associated with allelic variation of 5HTT function.

Cortical Development in 5HTT-Deficient Mice

Developmental genetics has made considerable progress in the elucidation of the molecular events regulating the generation of distinct neuronal cell types at precise locations and in appropriate numbers in the neural tube that is separated into transverse and longitudinal domains at early embryonic stages. The differentiation of cortical areas has been suggested to be controlled by an interplay of intrinsic genetic programs (e.g., sequential activation of transcription factor and expression of cell adhesion molecules) and extrinsic mechanisms including those mediated by glutamatergic thalamocortical afferents (Nakagawa, Johnson, & O'Leary, 1999).

Investigations of 5HT participation in neocortical development and plasticity have been concentrated on the rodent somatosensory cortex (SSC; see Figure 20.2), due to its one-to-one correspondence between each whisker and its cortical barrel-like projection area. The timing of serotonergic innervation coincides with pronounced growth, and synaptogenesis in the SSC and perinatal manipulations of 5HT affects cortical 5HT receptors. The processes underlying patterning of projections in the SSC have been intensively studied with a widely held view that the formation of somatotopic maps does not depend on neural activity (Katz & Shatz, 1996). Whereas pharmacologically induced 5HT depletion at birth yields smaller barrels but does not prevent the formation of the barrel pattern itself (Bennett-Clarke, Leslie, Lane, & Rhoades, 1994; Osterheld-Haas, Van der Loos, & Hornung, 1994), excess of extracellular 5HT, as demonstrated in MAOA knock out, results in a complete absence of cortical barrel patterns (Cases et al., 1996). Additional evidence for a role of 5HT in the development of neonatal rodent SSC derives from the transient barrel-like distribution of 5HT, $5HT_{1B}$, and $5HT_{2A}$ receptors and of the 5HTT (Lebrand et al., 1996; Mansour-Robaey, Mechawar, Radja, Beaulieu, & Descarries, 1998). The transient barrel-like 5HT pattern visualized in layer IV of the SSC of neonatal rodents apparently stems from 5HT uptake and vesicular

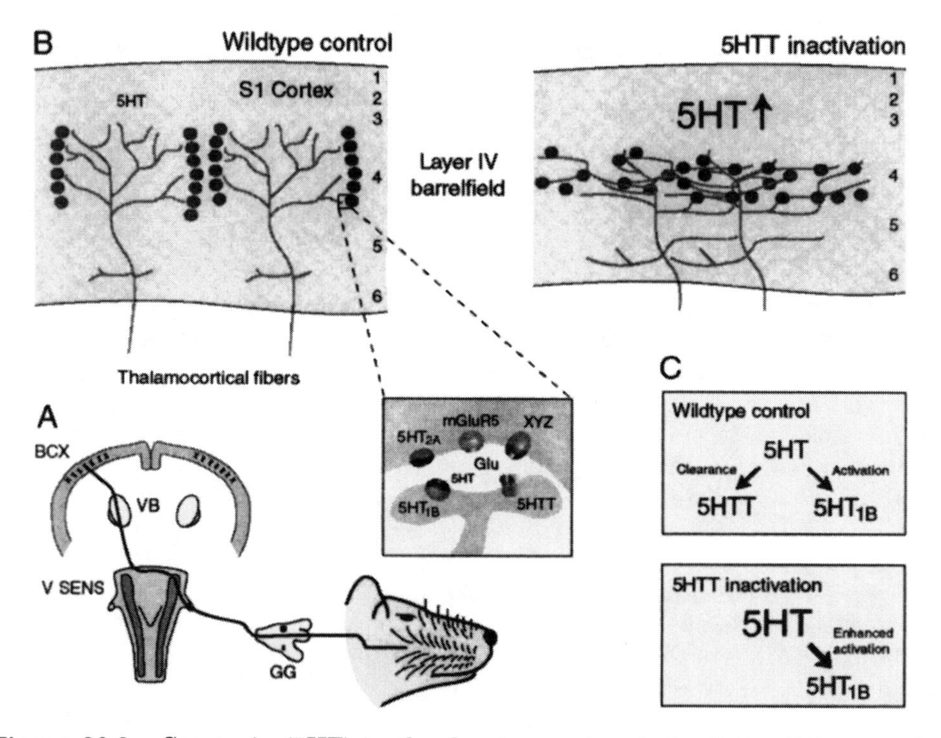

Figure 20.2. Serotonin (5HT) in the development and plasticity of the somatosensory cortex (SSC) in rodents. A: One-to-one correspondence between each whisker and its cortical barrel-like projection area (GG = ganglion gasseri; V SENS = sensory nerve V; VB = ventrobasal thalamus; BCX = barrelfield cortex). B: SSC in serotonin transporter (5HTT) knock-out mice versus wild-type controls ($5HT_{1B/2A}$ = serotonin receptor subtypes; Glu = glutamate; mGluR5 = meatobotropic glutamate receptor 5; XYZ = unknown receptors). C: Two key players mediate the deleterious effects of excess 5HT: the 5HTT and the $5HT_{1B}$ receptor. Both molecules are expressed in primary sensory thalamic nuclei during the period when the segregation of thalamocortical projections occurs. 5HT is internalized via 5HTT in thalamic neurons, stored in axon terminals, and used as a cotransmitter of glutamate. Severe impairment of 5HT clearance in mice with an inactivation of the 5HTT results in an accumulation of 5HT and overstimulation of 5HT receptors all along thalamic neurons. Because $5HT_{1B}$ receptors inhibit the release of glutamate in the thalamocortical somatosensory pathway, excessive activation of $5HT_{1B}$ receptors prevents activity-dependent processes involved in the patterning of afferents and barrel structures.

storage in thalamocortical neurons, which express both the 5HTT and the vesicular monoamine transporter (VMAT2) at this developmental stage (Lebrand et al., 1996).

Inactivation of the 5HTT gene profoundly disturbs formation of the SSC with altered cytoarchitecture of cortical layer IV, the layer that contains synapses between thalamocortical terminals and their postsynaptic target neurons (see Figure 20.2; Persico et al., 2001). Brains of 5HTT knock-out mice display no or only very few barrels. Cell bodies as well as terminals, typically more dense in barrel septa, appear homogeneously distributed in layer IV of adult 5HTT knock-out brains. Injections of a 5HT synthesis inhibitor within a narrow time window of 2 days postnatally completely rescued formation of SSC barrel fields. Of note, heterozygous knock-out mice develop all SSC barrel fields but frequently present irregularly shaped barrels and less defined cell gradients between septa and barrel hollows. These findings demonstrate that excessive concentrations of extracellular 5HT are deleterious to SSC development and suggest that transient 5HTT expression in thalamocortical neurons is responsible for barrel patterns in neonatal rodents, and its permissive action is required for normal barrel pattern formation, presumably by maintaining extracellular 5HT concentrations below a critical threshold. Because normal synaptic density in SSC layer IV of 5HTT knock-out mice was shown, it is more likely that 5HT affects SSC cytoarchitecture by promoting dendritic growth toward the barrel hollows as well as by modulating cytokinetic movements of cortical granule cells, similar to concentration-dependent 5HT modulation of cell migration described in other tissues. Because the gene–dose dependent reduction in 5HTT availability in heterozygous knock-out mice, which leads to a modest delay in 5HT uptake but distinctive irregularities in barrel and septum shape, is similar to that reported in humans carrying low-activity alleles of the 5HTTLPR, allelic variation in 5HTT function also affects the human brain during development, with subsequent consequences for disease liability and therapeutic response.

These findings demonstrate that excessive amounts of extracellular 5HT are detrimental to SSC development and suggest that transient 5HTT expression and its permissive action are required for barrel pattern formation, presumably by maintaining extracellular 5HT concentrations below a critical threshold. Two key players of serotonergic neurotransmission appear to mediate the deleterious effects of excess 5HT: the 5HTT and the $5HT_{1B}$ receptor. Both molecules are expressed in primary sensory thalamic nuclei during the period when the segregation of thalamocortical projections occurs (Bennett-Clarke, Chiaia, & Rhoades, 1996; Hansson et al., 1998; Lebrand et al., 1996). 5HT is internalized through 5HTT in thalamic neurons and is detectable in axon terminals (Cases et al., 1998; Lebrand et al., 1996). The presence of VMAT2 within the same neurons allows internalized 5HT to be stored in vesicles and used as a cotransmitter of glutamate. Lack of 5HT degradation in MAOA knock-out mice as well as severe impairment of 5HT clearance in mice with an inactivation of the 5HTT results in an accumulation of 5HT and overstimulation of 5HT receptors all along thalamic neurons (see Figure 20.2; Cases et al., 1998). Because $5HT_{1B}$ receptors are known to inhibit the release of glutamate in the thalamocortical somatosensory pathway, excessive activation of $5HT_{1B}$ receptors could prevent activity-dependent processes involved in the patterning of afferents and barrel structures. This hypothesis is supported by a recent study using a strategy of combined knock out of MAOA, 5HTT, and $5HT_{1B}$ receptor genes (Salichon et al., 2001). Whereas only partial disruption of the patterning of somatosensory thalamocortical projections was observed in 5HTT knock out, MAOA/5HTT double knock-out mice showed that 5HT accumulation in the extracellular space causes total disruption of the patterning of these projections. Moreover, the removal of $5HT_{1B}$ receptors in MAOA and 5HTT knock-out as well as in MAOA/5HTT double knock-out mice allows a normal segregation of the somatosensory projections. These findings point to an essential role of the $5HT_{1B}$ receptor in mediating the deleterious effects of excess 5HT in the somatosensory system.

The evidence that changes in 5HT system homeostasis exert long-term effects on cortical development and adult brain plasticity may be an important step forward in establishing the psychobiological groundwork for a neurodevelopmental hypothesis of neuroticism, negative emotionality, and anxiety disorders. Although there is converging evidence that serotonergic dysfunction contributes to anxiety-related behavior, the precise mechanism that renders $5HT_{1A}$ receptor and 5HTT-deficient mice more anxious remains to be elucidated. While increased 5HT availability and activation of other serotonergic receptor subtypes that have been shown to mediate anxiety (e.g., $5HT_{2C}$ receptor) may contribute to increased anxiety in rodent models, multiple downstream cellular pathways or neurocircuits, including noradrenergic, GABAergic, glutamatergic, and peptidergic transmission, as suggested by overexpression or targeted inactivation of critical genes within these systems, have been implicated as participants in the processing of this complex behavioral trait. Recent work has therefore been focused on a large number of genes that have known relevance in the neurocircuitries of anxiety, although the knock out of some genes that appear not directly to be involved in anxiety may also lead to an anxiety-related phenotype (for a review, see Lesch, in press-a).

5HTT Gene–Gene Interaction

Because a considerable number of genes appear to contribute to the phenotypic expression of anxiety-related behavior, dissection of gene–gene interaction (i.e., epistasis) in the development of personality and behavioral traits is a pertinent and exciting avenue of research (see Figure 20.3). Moreover, in the evaluation of complex genetic effects, it seems to be essential to control for environmental factors. Recent studies have therefore focused on the human neonatal period, a time in early development when environmental influences may be minimal and least likely to confound associations between temperament and genes. In this context, the term *temperament* is used to refer to the psychological qualities of infants that display considerable variation and have a relatively stable biological basis in the individual's genotype, even though different phenotypes may emerge as the child grows (Kagan, 1989; Kagan, Reznick, & Snidman, 1988).

Ebstein and colleagues (Ebstein et al., 1998) investigated the behavioral effects of the VNTR polymorphism in exon 3 of the dopamine D4 receptor (DRD4), which had previously been linked to the Tridimensional Personality Questionnaire (TPQ) personality trait of Novelty Seeking

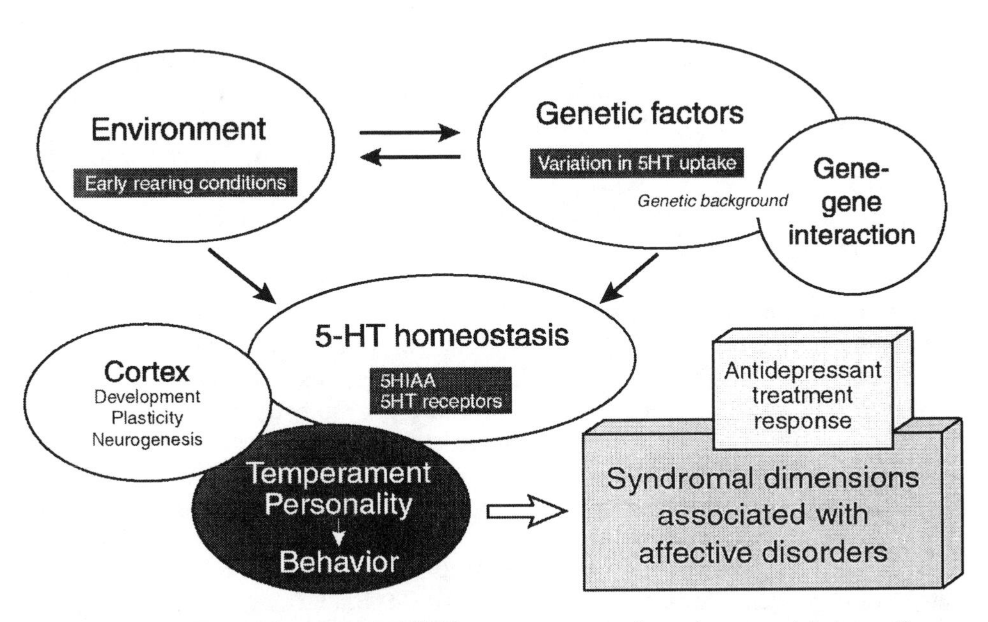

Figure 20.3. The influence of 5HTT gene–gene and environment interaction on central 5HT homeostasis in humans and nonhuman primates and its consequences for cortical plasticity, personality, behavior, and the syndromal dimensions as well as the treatment of affective disorders.

(Benjamin et al., 1996; Ebstein et al., 1996) and of 5HTTLPR, which seems to influence NEO Neuroticism and TPQ Harm Avoidance, in 2-week-old neonates. Neonate temperament and behavior were assessed using the Brazelton Neonatal Behavior Assessment Scale (NBAS). In addition to a significant association of the DRD4 polymorphism across four behavioral clusters relevant to temperament (orientation, motor organization, range of state, and regulation of state), an interaction was observed between the DRD4 polymorphism and 5HTTLPR. The presence of the 5HTTLPR s/s genotype decreased the orientation score for the group of neonates lacking the long variant of DRD4. The DRD4 polymorphism-5HTTLPR interaction was also assessed in a sample of adult subjects. There was no significant effect of the L-DRD4 genotype in those subjects homozygous for 5HTTLPR, whereas in the group without the homozygous genotype the effect of L-DRD4 was significant and represented 13% of the variance in novelty seeking scores between groups.

Temperament and behavior of the infants were reexamined at 2 months using Rothbart's Infant Behavior Questionnaire (IBQ) as well as at 1 year using a structured play situation and an information-processing task (Auerbach et al., 1999; Auerbach, Benjamin, Faroy, Geller, & Ebstein, 2001). There were significant negative correlations between neonatal orientation and motor organization as measured by the NBAS at 2 weeks and negative emotionality, especially distress in daily situations, at 2 months of age. Furthermore, grouping of the infants by DRD4 polymorphism and 5HTTLPR revealed significant main effects for negative emotionality and distress. Infants with long DRD4 alleles had lower scores on negative emotionality and distress to limitations than infants with short DRD4 alleles. In contrast, infants homozygous for the 5HTTLPR s allele had higher scores on negative emotionality and distress than infants with the l/s or l/l genotypes. Infants with the s/s genotype who also were lacking the novelty seeking-associated long DRD4 alleles showed most negative emotionality and distress, traits that possibly contribute to the predisposition for adult neuroticism.

In addition, an interaction between the 5HTT gene and the catechol-O-methyltransferase (COMT) gene has been shown to influence the trait of Persistence (RD2), a subscale of TPQ major personality factor Reward Dependence. Persistence is considered to be highly adaptive in the presence of stable intermittent reward patterns. Persistent individuals are eager, ambitious, and determined overachievers (Cloninger, 1994). In the presence of COMT val/val or met/met homozygosity, the low-function 5HTTLPR s allele significantly raises TPQ RD2 scores, including perfectionism (Benjamin, Osher, Lichtenberg, et al., 2000). Benjamin and associates (Benjamin, Osher, Kotler, et al., 2000) also investigated the interaction of three different gene variants (DRD4, 5HTT, and COMT) and TPQ personality factors. In the absence of the 5HTTLPR s allele and in the presence of the high-enzyme activity COMT val/val genotype, novelty seeking scores were higher in the presence of the DRD4 7-repeat allele. Because analysis of a large number of epistatic interactions is anticipated to bear a considerable potential of false-positive findings that may lead to meaningless conclusions, gene–gene interaction analyses should at this stage be limited to polymorphisms known to be functional, preferentially within a single neural circuit, such as the 5HT system (see Figure 20.4), or demonstrated to be associated with the behavioral phenotype of interest.

Gene–Environment Interaction in a Nonhuman Primate Model

Human and nonhuman primate behavior are similarly affected by deficits in 5HT function. In rhesus monkeys (*Macacca mulatta*), 5HT turnover, as measured by cerebrospinal (CSF) 5-hydroxyindoleacetic acid (5HIAA) concentrations, has a strong heritable component and is traitlike, with demonstrated stability over an individual's life span (Higley, Suomi, & Linnoila, 1991, 1992; Higley et al., 1993; Kraemer, Ebert, Schmidt, & McKinney, 1989). Moreover, CSF 5-HIAA concentrations are subject to the long-lasting influence of deleterious events early in life as well as by situational stressors. Monkeys separated from their mother and reared in absence of conspecific adults (peer-reared) have altered serotonergic function and exhibit behavioral deficits throughout their lifetimes when compared with their mother-reared counterparts.

Comparison of different mammalian species indicates that the 5HTTLPR is unique to humans and simian primates. In hominids, all alleles originate from variation at a single site (polymorphic locus 1, PL1), whereas an alternative location for a 21-bp length variation (PL2) was

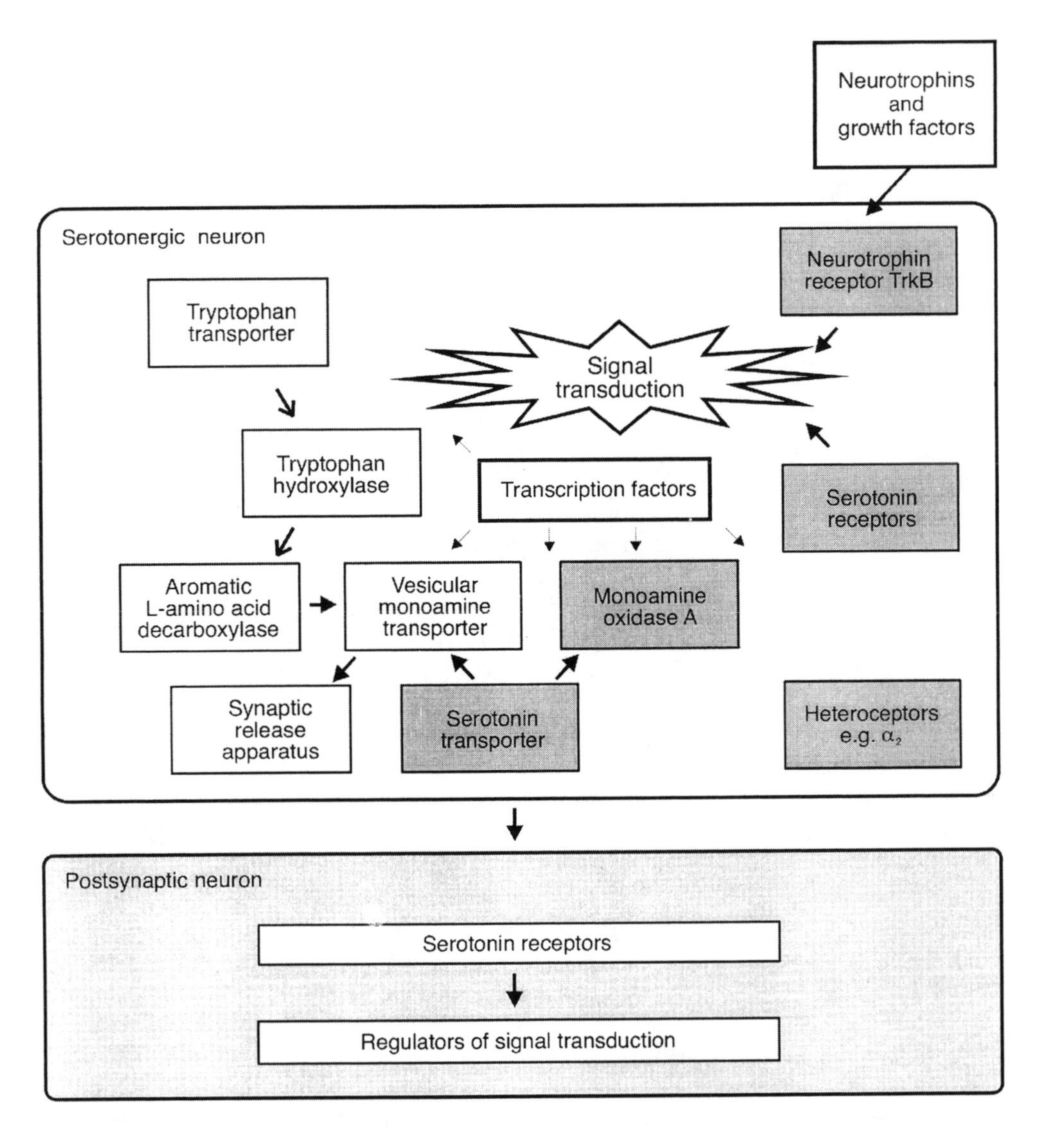

Figure 20.4. Serotonergic gene pathways, including different components of serotonergic system development, plasticity, and function currently under behavioral genetic investigation.

found in the 5HTTLPR of rhesus monkeys (rh5HTTLPR; Lesch et al., 1997). The presence of an analogous rh5HTTTLPR and resulting allelic variation of 5HTT activity in rhesus monkeys provides a unique model to dissect the relative contribution of genes and environmental sources to central serotonergic function and related behavioral outcomes.

To study genotype–environment interaction, researchers tested the association between central 5HT turnover and rh5HTTLPR genotype in rhesus monkeys with well-characterized environmental histories (Higley et al., 1998). This sample of rhesus monkeys (N = 177) showed allele frequencies of 83% for the 24-repeat allele (long, l) and 16% for the 23-repeat (short, s) rh5HTTLPR variant with a genotype distribution of 68% l/l, 31% l/s, and 1% s/s; one individual was heterozygous for a rare allele with an additional repeat unit (xl) similar to longer alleles found occasionally in humans (Lesch et al., 1997). The monkeys' rearing fell into one of the following categories: mother-reared (either reared with the biological mother or cross-fostered) or peer-reared (either with a peer group of 3–4 monkeys or with an inanimate surrogate and daily contact with a play-group of peers). Peer-reared monkeys were separated from their mothers, placed in the nursery at birth, and given access to peers at 30 days of age either continuously or during daily play sessions. Mother-reared and cross-fostered monkeys remained with the mother, typically within a social group. At roughly 7 months of age, mother-reared monkeys were weaned and placed together with their peer-reared cohort in large, mixed-gender social groups.

Because the monkey population constituted two groups that received dramatically different social and rearing experience early in life, the interactive effects of environmental experience and the rh5HTTLPR on cisternal CSF 5HIAA levels and 5HT-related behavior were assessed. CSF 5HIAA concentrations were significantly influenced by genotype for peer-reared but not for mother-reared subjects (Bennett et al., 2002). Peer-reared rhesus monkeys with the low-activity rh5HTTLPR s allele had significantly lower concentrations of CSF 5HIAA than their homozygous l/l counterparts. Low 5HT turnover in monkeys with the s allele is congruent with in vitro studies that show reduced binding and transcriptional efficiency of the 5HTT gene associated with the 5HTTLPR s allele (Heils et al., 1996; Lesch et al., 1996). This suggests that the rh5HTTLPR genotype is predictive of CSF 5HIAA concentrations but that early experiences make unique contributions to variation in later 5HT functioning. This finding provides evidence for an environment-dependent association between a 5HTT gene variation and a direct measure of 5HT functioning, cisternal CSF 5HIAA concentration, thus revealing an interaction between rearing environment and rhHTTLPR genotype. Similar to the 5HTTLPR's influence on NEO Neuroticism in humans, however, the effect size is small, with 4.7% of variance in CSF 5HIAA accounted for by the rh5HTTLPR and rearing environment interaction.

Previous work has shown that monkeys' early experiences have long-term consequences for the functioning of the central 5HT system, as indicated by robustly altered CSF 5HIAA levels, as well as anxiety and depression-related behavior, in monkeys deprived of their parents at birth and raised only with peers (Higley et al., 1991; Higley, Suomi, et al., 1992; Kraemer et al., 1989). Intriguingly, the biobehavioral results of deleterious early experiences of social separation are consistent with the notion that the 5HTTLPR may influence the risk for affective spectrum disorders. Evolutionary preservation of two prevalent 5HTTLPR variants and the resulting allelic variation in 5HTT expression may be part of the genetic mechanism resulting in the emergence of temperamental traits that facilitate adaptive functioning in the complex social worlds most primates inhabit. The uniqueness of the 5HTTLPR among humans and simian nonhuman primates, but not among prosimians or other mammals, along with the role 5HT plays in complex primate sociality, form the basis for the hypothesized relationship between the 5HTT function and personality traits that mediate individual differences in social behavior. Accumulating evidence demonstrates the complex interplay between individual differences in the central 5HT system and social success. In monkeys, lowered 5HT functioning, as indicated by decreased CSF 5HIAA levels, is associated with lower rank within a social group, less competent social behavior, and greater impulsive aggression (Higley et al., 1996; Higley, Mehlman, et al., 1992; Mehlman et al., 1994, 1995). It is well established that although subjects with low CSF 5HIAA concentrations are no more likely to engage in competitive aggression than other monkeys, when they do engage in aggression it frequently escalates to violent and hazardous levels. A presumed role of s allele-dependent low 5HTT function in nonhuman primate aggressive behavior would be in agreement with the association in humans between NEO subscales Neuroticism (increased angry hostility) and Agreeableness (decreased compliance = increased aggressiveness and hostility) and 5HTTLPR S genotypes.

As the scope of human studies has been extended to the neonatal period, a time in early development when environmental influences are modest and least likely to confound gene–temperament associations, complementary approaches have recently been applied to nonhuman primates. Rhesus macaque infants heterozygous for the s and l variant of the rh-HTTLPR (l/s) displayed higher behavioral stress reactivity compared with infants homozygous for the long variant of the allele (l/l; Champoux et al., 1999). Mother-reared and peer-reared monkeys were assessed on Days 7, 14, 21, and 30 of life, on a standardized primate neurobehavioral test designed to measure orientation, motor maturity, reflex functioning, and temperament. Main effects of genotype, and, in some cases, interactions between rearing condition and genotype, were demonstrated for items indicative of orienting, attention, and temperament. In general, heterozygote animals demonstrated diminished orientation, lower attentional capabilities, and increased affective responding relative to l/l homozygotes. However, the genotype effects were more pronounced for animals raised in the neonatal nursery than for animals reared by their mothers. These results demonstrate the contributions of rearing environment and genetic background, and their interaction, in a nonhuman primate model of behavioral development.

Another well-defined behavioral pattern among nonhuman primates that seems to be directly related to CSF 5HIAA concentrations is dispersal from the natal group (Mehlman et al., 1995). Most male rhesus monkeys leave their natal group and either visit or join other social groups or form small transient all-male groups before returning to their birth group. Although natal dispersal is occurring at a highly variable age and is almost always associated with loss of social status and an increase in stress, injury, and mortality, the cause and intention remain controversial. Interestingly, a recent study by Trefilov and colleagues (Trefilov, Berard, Krawczak, & Schmidtke, 2000) showed a gene–dose effect of the rh5HTTLPR s variant on the age of dispersal, with s/s homozygotes leaving earlier than carriers of the l allele. This finding further supports the notion that impaired 5HTT function resulting in low 5HT turnover is associated with impulsive behavior together with a high tendency toward risk-taking activity that leads to early dispersal.

Taken together, these findings provide evidence of an environment-dependent association between allelic variation of 5HTT expression and central 5HT function (see Figure 20.3) and illustrate the possibility that specific genetic factors play a role in 5HT-mediated social competence in higher order primates. The objective of further studies will be the elucidation of the relationship between the rh5HTTLPR genotype and sociability in monkeys, as this behavior is expressed with characteristic individual differences both in daily life and in response to challenge. Because rhesus monkeys exhibit temperamental and behavioral traits that parallel anxiety, depression, and aggression-related personality dimensions associated in humans with the low-activity 5HTTLPR variant, it may be possible to search for evolutionary continuity in this genetic mechanism for individual differences. Nonhuman primate studies also may be useful to help identify environmental factors that either compound the vulnerability conferred by a particular genetic makeup or, conversely, act to improve the behavioral outcome associated with that genotype.

5HTTLPR and Syndromal Dimensions of Mood Disorders

A notable number of studies have explored the role of allelic variation in 5HTT function in various neuropsychiatric disorders, including mood disorders, panic disorder, obsessive–compulsive disorder, attention deficit hyperactivity disorder, autism, schizophrenia, Alzheimer's disease, and substance abuse including alcoholism (for a review, see Lesch, 2001).

A growing body of evidence implicates personality traits, such as neuroticism or negative emotionality, in the comorbidity of affective spectrum disorders (Kendler, Kessler, Neale, Heath, & Eaves, 1993; Livesley, Jang, & Vernon, 1998). The paradigmatic separation of mental illness from personality disorders in current consensual diagnostic systems remarkably enhanced interest in the link between temperament, personality, and psychiatric disorders as well as in the involvement of this interrelationship to the heterogeneity within diagnostic entities, prediction of long-term course, and treatment response (Mulder, Joyce, & Cloninger, 1994). Based on multivariate genetic analyses of comorbidity, generalized anxiety disorder and major depression have common genetic origins, and the phenotypic differences between anxiety and depression are dependent on the environment (Kendler, 1996; Kendler, Neale, Kessler, Heath, & Eaves, 1992a, 1992b). Moreover, indexed by the personality scale of neuroticism, general vulnerability overlaps genetically to a substantial extent with both anxiety and depression (Kendler, Kessler, et al., 1993; Kendler, Neale, Kessler, Heath, & Eaves, 1993). These results predicted that when a QTL, such as the 5HTTLPR, is found for neuroticism, the same QTL should be associated with symptoms of anxiety and depression. Anxiety and mood disorders are therefore likely to represent the extreme end of variation in negative emotionality (Eley & Plomin, 1997; Eley & Stevenson, 1999). The genetic factor contributing to the extreme ends of dimensions of variation commonly recognized as a disorder may be quantitatively, not qualitatively, different from the rest of the distribution. Although it has been challenged recently (Sirota et al., 1999), this vista has important implications for identifying genes for complex traits related to a distinct disorder.

An association of the 5HTTLPR and affective illness, including unipolar depression and bipolar disorder, has been reported by Collier et al. (1996); however, subsequent studies did not consistently replicate these results (Tables 20.2 and 20.3). Among likely reasons for these conflicting results are the traditional approach of studying individuals with categorically defined psychiatric disorders, inadequate statistical power to replicate an association between 5HTTLPR genotype and affective illness, difficulties in interpretation of population-based association studies due to ethnic differences in 5HTTLPR allele frequencies, and the lack of within-family designs, raising the chance of false-positive (and false-negative) findings due to ethnic admixture. Using categorically defined psychiatric diagnoses does not guarantee biological homogeneity; in fact, diagnostic categories include a high degree of heterogeneity (e.g., anxiety, melancholic features, delusions, variable time course, and drug response). Serretti et al. (1999) reported an association of the 5HTTLPR s variant and anxiety scores in patients with affective illness, whereas no major influence on the remaining symptom clusters of depressive symptomatology was found. Thus, genetic studies based on diagnostic criteria may furnish conflicting results owing to different frequencies of various subgroups within these populations.

Among various subtypes of affective spectrum disorders, distinct forms of 5HT system-specific depression with anxiety and aggression as "pacemaking" features have been described by van Praag (1996, 1998). Particularly, seasonal variations in mood and behavior (seasonality) and seasonal affective disorder (SAD) have been attributed to seasonal fluctuations in brain 5HT functioning. Comparison of SAD patients and nonseasonal healthy controls with low seasonality levels showed that patients with SAD were more likely to have the low-activity 5HTTLPR s allele (Rosenthal et al., 1998). Moreover, SAD patients with the l/l genotype had a lower mean seasonality score than did patients with the l/s and s/s genotypes, indicating that the 5HTTLPR contributes to the trait of seasonality and the risk of developing SAD. Interestingly, the 5HTTLPR also seems to influence the development of seasonality in a general sense (Sher et al., 1999).

In addition to anxiety, depression, externally directed hostility, and aggression, disturbances of 5HT function have been implicated in self-directed injurious behavior, including suicidality. A genetic contribution to suicidality is supported by familial, twin, and adoption studies, and at the same time a growing body of evidence indicates that genetic factors underlying suicidal behavior are independent of the genetic transmission of

Table 20.2. 5HTTLPR Alleles in Major Depression

Study	Size of study	Δ	C	R
Collier et al. (1996)	275 MDD vs. 739 controls	MDD		s/l
Menza et al. (1999)	32 patients	HDRS scores	Parkinson patients	s/l
Mössner et al. (2001)	72 patients	HDRS scores	Parkinson patients	s/l
Rosenthal et al. (1998)	97 SAD vs. 71 controls	SAD	74 unipolar, 25 bipolar	s/l
Serretti et al. (1999)	67 MDD	HDRS	Inpatients	*ns*
Furlong et al. (1998)	125 MDD vs. 169 controls	MDD		*ns*
Kunugi et al. (1997)	49 MDD vs. 207 controls	MDD	Asian study	*ns*
Rees et al. (1997)	80 MDD vs. 118 controls	MDD		*ns*
Bellivier et al. (1998)	45 MDD vs. 99 controls	MDD		*ns*
Kim et al. (2000)	120 MDD vs. 252 controls	MDD	Asian study	*ns*
Kunugi et al. (1996)	36 MDD vs. 281 controls	MDD	Asian study	*ns*

Note. MDD = major depressive disorder; HDRS = Hamilton Depression Rating Scale; Δ = diagnosis; C = comment; R = results; s/l = short/long variant of the 5HTT gene-linked polymorphic region (5HTTLPR).

Table 20.3. 5HTTLPR Alleles in Bipolar Disorder

Study	Size of study	Δ	C	R
Collier et al. (1996)	304 bipolar vs. 570 controls	Bipolar		s/l
Bellivier et al. (1998)	204 bipolar vs. 99 controls	Bipolar		s/l
Mundo et al. (2001)	27 bipolar IM+ vs. 29 bipolar IM−	Bipolar I and II	SRI-induced mania	s/l
Furlong et al. (1998)	87 bipolar vs. 169 controls	Bipolar		*ns*
Kunugi et al. (1997)	140 bipolar vs. 207 controls	Bipolar	Japanese study	*ns*
Rees et al. (1997)	168 bipolar I vs. 118 controls	Bipolar I		*ns*
Hoehe et al. (1998)	79 bipolar I vs. 281 controls	Bipolar I		*ns*
Oruc et al. (1997)	42 bipolar I vs. 40 controls	Bipolar		*ns*
Gutierrez et al. (1998)	88 bipolar vs. 113 controls	Bipolar		*ns*
Vincent et al. (1999)	108 bipolar vs. 108 controls	Bipolar	Nonreplication, initial results positive	*ns*
Family design studies				
Mynett-Johnson et al. (2000)	92 families	Bipolar I	TDT	s/l
Esterling et al. (1998)	22 families	Bipolar	TDT	*ns*
Kirov et al. (1999)	120 families	Bipolar I	TDT	*ns*
Mundo et al. (2000)	133 bipolar patients	Bipolar	TDT	*ns*

Note. SRI = serotonin reuptake inhibitor; TDT = transmission disequilibrium test; IM = SRI-induced mania; Δ = diagnosis; C = comment; R = results; s/l = short/long variant of the 5HTT gene-linked polymorphic region (5HTTLPR).

psychiatric disorder comorbid with suicidal behavior (Mann, 1998). In line with this view, several but not all reports show an association between the low 5HTT activity 5HTTLPR s variant and violent suicidal behavior but not with diagnostic entities (e.g., mood disorders, alcohol dependence; Bellivier et al., 1997, 1998; Bondy, Erfurth, de Jonge, Kruger, & Meyer, 2000; Courtet et al., 2001; Du et al., 1999; Fitch et al., 2001; Geijer et al., 2000; Ohara, Nagai, Tsukamoto, Tani, & Suzuki, 1998; Zalsman et al., 2001).

Based on preclinical data, a complex interaction among genotype, syndromal dimensions, and drug response has been hypothesized (Catalano, 1999; see Figure 20.3). A given genetic predisposition, such as allelic variation in 5HTT function, may confer susceptibility to anxious or depressive features and a less favorable antidepressant response in patients affected by mood disorders. Impaired 5HTT function apparently confers, if any, only a very modest susceptibility to depressed states, because adaptive mechanisms are likely to compensate for the deficiency, whereas more robust alterations of 5HT turnover observed during antidepressant treatment may reveal 5HTTLPR genotype effects that lead to variable SRI efficacy. Smeraldi and associates (Smeraldi et al., 1998) investigated whether the 5HTTLPR genotype is related to the antidepressant response to the SRI fluvoxamine or augmentation with the $5HT_{1A}$ receptor antagonist pindolol in patients with major depression with psychotic features who had been randomly assigned to treatment with a fixed dose of fluvoxamine and either placebo or pindolol for 6 weeks. Both homozygotes for the long variant (l/l) of the 5HTT promoter and heterozygotes (l/s) showed a better response to fluvoxamine than homozygotes for the short variant (s/s). In the group treated with fluvoxamine plus pindolol, all the genotypes acted like l/l treated with fluvoxamine alone. Fluvoxamine efficacy in delusional depression seems to be related, in part, to allelic variation within the promoter of the 5HTT gene. This 5HTTLPR genotype effect on antidepressant response has been replicated recently by several other studies (Table 20.4). Furthermore, interactions between 5HTTLPR genotype and therapeutic efficacy of the antimanic/antibipolar agent lithium, and the antipsychotic drug clozapine, both of which are assumed to act

Table 20.4. 5HTTLPR Alleles in Treatment Response to Selective 5HT Transporter Inhibitors

Study	Medication	*N*	Result
Smeraldi et al. (1998)	Fluvoxamine	53	s/l
Zanardi et al. (2000)	Paroxetine	58	s/l
Pollock, Ferrell, et al. (2000)	Paroxetine	51	s/l
Arias et al. (2000)	Citalopram	71	s/l
Rausch et al. (2000)	Fluoxetine	51	s/l
Benedetti et al. (1999)	Sleep deprivation	68	s/l
Pollock, Laghrissi-Thode, et al. (2000)	Nortriptyline	45	*ns*
Kim et al. (2000)	Fluoxetine or paroxetine	120	s/l

Note. s/l = short/long variant of the 5HTT gene-linked polymorphic region (5HTTLPR).

through serotonergic mechanisms, have been demonstrated (Arranz et al., 2000; Del Zompo et al., 1999). Finally, drug-free patients with bipolar depression who were homozygotic for the l variant of 5HTTLPR showed better mood improvement after total sleep deprivation than those with the s/l and s/s genotype (Benedetti et al., 1998). These findings support the notion that 5HTTLPR genotyping may represent a useful pharmacogenetic tool to individualize treatment of depression, and that 5HTT function is critical for the antidepressant mechanism of action of sleep deprivation.

Conclusion and Outlook

A considerable body of evidence indicates that neuroticism, which comprises a wide spectrum of anxiety-related traits and behavior, is influenced by genetic factors and that the genetic component is highly complex, polygenic, and epistatic. Although several genes that may contribute to the genetic variance of anxiety or may modify the phenotypic expression of pathologic anxiety are currently under investigation, molecular genetics has so far failed to identify a genomic variation that consistently contributes to the expression of anxiety-related traits.

Converging lines of evidence suggest that allelic variation in functional 5HTT expression plays a critical role in neurodevelopment and synaptic plasticity, thus setting the stage for expression of anxiety-related traits and their associated behavior throughout adult life. Moreover, genetically driven variation of 5HTT function, in conjunction with other predisposing genetic factors and with inadequate adaptive responses to environmental stressors, also is likely to contribute to syndromal dimensions and treatment response of anxiety and mood disorders. Integration of emerging technologies for genetic analysis will further fertilize the ground for an advanced stage of gene identification and functional studies in behavioral genetics. The documented heterogeneity of both genetic and environmental constituents of brain development and function suggests the futility of searching for unitary determinants of anxiety-related behavior.

Several refined concepts should therefore be adopted regarding future behavioral genetic research. First, future studies will require control for nonindependence within cohorts. To minimize the risk of population stratification bias, rigorous methods of “genomic control” have been designed. These statistical strategies are based on the assessment of 60 single nucleotide polymorphisms (SNPs) or genotypes of 100 unlinked microsatellite markers spread throughout the genome to adjust the significance level of a candidate gene polymorphism (Bacanu, Devlin, & Roeder, 2000; Pritchard, Stephens, Rosenberg, & Donnelly, 2000). With recent advances in molecular genetics, the rate-limiting step in identifying candidate genes has nevertheless become the definition of phenotype.

Second, more functionally relevant polymorphisms in genes within a single neurotransmitter system, or in genes that comprise a developmental and functional unit in their concerted actions, need to be identified and investigated in both large association studies to avoid stratification artifacts and to elucidate complex epistatic and epigenetic interactions of multiple loci with the environment. Not only will DNA variants in coding and regulatory regions of genes be useful for systematic genome scans for identifying genes associated with personality and behavior, but they also will make it possible to study integrated systems of gene pathways as an important step on the route to behavioral genomics (see Figure 20.3; Lesch, in press-b; see also Craig & McClay, chapter 2, this volume). Although great strides have been made in understanding the diversity of the human genome, such as the frequency, distribution, and type of genetic variation that exists, the feasibility of applying this information to uncover useful genomic markers of behavioral traits remains uncertain. Based on the first draft sequence of the human genome, several million SNPs in addition to microdeletions (or insertions), and polymorphic simple sequence repeats (SSRs) of 2 to 50+ nucleotides in length, have been identified in the human genome, yet no more than roughly 30,000–60,000 common polymorphisms are located in coding and regulatory regions of genes that are the ultimate causes of the heritability of complex traits (McPherson et al., 2001; Sachidanandam et al., 2001). Although the industry is relying heavily on commercialized access to SNP databases for use in research, rarely addressed are issues related to the reality of using SNPs to uncover markers for behavioral traits and disorders as well as treatment response, such as population and patient sample size, SNP density and genome coverage, SNP functionality, and data interpretation that will be important for determining the suitability of genomic information. Success will depend on the availability of SNPs in the coding or regulatory regions (cSNPs or rSNPs, respectively) of a large number of candidate genes as well as knowledge of the average extent of linkage disequilibrium between SNPs, the development of high-throughput technologies for genotyping SNPs, identification of protein-altering SNPs by DNA and protein microarray-assisted expression analysis, and collection of DNA from well-assessed cohorts. As more and more appreciation of the potential for polymorphisms in gene regulatory regions to impact gene expression is gained, knowledge of novel functional variants is likely to emerge.

Third, genetic influences are not the only pathway that leads to individual differences in personality dimensions, behavior, psychopathology, and drug response. Complex traits are most likely to be generated by a complex interaction of environmental and experimental factors with a number of genes and their products. Even pivotal regulatory proteins of cellular pathways and neurocircuits are most likely to have only a very modest, if not minimal, impact, whereas noise from nongenetic mechanisms obstructs identification of relevant gene variants. Although current methods for the detection of gene–gene and gene–environment interaction in behavioral genetics are largely indirect, the most relevant consequence of gene identification for behavioral traits may be that it will provide the tools required to systematically clarify the effects of gene–environment interaction on brain development and plasticity.

Finally, future benefits will stem from novel techniques in molecular cell biology, transgenics, and gene transfer technologies. However, in the postgenomic era, behavioral genetic research will require integration of the entire spectrum of genomics, DNA variants, gene expression, proteomics, brain development, structure, and function, as well as behavior in a wide spectrum of species. Although bioinformatics resources are evolving in most of these areas, incorporation of these resources from the perspective of functional genomics of personality and behavior will greatly facilitate research.

References

Arias, B., Gasto, C., & Catalan, R. (2000). Variation in the serotonin transporter gene and clinical response to citalopram in major depression. *American Journal of Medical Genetics, 96*, 536.

Arranz, M., Nunro, J., Birkett, J., Bolonna, A., Manacama, D., Sodhi, M., et al. (2000). Pharmacogenetic prediction of clozapine response. *Lancet, 355*, 1615–1616.

Auerbach, J. G., Benjamin, J., Faroy, M., Geller, V., & Ebstein, R. (2001). DRD4 related to infant attention and information processing: A developmental link to ADHD? *Psychiatric Genetics, 11*, 31–35.

Auerbach, J. G., Geller, V., Lezer, S., Shinwell, E., Belmaker, R., Levine, J., & Ebstein, R. (1999). Dopamine D4 receptor (D4DR) and serotonin transporter promoter (5-HTTLPR) polymorphisms in the determination of temperament in 2-month-old infants. *Molecular Psychiatry, 4*, 369–373.

Azmitia, E. C., & Whitaker-Azmitia, P. M. (1997). Development and adult plasticity of serotonergic neurons and their target cells. In H. G. Baumgarten & M. Göthert (Eds.), *Serotonergic neurons and 5-HT receptors in the CNS* (Vol. 129, pp. 1–39). New York: Springer.

Bacanu, S. A., Devlin, B., & Roeder, K. (2000). The power of genomic control. *American Journal of Human Genetics, 66*, 1933–1944.

Ball, D., Hill, L., Freeman, B., Eley, T. C., Strelau, J., Riemann, R., et al. (1997). The serotonin transporter gene and peer-rated neuroticism. *Neuroreport, 8*, 1301–1304.

Battersby, S., Ogilvie, A. D., Blackwood, D. H., Shen, S., Muqit, M. M., Muir, W. J., et al. (1999). Presence of multiple functional polyadenylation signals and a single nucleotide polymorphism in the 3′ untranslated region of the human serotonin transporter gene. *Journal of Neurochemistry, 72*, 1384–1388.

Baumeister, R. F., & Tice, D. M. (1990). Anxiety and social exclusion. *Journal of Social and Clinical Psychology, 9*, 165–195.

Bellivier, F., Henry, C., Szoke, A., Schurhoff, F., Nosten-Bertrand, M., Feingold, J., et al. (1998). Serotonin transporter gene polymorphisms in patients with unipolar or bipolar depression. *Neuroscience Letters, 255*, 143–146.

Bellivier, F., Laplanche, J. L., Leboyer, M., Feingold, J., Bottos, C., Allilaire, J. F., & Launay, J. M. (1997). Serotonin transporter gene and manic depressive illness: An association study. *Biological Psychiatry, 41*, 750–752.

Benedetti, F., Colombo, C., Barbini, B., Campori, E., & Smeraldi, E. (1999). Ongoing lithium treatment prevents relapse after total sleep deprivation. *Journal of Clinical Psychopharmacology, 19*, 240–245.

Benedetti, F., Serretti, A., Colombo, C., Campori, E., Barbini, B., di Bella, D., & Smeraldi, E. (1998). Influence of a functional polymorphism within the promoter of the serotonin transporter gene on the effects of total sleep deprivation in bipolar depression. *American Journal of Psychiatry, 156*, 1450–1452.

Bengel, D., Murphy, D. L., Andrews, A. M., Wichems, C. H., Feltner, D., Heils, A., et al. (1998). Altered brain serotonin homeostasis and locomotor insensitivity to 3,4-methylenedioxymethamphetamine ("Ecstasy") in serotonin transporter-deficient mice. *Molecular Pharmacology, 53*, 649–655.

Benjamin, J., Li, L., Patterson, C., Greenberg, B. D., Murphy, D. L., & Hamer, D. H. (1996).

Population and familial association between the D4 dopamine receptor gene and measures of novelty seeking. *Nature Genetics, 12*, 81–84.

Benjamin, J., Osher, Y., Kotler, M., Gritsenko, I., Nemanov, L., Belmaker, R. H., & Ebstein, R. P. (2000). Association between Tridimensional Personality Questionnaire (TPQ) traits and three functional polymorphisms: Dopamine receptor D4 (DRD4), serotonin transporter promoter region (5-HTTLPR) and catechol O-methyltransferase (COMT). *Molecular Psychiatry, 5*, 96–100.

Benjamin, J., Osher, Y., Lichtenberg, P., Bachner-Melman, R., Gritsenko, I., Kotler, M., et al. (2000). An interaction between the catechol O-methyltransferase and serotonin transporter promoter region polymorphisms contributes to Tridimensional Personality Questionnaire persistence scores in normal subjects. *Neuropsychobiology, 41*, 48–53.

Bennett, A. J., Lesch, K. P., Heils, A., Long, J., Lorenz, J., Shoaf, S. E., et al. (2002). Early experience and serotonin transporter gene variation interact to influence primate CNS function. *Molecular Psychiatry, 7*, 118–122.

Bennett-Clarke, C. A., Chiaia, N. L., & Rhoades, R. W. (1996). Thalamocortical afferents in rat transiently express high-affinity serotonin uptake sites. *Brain Research, 733*, 301–306.

Bennett-Clarke, C. A., Leslie, M. J., Lane, R. D., & Rhoades, R. W. (1994). Effect of serotonin depletion on vibrissa-related patterns of thalamic afferents in the rat's somatosensory cortex. *Journal of Neuroscience, 14*, 7594–7607.

Bondy, B., Erfurth, A., de Jonge, S., Kruger, M., & Meyer, H. (2000). Possible association of the short allele of the serotonin transporter promoter gene polymorphism (5-HTTLPR) with violent suicide. *Molecular Psychiatry, 5*, 193–195.

Bradley, C. C., & Blakely, R. D. (1997). Alternative splicing of the human serotonin transporter gene. *Journal of Neurochemistry, 69*, 1356–1367.

Buss, D. M. (1991). Evolutionary personality psychology. *Annual Review of Psychology, 42*, 459–491.

Cases, O., Lebrand, C., Giros, B., Vitalis, T., De Maeyer, E., Caron, M. G., et al. (1998). Plasma membrane transporters of serotonin, dopamine, and norepinephrine mediate serotonin accumulation in atypical locations in the developing brain of monoamine oxidase A knock-outs. *Journal of Neuroscience, 18*, 6914–6927.

Cases, O., Vitalis, T., Seif, I., De Maeyer, E., Sotelo, C., & Gaspar, P. (1996). Lack of barrels in the somatosensory cortex of monoamine oxidase A-deficient mice: Role of a serotonin excess during the critical period. *Neuron, 16*, 297–307.

Catalano, M. (1999). The challenges of psychopharmacogenetics. *American Journal of Human Genetics, 65*, 606–610.

Champoux, M., Bennett, A., Lesch, K. P., Heils, A., Nielsen, D. A., Higley, J. D., et al. (1999). Serotonin transporter gene polymorphism and neurobehavioral development in rhesus monkey neonates. *Society of Neuroscience Abstracts, 25*, 69.

Cloninger, C. (1994). Temperament and personality. *Current Opinion in Neurobiology, 4*, 266–273.

Collier, D. A., Stöber, G., Li, T., Heils, A., Catalano, M., Di Bella, D., et al. (1996). A novel functional polymorphism within the promoter of the serotonin transporter gene: Possible role in susceptibility to affective disorders. *Molecular Psychiatry, 1*, 453–460.

Costa, P., & McCrae, R. (1992). *Revised NEO Personality Inventory (NEO–PI–R) and NEO Five Inventory (NEO–FFI) professional manual*. Odessa, FL: Psychological Assessment Resources.

Courtet, P., Baud, P., Abbar, M., Boulenger, J. P., Castelnau, D., Mouthon, D., et al. (2001). Association between violent suicidal behavior and the low activity allele of the serotonin transporter gene. *Molecular Psychiatry, 6*, 338–341.

Deary, I. J., Battersby, S., Whiteman, M. C., Connor, J. M., Fowkes, F. G., & Harmar, A. (1999). Neuroticism and polymorphisms in the serotonin transporter gene. *Psychological Medicine, 29*, 735–739.

Del Zompo, M., Ardau, R., Palmas, M., Bocchetta, A., Reina, A., & Piccardi, M. (1999). Lithium response: Association study with two candidate genes. *Molecular Psychiatry, 4*(Suppl.), 66–67.

Du, L., Bakish, D., & Hrdina, P. D. (2000). Gender differences in association between serotonin transporter gene polymorphism and personality traits. *Psychiatric Genetics, 10*, 159–164.

Du, L., Faludi, G., Palkovits, M., Demeter, E., Bakish, D., Lapierre, Y. D., et al. (1999). Frequency of long allele in serotonin transporter gene is increased in depressed suicide victims. *Biological Psychiatry, 46*, 196–201.

Ebstein, R. P., Gritsenko, I., Nemanov, L., Frisch, A., Osher, Y., & Belmaker, R. H. (1997). No association between the serotonin transporter gene regulatory region polymorphism and the Tridimensional Personality Questionnaire (TPQ) temperament of harm avoidance. *Molecular Psychiatry, 2*, 224–226.

Ebstein, R. P., Levine, J., Geller, V., Auerbach, J., Gritsenko, I., & Belmaker, R. H. (1998). Dopamine D4 receptor and serotonin transporter promoter in the determination of neonatal temperament. *Molecular Psychiatry, 3*, 238–246.

Ebstein, R. P., Novick, O., Umansky, R., Priel, B., Osher, Y., Blaine, D., et al. (1996). Dopamine D4 receptor (D4DR) exon III polymorphism associated with the human personality trait of novelty seeking. *Nature Genetics, 12*, 78–80.

Eley, T. C., & Plomin, R. (1997). Genetic analyses of emotionality. *Current Opinion in Neurobiology, 7*, 279–284.

Eley, T. C., & Stevenson, J. (1999). Exploring the covariation between anxiety and depression symptoms: A genetic analysis of the effects of age and sex. *Journal of Child Psychology and Psychiatry, 40*, 1273–1282.

Esterling, L. E., Yoshikawa, T., Turner, G., Badner, J. A., Bengel, D., Gershon, E. S., et al. (1998). Serotonin transporter (5-HTT) gene and bipolar affective disorder. *American Journal of Medical Genetics, 81*, 37–40.

Fink, G., Sumner, B., Rosie, R., Wilson, H., & McQueen, J. (1999). Androgen actions on central serotonin neurotransmission: Relevance for mood, mental state and memory. *Behavioral Brain Research, 105*, 53–68.

Fitch, D., Lesage, A., Seguin, M., Trousignant, M., Bankelfat, C., Rouleau, G. A., et al. (2001). Suicide and the serotonin transporter gene. *Molecular Psychiatry, 6*, 127–128.

Flory, J. D., Manuck, S. B., Ferrell, R. E., Dent, K. M., Peters, D. G., & Muldoon, M. F. (1999). Neuroticism is not associated with the serotonin transporter (5-HTTLPR) polymorphism. *Molecular Psychiatry, 4*, 93–96.

Furlong, R. A., Ho, L., Walsh, C., Rubinsztein, J. S., Jain, S., Paykel, E. S., et al. (1998). Analysis and meta-analysis of two serotonin transporter gene polymorphisms in bipolar and unipolar affective disorders. *American Journal of Medical Genetics, 81*, 58–63.

Geijer, T., Frisch, A., Persson, M. L., Wasserman, D., Rockah, R., Michaelovsky, E., et al. (2000). Search for association between suicide attempt and serotonergic polymorphisms. *Psychiatric Genetics, 10*, 19–26.

Gelernter, J., Kranzler, H., Coccaro, E. F., Siever, L. J., & New, A. S. (1998). Serotonin transporter protein gene polymorphism and personality measures in African American and European American subjects. *American Journal of Psychiatry, 155*, 1332–1338.

Gelernter, J., Kranzler, H., & Cubells, J. F. (1997). Serotonin transporter protein (SLC6A4) allele and haplotype frequencies and linkage disequilibria in African- and European-American and Japanese populations and in alcohol-dependent subjects. *Human Genetics, 101*, 243–246.

Gould, E. (1999). Serotonin hippocampal neurogenesis. *Neuropsychopharmacology, 21*, 46S–51S.

Greenberg, B. D., Li, Q., Lucas, F. R., Hu, S., Sirota, L. A., Benjamin, J., et al. (2000). Association between the serotonin transporter promoter polymorphism and personality traits in a primarily female population sample. *American Journal of Medical Genetics, 96*, 202–216.

Greenberg, B. D., Tolliver, T. J., Huang, S. J., Li, Q., Bengel, D., & Murphy, D. L. (1999). Genetic variation in the serotonin transporter promoter region affects serotonin uptake in human blood platelets. *American Journal of Medical Genetics, 88*, 83–87.

Gustavsson, J. P., Nothen, M. M., Jonsson, E. G., Neidt, H., Forslund, K., Rylander, G., et al. (1999). No association between serotonin transporter gene polymorphisms and personality traits. *American Journal of Medical Genetics, 88*, 430–436.

Gutierrez, B., Arranz, M. J., Collier, D. A., Valles, V., Guillamat, R., Bertranpetit, J., et al.

(1998). Serotonin transporter gene and risk for bipolar affective disorder: An association study in Spanish population. *Biological Psychiatry, 43*, 843–847.

Hanna, G. L., Himle, J. A., Curtis, G. C., Koram, D. Q., Veenstra-Vander Weele, J., Leventhal, B. L., et al. (1998). Serotonin transporter and seasonal variation in blood serotonin in families with obsessive–compulsive disorder. *Neuropsychopharmacology, 18*, 102–111.

Hansson, S. R., Mezey, E., & Hoffman, B. J. (1998). Serotonin transporter messenger RNA in the developing rat brain: Early expression in serotonergic neurons and transient expression in non-serotonergic neurons. *Neuroscience, 83*, 1185–1201.

Hansson, S. R., Mezey, E., & Hoffman, B. J. (1999). Serotonin transporter messenger RNA expression in neural crest-derived structures and sensory pathways of the developing rat embryo. *Neuroscience, 89*, 243–265.

Heils, A., Teufel, A., Petri, S., Stöber, G., Riederer, P., Bengel, B., et al. (1996). Allelic variation of human serotonin transporter gene expression. *Journal of Neurochemistry, 6*, 2621–2624.

Heinz, A., Jones, D. W., Mazzanti, C., Goldman, D., Ragan, P., Hommer, D., et al. (1999). A relationship between serotonin transporter genotype and in vivo protein expression and alcohol neurotoxicity. *Biological Psychiatry, 47*, 643–649.

Heisler, L. K., Chu, H. M., Brennan, T. J., Danao, J. A., Bajwa, P., Parsons, L. H., et al. (1998). Elevated anxiety and antidepressant-like responses in serotonin 5-HT1A receptor mutant mice. *Proceedings of the National Academy of Sciences USA, 95*, 15049–15054.

Hensler, J. G., Ferry, R. C., Labow, D. M., Kovachich, G. B., & Frazer, A. (1994). Quantitative autoradiography of the serotonin transporter to assess the distribution of serotonergic projections from the dorsal raphe nucleus. *Synapse, 17*, 1–15.

Higley, J., Bennett, A., Heils, A., Lesch, K., Shoaf, S., White, I., et al. (1998). Serotonin transporter gene variation is associated with CSF 5-HIAA concentrations in rhesus monkeys. *Society of Neuroscience Abstracts, 24*, 1113.

Higley, J. D., King, S. T., Jr., Hasert, M. F., Champoux, M., Suomi, S. J., & Linnoila, M. (1996). Stability of interindividual differences in serotonin function and its relationship to severe aggression and competent social behavior in rhesus macaque females. *Neuropsychopharmacology, 14*, 67–76.

Higley, J. D., Mehlman, P. T., Taub, D. M., Higley, S. B., Suomi, S. J., Vickers, J. H., et al. (1992). Cerebrospinal fluid monoamine and adrenal correlates of aggression in free-ranging rhesus monkeys. *Archives of General Psychiatry, 49*, 436–441.

Higley, J. D., Suomi, S. J., & Linnoila, M. (1991). CSF monoamine metabolite concentrations vary according to age, rearing, and sex, and are influenced by the stressor of social separation in rhesus monkeys. *Psychopharmacology, 103*, 551–556.

Higley, J. D., Suomi, S. J., & Linnoila, M. (1992). A longitudinal assessment of CSF monoamine metabolite and plasma cortisol concentrations in young rhesus monkeys. *Biological Psychiatry, 32*, 127–145.

Higley, J. D., Thompson, W. W., Champoux, M., Goldman, D., Hasert, M. F., Kraemer, G. W., et al. (1993). Paternal and maternal genetic and environmental contributions to cerebrospinal fluid monoamine metabolites in rhesus monkeys (*Macaca mulatta*). *Archives of General Psychiatry, 50*, 615–623.

Hoehe, M. R., Wendel, B., Grunewald, I., Chiaroni, P., Levy, N., Morris-Rosendahl, D., et al. (1998). Serotonin transporter (5-HTT) gene polymorphisms are not associated with susceptibility to mood disorders. *American Journal of Medical Genetics, 81*, 1–3.

Ishiguro, H., Arinami, T., Yamada, K., Otsuka, Y., Toru, M., & Shibuya, H. (1997). An association study between a transcriptional polymorphism in the serotonin transporter gene and panic disorder in a Japanese population. *Psychiatry and Clinical Neuroscience, 51*, 333–335.

Jorm, A. F., Henderson, A. S., Jacomb, P. A., Christensen, H., Korten, A. E., Rodgers, B., et al. (1998). An association study of a functional polymorphism of the serotonin transporter gene with personality and psychiatric symptoms. *Molecular Psychiatry, 3*, 449–451.

Kagan, J. (1989). Temperamental contributions to social behavior. *American Psychologist, 44*, 664–668.

Kagan, J., Reznick, J. S., & Snidman, N. (1988). Biological bases of childhood shyness. *Science, 240*, 167–171.

Katsuragi, S., Kunugi, H., Sano, A., Tsutsumi, T., Isogawa, K., Nanko, S., et al. (1999). Association between serotonin transporter gene polymorphism and anxiety-related traits. *Biological Psychiatry, 45*, 368–370.

Katz, L. C., & Shatz, C. J. (1996). Synaptic activity and the construction of cortical circuits. *Science, 274*, 1133–1138.

Kendler, K. S. (1996). Major depression and generalised anxiety disorder: Same genes, (partly) different environments—revisited. *British Journal of Psychiatry (Suppl.)*, 30, 68–75.

Kendler, K. S., Kessler, R. C., Neale, M. C., Heath, A. C., & Eaves, L. J. (1993). The prediction of major depression in women: Toward an integrated etiologic model. *American Journal of Psychiatry, 150*, 1139–1148.

Kendler, K. S., Neale, M. C., Kessler, R. C., Heath, A. C., & Eaves, L. J. (1992a). Generalized anxiety disorder in women: A population-based twin study. *Archives of General Psychiatry, 49*, 267–272.

Kendler, K. S., Neale, M. C., Kessler, R. C., Heath, A. C., & Eaves, L. J. (1992b). Major depression and generalized anxiety disorder: Same genes, (partly) different environments? *Archives of General Psychiatry, 49*, 716–722.

Kendler, K. S., Neale, M. C., Kessler, R. C., Heath, A. C., & Eaves, L. J. (1993). A longitudinal twin study of personality and major depression in women. *Archives of General Psychiatry, 50*, 853–862.

Kim, D. K., Lim, S. W., Lee, S., Sohn, S. E., Kim, S., Hahn, C. G., et al. (2000). Serotonin transporter gene polymorphism and antidepressant response. *Neuroreport, 11*, 215–219.

Kirov, G., Rees, M., Jones, I., MacCandless, F., Owen, M. J., & Craddock, N. (1999). Bipolar disorder and the serotonin transporter gene: A family-based association study. *Psychological Medicine, 29*, 1249–1254.

Knutson, B., Wolkowitz, O. M., Cole, S. W., Chan, T., Moore, E. A., Johnson, R. C., et al. (1998). Selective alteration of personality and social behavior by serotonergic intervention. *American Journal of Psychiatry, 155*, 373–379.

Kraemer, G. W., Ebert, M. H., Schmidt, D. E., & McKinney, W. T. (1989). A longitudinal study of the effect of different social rearing conditions on cerebrospinal fluid norepinephrine and biogenic amine metabolites in rhesus monkeys. *Neuropsychopharmacology, 2*, 175–189.

Kumakiri, C., Kodama, K., Shimizu, E., Yamanouchi, N., Okada, S., Noda, S., et al. (1999). Study of the association between the serotonin transporter gene regulatory region polymorphism and personality traits in a Japanese population. *Neuroscience Letters, 263*, 205–207.

Kunugi, H., Hattori, M., Kato, T., Tatsumi, M., Sakai, T., Sasaki, T., et al. (1997). Serotonin transporter gene polymorphisms: Ethnic difference and possible association with bipolar affective disorder. *Molecular Psychiatry, 2*, 457–462.

Kunugi, H., Tatsumi, M., Sakai, T., Hattori, M., & Nanko, S. (1996). Serotonin transporter gene polymorphism and affective disorder. *Lancet, 347*, 1340.

Lauder, J. M. (1990). Ontogeny of the serotonergic system in the rat: Serotonin as a developmental signal. *Annals of the New York Academy of Sciences, 600*, 297–314.

Lauder, J. M. (1993). Neurotransmitters as growth regulatory signals: Role of receptors and second messengers. *Trends in Neuroscience, 16*, 233–240.

Lebrand, C., Cases, O., Adelbrecht, C., Doye, A., Alvarez, C., Elmestikawy, S., et al. (1996). Transient uptake and storage of serotonin in developing thalamic neurons. *Neuron, 17*, 823–835.

Lesch, K. P. (1997). Molecular biology, pharmacology, and genetics of the serotonin transporter: Psychobiological and clinical implications. In H. G. Baumgarten & M. Göthert (Eds.), *Serotonergic neurons and 5-HT receptors in the CNS* (Vol. 129, pp. 671–705). New York: Springer.

Lesch, K. P. (in press-a). Genetic dissection of anxiety and related disorders. In D. Nutt & T. Ballenger (Eds.), *Anxiety disorders*. Oxford, England: Blackwell.

Lesch, K. P. (in press-b). The psychopharmacogenetic–neurodevelopmental interface: Focus on serotonergic gene pathways. In B. Lerer (Ed.), *Pharmacogenetics of psychotropic drugs*. Cambridge, England: Cambridge University Press.

Lesch, K. P. (2001). Serotonin transporter: From genomics and knockouts to behavioral traits and psychiatric disorders. In M. Briley & F. Sulser (Eds.), *Molecular genetics of mental disorders* (pp. 221–267). London: Martin Dunitz.

Lesch, K. P., Balling, U., Gross, J., Strauss, K., Wolozin, B. L., Murphy, D. L., et al. (1994). Organization of the human serotonin transporter gene. *Journal of Neural Transmission, 95*, 157–162.

Lesch, K. P., Bengel, D., Heils, A., Sabol, S. Z., Greenberg, B. D., Petri, S., et al. (1996). Association of anxiety-related traits with a polymorphism in the serotonin transporter gene regulatory region. *Science, 274*, 1527–1531.

Lesch, K. P., Meyer, J., Glatz, K., Flugge, G., Hinney, A., Hebebrand, J., et al. (1997). The 5-HT transporter gene-linked polymorphic region (5-HTTLPR) in evolutionary perspective: Alternative biallelic variation in rhesus monkeys [Rapid communication]. *Journal of Neural Transmission, 104*, 1259–1266.

Lesch, K. P., & Mössner, R. (1998). Genetically driven variation in serotonin uptake: Is there a link to affective spectrum, neurodevelopmental, and neurodegenerative disorders? *Biological Psychiatry, 44*, 179–192.

Lesch, K. P., & Mössner, R. (1999). 5-HT1A receptor inactivation: Anxiety or depression as a murine experience. *International Journal of Neuropsychopharmacology, 2*, 327–331.

Li, Q., Wichems, C., Heils, A., Lesch, K. P., & Murphy, D. L. (2000). Reduction in the density and expression, but not G-protein coupling, of serotonin receptors (5-HT1A) in 5-HT transporter knock-out mice: Gender and brain region differences. *Journal of Neuroscience, 20*, 7888–7895.

Little, K. Y., McLaughlin, D. P., Zhang, L., Livermore, C. S., Dalack, G. W., McFinton, P. R., et al. (1998). Cocaine, ethanol, and genotype effects on human midbrain serotonin transporter binding sites and mRNA levels. *American Journal of Psychiatry, 155*, 207–213.

Livesley, W. J., Jang, K. L., & Vernon, P. A. (1998). Phenotypic and genetic structure of traits delineating personality disorder. *Archives of General Psychiatry, 55*, 941–948.

Mann, J. J. (1998). The neurobiology of suicide. *Nature Medicine, 4*, 25–30.

Mansour-Robaey, S., Mechawar, N., Radja, F., Beaulieu, C., & Descarries, L. (1998). Quantified distribution of serotonin transporter and receptors during the postnatal development of the rat barrel field cortex. *Developmental Brain Research, 107*, 159–163.

Mazzanti, C. M., Lappalainen, J., Long, J. C., Bengel, D., Naukkarinen, H., Eggert, M., et al. (1998). Role of the serotonin transporter promoter polymorphism in anxiety-related traits. *Archives of General Psychiatry, 55*, 936–940.

McPherson, J. D., Marra, M., Hillier, L., Waterston, R. H., Chinwalla, A., Wallis, J., et al. (2001). A physical map of the human genome. *Nature, 409*, 934–941.

McQueen, J. K., Wilson, H., & Fink, G. (1997). Estradiol-17 beta increases serotonin transporter (SERT) mRNA levels and the density of SERT-binding sites in female rat brain. *Molecular Brain Research, 45*, 13–23.

Mehlman, P. T., Higley, J. D., Faucher, I., Lilly, A. A., Taub, D. M., Vickers, J., et al. (1994). Low CSF 5-HIAA concentrations and severe aggression and impaired impulse control in nonhuman primates. *American Journal of Psychiatry, 151*, 1485–1491.

Mehlman, P. T., Higley, J. D., Faucher, I., Lilly, A. A., Taub, D. M., Vickers, J., et al. (1995). Correlation of CSF 5-HIAA concentration with sociality and the timing of emigration in free-ranging primates. *American Journal of Psychiatry, 152*, 907–913.

Melke, J., Landen, M., Baghei, F., Rosmond, R., Holm, G., Bjorntorp, P., et al. (2001). Serotonin transporter gene polymorphisms are associated with anxiety-related personality traits in women. *American Journal of Medical Genetics, 105*, 458–463.

Menza, M. A., Palermo, B., DiPaola, R., Sage, J. I., & Ricketts, M. H. (1999). Depression and anxiety in Parkinson's disease: Possible effect of genetic variation in the serotonin transporter. *Journal of Geriatric Psychiatry and Neurology, 12*, 49–52.

Mortensen, O. V., Thomassen, M., Larsen, M. B., Whittemore, S. R., & Wiborg, O. (1999). Functional analysis of a novel human serotonin transporter gene promoter in immortalized raphe cells. *Molecular Brain Research, 68*, 141–148.

Mössner, R., Henneberg, A., Schmitt, A., Syagailo, Y. V., Grassle, M., Hennig, T., et al. (2001).

Allelic variation of serotonin transporter expression is associated with depression in Parkinson's disease. *Molecular Psychiatry, 6*, 350–352.

Mulder, R., Joyce, P., & Cloninger, C. (1994). Temperament and early environment influence comorbidity and personality disorders in major depression. *Comprehensive Psychiatry, 35*, 225–233.

Mundo, E., Walker, M., Cate, T., Macciardi, F., & Kennedy, J. L. (2001). The role of serotonin transporter protein gene in antidepressant-induced mania in bipolar disorder: Preliminary findings. *Archives of General Psychiatry, 58*, 539–544.

Mundo, E., Walker, M., Tims, H., Macciardi, F., & Kennedy, J. L. (2000). Lack of linkage disequilibrium between serotonin transporter protein gene (SLC6A4) and bipolar disorder. *American Journal of Medical Genetics, 96*, 379–383.

Murakami, F., Shimomura, T., Kotani, K., Ikawa, S., Nanba, E., & Adachi, K. (1999). Anxiety traits associated with a polymorphism in the serotonin transporter gene regulatory region in the Japanese. *Journal of Human Genetics, 44*, 15–17.

Mynett-Johnson, L., Kealey, C., Claffey, E., Curtis, D., Bouchier-Hayes, L., Powell, C., & McKeon, P. (2000). Multimarker haplotypes within the serotonin transporter gene suggest evidence of an association with bipolar disorder. *American Journal of Medical Genetics, 96*, 845–849.

Nakagawa, Y., Johnson, J. E., & O'Leary, D. D. (1999). Graded and areal expression patterns of regulatory genes and cadherins in embryonic neocortex independent of thalamocortical input. *Journal of Neuroscience, 19*, 10877–10885.

Nakamura, T., Muramatsu, T., Ono, Y., Matsushita, S., Higuchi, S., Mizushima, H., et al. (1997). Serotonin transporter gene regulatory region polymorphism and anxiety-related traits in the Japanese. *American Journal of Medical Genetics, 74*, 544–545.

Neumeister, A., Konstantinidis, A., Stastny, J., Schwarz, M., Vitouch, O., Willeit, M., et al. (in press). An association between serotonin transporter gene promoter polymorphism (5HTTLPR) and behavioral responses to tryptophan depletion in healthy female subjects with and without family history of depression. *Archives of General Psychiatry*.

Nobile, M., Begni, B., Giorda, R., Frigerio, A., Marino, C., Molteni, M., et al. (1999). Effects of serotonin transporter promoter genotype on platelet serotonin transporter functionality in depressed children and adolescents. *Journal of the American Academy of Child and Adolescent Psychiatry, 38*, 1396–1402.

Ohara, K., Nagai, M., Tsukamoto, T., Tani, K., & Suzuki, Y. (1998). Functional polymorphism in the serotonin transporter promoter at the SLC6A4 locus and mood disorders. *Biological Psychiatry, 44*, 550–554.

Ono, Y., Yoshimura, K., Sueoka, R., Yamauchi, K., Mizushima, H., Momose, T., et al. (1996). Avoidant personality disorder and taijin kyoufu: Sociocultural implications of the WHO/ADAMHA international study of personality disorders in Japan. *Acta Psychiatrica Scandinavia, 93*, 172–176.

Oruc, L., Verheyen, G. R., Furac, I., Jakovljevic, M., Ivezic, S., Raeymaekers, P., et al. (1997). Association analysis of the 5-HT2C receptor and 5-HT transporter genes in bipolar disorder. *American Journal of Medical Genetics, 74*, 504–506.

Osher, Y., Hamer, D., & Benjamin, J. (1999). Association and linkage of anxiety-related traits with a functional polymorphism of the serotonin transporter gene in Israeli siblings pairs. *Molecular Psychiatry, 4*, 24–25.

Osterheld-Haas, M. C., Van der Loos, H., & Hornung, J. P. (1994). Monoaminergic afferents to cortex modulate structural plasticity in the barrelfield of the mouse. *Developmental Brain Research, 77*, 189–202.

Parks, C. L., Robinson, P. S., Sibille, E., Shenk, T., & Toth, M. (1998). Increased anxiety of mice lacking the serotonin1A receptor. *Proceedings of the National Academy of Sciences USA, 95*, 10734–10739.

Persico, A. M., Revay, R. S., Mössner, R., Conciatori, M., Marino, R., Baldi, A., et al. (2001). Barrel pattern formation in somatosensory cortical layer IV requires serotonin uptake by thalamocortical endings, while vesicular monoamine release is necessary for development of supragranular layers. *Journal of Neuroscience, 21*, 6862–6873.

Plomin, R., Owen, M. J., & McGuffin, P. (1994). The genetic basis of complex human behaviors. *Science, 264*, 1733–1739.

Pollock, B. G., Ferrell, R. E., Mulsant, B. H., Mazumdar, S., Miller, M., Sweet, R. A., et al.

(2000). Allelic variation in the serotonin transporter promoter affects onset of paroxetine treatment response in late-life depression. *Neuropsychopharmacology, 23*, 587–590.

Pollock, B. G., Laghrissi-Thode, F., & Wagner, W. R. (2000). Evaluation of platelet activation in depressed patients with ischemic heart disease after paroxetine or nortriptyline treatment. *Journal of Clinical Psychopharmacology, 20*, 137–140.

Pritchard, J. K., Stephens, M., Rosenberg, N. A., & Donnelly, P. (2000). Association mapping in structured populations. *American Journal of Human Genetics, 67*, 170–181.

Ramboz, S., Oosting, R., Amara, D. A., Kung, H. F., Blier, P., Mendelsohn, M., et al. (1998). Serotonin receptor 1A knockout: An animal model of anxiety-related disorder. *Proceedings of the National Academy of Sciences USA, 95*, 14476–14481.

Rausch, J., Fei, Y., & Li, J. (2000). SERT genotype influence in fluoxetine antidepressant response: A study of initial conditions. *Biological Psychiatry, 47*, 92S–93S.

Rees, M., Norton, N., Jones, I., McCandless, F., Scourfield, J., Holmans, P., et al. (1997). Association studies of bipolar disorder at the human serotonin transporter gene (hSERT; 5HTT). *Molecular Psychiatry, 2*, 398–402.

Reist, C., Mazzanti, C., Vu, R., Tran, D., & Goldman, D. (2001). Serotonin transporter promoter polymorphism is associated with attenuated prolactin response to fenfluramine. *American Journal of Medical Genetics, 105*, 363–368.

Ricketts, M. H., Hamer, R. M., Sage, J. I., Manowitz, P., Feng, F., & Menza, M. A. (1998). Association of a serotonin transporter gene promoter polymorphism with harm avoidance behaviour in an elderly population. *Psychiatric Genetics, 8*, 41–44.

Rosenthal, N. E., Mazzanti, C. M., Barnett, R. L., Hardin, T. A., Turner, E. H., Lam, G. K., et al. (1998). Role of serotonin transporter promoter repeat length polymorphism (5-HTTLPR) in seasonality and seasonal affective disorder. *Molecular Psychiatry, 3*, 175–177.

Rudolph, U., & Mohler, H. (1999). Genetically modified animals in pharmacological research: Future trends. *European Journal of Pharmacology, 375*, 327–337.

Sachidanandam, R., Weissman, D., Schmidt, S. C., Kakol, J. M., Stein, L. D., Marth, G., et al. (2001). A map of human genome sequence variation containing 1.42 million single nucleotide polymorphisms. *Nature, 409*, 928–933.

Salichon, N., Gaspar, P., Upton, A. L., Picaud, S., Hanoun, N., Hamon, M., et al. (2001). Excessive activation of serotonin (5-HT) 1B receptors disrupts the formation of sensory maps in monoamine oxidase a and 5-ht transporter knock-out mice. *Journal of Neuroscience, 21*, 884–896.

Serretti, A., Cusin, C., Lattuada, E., Di Bella, D., Catalano, M., & Smeraldi, E. (1999). Serotonin transporter gene (5-HTTLPR) is not associated with depressive symptomatology in mood disorders. *Molecular Psychiatry, 4*, 280–283.

Shen, S., Battersby, S., Weaver, M., Clark, E., Stephens, K., & Harmar, A. J. (2000). Refined mapping of the human serotonin transporter (SLC6A4) gene within 17q11 adjacent to the CPD and NF1 genes. *European Journal of Human Genetics, 8*, 75–78.

Sher, L., Hardin, T. A., Greenberg, B. D., Murphy, D. L., Li, Q., & Rosenthal, N. E. (1999). Seasonality associated with the serotonin transporter promoter repeat length polymorphism [Letter]. *American Journal of Psychiatry, 156*, 1837.

Sirota, L. A., Greenberg, B. D., Murphy, D. L., & Hamer, D. H. (1999). Non-linear association between the serotonin transporter promoter polymorphism and neuroticism: A caution against using extreme samples to identify quantitative trait loci. *Psychiatric Genetics, 9*, 35–38.

Smeraldi, E., Zanardi, R., Benedetti, F., Di Bella, D., Perez, J., & Catalano, M. (1998). Polymorphism within the promoter of the serotonin transporter gene and antidepressant efficacy of fluvoxamine. *Molecular Psychiatry, 3*, 508–511.

Trefilov, A., Berard, J., Krawczak, M., & Schmidtke, J. (2000). Natal dispersal in rhesus macaques is related to serotonin transporter gene promoter variation. *Behavioral Genetics, 30*, 295–301.

van Praag, H. M. (1996). Faulty cortisol/serotonin interplay: Psychopathological and biological characterisation of a new, hypothetical depression subtype. *Psychiatry Research, 60*, 143–157.

van Praag, H. M. (1998). Anxiety and increased aggression as pacemakers of depression. *Acta Psychiatrica Scandinavia (Suppl.), 393*, 81–88.

Vincent, J. B., Masellis, M., Lawrence, J., Choi, V., Gurling, H. M., Parikh, S. V., & Kennedy, J. L. (1999). Genetic association analysis of serotonin system genes in bipolar affective disorder. *American Journal of Psychiatry, 156*, 136–138.

Whale, R., Quested, D. J., Laver, D., Harrison, P. J., & Cowen, P. J. (2000). Serotonin transporter (5-HTT) promoter genotype may influence the prolactin response to clomipramine. *Psychopharmacology (Berlin), 150*, 120–122.

Whitaker-Azmitia, P. M., & Peroutka, S. J. (1990). The neuropharmacology of serotonin. *Annals of the New York Academy of Sciences USA, 600*, 1–718.

Wichems, C. H., Li, Q., Holmes, A., Crawley, J. N., Tjurmina, O., Goldstein, D., et al. (2000). Mechanisms mediating the increased anxiety-like behavior and excessive responses to stress in mice lacking the serotonin transporter. *Society of Neuroscience Abstracts, 26*, 400.

Zalsman, G., Frisch, A., Bromberg, M., Gelernter, J., Michaelovsky, E., Campino, A., et al. (2001). Family-based association study of serotonin transporter promoter in suicidal adolescents: No association with suicidality but possible role in violence traits. *American Journal of Medical Genetics, 105*, 239–245.

Zanardi, R., Benedetti, F., Di Bella, D., Catalano, M., & Smeraldi, E. (2000). Efficacy of paroxetine in depression is influenced by a functional polymorphism within the promoter of the serotonin transporter gene. *Journal of Clinical Psychopharmacology, 20*, 105–107.

21

Animal Models of Anxiety

Jonathan Flint

Over the past 70 years, artificial selection has been used to create animal strains with heritable differences in behavior. Because most human behaviors are known to have an inherited component, one of the interests of these strains is that they may model susceptibility to human behavioral disorders and hence enable experimental investigation, impossible in human subjects, that would expose the biological foundations of psychiatric illnesses. However, the extent to which heritable differences in animal behavior model human susceptibility to psychiatric illness is not clear. While some disorders, such as abnormalities of language, cannot be effectively modeled, others, such as certain emotional disorders, probably do have parallels in other species. For example, some of the neuroanatomical and neurophysiological bases of fear-related behaviors are common across species, leading to the possibility that anxiety can be modeled effectively in animals. This chapter reviews the rodent models of anxiety and what is known about their genetic basis. After introducing the behavioral measures that are used, I discuss the information about genetic architecture inferred first from genetic selection experiments and then from the more recent molecular mapping experiments. I do not consider here the evidence from genetically manipulated rodents.

Tests of Anxiety in Rats and Mice

Table 21.1 lists the tests most commonly used to measure fearfulness, or susceptibility to anxiety in rodents. The table is divided into two sections, depending on whether the measures are of conditioned or unconditioned behavior. This distinction is important because conditioned tests are amenable to experimental manipulation to a degree that is impossible with unconditioned (or ethological) tests. Consequently, we know more about the reasons for the behavior in conditioned than unconditioned tests (Gray, 1982). For instance, in the case of unconditioned tests that involve behavior in a novel arena (the open-field arena, the elevated plus maze, and the light–dark box), while it is supposed that exploration as well as fearfulness is determining the animals' activity, we have no easy way of testing this hypothesis nor a clear formulation of the relationship between exploration and fearfulness. By contrast, theoretical formulations of behavior

Table 21.1. Animal Tests of Anxiety and Depression

Test of anxiety	Response
Unconditioned responses	
Open-field arena	Activity, defecation, thigmotaxis
Elevated plus maze	Ratio of entries to open and closed arms
Light–dark box	Emergence time
Hyponeophagia	Latency to start eating
Holeboard	Head dipping
Social interaction	Frequency of social interaction in a novel environment
Conditioned responses	
Two-way avoidance conditioning	Rate of acquisition of response
Acoustic startle	Contraction in response to loud noise
Footshock-induced freezing	Contraction in response to conditioned stimulus
Fear-potentiated startle	Contraction in response to loud noise in conjunction with conditioned stimulus
Geller-Seifter	Frequency of conditioned response coincidental to an electrical shock
Vogel conflict	Frequency of conditioned licking coincidental to an electrical shock

in conditioned tasks are more sophisticated and the experimental literature much richer (reviewed in Gray, 1982).

Unconditioned tests typically involve assessing the animal's behavior in a mildly stressful environment. Rats tend to avoid open elevated alleys and prefer closed alleys, a preference that is driven at least in part by fearfulness. The open-field arena exploits the aversive nature of an open space, a plain brightly lit area into which the animal is introduced. Reduced activity, avoidance of the central area, increased defecation and urination, and decreased rearing are taken as indices of fearfulness (Hall, 1934; Rodgers, Cao, Dalvi, & Holmes, 1997; Willner, 1990). Consistent with the view that frightened animals freeze, urinate, and defecate, a negative correlation was found between the animal's level of activity in the open field and its defecation rate, suggesting that the same brain system mediated the fearful behaviors. Other open-field measures have also been used: Frightened animals are considered to spend less time in the center of the arena (thigmotaxis) and to rise more often on their hind feet. The elevated plus maze, one of the most popular models of anxiety, has four arms, positioned like a cross, two of which are covered and two exposed. The animal is placed in the central platform, and the total number of arm entries is recorded. A fearful animal is expected to show reduced open-arm entries compared with closed-arm entries. The light–dark box (or black–white box) is an emergence test: Fearful animals are expected to avoid the brightly lit compartment, preferring to remain in the enclosed dark box. *Hyponeophagia* refers to the inhibition of eating in a novel situation: A measure of fearfulness is thus the time taken to approach and taste food in a novel environment (Rodgers et al., 1997; Willner, 1990). In the hole-board apparatus, rodents explore a square board with up to 16 holes into which they will tend to put their heads, a behavior that can be encouraged by placing novel objects beneath the holes. In the social interaction test, the effect of a novel environment on the duration of social interaction with conspecific animals is measured (File & Hyde, 1978).

The grounds for believing that the unconditioned tests measure fearfulness are twofold. First, there is a large literature on the behavioral effects of anxiolytic and anxiogenic drugs (Rodgers et al., 1997; Willner, 1990). The argument for pharmacological validation is that if a test measures fearfulness, then it should be possible to manipulate it specifically with drugs that affect human fearfulness (or anxiety) and not with drugs without such effects. For example, compounds that cause anxiety reduce the percentage of entries into and time spent on the open arms of the elevated plus maze, whereas antidepressants (e.g., imipramine and mianserin) and antipsychotics (e.g., haloperidol) do not. In general, the pharmacological validation holds up: Anxiolytics tend to increase locomotion in the open-field arena (Simon & Soubrie, 1979), increase entries into the open arms of the elevated plus maze (Pellow, Chopin, File, & Briley, 1985; Pellow & File, 1986), consistently and selectively increase light–dark crossings in the black–white box (Crawley, 1981; Crawley et al., 1984), and increase social interaction. Second, tests can be validated by correlated behaviors. Thus we expect a frightened animal to defecate more and ambulate less in the open field, show a longer latency to emerge in the black–white box, and visit the open arms of the elevated plus maze less often. Factorial derived constructs should then confirm the observed correlations.

However, the motivational structure of the unconditioned tests is much more complex than this simple exposition suggests. For example, the anxiolytic effect of benzodiazepines in the elevated plus maze is reduced or abolished after a single prior exposure, although prior test experience increases baseline open-arm avoidance. Dawson and colleagues provided evidence that the anxiolytic effects of chlordiazepoxide in the elevated plus maze are confounded by increases in motor activity: For instance, psychostimulants such as amphetamine at doses that increase locomotor activity have an anxiolytic profile in the apparatus (Dawson, Crawford, Collinson, Iversen, & Tricklebank, 1995). Factor analysis also has been used to dissect out additional components from fearfulness: In an analysis of five tests of fearfulness, three independent factors were found to account for 85% of the variance (Ramos, Berton, Mormede, & Chaouloff, 1997). Factor 1 loaded primarily on latency measures (time to approach the center of an open field and emerge from a black–white box), whereas Factor 2 loaded on activity measures in novel environments. These findings suggest a more complex explanation of unconditioned behavior, though not necessarily one that excludes a fearfulness component (Ramos & Mormede, 1998).

Conditioned tests offer more experimental control over potential confounds, and they include the best animal models we have for anxiety. In footshock-induced freezing, the animal is placed in a conditioning chamber and a shock, paired with a tone (or light), is administered twice. Animals are returned to the home cage and, on the following day, the amount of time the animal freezes is measured with and without exposure to the tone (or light), the latter now considered to be a conditioned stimulus. Freezing in the chamber without the tone is called *freezing to the context* (the animals associate the surroundings with the unpleasant experience), and freezing to the tone is called *freezing to the conditioned stimulus*. In the conditioned emotional response, food-deprived animals are trained to press a lever for food. Once the lever pressing is stable, the animal is taught to associate a light with a footshock. The light is then found to suppress bar-pressing. In the Geller-Seifter conflict procedure (Geller & Seifter, 1960), a light or tone introduces footshock in the operant chamber (i.e., where the animal has learned to bar-press for a food reward); in the Vogel water-lick conflict test (Vogel & Beer, 1971), thirsty animals are first trained to lick from a tube, and then electric shocks are delivered either through the tube or the floor.

Avoidance learning paradigms are of two sorts. In passive avoidance, a response is punished so that the probability of its occurrence diminishes; in active avoidance, the animal has to make a specified response to avoid punishment. The distinction between these two responses to punishment was extensively investigated during the 1960s and 1970s but frequently is ignored in current discussions of emotional behavior (reviewed in Gray, 1982). Although both types of avoidance responses can be, and often are, described as measures of fear or emotionality, brain lesions and antianxiety drugs differentiate between them: Antianxiety drugs, typified by the benzodiazepines, impair passive avoidance but improve active avoidance. Thus an anxious animal can be defined as one that shows relatively poor shuttle box performance and relatively high rates of freezing to a conditioned stimulus. In one-way active avoidance, the shock is given in one area only, so that the animal learns to associate shock with that compartment; in two-way active avoidance, the shock is given in both compartments so that the animal has to shuttle in order to avoid the shock. Two-way active avoidance thus involves more complex learning because the animal has to learn to approach cues that had been associated with shock.

In fear-potentiated startle, the animal's fearfulness is measured by how much it jumps in response to a conditioned stimulus to footshock. First, light is associated with a footshock. Startle is then elicited by another stimulus, such as a loud noise. The amount of startle is measured when the conditioned stimulus (the light) is given before the loud noise. Any additional startle (or potentiated startle) is then regarded as a measure of fearfulness. Fear-potentiated startle correlates highly with freezing, but unlike freezing it is open to experimental manipulation. Potentiated startle is suppressed by anxiolytics, which do not affect the unconditioned startle to the large noise, as expected if only the potentiated startle represents fearfulness. The model has been investigated extensively at the anatomical, cellular, and molecular level. The brain pathways underlying the startle response have been mapped (Davis, 1991), and the amygdala has been implicated in the establishment of conditioned fear.

Animals Selected for Different Tests of Anxiety

Selection experiments provide evidence that individual differences in test performance reflect genetic variation and hence demonstrate the importance of genetic effects on behavior. But evidence from selection experiments also can be used to validate animal models of susceptibility to anxiety. For more than half a century, considerable research effort has been expended in the search for an equivalent in animal species (usually termed *emotionality*) of the human personality factor of neuroticism or trait anxiety (Gray, 1987). If emotionality is a unitary construct, equivalent to susceptibility to anxiety, then animals that have been selected for extremes of behavior in one test should show correlated changes in another test.

Mice have been bred selectively for high and low activity scores in the open-field test, and rats for high and low defecation scores in the open field. In both cases, selection not only resulted in stable differences in behavior between the high and low lines but also produced correlated changes in other measures: Selection for high activity in the open field results in low defecation scores (DeFries, Gervais, & Thomas, 1978); selection for low defecation results in high activity (Broadhurst, 1960). However, as explained above, behavior in the open field is influenced by many factors, some of which may have no relevance to susceptibility to anxiety. For this reason, selection for conditioned behaviors is likely to provide more convincing models of anxiety.

One good example is the Roman high- and low-avoidance rats (RHA and RLA, respectively), the product of bidirectional selection for two-way active avoidance acquisition in a shuttle box (Bignami, 1965). The behavioral differences of the Roman rat strains are consistent with an interstrain variation in responses to fear stimuli. In the shuttle box, RHA/Verh rats quickly acquire the active avoidance response, whereas RLA/Verh rats display much freezing and escape responses during the acquisition phase (Driscoll & Battig, 1982; Escorihuela, Tobena, Driscoll, & Fernández-Teruel, 1995; Fernández-Teruel, Escorihuela, Castellano, Gonzalez, & Tobena, 1997). Results from other models of anxiety (the open field, elevated plus maze, and light–dark box and freezing to a conditioned stimulus) concur: Both inbred and outbred RHA/Verh rats are less anxious than their inbred or outbred RLA/Verh counterparts (Driscoll et al., 1998; Escorihuela, et al., 1999; Steimer, Driscoll, & Schulz, 1997). Furthermore, differences in neuroendocrine responses support the view that the RLA/Verh rats are more susceptible to environmental stressors than the RHA/Verh rats, as would be expected from the strain that is more responsive to fear-provoking stimuli (Driscoll et al., 1998).

Selection also has been carried out in other paradigms, including ultrasonic vocalization rates in infancy (Dichter, Brunelli, & Hofer, 1996), low and high activity in runways (Cerbone, Pellicano, & Sadile, 1993; Fujita, Annen, & Kitaoka, 1994), and behavior in elevated mazes (Lapin & Mirzaev, 1996; Liebsch, Linthorst, et al., 1998; Liebsch, Montkowski, Holsboer, & Landgraf, 1998). In general, these results, together with the more detailed investigations carried out on earlier selected lines, indicate that selection on one measure of emotionality produces correlated changes in other measures. The critical question is whether the same set of genes acts on all behavioral tests, as required by the hypothesis that consistency across tests indicates the presence of a trait susceptibility to anxiety, or whether genetic variants act independently on separate tests. Emotionality may consist of a large number of independent traits, each with a limited domain of action (e.g., Ramos & Mormede, 1998), or, alternatively, derive from a small number of separable, but interacting, brain systems that are engaged to varying degrees depending on the specific demands imposed by the current aversive situation (Gray, 1987).

QTL Analysis of Emotionality

Factor-analytic approaches to this problem in general concluded that different animal tests do not measure the same phenomena. A fourfold structure emerged from the analysis of open-field emergence, plus maze, and contextual fear conditioning (Flaherty, Greenwood, Martin, & Leszczuk, 1998) and three factors from analysis of holeboard, plus maze, and sexual preference/social interest tests (Fernándes, González, Wilson, & File, 1999). Factor analysis has also been applied to the different components of each test. For example, a threefold factor structure accounts for measures in the elevated plus maze (Lister, 1987; Rodgers & Johnson, 1995); five factors emerge from analyzing behavior in the murine defense test (Griebel, Blanchard, & Blanchard, 1996). In short, factor analyses have led some to the conclusion that animal models of anxiety measure different phenomena (Ramos & Mormede, 1998).

However, intertest correlations are typically small (less than .4), limiting the power of factor analysis to detect common effects. Genetic mapping provides an alternative way to detect small common effects, because it is within the capacity of most laboratories to phenotype and genotype the several hundred animals needed to detect loci contributing to less than 5% of the phenotypic variance. Because the traits are scored quantitatively, the underlying loci are referred to as quantitative trait loci (QTLs).

Many studies have shown that it is possible to map QTLs that influence behavior in rodent tests of anxiety. Genetic effects on a variety of anxiety tests have been mapped in mice strains derived from open-field activity selection experiments, carried out by John DeFries 30 years ago (DeFries et al., 1978; DeFries, Wilson, & McClearn, 1970). DeFries's exemplary selection experiment began with an F3 generation derived from a cross between two inbred strains, BALBc/J and C57BL/6J. The most active male and female within each litter were selected to become the progenitors of one selected line, termed H1. The least active male and least active female from each of these same 10 litters were selected to serve as progenitors of a low-activity line, L1. The same selection procedure was applied to members of a second randomly chosen group of 10 F3 litters, thereby providing 10 pairs of parents for replicate selected lines H2 and L2. Following 30 generations of bidirectional selection, the lines were random mated and then inbred using full-sib matings (see Figure 21.1; Toye & Cox, 2001). The existence of the two pairs of strains provided an opportunity to investigate thoroughly the genetic architecture of the response to behavioral selection for an anxiety phenotype.

In an initial experiment, measuring activity and defecation in the open field and entries into the open and closed arms of the elevated plus maze, three loci, on chromosomes 1, 12, and 15, were found that appeared to influence all measures relevant to anxiety. The QTL on chromosome 1 accounted for almost half the genetic variance of open-field activity (OFA; Flint et al., 1995). It appeared therefore that the genetic basis of anxiety phenotypes was relatively simple in these selected lines. Subsequent QTL mapping experiments added weight to this view. Gershenfeld and Paul mapped thigmotaxis in the open-field and light–dark transitions (Gershenfeld et al., 1997; Gershenfeld & Paul, 1997). Thigmotaxis mapped to chromosome 1, though more centromeric than the previously identified locus influencing activity. Using the number of transitions between light and dark boxes as a measure of exploratory behavior, Gershenfeld and Paul mapped another QTL to chromosome 10 (Gershenfeld et al., 1997; Gershenfeld & Paul, 1997). One study of hyperactivity in the rat identified a locus on chromosome 8 that contributed substantially to the variation in spontaneous activity, rearing, and activity in the open field (Moisan et al., 1996). However, this locus is not homologous to any identified in the mouse. QTLs influencing variation in fear conditioning have also been mapped, providing an opportunity to test the hypothesis that fear conditioning and open-field behaviors have a common genetic basis. Two F2 intercrosses and one study of recombinant inbred strains identified a locus on chromosome 1 that lies within the region identified by the work on open-field behaviors (Caldarone et al., 1997; Owen, Christensen, Paylor, & Wehner, 1997; Wehner et al., 1997; see also Wehner & Balogh, chapter 7, this volume).

Although the QTL studies suggest that there is a relatively simple genetic architecture for each phenotype, they do not tell us whether a locus has a joint action on several behavioral measures or whether fear is multidimensional, consisting of independent traits, each with a limited domain of action (Ramos & Mormede, 1998). Recent QTL analyses have attempted to address this problem and have revealed a substantially more complicated picture of how genes operate.

By using both replicates of the DeFries selected lines, it has been possible to replicate the findings for open-field behaviors. QTLs on chromosomes 1, 4, 7, 12, 15, 18, and X influence open-field activity (Turri, Datta, DeFries, Henderson, & Flint, 2001), and open-field defecation mapped to four loci (1, 7, 14, and X; Turri, Henderson, DeFries, & Flint, 2001; see Figure 21.1). Selection appears to have operated on the same relatively small number of loci in the DeFries selection experiments. Of

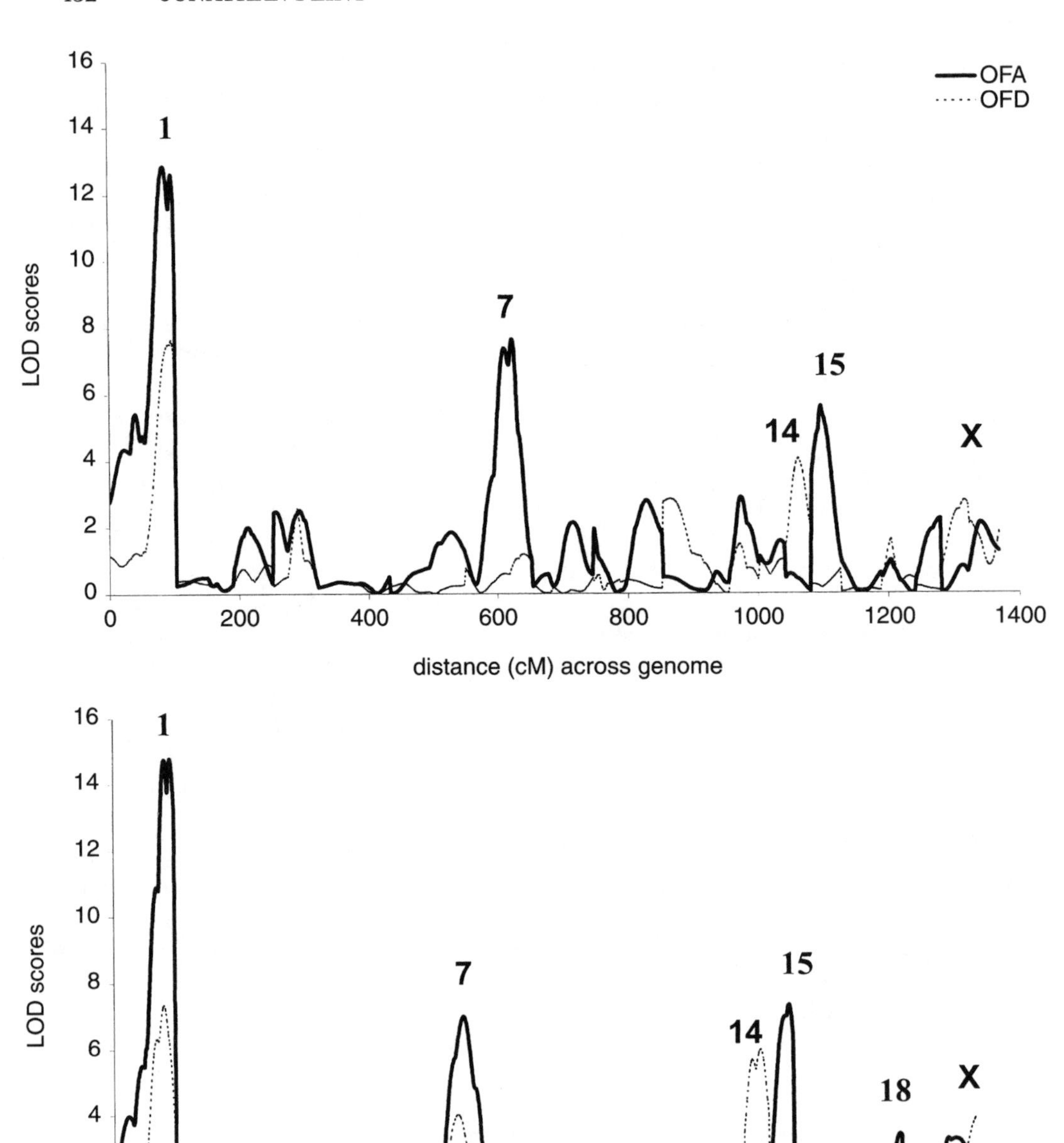

Figure 21.1. LOD scores (vertical axis) for measures from open-field activity (OFA; continuous line) and open-field defecation (OFD; dotted line) in a replicated quantitative trait loci mapping experiment. The horizontal axis represents the distance in centimorgans across the whole genome, from chromosome 1 on the left to the X chromosome on the right. The top panel shows the LOD scores for a cross between H1 and L1 strains, and the bottom panel the results from the H2 and L2 strains. Chromosomes with LOD scores exceeding a 5% significance threshold are shown as numbers on the graph.

five QTL for OFA that exceed the 5% significance threshold, three were found in crosses between H1 × L1 and in H2 × L2.

However, when other anxiety phenotypes were mapped, although a small number of QTLs influenced each additional measure, the pattern was much more complex (Turri, Datta, et al., 2001). Five unconditioned tests of anxiety were used: the open-field arena, the elevated plus maze (Pellow et al., 1985), the square maze (Shepherd, Grewal, Fletcher, Bill, & Dourish, 1994), the light-dark box (Costall, Jones, Kelly, Naylor, & Tomkins, 1989), and the mirror chamber (Toubas, Abla, Cao, Logan, & Seale, 1990). QTLs on chromosomes 1, 4, 15, and 18 influenced at least one measure obtained in all five tests. Other loci, on chromosomes 7, 8, 12, and X, influenced behaviors in some but not all tests, and QTLs on chromosomes 11 and 14 were found to be test specific. A small number of QTLs influenced a single measure taken in a single test (e.g., a QTL on chromosome 6 influenced time spent in the closed arms of the elevated plus maze).

To determine whether these QTLs have a common influence on anxiety, the effects of each QTL were examined on different components of the apparatus used to measure anxiety. Then the effect of some QTLs could be explained with respect to the phenotypes they influenced.

In each test, the animal is presented with a choice between threatening and nonthreatening environments. In the elevated plus maze and square maze, mice have a choice between two relatively anxiogenic regions (the open arms) and two relatively safe regions (the closed arms; Hogg, 1996; Pellow et al., 1985; Rodgers & Dalvi, 1997). The light–dark box and mirror chamber also permit the animal to explore a novel situation or remain in a relatively nonthreatening dark enclosure. The difference lies in the nature of the novel environment: either a well-lit exposed area or a mirrored chamber (Lamberty, 1998). Even within the open field, there are thought to be distinctions in the level of threat the exposed area provides: The periphery is safer than the center, and latency to reach the center of the open field is thought to measure the same behavior as latency to emerge from the light–dark box or enter a mirrored chamber (Lister, 1990).

QTL effects were interpreted within this framework. In the inbred cross experiments, each QTL has just two alleles. Alleles of a QTL with an influence on anxiety should have opposing effects on the animal's exposure to threatening environments and should have little or no effect in emotionally neutral measures, such as spontaneous activity. By categorizing QTLs in this way, a QTL on chromosome 4 was found to influence locomotor activity, not anxiety: It acted on activity in the home cage (a nonthreatening environment) and on activity in tests of anxiety, regardless of the supposed anxiogenic potential of the region in which activity is measured. For instance, it had no effect on the time spent in anxiogenic regions. By contrast, QTLs on chromosomes 1 and 15 did not influence spontaneous activity but did have effects on time measures in tests of anxiety, so that the allele that decreases activity in an anxiogenic component also decreases the time spent there.

More detailed examination of the phenotypes on which the latter two loci act showed that their effects could be dissociated in three ways. First, the locus on chromosome 15 had little influence on behavior in nonthreatening regions (e.g., the enclosed arms of the elevated plus maze and dark compartment of the light–dark box), whereas the locus on chromosome 1 had equal if not greater effects in nonthreatening regions. Second, the QTL on chromosome 15 was the only locus with consistent and large effects on the time taken to enter aversive regions of the test apparatus. Third, prior exposure to the apparatus, which should reduce the anxiogenic potential of the test, diminished the chromosome 15 QTL's effect.

Multivariate analyses have also been used to explore the genetic architecture of a battery of behavioral tests, all of which are used as animal models of anxiety (Fernández-Teruel et al., 2002). In a study in rats, eight QTLs were found that influenced a range of tests, of which three, on chromosomes 5, 10, and 15, influence more than one behavioral measure of fear (Fernández-Teruel et al., 2001). Multivariate approaches were used to establish the significance of the contribution to the LOD score of each trait. These analyses showed that loci on chromosomes 1, 3, 6, 19, and X had effects inconsistent with an influence on fear responses (e.g., the QTL on chromosome 6 affects spontaneous activity and defecation in a novel environment), whereas loci on chromosomes 5, 10, and 15 influenced a broad range of measures of fear, as would be expected if they contain genes involved in determining a response to fear-provoking stimuli. As was found in the mouse, most loci had relatively specific effects.

Again, examination of the direction of QTL effects provided support for the QTL's influence on anxiety. On chromosome 5, an allele that increased two-way active avoidance also decreased cue and contextual fear conditioning and grooming, while increasing time in the open arms of the elevated plus maze and activity in the open field, as well as rearing. The QTL had no discernible influence on spontaneous activity, the acoustic-startle response, or defecation, a pattern consistent with the action of a gene influencing an animal's reaction to a fear stimulus, and paralleling the effects of drugs used to treat anxiety disorders in humans (Fernández-Teruel et al., 1991; Gray, 1977; Simon & Soubrie, 1979). Neither anxiolytic drugs nor the QTL affects the acoustic startle response or defecation. The finding that genetic effects on defecation can be dissociated from other tests of fear is supported by an analysis of Maudsley rat strains, derived by selection for differences in open-field defecation (Paterson, Whiting, Gray, Flint, & Dawson, 2001).

Comparison with the genetic mapping carried out in mice shows no apparent overlap in the locations of QTL. In the mouse, QTLs influencing behavior in the open field, elevated plus maze, light–dark box, and fear conditioning have been mapped to chromosomes 1, 3, 10, 12, and 15 (Caldarone et al., 1997; Gershenfeld & Paul, 1997; Turri, Datta, et al., 2001; Wehner et al., 1997). The syntenic regions in rat are chromosomes 13, 2, 7, 6, and 7, respectively (Blake et al., 2001), chromosomes devoid of QTLs that have an effect on the same behaviors in rats. Crosses between inbred mouse and rat strains sample only a fraction of the genetic variation in the two species, so one cannot at this point say whether the failure to detect syntenic locations represents a difference in the genetic architecture of anxiety in mouse and rat or merely a failure to choose the right strain combination.

In summary, it is possible to detect QTLs that have a consistent pattern of effects across a broad range of relevant tests of animal behavior. This is important because, despite many years of research, evidence that genetic effects operate in this way has been lacking. Using a large battery of tests of anxiety, including both ethological and conditioned measures, and using multivariate methods for the analysis of QTL data, together with predictions of gene action, it has been possible to dissect the effects of each QTL and determine which are most likely to represent loci with an effect on anxiety. These analyses have identified two loci that influence variation in anxiety: In the mouse, a QTL on chromosome 15, and in the rat, a QTL on chromosome 5 are the best candidates at present.

Finding the Genes That Influence Emotionality

The results discussed so far have been obtained using crosses between inbred animal strains. While mapping QTLs is relatively straightforward using inbred crosses, identifying the responsible molecular variant is very taxing, largely because of the difficulties of resolving loci into intervals small enough to identify genes. Most QTL mapping experiments in animals start with a detection stage and then proceed to localize the QTL more precisely. The problem has been that QTL detection methodologies using inbred crosses have very poor resolving power: The 95% confidence intervals for moderate effect QTLs (accounting for 5%–10% of the phenotypic variance) usually encompass half a chromosome (Darvasi & Soller, 1997). A number of approaches to fine-map single loci have been introduced, but they founder when a single QTL of apparent large effect turns out to be due to multiple loci of small effect (Legare, Bartlett, & Frankel, 2000). Alternative approaches using inbred crosses require either very large numbers of animals in a single cross (tens of thousands) or many generations of intercrossing (which may take years) to achieve subcentimorgan resolution (Darvasi, 1998; Darvasi & Soller, 1995).

My colleagues and I have shown that outbred animals can be used for very high-resolution QTL mapping. We mapped QTLs accounting for less than 5% of the phenotypic variance into intervals of a centimorgan or less in a genetically heterogeneous stock (HS) of mice (Talbot et al., 1999). The strategy uses an eight-way cross that had been randomly bred for 60 generations, and the extent of linkage disequilibrium is known to be 1.7 cM (Talbot et al., 1999). Genetic effects should therefore be detectable by allelic association with markers spaced at 1 cM intervals. Using this strategy, we detected one QTL in the HS on chromosome 1, the same as the QTL segregating in an A/J and C57BL/6 F2 intercross (Gershenfeld et al., 1997; Gershenfeld & Paul, 1997). However, we could not find the QTL that we had first detected segregating in the DeFries strains, which were derived from C57BL/6 and BALBc/J (Turri, Talbot, Radcliffe, Wehner, & Flint, 1999).

One possible reason for the failure is that QTLs with opposing effects have identical alleles at nearby marker loci, thus avoiding detection by allelic association. For example, consider two markers (D1Mit100 and D1Mit496), less than 500 kb apart, near the QTL at position 64 cM on chromosome 1. Single-point analysis gave a highly significant result at one marker (D1Mit100) but failed to detect the QTL at the neighboring marker D1Mit496, situated less than 500 kb away. The proximity of the markers rules out recombination as a reason for the difference in significance. In fact, the explanation is that alleles of the same size are descended from different strains. The HS is derived from eight inbred strains (C57BL/6, BALB/c, RIII, AKR, DBA/2, I, A, and C3H; McClearn, Wilson, & Meredith, 1970), but we cannot unambiguously determine the strain from which an allele descends, because at most markers there are only two or three alleles. The important difference between D1Mit496 and D1Mit100 is that strain RIII can be distinguished from A/J, C3H, and I at D1Mit100. Markers with the same grouping of strains as D1Mit100 gave significant results; markers with other strain groupings show no significant association.

To overcome this problem, we developed a multipoint method that determines the likely progenitor of each allele in the HS, exploiting the fact that progenitor haplotypes are known. QTLs are then detected by testing for differences between the genetic effects of the progenitor haplotypes rather than by association at each locus. Note that it would not help to reconstruct the haplotypes of the HS at the generation we tested, as this would not determine whether (in the example) an allele at D1Mit496 was derived from RIII or from one of the other strains. The critical issue is to calculate the probability that an allele is descended from one of the eight progenitors. Because the number of possible ancestral haplotype reconstructions increases exponentially as we move from one end of a chromosome to the other, it is impossible to calculate the probability of each haplotype separately. However, a multipoint analysis, which uses a dynamic programming algorithm, reduces the complexity to manageable proportions. Therefore, a two-stage analysis was carried out: haplotype probability reconstruction using dynamic programming followed by hypothesis testing using linear regression. With this method, all QTLs found in the inbred crosses were detected in the HS and mapped to a high resolution. This approach opens the road to cloning a large number of variants; it also confirms suspicions that the genetics of anxiety has a complex architecture.

The development and application of the dynamic programming analysis overcomes the problem of phase association and makes the use of the outbred strains a robust and efficient strategy for high-resolution mapping. Mapping resolution in the HS is sufficiently high to restrict the search for candidate genes into a region of between one and two megabases. While this is still a large region to search, the task will become tractable with the availability of complete mouse and human genome sequences within the next few years. Confirming candidates as QTLs remains a challenge but is much easier than in human studies (Horikawa et al., 2000; Risch, 2000). The use of laboratory animals simplifies the task, as do advances in genomics.

Many possibilities are now available for confirming the identity of a gene as a QTL (Flint & Mott, 2001). First, it is possible to identify all genes in a given chromosomal region by cross-species comparison, a method whose power to identify both expressed sequences and noncoding regulatory elements is well documented (Flint et al., 2001; Hardison, 2000; Hardison, Oeltjen, & Miller, 1997; Loots et al., 2000). Second, it will be possible to resequence regions of interest to find all variants segregating in the outbred animals and identify the direction of effect of each QTL allele: To be considered a candidate QTL, a sequence variant must have a strain distribution pattern consistent with the pattern of effects observed in the QTL detection experiment (e.g., if the increasing QTL effect is found on an A/J allele and the decreasing allele on a C57BL/6 allele, then the QTL must be polymorphic between these two strains). Third, it is possible to test candidate genes by transgenic analysis (as was recently achieved with the successful identification of a QTL affecting tomato fruit weight; Frary et al., 2000). Finally, QTL mapping will complement the mutagenesis programs as a way to identify genes (Nolan et al., 2000). Mutations that map to the same locations as QTLs for the same phenotype may provide an additional way of identifying genes.

Conclusion

Access to the complete genome of the human species, and soon also of the mouse, coupled with complete transcript maps and high-density single nucleotide polymorphism maps, has raised expectations that it will soon become relatively straightforward to identify the molecular determinants of complex phenotypes, including behavior. Because so many of the genes discovered during the Human Genome Project are novel, it is likely that many of the genetic variants influencing behavioral phenotypes will lie in genes of unknown function. In the coming years, the challenge will be to understand the function of these genes. In this task, mouse models will play an essential role. Studies in the mouse will provide an unparalleled ability to relate genetic variants to behavior.

Perhaps the most critical skill that will be needed for postgenomic behavioral genetics will be that of the experimental psychologist, in devising the most appropriate behavioral tests. There is certainly room for innovation here: As I have emphasized, the list of behavioral tasks in mice is still relatively limited, but it is also clear that novel, useful tasks can be devised (Humby, Laird, Davies, & Wilkinson, 1999). We can expect further developments in this area. There is little doubt that animal models will remain a critical resource for genetic studies for a long while to come.

References

Bignami, G. (1965). Selection for high rates and low rates of avoidance conditioning in the rat. *Animal Behaviour, 13,* 221–227.

Blake, J. A., Eppig, J. T., Richardson, J. E., Bult, C. J., Kadin, J. A., & Group MGD. (2001). The Mouse Genome Database (MGD): Integration nexus for the laboratory mouse. *Nucleic Acids Research, 29,* 91–94.

Broadhurst, P. L. (1960). Application of biometrical genetics to the inheritance of behaviour. In H. J. Eysenck (Eds.), *Experiments in personality* (pp. 3–102). London: Routledge & Kegan Paul.

Caldarone, B., Saavedra, C., Tartaglia, K., Wehner, J. M., Dudek, B. C., & Flaherty, L. (1997). Quantitative trait loci analysis affecting contextual conditioning in mice. *Nature Genetics, 17,* 335–337.

Cerbone, A., Pellicano, M. P., & Sadile, A. G. (1993). Evidence for and against the Naples high-excitability and low-excitability rats as genetic model to study hippocampal functions. *Neuroscience and Biobehavioral Reviews, 17,* 295–303.

Costall, B., Jones, B. J., Kelly, M. E., Naylor, R. J., & Tomkins, D. M. (1989). Exploration of mice in a black and white test box: Validation as a model of anxiety. *Pharmacology, Biochemistry and Behavior, 32,* 777–785.

Crawley, J. N. (1981). Neuropharmacologic specificity of a simple animal-model for the behavioral actions of benzodiazepines. *Pharmacology, Biochemistry and Behavior, 15,* 695–699.

Crawley, J. N., Ninan, P. T., Pickar, D., Chrousos, G. P., Skolnick, P., & Paul, S. M. (1984). Behavioral and physiological-responses to benzodiazepine receptor antagonists. *Psychopharmacology Bulletin, 20,* 403–407.

Darvasi, A. (1998). Experimental strategies for the genetic dissection of complex traits in animal models. *Nature Genetics, 18,* 19–24.

Darvasi, A., & Soller, M. (1995). Advanced intercross lines, an experimental population for fine genetic mapping. *Genetics, 141,* 1199–1207.

Darvasi, A., & Soller, M. (1997). A simple method to calculate resolving power and confidence interval of QTL map location. *Behavior Genetics, 27,* 125–132.

Davis, M. (1991). Animal models of anxiety based on classical conditioning: The conditioned emotional response and fear-potentiated startle effect. In S. E. File (Ed.), *Psychopharmacology of anxiolytics and antidepressants* (pp. 187–212). New York: Pergamon Press.

Dawson, G. R., Crawford, S. P., Collinson, N., Iversen, S. D., & Tricklebank, M. D. (1995). Evidence that the anxiolytic-like effects of chlordiazepoxide on the elevated plus-maze are confounded by increases in locomotor-activity. *Psychopharmacology, 118,* 316–323.

DeFries, J. C., Gervais, M. C., & Thomas, E. A. (1978). Response to 30 generations of selection for open field activity in laboratory mice. *Behavior Genetics, 8,* 3–13.

DeFries, J. C., Wilson, J. R., & McClearn, G. E. (1970). Open-field behavior in mice: Selection response and situational generality. *Behavior Genetics, 1,* 195–211.

Dichter, G. S., Brunelli, S. A., & Hofer, M. A. (1996). Elevated plus-maze behavior in adult offspring of selectively bred rats. *Physiology and Behavior, 60,* 299–304.

Driscoll, P., & Battig, K. (1982). Behavioral, emotional and neurochemical profiles of rats selected for extreme differences in active two way avoidance. In I. Lieblich (Ed.), *Genetics of the brain* (pp. 95–123). Amsterdam: Elsevier.

Driscoll, P., Escorihuela, R. M., Fernández-Teruel, A., Giorgi, O., Schwegler, H., Steimer, T., et al. (1998). Genetic selection and differential stress responses: The Roman lines/strains of rats. *Annals of the New York Academy of Sciences, 851,* 501–510.

Escorihuela, R. M., Fernández-Teruel, A., Gil, L., Aguilar, R., Tobena, A., & Driscoll, P. (1999). Inbred Roman high- and low-avoidance rats: Differences in anxiety, novelty-seeking, and shuttlebox behaviors. *Physiology and Behavior, 67,* 19–26.

Escorihuela, R. M., Tobena, A., Driscoll, P., & Fernández-Teruel, A. (1995). Effects of training, early handling, and perinatal flumazenil on shuttle box acquisition in Roman low-avoidance rats: Toward overcoming a genetic deficit. *Neuroscience and Biobehavioral Reviews, 19,* 353–367.

Fernándes, C., González, M. I., Wilson, C. A., & File, S. E. (1999). Factor analysis shows that female rat behavior is characterised primarily by activity, male rats are driven by sex and anxiety. *Pharmacology, Biochemistry and Behavior, 64,* 731–738.

Fernández-Teruel, A., Escorihuela, R. M., Castellano, B., Gonzalez, B., & Tobena, A. (1997). Neonatal handling and environmental enrichment effects on emotionality, novelty/reward seeking, and age-related cognitive and hippocampal impairments: Focus on the Roman rat lines. *Behavior Genetics, 27,* 513–526.

Fernández-Teruel, A., Escorihuela, R. M., Gray, J. A., Aguilar, R., Gil, L., Gimenez-Llort, L., et al. (2002). A quantitative trait locus influencing emotionality in the laboratory rat. *Genome Research, 12,* 618–626.

Fernández-Teruel, A., Escorihuela, R. M., Nunez, J. F., Zapata, A., Boix, F., Salazar, W., & Tobena, A. (1991). The early acquisition of two-way (shuttle-box) avoidance as an anxiety-mediated behavior: Psychopharmacological validation. *Brain Research Bulletin, 26,* 173–176.

File, S. E., & Hyde, J. R. G. (1978). Can social interaction be used to measure anxiety? *British Journal of Pharmacology, 62,* 19–24.

Flaherty, C. F., Greenwood, A., Martin, J., & Leszczuk, M. (1998). Relationship of negative contrast to animal models of fear and anxiety. *Animal Learning and Behavior, 26,* 397–407.

Flint, J., Corley, R., DeFries, J. C., Fulker, D. W., Gray, J. A., Miller, S., & Collins, A. C. (1995). A simple genetic basis for a complex psychological trait in laboratory mice. *Science, 269,* 1432–1435.

Flint, J., & Mott, R. (2001). Finding the molecular basis of quantitative traits: Successes and pitfalls. *Nature Reviews Genetics, 2,* 438–445.

Flint, J., Tufarelli, C., Peden, J., Clark, K., Daniels, R. J., Hardison, R., et al. (2001). Comparative genome analysis delimits a chromosomal domain and identifies key regulatory elements in the alpha globin cluster. *Human Molecular Genetics, 10,* 371–382.

Frary, A., Nesbitt, T. C., Grandillo, S., Knaap, E., Cong, B., Liu, J., et al. (2000). fw2.2: A quantitative trait locus key to the evolution of tomato fruit size. *Science, 289,* 85–88.

Fujita, O., Annen, Y., & Kitaoka, A. (1994). Tsukuba high-emotional and low-emotional strains of rats (*Rattus-norvegicus*): An overview. *Behavior Genetics, 24,* 389–415.

Geller, I. & Seifter, J. (1960). The effect of meprobamate, barbiturates, *d*-amphetamine and promazine on experimentally induced conflict in the rat. *Psychopharmacology, 1,* 482–492.

Gershenfeld, H. K., Neumann, P. E., Mathis, C., Crawley, J. N., Li, X., & Paul, S. M. (1997). Mapping quantitative trait loci for open-field behavior in mice. *Behavior Genetics, 27,* 201–210.

Gershenfeld, H. K., & Paul, S. M. (1997). Mapping quantitative trait loci for fear-like behaviors in mice. *Genomics, 46,* 1–8.

Gray, J. A. (1977). Drug effects on fear and frustration: Possible limbic sites of action of minor tranquilizers. In L. L. Iversen, S. D. Iversen, & S. H. Snyder (Eds.), *Handbook of psychopharmacology* (pp. 433–529). New York: Plenum.

Gray, J. A. (1982). *The neuropsychology of anxiety: An enquiry into the function of the septo-hippocampal system.* Oxford, England: Oxford University Press.

Gray, J. A. (1987). *The psychology of fear and stress* (2nd ed.). Cambridge, England: Cambridge University Press.

Griebel, G., Blanchard, C., & Blanchard, R. (1996). Evidence that behaviors in the mouse defense test battery relate to different emotional states: A factor analytic study. *Physiology and Behavior, 60,* 1255–1260.

Hall, C. S. (1934). Emotional behaviour in the rat. I. Defecation and urination as measures of individual differences in emotionality. *Journal of Comparative Psychology, 22,* 345–352.

Hardison, R. C. (2000). Conserved noncoding sequences are reliable guides to regulatory elements. *Trends in Genetics, 16,* 369–372.

Hardison, R. C., Oeltjen, J., & Miller, W. (1997). Long human-mouse sequence alignments reveal novel regulatory elements: A reason to sequence the mouse genome. *Genome Research, 7,* 959–966.

Hogg, S. (1996). A review of the validity and variability of the elevated plus maze as an animal model of anxiety. *Pharmacology, Biochemistry and Behavior, 54,* 21–30.

Horikawa, Y., Oda, N., Cox, N. J., Li, X., Orho-Melander, M., Hara, M., et al. (2000). Genetic variation in the gene encoding calpain-10 is associated with Type 2 diabetes mellitus. *Nature Genetics, 26,* 163–175.

Humby, T., Laird, F. M., Davies, W., & Wilkinson, L. S. (1999). Visuospatial attentional functioning in mice: Interactions between cholinergic manipulations and genotype. *European Journal of Neuroscience, 11,* 2813–2823.

Lamberty, Y. (1998). The mirror chamber test for testing anxiolytics: Is there a mirror-induced stimulation? *Physiology and Behavior, 64,* 703–705.

Lapin, I. P., & Mirzaev, S. M. (1996). Selection of anxious mice by a 3-stage test in an elevated T-maze. *Bulletin of Experimental Biology and Medicine, 121,* 325–326.

Legare, M. E., Bartlett, F. S., II., & Frankel, W. N. (2000). A major effect QTL determined by multiple genes in epileptic EL mice. *Genome Research, 10,* 42–48.

Liebsch, G., Linthorst, A. C. E., Neumann, I. D., Reul, J., Holsboer, F., & Landgraf, R. (1998). Behavioral, physiological, and neuroendocrine stress responses and differential sensitivity to diazepam in two Wistar rat lines selectively bred for high- and low-anxiety-related behavior. *Neuropsychopharmacology, 19,* 381–396.

Liebsch, G., Montkowski, A., Holsboer, F., & Landgraf, R. (1998). Behavioural profiles of two Wistar rat lines selectively bred for high or low anxiety-related behaviour. *Behavioural Brain Research, 94,* 301–310.

Lister, R. G. (1987). The use of a plus-maze to measure anxiety in the mouse. *Psychopharmacology, 92,* 180–185.

Lister, R. G. (1990). Ethologically-based animal models of anxiety disorders. *Pharmacology and Therapeutics, 46,* 321–340.

Loots, G. G., Locksley, R. M., Blankespoor, C. M., Wang, Z. E., Miller, W., Rubin, E. M., & Frazer, K. A. (2000). Identification of a coordinate regulator of interleukins 4, 13, and 5 by cross-species sequence comparisons. *Science, 288,* 136–140.

McClearn, G. E., Wilson, J. R., & Meredith, W. (1970). The use of isogenic and heterogenic mouse stocks in behavioral research. In G. Lindzey & D. Thiessen (Eds.), *Contributions to behavior-genetic analysis: The mouse as a prototype* (pp. 3–22). New York: Appleton Century Crofts.

Moisan, M. P., Courvoisier, H., Bihoreau, M. T., Gauguier, D., Hendley, E. D., Lathrop, M., et al. (1996). A major quantitative trait locus influences hyperactivity in the Wkha rat. *Nature Genetics, 14,* 471–473.

Nolan, P. M., Peters, J., Strivens, M., Rogers, D., Hagan, J., Spurr, N., et al. (2000). A systematic, genome-wide, phenotype-driven mutagenesis programme for gene function studies in the mouse. *Nature Genetics, 25,* 440–443.

Owen, E. H., Christensen, S. C., Paylor, R., & Wehner, J. M. (1997). Identification of quantitative trait loci involved in contextual and auditory-cued fear conditioning in BXD recombinant inbred strains. *Behavioral Neuroscience, 111,* 292–300.

Paterson, A., Whiting, P. J., Gray, J. A., Flint, J., & Dawson, G. R. (2001). Lack of consistent behavioural effects of Maudsley reactive and non-reactive rats in a number of animal tests of anxiety and activity. *Psychopharmacology, 154,* 336–342.

Pellow, S., Chopin, P., File, S., & Briley, M. (1985). Validation of open:closed arms entries in an elevated plus maze as a measure of anxiety in the rat. *Journal of Neuroscience Methods, 14,* 149–167.

Pellow, S., & File, S. E. (1986). Anxiolytic and anxiogenic drug effects in exploratory activity in an elevated plus-maze: A novel test of anxiety in the rat. *Pharmacology, Biochemistry and Behavior, 24,* 525.

Ramos, A., Berton, O., Mormede, P., & Chaouloff, F. (1997). A multiple-test study of anxiety-related behaviours in six inbred rat strains. *Behavioural Brain Research, 85,* 57–69.

Ramos, A., & Mormede, P. (1998). Stress and emotionality: A multidimensional and genetic approach. *Neuroscience and Biobehavioral Reviews, 22,* 33–57.

Risch, N. J. (2000). Searching for genetic determinants in the new millennium. *Nature, 405,* 847–856.

Rodgers, R. J., Cao, B. J., Dalvi, A., & Holmes, A. (1997). Animal models of anxiety: An ethological perspective. *Brazilian Journal of Medical and Biological Research, 30,* 289–304.

Rodgers, R. J., & Dalvi, A. (1997). Anxiety, defence and the elevated plus-maze. *Neuroscience and Biobehavioral Reviews, 21,* 801–810.

Rodgers, R. J., & Johnson, J. T. (1995). Factor analysis of spatiotemporal and ethological measures in the murine elevated plus-maze of anxiety. *Pharmacology, Biochemistry and Behavior, 52,* 297–303.

Shepherd, J. K., Grewal, S. S., Fletcher, A., Bill, D. J., & Dourish, C. T. (1994). Behavioural and pharmacological characterisation of the elevated "zero-maze" as an animal model of anxiety. *Psychopharmacology (Berlin), 116,* 56–64.

Simon, P., & Soubrie, P. (1979). Behavioral studies to differentiate anxiolytic and sedative activity of the tranquilizing drugs. In J. R. Boisser (Eds.), *Modern problems of pharmacopsychiatry: Vol. 14. Differential psychopharmacology of anxiolytics and sedatives* (pp. 1188–2102). Basel, Switzerland: Karger.

Steimer, T., Driscoll, P., & Schulz, P. E. (1997). Brain metabolism of progesterone, coping behaviour and emotional reactivity in male rats from two psychogenetically selected lines. *Journal of Neuroendocrinology, 9,* 169–175.

Talbot, C. J., Nicod, A., Cherny, S. S., Fulker, D. W., Collins, A. C., & Flint, J. (1999). High-resolution mapping of quantitative trait loci in outbred mice. *Nature Genetics, 21,* 305–308.

Toubas, P. L., Abla, K. A., Cao, W., Logan, L. G., & Seale, T. W. (1990). Latency to enter a mirrored chamber: A novel behavioral assay for anxiolytic agents. *Pharmacology, Biochemistry and Behavior, 35,* 121–126.

Toye, A. A., & Cox R. (2001). Behavioral genetics: Anxiety under interrogation. *Current Biology, 11,* 473–476.

Turri, M. G., Datta, S. R., DeFries, J., Henderson, N. D., & Flint, J. (2001). QTL analysis identifies multiple behavioral dimensions in ethological tests of anxiety in laboratory mice. *Current Biology, 11,* 725–734.

Turri, M. G., Henderson, N., DeFries, J. C., & Flint, J. (2001). Quantitative trait locus (QTL) mapping in laboratory mice derived from a replicated selection experiment for open-field activity. *Genetics, 158,* 1217–1226.

Turri, M. G., Talbot, C. J., Radcliffe, R. A., Wehner, J. M., & Flint, J. (1999). High-resolution mapping of quantitative trait loci for emotionality in selected strains of mice. *Mammalian Genome, 10,* 1098–1101.

Vogel, J. R., & Beer, B. D. E. C. (1971). A simple and reliable conflict procedure for testing antianxiety agents. *Psychopharmacology, 100,* 138–140.

Wehner, J. M., Radcliffe, R. A., Rosmann, S. T., Christensen, S. C., Rasmussen, D. L., Fulker, D. W., & Wiles, M. (1997). Quantitative trait locus analysis of contextual fear conditioning in mice. *Nature Genetics, 17,* 331–334.

Willner, P. (1990). Animal models for clinical psychopharmacology: Depression, anxiety, schizophrenia. *International Review of Psychiatry, 2,* 253–276.

Part VIII

Psychopathology

22

Attention Deficit Hyperactivity Disorder: New Genetic Findings, New Directions

Anita Thapar

Attention deficit hyperactivity disorder (ADHD) is a condition characterized by severe hyperactivity, impulsiveness, and inattention with an onset before age 7. The disorder affects between 2% and 6% of children and commonly co-occurs with other disorders, for example, oppositional defiant disorder (ODD), conduct disorder (CD), and reading disability (Barkley, 1998). ADHD is known to be associated with major difficulties in childhood, such as educational failure and problems with peer and family relationships (Taylor, 1994). Moreover, there is increasing evidence that ADHD can persist into adolescence and even adult life and that it is associated with an increased risk of antisocial behavior and drug and alcohol misuse (Barkley, 1998). Although ADHD is defined categorically for clinical purposes, much research has also been undertaken in which a score based on ADHD-type symptoms is considered as a continuously distributed trait measure.

There is now a wealth of evidence demonstrating the importance of genetic factors in the etiology of ADHD. In this chapter, I begin by summarizing the evidence from family, twin, and adoption studies. These genetic epidemiological studies have also highlighted phenotype issues that are likely to be of importance for molecular genetic studies. The most consistent and relevant findings are discussed. I then move on to consider molecular genetic findings that have been replicated, before concluding with thoughts on new directions and strategies. Finally, research on the genetic basis of ADHD is flourishing and new studies are being published every week. The intention in this chapter is not to describe all of these findings but rather to focus on results that have been replicated or on findings that have converged across different study designs and also to discuss new evidence that seems likely to be of particular importance or significance.

This research was supported by the Wellcome Trust.

Evidence From Family, Twin, and Adoption Studies

Most family studies have focused on children who have been referred to clinics and who fulfill diagnostic criteria for ADHD. There is consistent evidence that the rate of ADHD in first-degree relatives is higher than among relatives of controls, although estimates of relative risk (λr) are modest with an overall reported λr of 4–5.4 (Biederman et al., 1992; Faraone, Biederman, & Monuteaux, 2000).

There have so far been at least 12 published twin studies from the United States, United Kingdom, Australia, and Norway, nearly all of which have examined ADHD trait measures (Eaves et al., 1997; Edlebrock, Rende, Plomin, & Thompson, 1995; Gjone, Stevenson, & Sundet, 1996; Goodman & Stevenson, 1989; Hudziak, Rudyer, Neale, Heath, & Todd, 2000; Kuntsi & Stevenson, 2001; Levy, Hay, McStephen, Wood, & Waldman, 1997; Neuman et al., 1997; Schmitz, Fulker, & Mrazek, 1995; Sherman, Iacono, & McGue, 1997; Thapar, Harrington, Ross, & McGuffin, 2000; Thapar, Hervas, & McGuffin, 1995). These studies have been based on general population samples, and most have used questionnaire measures of ADHD, with the Rutter A scale and the Child Behavior Checklist (CBCL) being two of the most commonly used instruments. Findings have been remarkably consistent with all these studies showing that ADHD symptom scores are highly heritable when symptoms are rated by parents (heritability estimate h^2 = 60%–88%) and by teachers (h^2 = 39%–72%; Thapar, Holmes, Poulton, & Harrington, 1999). All of these studies examined variation in scores across the entire range, and this raises the question of whether the etiology of symptoms at the upper end of the distribution, which may equate to the clinical range, is the same. Several groups have investigated this issue by using the DeFries and Fulker extremes (DF) method of analysis (Gillis, Gilger, Pennington, & DeFries, 1992; Kuntsi & Stevenson, 2001; Levy et al., 1997; Stevenson, 1992). This approach allows a test of whether heritability is the same at the extreme as in the rest of the range of scores, and with respect to ADHD this appears to be so; genetic influences also contribute to extreme scores (h_g^2 = 42%–91%). Similarly, there is also evidence of a genetic contribution to ADHD when it is defined as a broad category (h^2 = 64%–89%; Goodman & Stevenson, 1989; Sherman, McGue, & Iacono, 1997; Thapar et al., 2000). Twin study findings suggest that ADHD can be viewed as a continuously distributed trait, a finding also supported by epidemiological studies that have examined the risk effects and predictive validity of ADHD symptoms (Rutter, Giller, & Hagell, 1998).

Adoption studies of ADHD also have allowed examination of the contribution of genes and environment (Alberts-Corush, Firestone, & Goodman, 1986; Cantwell, 1975; Morrison & Stewart, 1975; Safer, 1973; Sprich, Biederman, Crawford, Mundy, & Faraone, 2000). These studies have tended to be small and based on highly selected samples; the earliest work did not use modern diagnostic criteria. Nevertheless, again there is consistent evidence from all of these studies that the rate of ADHD and attentional difficulties is much higher in the biological relatives of affected individuals than among adoptive relatives. Overall, although family, twin, and adoption studies have different methodological strengths and weaknesses, there are consistent and convergent findings from all of these types of study showing that genetic factors have a moderate to high contribution to ADHD.

Phenotype Issues of Importance for Molecular Genetic Studies

Comorbidity

It is well recognized that ADHD is commonly accompanied by antisocial behavior. There is evidence from epidemiological and clinical studies that symptoms of ADHD and CD covary and that there is a high degree of co-occurrence of ADHD and ODD and CD. Family and twin studies have examined the issue of comorbidity in two ways. The first is essentially dimensional, examining the extent to which a common set of genes and environmental influences contribute to the overlap of ADHD symptoms and antisocial behaviors; the second is categorical, investigating the subgroup of children who show both ADHD and CD problems (ADHD+CD).

Twin studies that have examined the overlap of ADHD and antisocial symptoms (Kuntsi & Stevenson, 2001; Nadder, Silberg, Eaves, Maes, & Meyer, 1998; Silberg et al., 1996; Thapar, Harrington, & McGuffin, 2001) have all shown that a common genetic influence accounts for much of the covariation. These findings suggest that the genetic factors that influence antisocial behavior are the same ones that contribute to ADHD symptoms.

The second issue that has been examined relates to the subgroup of children with ADHD+CD. Clinical and follow-up studies have shown consistently that those with ADHD+CD represent a more severely affected group in that they show more severe symptoms and a poorer outcome than children with pure ADHD or pure CD. In a series of family studies of ADHD (Biederman et al., 1992; Faraone, Biederman, Keenan, & Tsuang, 1991; Faraone, Biederman, Mennin, Russell, & Tsuang, 1998), ADHD+CD appeared to represent a qualitatively or quantitatively distinct group. Furthermore, the relative risk, λr, of ADHD among relatives of those with ADHD+CD was estimated as between 4.8 and 9.5, which is higher than the overall λr for ADHD (Faraone et al., 2000). Increased familial loading could, however, be explained by shared environmental adversity as well as by genetic influences. Findings from one twin study (Thapar et al., 2001) also have suggested that ADHD+CD represents a subgroup for which there is higher genetic loading, although here ADHD+CD had to be defined broadly using cutoffs on questionnaire measures. Thus family and twin study findings raise the possibility that those with ADHD+CD may represent a subgroup of interest for molecular genetic studies.

ADHD also commonly occurs with reading disability, and twin study findings have suggested a common genetic influence on symptoms of inattentiveness and reading disability (Willcutt, Pennington, & DeFries,

2000). This topic is covered in more detail in Willcutt et al. (chapter 13, this volume).

Rater Issues

ADHD symptoms in research studies of children are most commonly rated by parents and teachers. For clinical purposes, a clinical history generally is required from both the parents and the school given that the *International Classification of Diseases (ICD–10;* World Health Organization, 1992) criteria for hyperkinetic disorder and the *Diagnostic and Statistical Manual of Mental Disorders* (4th ed. [*DSM–IV*]; American Psychiatric Association, 1994) criteria for ADHD require the manifestation of symptoms in more than one setting. Twin studies of ADHD have repeatedly revealed a puzzling finding: A low and in many instances zero or negative dizygotic (DZ) twin correlation. Although nonadditive effects such as genetic dominance, epistasis, and gene–environment interaction could account for the very much higher monozygotic (MZ) twin correlations relative to the DZ twin correlations, there is clearly no plausible biological explanation for a negative DZ twin correlation.

At least five separate twin studies have found a negative DZ twin correlation (Eaves et al., 1997; Goodman & Stevenson, 1989; Kuntsi & Stevenson, 2001; Nadder et al., 1998; Thapar et al., 1995), and although many of these studies used the Rutter A questionnaire (Rutter, Tizard, & Whitmore, 1970), one used the Conners scale (Kuntsi & Stevenson, 2001), another used a telephone interview (Nadder et al., 1998), and the other included a comprehensive research diagnostic interview (Eaves et al., 1997), the Child and Adolescent Psychiatric Assessment (Angold et al., 1995), an instrument that is now being widely used in molecular genetic studies of ADHD. These findings could be explained by sibling competition (increased ADHD scores in one twin leading to decreased scores in the other twin) or rater contrast effects, or they may be attributable to nonlinear chaotic influences (Eaves, Kirk, Martin, & Russell, 1999). So far, analyses of parent and teacher data have suggested that these effects could be explained by rater contrast effects whereby mothers of twins tend to exaggerate differences in symptom scores between the twins (Simonoff et al., 1998).

This rater effect would clearly pose a potential confounder for studies attempting to find linkage with quantitative trait loci (QTLs) in sibling pairs and raises the question of whether mothers are the best informants for rating symptoms in such studies. However, findings have not been entirely consistent. Other twin studies (e.g., Hudziak et al., 2000; Levy et al., 1997; Sherman, Iacano, & McGue, 1997; Thapar et al., 2000) have failed to show a negative DZ twin correlation, with reported DZ twin correlations that vary between 0.3 and 0.5. This suggests that rater contrast effects appear to vary depending on the measure used. For example, reasonable DZ twin correlations have been reported in some studies that used the CBCL (Hudziak et al., 2000), the DuPaul ADHD rating scale (Thapar et al., 2000), and DSM-based rating scales (e.g., Levy et al., 1997). Thus the choice of measure used for molecular genetic studies that include siblings clearly is critical. Nevertheless, it remains unclear as to what extent other factors also contribute to the low DZ twin correlations.

Related questions are whether teachers should be used as informants instead of parents and do teachers rate a similar phenotype? Overall, twin studies suggest that genetic influences contribute to teacher ratings of ADHD, although some (Sherman, Iacono, & McGue, 1997; Thapar et al., 2000) but not all studies (e.g., Eaves et al., 1997) have found that shared environmental factors also contribute substantially. In part, this could be attributable to rater biases, with teachers rating twins more similarly. DF analyses of extreme teacher-rated ADHD scores have yielded relatively low heritability estimates for high scores (16%–20%; Kuntsi & Stevenson, 2001; Stevenson, 1992), although these have been based on smaller samples. Nevertheless, there is still a considerable overlap of teacher and parent symptom ratings, and a number of twin studies (e.g., Sherman, Iacono, & McGue, 1997; Thapar et al., 2000) have shown common genetic influences on parent- and teacher-rated symptoms. Although for clinical purposes there is much advantage in using parent and teacher reports to obtain a measure of pervasiveness of symptoms, so far there has been no evidence from twin studies to suggest that pervasive ADHD necessarily represents an improved phenotype definition for genetic studies (Sherman, McGue, & Iacono, 1997; Thapar et al., 2000). Nevertheless, this issue has not yet been examined fully in genetic studies, and thus it is premature to conclude that teacher as well as parent reports are not required for molecular genetic studies of ADHD. The use of different informants and multiple indices of a phenotype seems to represent a sensible way forward. However, again, so far there is little evidence to suggest that these approaches necessarily yield clearer findings from genetic studies, and clinicians would argue that it may prove difficult to then apply results from studies using such phenotypic definitions to clinically defined cases.

Related Measures / Endophenotypes

A difficulty for those investigating the genetic basis of nearly all psychiatric disorders is that phenotypes are defined on the basis of reported symptoms with no underlying biological measures available. Given this difficulty, there is clearly considerable appeal in examining whether phenotypes such as ADHD can be defined more "objectively." There has been a wealth of research examining the utility of tasks measuring traits and processes such as sustained attention, executive functioning processes, and response inhibition (Barkley, 1998). Although none of these tasks reliably distinguish children with ADHD, they should at least avoid the problem of rater bias. Nevertheless, before these measures are adopted for molecular genetic studies, it is clearly essential to demonstrate that the measures are strongly related to ADHD, that they are influenced genetically, and that genes that influence these task measures also are likely to be involved in the etiology of ADHD. Unfortunately, there have been relatively few family, twin, or adoption studies examining these issues.

There has been one adoption study of ADHD (Alberts-Corush et al., 1986) in which the biological parents of children with hyperactivity were found to show greater attentional difficulties than adoptive parents and control parents, but there was no evidence of greater impulsivity. There have been three twin studies in which objective measures of attention have been examined in children (Goodman & Stevenson, 1989; Holmes, Hever, et al., in press; Kuntsi & Stevenson, 2001). In Goodman and Stevenson's study, two indices of attentiveness (freedom from distractibility and "E" scan attentiveness) were used. These measures were found to be influenced moderately by genetic factors (32%–42%). In a smaller study of 40 twin pairs (Holmes, Hever, et al., in press), genetic influences were not found to contribute to measures of sustained attention using a computerized version of the continuous performance test. Although measures of impulsiveness using the Matching Familiar Figures Test appeared to be influenced by genetic factors (Holmes, Hever, et al., in press), surprisingly and unaccountably the DZ twin correlations were found to be zero or negative. In a more recent larger study (Kuntsi & Stevenson, 2001), results were again inconclusive. Bivariate twin analysis showed no evidence of a significant common genetic influence on reported hyperactivity symptoms and neuropsychological task measures of delay aversion and working memory. The only significant shared genetic effects detected were on hyperactivity and variability of reaction times on the Stroop task.

Thus, overall, although the use of neuropsychological tasks to define phenotypes for molecular genetic studies is appealing, the evidence so far is not at all compelling that they provide a superior measure compared with reported clinical symptoms. It is clear that further family, twin, and adoption work is needed to provide evidence that measures derived from these tasks are familial, genetically influenced, and genetically related to ADHD and for informing researchers which measures are likely to be the most useful and whether they are really likely to provide significant advantages over a phenotype defined by clinical symptoms.

ADHD Subtypes

The *DSM–IV* classification system (American Psychiatric Association, 1994) allows for defining subtypes of ADHD. The combined subtype requires the presence of both hyperactive–impulsive and inattentive symptoms, whereas the inattentive and hyperactive–impulsive subtypes only require the necessary number of symptoms from each of those separate symptom groups.

Although there has been much clinical research on these subgroups, unfortunately, family and twin studies yield conflicting findings. Findings based on a family study of ADHD (Faraone et al., 2000) and analysis of a sample of affected sibpairs collected for molecular genetic studies (Smalley et al., 2000) suggest that there is no familial distinction between the subtypes. That is, they do not appear to breed true, and nonfamilial factors seem to contribute to the subtype of ADHD. However, findings from two twin studies suggested somewhat distinct genetic (and environmental) influences on inattentive and hyperactive–impulsive symptoms (Hudziak et al., 1998; Sherman, Iacono, & McGue, 1997).

These conflicting sets of findings simply may reflect sampling differences. The family studies included selected clinical samples and used diagnostic measures of ADHD, whereas the twin studies were based on ADHD symptom scores in population samples.

Persistent/Adult ADHD

It is being recognized increasingly that ADHD can persist into adolescence and adult life. Although the concept of adult ADHD has been considered by some as somewhat controversial, there is increasing evidence that many children with ADHD continue to show significant symptoms and associated impairment in adulthood. However, follow-up studies have varied considerably in their estimates of persistence (10%–70%; Spencer, Biederman, Wilens, & Faraone, 1998). Nevertheless, there is an increasing number of studies reporting the benefits of stimulant medication for adult ADHD and showing evidence of abnormalities on neuropsychological testing and with neuroimaging studies (Spencer et al., 1998).

There have been two family studies (Biederman et al., 1995; Manshadi, Lippmann, O'Daniel, & Blackman, 1983), both of which suggest that adult ADHD is highly familial. Moreover, follow-up findings indicate that ADHD that persists over time may represent a much more strongly familial variant than childhood ADHD, with estimated λr of 17.2–19.7 (compared with about 4–5.4 when the probands were children with ADHD diagnosed at one time point). Should we be commencing preferentially molecular genetic studies of adult ADHD? A degree of caution is warranted. There have been no twin or adoption studies of adult ADHD that are needed to demonstrate the importance of a genetic influence, but, more importantly, there is much controversy regarding how adult ADHD should be defined. It is not clear whether adult ADHD should be defined strictly according to *DSM–IV* diagnostic criteria or whether a partial syndrome, in which there is a clear history of ADHD in childhood with fewer symptoms but associated impairment in adulthood, is preferable or whether the phenotype should be broadened considerably. Current diagnostic criteria were devised for children and may not be appropriate developmentally for an adult population. Some scales devised to assess adult ADHD (e.g., the Wender Utah scale; Ward, Wender, & Reimherr, 1993) include a very broad range of symptoms such as being "hot tempered," and it is not clear whether such symptoms should be included as part of the adult ADHD phenotype.

Another issue is the choice of informant for adult ADHD. Self-reports are usual in defining adult psychopathology, yet the evidence that is available suggests that self-reports of ADHD symptoms, at least in adolescents, show extremely low heritability (Martin, 2000) and may result in underestimates of symptoms (Barkley, 1998).

Molecular Genetic Findings

Despite the difficulties and unanswered questions, the wealth of family, twin, and adoption studies all suggesting the important role of genetic factors in the etiology of ADHD provide fairly compelling evidence for conducting molecular genetic studies. The search for susceptibility genes for ADHD has begun relatively recently, but preliminary findings have so far been encouraging. In this chapter, I focus on findings that have been independently replicated. To date, this includes results on the dopamine (DRD4) receptor gene and the dopamine transporter gene (DAT1).

Most studies to date have adopted a candidate gene association approach and used both family-based and case control designs. This seems reasonable given the likelihood that susceptibility genes are likely to be of small effect, particularly given that the λr for ADHD is only modest. The choice of candidate genes has been influenced by pharmacological, animal, and neuroimaging study findings.

There is particularly persuasive evidence supporting the involvement of the dopaminergic system (Vallone, Picetti, & Borrelli, 2000; Zametkin & Liotta, 1998). Seventy to 80% of children with ADHD show an immediate improvement in ADHD symptoms when given stimulant medication such as methylphenidate (e.g., Ritalin and Equasym) or dexamphetamine (e.g., dexedrine). These drugs are known to inhibit reuptake via the dopamine transporter (Amara & Kuhar, 1993) and increase synaptic levels of dopamine, although the mechanism of action is more complex than commonly assumed and involves other neurotransmitter systems (Solanto, 1998). Animal studies have also implicated the dopamine system. For example, dopamine transporter knock-out mice, that is, animals genetically engineered so that they do not have active dopamine transporter genes (Caron, 1996), show motor overactivity that is reduced when treated with methylphenidate. More recent findings also showed that fluoxetine (Prozac), a selective serotonin reuptake inhibitor (SSRI), has a calming effect on DAT1 knock-out mice (Gainetdinov et al., 1999). However, so far there is no consistent evidence that SSRIs are an effective treatment for ADHD in humans. Another interesting animal model is the DRD4 knock-out mouse that shows decreased novelty seeking (Dulawa, Grandy, Low, Paulus & Geyer, 1999), a behavior thought to be related to ADHD. Neuroimaging studies also implicate involvement of the dopamine system and more specifically the dopamine transporter, at least in adults (Krause, Dresel, Krause, Kung, & Tatsch, 2000).

Although interest has primarily focused on a dopamine hypothesis in the genetic etiology of ADHD and more recently the potential role of 5HT has been highlighted, involvement of the noradrenergic system has also been implicated. This has been suggested by findings from animal studies and the effect that stimulant medication (Solanto, 1998) and second-line therapeutic drugs for ADHD (e.g., clonidine, an alpha 2a receptor agonist and tricyclic antidepressants, particularly those such as desipramine and bupropion) have on noradrenergic pathways.

Dopamine Receptor DRD4 Gene

To date, there have been at least 15 published studies that have examined for association of a genetic variant/polymorphism (the 48bp VNTR in exon 3) of the DRD4 gene with ADHD (summarized in Table 22.1). Eight independent U.S., Canadian, and Turkish studies (Comings et al., 1999; Faraone et al., 1999; LaHoste et al., 1996; Muglia, Jain, Macciardi, & Kennedy, 2000; Smalley et al., 1998; Sunohara et al., 2000; Swanson et al., 1998; Tahir et al., 2000) have found evidence of association of the DRD4 7-repeat allele with ADHD. Three additional studies that examined unrelated cases compared with controls and also used a family-based method called the transmission disequilibrium test (TDT) analyses reported positive case control findings but negative TDT results (Holmes et al., 2000; Mill et al., 2001; Rowe et al., 1999). Four other studies to date from Ireland, Israel, the United States (three family-based and one case control) have failed to show any association with the DRD4 7-repeat allele (Castellanos et al., 1998; Eisenberg et al., 2000; Hawi et al., 2000; Kotler et al., 2000). A recent meta-analysis has used data from all these studies (Faraone, Doyle, Mick, & Biederman, 2001) and demonstrated significant association (and linkage) of the DRD4 7 repeat allele with ADHD from both family-based studies (odds ratio [OR] = 1.4; confidence intervals = 1.1–1.6) and case control studies (OR = 1.9; confidence intervals = 1.5–2.2). Thus overall, so far the evidence suggests that there is an association between the DRD4 7-repeat allele and ADHD, although the effect size is small.

Phenotype Issues and DRD4

Molecular genetic studies of ADHD that have specifically focused on DRD4 also have highlighted some potentially important phenotype issues. At least three of the studies further examined the effects of treatment response to stimulant medication (Holmes et al., 2000; Swanson et al., 1998; Tahir et al., 2000). One of these studies used a refined ADHD phenotype (no comorbidity, responders to stimulants) and showed a robust association with the 7-repeat allele (Swanson et al., 1998). The other study also found increased evidence of association in the subgroup of methylphenidate responders (Tahir et al., 2000). However, in the British study (Holmes et al., 2000), no effects of stimulant response were found.

The second area of interest has been in further examining the subgroup of those with ADHD and conduct problems given findings from family and twin studies that suggest that ADHD+CD may represent a distinct group. In the U.K. and Irish groups, in which association had either been

Table 22.1. Molecular Genetic Studies of DRD4[a] and ADHD

Authors	Sample (*N*) and origin	Design	Association and linkage finding
LaHoste et al., 1996	39, U.S.	case control	positive
Castellanos et al., 1998	41, U.S.	case control	negative
Swanson et al., 1998	52, U.S.	family based, HRR	positive
Rowe et al., 1998	70, U.S.	case control	positive
		family-based, TDT	negative
Smalley et al., 1998	133, U.S.	family-based, TDT	positive
Comings et al., 1999	52, U.S.	case control	positive[b]
Faraone et al., 1999	27, adults U.S.	family-based, TDT	positive
Tahir et al., 2000	111, Turkey	family-based, TDT	positive
Holmes et al., 2000	129, U.K.	case control	positive
	110,	family-based, TDT	negative
Hawi et al., 2000	99, Ireland	family-based, HHRR	negative
Eisenberg et al., 2000	46, Israel	family-based, HRR	negative
Muglia et al., 2000	66, adults, Canada	case control	positive
	44,	family-based, TDT	negative (positive trend)
Kotler et al., 2000	49, Israel	family-based, HRR	negative
Mill et al., 2001	132, U.K.	case control	positive
	85,	family-based, TDT	negative
Sunohara et al., 2000	88, Canada	family-based, TDT	positive
	59, U.S.—new cases plus 52 from Swanson et al. 1998, U.S.		negative (new U.S. cases alone)

Notes. Data published or known to be in press before January 1, 2001. See text for pooled odds ratios for DRD4 replicated; sample characteristics obtained from original publishers' papers. TDT = transmission disequilibrium test, HRR = haplotype relative risk method, HHRR = haplotype-based haplotype relative risk.

[a]Candidate allele: exon 4 VNTR 48 bp, −7 repeat allele.

[b]Positive findings when alleles 5, 6, 7, and 8 were pooled.

found using case control analysis but not with the TDT (Holmes et al., 2000; Mill et al., 2001) or not found at all using the TDT (Hawi et al., 2000), further TDT analysis of the subgroup of children with ADHD and conduct problems was performed. Separate analysis of this subgroup provided evidence of association and linkage with the 7-repeat allele (Holmes, Payton, Barrett, et al., 2002). However, there were too few cases to specifically examine those who met full criteria for CD. Although these findings support those suggested by family and twin studies, replication is needed. Furthermore, although heterogeneity according to stimulant response and comorbidity are clearly important and initial results appear promising, so far the sample sizes of subgroups have in general been too small, and consistent findings have yet to be reported before firm conclusions can be drawn.

The final issue of interest is how important is the diagnostic boundary of ADHD for molecular genetic studies? There are clearly compelling clinical reasons for maintaining a careful and stringent approach to making the diagnosis of ADHD. There is also a long tradition of painstaking care being taken over making psychiatric diagnoses for molecular genetic studies. This is clearly important and indeed essential, given the need for a high degree of reliability. There is also the assumption that selecting as homogeneous a group as possible (i.e., the most severely affected) will be more likely to successfully yield positive findings. It is therefore perhaps somewhat surprising yet interesting that two studies that used questionnaire measure screens on general population samples have shown or suggested association of the DRD4 7-repeat allele with high trait scores (Curran et al., 2001; Payton, Holmes, Barrett, Sham, et al., 2001). One study selected a sample of high-scoring concordant and low-scoring concordant MZ twins and showed a trend for association with the 7-repeat allele (OR = 1.4; 95% confidence interval = 0.6–2.9; Payton, Holmes, Barrett, Sham, et al., 2001). In the other study (Curran et al., 2001), high scorers were selected from a general population sample of children, and here significant association was found with the 7-repeat allele (OR = 2.09; 95% confidence interval = 1.24–3.54).

Dopamine Transporter Gene (DAT1)

There have been at least eight published studies of DAT1 and ADHD, all of which have examined the same genetic variant (a VNTR at the 3′ region). Four independent family-based studies based on samples from Ireland, the United States, and the United Kingdom have shown significant association and linkage with allele 10, the 480 bp repeat (Cook et al., 1995; Curran et al., 2001; Daly, Hawi, Fitzgerald, & Gill, 1999 [extension of Gill, Daly, Heron, Hawi, & Fitzgerald, 1997]; Waldman et al., 1998), and one study found evidence of a trend for association (Barr et al., 2001). However, three independent family-based studies all of which used reasonable sample sizes failed to show association (Holmes et al., 2000; Palmer et al.,

1999; Swanson et al., 1998). Examination of these findings suggests heterogeneity (Curran et al., 2001). A meta-analysis is currently under way.

Phenotype Issues and DAT1

Again, so far, there has been little specific subgroup analysis and thus no consistent findings reported yet. First, the Irish study found stronger evidence of association and linkage with the DAT1 480bp allele in those with a positive family history of ADHD (Daly et al., 1999; defined using the cut point on the Wender Utah rating scale). Second, Waldman et al. (1998) found evidence of association of DAT1 with hyperactive–impulsive but not inattentive symptoms. This was also shown when using categorical measures.

Future Directions

Molecular genetic research in ADHD is at a relatively early stage. In contrast to other disorders covered in this book, there have so far been no published genome scans in ADHD. However, at least one group has adopted a sibpair linkage approach, and results of this work are awaited (S. L. Smalley, personal communication, February, 1, 2001). Thus far, results have all been based on candidate gene association studies. Nevertheless, preliminary findings appear promising, and this leads to the question of what directions need to be considered in the future. Already there has been much cooperation between different research groups that facilitated the meta-analysis of the DRD4 results. Undoubtedly, many of the same polymorphisms in the same candidate genes will be examined by different research groups, and meta-analyses of these data represent a first key step in terms of integrating findings for specific candidate genes. A meta-analysis of findings for the DAT1 480bp VNTR is already planned. Another possible candidate for meta-analysis in the future is the DRD5 gene, given that one group has detected significant linkage and association with the 148bp allele of a CA repeat polymorphism (Daly et al., 1999), and three groups so far have found a trend for association with the same allele (Barr et al., 2000; Payton et al., 2000; Tahir et al., 2000).

There is also clearly a need to examine new functional polymorphisms in candidate genes and understand more about how they affect function, given that one of the major goals of molecular genetics research is to understand more about disease pathogenesis and the mechanisms through which associated gene variants influence disorder. For example, although several groups have found an association with the DAT1 VNTR, there is as yet no plausible hypothesis of how this finding affects DAT functioning and is related to ADHD. Similarly, work on DRD4 is needed to further determine mechanisms through which the 7-repeat allele contributes to disorder before concluding that this is a true susceptibility allele and also to consider the role of other DRD4 polymorphisms. This molecular work must go hand in hand with further examination of the phenotype. For example, what particular aspects of the ADHD phenotype are related to specific polymorphisms that are found to be associated with ADHD? Similarly, as discussed in Rowe (chapter 5, this volume), careful consideration must be taken of gene–environment correlation and interaction. Even where specific susceptibility genes are detected, how are effects moderated by environmental risk factors, and do environmental mechanisms play a role in mediating the effects of specific genetic variants?

New molecular genetic as well as twin study findings suggest that ADHD appears to function as a dimensional attribute. It also appears that studies based on population-based samples using fairly broad phenotype definitions of ADHD can yield similar findings to those from clinical studies. This clearly challenges the traditional psychiatric genetic approach, in which much emphasis is placed on diagnostic boundaries. So far, findings on ADHD at least suggest that studies using large population-based samples may yield useful results and have the advantage of being able to adopt an epidemiological framework (e.g., including measures of environmental risk). Nevertheless, the collection of large, well-characterized clinical samples remains essential. It ultimately will still be necessary to test whether molecular genetic findings are relevant in patient populations and to examine important clinical factors such as the effects of comorbidity and the influence on treatment response.

A key question for molecular genetic researchers is how important is heterogeneity, and should we be examining subgroups? At present, there probably is sufficient genetic epidemiological, neurobiological, or clinical evidence to warrant further examining heterogeneity at a clinical level, specifically stimulant treatment response and comorbidity with other disorders. The evidence to support subdividing according to ADHD subtypes is more equivocal. Nevertheless, given the necessity for large sample sizes to be able to detect susceptibility genes of small effect size, at present the main initial focus has to be on the ADHD phenotype rather than on subgroups. Nevertheless, provided the phenotype is carved meaningfully on the basis of evidence and is hypothesis driven, further examination of subgroups may be helpful with the proviso that findings need to be easily replicable.

The next area of rapidly increasing interest is adult ADHD. Given that the concept of ADHD in adult life is relatively new, clearly much more clinical and longitudinal research is needed to examine the phenotype in adults. Similarly, there is also a need for genetic epidemiological studies of adult ADHD, including further family and twin studies, particularly given that there have been no twin studies of adult ADHD. It seems likely that with an increasing knowledge base, adult ADHD is likely to become an increasingly fruitful area for research over the next decade.

Finally, so far, genetic research of ADHD has been characterized by a high degree of cooperation and collaboration, exemplified by the establishment of the International ADHD Molecular Genetics Network. Collaboration will undoubtedly prove to be a major facilitating factor for progress in the field and indeed is necessary if we are to address many if not most of the key questions and issues highlighted in this chapter.

References

Alberts-Corush, J., Firestone, P., & Goodman, J. J. (1986). Attention and impulsivity characteristics of the biological and adoptive parents of hyperactive and normal control children. *American Journal of Orthopsychiatry, 56,* 413–423.

Amara, S. G., & Kuhar, M. J. (1993). Neurotransmitter transporters: Recent progress. *Annual Review of Neuroscience, 16,* 73–93.

American Psychiatric Association. (1994). *Diagnostic and statistical manual of mental disorders* (4th ed.). Washington, DC: Author.

Angold, A., Prendergast, M., Cox, A., Harrington, R., Simonoff, E., & Rutter, M. (1995). The Child and Adolescent Psychiatric Assessment (CAPA). *Psychological Medicine, 25,* 739–753.

Barkley, R. A. (1998). *Attention-deficit hyperactivity disorder* (2nd ed.). New York: Guilford Press.

Barr, C. L., Wigg, K. G., Feng, Y., Zai, G., Malone, M., Roberts, W., et al. (2000). Attention deficit hyperactivity disorder and the gene for the dopamine D5 receptor. *Molecular Psychiatry, 5,* 548–551.

Barr, C. L., Xu, C., Kroft, J., Feng, Y., Wigg, K., Zai, G., et al. (2001). Haplotype study of 3 polymorphisms at the dopamine transporter locus confirm linkage to attention deficit hyperactivity disorder. *Biological Psychiatry, 49,* 333–339.

Biederman, J., Faraone, S. V., Keegan, K., Benjamin, J., Krifcher, B., Moore, C., et al. (1992). Further evidence for family-genetic risk factors in attention deficit hyperactivity disorder (ADHD): Patterns of comorbidity in probands and relatives in psychiatrically and pediatrically referred samples. *Archives of General Psychiatry, 49,* 728–738.

Biederman, J., Faraone, S. V., Mick, E., Spencer, T., Wilens, T., Kiely, K., et al. (1995). High-risk for attention-deficit hyperactivity disorder among children of parents with childhood-onset of the disorder: A pilot-study. *American Journal of Psychiatry, 152,* 431–435.

Cantwell, D. P. (1975). Genetics of hyperactivity. *Journal of Child Psychology and Psychiatry, 16,* 261–264.

Caron, M. G. (1996). Molecular biology: 2. A dopamine transporter mouse knockout. *American Journal of Psychiatry, 153,* 1515.

Castellanos, F. X., Lau, E., Tayebi, N., Lee, P., Long, R. E., Giedd, J. N., et al. (1998). Lack of an association between a dopamine-4 receptor polymorphism and attention-deficit/hyperactivity disorder: Genetic and brain morphometric analysis. *Molecular Psychiatry, 3,* 431–434.

Comings, D. E., Gonzalez, N., Wu, S., Gade, R., Muhleman, D., Saucier, G., et al. (1999). Studies of the 48 bp repeat polymorphism of the DRD4 gene in impulsive, compulsive, addictive behaviors: Tourette syndrome, ADHD, pathological gambling, and substance abuse. *American Journal of Medical Genetics, 88,* 358–368.

Cook, E. H., Stein, M. A., Krasowski, M. D., Cox, N. J., Olkon, D. M., Kieffer, J. E., et al. (1995). Association of attention deficit hyperactivity disorder and the dopamine transporter gene. *American Journal of Human Genetics, 56,* 993–998.

Curran, S., Mill, J., Tahir, E., Kent, L., Richards, S., Gould, A., et al. (2001). Association study of a dopamine transporter polymorphism and attention deficit hyperactivity disorder in UK and Turkish samples. *American Journal of Medical Genetics, 105,* 387–393.

Daly, G., Hawi, Z., Fitzgerald, M., & Gill, M. (1999). Mapping susceptibility loci in attention deficit hyperactivity disorder: Preferential transmission of parental alleles at DAT1, DBH and DRD5 to affected children. *Molecular Psychiatry, 4,* 192–196.

Dulawa, S. C., Grandy, D. K., Low, M. J., Paulus, M. P., & Geyer, M. A. (1999). Dopamine D4 receptor-knock-out mice exhibit reduced exploration of novel stimuli. *Journal of Neuroscience, 19,* 9550–9556.

Eaves, L. J., Kirk, K. M., Martin, N. G., & Russell, R. J. (1999). Some implications of chaos theory for the genetic analysis of human development and variation. *Twin Research, 2,* 49–52.

Eaves, L. J., Silberg, J. L., Meyer, J. M., Maes, H. H., Simonoff, E., Pickles, A., et al. (1997). Genetics and development psychopathology: 2. The main effects of genes and environment on behavioral problems in the Virginia twin study of adolescent behavioral development. *Journal of Child Psychology and Psychiatry, 38,* 965–980.

Edlebrock, C., Rende, R., Plomin, R., & Thompson, L. A. (1995). A twin study of competence and problem behavior in childhood and early adolescence. *Journal of Child Psychology and Psychiatry, 36,* 775–785.

Eisenberg, J., Zchar, A., Mei-Tal, G., Steinberg, A., Tartakovsky, E., Gritsenko, I., et al. (2000). A haplotype relative risk study of the dopamine D4 receptor (DRD4) exon III repeat polymorphism and attention deficit hyperactivity disorder (ADHD). *American Journal of Medical Genetics, 96,* 258–261.

Faraone, S. V., Biederman, J., Keenan, K., & Tsuang, M. T. (1991). Separation of *DSM–III* attention deficit disorder and conduct disorder: Evidence from a family-genetic study of American child psychiatric patients. *Psychological Medicine, 21,* 109–121.

Faraone, S. V., Biederman, J., Mennin, D., Russell, R., & Tsuang, M. T. (1998). Familial subtypes of attention deficit hyperactivity disorder: A 4 year follow-up study of children from antisocial-ADHD families. *Journal of Child Psychology and Psychiatry, 39,* 1045–1053.

Faraone, S. V., Biederman, J., & Monuteaux, M. C. (2000). Toward guidelines for pedigree selection in genetic studies of attention deficit hyperactivity disorder. *Genetic Epidemiology, 18,* 1–16.

Faraone, S. V., Biederman, J., Weiffenbach, B., Keith, T., Chu, M. P., Weaver, A., et al. (1999). Dopamine D4 gene 7-repeat allele and attention deficit hyperactivity disorder. *American Journal of Psychiatry, 156,* 768–770.

Faraone, S. V., Doyle, A. E., Mick, E., & Biederman, J. (2001). Meta-analysis of the association between the dopamine D4 gene 7-repeat allele and attention deficit hyperactivity disorder. *American Journal of Psychiatry, 158,* 1052–1057.

Gainetdinov, R. R., Wetsel, W. C., Jones, S. R., Levin, E. D., Jaber, M., & Caron, M. G. (1999). Role of serotonin in the paradoxical calming effects of psychostimulants on hyperactivity. *Science, 283,* 397–401.

Gill, M., Daly, G., Heron, S., Hawi, Z., & Fitzgerald, M. (1997). Confirmation of association between attention deficit hyperactivity disorder and a dopamine transporter polymorphism. *Molecular Psychiatry, 2,* 311–313.

Gillis, J. J., Gilger, J. W., Pennington, B. F., & DeFries, J. C. (1992). Attention deficit disorders in reading disabled twins: Evidence for a genetic aetiology. *Journal of Abnormal Child Psychology, 20,* 303–315.

Gjone, H., Stevenson, J., & Sundet, J. (1996). Genetic influence on parent-reported attention-related problems in a Norwegian general population twin sample. *Journal of the American Academy of Child and Adolescent Psychiatry, 35,* 588–598.

Goodman, R., & Stevenson, J. (1989). A twin study of hyperactivity: II. The aetiological role of genes, family relationships and perinatal adversity. *Journal of Child Psychology and Psychiatry, 30,* 691–709.

Hawi, Z., McCarron, M., Kirley, A., Daly, G., Fitzgerald, M., & Gill, M. (2000). No association of the dopamine DRD4 receptor (DRD4) gene polymorphism with attention deficit hyperactivity disorder (ADHD) in the Irish population. *American Journal of Medical Genetics, 96,* 268–272.

Holmes, J., Hever, T., Hewitt, L., Ball, C., Taylor, E., Rubia, K., & Thapar, A. (in press). A pilot study of psychological measures of attention deficit hyperactivity disorder. *Behavior Genetics.*

Holmes, J., Payton, A., Barrett, J. H., Harrington, R., McGuffin, P., Owen, M., et al. (2000). Association of DRD4 in children with ADHD and comorbid conduct problems. *American Journal of Medical Genetics, 114,* 150–153.

Hudziak, J. J., Heath, A. C., Madden, P. F., Reich, W., Bucholz, K. K., Slutske, W., et al. (1998). Latent class and factor analysis of *DSM–IV*: A twin study of female adolescents. *Journal of the American Academy of Child and Adolescent Psychiatry, 37,* 848–857.

Hudziak, J. J., Rudyer, L. P., Neale, M. C., Heath, A. C., & Todd, R. D. (2000). A twin study of inattentive, aggressive and anxious/depressed behaviors. *Journal of the American Academy of Child and Adolescent Psychiatry, 39,* 469–476.

Kotler, M., Manor, I., Sever, Y., Eisenberg, J., Cohen, H., Ebstein, R. P., & Tyano, S. (2000). Failure to replicate an excess of the long dopamine D4 exon III repeat polymorphism in ADHD in a family-based study. *American Journal of Medical Genetics, 96,* 278–281.

Krause, K. H., Dresel, S. H., Krause, J., Kung, H. F., & Tatsch, K. (2000). Increased striatal dopamine transporter in adult patients with attention deficit hyperactivity disorder: Effects of methylphenidate as measured by single photon emission computed tomography. *Neuroscience Letters, 285,* 107–110.

Kuntsi, J., & Stevenson, J. (2001). Psychological mechanisms in hyperactivity: II. The role of genetic factors. *Journal of Child Psychology and Psychiatry, 42,* 211–219.

LaHoste, G. J., Swanson, J. M., Wigal, S. B., Glabe, C., Wigal, T., King, N., et al. (1996). Dopamine D4 receptor gene polymorphism is associated with attention deficit hyperactivity disorder. *Molecular Psychiatry, 1,* 121–124.

Levy, F., Hay, D. A., McStephen, M., Wood, C., & Waldman, I. (1997). Attention-deficit hyperactivity disorder: A category or a continuum? Genetic analysis of a large-scale twin study. *Journal of the American Academy of Child and Adolescent Psychiatry, 36,* 737–744.

Manshadi, M., Lippmann, S., O'Daniel, R. G., & Blackman, A. (1983). Alcohol abuse and attention deficit hyperactivity disorder. *Journal of Clinical Psychiatry, 44,* 379–380.

Martin, N. (2000). *A twin study of the genetics of childhood attention deficit hyperactivity disorder and its comorbidity with conduct disorder*. Unpublished doctoral dissertation, University of Wales College of Medicine.

Mill, J., Curran, S., Kent, L., Richards, S., Gould, A., Virdee, V., et al. (2001). Attention deficit hyperactivity disorder (ADHD) and the dopamine D4 receptor gene: Evidence of association but no linkage in a UK sample. *Molecular Psychiatry, 6,* 440–444.

Morrison, J. R., & Stewart, M. A. (1975). The psychiatric status of the legal families of adopted hyperactive children. *Archives of General Psychiatry, 28,* 888–891.

Muglia, P., Jain, U., Macciardi, F., & Kennedy, J. L. (2000). Adult attention deficit hyperactivity disorder and the dopamine D4 receptor gene. *American Journal of Medical Genetics, 96,* 273–277.

Nadder, N. S., Silberg, J. L., Eaves, L. J., Maes, H. H., & Meyer, J. M. (1998). Genetic effects on ADHD symptomatology in 7-to-13-year-old twins: Results from a telephone survey. *Behavior Genetics, 28,* 83–99.

Neuman, R. J., Todd, R. D., Heath, A. C., Bucholz, K., Reich, W., Begleiter, H., et al. (1997). Latent class analysis deficit/hyperactivity symptoms in three populations. *American Journal of Medical Genetics, 74,* 570–571.

Palmer, C. G. S., Bailey, J. N., Ramsey, C., Cantwell, D., Sinsheimer, J. S., Del'Homme, M., et al. (1999). No evidence of linkage disequilibrium between DAT1 and attention deficit hyperactivity disorder in a large sample. *Psychiatric Genetics, 9,* 157–160.

Payton, A., Holmes, J., Barrett, J., Hever, T., Fitzpatrick, H., Trumper, A., et al. (2000). A family based candidate gene association study of attention deficit hyperactivity disorder. *American Journal of Medical Genetics, 96,* 6.

Payton, A., Holmes, J., Barrett, J. H., Sham, P., Harrington, R., McGuffin, P., et al. (2001). Susceptibility genes for a trait measure of attention deficit hyperactivity disorder: A pilot study in a non-clinical sample of twins. *Psychiatry Research, 105,* 273–278.

Rowe, D. C., Stever, C., Giedinghagen, L. N., Gard, J. M. C., Cleveland, H. H., Terris, S. T., et al. (1999). Dopamine DRD4 receptor polymorphism and attention deficit hyperactivity disorder. *Molecular Psychiatry, 3,* 419–426.

Rutter, M., Giller, H., & Hagell, A. (1998). *Antisocial behaviour by young people.* Cambridge, England: Cambridge University Press.

Rutter, M., Tizard, J., & Whitmore, K. (1970). *Education, health, and behavior.* London: Longman.

Safer, D. J. (1973). A familial factor in minimal brain dysfunction. *Behavioral Genetics, 3,* 175–186.

Schmitz, S., Fulker, D. W., & Mrazek, D. A. (1995). Problem behavior in early and middle childhood: An initial behavior genetic analysis. *Journal of Child Psychology and Psychiatry, 32*, 1443–1458.

Sherman, D. K., Iacono, W. G., & McGue, M. K. (1997). Attention-deficit hyperactivity disorder dimensions: A twin study of inattention and impulsivity-hyperactivity. *Journal of the American Academy of Child and Adolescent Psychiatry, 36*, 745–753.

Sherman, D. K., McGue, M. K., & Iacono, W. G. (1997). Twin concordance for attention deficit hyperactivity disorder: A comparison of teachers' and mothers' reports. *American Journal of Psychiatry, 154*, 532–535.

Silberg, J., Rutter, M., Meyer, J., Maes, H., Hewitt, J., Simonoff, E., et al. (1996). Genetic and environmental influences on the covariation between hyperactivity and conduct disturbance in juvenile twins. *Journal of Child Psychology and Psychiatry, 37*, 803–816.

Simonoff, E., Pickles, A., Hervas, A., Silberg, J., Rutter, M., & Eaves, L. (1998). Genetics influences on childhood hyperactivity: Contrast effects imply parental rating bias, not sibling interaction. *Psychological Medicine, 28*, 825–837.

Smalley, S. L., Bailey, J. N., Palmer, C. G., Cantwell, D. P., McGough, J. J., Del'Homme, M. A., et al. (1998). Evidence that the dopamine D4 receptor is a susceptibility gene in attention deficit hyperactivity disorder. *Molecular Psychiatry, 3*, 427–430.

Smalley, S. L., McGough, J. J., Del'Homme, M., New Delman, J., Gordon, E., Kim, T., et al. (2000). Familial clustering of symptoms and disruptive behaviours in multiplex families with Attention Deficit Hyperactivity Disorder. *Journal of the American Academy of Child and Adolescent Psychiatry, 39*, 1135–1143.

Solanto, M. V. (1998). Neuropsychopharmacological mechanisms of stimulant drug addiction in attention deficit hyperactivity disorder: A review and integration. *Behavioural Brain Research, 94*, 127–152.

Spencer, T., Biederman, J., Wilens, T. E., & Faraone, S. V. (1998). Adults with attention-deficit/hyperactivity disorder: A controversial diagnosis. *Journal of Clinical Psychiatry, 59*, 59–68.

Sprich, S., Biederman, J., Crawford, M. H., Mundy, E., & Faraone, S. V. (2000). Adoptive and biological families of children and adolescents with ADHD. *Journal of the American Academy of Child and Adolescent Psychiatry, 39*, 1432–1437.

Stevenson, J. (1992). Evidence for a genetic aetiology in hyperactivity in children. *Behavioral Genetics, 22*, 337–344.

Sunohara, G. A., Roberts, W., Malone, M., Schachar, R. J., Tannock, R., Basile, V. S., et al. (2000). Linkage of the dopamine D4 receptor gene and attention deficit/hyperactivity disorder. *Journal of the American Academy of Child and Adolescent Psychiatry, 39*, 1537–1542.

Swanson, J. M., Sunohara, G. A., Kennedy, J. L., Regino, R., Fineberg, E., Wigal, T., et al. (1998). Association of the dopamine receptor D4 (DRD4) gene with a refined phenotype of attention deficit hyperactivity disorder (ADHD): A family-based approach. *Molecular Psychiatry, 3*, 38–41.

Tahir, E., Curran, S., Yazgan, Y., Ozbay, F., Cirakoglu, B., & Asherson, P. J. (2000). No association between low and high-activity catecholamine-methyl-transferase (COMT) and attention deficit hyperactivity disorder (ADHD) in a sample of Turkish children. *American Journal of Medical Genetics, 96*, 285–288.

Taylor, E. (1994). Syndromes of attention deficit and overactivity. In M. Rutter, E. Taylor, & L. Hersov (Eds.), *Child and adolescent psychiatry: Modern approaches.* Oxford: Blackwell Scientific Publications.

Thapar, A., Harrington, R., & McGuffin, P. (2001). Examining the comorbidity of ADHD related behaviours and conduct problems using a twin study design. *British Journal of Psychiatry*, 179, 224–229.

Thapar, A., Harrington, R., Ross, K., & McGuffin, P. (2000). Does the definition of ADHD affect heritability? *Journal of the American Academy of Child and Adolescent Psychiatry, 39*, 1528–1536.

Thapar, A., Hervas, A., & McGuffin, P. (1995). Childhood hyperactivity scores are highly heritable and show sibling competition effects: Twin study evidence. *Behavior Genetics, 25*, 537–544.

Thapar, A., Holmes, J., Poulton, K., & Harrington, R. (1999). Genetic basis of attention deficit and hyperactivity. *British Journal of Psychiatry, 174*, 105–111.

Vallone, D., Picetti, R., & Borrelli, E. (2000). Structure and function of dopamine receptors. *Neuroscience and Biobehavioral Reviews, 24,* 125–132.

Waldman, I. D., Rowe, D. C., Abramowitz, A., Kozel, S. T., Mohr, J. H., Sherman, S. L., et al. (1998). Association of the dopamine transporter (DAT1) and attention deficit hyperactivity disorder in children: Heterogeneity owing to diagnostic subtype and severity. *American Journal of Human Genetics, 63,* 1767–1776.

Ward, M. F., Wender, P. H., & Reimherr, F. W. (1993). The Wender Utah Rating Scale: An aid in the retrospective diagnosis of childhood attention deficit hyperactivity disorder. *American Journal of Psychiatry, 150,* 885–890.

Willcutt, E. G., Pennington, B. F., & DeFries, J. C. (2000). Twin study of etiology of comorbidity between reading disability and attention-deficit/hyperactivity disorder. *American Journal of Medical Genetics, 96,* 293–301.

World Health Organization. (1992). *The ICD–10 classification of mental and behavioral disorders.* Geneva.

Zametkin, A. J., & Liotta, W. (1998). The neurobiology of attention-deficit/hyperactivity disorder. *Journal of Clinical Psychiatry, 59,* 17–23.

23

Schizophrenia and Genetics

Michael J. Owen and Michael C. O'Donovan

Schizophrenia is the heartland of psychiatry and the core of its clinical practice. (Kendell, 1983, p. 275)

Schizophrenia is a severe and debilitating mental disorder that, despite the move toward community care, still accounts for more hospital bed occupancy than any other disease (Gottesman, 1991). It is a disorder with a relatively low population incidence (0.17–0.57 per 1,000), but because it tends to run a chronic course, there is a relatively high point prevalence (2.4–6.7 per 1,000). The lifetime morbid risk is slightly less than 1% (Gottesman, 1991; Jablensky, 1999). The incidence of schizophrenia is surprisingly uniform across a variety of different populations with different cultures, and the evidence for pockets of particularly high and low incidence and prevalence is not strong (Jablensky, 1999).

Most individuals with schizophrenia have an onset in early adulthood, have long periods of illness, are unable to work, and have difficulty in sustaining family relationships because of their illness. Schizophrenia also has an enormous impact on families, including the necessity of providing informal care. Epidemiological studies have identified a number of putative associations. These include certain pregnancy and delivery complications (Hultman, Sparen, Takei, Murray, & Cnattingius, 1999; Jones, Rantakallio, Hartikainen, Isohanni, & Sipilä, 1998), delayed developmental milestones (Jones, Rodgers, Murray, & Marmot, 1994), low IQ score (David, Malmberg, Brandt, Allebeck, & Lewis, 1997), personality characteristics concerned with social relations (Malmberg, Lewis, David, & Allebeck, 1998), urban upbringing (Lewis, David, Andreasson, & Allebeck, 1992), immigration (Hutchinson et al., 1996), and the use of illegal drugs, especially cannabis (Andreasson, Allebeck, Engstrom, & Rydberg, 1987). However, in each case, the relative risks are small compared with those associated with having an affected first-degree relative.

The main objective of this chapter is to provide an overview of the current status of molecular genetic studies of schizophrenia and to suggest how this may progress over the next few years. However, to allow readers to appreciate why schizophrenia research is now where it is, and why we expect it to move where we think it will move, we start by briefly discussing the leading hypotheses of the etiology of schizophrenia and the genetic epidemiology that should be a prerequisite to molecular genetic investigation.

Pathogenic theories for many years have been dominated by ideas based on neuropharmacological studies suggesting that abnormalities in monoamine neurotransmission, in particular of the dopaminergic and serotonergic systems, underlie schizophrenia. The predominant hypothesis for more than three decades was that schizophrenia is caused by excessive dopaminergic neurotransmission. This view is based on two observations. First, most conventional antipsychotic drugs block dopamine receptors. Second, dopaminergic drugs such as amphetamine are psychotomimetic. However, firm evidence for a primary dopaminergic abnormality in schizophrenia is yet to be established. Moreover, the clinical efficacy of newer antipsychotic drugs appears to be partly associated with actions on other neurotransmitter systems, and there is evidence to support hypotheses of serotonin, glutamate, and gamma-aminobutyric acid (GABA) dysfunction in schizophrenia (Bray & Owen, 2001). As neurotransmitter systems do not function in isolation, it seems likely that more than one will be disturbed in schizophrenia, and more recent neurochemical theories have centered on the complex interactions between these various systems. However, at present, it remains unclear to what extent any neurochemical findings reflect primary rather than downstream pathology, compensatory mechanisms, or environmental influences.

In recent years, schizophrenia increasingly has been regarded as a neurodevelopmental disorder (Murray & Lewis, 1987; Weinberger, 1987). Views differ as to the timing and nature of the supposed developmental disturbance(s), but the view that schizophrenia has its origins in early, perhaps even prenatal, brain development is supported by epidemiological evidence of increased obstetric complications and childhood neuropsychological deficits in individuals who subsequently develop the disorder, together with a pattern of nonspecific neuropathological anomalies (Weinberger, 1995). Neuroimaging studies have shown increased lateral ventricle size in people with schizophrenia, and this is present at onset of symptoms (Degreef et al., 1992) and in currently unaffected adolescents who are of a high genetic risk for the disorder (Cannon et al., 1993). Volumetric studies also have shown a small but significant reduction in brain size, particularly within the temporal lobes (McCarley et al., 1999), for which there may be a significant genetic component (Lawrie et al., 1999). Histological studies, though prone to methodological problems, have provided evidence for subtle cytoarchitectural anomalies of putatively developmental origin within the frontal lobes and temporo-limbic structures such as the hippocampus (Harrison, 1999). Although reports of neuronal displacement have proved difficult to replicate, a relatively consistent finding is of a reduction in axonal and dendritic markers within these brain regions (Eastwood & Harrison, 2000). In the absence of classical degenerative changes, such findings potentially could reflect a defect in the development or maintenance of synaptic connectivity. The development of mature synaptic networks is, however, extremely complex, involving a dynamic interplay of genetic, environmental, and stochastic factors. Thus, despite some broad clues, the precise biological mechanisms underlying these putative neurodevelopmental disturbances remain largely speculative. This means that at the present time it is difficult to base rational selection of candidate genes and pathways on clear evidence for disturbed neurodevelopmental processes.

Ignorance of pathogenesis has meant that, despite considerable effort and ingenuity, it has not been possible to delineate etiologically distinct subgroups. The diagnosis of schizophrenia therefore still embraces conditions with a variety of symptoms, courses, outcomes, responses to treatment, and probably etiologies. This means that the disorder is effectively syndromic, and diagnosis has to rely on a combination of heterophenomenology and clinical observation. In spite of this, the use of structured and semistructured interviews together with explicit operational diagnostic criteria means that it is possible to achieve high degrees of diagnostic reliability. Moreover, the syndrome so defined has a high heritability. However, we also have to accept that the disorder, as defined by current diagnostic criteria, may include a number of heterogeneous disease processes and that this will hinder any attempts at identifying etiological factors.

Genetic Epidemiology

Familiality and Genetic Risk

The extensive literature on the genetic epidemiology of schizophrenia provides strong support for genetic factors in etiology. Gottesman and Shields (1982), using data from around 40 family studies spanning much of the 20th century, were able to show conclusively that the risk of schizophrenia is increased in relatives of probands with the disorder. The lifetime risk for schizophrenia in the general population worldwide is generally reported to be around 1%. This contrasts with the risk to siblings (10%) and offspring of affected probands (13%), and although the risk to parents is somewhat lower (6%), this reflects the reproductive disadvantage conferred by schizophrenia. Many of the studies in this analysis predate up-to-date methods, but in a review of seven methodologically modern studies, the finding of an approximately 10-fold increase in risk of schizophrenia in first-degree relatives remains (Kendler & Diehl, 1993).

The results of twin and adoption studies of schizophrenia permit genetic factors to be differentiated from infective, cultural, socioeconomic, or other environmental factors that could potentially cause familiality. Adoption studies of the three main study designs (adoptee studies, cross-fostering studies, and adoptee family studies) provide evidence that there is an increased risk of schizophrenia in biological first-degree relatives of probands, though not in nonbiologically related adopted or adoptive family members who share the same environment as probands (Heston, 1966; Kety et al., 1994; Rosenthal, Wender, Kety, Welner, & Schulsinger, 1971;

Wender, Rosenthal, Kety, Schulsinger, & Welner, 1974). The data of Kety et al. (1994) from Denmark have been reanalyzed applying *Diagnostic and Statistical Manual of Mental Disorders* (3rd ed. [*DSM–III*]; American Psychiatric Association, 1980) criteria, with consistent results (Kendler, Gruenberg, & Kinney, 1994): 7.9% of first-degree biological relatives of proband adoptees had *DSM–III* schizophrenia, compared with 0.9% of first-degree relatives of control adoptees. A further Finnish study (Tienari et al., 1994) found a higher prevalence of *DSM–III–R* (3rd ed., Revised; American Psychiatric Association, 1987) schizophrenia in the adoptive offspring of mothers with schizophrenia or a related disorder, compared with control offspring (4.4% vs. 0.5%).

There is also overwhelming evidence from twin studies that shared genes rather than shared environments underlie the increased risk in relatives (McGuffin, Owen, O'Donovan, Thapar, & Gottesman, 1994). Based on the five most recent systematically ascertained twin studies, the probandwise concordance rate for schizophrenia in monozygotic (MZ) twin pairs is 41%–65% compared with 0%–28% for dizygotic (DZ) twin pairs, corresponding to heritability estimates of approximately 80%–85% (Cardno & Gottesman, 2000).

Mode of Inheritance

Although it is clear that there is a genetic contribution to schizophrenia, the frequency with which MZ twins can be discordant for schizophrenia strongly suggests that what is inherited is not the certainty of disease accompanying a particular genotype but rather a predisposition or liability to develop the disorder. Studies on the recurrence risk in various classes of relative allow us to exclude the possibility that schizophrenia is a single-gene disorder or collection of single-gene disorders even when incomplete penetrance is taken into account. Rather, the mode of transmission, like that of other common disorders, is complex and non-Mendelian (McGue & Gottesman, 1989). The most common mode of transmission is probably *oligogenic* (a small number of genes of moderate effect) or *polygenic* (many genes of small effect) or a mixture of the two (McGuffin et al., 1994). However, the number of susceptibility loci, the disease risk conferred by each locus, and the degree of interaction between loci all remain unknown. Risch (1990) calculated that the data for recurrence risks in the relatives of probands with schizophrenia are incompatible with the existence of a single locus conferring a relative risk (λs) of more than 3 and, unless extreme epistasis exists, models with two or three loci of $\lambda s \leq 2$ are more plausible. It should be emphasized that these calculations are based on a homogeneous population, and under a model of heterogeneity, it is quite possible that alleles of larger effect are operating in some subpopulations of patients, for example, families chosen for having a high density of illness. However, it also has been demonstrated that such families would be expected to occur even under polygenic inheritance, and their existence does not force the conclusion that alleles of large effect exist (McGue & Gottesman, 1989).

Defining the Phenotype

Although it is clear that genotype confers liability to schizophrenia, we have no clear idea what form this "liability" actually takes. That is to say, we do not understand the nature of the inherited intermediate phenotypes that predispose to schizophrenia. It follows then that unless it is manifest as disorder, we do not know who has inherited high liability, or more accurately, how liability should be measured and quantified.

The situation is made more complex by the fact that we are unable to demarcate the clinical syndrome to which liability may lead. Family, twin, and adoption studies have shown that the phenotype extends beyond the core diagnosis of schizophrenia to include a spectrum of disorders, including mainly schizophrenic schizoaffective disorder and schizotypal personality disorder (Farmer, McGuffin, & Gottesman, 1987; Kendler et al., 1994). However, the limits of this spectrum and its relationship to other psychoses, affective and nonaffective, and nonpsychotic affective disorders remain uncertain (Cardno, Rijsdijk, Sham, Murray, & McGuffin, 2002; Kendler, Karkowski, Walsh, & Crow, 1998; Kendler, Neale, & Walsh, 1995).

However, in spite of the problems and uncertainties we have described, and the difficulties that ensue, schizophrenia seems a compelling candidate for molecular genetic approaches. This is because genes make such a strong contribution to its etiology, and notwithstanding improvements in neuroimaging, the brain is still difficult to study directly, and the interpretation of cause and effect is problematic.

Molecular Genetics

There are two main approaches for identifying disease genes in human populations: linkage and association. Both approaches have been widely used in studies of schizophrenia.

Linkage

The first wave of systematic molecular genetic studies of schizophrenia in the late 1980s and early 1990s focused on linkage analysis of large pedigrees containing multiple effects. This approach had been highly effective for identifying genes of large effect in single-gene disorders. The proponents of this method were aware it was unlikely that such genes account for most of the genetic risk for schizophrenia. Instead, the hope was that under a mixed genetic model, the multiplex families, or at least a sufficient proportion of them, were the result of segregation of rare alleles of large effect.

This optimistic view proved to be well founded in a number of complex disorders, for example, Alzheimer's disease, in which mutations in three genes (APP, PS1, and PS2) are now known to cause rare forms of the disorder. In such cases, the disease is of unusually early onset and is transmitted through multiplex pedigrees as an autosomal dominant (Goate et al., 1991; Levy-Lahad et al., 1995; Sherrington et al., 1995). Early studies of large schizophrenia pedigrees also initially produced a strongly positive finding (Sherrington et al., 1988). Unfortunately, this could not be replicated. However, modest evidence for several linked loci has been reported, some of which have received supportive evidence from international collaborative studies. These include chromosome 22q11-12, 6p24-22, 8p22-21, and 6q (Gill et al., 1996; Levinson et al., 2000; Schizophrenia Linkage Collaborative, 1996). There are also many other promising areas of putative linkage, which have been supported by other individual groups although not from larger international consortia. These include 13q14.1-q32 (Blouin et al., 1998; Brzustowicz et al., 1999; Lin et al., 1995), 5q21-q31 (Schwab et al., 2000; Straub, MacLean, Oneill, Walsh, & Kendler, 1997), and 10p15-p11 (Faroane et al., 1998; Schwab et al., 1997; Straub et al., 1998). One other region that is notable for the strength of evidence (maximum heterogeneity log of the odds, LOD, score of 6.5) from a single study, rather than the more important convergence of evidence from different sources, is 1q21-q22 (Brzustowicz, Hodgkinson, Chow, Honer, & Bassett, 2000). The finding on 1q21-q22 more than meets standard criteria for claiming "significant" linkage. Early experience in psychiatric genetics has, however, impressed the need for replication firmly on the minds of most researchers, and although there have been other reports of linkage to markers on chromosome 1 (Ekelund et al., 2000; Hovatta et al., 2000), these are distal to 1q21-q22, and it is not yet clear that these findings are consistent with the existence of a single susceptibility locus on 1q. Not surprisingly, this region is now under intense investigation by an international linkage consortium.

While the linkage data, based on multiplex families, are not definitive, it seems likely that one or more of the above loci are true positives. However, in every case there are negative as well as positive findings, and in only two cases, 13q14.1-q32 and 1q21-q22, did any single study achieve genomewide significance at $p < .05$, that is, a linkage value that is expected less than once by chance in 20 complete genome scans.

As more linkage data from genome scans of schizophrenia and bipolar disorder have emerged, putative linkages to both disorders have been reported in the same regions of the genome, for example, 13q22 and 22q11-13 (Berrettini, 2000). The observation of common linked loci may indicate at least a partial degree of overlap in the etiologies of these two disorders. However, although we are inclined to be sympathetic to this view, there are several reasons for caution. First, none of the linkages are as yet definitive, and therefore the strength of the evidence for joint areas of linkage is not conclusive. Second, similarity is more noticeable than dissimilarity. Until the apparent clustering has been subjected to rigorous statistical analysis, the pattern may be coincidence. Third, even if correct, the regions of linkage contain hundreds or even thousands of genes. It follows that common linked loci does not prove common genes. Fourth, it is possible that the joint linkages are to loci that modify the phenotypes in such a way as to influence measurement of affected status. The prognosis for both schizophrenia and bipolar disorder is generally worse in individuals who are subjected to chronic stress. Therefore, loci that modulate the stress response might appear to be linked to both disorders simply because of an effect on chronicity, and thereby ascertainment.

As an alternative to linkage based on multiplex families, some groups have advocated the use of smaller families with two or more affected members in a single sibship. This is in the belief that such families may be more suited for the analysis of complex traits, and being much more common, also may be more representative of disease in the general population. This approach has been successful in detecting linkage to several complex diseases. Of most relevance to this readership, these include late-onset Alzheimer's disease (Myers et al., 2000). We have recently reported the largest systematic search for linkage yet published using a sample of 196 affected sibling pairs (N. M. Williams et al., 1999). This study was designed to have power >.95 to detect a susceptibility locus of $\lambda s = 3$ (the maximum effect size estimated by Risch, 1990) with genomewide significance of .05. However, no regions of the genome gave linkage results approaching genomewide significance, but we did find modest evidence (at the level expected by chance once per genome scan) for loci on chromosomes 4p, 18p, and the centromeric region of X.

Because, arguably, no definitive linkages have emerged to date from linkage analysis of schizophrenia, the results must be viewed as disappointing. However, as some of the more promising findings described above have only recently emerged, we believe that persistence with the linkage approach is justified. We must continue to explore different models of disease definition but also tailor our sample sizes appropriately. For example, our own genome scan (N. M. Williams et al., 1999) based on sibpairs was powered predicated on a locus of $\lambda s = 3$ but only had power of 0.7 to detect significant linkage to loci of $\lambda s = 2$. Such loci are still within the detection limits of linkage, with sample sizes of 600–800 sibling pairs (Hauser, Boehnke, Guo, & Risch, 1996; Scott, Pericak-Vance, & Haines, 1997), and with colleagues in Sweden, we are currently engaged in such a study. Those engaged in linkage analysis should now give priority to collecting such samples (or equivalently powered samples of multiplex families). Furthermore, clinical methodology must be, as far as is possible, comparable across all interested research groups if "failure to replicate" is to have much meaning. However, if liability to schizophrenia is entirely due to the operation of many genes of small effect, then even these large-scale studies will be unsuccessful. To detect genetic risk factors of this magnitude, we must turn to the second main gene identification strategy based on association designs.

Association Studies

Association studies offer a powerful means of identifying genes of small effect. Screening the whole genome by association presents difficulties because of the large, and as yet unknown, number of markers required to achieve comprehensive coverage. Although searches of whole chromosomes (e.g., Fisher et al., 1999) and whole genome association screens (Plomin et al., 2001) have been carried out by ourselves and colleagues in other traits, such methods remain exploratory. To date, therefore, most researchers have chosen to select specific genes or loci to be tested by association. This may be on the grounds that a gene is a functional candidate gene (i.e., encodes a protein that may be involved in pathogenesis), a positional candidate gene (i.e., maps to a region implicated by previous linkage studies), or a combination of the two. Because of doubts concerning the robustness of linkage findings in schizophrenia, most studies to date have focused on functional candidate genes.

There are many potential problems with association studies that have been reviewed elsewhere in this volume (see Cardon, chapter 4). It is also worth emphasizing that, for disorders of unknown etiology such as schizophrenia, the potential choice of candidate genes is limited largely by the number of genes in the human genome (probably around 30,000–40,000). This places a heavy burden of proof on positive results because, with very few exceptions, each gene has an extremely low prior probability that it is involved in the disorder (Owen, Holmans, & McGuffin, 1997). It is also worth pointing out that many studies to date have not had sufficient power to detect or confirm the presence of alleles of small effect.

The most obvious candidate genes derive from the neuropharmacological literature. Thus, genes involved in dopaminergic and serotonergic neurotransmission have received a great deal of attention. With recent developments in genome analysis technology and the recent publication of the (almost) complete sequence of the human genome (Lander et al., 2001), these data are being rapidly extended to other systems including the glutamatergic, GABAergic, and genes involved in neuromodulation and neurodevelopment. It is impossible for us to present the results of the hundreds, possibly thousands, of candidate gene analyses that have been published or reported at meetings. Fortunately for the reader, this can be avoided by noting that, for the reasons outlined, the overwhelmingly negative reports cannot be considered as definitive exclusions of small genetic effects. Moreover, with perhaps the two exceptions we discuss below, no positive finding has received sufficient support from more than one group to suggest a true association with even a modest degree of probability.

Serotonin 5HT2A receptor. The serotonergic system is a therapeutic target for several antipsychotic drugs. The first genetic evidence for its involvement in schizophrenia was a report of association with a T > C polymorphism at nucleotide 102 in the HTR2A gene that encodes the 5HT2a receptor (Inayama et al., 1996). This association was based on a small sample of Japanese subjects. Firmer evidence for association subsequently emerged from a large multicenter European consortium (J. Williams et al., 1996) and a meta-analysis of all the data available in 1997 based on more than 3,000 subjects (J. Williams et al., 1997).

Since the meta-analysis was undertaken, a few further negative reports have followed, none with the sample sizes required. If we assume homogeneity and that the association is correct, the odds ratio for the C allele is approximately 1.2. Sample sizes of 1,000 cases and 1,000 controls are required for 80% power to detect an effect of this size even at $p = .05$. However, although the negative studies do not refute the putative association, the evidence presented even in the meta-analysis ($p = .0009$) is well short of genomewide significance (estimated at around $p = 5 \times 10^{-8}$ by Risch & Merikangas, 1996). Admittedly, demanding genomewide significance is inappropriate for such a strong candidate gene because the prior probability that it is involved in schizophrenia is greater than a gene selected at random. However, as there is no way of quantifying how much greater the prior probability is, we have no compelling alternative to genomewide significance. Consequently, we believe that the evidence tends to indicate association between HTR2A and schizophrenia but that the burden of proof has not yet been met. Another reason to be skeptical about the putative association is that there are no variants yet detected in the gene that clearly alter receptor function or expression. However, there is some indirect evidence for the existence of sequence variation elsewhere that alters HTR2A expression in some regions of the brain (Bunzel et al., 1998).

Dopamine receptor genes. The dominant neurochemical hypothesis of schizophrenia involves dysregulation of the dopaminergic system, with disordered transmission at the dopamine D2 receptor widely favored. However, association analyses of DRD2 to date have essentially been negative, although recently, three groups—two Japanese (Arinami, Gao, Hamaguchi, & Toru, 1997; Ohara et al., 1998) and one Swedish (Jonsson et al., 1999)—have implicated a polymorphism in the promoter that alters DRD2 expression in vitro. Unfortunately, the biggest single study to date found no evidence for association (Li et al., 1998). Furthermore when the U.K. data from this study were combined with a large dataset from Scotland, the allele that was associated with schizophrenia in the other three studies was actually significantly less common in the patient group (Breen et al., 1999). This finding has been interpreted as suggestive that the polymorphism tested in these studies is simply a marker for the true susceptibility variant (i.e., association due to linkage disequilibrium, LD). While it is reasonable to attribute reversal of allelic association to LD differences between samples of different ethnic origins, the explanation would be more convincing if the same allele was associated in both U.K. and Swedish samples. Furthermore, if LD is the explanation, then the fact that this particular polymorphism is functional is no longer relevant. Data from more groups are required before a sensible discussion of the role of this variant can be undertaken.

The dopamine hypothesis has received somewhat stronger, although not robust, support from studies of DRD3 that encodes the dopamine D3 receptor. Association has been reported between schizophrenia and homozygosity for a Ser9Gly polymorphism in exon 1 of this gene (Crocq et al., 1992). As expected, positive and negative data have emerged, but unlike DRD2, sufficient data are available for meta-analysis. On the basis of all available data from over 5,000 individuals, Williams and colleagues (J. Williams et al., 1998) concluded in favor of association although the effect size was small (odds ratio = 1.23) but nominally significant ($p = .0002$). As for HTR2A, for the reasons outlined above, level of statistical evidence, while supportive, is not conclusive.

Anticipation and Trinucleotide Repeats

Anticipation is the phenomenon whereby a disease becomes more severe or has an earlier age at onset in successive generations. Numerous studies have now suggested that the pattern of illness in multiplex pedigrees segregating schizophrenia and other psychotic illnesses is at least consistent with anticipation, although ascertainment biases offer an alternative explanation (O'Donovan et al., 1995).

True genetic anticipation is a hallmark of a particular class of mutation called *expanded trinucleotide repeats*. The observations in families have therefore been taken as suggestive that such mutations might be involved in the etiology of psychosis (Petronis & Kennedy, 1996). This hypothesis has received support from a number of studies reporting that large but unidentified CAG/CTG trinucleotide repeats are more common in patients with schizophrenia and other psychoses than in unaffected controls (e.g., Morris, Gaitonde, McKenna, Mollon, & Hunt, 1995; O'Donovan & Owen, 1996). Unfortunately, these early findings have been followed by numerous failures of replication. Also, although two autosomal loci, one at 18q21.1 and the other at 17q21.3, account for around 50% of large CAG/CTG repeats detected by repeat expansion detection (RED), neither is associated with psychosis (e.g., Bowen et al., 2000; Vincent et al., 1999). Although these data have tempered much of the enthusiasm for the CAG/CTG hypothesis, it is still premature to discard it. Indeed, two independent groups have recently detected the presence of proteins in a small number of schizophrenics that react with an antibody against moderate-large polyglutamine sequences (see Ross, 1999). This is of relevance here because polyglutamine sequences are often encoded in DNA by CAG repeats. It remains to be seen whether these proteins are involved in the disease rather than simply chance findings.

Cytogenetic Abnormalities

Apart from linkage and association, a third approach by which investigators have sought to locate susceptibility genes for schizophrenia has been to identify chromosomal abnormalities in affected individuals. Cytogenetic anomalies, such as translocations and deletions, may be pathogenic through several mechanisms: direct disruption of a gene or genes, the formation of a new gene comprised of a fusion of two or more genes that are normally spatially separated, indirect disruption of the function of neighboring genes by a so-called position effect, or an alteration of gene dosage in the case of deletions, duplications, and unbalanced translocations. It also is possible for an abnormality merely to be linked with a susceptibility variant in a particular family. Owing to the high prevalence of schizophrenia, a single incidence of a cytogenetic abnormality is insufficient to suggest causality. To warrant further investigation, a cytogenetic abnormality should be shown to exist in greater frequencies in affected individuals, to disrupt a region already implicated by genetic analyses, or to show cosegregation with the condition in affected families.

There have been numerous reports of associations between schizophrenia and chromosomal abnormalities (Baron, 2001; Bassett, 1992), but, with two exceptions, none have as yet provided convincing evidence to support the location of a gene conferring risk to schizophrenia. The first strongly suggestive finding is a t(1;11) balanced reciprocal translocation found to cosegregate with schizophrenia in a large Scottish family (St. Clair et al., 1990). The term t(1;11) reciprocal translocation implies an exchange of chromosomal material between chromosomes 1 and 11 resulting in two abnormal chromosomes. One of the abnormal chromosomes comprises most of chromosome 11 and a small fragment originating from chromosome 1, the other comprises most of chromosome 1 and the missing fragment of chromosome 11. Individuals with a balanced translocation carry both and therefore a complete complement of chromosomes 1 and 11. Recently, the translocation has been reported to directly disrupt three genes of unknown function on chromosome 1 (Millar et al., 2000). These are located close to the chromosome 1 markers implicated in the two Finnish linkage studies mentioned previously (Ekelund et al., 2000; Hovatta et al., 2000). However, until mutations have been identified in other families or convincing biological evidence implicating this locus has been obtained, the mechanism by which the translocation confers risk to mental illness remains obscure.

The second finding of interest is the association between velo-cardio-facial syndrome (VCFS) and schizophrenia. VCFS, also known as DiGeorge or Shprintzen syndrome, is associated with small interstitial deletions of chromosome 22q11. The phenotype of VCFS is variable, but in addition to characteristic core features of dysmorphology and congenital heart disease, there is strong evidence that individuals with VCFS have a dramatic increase in the risk of psychosis (Papolos et al., 1996; Pulver et al., 1994; Shprintzen, Goldberg, & Golding-Kushner, 1992). There is still some dispute about the nature of the psychotic phenotype associated with VCFS. Thus, the first two of these studies (Pulver et al., 1994; Shprintzen et al., 1992) suggested the association to be with schizophrenia and schizoaffective disorders, whereas a third study demonstrated a high prevalence of bipolar spectrum disorders (Papolos et al., 1996). At least some of the differences can be attributed to differing methods of proband ascertainment, the proband age group, and the diagnostic methodology used.

To characterize the psychosis phenotype in VCFS further, we recently evaluated 50 VCFS cases with a structured clinical interview to establish *DSM–IV* (American Psychiatric Association, 1994) diagnosis (Murphy,

Jones, & Owen, 1999). Fifteen individuals with VCFS (30%) had a psychotic disorder, 12 of which met *DSM–IV* criteria for schizophrenia. These data strongly suggest that it is schizophrenia and allied disorders that explain the high rates of psychosis in VCFS, but studies based on sounder epidemiological principles are now required to demonstrate this beyond all doubt.

It is clear that, with an estimated prevalence of 1 in 4,000 live births, one can estimate that VCFS cannot be responsible for more than a small fraction (~1%) of cases of schizophrenia, and this estimate is in keeping with empirical data (Karayiorgou et al., 1995). From the practical perspective, clinicians should be vigilant for VCFS, especially when psychosis occurs in the presence of other features suggestive of the syndrome, such as dysmorphology, mild learning disability, or a history of cleft palate or congenital heart disease (Bassett et al., 1998; Gothelf et al., 1997). However, from the perspective of the genetic researcher, the most pressing question is whether the high rate of psychosis in VCFS provides a shortcut to a gene within the deleted region that is involved in susceptibility to schizophrenia in cases without a deletion. To answer this, we are currently screening all the genes that map to the deleted region. There are reasons to be optimistic that this will be a route to a general susceptibility gene rather than a blind alley. As we have already seen, some linkage studies suggest the presence of a general schizophrenia susceptibility locus on 22q. While most suggest that this maps outside the VCFS region, linkage mapping in complex diseases is imprecise, and modest evidence for linkage within the VCFS region also has been reported (Blouin et al., 1998; Lasseter et al., 1995; Shaw et al., 1998).

Future Directions

The linkage studies to date have shown that genes of major effect are not common causes of schizophrenia and allied psychoses. They also have provided information concerning the *possible* location of some genes of moderate effect. Association studies have provided further mapping information, with VCFS providing fairly strong evidence for a susceptibility locus in 22q11. To a lesser degree, with the proviso that the associations in HTR2A and DRD3 are weak and still provisional, they also suggest that the dominant neurochemical hypotheses are at least partially correct. It is clear, however, that the findings from molecular genetics are as fragile as those from genetic epidemiology are robust. How are we to make progress?

As in other common diseases, it is hoped that advances in the genetics of schizophrenia will come through the application of detailed information on the anatomy and sequence of the human genome combined with new methods of genotyping and statistical analysis. To take advantage of these developments, collection of large, well-characterized samples is a priority (Owen, Cardno, & O'Donovan, 2000). Samples for genetic studies of schizophrenia are generally ascertained in differing and unsystematic ways, and this may be one of the explanations of failures to replicate. For example, if a positive association is detected in one patient group that, by virtue of the method of ascertainment, has been inadvertently selected for chronicity, the putative "risk allele" might actually be an allele that modifies treatment response rather than predisposing to the disorder itself. Unless this is recognized, replication will be difficult except in samples with a bias toward chronicity.

It is possible that new genes and pathways might be implicated by emerging transcriptomic and proteomic approaches, and preliminary findings using both approaches have already been reported (Edgar et al., 1999; Hakak et al., 2001). The problem here is that human studies will be hampered by the many confounding variables associated with postmortem studies of brain, whereas animal studies suffer from the difficulties inherent in extrapolating from animal behavior to a complex human psychiatric disorder such as schizophrenia.

Finally, it is worth reminding ourselves that the effectiveness of molecular genetic studies depends on the genetic validity of the phenotypes studied. We therefore need to focus research on the development and refinement of phenotypic measures and biological markers, which might simplify the task of finding genes. It is possible that genetic validity may be improved by focusing on quantitative aspects of clinical variation such as symptom profiles (e.g., Cardno et al., 1996) or by identifying biological markers that predict degree of genetic risk or define more homogeneous subgroups. However, it seems unlikely that these phenotypes will provide a rapid solution to the problem. First, we will need measures that can be practicably applied to a sufficient number of families or unrelated patients and, second, we will need to ensure that the traits identified are highly heritable, which will itself require a return to classic genetic epidemiology and model fitting.

References

American Psychiatric Association. (1980). *Diagnostic and statistical manual of mental disorders* (3rd ed.). Washington, DC: Author.

American Psychiatric Association. (1987). *Diagnostic and statistical manual of mental disorders* (3rd ed., Rev.). Washington, DC: Author.

American Psychiatric Association. (1994). *Diagnostic and statistical manual of mental disorders* (4th ed., Rev.). Washington, DC: Author.

Andreasson, S., Allebeck, A., Engstrom, A., & Rydberg, U. (1987). Cannabis and schizophrenia: A longitudinal study of Swedish conscripts. *Lancet, 2,* 1483–1486.

Arinami, T., Gao, M., Hamaguchi, H., & Toru, M. (1997). A functional polymorphism in the promoter region of the dopamine D2 receptor gene is associated with schizophrenia. *Human Molecular Genetics, 6,* 577–582.

Baron, M. (2001). Genetics of schizophrenia and the new millennium: Progress and pitfalls. *American Journal of Human Genetics, 68,* 299–312.

Bassett, A. S. (1992). Chromosomal aberrations and schizophrenia: Autosomes. *British Journal of Psychiatry, 161,* 323–334.

Bassett, A. S., Hodgkinson, K., Chow, E. W., Correia, S., Scutt, L. E., & Weksberg, R. (1998). 22q11 deletion syndrome in adults with schizophrenia. *American Journal of Medical Genetics, 81,* 328–337.

Berrettini, W. H. (2000). Susceptibility loci for bipolar disorder: Overlap with inherited vulnerability to schizophrenia. *Biological Psychiatry, 47,* 245–251.

Blouin, J. L., Dombroski, B. A., Nath, S. K., Lassetter, V. K., Wolniec, P. S., Nestadt, G., et al. (1998). Schizophrenia susceptibility loci on chromosomes 12q32 and 8p21. *Nature Genetics, 20,* 70–73.

Bowen, T., Guy, C. A., Cardno, A. G., Vincent, J. B., Kennedy, J. L., Jones, L. A., et al. (2000). Repeat sizes at CAG/CTG loci CTG18.1, ERDA1 and TGC13-7a in schizophrenia. *Psychiatric Genetics, 10,* 33–37.

Bray, N. J., & Owen, M. J. (2001). Searching for schizophrenia genes. *Trends in Molecular Medicine, 7,* 169–175.

Breen, G., Brom, J., Maude, S., Fox, H, Collier, D., Li, T., et al. (1999). −141C del/ins polymorphism of the dopamine receptor 2 gene is associated with schizophrenia in a British population. *American Journal of Medical Genetics, 88,* 407–410.

Brzustowicz, L. M., Hodgkinson, K. A., Chow, E. W. C., Honer, W. G., & Bassett, A. S. (2000). Location of a major susceptibility locus for familial schizophrenia on chromosome 1q21–22. *Science, 288,* 678–682.

Brzustowicz, L. M., Honer, W. G., Chow, E. W. C., Little, D., Hogan, J., Hodgkinson, K., et al. (1999). Linkage of familial schizophrenia to chromosome 13q32. *American Journal of Human Genetics, 65,* 1096–1103.

Bunzel, R., Blumucke, I., Cichon, S., Normann, S., Schramm, J., Propping, P., & Nothern, M. M. (1998). Polymorphic imprinting of the serotonin-2A (5-HT2A) receptor gene in human adult brain. *Molecular Brain Research, 59,* 90–92.

Cannon, T. D., Mednich, S. A., Parnas, J., Schulsinger, F., Schulsinger, F., Mednich, S. A., et al. (1993). Developmental brain abnormalities in the offspring of schizophrenic mothers: I. Contributions of genetic and perinatal factors. *Archives of General Psychiatry, 49,* 531–564.

Cardno, A. G., & Gottesman, I. I. (2000). Twin studies of schizophrenia: From bow-and-arrow concordances to star wars Mx and functional genomics. *American Journal of Medical Genetics, 97,* 12–17.

Cardno, A. G., Jones, L. A., Murphy, K. C., Asherson, P., Scott, L. C., Williams, J., et al. (1996). Factor analysis of schizophrenic symptoms using the OPCRIT checklist. *Schizophrenia Research, 22,* 233–239.

Cardno, A. G., Rijsdijk, F. V., Sham, P. C., Murray, R. M., & McGuffin, P. (2002). A twin study of genetic relationships between psychotic symptoms. *American Journal of Psychiatry, 159,* 539–545.

Crocq, M. A., Mant, R., Asherson, P., Williams, J., Hode, Y., Mayerova, A., et al. (1992). Association between schizophrenia and homozygosity at the dopamine-d3 receptor gene. *Journal of Medical Genetics, 29,* 858–860.

David, A. S., Malmberg, A., Brandt, L., Allebeck, A., & Lewis, G. (1997). IQ and risk for schizophrenia: A population-based cohort study. *Psychological Medicine, 27,* 1311–1323.

Degreef, G., Ashtari, M., Bogerts, B., Bilder, R. M., Jody, D. N., Alvir, J. M., et al. (1992). Volumes of ventricular system subdivisions measured from magnetic resonance images in first-episode schizophrenia patients. *Archives of General Psychiatry, 49,* 531–537.

Eastwood, S. L., & Harrison, P. J. (2000). Hippocampal synaptic pathology in schizophrenia, bipolar disorder and major depression: A study of complexin mRNAs. *Molecular Psychiatry 5,* 425–432.

Edgar, P. F., Schonberger, S. J., Dean, B., Faull, R. L. M., Kydd, R., & Cooper, G. J. S. (1999). A comparative proteome analysis of hippocampal tissue from schizophrenic and Alzheimer's disease individuals. *Molecular Psychiatry, 4,* 173–178.

Ekelund, J., Lichtermann, D., Hovatta, I., Ellonen, P., Suvisaari, J., Terwilliger, J. D., et al. (2000). Genome-wide scan for schizophrenia in the Finnish population: Evidence for a locus on chromosome 7q22. *Human Molecular Genetics, 9,* 1049–1057.

Faraone, S. V., Matise, T., Svrakic, D., Pepple, J., Malaspina, D., Suarez, B., et al. (1998). Genome scan of European-American schizophrenia pedigrees: Results of the NIMH genetics initiative and millennium consortium. *American Journal of Medical Genetics, 81,* 290–295.

Farmer, A. E., McGuffin, P., & Gottesman, I. I. (1987). Twin concordance for *DSM–III* schizophrenia: Scrutinizing the validity of the definition. *Archives of General Psychiatry, 44,* 634–641.

Fisher, P. J., Turic, D., Williams, N. M., McGuffin, P., Asherson, P., Ball, D., et al. (1999). DNA pooling identifies QTLs for general cognitive ability in children on chromosome 4. *Human Molecular Genetics, 8,* 915–922.

Gill, M., Vallada, H., Collier, D., Sham, P., Holmans, P., Murray, R., et al. (1996). A combined analysis of D22s278 marker alleles in affected sib-pairs: Support for a susceptibility locus for schizophrenia at chromosome 22Q12. *American Journal of Medical Genetics, 67,* 40–45.

Goate, A., Chartierharlin, M. C., Mullan, M., Brown, J., Crawford, F., Fidani, L., et al. (1991). Segregation of a missense mutation in the amyloid precursor protein gene with familial Alzheimers-disease. *Nature, 349,* 704–706.

Gothelf, D., Frisch, A., Munitz, H., Rockah, R., Aviram, T., Mozes, A., et al. (1997). Velocardiofacial manifestations and microdeletions in schizophrenic patients. *American Journal of Medical Genetics, 72,* 455–461.

Gottesman, I. I. (1991). *Schizophrenia genesis: The origins of madness.* New York: Freeman.

Gottesman, I. I., & Shields, J. (1982). *Schizophrenia: The epigenetic puzzle.* Cambridge, England: Cambridge University Press.

Hakak, Y., Walker, J. R., Li, C., Wong, W. H., Davis, K. L., Buxbaum, J. D., et al. (2001). Genome-wide expression analysis reveals dysregulation of myelination-related genes in chronic schizophrenia. *Proceedings of the Academy of Medical Sciences, 98,* 4746–4751.

Harrison, P. J. (1999). The neuropathology of schizophrenia: A critical review of the data and their interpretation. *Brain, 122,* 593–624.

Hauser, E. R., Boehnke, M., Guo, S. W., & Risch, N. J. (1996). Affected sib pair interval mapping and exclusion for complex traits: Sampling considerations. *Genetic Epidemiology, 13,* 117–137.

Heston, L. L. (1966). Psychiatric disorders in foster home reared children of schizophrenic mothers. *British Journal of Psychiatry, 112,* 819–825.

Hovatta, I., Varilo, T., Suvisaari, J., Terwilliger, J. D., Ollikainen, V., Arajärvi, R., et al. (2000). Screen for schizophrenia genes in an isolated Finnish subpopulation, suggesting multiple susceptibility loci. *American Journal of Human Genetics, 65,* 1114–1125.

Hultman, C. M., Sparen, P., Takei, N., Murray, R. M., & Cnattingius, S. (1999). Prenatal and perinatal risk factors for schizophrenia, affective psychosis and reactive psychosis: Case-control study. *British Medical Journal, 318,* 421–426.

Hutchinson, G., Takei, N., Fahy, T., Bhugra, D., Gilvarry, K., Moran, P., et al. (1996). Morbid risk of psychotic illness in first degree relatives of White and African-Caribbean patients with psychosis. *British Journal of Psychiatry, 169,* 776–780.

Inayama, Y., Yoneda, H., Sakai, T., Ishida, Y., Nonomura, Y., Kono, R., et al. (1996). Positive association between a DNA sequence variant in the serotonin 2A receptor gene and schizophrenia. *American Journal of Medical Genetics, 67,* 103–105.

Jablensky, A. (1999). The 100-year epidemiology of schizophrenia. *Schizophrenia Research 28,* 111–125.

Jones, P., Rantakallio, P., Hartikainen, A.-L., Isohanni, M., & Sipilä, P. (1998). Schizophrenia as long-term outcome of pregnancy, delivery and perinatal complications: A 28-year follow-up of the 1966 North Finland general population birth cohort. *American Journal of Psychiatry, 155,* 355–364.

Jones, P., Rodgers, B., Murray, R., & Marmot, M. (1994). Child development risk factors for adult schizophrenia in the British 1946 birth cohort. *Lancet, 344,* 1398–1402.

Jonsson, E. G., Nothen, M. M., Neidt, H., Forslund, K., Rylander, G., Mattila-Evenden, M., et al. (1999). Association between a promoter polymorphism in the dopamine D2 receptor gene and schizophrenia. *Schizophrenia Research, 9, 40,* 31–36.

Karayiorgou, M., Morris, M. A., Morrow, B., Shprintzen, R. J., Goldberg, R., Borrow, J., et al. (1995). Schizophrenia susceptibility associated with interstitial deletions of chromosome 22q11. *Proceedings of the National Academy of Sciences USA, 92,* 7612–7616.

Kendell, R. E. (1983). Schizophrenia. In R. E. Kendell & A. K. Zeally (Eds.), *Companion to psychiatric studies* (pp. 275–296). Edinburgh, Scotland: Churchill Livingstone.

Kendler, K. S., & Diehl, S. R. (1993). The genetics of schizophrenia: A current, genetic–epidemiologic perspective. *Schizophrenia Bulletin, 19,* 261–285.

Kendler, K. S., Gruenberg, A. M., & Kinney, D. K. (1994). Independent diagnoses of adoptees and relatives as defined by *DSM–III* in the provincial and national samples of the Danish Adoption Study of Schizophrenia. *Archives of General Psychiatry, 51,* 456–468.

Kendler, K. S., Karkowski, L. M., Walsh, D., & Crow, T. J. (1998). The structure of psychosis: Latent class analysis of probands from the Roscommon family study. *Archives of General Psychiatry, 55,* 492–509.

Kendler, K. S., Neale, M. C., & Walsh, D. (1995). Evaluating the spectrum concept of schizophrenia in the Roscommon Family Study. *American Journal of Psychiatry, 152,* 749–754.

Kety, S. S., Wender, P. H., Jacobsen, B., Ingraham, L. J., Jansson, L., Faber, B., & Kinney, D. K. (1994). Mental illness in the biological and adoptive relatives of schizophrenic adoptees: Replication of the Copenhagen study in the rest of Denmark. *Archives of General Psychiatry, 51,* 442–455.

Lander, E. S., Linton, L. M., Birren, B., Nasbaum, C., Zody, M. C., Baldwin, J., et al. (2001). Initial sequencing and analysis of the human genome. *Nature, 409,* 860–921.

Lasseter, V. K., Pulver, A. E., Wolyniec, P. S., Nestadt, G., Meyers, D., Karayiorgou, M., et al. (1995). Follow-up report of potential linkage for schizophrenia on chromosome 22q.3. *American Journal of Medical Genetics, 60,* 172–173.

Lawrie, S. M., Whalley, H., Kastelman, H., Abukmeils, S., Byrne, M., Hodges, A., et al. (1999). Magnetic resonance imaging of brain in people at high risk of developing schizophrenia. *Lancet, 353,* 30–33.

Levinson, D. F., Holmans, P., Straub, R. E., Owen, M. J., Wildenauer, D. B., Gejman, P. V., et al. (2000). Multicenter linkage study of schizophrenia candidate regions on chromosomes 5q, 6q, 10p, and 13q: Schizophrenia Linkage Collaborative Group III. *American Journal of Human Genetics, 67,* 652–663.

Levy-Lahad, E., Wasco, W., Poorkaj, P., Romano, D. M., Oshima, J., Pettingell, W. H., et al. (1995). Candidate gene for the chromosome-1 familial Alzheimers-disease locus. *Science, 269,* 973–977.

Lewis, G., David, A., Andreasson, S., & Allebeck, P. (1992). Schizophrenia and city life. *Lancet, 340,* 137–140.

Li, T., Hu, X., Chandy, K. G., Fantino, E., Kalman, K., Gutman, G., et al. (1998). Transmission disequilibrium analysis of a triplet repeat within the HkCa3 gene using family trios with schizophrenia. *Biochemical and Biophysical Research Communications, 251,* 662–665.

Lin, M. W., Curtis, D., Williams, N., Arranz, M., Nanko, S., Collier, D., et al. (1995). Suggestive evidence for linkage of schizophrenia to markers on chromosome 13q14.1-q32. *Psychiatric Genetics, 5,* 117–126.

Malmberg, A., Lewis, G., David, A. S., & Allebeck, A. (1998). Premorbid adjustment and personality in schizophrenia. *British Journal of Psychiatry, 172,* 308–313.

McCarley, R. W., Wible, C. G., Frumin, M., Hirayasu, Y., Levitt, J. J., Fischer, I. A., et al. (1995). MRI anatomy of schizophrenia. *Biological Psychiatry, 45,* 1099–1119.

McGue, M., & Gottesman, I. I. (1989). A single dominant gene still cannot account for the transmission of schizophrenia. *Archives of General Psychiatry, 46,* 478–479.

McGuffin, P., Owen, M. J., O'Donovan, M. C., Thapar, A., & Gottesman, I. I. (1994). *Seminars in psychiatric genetics.* London: Gaskell Press.

Millar, J. K., Wilson-Annan, J. C., Anderson, S., Christie, S., Taylor, M. S., Semple, C. A., et al. (2000). Disruption of two novel genes by a translocation co-segregating with schizophrenia. *Human Molecular Genetics,* 9, 1415–1423.

Morris, A. G., Gaitonde, E., Mckenna, P. J., Mollon, J. D., & Hunt, D. M. (1995). CAG repeat expansions and schizophrenia: Association with disease in females and with early age at onset. *Human Molecular Genetics, 4,* 1957–1961.

Murphy, K. C., Jones, L. A., & Owen, M. J. (1999). High rates of schizophrenia in adults with velo-cardio-facial syndrome. *Archives of General Psychiatry, 56,* 940–945.

Murray, R. M., & Lewis, S. W. (1987). Is schizophrenia a neurodevelopmental disorder? *British Medical Journal (Clinical Research), 295,* 681–682.

Myers, A., Holmans, P., Marshall, H., Kwon, J., Meyer, D., Ramic, D., et al. (2000). Susceptibility locus for Alzheimer's disease on chromosome 10. *Science, 290,* 2304–2305.

O'Donovan, M. C., Guy, C., Craddock, N., Murphy, K. C., Cardno, A. G., Jones, L., et al.

(1995). Expanded CAG repeats in schizophrenia and bipolar disorder. *Nature Genetics, 10,* 380–381.
O'Donovan, M. C., & Owen, M. J. (1996). Dynamic mutations and psychiatric genetics. *Psychological Medicine, 28,* 1–6.
Ohara, K., Nagai, M., Tani, K., Nakamura, Y., Ino, A., & Ohara, K. (1998). Functional polymorphism of −141c Ins/Del in the dopamine D2 receptor gene promoter and schizophrenia. *Psychiatry Research, 81,* 117–123.
Owen, M. J., Cardno, A. G., & O'Donovan, M. C. (2000). Psychiatric genetics: Back to the future. *Molecular Psychiatry, 5,* 22–31.
Owen, M. J., Holmans, P., & McGuffin, P. (1997). Association studies in psychiatric genetics. *Molecular Psychiatry, 2,* 270–273.
Papolos, D. F., Faedda, G. I., Veit, S., Goldberg, R. B., Morrow, B., Kucherlapati, R., et al. (1996). Bipolar spectrum disorders in patients diagnosed with velo-cardio-facial syndrome: Does a hemizygous deletion of chromosome 22q11 result in bipolar affective disorder? *American Journal of Psychiatry, 153,* 1541–1547.
Petronis, A., & Kennedy, J. L. (1996). Unstable genes—unstable mind? *American Journal of Human Genetics, 152,* 164–172.
Plomin, R., Hill, L., Craig, I. W., McGuffin, P., Purcell, S., Sham, P., et al. (2001). A genome-wide scan of 1842 DNA markers for allelic associations with general cognitive ability: A five-stage design using DNA pooling and extreme selected groups. *Behaviour Genetics, 6,* 497–509.
Pulver, A. E., Nestadt, G., Goldberg, R., Shprintzen, R. J., Lamacz, M., Wolyniec, P. S., et al. (1994). Psychotic illness in patients diagnosed with velo-cardio-facial syndrome and their relatives. *Journal of Nervous and Mental Disease, 182,* 476–478.
Risch, N. (1990). Linkage strategies for genetically complex traits: 2. The power of affected relative pairs. *American Journal of Human Genetics, 46,* 229–241.
Risch, N., & Merikangas, K. (1996). The future of genetic studies of complex human diseases. *Science, 273,* 1516–1517.
Rosenthal, D., Wender, P. H., Kety, S. S., Welner, J., & Schulsinger, F. (1971). The adopted-away offspring of schizophrenics. *American Journal of Psychiatry, 128,* 307–311.
Ross, C. A. (1999). Schizophrenia genetics: Expansion of knowledge? *Molecular Psychiatry, 4,* 4–5.
Schizophrenia Linkage Collaborative Group for Chromosomes 3, 6, and 8. (1996). Additional support for schizophrenia linkage on chromosomes 6 and 8: A multicenter study. *American Journal of Medical Genetics, 67,* 580–594.
Schwab, S. G., Eckstein, G. N., Hallmayer, J., Lerer, B., Albus, M., Borrmann, M., et al. (1997). Evidence suggestive of a locus on chromosome 5q31 contributiong to susceptibility for schizophrenia in German and Israeli families by multipoint affected sib-pair linkage analysis. *Molecular Psychiatry, 2,* 156–160.
Schwab, S. G., Hallmayer, J., Albus, M., Lerer, B., Eckstein, G. N., Borrmann, M., et al. (2000). A genome-wide autosomal screen for schizophrenia susceptibility loci in 71 families with affected siblings: Support for loci on chromosome 10p and 6. *Molecular Psychiatry, 5,* 638–649.
Scott, W., Pericak-Vance, M., & Haines, J. (1997). Genetic analysis of complex diseases. *Science, 275,* 1327.
Shaw, S. H., Kelly, M., Smith, A. B., Shields, G., Hopkins, P. J., Loftus, J., et al. (1998). A genome wide search for schizophrenia susceptibility genes. *American Journal of Medical Genetics, 81,* 364–376.
Sherrington, R., Brynjolfsson, J., Petursson, H., Potter, M., Dudleston, K., Barraclough, B., et al. (1988). Localization of a susceptibility locus for schizophrenia on chromosome-5. *Nature, 336,* 164–167.
Sherrington, R., Rogaev, E. I., Liang, Y., Rogaeva, E. A., Levesque, G., Ikeda, M., et al. (1995). Cloning of a gene bearing missense mutations in early-onset familial Alzheimers-disease. *Nature, 375,* 754–760.
Shprintzen, R. J., Goldberg, R. B., & Golding-Kushner, K. J. (1992). Late-onset psychosis in the velo-cardio-facial syndrome. *American Journal of Medical Genetics, 42,* 141–142.
St. Clair, D., Blackwood, D., Muir, W., Carothers, A., Walker, M., Spowart, G., et al. (1990).

Association within a family of a balanced autosomal translocation with major mental illness. *Lancet, 336,* 13–16.

Straub, R. E., MacLean, C. J., Martin, R. B., Ma, Y., Myakishev, M. V., Harris-Kerr, C., et al. (1998). A schizophrenia locus may be located in region 10p15-p11. *American Journal of Medical Genetics, 81,* 296–301.

Straub, R. E., MacLean, C. J., O'Neill, F. A., Walsh, D., & Kendler, K. S. (1997). Support for a possible schizophrenia vulnerability locus in region 5q22-31 in Irish families. *Molecular Psychiatry, 2,* 148–155.

Tienari, P., Wynne, L. C., Moring, J., Lahti, I., Naarala, M., Sorri, A., et al. (1994). The Finnish Adoptive Family Study of Schizophrenia: Implications for family research. *British Journal of Psychiatry, 164*(Suppl.), 20–26.

Vincent, J. B., Petronis, A., Strong, E., Parikh, S. V., Meltzer, H. Y., Lieberman, J., & Kennedy, J. L. (1999). Analysis of genome-wide CAG/CTG repeats, and at SEF2-1B and ERDA1 in schizophrenia and bipolar affective disorder. *Molecular Psychiatry, 4,* 229–234.

Wender, P. H., Rosenthal, D., Kety, S. S., Schulsinger, F., & Welner, J. (1974). Crossfostering: A research strategy for clarifying the role of genetic and experiential factors in the etiology of schizophrenia. *Archives of General Psychiatry, 30,* 121–128.

Williams, J., McGuffin, P., Nothen, M., Owen, M. J., & the EMASS Collaborative Group. (1997). Meta-analysis of association between the 5-HT2a receptor T102C polymorphism and schizophrenia. *Lancet, 349,* 1221.

Williams, J., Spurlock, G., Holmans, P., Mant, R., Murphy, K., Jones, L., et al. (1998). A meta-analysis and transmission disequilibrium study of association between the dopamine D3 receptor gene and schizophrenia. *Molecular Psychiatry, 3,* 141–149.

Williams, J., Spurlock, G., McGuffin, P., Mallet, J., Nothen, M. M., Gill, M., et al. (1996). Association between schizophrenia and T102C polymorphism of the 5-Hydroxytryptamine type 2A-Receptor gene. *Lancet, 347,* 1294–1296.

Williams, N. M., Rees, M. I., Holmans, P., Norton, N., Cardno, A. G., Jones, L. A., et al. (1999). A two-stage genome scan for schizophrenia susceptibility genes in 196 affected sibling pairs. *Human Molecular Genetics, 8,* 1729–1739.

Weinberger, D. R. (1987). Implications of normal brain development for the pathogenesis of schizophrenia. *Archives of General Psychiatry, 44,* 660–669.

Weinberger, D. R. (1995). From neuropathology to neurodevelopment. *Lancet, 346,* 552–557.

24

The Genetics of Affective Disorders: Present and Future

Sridevi Kalidindi and Peter McGuffin

Mood disorders are of enormous public health importance. The World Bank's Global Burden of Disease study (Murray & Lopez, 1997) predicted that, in 2020, in economic terms, among adult diseases, the burden imposed worldwide by mood disorders will only be exceeded by ischemic heart disease. Depression also was estimated to lead to more years of healthy function lost through illness than any other disorders except for perinatal and communicable diseases of childhood. As with other common diseases, however, there can be problems of definition and nomenclature. The terms *mood disorder* and *affective disorder* incorporate a spectrum of illnesses, and this has led to some of the difficulties for genetic studies as the phenotypes have not always been clearly demarcated. This chapter concentrates on conditions at the more severe end of the spectrum, those that are classified in the *Diagnostic and Statistical Manual of Mental Disorders* (4th ed., *DSM–IV*; American Psychiatric Association, 1994) as major depression and bipolar affective disorder. We also follow the convention of distinguishing between unipolar disorder, characterized by episodes of depression, and bipolar disorder, in which there are episodes of both mania and depression (or more rarely recurrent episodes of mania alone).

Family Studies

Compared with the general population, the relatives of depressed probands have between a threefold (McGuffin & Katz, 1986) and ninefold (Farmer et al., 2000) increase in their risk of developing major depressive disorder. The high relative risk in the study by Farmer et al. may have been partly due to the application of a more stringent *International Classification of Diseases* (*ICD–10;* World Health Organization, 1993) definition of depression together with use of a control group who were siblings of "super healthy" volunteers. However, the consistent pattern that emerges is that whatever definition is applied, depression appears to be familial (Sullivan, Neale, & Kendler, 2000).

Some studies have suggested that those who develop a depressive episode early, before the age of 20, carry a greater genetic loading as shown by the greater prevalence of major depression in their relatives (Kupfer, Frank, Carpenter, & Neiswanger, 1989; Weissman, Warner, Wickramaratne, & Prusoff, 1988). However, the data are not consistent because some studies show little or no relationship between age of onset and the proportion of relatives affected. A more solid predictor of strong familiality appears to be recurrent illness rather than a single episode of depression in the proband or index case (Sullivan et al., 2000).

Studies of the relatives of unipolar probands also are consistent in finding an excess just of unipolar disorder with no evidence of an increased risk of bipolar illness. Family studies of bipolar disorder, in contrast, find an excess of both unipolar and bipolar disorder among relatives.

In a meta-analysis of the family studies that included all studies that used the modern concept of bipolar disorder and had their own control groups, the relative risk in first-degree relatives was between 3% and 19% (Craddock & Jones, 1999), and the mean estimate of risk of bipolar disorder in first-degree relatives of bipolar probands was 7% (confidence interval [CI] = 5%–10%). In 13 other studies using similar diagnostic concepts but in which comparisons were made with assumed population risk figures rather than a sample of controls, there was a broadly similar pattern. However, as had been noted in an earlier survey of published data (McGuffin & Katz, 1989), the range of the estimated recurrence risks in relatives was broad. The risk of bipolar disorder was between 5% and 10% and that of unipolar disorder was between 10% and 20%, giving a combined rate of any form of major affective disorder in first-degree relatives of bipolar probands of 15% to 30%.

So-called Bipolar II disorder (in which there is a milder form of elevated mood, hypomania) also is increased in the families of Bipolar I probands compared with the general population (Gershon et al., 1982; Heun & Maier, 1993; Rice et al., 1987), indicating a genetic component to the milder Bipolar II disorder also. In addition, schizoaffective disorder with manic symptoms has been shown to be increased in some studies of families of bipolar probands (Gershon et al., 1982; Kendler, Karkowski, & Walsh, 1998; Maier et al., 1993; Rice et al., 1987), suggesting that there may be genetic or other familial factors that are shared across the functional psychoses.

In relatives of an affected bipolar proband, the lifetime risk does not appear to be altered by the sex of the proband or the relative (Heun & Maier, 1993; Pauls, Morton, & Egeland, 1992; Rice et al., 1987). An increase of the lifetime risk of affective disorder in relatives has again been suggested where the proband has an early age of onset (Strober, 1992) and where there are more than one family member already affected (Gershon et al., 1982).

Twin Studies

The great majority of studies of twins ascertained through unipolar probands (Bertelsen, Harvald, & Hauge, 1977; Beirut et al., 1999; Kendler,

Pedersen, Neale, & Mathe, 1995; Kendler & Prescott, 1999; Lyons et al., 1998; McGuffin, Katz, Watkins, & Rutherford, 1996; Torgersen, 1986) shows a genetic effect with significantly higher probandwise concordance rates in monozygtic (MZ) twins than in dizygotic (DZ) twins. Only one study, by Andrews, Stewart, Allen, and Henderson (1990), failed to find such a pattern.

In a meta-analysis of twin studies of unipolar depression, Sullivan et al. (2000) found that the variance in liability to depression could be explained by unique environmental effects (58% to 67%) and additive genetic effects (31% to 42%), with very little contribution by shared environmental factors (0% to 5%). However, some studies suggest that more severe unipolar disorder may differ etiologically from milder forms (Lyons et al., 1998). Indeed, the one recent study that was entirely based on a clinically ascertained sample, at the Maudsley Hospital in London, found that major depression was substantially heritable, with estimates, depending on assumptions about the population frequency, in the region of 40% to 70% (McGuffin et al., 1996). Again all of the environmental effects were of the nonshared kind.

The authors of the Maudsley study found no difference in the heritability of major depression between the sexes but only had same-sex DZ twins in their sample and were therefore limited in what inferences they could make concerning whether the *same* genes and environmental effects operate in causing depression in men and women (McGuffin et al., 1996). Kendler and Prescott (1999) carried out an analysis of population-based twin pairs that examined this issue. In both men and women, depression showed substantial and significant heritability, but there was evidence of both common genetic effects and genetic effects that differ between the sexes.

There is significant comorbidity between affective disorders and other psychiatric conditions. Using multivariate twin models, one can assess the correlation between disorders in terms of the contribution of additive genes, of familial environment, and of unique environmental influences to the illness. Kendler, Neale, Kessler, Heath, and Eaves (1992) showed that shared genetic factors appear to contribute to susceptibility to both unipolar depression and generalized anxiety disorder. In the same sample, the genetic factors affecting susceptibility to major depression and alcoholism were distinct (Kendler et al., 1995). A marked overlap of the genes contributing to depressive and anxiety symptoms was also found in a twin sample of children and adolescents, but there also was evidence of a significant specific genetic effect on depressive symptoms (Eley & Stevenson, 1999; Thapar & McGuffin, 1997).

Jones, Kent, and Craddock (in press) pooled the data from six studies using current diagnostic concepts of bipolar disorder in adults (Allen, Cohen, Pollin, & Greenspan, 1974; Bertelsen et al., 1977; Cardno et al., 1999; Kendler, Neale, Kessler, Heath, & Eaves, 1993; Kringlen, 1967; Torgersen, 1986). A weighted average MZ concordance for bipolar disorder of 50% was found compared with an average DZ concordance of 7%. The more recent studies may have underestimated concordances by being derived from indirect sources, such as questionnaires (Kendler et al., 1993) or hospital case notes (Cardno et al., 1999). Notably, an earlier study with more complete cotwin information (Bertelsen et al., 1977) found an MZ concordance for any affective disorder (unipolar or bipolar disorder in cotwins of bipolar probands) of 69% compared with 19% in DZ cotwins.

The incomplete concordance in MZ twins indicates that nongenetic factors play a role in the liability to bipolar disorder. Thus, discordant MZ pairs might arise because the affected proband has a nongenetic type of disorder. Gottesman and Bertelsen (1989) tested this by studying the offspring of discordant bipolar twins. They found that in the offspring of the unaffected MZ twins, the risk of bipolar disorder was increased to the same degree as in the offspring of the affected proband. That is, even though they themselves did not express the phenotype, the unaffected cotwins had the same genotypic susceptibility to bipolar disorder as the probands, and therefore the observed MZ twin discordance could be attributed to nongenetic factors.

Some overlap in bipolar susceptibility genes with those conferring liability to the spectrum of schizophrenic disorder is again apparent in studies of MZ twins, who are genetically identical. Schizoaffective disorder (Bertelsen et al., 1977), and rarely schizophrenia (McGuffin, Reveley, & Holland, 1982), have been reported in MZ cotwins of a bipolar proband. More recently, an analysis in which the authors relaxed conventional diagnostic hierarchies (e.g., when a diagnosis of schizophrenia excludes a diagnosis of bipolar disorder) and allowed subjects to have more than one diagnosis, supported the existence of an overlap between the genes causing liability to bipolar disorder and schizophrenia (Cardno, Rijsdijk, Sham, Murray, & McGuffin, 2002).

Adoption Studies

Of the three adoption studies of unipolar disorder reviewed by Sullivan et al. (2000), the largest study showed a genetic effect, with an odds ratio of more than 7 for occurrence of depression in those biologically related versus those not biologically related to a depressed proband (Wender et al., 1986). There were significant methodological problems such as indirect methods of diagnosis and small sample size in all three studies (Sullivan et al., 2000), and this is likely to have contributed to the failure of two studies to support the evidence for a genetic contribution. In the only study of bipolar disorder, probands who were adopted were found to have a significantly higher rate of affective disorder in their biological than their adoptive parents. The frequency of affective disorder in biological parents did not differ from that in the biological parents of bipolar nonadoptees (Mendlewicz & Rainer, 1977).

Molecular Genetic Studies

Despite the comparative paucity of adoption data, the combined evidence from family, twin, and adoption studies suggests an important genetic contribution to both unipolar and bipolar affective disorder. Therefore, the question arises, can the genes be located and identified? There are broadly two alternative starting points to the process of finding the genes involved in complex diseases: linkage analysis and allelic association analysis.

Linkage Studies

The classic approach to linkage studies of disease is to focus on multiplex families, that is, multigenerational or extended pedigrees containing multiple affected members in which the aim is to detect one or more genetic markers that do not assort independently of the disease. Significant departure from independent assortment indicates that the marker and a disease susceptibility locus are in close proximity on the same chromosome. The definition of a so-called region of interest on a particular chromosome is then the first step in the process of positional cloning, which involves narrowing down the region, identifying the genes within it, and finally cloning the susceptibility gene itself. One of the attractive aspects of carrying out linkage studies is that the marker need not be close to the disease locus. The whole genome, averaged across the sexes, is about 3,500 centimorgans (cM) in length, where 1 cM corresponds to crossing over, or recombination, during meiosis one time in a hundred. For most practical purposes, linkage is detectable over distances of 10 cM or even up to 20 cM. Therefore, given the right conditions, including an adequate sample size and a disorder in which there is a major gene effect and a known mode of transmission, it should be possible to mount a whole genome scan to detect linkage with just a few hundred evenly spaced markers. Because there is now an abundance of DNA-based polymorphisms and access to semiautomated high-throughput genotyping is now widely available, this makes the linkage approach to finding affective disorder genes more attractive still.

There are, however, several obstacles that need to be overcome. First, we do not know the mode of transmission of unipolar or bipolar disorder and have no evidence of major gene effects. Indeed a single-gene model of bipolar disorder can be effectively ruled out mathematically (Craddock, Van Eerdewegh, & Reich, 1995). Second, there could be genetic heterogeneity in which different genes are involved in predisposing to the disorder in different families. Third, there is the difficulty of how to define family members who have mild or borderline forms of disorder. These problems typically have been dealt with by focusing on large multiplex families and by making some simplifying assumptions. The assumptions are that major genes segregate in at least some families, that the mode of transmission can be inferred at least approximately, and that even though the disorder as a whole might be heterogeneous, there should be homogeneity within pedigrees. The problem of borderline cases is particularly difficult for behavioral phenotypes but can be tackled by applying explicit operational diagnostic criteria and by allowing more than one diagnostic model (e.g., classifying only those with bipolar disorder as affected under a more restricted model and allowing both unipolar and bipolar cases to be classed as affected under a broader model).

The psychiatric disorder in which taking this set of simplifying assumptions has worked best is early onset Alzheimer's disease (see Williams, chapter 25, this volume), in which three different major genes with dominant patterns of transmission have been identified. Unfortunately, in affective disorder, the situation is more complicated and more difficult to interpret.

Studies indicating evidence of X linkage, using markers that show cosegregation with the X chromosome and bipolar disorder (Baron et al., 1987; Mendlewicz, Fleiss, & Fieve, 1972; Reich, Clayton, & Winokur, 1969), have a long history but have been controversial because of methodological problems. Furthermore, inheritance through a major X-linked gene cannot apply in bipolar disorder as a whole because numerous instances of father to son transmission have been reported (Hebebrand, 1992). However, there may be a small number of instances in which X-linked inheritance is the main mode of transmission (Bucher et al., 1981).

There also have been reports of single large multiplex pedigrees in which bipolar disorder follows an autosomal dominant like pattern and in which there is significant evidence of linkage, most notably with a set of markers on chromosome 4p (Blackwood et al., 1996). However, a major gene in this region has not been widely replicated in other families. Four other regions offer more general promise, having been implicated in more than one linkage study of bipolar disorder (Craddock & Jones, 1999). These are 12q23-q24, chromosome 18 (Berrettini et al., 1994), 16p13 (Ewald et al., 1995), Xq24-q26 (Pekkarinen, Bredbacka, Loonqvist, & Peltonen, 1995), and 15q11-q13 (Craddock & Lendon, 1999). The chromosome 18 studies, although there have been several reports of positive findings, cannot be regarded as mutually supportive because "hits" have been found by various groups virtually throughout the chromosome (Van Broeckhoven & Verheyen, 1999). A region on chromosome 21 is also of interest. In addition to positive linkage results in some families, Craddock and Owen (1994) suggested that the reduced incidence of mania in trisomy 21 is consistent with a chromosome 21 locus.

Linkage studies in unipolar depression are only just beginning, and, currently, the findings are unclear (Malhi, Moore, & McGuffin, 2000). Older studies, carried out before the advent of DNA markers, used classical genetic markers such as protein polymorphisms and blood group antigens with equivocal results. Using DNA polymorphisms, researchers have screened candidate genes for neuroendocrine and neurotransmitter systems linked with unipolar disorder (Malhi et al., 2000; Neiswanger et al., 1998) or those implicated in bipolar disorder (Balciuniene et al., 1998), but the findings have been negative.

Association Studies

The advantage of the allelic association approach is that it enables genes of a smaller effect to be identified. The main drawback is that allelic association either results from the marker locus itself conferring susceptibility or from very tight linkage between the marker and a disease susceptibility locus resulting in linkage disequilibrium (LD), the phenomenon in which a pair of alleles at adjacent loci continue to be found together over many generations of recombination. In outbred populations, LD occurs only over very short distances. How big this distance might be has been the subject of much debate, but it is certainly less than 1 cM, varies throughout the genome, and for some pairs of polymorphisms may only extend for a few hundred base pairs (Pritchard & Przeworski, 2001). Consequently, whole genome scans for association require many thousands rather than hundreds of evenly spaced markers (McGuffin & Martin, 1999). The technology is developing whereby this will soon be feasible on a large enough scale and whole chromosome searches have already been mounted. However, to date, the main focus has been on candidate genes, that is, genes that encode for a protein that might plausibly be involved in the pathophysiology of affective disorder. The prime candidates are variations, or polymorphisms that affect the structure of the protein or the level at which the gene is expressed: In affective disorders, the most obvious candidates are the genes associated with neurotransmitter systems. However, candidate gene approaches are limited as they are linked to the current neurobiological understanding of the disease (Craddock & Owen, 1996). Because of what is known about the mechanism of action of drugs that affect mood, the serotonergic, dopaminergic, noradrenergic systems as well as the glutaminergic and gamma-aminobutyric acid (GABA) pathway have been targeted. Elsewhere in this volume, Lesch (chapter 20, this volume) has focused in detail on the genes involved in serotonergic neurotransmission, so that here we review the results of studies involving genes that have a role in dopaminergic, noradrenergic, GABAergic, and glutaminergic pathways.

Dopaminergic Neurotransmitter System

Disturbances in dopaminergic systems have been implicated in the etiology of mood disorders (Sokoloff et al., 1992; Willner, 1995). This hypothesis is based on the observation that drugs used in the management of affective disorders affect dopaminergic activity. Among the mood stabilizers, lithium reduces dopaminergic turnover and carbamazepine affects the GH response to apomorphine, whereas the antipsychotics used in the treatment of mania and psychotic depression mainly exert their action through the blocking of dopamine receptors (Avissar & Schreiber, 1992; Elphick, Yang, & Cowen, 1990). Psychostimulants such as amphetamine and methylphenidate, which enhance the activity of dopaminergic synapses, cause a transient elevation of mood in patients with depression (Little, 1988). A depressive syndrome is frequently encountered in subjects affected by Parkinson's disease, in which dopamine depletion is observed (Taylor & Saint-Cyr, 1990). Also, dopamine receptor antagonists are a frequent cause of pharmacologically induced depression (Siris, 1991). Abnormalities in levels of the dopamine metabolite homovanillic acid in cerebrospinal fluid in depressed subjects with motor retardation indicate reduced dopamine turnover (Willner, 1983).

Thus, the dopamine receptor and transporter genes may be considered good candidates. The dopamine transporter that functions in the reuptake of released dopamine into presynaptic terminals plays an important role in dopaminergic systems (Giros et al., 1992). The D2, D3, and D4 receptor gene (DRD2, DRD3, and DRD4, respectively) polymorphisms that cause substitution in the amino acid sequence of these proteins have been identified (Gejman et al., 1994; Itokawa, Arinami, Futamura, Hamaguchi, & Toru, 1993; Lannfelt et al., 1992; Van Tol et al., 1992). Although no structural polymorphism has been reported for the dopamine transporter gene (DAT1), a 40 bp-repeat polymorphism has been found in the 3′-untranslated region of this gene (Vandenbergh et al., 1992).

Manki et al. (1996) investigated dopamine D2, D3, and D4 receptor and transporter gene polymorphisms and compared bipolar and unipolar patients with controls. They found no association with the DRD2, DRD3, or DAT1 gene polymorphisms and mood disorder. However, there was a significant association between the 48 bp-repeat polymorphism short allele (the 5-repeat allele) of the dopamine D4 receptor gene and major depression. It also was found that the 4-repeat allele was significantly lower in mood-disordered patients than controls, possibly indicating a protective role. The polymorphism involves 48 base pairs in the third exon of the coding sequence that repeats 2 to 10 times (Van Tol et al., 1991). Within this polymorphism, it appears that the potency of inhibition of cAMP formation by dopamine is reduced twofold in vitro for the DRD4 7-repeat variant as compared with the short allele variants (Asghari et al., 1995).

Serretti, Macciardi, Cusin, Lattuada, Lilli, and Smeraldi (1998) found that delusional symptoms were significantly associated with long DRD4 variant, 7-repeat allele, containing genotypes DRD4 2/7 and DRD4 4/7, in bipolar and unipolar affective disorder. However, the variance in liability explained was small at 5.3%.

An association with the 48 bp-repeat polymorphism of the DRD4 gene and unipolar depression was also reported by Macciardi et al. (1994), although Lim et al. (1994) and Perez de Castro, Torres, Fernandez, Saiz, and Llinares (1994) found no evidence of an association between this polymorphism and bipolar disorder. This suggests that the association may be specific to unipolar depression.

The 48 bp-repeat polymorphism of DRD4 has been reported to produce altered receptor function; it confers differential neuroleptic binding properties in cell culture (Van Tol et al., 1992). This polymorphism is located at the third cytoplasmic loop of the receptor protein, the site believed to interact with G protein in this family of receptors (Ross, 1989). It is possible that the size change in the cytoplasmic loop affects the conformation of the protein and receptor function, thereby leading to susceptibility to mood disorders. Alternatively, there may again be linkage disequilibrium with unknown loci affecting the risk of developing major depression.

There also is some evidence to implicate the DRD3 gene in unipolar depression (Dikeos et al., 1999). The D3 receptor was identified by Sokoloff, Giros, Martres, Bouthenet, and Schwartz (1990) and localized to chromosome 3q13.3 (Le Coniat et al., 1991). The receptor is almost exclusively present in the limbic region of the brain known to be associated with the generation of emotions and the functional anatomy of depression (Drevets et al., 1992; Sokoloff et al., 1992). A relationship has been observed between D3 stimulation and changes in frontal lobe cortex serotonin (Lynch, 1997).

The polymorphism of the DRD3 gene studied by Dikeos et al. (1999) was the Bal I restriction site polymorphism. This polymorphism is located in the first exon of the DRD4 gene and gives rise to a substitution of serine by glycine in the ninth amino- acid of the protein (Lannfelt et al., 1992), and there also have been many studies of association with schizophrenia with a meta-analysis suggesting a small but significant effect (see Owen & O'Donovan, chapter 23, this volume). However, overall studies in unipolar disorder have been inconsistent. In some, no association has been detected (Lannfelt et al., 1992; Sokoloff et al., 1992), and in another, the two different forms of the receptor appeared to differ in function as they result in varying levels of the serotonin metabolite 5-HIAA in the CSF (Jonsson et al., 1996).

The inconsistent results regarding the relationship of DRD3 with unipolar depression and the relatively low odds ratios observed in the study with positive results indicate that the effect of the DRD3 gene in the pathogenesis of this disorder must be small. Once more, the Bal I polymorphism may in fact be in linkage disequilibrium with another functional part of the DRD3 gene or another gene in the proximity directly influencing pathogenesis.

In summary, although there are good theoretical reasons for suspecting involvement in affective disorder of dopamine receptor genes, the results so far are inconclusive. At best we can say that there is some support for a small role of functional polymorphisms in the DRD3 and DRD4 genes.

Tyrosine hydroxylase (TH) gene. Polymorphisms in this gene have been of interest because TH is the enzyme that is responsible for the rate-limiting step in the catecholamine synthetic pathway (i.e., the pathway for both dopamine and norepinephrine synthesis). The TH gene maps to chromosome 11p15, which is also the region in which linkage was reported by Egeland et al. (1987) in the Old Order Amish. An association has been reported with two restriction fragment length polymorphisms (Leboyer et al., 1990), but following further work by other groups and a meta-analysis of all the studies, there appears to be no consistent evidence of an association with bipolar disorder (Turecki, Rouleau, Mari, Joober, & Morgan, 1997). Furthermore, the 11p15 linkage finding was subsequently not supported by further work on the same Amish pedigree or by other independent groups (Jones, Kent, & Craddock, in press).

Catecholamine-O-methyltransferase (COMT) gene. This enzyme is responsible for the degradation of monoamines, and hence, like TH, is part of both the dopamine and norepinephrine pathway. There is a common functional polymorphism in the gene influencing its enzyme activity. Craddock et al. (1997) found no evidence to associate it with bipolar disorder as a whole. However, three other studies suggest that an allele that confers reduced enzyme activity may be associated with a modest increase in susceptibility to rapid cycling bipolar disorder (Kirov et al., 1998; Lachman et al., 1996; Papolos, Veit, Faedda, Saito, & Lachman, 1998).

Norepinephrine (or Noradrenaline) Neurotransmitter System

Interest in the norepinephrine transporter (NET) gene has arisen partly as a result of the mechanism of action of certain antidepressant drugs such as venlafaxine, which inhibit the uptake of norepinephrine as well as serotonin in depression. The transporter reaccumulates and reuses neurotransmitters into presynaptic terminals, aiding termination of synaptic transmission. The gene encoding the NET is found on the long arm of chromosome 16 (Gelernter et al., 1993). Five missense mutations have been identified: Substitutions Val69Ile, Thr99Ile, Val245Ile, Val449Ile, and Gly478Ser are located at putative transmembrane domains 1, 2, 4, 9, and 10, respectively. No association has been found between a silent 1287G/A polymorphism (NET-8) in exon 9 of the NET gene or within the Thr99Ile in exon 2 of the NET gene (NET-1) and unipolar depression (Owen, Du, Bakish, Lapierre, & Hrdina, 1999). No association of the five substitutions was reported in a case control study of bipolar subjects (Stober et al., 1996).

GABA Neurotransmitter System

Changes in the activity of GABA, the major inhibitory neurotransmitter in the cortex, can potentially disturb other neurotransmitter pathways such as the serotonin and dopamine systems, which are involved in the neurobiological etiology of affective disorders (Goodwin, 1990; Petty, 1995). GABA-A receptors consist of at least 15 different receptor subunits, classified into five families, (α, β, γ, δ, and ρ). Neural inhibition is achieved through GABA-A receptors (Zorumski, 1991). Relatively high expression, accounting for around 20% of the GABA-A receptors that contain α5 subunits, is found in the hippocampus (McKernan, 1996), an area that has been implicated in the neurobiology of affective disorders. Indeed, a positive association with allele 4 of a dinucleotide repeat polymorphism in the GABA receptor gene, GABRA5, in subjects with unipolar depression (Oruc, Verheyen, et al., 1997), has been reported. In a linkage study in bipolar subjects, an association with the 282 bp allele of the GABRA5 locus has been shown, which appears to confer a preponderance of manic episodes in the course of the illness but not to influence the age of onset or the overall clinical severity (Papadimitriou et al., 1998). However, negative findings in bipolar subjects in both linkage studies (Coon, Hicks, & Bailey,

1994; De Bruyn, Souery, & Mendelbaum, 1994; Ewald et al., 1995) and association studies (Oruc, Lindblad, et al., 1997) have been reported.

A link with the GABRAB1 loci could not be ruled out in bipolar subjects (De Bruyn, Souery, Mendelbaum, Mendlewicz, & Van Broeck, 1996) but did not show a positive association in another study in unipolar or bipolar disorder (Serretti, Macciardi, Cusin, Lattuada, Lilli, Di Bella, & Smeraldi, 1998). Further work is required in this neurotransmitter system to gain a consensus on whether genetic polymorphisms play a role in the etiology of mood disorders.

Glutaminergic System

The *N*-methyl D-aspartate (NMDA) class of glutamate receptors would appear to play a role in the pathophysiology of affective disorders (Skolnick et al., 1996). Evidence from animal model studies of depression shows NMDA receptor antagonists such as ketamine acting effectively (Berman et al., 2000; Layer, Popik, Olds, & Skolnick, 1995; Meloni et al., 1993; Moryl, Danysz, & Quack, 1993; Trullas & Skolnick, 1990). An interesting finding in mice is that repeated antidepressant administration alters mRNA expression encoding the NMDA receptor subunits (Boyer, Skolnick, & Fossom, 1998). Levine et al. (2000) showed an abnormal glial-neuronal glutamine/glutamate cycle associated with the NMDA receptor system in patients with depression. Polymorphisms have been identified in the NMDA receptor genes, such as in the 2B subunit (Nishiguchi, Shirakawa, Ono, Hashimoto, & Maeda, 2000), but research as to links with the affective disorders have yet to be undertaken, although such work has commenced in subjects with schizophrenia (Nishiguchi et al., 2000).

Trinucleotide Repeat Expansion (Dynamic Mutation)

As with schizophrenia (see Lesch, chapter 20, this volume), it has been suggested that bipolar disorder may show anticipation in families (Grigoroiu-Serbanescu, Wickramaratne, Hodge, Milea, & Mihailescu, 1997). This is a process whereby successive generations suffer with progressively earlier age of onset or more severe disease. This phenomenon is apparent in certain single-gene neurological conditions such as fragile X syndrome and Huntington's disease, and it is explained by the presence of a mutation involving trinucleotide repeats whereby having more than a critical number of repeats is associated with expression of the disease. Such mutations tend to be unstable or dynamic such that each generation has an increased number of repeats, which in turn is associated with increased severity.

A method known as repeat expansion detection (RED) analysis (Schalling, Hudson, Buetow, & Houseman, 1993) is a way of searching for trinucleotide repeats occurring anywhere in the genome; with this approach, there have been positive reports of increased trinucleotide repeat sequences in bipolar disorder (Lindblad et al., 1995; O'Donovan et al., 1995; O'Donovan et al., 1996; Oruc, Lindblad, et al., 1997) but also a negative one showing no difference in the number of repeats when comparing patients with bipolar disorder and controls (Vincent et al., 1996). So far, the gene or genes responsible for the RED analysis findings have not been identified.

Conclusion and a Look to the Future

There is clear evidence from family, twin, and adoption studies of a genetic component to affective disorders as well as a substantial contribution from the environment. The tasks ahead now are, first, to gain a better understanding of gene–environment interplay and, second, to locate and identify the genes involved and discover how they exert their effects.

Although a relationship between recent adversity and the onset of depression is generally accepted as proven (Brown & Harris, 1978), the relationship between genetic liability to depression and reaction to unpleasant happenings appears to be complicated. There is evidence dating back to the 1980s that relatives of depressed subjects not only show increased rates of depression but also increased rates of experiencing or reporting threatening life events (McGuffin, Katz, & Bebbington, 1988). At least some of this familiality results from the same event impinging on two or more members of the family (Farmer, Harris, et al., 2000), but this does not offer a complete explanation. Familial clustering of life events also has been supported by twin studies that have produced the surprising finding that some categories of events appear to be influenced by genes (Plomin & Bergeman, 1991; Thapar & McGuffin, 1996). This has led to the conjecture that familial factors may influence the liability to depression indirectly by predisposing individuals to select more aversive environments and a "hazard-prone" lifestyle. Alternatively, it has been hypothesized that an inherited cognitive schema, a tendency toward "threat perception," may predispose both to depressive symptomatology and reported unpleasant happenings (McGuffin, Katz, & Bebbington, 1988). The results of a recent sibpair study (Farmer, Harris, et al., 2001) suggested that depressed subjects have high levels of dysfunctional attitudes and neuroticism but actually score low on measures of hazard-prone behavior as reflected in sensation seeking. However, one of the major difficulties in teasing out the types of personality *trait* that characterize vulnerability to depression is that the trait measures are inevitably contaminated by mood *state*.

So far, twin and family studies investigating life events and depression in the same sample have taken a cross-sectional approach. This means that it is difficult to separate trait from state and to take into account such phenomena as life events being more strongly associated with first onsets of depression than with recurrences. One way forward would be to take a lifetime approach not just to eliciting depressive symptoms but to charting life happenings. Ideally, this would be done prospectively, but retrospective studies are more feasible and affordable. Such an approach has been shown to be practicable as well as reliable and valid using a life history calendar method (Caspi et al., 1996). It also has been shown that a relatively small number of events categories, which can be captured by a 12-item checklist (Brugha, Bebbington, Tennant, & Hurry, 1985), account for the bulk of events associated with depressive onsets. Studies putting together this type of methodology with a genetic design may hold considerable promise.

With regard to linkage and association studies, larger samples need to be collected. The majority of recent studies have been underpowered to find genes of small effect (see Cardon, chapter 4, this volume). In particular, the apparent nonreplication of linkage study findings has, for many, been perplexing. However, it is important to realize that if detection of linkage in multiple gene disorders is difficult, replication is even more difficult. This is because initial genome scans of bipolar disorder so far have typically each reported one or two "hits" out of, we assume, several, perhaps many potential targets. The task of a replication study by contrast is to confirm the existence of linkage in precisely those one or two target regions, not just in *any* one or two of several regions. Consequently, the sample size required to replicate linkage in multiple gene disorders will tend to be substantially larger than the sample size for initial detection. Therefore, we are likely to see future linkage studies involving larger numbers of families, and multicenter collaborations will inevitably become even more important.

The application of association studies has, as we have mentioned and as is discussed elsewhere in this volume (see Sham, chapter 3; Cardon, chapter 4; Plomin, chapter 11; Owen & O'Donovan, chapter 23, this volume), been limited by the practical difficulties of carrying out systematic searches using a very dense grid of thousands of markers. The technical difficulties here are likely to be overcome, but the identification of all genes expressed in the brain together with identification of functional polymorphisms in these and their regulatory regions will enable a different kind of systematic search, essentially a whole genome scan focusing on all possible candidates (see also Craig & McClay, chapter 2, this volume).

In the meantime, there is work to be done on existing candidate genes. Although differences in receptor number have been associated with some of the genetic polymorphisms that may be involved in depression, there has been very little research into the clinical relevance of functional differences conferred by these polymorphisms.

It is likely that the various polymorphisms do result in altered neurotransmitter receptor numbers and function, giving rise to changes in neurotransmitter levels in the relevant brain regions, which in turn could confer an increased susceptibility to psychiatric illness in conjunction with other psychosocial factors. It may be that the presence of several polymorphisms increases the vulnerability of an individual to cross the liability threshold and become mentally ill.

Many abnormalities have been identified using neuroimaging in patients with depression (for reviews, Grasby, 1999; Kennedy, Javanmard, &

Vaccarino, 1997; Mann, 1998). It is yet to be seen whether any of these are a result of particular genetic polymorphisms.

As discussed in greater detail by Aitchison and Gill (chapter 18, this volume), identification of the genes and gene products involved in psychiatric disorders should have important implications for pharmacotherapy in two ways. The first of these is in directing drugs discovery (Roses, 2000). At present, the drugs used in treatment of psychiatric or other central nervous system diseases have their actions at a limited number of target sites that include cell surface receptors, nuclear receptors, iron channels, and enzymes. It is likely that identifying the genes involved in the pathogenesis of psychiatric disorders will identify new targets, some of which fall within these categories but will include others that have thus far not been included as target sites of drug action. In addition to targeting of treatments, advances in genomics will allow tailoring of treatments. As reviewed by Aitchison and Gill, there is already some evidence that response to antidepressants or neuroleptics is influenced by receptor genotypes. It is also well established that the rate at which most psychoactive drugs are metabolized is influenced by genetic factors, in particular genes in the cytochrome p450 system of which variation in the CYP2D6 gene have particular importance. This may be relevant to the development of side effects as well as to treatment response (Wolf, Smith, & Smith, 2000). This whole area of pharmacogenetics is still comparatively novel but holds enormous potential for individual tailoring of treatments that will be a major advance on current clinical trial-and-error approaches.

Finally, there has been a degree of public concern that the current pace of advance will tend to "geneticize" common diseases and encourage deterministic attitudes. In particular, worries have been expressed that insurance companies may wish to force DNA testing on individuals thought to be at genetic high risk of disorder. While prediction is already possible for rare early onset dementia such as Huntington's disease or the single-gene forms of Alzheimer's disease, the ApoE4 genotype, which has a proven association with later onset Alzheimer's disease in the general population is of limited value as a predictor at an individual level (see also Williams, chapter 25). The situation is likely to prove even more complicated with disorders such as bipolar or unipolar affective disorder, and, at best, DNA-based tests may be used to modify the predicted risk in individuals who are already at high risk because of having an affected close relative. However, it is unlikely that risk prediction will ever be better than about 40% to 70% accurate because genetically identical individuals (MZ twins) are discordant for schizophrenia 30% to 60% of the time. This means that DNA-based population screening for complex psychiatric disorders will never become a reality, even if the option of offering screening to high-risk individuals probably will. Given that most common diseases, and not just psychiatric ones, will depend on the combined action of multiple common gene variations together with environmental risk factors, this means that batteries of genetic tests will be of limited usefulness to insurance companies and are not likely to be widely used.

The other potentially worrying aspect of psychiatric disorders becoming "geneticized" is, it has been suggested, an increase in stigma. Of course, just the opposite could be the case, so that increasing knowledge about the causation of disorders may serve to demystify them and therefore make them, in the public eye, something that is more tangible and acceptable. The postgenomic impact on disorders such as bipolar affective disorder and depression might therefore be to legitimize them as "real" diseases rather than, as is all too often the case, their being perceived as a result of personal failing or weakness.

References

Allen, M. G., Cohen, S., Pollin, W., & Greenspan, S. I. (1974). Affective illness in veteran twins: A diagnostic review. *American Journal of Psychiatry, 131,* 1234–1239.

American Psychiatric Association. (1994). *Diagnostic and statistical manual of mental disorders* (4th ed.). Washington, DC: Author.

Andrews, G., Stewart, G., Allen, R., & Henderson, A. S. (1990). The genetics of six neurotic disorders: A twin study. *Journal of Affective Disorder, 19,* 23–29.

Asghari, V., Sanyal, S., Buchwaldt, S., Paterson, A., Jovanovic, V., & Van Tol, H. (1995). Modulation of intracellular cyclic AMP levels by different human dopamine D4 receptor variants. *Journal of Neurochemistry, 65,* 1157–1165.

Avissar, S., & Schreiber, G. (1992). The involvement of guanine nucleotide binding proteins in the pathogenesis and treatment of affective disorders. *Biological Psychiatry, 31,* 435–459.

Balciuniene, J., Yuan, Q. P., Engstrom, C., Lindblad, K., Nylander, P. O., Sundvall, M., et al. (1998). Linkage analysis of candidate loci in families with recurrent major depression. *Molecular Psychiatry, 3,* 162–168.

Baron, M., Risch, N., Hamburger, R., Mandel, B., Kushner, S., Newma, M., et al. (1987). Genetic linkage between X-chromosome markers and bipolar affective illness. *Nature, 326,* 289–292.

Beirut, L. J., Heath, A. C., Bucholz, K. K., Dinwiddie, S. H., Madden, P. A., Statham, D. J., et al. (1999). Major depressive disorder in a community based sample: Are there different genetic and environmental contributions for men and women? *Archives of General Psychiatry, 56,* 557–563.

Berman, R. M., Cappiello, A., Anand, A., Oren, D. A., Heninger, G. R., Charney, D. S., & Krystal, J. H. (2000). Antidepressant effects of ketamine in depressed patients. *Biological Psychiatry, 47,* 351–354.

Berrettini, W. H., Ferraro, T. N., Goldin, L. R., Weeks, D. E., Detera-Wadleigh, S., Nurnberger, J. I., Jr., et al. (1994). Chromosome 18 DNA markers and manic-depressive illness: Evidence for a susceptibility gene. *Proceedings of the National Academy of Sciences, 91,* 5918–5921.

Bertelsen, A., Harvald, B., & Hauge, M. (1977). A Danish twin study of manic-depressive illness. *British Journal of Psychiatry, 130,* 330–351.

Blackwood, D. H. R., He, L., Morris, S. W., Whitton, C., Thomson, M., Walker, M. T., et al. (1996). A locus for bipolar affective disorder on chromosome 4p. *Nature Genetics, 12,* 427–430.

Boyer, P. A., Skolnick, P., & Fossom, L. H. (1998). Chronic administration of imipramine and citalopram alters the expression of NMDA receptor subunit mRNAs in mouse brain: A quantitative in situ hybridisation study. *Journal of Molecular Neuroscience, 10,* 219–233.

Brown, G. W., & Harris, T. O. (1978). *The social origins of depression: A study of psychiatric disorder in women.* London: Tavistock Press.

Brugha, T., Bebbington, P., Tennant, C., & Hurry, J. (1985). The list of threatening experiences: A subset of 12 life event categories with considerable long-term contextual threat. *Psychological Medicine, 15,* 189–194.

Bucher, K. D., Elston, R. C., Green, R., Whybrow, P., Helzer, J., Reich, T., et al. (1981). The transmission of manic depressive illness: II. Segregation analysis of three sets of family data. *Journal of Psychiatric Research, 16,* 65–78.

Cardno, A. G., Marshall, E. J., Coid, B., Macdonald, A. M., Ribchester, T. R., Davies, N. J., et al. (1999). Heritability estimates for psychotic disorders. *Archives of General Psychiatry, 56,* 162–168.

Cardno, A. G., Rijsdijk, F. V., Sham, P. C., Murray, R. M., & McGuffin, P. (2002). A twin study of genetic relationships between psychotic symptoms. *American Journal of Psyciatry, 159,* 539–545.

Caspi, A., Moffitt, T. E., Thorton, A., Fredman, D., Amell, J. W., Harrington, H., et al. (1996). The life history calendar: A research and clinical assessment method for collecting retrospective event-history data. *International Journal of Methods in Psychiatric Research, 6,* 101–114.

Coon, H., Hicks, A. A., & Bailey, M. E. S. (1994). Analysis of GABA-A receptor subunit genes in multiplex pedigrees with manic depression. *Psychiatric Genetics, 4,* 185–191.

Craddock, N., & Jones, I. (1999). Genetics of bipolar disorder. *Journal of Medical Genetics, 36,* 585–594.

Craddock, N., & Lendon, C. (1999). Chromosome workshop: Chromosomes 11, 14 and 15. *American Journal of Medical Genetics (Neuropsychiatric Genetics), 88,* 244–254.

Craddock, N., & Owen, M. (1994). Is there an inverse relationship between Down's syndrome and bipolar affective disorder? Literature review and genetic implications. *Journal of Intellectual Disability Research, 38,* 613–620.

Craddock, N., & Owen, M. J. (1996). Modern molecular genetic approaches to psychiatric disease. *British Medical Bulletin, 52,* 434–453.

Craddock, N., Spurlock, G., McGuffin, P., Owen, M. J., Nosten-Bertrand, M., Bellivier, F., et al. (1997). No association between bipolar disorder and alleles at a functional polymorphism in the COMT gene. *British Journal of Psychiatry, 170,* 526–528.

Craddock, N., Van Eerdewegh, P., & Reich, T. (1995). Mathematical limits of multilocus models: The genetic transmission of bipolar disorder. *American Journal of Medical Genetics, 57,* 690–702.

De Bruyn, A., Souery, D., & Mendelbaum, K. (1994). A linkage study between bipolar disorder and candidate genes involved in dopaminergic and GABAergic neurotransmission. *Psychiatric Genetics, 6,* 67–73.

De Bruyn, A., Souery, D., Mendelbaum, K., Mendlewicz, J., & Van Broeckhoven, C. (1996). A linkage study between bipolar disorder genes involved in dopaminergic and GABAergic neurotransmission. *Psychiatric Genetics, 6*(2), 67–73.

Dikeos, D. G., Papadimitriou, G. N., Avramopoulos, D., Karadima, G., Daskalopoulou, E. G., Sourey, D., et al. (1999). Association between the dopamine D3 receptor gene locus (DRD3) and unipolar affective disorder. *Psychiatric Genetics, 9,* 189–195.

Drevets, W. C., Videen, T. O., Price, J. L., Prescorn, S. H., Carmichael, S. T., & Raichle, M. E. (1992). A functional anatomical study of unipolar depression. *Journal of Neuroscience, 12,* 3628–3641.

Egeland, J. A., Gerhard, D. S., Pauls, D. L., Sussex, J. N., Kidd, K. K., Allen, C. R., et al. (1987). Bipolar affective disorders linked to DNA markers on chromosome 11. *Nature, 325,* 783–787.

Elphick, M., Yang, J. D., & Cowen, P. J. (1990). Effects of carbamazepine on dopamine and serotonin-mediated neuroendocrine responses. *Archives of General Psychiatry, 47,* 135–140.

Ewald, H., Mors, O., Flint, T., Koed, K., Eiberg, H., & Kruse, T. A. (1995). A possible locus for manic depressive illness on chromosome 16p13. *Psychiatric Genetics, 5,* 71–81.

Farmer, A., Harris, T., Redman, K., Sadler, S., Mahmood, A., & McGuffin, P. (2000). The Cardiff depression study: A sib pair study of life events and familiality in major depression. *British Journal of Psychiatry, 176,* 150–155.

Farmer, A., Harris, T., Redman, K., Sadler, S., Mahmood, A., & McGuffin, P. (2001). The Cardiff depression study: A sib-pair study of dysfunctional attitudes in depressed probands and healthy control subjects. *Psychological Medicine, 31,* 627–633.

Gejman, P. V., Ram, A., Gelernter, J., Friedman, E., Cao, Q., Pickar, D., et al. (1994). No structural mutation in the dopamine D2 receptor gene in alcoholism or schizophrenia. Analysis using denaturing gradient gel electrophoresis. *Journal of American Medical Assessment, 271,* 204–208.

Gelernter, J., Kruger, S., Pakstis, A. J., Pacholczyk, T., Sparkes, R. S., Kidd, K. K., & Amara, S. G. (1993). Assignment of the norepinephrine transporter protein (NET 1) locus chromosome 16. *Genomics, 18,* 690–692.

Gershon, E. S., Hamovit, J., Guroff, J. J., Dibble, E., Leckman, J. F., Sceery, W., et al. (1982). A family study of schizoaffective, Bipolar I, Bipolar II, unipolar, and normal control probands. *Archives of General Psychiatry, 39,* 1157–1167.

Giros, B., El Mestikawy, S., Godinot, N., Zheng, K., Hans, H., Yang, F. T., & Caron, M. G. (1992). Cloning, pharmacological characterization, and chromosome assignment of the human dopamine transporter. *Molecular Pharmacology, 42,* 383–390.

Goodwin, F. K. J. K. (1990). *Manic depressive illness*. New York: Oxford University Press.

Gottesman, I. I., & Bertelsen, A. (1989). Confirming unexpressed genotypes for schizophrenia. *Archives of General Psychiatry, 46,* 274–287.

Grasby, P. M. (1999). Imaging strategies in depression. *Journal of Psychopharmacology, 13,* 346–351.

Grigoroiu-Serbanescu, M., Wickramaratne, P. J., Hodge, S. E., Milea, S., & Mihailescu, R. (1997). Genetic anticipation and imprinting in Bipolar I illness. *British Journal of Psychiatry, 170,* 162–166.

Hebebrand, J. (1992). A critical appraisal of X-linked bipolar illness. Evidence for the assumed mode of inheritance is lacking. *British Journal of Psychiatry, 160,* 7–11.

Heun, R., & Maier, W. (1993). The distinction of Bipolar II disorder from Bipolar I and recurrent unipolar depression: Results of a controlled family study. *Acta Psychiatrica Scandinavica, 87,* 279–284.

Itokawa, M., Arinami, T., Futamura, N., Hamaguchi, H., & Toru, M. (1993). A structural polymorphism of human dopamine D2 receptor. D2 (Ser311-Cys). *Biochemical and Biophysical Research Communications, 196,* 1369–1375.

Jones, I., Kent, L., & Craddock, N. (in press). The genetics of psychiatric disorders. In P. McGuffin, I. Gottesman, & M. Owen (Eds.), *Psychiatric genetics and genomics.* New York: Oxford University Press.

Jonsson, E., Sedvall, G., Brene, S., Gustavsson, J. P., Geijer, T., Terenius, L., et al. (1996). Dopamine-related genes and their relationships to monoamine metabolites in CSF. *Biological Psychiatry, 40,* 1032–1043.

Kendler, K. S., Karkowski, L. M., & Walsh, D. (1998). The structure of psychosis. *Archives of General Psychiatry, 54,* 299–304.

Kendler, K. S., Neale, M. C., Kessler, R. C., Heath, A. C., & Eaves, L. J. (1992). Major depression and generalized anxiety disorder: Same genes, (partly) different environments? *Archives of General Psychiatry, 49,* 716–722.

Kendler, K. S., Neale, M., Kessler, R., Heath, A., & Eaves, L. (1993). A twin study of recent life events and difficulties. *Archives of General Psychiatry, 50,* 789–796.

Kendler, K. S., Pedersen, N. L., Neale, M. C., & Mathe, A. A. (1995). A pilot Swedish twin study of affective illness including hospital and population ascertained subsamples: Results of model fitting. *Behavior Genetics, 25,* 217–232.

Kendler, K. S., & Prescott, C. A. (1999). A population based twin study of lifetime major depression. *Archives of General Psychiatry, 56,* 39–44.

Kennedy, S. H., Javanmard, M., & Vaccarino, F. J. (1997). A review of functional neuroimaging in mood disorders: Positron emission tomography and depression. *Canadian Journal of Psychiatry, 42,* 467–475.

Kirov, G., Murphy, K. C., Arranz, M. J., Jones, I., McCandless, F., Kunigi, H., et al. (1998). Low activity allele of catechol-O-methyltransferase gene associated with rapid cycling bipolar disorder. *Molecular Psychiatry, 3,* 342–345.

Kringlen, E. (1967). *Hereditary and environment in the functional psychoses.* London: William Heinemann.

Kupfer, D. J., Frank, E., Carpenter, L. L., & Neiswanger, K. (1989). Family history in recurrent depression. *Journal of Affective Disorders, 17,* 113–119.

Lachman, H. M., Morrow, B., Shprintzen, R., Veit, S., Parsia, S. S., Faedda, G., et al. (1996).

Association of Codon 108/158 catechol-O-methyltransferase gene polymorphism with the psychiatric manifestations of velo-cardio-facial syndrome. *American Journal of Medical Genetics (Neuropsychiatric Genetics), 67,* 468–472.

Lannfelt, L., Sokoloff, P., Martres, M. P., Pilon, C., Giros, B., Jonnson, E., et al. (1992). Amino acid substitution in the dopamine D3 receptor as a useful polymorphism for investigating psychiatric disorders. *Psychiatric Genetics, 2,* 249–256.

Layer, R. T., Popik, P., Olds, T., & Skolnick, P. (1995). Antidepressant-like actions of the polyamine site NMDA antagonist, eliprodil. *Pharmacology, Biochemistry and Behavior, 52,* 621–627.

Leboyer, M., Malafosse, A., Boularand, S., Campion, D., Gheysen, F., Samolyk, D., et al. (1990). Tyrosine hydroxylase polymorphisms associated with manic-depressive illness. *Lancet, 335,* 1219.

Le Coniat, M., Sokoloff, P., Hillion, J., Martres, M. P., Giros, B., & Pilon, C. (1991). Chromosomal localization of the human D3 dopamine receptor gene. *Human Genetics, 87,* 618–620.

Levine, J., Panchalingam, K., Rapoport, A., Gershon, S., McClure, M. J., & Pettegrew, J. W. (2000). Increased cerebrospinal fluid glutamate levels in depressed patients. *Biological Psychiatry, 47,* 586–593.

Lim, L. C., Nothen, M. M., Korner, J., Rietschel, M., Castle, D., Hunt, N., et al. (1994). No evidence of association between dopamine D4 receptor variants and bipolar affective disorder. *American Journal of Medical Genetics, 54,* 259–263.

Lindblad, K., Nylander, P.-O., De Bruyn, A., Sourey, D., Zander, C., Engstrom, C., et al. (1995). Detection of expanded CAG repeats in bipolar affective disorder using the repeat expansion detection (RED) method. *Neurobiological Disorders, 2,* 55–62.

Little, K. Y. (1988). Amphetamine, but not methylphenidate, predicts antidepressant response. *Journal of Clinical Pharmacology, 8,* 177–183.

Lynch, M. R. (1997). Selective effects on prefrontal cortex serotonin by dopamine D3 receptor agonism: Interaction with low-dose haloperidol. *Progress in Neuropsychopharmacological and Biological Psychiatry, 21,* 1141–1153.

Lyons, M. J., Eisen, S. A., Goldberg, J., True, W., Lin, N., Meyer, J. M., et al. (1998). A registry-based twin study of depression in men. *Archives of General Psychiatry, 55,* 468–472.

Macciardi, F., Cavallini, M., Petronis, A., Marino, C., Verga, M., Cauli, G., et al. (1994). The D4 receptor gene and mood disorders: An association study [Abstract]. *American Journal of Human Genetics, 55,* A336.

Maier, W., Lichtermann, D., Minges, J., Hallmayer, J., Heun, R., Benkert, O., & Levinson, D. F. (1993). Continuity and discontinuity of affective disorders and schizophrenia: Results of a controlled family. *Archives of General Psychiatry, 50,* 871–883.

Malhi, G. S., Moore, J., & McGuffin, P. (2000). The genetics of major depression. *Current Psychiatric Reports, 2,* 165–169.

Manki, H., Kanba, S., Muramatsu, T., Higuchi, S., Suzuki, E., Matsushita, S., et al. (1996). Dopamine D2, D3 and D4 receptor and transporter gene polymorphisms and mood disorders. *Journal of Affective Disorders, 40,* 7–13.

Mann, J. J. (1998). The role of in vivo neurotransmitter system imaging studies in understanding major depression. *Biological Psychiatry, 44,* 1077–1078.

McGuffin, P., & Katz, R. (1986). Nature, nurture, and affective disorder. In J. W. F. Deakin (Ed.), *The biology of depression* (pp. 26–51). London: Gaskell.

McGuffin, P., & Katz, E. (1989). The genetics of depression and manic-depressive disorder. *British Journal of Psychiatry, 155,* 294–304.

McGuffin, P., Katz, E., & Bebbington, P. (1988). The Camberwell collaborative Study. III. Depression and adversity in the relatives of depressed probands. *British Journal of Psychiatry, 152,* 775–782.

McGuffin, P., Katz, R., Watkins, S., & Rutherford, J. (1996). A hospital-based twin register of the hereditability of *DSM IV* unipolar depression. *Archives of General Psychiatry, 53,* 129–136.

McGuffin, P., & Martin, N. (1999). Behaviour and genes. *British Medical Journal, 319,* 37–40.

McGuffin, P., Reveley, A., & Holland, A. (1982). Identical triplets: Non-identical psychosis? *British Journal of Psychiatry, 140,* 1–6.

McKernan, R. M. W. P. (1996). Which GABA-A-receptor subtypes really occur in the brain. *Trends in Neuroscience, 4,* 139–143.

Meloni, D., Gambara C., Graziella de Montis, M., Pra, P., Taddei, I., & Tagliamonte, A. (1993). Dizocilpine antagonizes the effect of chronic imipramine on learned helplessness in rats. *Pharmacology, Biochemistry and Behavior, 46,* 423–426.

Mendlewicz, J., Fleiss, J. L., & Fieve, R. R. (1972). Evidence for X-linkage in the transmission of manic-depressive illness. *Journal of the American Medical Association, 222,* 1624.

Mendlewicz, J., & Rainer, J. D. (1977). Adoption study supporting genetic transmission in manic-depressive illness. *Nature, 268,* 327–329.

Moryl, E., Danysz, W., & Quack, G. (1993). Potential antidepressive properties of amantadine, memantine and bifemelane. *Pharmacology and Toxicology, 72,* 394–397.

Murray, C. J. L., & Lopez, A. D. (1997). Global mortality, disability, and the contribution of risk factors: Global burden of disease study. *Lancet, 349,* 1436–1442.

Neiswanger, K., Zubenko, G. S., Giles, D. E., Frank, E., Kupfer, D. J., & Kaplan, B. B. (1998). Linkage and association analysis of chromosomal regions containing genes related to neuroendocrine or serotonin function in families with early-onset recurrent major depression. *American Journal of Medical Genetics, 81,* 443–439.

Nishiguchi, N., Shirakawa, O., Ono, H., Hashimoto, T., & Maeda, K. (2000). Novel polymorphism in the gene region encoding the carboxyl-terminal intracellular domain of the NMDA receptor 2B subunit: Analysis of association with schizophrenia. *American Journal of Psychiatry, 157,* 1329–1331.

O'Donovan, M. C., Guy, C., Craddock, N., Bowen, T., McKeon, P., Macedo, A., et al. (1996). Confirmation of association between expanded CAG/CTG repeats and both schizophrenia and bipolar disorder. *Psychological Medicine, 26,* 1145–1153.

O'Donovan, M. C., Guy, C., Craddock, N., Murphy, K. C., Cardno, A. G., Jones, L. A., et al. (1995). Expanded CAG repeats in schizophrenia and bipolar disorder. *Nature Genetics, 10,* 380–381.

Oruc, L., Lindblad, K., Verheyen, G. R., Ahlberg, S., Jakovljevic, M., Ivezic, S., et al. (1997). CAG repeat expansions in bipolar and unipolar disorders. *American Journal of Human Genetics, 60,* 730–732.

Oruc, L., Verheyen, G. R., Furac, I., Evezic, S., Jakovljevic, M., Raeymaekers, & Van Broeckhoven, C. (1997). Positive association between the GABRA5 gene and unipolar recurrent major depression. *Neuropsychobiology, 36,* 62–64.

Owen, D., Du, L., Bakish, D., Lapierre, Y. D., & Hrdina, P. D. (1999). Norepinephrine transporter gene polymorphism is not associated with susceptibility to major depression. *Psychiatry Research, 87,* 1–5.

Papadimitriou, G. N., Dikeos, D. G., Karadima, G., Avramopoulos, D., Daskalopoulou, E. G., Vassilopoulos, D., & Stefanis, C. N. (1998). Association between the GABA(A) receptor alpha5 subunit gene locus (GABRA5) and bipolar disorder. *American Journal of Medical Genetics, 81,* 73–80.

Papolos, D. F., Veit, S., Faedda, G. L., Saito, T., & Lachman, H. M. (1998). Ultra-ultra rapid cycling bipolar disorder is associated with the low activity catechol-O-methyltransferase allele. *Molecular Psychiatry, 3,* 346–349.

Pauls, D. L., Morton, L. A., & Egeland, J. A. (1992). Risks of affective illness among first-degree relatives of Bipolar I Old-Order Amish probands. *Archives of General Psychiatry, 49,* 703–708.

Pekkarinen, P. J., Bredbacka, P. E., Loonqvist, J., & Peltonen, L. (1995). Evidence of a predisposing locus to bipolar disorder on Xq24-q27.1 in an extended Finnish pedigree. *Genome Research, 5,* 105–115.

Perez de Castro, I., Torres, P., Fernandez, P. J., Saiz, R. J., & Llinares, C. (1994). No association between dopamine D4 receptor polymorphism and manic depressive illness. *Journal of Medical Genetics, 31,* 897–898.

Petty F. (1995). GABA and mood disorders: A brief review and hypothesis. *Journal of Affective Disorders, 34,* 275–281.

Plomin, R., & Bergeman, C. S. (1991). The nature of nurture: Genetic influence on 'environmental' measures. *Behaviour and Brain Sciences, 14,* 373–427.
Pritchard, J. K., & Przeworski, M. (2001). Linkage disequilibrium in humans: Models and data. *American Journal of Human Genetics, 69,* 1–14.
Reich, T., Clayton, P., & Winokur, G. (1969). Family history studies: V. The genetics of mania. *American Journal of Psychiatry, 125,* 1358–1368.
Rice, J., Reich, T., Andreasen, N. C., Endicott, J., Van Eerdewegh, M., Fishman, R., et al. (1987). The familial transmission of bipolar illness. *Archives of General Psychiatry, 44,* 441–447.
Roses, A. D. (2000). Pharmacogenetics and the practice of medicine. *Nature, 405,* 857–865.
Ross, E. M. (1989). Signal sorting and amplification through G protein-coupled receptors. *Neuron, 3,* 141–152.
Schalling, M., Hudson, T. J., Buetow, K. H., & Houseman, D. E. (1993). Direct detection of novel expanded trinucleotide repeats in the human genome. *Nature Genetics, 4,* 135–139.
Serretti, A., Macciardi, F., Cusin, C., Lattuada, E., Lilli, R., Di Bella, D. C. M., & Smeraldi, E. (1998). GABAA alpha-1 subunit gene not associated with depressive symptomatology in mood disorders. *Psychiatric Genetics, 8,* 251–254.
Serretti, A., Macciardi, F., Cusin, C., Lattuada, E., Lilli, R., & Smeraldi, E. (1998). Dopamine receptor D4 gene is associated with delusional symptomatology in mood disorders. *Psychiatry Research, 80,* 129–136.
Siris, S. (1991). Diagnosis of secondary depression in schizophrenia: Implications for *DSM–IV. Schizophrenia Bulletin, 17,* 75–98.
Skolnick, P., Layer, R. T., Popik, P., Nowak, G., Paul, I. A., & Trullas, A. (1996). Adaptation of *N* methyl-D-aspartate (NMDA) receptors following antidepressant treatment: Implications for the pharmacotherapy of depression. *Pharmacopsychiatry, 29,* 23–26.
Sokoloff, P., Giros, B., Martres, M. P., Bouthenet, M. L., & Schwartz, J. C. (1990). Molecular cloning and characterization of a novel dopamine receptor (D3) as a target for neuroleptics. *Nature, 346,* 146–151.
Sokoloff, P., Lannfelt, L., Martres, M. P., Giros, B., Bouthenet, M. L., & Schwartz, J. C. (1992). The D3 dopamine receptor gene as a candidate gene for genetic linkage studies. In J. Mendlewicz, H. Hippius, B. Bondy, M. Ackenheil, & M. Sandler (Eds.), *Genetic research in psychiatry* (pp. 219–255). Berlin: Springer-Verlag.
Stober, G., Nothen, M. M., Porzgen, P., Bruss, M., Bonisch, H., Knapp, M., et al. (1996). Systematic search for variation in the human norepinephrine transporter gene: Identification of five naturally occurring missense mutations and study of associations with major psychiatric disorders. *American Journal of Medical Genetics (Neuropsychiatric Genetics), 67,* 523–532.
Strober, M. (1992). Relevance of early age-of-onset in genetic studies of bipolar affective disorder. *Journal of American Academic Child and Adolescent Psychiatry, 31,* 606–610.
Sullivan, P., Neale, M., & Kendler, K. (2000). General epidemiology of major depression: Review and meta-analysis. *American Journal of Psychiatry, 157,* 1552–1562.
Taylor, A. E., & Saint-Cyr, A. (1990). Depression in Parkinson's disease reconciling physiological and psychological perspectives. *Neuropsychiatric Practical Opinions, 2,* 92–98.
Thapar, A., & McGuffin, P. (1996). Genetic influences on life events in childhood. *Psychological Medicine, 26,* 813–820.
Thapar, A., & McGuffin, P. (1997). Anxiety and depressive symptoms in childhood: A genetic study of comorbidity. *Journal of Child Psychology, 38,* 651–656.
Torgersen, S. (1986). Genetic factors in moderately severe and mild affective disorders. *Archives of General Psychiatry, 43,* 222–226.
Trullas, R., & Skolnick, P. (1990). Functional antagonists at the NMDA receptor complex exhibit antidepressant actions. *European Journal of Pharmacology, 185,* 1–10.
Turecki, G., Rouleau, G. A., Mari, J., Joober, R., & Morgan, K. (1997). Lack of association between bipolar disorder and tyrosine hydroxylase: A meta-analysis. *American Journal of Medical Genetics (Neuropsychiatric Genetics), 74,* 348–352.
Van Broeckhoven, D., & Verheyen, G. (1999). Report of the chromosome 18 workshop. *American Journal of Medical Genetics (Neuropsychiatric Genetics), 88,* 263–270.
Vandenbergh, D. J., Persico, A. M., Hawkins, A. L., Griffin, C. A., Li, X., Jabs, E. W., & Uhl,

G. R. (1992). Human dopamine transporter gene (DAT1) maps to chromosome 5p15.3 and displays a VNTR. *Genomics, 14,* 1104–1106.

Van Tol, H. H., Bunzow, J., Guan, H., Sunahara, R., Seeman, P., Niznik, H., & Civelli, O. (1991). Cloning of the gene for a human dopamine D4 receptor with high affinity for the antipsychotic clozapine. *Nature, 350,* 610–614.

Van Tol, H. H., Caren, M., Guan, H., Ohara, K., Bunzow, J., Civelli, O., et al. (1992). Multiple dopamine D4 receptor variants in the human population. *Nature, 358,* 149–152.

Vincent, J. B., Klempan, T., Parikh, S. S., Sasaki, T., Meltzer, H. Y., Sirugo, G., et al. (1996). Frequency analysis of large CAG/CTG trinucleotide repeats in schizophrenia and bipolar affective disorder. *Molecular Psychiatry, 1,* 141–148.

Wender, P. H., Kety, S. S., Rosenthal, D., Schulsinger, F., Ortmann, J., & Lunde, I. (1986). Psychiatric disorders in the biological and adoptive families of adopted individuals with affective disorders. *Archives of General Psychiatry,* 43, 923–929.

Weissman, M. M., Warner, V., Wickramaratne, P., & Prusoff, B. A. (1998). Early-onset major depression in parents and their children. *Journal of Affective Disorder, 15,* 269–277.

Willner, P. (1983). Dopamine and depression: A review of recent evidence. *Brain Research Review, 6,* 211–246.

Willner, P. (1995). Dopamine mechanisms in depression and mania. In F. E. Bloom & D. J. Kupfer (Eds.), *Psychopharmacology: The fourth generation of progress* (pp. 921–931). New York: Raven Press.

Wolf, R., Smith, G., & Smith, R. L. (2000). Science, medicine, and the future: Pharmacogenetics. *British Medical Journal, 320,* 987–990.

World Health Organization. (1993). *Diagnostic criteria for research: International classification of diseases* (10th ed.). Geneva: Author.

Zorumski, C. F. I. K. (1991). Insights into the structure and function of GABA-benzodiazepine receptors: Ion channels and psychiatry. *American Journal of Psychiatry, 148,* 162–173.

25

Dementia and Genetics

Julie Williams

At least 55 diseases result in some form of dementia, and we are just beginning to understand their causes. Recently, many advances have stemmed from the identification of genes contributing to disease development. Indeed, research into the molecular genetics of dementia is probably the most successful in the field of behavioral genetics. Twelve genes and over 80 mutations known to contribute to dementia have already been identified. These genes have a variety of effects. In some cases, the possession of a single mutation will result in the certain development of dementia, the onset of which can be predicted almost to the year, whereas other genetic changes increase the risk of dementia development but are neither necessary nor sufficient to cause the disease. In some genes, only a single mutation predisposes to disease, whereas in others multiple mutations are found. There also are complex relationships between genetic mutations and the behavioral phenotypes with both pleiotrophy and heterogeneity observed. It is encouraging that the progress made in identifying genes contributing to dementia has and will continue to improve our knowledge of disease development. The effects of the susceptibility alleles identified and the knowledge we gain about the mechanisms involved in disease development may also help us identify genes for other complex disorders. Consequently, the focus of this chapter is on a selection of research findings that reflect some of the major areas of success to date.

Among the most common forms of dementia are Alzheimer's disease, vascular dementia, Parkinson's disease, frontotemporal dementia, Huntington's disease, Pick's disease, progressive supranuclear palsy, and Creutzfeldt-Jakob disease. Table 25.1 illustrates the genes already identified that contribute to the development of these disorders: Creutzfeldt-Jakob disease (Rossi et al., 2000), familial British dementia (Ghiso et al., 2000), vascular dementia–CADASIL (Joutel et al., 1996), Huntington's disease (Huntington's Disease Collaborative Research Group, 1993), and Parkinson's disease (Dunnett & Bjorklund, 1999; Gasser, 1998). This chapter focuses on two in which dementia is the prominent symptom and in

I acknowledge the Medical Research Council, the Alzheimer's Research Trust, and the Alzheimer's Society for funding the work of our group and thank Professor Mike Owen and my other colleagues for their support. I also thank Professors John Hardy and Mike Hutton for contributing figures.

Table 25.1. Genes Contributing to Dementia

Dementia	Gene
Alzheimer's disease	APP
	PS1&PS2
	ApoE
Frontotemporal dementia	TAU
Pick's disease	
Progressive supranuclear palsy	
Parkinson's disease	a-Syn
	PARKIN
	UCH-L1
Huntington's disease	Huntington
Creutzfeldt-Jacob disease	PRP
Vascular dementia-CADASIL	NOTCH 3
Family British/Danish dementia	BR1

Note. APP = amyloid precursor protein; PS1 and PS2 = presenilin genes; ApoE = apolipoprotein E; A-Syn = alpha-synuclein; BR1 = Familial British Dementia Gene–1; UCH-L1 = ubiquitin carboxyl-terminal esterase L1 (ubiquitin thiolesterase); PRP = prion protein (Creutzfeldt-Jakob disease, Gerstmann-Strausler-Scheinker syndrome, fatal familial insomnia)

which a number of genes have been found: Alzheimer's disease and frontotemporal dementia.

Alzheimer's Disease

Alzheimer's disease is the main cause of dementia in the elderly and the fourth leading cause of death after heart disease, cancer, and stroke (McKahnn, Drachman, Folstein, Katzman, Price, & Stadlan, 1984). Alzheimer's disease is common, affecting 5% of individuals over the age of 65, rising to over 40% of individuals over the age of 90. Alzheimer's disease is a progressive neurodegenerative disorder, which begins typically with subtle memory loss and progresses to global loss of cognitive function (McKahnn et al., 1984). Following diagnosis, Alzheimer's disease sufferers survive for an average of 8 years, with death usually resulting from a secondary disease. Alzheimer's disease results in definitive neuropathological change, which is characterized by extensive neuronal cell loss and the deposition of extracellular senile plaques (made up of mostly β amyloid Aβ) and intracellular neurofibrillary tangles (formed from hyperphosphorylated tau) in the cerebral cortex. Epidemiological studies consistently show increasing age and a positive family history for Alzheimer's disease to be the most important risk factors. In addition, women appear to be at increased risk for Alzheimer's disease compared with men, even controlling for differences in longevity (van Duijn, 1996). Limited years of education have also been noted as a risk factor for Alzheimer's disease (Stern, Gurland, Tatemichi, Tang, Wilder, & Mayeux, 1994), although this could be confounded by test bias, such as impaired nutrition. Alternatively, increased years of education may reflect greater cognitive capacity and consequently more cerebral reserve. Finally, head injury has been reported to double the risk of Alzheimer's disease (van Duijn et al., 1992).

Twin and Family Studies

Diseases of late onset, such as Alzheimer's disease, pose problems for twin and family studies because individuals from previous generations are difficult to ascertain. However, family studies using modern diagnostic criteria for Alzheimer's disease (McKahnn et al., 1984) have observed a high cumulative risk of dementia (50%) in first-degree relatives of Alzheimer's disease cases, by age 85–90 years (Brightner & Falsedeen, 1984; Huff, Auerbach, Chakravarti, & Boller, 1988; Martin, Gerteis, & Gabrielli, 1988; Mohs, Breitner, Silverman, & Davis, 1987). Farrer and colleagues (1989) found a similar cumulative incidence (44% ± 8%). However, when nonrandomly ascertained cases were excluded, the accumulative incidence of Alzheimer's type dementia in first-degree relatives with probable Alzheimer's disease was maximally 39% but probably closer to 24% when risk estimates were weighted for accuracy of diagnosis. Farrer and colleagues also divided their families into groups depending on age of onset (above and below 67 years). They found no difference between the cumulative risk of Alzheimer's disease in early- and late-onset relatives but observed an early age of onset in those related to early-onset Alzheimer's disease cases. They also found, when comparing relatives of cases and relatives of normal controls, that over the life span of 90 years, the proportion of relatives who became affected did not differ but those related to Alzheimer's disease cases became affected at an earlier age. However, this finding is based on a very small number of elderly survivors and needs to be replicated in larger samples (Owen, 1994).

The early evidence suggests significant genetic effects in Alzheimer's disease. For example, Callmann (1956) used a series of 54 monozygotic (MZ) and 54 dizygotic (DZ) pairs in which at least one was suffering from senile psychosis, a diagnostic category that included both Alzheimer's disease and multi-infarct dementia. He observed concordance rates of 43% in MZ and 8% in DZ twins. However, Nee and colleagues (1997), using modern diagnostic criteria but only 22 pairs (17 MZ and 5 DZ), showed no difference between concordance rates. The most recent twin study in Alzheimer's disease was undertaken by Bergem and colleagues (Bergem, Engedal, & Kringlen, 1997). They observed a pairwise concordance rate for Alzheimer's disease of 78% (7 out of 9) among MZ twins and 39% (9 out of 23) among DZ twin pairs. The probandwise concordance rate was 83% (10 out of 12) among MZ and 45% (12 out of 26) among DZ twin pairs. The estimated heritability was approximately 60%. The environmental factors shared by cotwins seemed to explain most of the remaining variance. They did not find a significant difference in the rate of ApoE e4 alleles between twin pairs concordant and discordant for Alzheimer's disease.

Molecular Genetics of Alzheimer's Disease

Four genes have been identified, which contribute to the development of Alzheimer's disease (see Table 25.1). Mutations in the amyloid precursor protein (APP) gene and the presenilin genes (PS1 and PS2) result in rare forms of early-onset familial Alzheimer's disease (onset before age 65). These three genes are highly penetrant, acting in an autosomal dominant fashion. A fourth gene, the apolipoprotein E (ApoE) gene, has quite a different pattern of effect and influences both early- and late-onset Alzheimer's disease but does so by increasing general risk and is neither necessary nor sufficient to cause disease.

Early-Onset Alzheimer's Disease

In 1984, Glenner and Wong identified the major constituent of senile plaques as a 4-kd peptide, which became known as β-amyloid (Aβ). Aβ is cleaved from a larger fragment known as β-amyloid precursor protein (APP), the gene for which maps to chromosome 21 (Gajdusek, 1987). At the time, it was also noted that individuals with Down syndrome, who have three copies of chromosome 21, develop Alzheimer's disease neuropathology, and most show cognitive decline beyond the age of 40 (Mann, Yates, Marcyniuk, & Ravindra, 1986). Even so, when linkage was reported between chromosome 21 and some Alzheimer's disease families, the APP gene was initially discounted because the linkage evidence was not wholly consistent (Goate et al., 1989; Pericak-Vance et al., 1998; Schellenberg et al., 1988; St George-Hyslop et al., 1987; St George-Hyslop, Haines, Farrer, et al., 1990; Van Broeckhoven et al., 1992). However, sequencing of the APP gene in some of these families identified six mutations that cosegregated with Alzheimer's disease (Chartier-Harlin et al., 1991; Goate et al., 1991; Hendriks et al., 1992; Mullan, Crawford, et al., 1992; Murrell, Farlow, Ghetti, & Benson, 1991). The age of onset of Alzheimer's disease produced by APP mutations varies between 40 and 60 years, with a mean in the early 50s (Mullan et al., 1993). All the mutations identified cluster around sites of proteolytic cleavage and thus appear to affect the processing of β-amyloid (see Figure 25.1). Evidence suggests that APP mutations seem to act by a gain-of-function mechanism, namely, the increased production of potentially cytotoxic Aβ fragment.

Combinations of α, γ, and β-secretase activity on the larger APP molecule result in the production of different forms of β-secretase ($A\beta_{-40}$ and $A\beta_{-42}$) and associated products (p3). The production of $A\beta_{-42}$ and its balance with $A\beta_{-40}$ appears to be significant in the development of Alzheimer's disease pathology. $A\beta_{-42}$ is more susceptible to aggregate and is the first form of Aβ to be deposited in diffuse plaques in Alzheimer's disease and

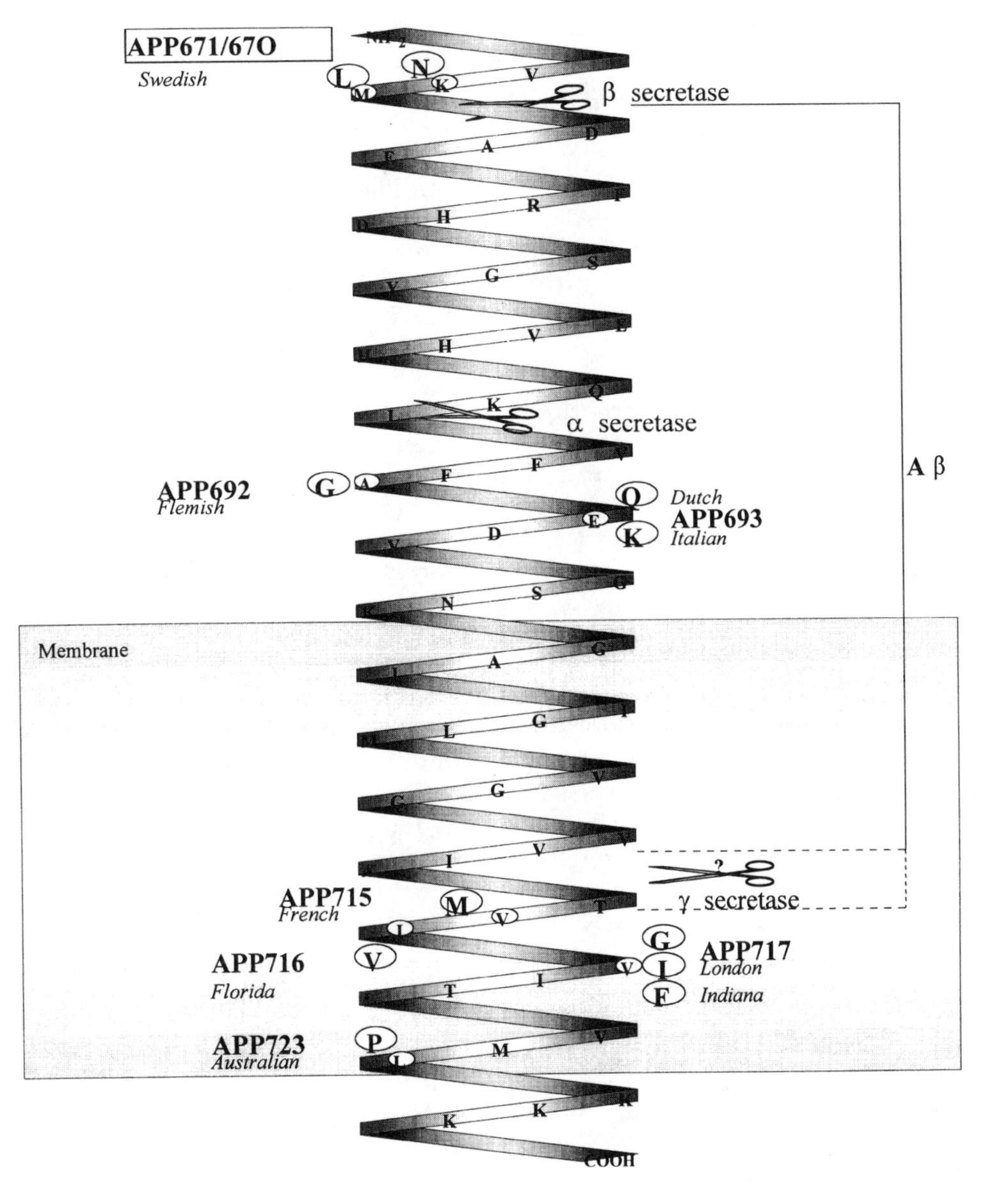

Figure 25.1. The Amyloid Precursor Protein (APP) Gene.

Down syndrome brains. For example, Aβ-$_{42}$ is observed in plaques from Down syndrome patients at the age of 12, whereas Aβ-$_{40}$ is first detected in the plaques some 20 years later (Lemere et al., 1996). In addition, overexpression of some of the APP mutations in transgenic mice increases Aβ deposition in the brain and also causes memory impairment (Games et al., 1995; Hsiao et al., 1996; Schellenberg et al., 1992). It therefore appears that changes in the APP gene could lead to Alzheimer's disease in a number of ways, which include the overexpression of forms of β-amyloid (Aβ-$_{40}$ and Aβ-$_{42}$) observed in Down syndrome and by missense mutations increasing the amyloidogenic cleavages of APP at either the β-secretase site (increasing production of Aβ-$_{40}$ and Aβ-$_{42}$) or the γ-secretase site (increasing production of Aβ-$_{42}$).

Only 20 families worldwide have been identified that have mutations in the APP gene that cause Alzheimer's disease and are estimated to account for between 5% and 20% of early-onset familial Alzheimer's disease cases but less than 0.1% of all Alzheimer's disease cases (Selkoe, 1999). In 1992, linkage to chromosome 14q was observed (Schellenberg et al., 1992), providing evidence for a second gene contributing to early-onset familial Alzheimer's disease (Mullan, Houlden, et al., 1992; St George-Hyslop et al., 1992; Van Broeckhoven et al., 1992). Subsequently, the presenilin gene (PS-1) was cloned in the region (see Figure 25.2), which carries numerous missense mutations (Sherrington et al., 1995). Currently, over 50 mutations have been identified in families who have a variety of ethnic origins (Selkoe, 1999). In 1995, linkage between Alzheimer's disease and chromosome 1 was detected in a series of families originating from Germany (Levy-Lahad, Wasco, et al., 1995). At about the same time, a DNA sequence was identified on chromosome 1, which bore homology to part of the PS-1 gene. As this was near to the region of linkage, the remainder of this gene was sequenced in individuals from several Volga-German families and a missense mutation identified (Levy-Lahad, Wasco, et al., 1995). Since then, only two additional Alzheimer's disease mutations have been identified in the presenilin-2 (PS-2) gene. The age of onset of Alzheimer's disease in families with PS-1 mutations ranges between 29 and 62 years, whereas the range for PS-2 mutations is between 40 and 88 years old. In addition, in contrast to PS-1, PS-2 mutations appear to show incomplete penetrance (i.e., some individuals carrying the mutation may not develop Alzheimer's disease).

The proteins produced by the PS-1 and PS-2 genes show 67% homology, and both proteins are thought to span the cell membrane eight times (see Figure 25.2). The mutations associated with Alzheimer's disease are in nucleotides coding for amino acids that are conserved between the two genes, which suggests that the two resulting proteins serve similar functions. The mutations are widely scattered through the molecule but tend to cluster within and adjacent to the transmembrane domains (Lendon, Ashall, & Goate, 1997). Mutations in both presenilin genes result in the elevation of Aβ-$_{42}$ (Scheumer et al., 1996). Indeed, crossing mice transgenic for mutant APP and mice expressing PS-1 mutations produces an accelerated Alzheimer's disease-like phenotype, involving Aβ-$_{42}$ plaques (diffuse and mature) occurring as early as age 3 months (Holcomb et al., 1998). The basis of this relationship has recently been illuminated by the finding that the presenilins are a component of γ-secretase, which acts to cleave APP producing Aβ-$_{42}$ (Esler et al., 2000; Herreman et al., 2000; Li et al., 2000; Zhang et al., 2000). Incidentally, the gene encoding β-secretase also has been identified (Vassar et al., 1999) on chromosome 12 but does not as yet show association with Alzheimer's disease development.

Figure 25.2. Presenilin 1.

Late-Onset Alzheimer's Disease

Although the identification of three genes contributing to early-onset Alzheimer's disease has been extremely important to our understanding of Alzheimer's disease development, they account for less than 2% of all Alzheimer's disease cases (Farrer et al., 1997). Alzheimer's disease is usually observed in individuals over the age of 65. The Apolipoprotein E gene (ApoE) is the first gene to show confirmed association with late-onset Alzheimer's disease (Strittmatter et al., 1993) and has been estimated to account for between 45% and 60% of the genetic variance in selected samples (Nalbantoglu et al., 1994), although this might be much less in the general population. Initially, some evidence of linkage was observed in the region of chromosome 19 to which ApoE maps. Association between Alzheimer's disease and a neighboring gene APOC II was observed and then, subsequently, to the ApoE gene. The ApoE gene has three common alleles (e2, e3, and e4). In a recent meta-analysis of over 40 studies (representing 30,000 ApoE alleles), Farrer and colleagues (1997) observed elevated frequencies of the ApoE 4 allele among Alzheimer's disease patients in every group, although the association was stronger among White people (odds ratios [OR] = 2.7–3.3 and 12.5–14.9 in e3–e4 and e4–e4 subjects, respectively) and the Japanese (ORs = 5.6 and 33.1, respectively) but weaker in African Americans (OR = 1.1 and 5.7, respectively). Allele e2 was observed to have a protective effect in all ethnic groups (OR = 0.6, e2–e3), and they also demonstrated ApoE 4 to be a risk factor for Alzheimer's disease in people as young as 40, but this risk diminished in those over the age of 70. There also appeared to be a greater risk of Alzheimer's disease in women attributable to increased risk in those within e3–e4 genotype, although it was noted that the increased risk of Alzheimer's disease in women is unlikely to be due to ApoE alone and was also not observed in the Japanese. However, unlike genes for early-onset Alzheimer's disease, which behave generally in an autosomal dominant manner, ApoE is a risk factor, which is neither necessary nor sufficient to cause Alzheimer's disease. For example, at least one third of individuals with Alzheimer's disease lack an e4 allele (Farrer, 1997; Strittmatter et al., 1993), and up to 50% of individuals who have a double dose of e4 survive to age 80 without developing Alzheimer's disease (Henderson et al., 1995; R. H. Myers et al., 1996). In addition, factors such as serious head injury, smoking, cholesterol level, and estrogen may also modify ApoE-related risk of Alzheimer's disease development (Farrer et al., 1995; Jarvik et al., 1995; Mayeux et al., 1995; Payami et al., 1994; van Duijn et al., 1994; van Duijn et al., 1996; van Duijn, Havekes, Van Broeckhoven, De Kniff, & Hofman, 1995).

ApoE is a major serum lipoprotein involved in cholesterol metabolism. Lipoproteins containing ApoE 4 are cleared more effectively from blood than those containing e3 and e2 isoforms. ApoE 4 is found in senile plaques and neurofibrillary tangles and binds to Aβ in cerebrospinal fluid (Namba, Tomonaga, Kawasaki, Otomo, & Okeda, 1991; Strittmatter et al., 1993). Coexpression of ApoE alleles with APP in cultured cells shows no differential change in the cleavage of APP to form Aβ (Biere et al., 1995). The disease-promoting effect of having two ApoE 4 alleles appears due to the enhanced aggregation or decreased clearance of Aβ (Evans, Berger, Cho, Weisbrager, & Lansbury, 1995; Rebeck, Reiter, Strickland, & Hyman, 1993; Schmechel et al., 1993). Other studies have shown that ApoE e3 but not ApoE e4 binds to Tau, preventing the aggregation of Tau and the formation of neurofibrillary tangles (Glenner & Wong, 1984; Goldgaber, Lerman, McBride, Saffiotti, & Gajdusek, 1987; Kang et al., 1987; McKhann et al., 1984; Robakis et al., 1987; Tanzi et al., 1987).

Finally, association with Alzheimer's disease has been reported between polymorphisms in the ApoE promoter at position −491 and polymorphisms within a potential Ch 1/E47 transcription factor binding consensus sequence (Bullido et al., 1998). Attempts to replicate these associations have resulted in some positive studies but also some negative (see Rebeck et al., 1999).

Other Susceptibility Genes

Recent simulation studies have estimated that at least four genes contribute to age of onset variance in Alzheimer's disease, with an equal effect or greater than ApoE (Warwick Daw et al., 2000). ApoE-observed contribution to the total age of onset variance was between 7% and 9%, whereas the contribution for the largest quantitative trait locus (QTLs) ranged from 5% to over 50%. Consequently, additional genes contributing to late-onset Alzheimer's disease remain to be identified. Schellenberg and colleagues (Schellenberg, D'Souza, & Poorkaj, 2000) reviewed evidence for a further 40 candidate genes that have been tested for association with Alzheimer's disease. Table 25.2 contains a selection of these genes, which show an initial positive association and demonstrate some evidence of replication. However, none of these genes can be said to show confirmed evidence of association. It is possible that a number will prove to be genuinely associated but may have a small or limited effect. It is unlikely that we will be able to identify these with great confidence until our sample sizes increase considerably, offering greater statistical power.

Further evidence for the existence of new genes for late-onset Alzheimer's disease comes from recent linkage studies by an Anglo American Alzheimer's disease collaborative group (Kehoe et al., 1999; A. Myers et al., 2002) and others (Pericak-Vance et al., 1997, 2000), which have implicated new chromosomal regions showing evidence of genes contributing to common Alzheimer's disease. This collaborative group (A. Myers et al., 2000, 2002) performed a two-stage genome screen involving a total of 451 affected sibling pairs (ASPs) and over 400 markers. Stage 1 included 292 ASPs and markers spaced at roughly 20 cM throughout the genome and identified 16 regions showing log of the odds (LOD) scores in excess of 1, taking account of ApoE genotype. In Stage 2, these 16 regions were tested using an extended sample of 451 ASPs and produced significant linkage to chromosome 10 (multiple LOD = 3.9) and four regions of suggestive

Table 25.2. Selected Candidate Gene Association Studies of Alzheimer's Disease

Gene/Locus	Chromosome	Positive studies	Negative studies
Butyrylcholinesterase	3q	2	12
HLA-A2	6p	5	4
HLA-DR	6p	2	1
VLDLR	9p	4	8
Estrogen receptor	11q	2	0
2-Macroglobulin	11q	4	10
LRP	12q	3	5
Presenilin-1 LOAD	14q	6	16
Dihydrolipoamide Succinyltransferase	14q	3	1
a-1-antichymotrypsin	14q	23	22
Serotonin transporter	17q	2	0
Angiotensin-converting enzyme	17q	3	1
CYP2D6	22q	2	3
IDE	10q	1	1

Note. LOAD = late onset Alzheimer's disease; UCH-L1 = ubiquitin carboxyl-terminal esterase L1 (ubiquitin thiolesterase); PRP = prion protein (p27-30) (Creutzfeldt-Jakob disease, Gerstmann-Strausler-Scheinker syndrome, fatal familial insomnia); BR1 = Familial British Dementia Gene–1; HLA-A2 = major histocompatibility complex, class I, A (a-2 alpha chain precursor); HLA-DR = major histocompatibility complex, class II, DR; VLDLR = very low density lipoprotein receptor; LRP1 = low density lipoprotein-related protein 1 (alpha-2-macroglobulin receptor); IDE = insulin-degrading enzyme.

linkage on chromosome 9 (44cM:MLS = 1.7, ApoE E4+, 105cM, MLS = 1.9), chromosome 1 (E4+, 149cM, MLS = 1.7), and chromosome 5 (E4+, 45cM, MLS = 1.6). Two groups also have found evidence of linkage to chromosome 10 (Bertram et al., 2000; Pericak-Vance et al., 1997, 2000), and another found evidence using a phenotype related to Alzheimer's disease (Ertekin-Taner et al., 2001). However, the former studies are not independent of the original observations, as there is considerable overlap between samples. Ertekin-Taner and colleagues tested for linkage between Aβ levels in families with above-average plasma Aβ-42 and observed an MLS of 3.9 in a subset of five families showing high Aβ levels (>90th centile). In addition, this peak maps to within 1.5 cM of the original finding.

Pericak-Vance and colleagues have detected significant evidence of linkage to chromosome 9 (MLS = 4.31 around marker D9S741) in a subset of autopsy-confirmed Alzheimer's disease families (199), which coincides with a region of suggestive linkage (A. Myers et al., 2002). Finally, early observations of suggestive linkage to chromosome 12 (Pericak-Vance et al., 1997; Scott et al., 2000) have not been consistently supported.

In summary, four genes have already been identified that contribute to Alzheimer's disease; at least another three of similar magnitude to ApoE remain to be detected. Linkage evidence strongly suggests that one of these resides on chromosome 10 and another on chromosome 9, but the genes remain to be cloned. The genes already identified act in a variety of ways, for example, as autosomal dominant causes, that is, possessing a mutation results in definite Alzheimer's disease, whereas others are common alleles that are neither necessary nor sufficient to cause disease but simply increase risk. Some genes contain numerous mutations, whereas others contain limited variation that contributes to susceptibility. However, there is no doubt that the identification of these genes has resulted in an explosion of basic research that has improved our understanding of Alzheimer's disease development, and with evidence for new genes already established, there is every hope that our knowledge will increase considerably in the future.

Frontotemporal Dementia

Frontotemporal dementia (FTD) may account for up to a quarter of dementia cases with onset before the age of 65 years (Snowden, Neary, & Mann, 1996) and is estimated to account for between 3% and 10% of total dementia (Brun, 1987; Foster et al., 1997; Gustafsen, 1987; The Lund and Manchester Groups, 1994; Neary, Snowden, Northen, & Goulding, 1988). In contrast to Alzheimer's disease, individuals with FTD retain good memory and visual–spatial orientation early in the disease. FTD is characterized by diverse alterations in personal and social conduct sufficient to render individual sufferers incapable of managing their own lives. These changes include inattention, disinhibition, stereotypic behaviors, antisocial acts, and economy of speech. The disease may present with affective and transient psychiatric symptoms, which are followed by behavioral and cognitive decline that leads to apathy, withdrawal, mutism, and late neurological signs (frontal release signs or striatal signs of akinesia and rigidity; Snowden et al., 1996). This is a disorder of executive function with sufferers showing severe impairment in the generation, monitoring, and organization of thought and skills.

Typically disease onset occurs between 45 and 65 years of age, although the youngest recorded patient is 21 years old. The duration of the illness is variable and can range from 2 to 20 years (average 8 years) and is observed with equal frequency in men and women (Mann, 2000). Approximately half of FTD cases report a positive family history for the disease (Brun, 1987; Gustafsen, 1987; The Lund and Manchester Groups, 1994; Neary et al., 1988; Stevens et al., 1998).

Structural imaging of FTD cases reveals atrophy in the anterior frontotemporal region of the brain, and functional imaging reveals a bilateral reduction in cerebral blood flow in the frontal and temporal lobes (Mann, 2000). Distinct histological changes are associated with the cortical atrophy observed in FTD, which include microvacuolar, glyotic with inclusion bodies (Pick-type), and degeneration of alba cranial nerve nuclei and anterior horn cells (Growdon, Cheung, Hyman, & Rebeck, 1999). Most individuals with FTD also show filamentous pathology made up of hyperphosphorylated tau protein in the absence of amyloid pathology (Foster et al., 1997; Spillantini, Murrell, et al., 1998).

Molecular Genetics of FTD

In 1998 Spillantini and colleagues found linkage for FTD with Parkinsonism to chromosome 17 (Spillantini, Bird, & Ghetti, 1998). Other families also showed linkage to FTD in this region (see Figure 25.3), and shortly after mutations in the Tau gene located within the region of overlapping linkage were identified (Clark et al., 1998; Goedert, Jakes, & Crowther, 1999; Hutton et al., 1998; Spillantini, Crowther, et al., 1998). It is now clear that a substantial proportion of familial FTD is caused by mutations in the Tau gene. For example, in a nationwide population-based study, Rizzu and colleagues (Rizzu et al., 1999) found pathogenic mutations in the Tau gene in 40% of familial cases. Others have reported frequencies between 13.6% and 29% (Dumanchin et al., 1998). However, Tau mutations do not appear to be a common cause of general dementia, as none have been identified in early-onset Alzheimer's disease, Parkinson's disease, or sporadic FTD (Rizzu et al., 1999; Roks et al., 1999; Zareparsi et al., 1999). Furthermore, the genetic defect remains unknown in the majority of familial FTD cases. Thus other genes that contribute to FTD remain to be detected. Indeed, linkage has already been reported to chromosome 3 in a single family (Brown et al., 1995).

Tau

In normal nerve cells, six isoforms of tau exist that result from the alternative splicing of a single messenger RNA (mRNA) transcript produced from the Tau gene (see Figure 25.4). These differ, first, in terms of the inclusion of certain N-terminal domains and, second, in the inclusion of either 3- or 4-imperfect repeat domains within the carboxy-terminal region of the molecule (Goedert, Spillantini, et al., 1999). These latter domains are important because they act as the microtubule-binding region and thus play a precise role in the maintenance of microtubule structure within the cell. A little more than half the tau molecules found in normal neurons are of the 3-repeat tau form with a remainder being of the 4-repeat form (Goedert, Spillantini, et al., 1999). If either the 3- or 4-repeat isoforms fail to function, or if they become unbalanced within the cell, microtubule formation and the stability of the existing microtubules become compromised. In addition, any excess of unused tau of either form can be bundled into a tangle (indigestible residue) that will choke the cell and impair its function (Mann, 2000).

Twenty mutations resulting in forms of FTD have been identified in the Tau gene, although only 14 have been published in detail (see Figure 25.4). All disease-causing mutations occur within the C-terminal region of the gene and most within or adjacent to the microtubule-binding domain.

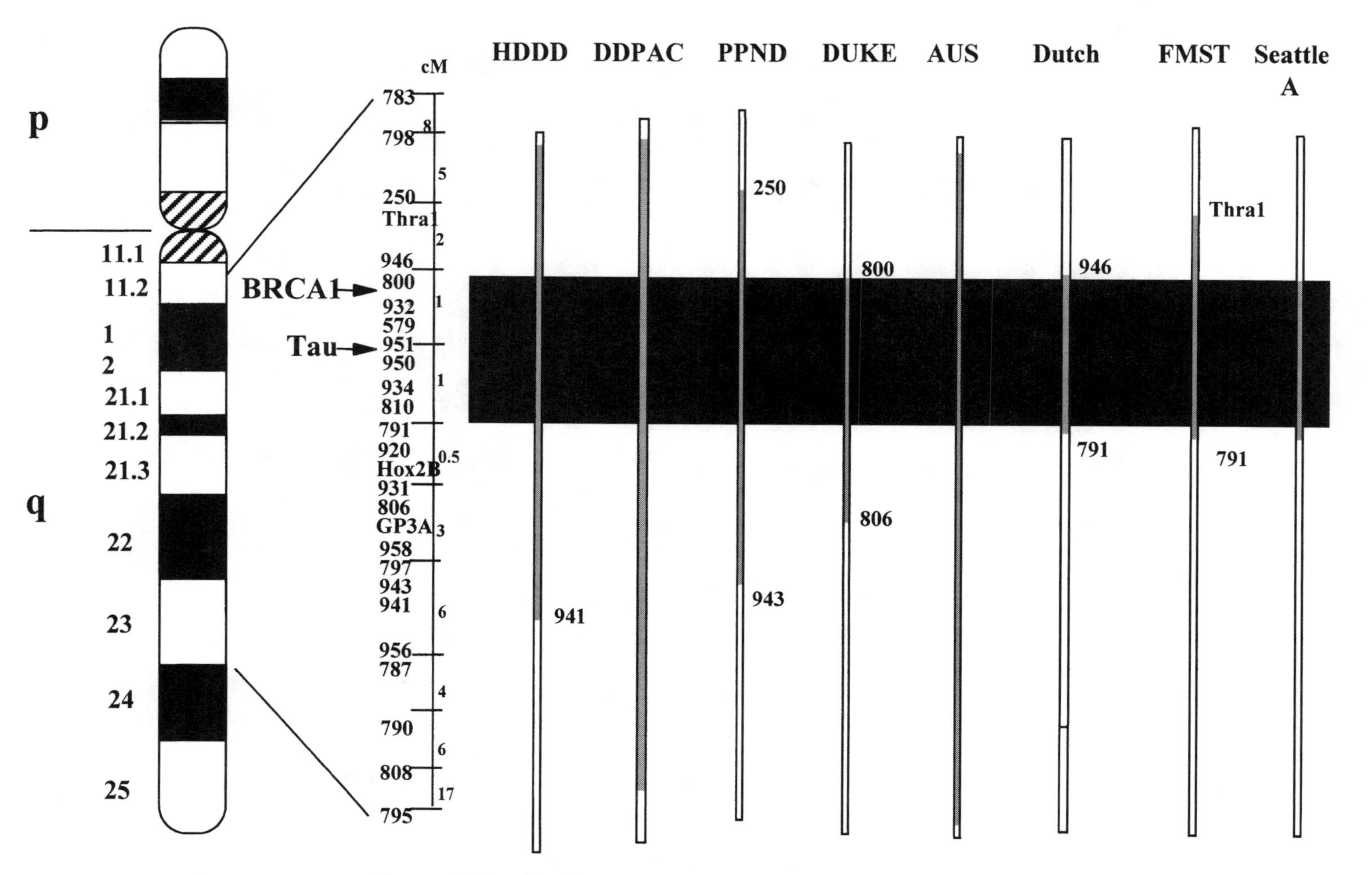

Figure 25.3. Families with linkage to the FTDP-17 locus.

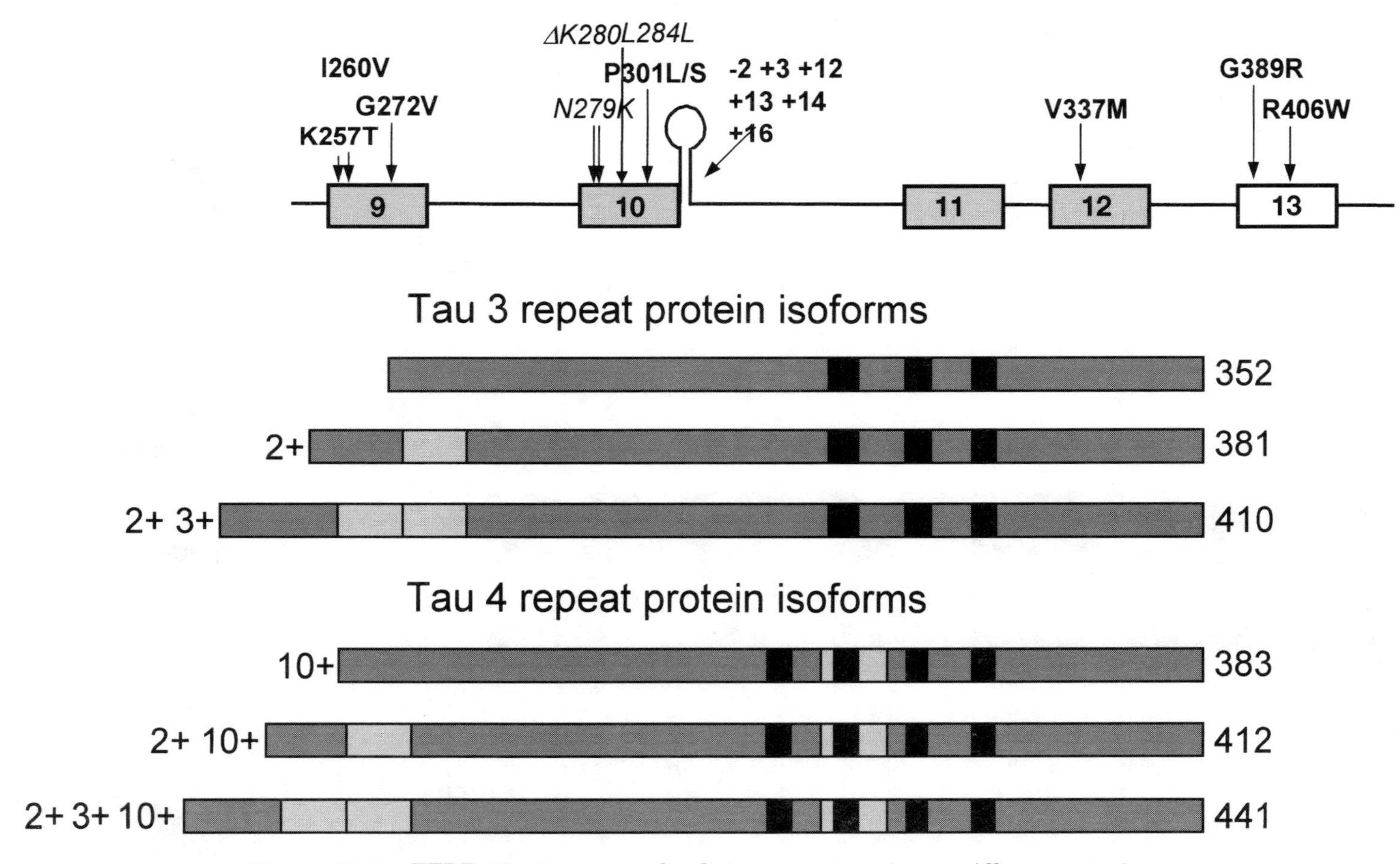

Figure 25.4. FTDP-17 missense and splicing mutations in tau. All tau mutations.

The mutations in exons 9, 10, 12, and 13 (i.e., ΔK280, G272V, P301L, P301S, V337M, and R406W) will result in tau proteins with conformational changes either within or near the microtubule-binding region, thus impairing the efficiency with which tau can interact with tubuline and thereby destabling microtubules (Hasegawa, Smith, & Goedert, 1998; Rizzu et al., 1999). The exonic mutations of V337M and R406W result in abnormal tau made up of a mixture of 3- and 4-repeat isoforms (Hong et al., 1998; Hutton et al., 1998; Poorkaj et al., 1998; Rizzu et al., 1999; Spillantini, Murrell, et al., 1998), which produces structures identical to the paired helical filaments observed in Alzheimer's disease (Reed et al., 1997). The intronic mutations identified (see Figure 25.4) are found within the stem loop structure of a splice accepter domain and act to increase splicing of exon 10 (Hutton et al., 1998; Poorkaj et al., 1998; Spillantini, Murrell, et al., 1998). This appears to result in an increased proportion of mRNA transcripts containing exon 10, producing an imbalance in the relative formation of tau isoforms in favor of those containing the 4-repeats. Similarly, the N279K and S305N mutation may enhance splicing of exon 10 (Clark et al., 1998; Hasegawa, Smith, Iijima, Tabira, & Goedert, 1999; Hong et al., 1998; Morris, Perez-Tur, et al., 1999), producing an excess of 4-repeat tau. These mutations as well as P301L and P301S and the intronic mutations produce tangles composed of predominantly 4-repeat tau (Clark et al., 1998; Hasegawa et al., 1999; Hong et al., 1998; Hutton et al., 1998; Spillantini, Murrell, et al., 1998), which occur as flat ribbons rather than the paired helical filaments observed in Alzheimer's disease. Mutations outside exon 10 appear to produce only neuronal pathology, whereas the intronic mutations produce pathology within glia as well as in neurons.

Finally, the mutations K257T and G389R produce a Pick-type histology (Pickering-Brown et al., 2000). They also may confer a reduced ability to bind microtubules and to promote their efficient assembly. Mutation G272V also causes the production of Pick-like bodies (Rizzu et al., 1999; Spillantini, Murrell, et al., 1998).

It has been postulated that all the mutations in some way result in a toxic gain of function caused by the excess deposition of hyperphosphorylated tau protein (Heutink et al., 1997). Three-repeat tau binds to microtubules at different sites to those bound by 4-repeat isoforms (Goode & Feinstein, 1994). Because the splicing mutations cause an excess of specific isoforms of the protein, the result could be a shortage of available binding sites for the overexpressed tau isoform, leading to an excess of free tau. In addition, the partial loss of function postulated to be the effect of the coding mutations may also result in excess free tau. The resulting free tau could then hyperphosphorylate or assemble into filaments or aggregates (Arrasate, Parez, Armas-Portela, & Avila, 1999; Goedert, Spillantini, et al., 1999; Nacharaju et al., 1999). There is evidence that tau aggregates become tagged for degradation by the ubiquitin degradation pathway, and it has been suggested that because these aggregates resist degradation they may interfere with normal ubiquitin recycling and disrupt the proteasome (Cummings et al., 1998). This would disrupt the normal functioning of the cell, making the neuron vulnerable to other stress factors. However, whether the tau mutations cause a loss of function or a toxic gain of function or a combination of both has yet to be fully established.

Phenotypic–Genotypic Relationships

Perhaps the most interesting aspect of the genetic findings in FTD is the variable relationship between specific mutations and the neuropathological and behavioral phenotypes observed. From the perspective of the researcher whose starting point is usually the behavioral phenotype with some pathological information and whose aim is to identify susceptibility genes, it is a sobering thought that mutations in one single gene can result in behavioral phenotypes with such wide variation. For example, the behavioral phenotype in one FTD individual is characterized by disinhibition, initiative loss, cognitive decline, and mutism, whereas others display antisocial and psychotic behavior, or Parkinsonism and epileptic seizures (see Table 25.3; see Heutink, 2000, for a detailed review). Although phenotype–genotype relationships do begin to emerge once the mutations are identified and their neuropathological and behavioral phenotypes are fully described, it does serve to illustrate the complexity of identifying genes in the first place, with the limited information that is available initially.

The P301L mutation is probably the best described in the FTD literature (Bird et al., 1999; Clarke et al., 1998; Dumanchin et al., 1998; Heutink, 2000; Heutink et al., 1997; Hutton et al., 1998; Mirra et al., 1999; Nasreddine et al., 1999; van Swieten et al., 1999). There is a considerable variation in the age of onset in families segregating this mutation, although most who appear to show disease before the age of 50 show a disease duration of roughly 8 years. Cases typically present with behavioral changes, which include disinhibition, language deficits, and mutism, but show preservation of visual–spatial skills and have no Parkinsonian symptoms. Tau pathology consists of intracytoplasmic tau deposits in neurons, glial cells, and neurites in the hypercampus, neocoptex, and substantia nigra, consisting predominantly of slender twisted filaments of 4-repeat tau. Although one small family experienced memory loss more prominently than behavioral disturbance (Clark et al., 1998) and other families had later ages of onset and shorter durations (Bird et al., 1999), most P301L cases had broadly similar clinical and pathological characteristics.

However, finding a phenotypic–genotypic pattern between mutations is less straightforward. The P301S mutation produces a more severe form of FTD characterized by a very early onset (Bugiani et al., 1999; Sperfeld et al., 1999), behavioral changes, memory loss, and Parkinsonism, with one family showing epileptic seizures and another showing hallucinations and delusions. This mutation is associated with extensive tau pathology in neurons and glial cells. It is possible that the severity of this phenotype may be attributable to an additional phosphorylation site within the

Table 25.3. The Relationships Between Tau Mutations and Behavioral Phenotypes

Mutation	Characteristic symptom	Mutation	Characteristic symptom
G272V (exon 9)	Disinhibition, initiative loss, cognitive decline, mutism	ΔK280	Initiative loss, disinhibition
P301L (exon 10)	Disinhibition, initiative loss, cognitive decline, mutism	284L	Word finding, visual spatial abilities, behavioral changes, Aβ pathology, executive function
	Behavioral changes	S305N	Personality change, cognitive function, memory loss
	Behavioral changes, disinhibition	N279K	Parkinsonism, personality change, memory loss
	Language deficits, mutism	+3	Memory, bradykinesia, rigidity
	Disinhibition, initiative loss, cognitive decline	+13	Not available
P301S (exon 10)	Depression, Parkinsonism, memory loss	+14	Disinhibition, dementia, Parkinsonism
	Behavioral changes, Parkinsonism, epileptic seizures	+16	Personality changes Behavioral changes, cognitive impairment
V337M (exon 12)	Anti-social, psychotic behavior		
R406W (exon 13)	Memory loss, personality change, Parkinsonism		
	Memory loss, personality change, Parkinsonism		

microtubule-binding domains of tau, making it more prone to the formation of insoluble filaments (Heutink, 2000).

Finally, the finding that tau mutations contribute to FTD has generated renewed interest in its possible role in other forms of dementia that show extensive tau pathology, such as Pick's disease (PiD) corticobasal degeneration (CBD), progressive supranuclear palsy (PSP), and the amyotrophic lateral sclerosis/Parkinsonism/dementia complex of Guam. Genetic association studies have found no evidence of a relationship between polymorphisms in the tau gene and these disorders with the exception of PSP. Linkage to chromosome 17 for familial forms of PSP has been excluded (Hoenika et al., 1999), but several studies have reported allelic association with the intronic polymorphism after exon 9 and sporadic PSP cases (Baker et al., 1999; Bennett et al., 1998; Bonifati et al., 1999; Conrad et al., 1997; Higgins, Litvan, Pho, Li, & Nee, 1998; Morris, Janssen, et al., 1999; Oliva et al., 1998). This association was found to an extended haplotype (H1/H1) spanning the whole tau gene including the promoter region

(Ezquerra et al., 1999). This has rendered the search for the biologically relevant defect difficult and so far unsuccessful. However, the finding that only the 4-repeat tau contributes to NFT (neurofibrillary tangle) tau pathology in PSP cases indicates that mutations involved with the splicing of exon 10 may be of importance (Sergeant, Wattez, & Delacourte, 1999). It also has been speculated (Heutink, 2000) that the relevant mutation may be within an as yet unknown element that regulates this splicing and that a similar mechanism might contribute to other dementias in which tau pathology consists mainly of 3-repeat (e.g., PiD) or 4-repeat (e.g., CBD) isoforms (Delacourte, 1999; Feany & Dickson, 1996; Feany, Mattiace, & Dickson, 1996).

Conclusion

Great progress has been made in identifying genes for many of the major dementia disorders. This chapter has focused on the molecular genetics of Alzheimer's disease and FTD and has highlighted two important issues. First, identifying genes really does help us understand the biological basis of these disorders and will therefore lay the platform for the development of novel treatments in the future. Second, the relationships between genotypes and phenotypes and the genetic characteristics of mutations detected (e.g., rare and dominant, or common risk contributors) are complex and variable and are difficult to predict beforehand. It is also likely that this complicated pattern of relationships may exist for other behavioral phenotypes. In conclusion, genes for complex behavioral phenotypes are important to identify, they can be found using current research strategies, and there are definitely many that remain to be detected.

References

Arrasate, M., Perez, M., Armas-Portela, R., & Avila, J. (1999). Polymerization of tau peptides into fibrillar structures: The effect of FTDP-17 mutations. *Federation of European Biochemical Societies Letters*, *446*, 199–202.

Baker, M., Litvian, I., Houlden, H., Adamson, J., Dickson, D., Perez-Tur, J., et al. (1999). Association of an extended haplotype in the tau gene with progressive supranucleur palsy. *Human Molecular Genetics*, *8*, 711–715.

Bennett, P., Bonifati, V., Bonnuccelli, U., Colosimo, C., De Mari, M., Fabbrini, G., et al. (1998). Direct genetic evidence for involvement of tau in progressive supranuclear palsy: European Study Group on Atypical Parkinsonism Consortium. *Neurology*, *51*, 982–985.

Bergem, A. L. M., Engedal, K., & Kringlen, E. (1997). The role of heredity in late-onset Alzheimer's disease and vascular dementia: A twin study. *Archives of General Psychiatry*, *54*, 264–270.

Bertram, L., Blacker, D., Mullin, K., Keeney, D., Jones, J., Basu, S., et al. (2000). Evidence for genetic linkage of Alzheimer's disease to chromosome 10q. *Science*, *290*, 2302.

Biere, A. L., Ostaszewski, B., Zhao, H., Gillespie, S., Younkin, S. G., & Selkoe, D. J. (1995).

Co-expression of β-amyloid precursor protein (βAPP) and apolipoprotein E in cell culture: Analysis of βAPP processing. *Neurobiological Disorders, 2,* 177–187.

Bird, T. D., Nochlin, D., Poorkaj, P., Cherrier, M., Kaye, J., Payami, H., et al. (1999). A clinical pathological comparison of three families with frontotemporal dementia and identical mutations in the tau gene (P301l). *Brain, 122,* 741–756.

Bonifati, V., Joosse, M., Nicholl, D. J., Vanacore, N., Bennett, P., Rizzu, P., et al. (1999). The tau gene in progressive supranuclear palsy: Exclusion of mutations in coding exons and exon 10 splice sites, and identification of a new intronic variant of the disease-associated H1 haplotype in Italian cases. *Neuroscience Letters, 274,* 61–65.

Brightner, J. C. S., & Falsedeen, M. (1984). Familial Alzheimer's dementia: A prevalent disorder with specific clinical features. *Psychological Medicine, 14,* 63–80.

Brown, J., Ashworth, A., Gydesen, S., Sorensen, A., Rossor, M., Hardy, J., et al. (1995). Familial non-specific dementia maps to chromosome 3. *Human Molecular Genetics, 4,* 1625–1628.

Brun, A. (1987). Frontal lobe degeneration of non-Alzheimer's type: I. Neuropathology. *Archives of Gerontology and Geriatrics, 6,* 193–208.

Bugiani, O., Murrell, J. R., Giaccone, G., Hasegawa, M., Ghigo, G., Tabaton, M., et al. (1999). Frontotemporal dementia and corticobasal degeneration in a family with a P301S mutation in tau. *Journal of Neuropathology and Neurology, 58,* 667–677.

Bullido, M. J., Artiga, M. J., Recuero, M., Sastre, I., Garcia, M. A., Aldudo, J., et al. (1998). A polymorphism in the regulatory region of APOE associated with risk for Alzheimer's dementia. *Nature Genetics, 18,* 69–71.

Callmann, F. J. (1956). Genetic aspects of mental disorders in later life. In O. J. Kaplan (Ed), *Mental disorders in later life* (2nd ed.). Stanford: Stanford University Press.

Chartier-Harlin, M.C., Crawford, F., Houlden, H., Warren, A., Hughes, D., Fidani, L., et al. (1991). Early onset Alzheimer's disease caused by mutations at codon 717 of the β-amyloid precursor protein gene. *Nature, 353,* 844–846.

Clark, L. N., Poorkaj, P., Wzolek, Z., Geschwind, D. H., Nasreddine, Z. S., Miller, B., et al. (1998). Pathogenic implications of mutation in the tau gene in pallido-ponto-nigral degeneration and related neurodegenerative disorders linked to chromosome 17. *Proceedings of the National Academy of Sciences USA, 95,* 13103–13107.

Conrad, C., Andreadis, A., Trojanaowski, J. Q., Dickson, D. W., Kang, D., Chen, X., et al. (1997). Genetic evidence for the involvement of tau in progressive supranuclear palsy. *Annals of Neurology, 41,* 277–281.

Cummings, C. J., Mancini, M. A., Antalffy, B., De Franco, D. B., Orr, H. T., & Zoghbi, H. Y. (1998). Chaperone suppression of aggregation and altered subcellular proteasome localization imply protein misfolding in SCA1. *Nature Genetics, 19,* 148–154.

Delacourte, A. (1999). Biochemical and molecular characterization of neurofibrillary degeneration in frontotemporal dementias. *Dementia and Geriatric Cognitive Disorders, 10,* 75–79.

Dumanchin, C., Camuzat, A., Campion, D., Verpillat, P., Hannequin, D., Dubois, B., et al. (1998). Segregation of a missense mutation in the microtubule-associated protein tau gene with familial frontotemporal dementia and Parkinsonism. *Human Molecular Genetics, 7,* 1825–1829.

Dunnett, S. B., & Bjorkland, A. (1999). Prospects for new restorative and neuroprotective treatments in Parkinson's disease. *Nature, 399,* A32-A39.

Ertekin-Taner, N., Graff-Radford, N., Younkin, L. H., Eckman, C., Adamson, J., Schaid, D. J., et al (2001). Heritability of plasma amyloid beta in typical late-onset Alzheimer's disease pedigrees. *Genetic Epidemiology, 21(1),* 19–30.

Esler, W. P., Kimberley, W. T., Ostaszewski, B. L., Diehl, T. S., Moore, C. L., Tsai, J. Y., et al. (2000). Transition-state analogue inhibitors of gamma-secretase bind directly to presenilin-1. *Natural Cell Biology, 2,* 428–434.

Evans, K. C., Berger, E. P., Cho, C.-G., Weisbrager, K. H., & Lansbury, P. T., Jr. (1995). Apolipoprotein E is a kinetic but not a thermodynamic inhibitor of amyloid formation: Implications for the pathogenesis and treatment of Alzheimer disease. *Proceedings of the National Academy of Sciences, 92,* 763–767.

Ezquerra, M., Pastor, P., Valldeoriola, F., Molinuevo, J. L., Blesa, R., Tolosa, E., et al. (1999). Identification of a novel polymorphism in the promoter region of the tau gene highly associated to progressive supranuclear palsy in humans. *Neuroscience Letters, 275*, 183–186.

Farrer, L. A. (1997). Genetics and the dementia patient. *Neurologist, 3*, 13–30.

Farrer, L. A., Cupples, L. A., Haines, J. L., Hyman, B., Kukull, W. A., Mayeux, R., et al. (1997). Effects of age, sex and ethnicity on the association between apolipoprotein E genotype and Alzheimer's disease: A meta-analysis. *Journal of the American Medical Association, 278*, 1349–1356.

Farrer, L. A., Cupples, L. A., van Duijn, C. M., Kurz, A., Zimmer, R., Muller, U., et al. (1995). Apolipoprotein E genotype in patients with Alzheimer disease. *Annals of Neurology, 38*, 797–808.

Farrer, L. A., O'Sullivan, D. M., Cupples, L. A., Growdon, J. H., & Myers, R. H. (1989). Assessment of genetic risk for Alzheimer's disease among first-degree relatives. *Annals of Neurology, 25*, 485–493.

Feany, M. B., & Dickson, D. W. (1996). Neurodegenerative disorders with extensive tau pathology: A comparative study and review. *Annals of Neurology, 40*, 139–148.

Feany, M. B., Mattiace, L. A., & Dickson, D. W. (1996). Neuropathologic overlap of progressive supranuclear palsy: Pick's disease and corticobasal degeneration. *Journal of Neuropathology and Experimental Neurology, 55*, 53–67.

Foster, N. L., Wilhelmsen, K., Sime, A. A., Jones, M. Z., D'Amato, C. J., & Gilman, S. (1997). Frontotemporal dementia and Parkinsonism linked to chromosome 17: A consensus conference. *Annals of Neurology, 41*, 706–715.

Gajdusek, D. C. (1987). Characterisation and chromosomal localization of cDNA encoding brain amyloid of Alzheimer's disease. *Science, 235*, 877–880.

Games, D., Adams, D., Alessandrini, R., Barbour, R., Berthelette, P., Blackwell, C., et al. (1995). Alzheimer-type neuropathology in transgenic mice over-expressing V717F β-amyloid precursor protein. *Nature, 373*, 523–527.

Gasser, T. (1998). Genetics of Parkinson's disease. *Annals of Neurology, 44*, s53-s57.

Ghiso, J., Vidla, R., Rostagno, A., Mead, S., Revesz, T., Plant, G., et al. (2000). A newly formed amyloidogenic fragment due to a stop codon mutation causes familial British Dementia. *Annals of New York Academy of Sciences, 903*, 129–137.

Glenner, G. G., & Wong, C. W. (1984). Alzheimer's disease: Initial report of the purification and characterization of a novel cerebrovascular amyloid protein. *Biochemical and Biophysical Research Communications, 120*, 885–890.

Goate, A., Chartier-Harlin, M. C., Mullan, M., Brown, J., Crawford, F., Fidani, L., et al. (1991). Segregation of a missense mutation in the amyloid precursor protein gene with familial Alzheimer's disease. *Nature, 349*, 704–706.

Goate, A. M., Haynes, A. R., Owen, M. J., Farrall, M., James, L. A., Lai, L. Y., et al. (1989). Predisposing locus for Alzheimer's disease on chromosome 21. *Lancet, 1*, 352–355.

Goedert, M., Jakes, R., & Crowther, R. A. (1999). Effects of frontotemporal dementia FTDP-17 mutations on heparin-induced assembly of tau filaments. *FEBS Letters, 450*, 306–311.

Goedert, M., Spillantini, M. G., Crowther, R. A., Chen, S. G., Parchi, P., Tabaton, M., et al. (1999). Tau gene mutation in familial progressive subcortical gliosis. *Nature Medicine, 5*, 454–457.

Goldgaber, D., Lerman, M. I., McBride, O. W., Saffiotti, U., & Gajdusek, D. C. (1987). Characterization of a cDNA encoding brain amyloid of Alzheimer's disease. *Science, 235*, 877–880.

Goode, B. L., & Feinstein, S. C. (1994). Identification of a novel microtubule binding and assembly domain in the developmentally regulated inter-repeat region of tau. *Journal of Cell Biology, 124*, 769–782.

Growdon, W. B., Cheung, B. S., Hyman, B. T., & Rebeck, G. W. (1999). Lack of allelic imbalance in APOE epsilon 3/4 brain mRNA expression in Alzheimer's disease. *Neuroscience Letters, 272*, 83–86.

Gustafsen, L. (1987). Frontal lobe degeneration of non-Alzheimer type: II. Clinical picture and differential diagnosis. *Archives of Gerontology and Geriatrics, 6*, 225–233.

Hasegawa, M., Smith, M. J., & Goedert, M. (1998). Tau proteins with FTDP-17 mutations have a reduced ability to promote microtubule assembly. *Federation of European Biochemical Societies Letters, 437*, 207–210.

Hasegawa, M., Smith, M. J., Iijima, M., Tabira, T., & Goedert, M. (1999). FTDP-17 mutations N279K and S305N in tau produce increased splicing of exon 10. *Federation of European Biochemical Societies Letters, 443*, 93–96.

Henderson, A. S., Easteal, S., Jorm, A. S., Mackinnon, A. J., Korten, A. E, Christensen, H., et al. (1995). Apolipoprotein E allele e4, dementia and cognitive decline in a population sample. *Lancet, 346*, 1387–1390.

Hendriks, L., van Duijn, C., Cras, P., Cruts, M., Van Hul, W., Van Harskamp, F., et al. (1992). Presenile dementia and cerebral haemorrhage linked to a mutation at codon 692 of the β-amyloid precursor protein gene. *Nature Genetics, 1*, 218–221.

Herreman, A., Serneels, L., Annaert, W., Collen, D., Schoonjans, L., & De Strooper, B. (2000). Total inactivation of gamma-secretase activity in presenilin-deficient embryonic stem cells. *Natural Cellular Biology, 2*, 461–462.

Heutink, P. (2000). Untangling tau-related dementia. *Human Molecular Genetics, 9*, 979–986.

Heutink, P., Stevens, M., Rizzu, P., Bakker, E., Kros, J. M., Tibben, A., et al. (1997). Hereditary frontotemporal dementia is linked to chromosome 17q21-q22: A genetic and clinicopathological study of three Dutch families. *Annals of Neurology, 41*, 150–159.

Higgins, J. J., Litvan, I., Pho., L. T., Li, W., & Nee, L. E. (1998). Progressive supranuclear gaze palsy is in linkage disequilibrium with the tau and not the alpha-synuclein gene. *Neurology, 50*, 270–273.

Hoenika, J., Perez, M., Perez-Tur, J., Barabash, A., Godoy, M., Vidal, L., et al. (1999). The tau gene A0 allele and progressive supranuclear palsy. *Neurology, 53*, 1219–1225.

Holcomb, L., Gordon, M. N., McGowan, E., Yu, X., Benkovic, S., Jantzen, P., et al. (1998). Accelerated Alzheimer-type phenotypes in transgenic mice carrying both mutant amyloid precursor protein and pre-senilin-1 transgenes. *Nature Medicine, 4*, 97–100.

Hong, M., Zhukareva, V., Vogelsberg-Ragaglia, V., Wszolek, Z., Reed, L., Miller, B. I., et al. (1998). Mutation-specific functional impairments in distinct tau isoforms of hereditary FTDP-17. *Science, 282*, 1914–1917.

Hsiao, K., Chapman, P., Nilsen, S., Eckman, C., Harigaya, Y., Younkin, S., et al. (1996). Correlative memory deficits, Aβ elevation and amyloid plaques in transgenic mice. *Science, 274*, 99–102.

Huff, J. F., Auerbach, J., Chakravarti, A., & Boller, F. (1988). Risk of dementia in relatives of patients with Alzheimer's disease. *Neurology, 38*, 786–790.

Huntington's Disease Collaborative Research Group. (1993). A novel gene containing a trinucleotide repeat that is expanded and unstable on Huntington's disease chromosomes: The Huntington's Disease Collaborative Research Group. *Cell, 72*, 971–983.

Hutton, M., Lendon, C. L., Rizzu, P., Baker, M., Froelich, S., Houlden, H., et al. (1998). Association of missense and 5'-splice-site mutations in tau with the inherited dementia FTDP-17. *Nature, 393*, 702–705.

Jarvik, G. P., Wijsman, E. M., Kukull, W. A., Schellenberg, G. D., Yu, C., & Larson, E. B. (1995). Interactions of apolipoprotein E genotype, total cholesterol level, age, and sex in prediction of Alzheimer's disease: A case-control study. *Neurology, 45*, 1092–1096.

Joutel, A., Corpechot, C., Ducros, A., Vahedi, K., Chabriat, H., Mouton, P., et al. (1996). Notch three mutations in CADASIL, a heredity adult-onset condition causing stroke and dementia. *Nature, 383*, 707–710.

Kang, J., Lemaire, H. G., Unterbeck, A., Salbaum, J. M., Masters, C. L., Grzeschik, K. H., et al. (1987). The precursor of Alzheimer's disease amyloid A4 protein resembles a cell-surface receptor. *Nature, 325*, 733–736.

Kehoe, P. G., Wavrant-De Vrieze, F., Crook, R., Wu, W. S., Holmans, P., Fenton, I., et al. (1999). A full genome scan for late onset Alzheimer's disease. *Human Molecular Genetics, 8*, 237–245.

Lemere, C. A., Blusztajn, J. K., Yamaguchi, H., Wisniewski, T., Saido, T. C., Selkoe, D. J. (1996). Sequence of deposition of heterogeneous amyloid β-peptides and APOE in Down's syndrome: Implications for initial events in amyloid plaque formation. *Neurobiological Disorders, 3*, 16–32.

Lendon, C. L., Ashall, S., & Goate, A. M. (1997). Exploring the aetiology of Alzheimer's disease using molecular genetics. *Journal of the American Medical Association, 277*, 825–831.

Levy-Lahad, E., Wasco, W., Poorkaj, P., Romano, D. M., Oshima, J., Pettingell, W. H., et al. (1995). Candidate gene for the chromosome 1 familial Alzheimer's disease locus. *Science, 269*, 973–977.

Levy-Lahad, E., Wijsman, E. M., Nemens, E., Anderson, L., Goddard, K. A., Weber, J. L., et al. (1995). A familial Alzheimer's disease locus on chromosome 1. *Science, 269*, 970–973.

Li, Y. M., Xu, M., Lai, M. T., Huang, Q., Castro, J. L., DiMuzio-Mower, J., et al. (2000). Photoactivated gamma-secretase inhibitors directed to the active site covalently label presenilin 1. *Nature, 405*, 689–694.

The Lund and Manchester Groups. (1994). Clinical and neuropathological criteria for frontotemporal dementia. *Journal of Neurology and Neurosurgery Psychiatry, 57*, 416–418.

Mann, D. (2000). Temporal dementia. *Genetics, 2*, 19–23.

Mann, D. M., Yates, P. O., Marcyniuk, B., & Ravindra, C. R. (1986). The topography of plaques and tangles in Down's syndrome patients of different ages. *Neuropathology and Applied Neurobiology, 12*, 447–457.

Martin, R. L., Gerteis, G., & Gabrielli, W. F. (1988). A family-genetic study of dementia of Alzheimer's type. *Archives of General Psychiatry, 45*, 894–900.

Mayeux, R., Ottman R., Maestre, G., Ngai, C., Tang, M. X., Ginsberg, H., et al. (1995). Synergistic effects of traumatic head injury and apolipoprotein-epsilon 4 in patients with Alzheimer's disease. *Neurology, 45*, 555–557.

McKhann, G., Drachman, D., Folstein, M., Katzman, R., Price, D., & Stadlan, E. M. (1984). Clinical diagnosis of Alzheimer's disease: Report of the NINCDS-ADRDA Work Group under the auspices of Department of Health and Human Services Task Force on Alzheimer's Disease. *Neurology, 34*, 939–944.

Mirra, S. S., Murrell, J. R., Gearing, M., Spillantini, M. G., Goedert, M., Crowther, R. A., et al. (1999). Tau pathology in a family with dementia and a P301L mutation in tau. *Journal of Neuropathology and Experimental Neurology, 58*, 335–345.

Mohs, R. C., Breitner, J. C., Silverman, J. M., Davis, K. L. (1987). Alzheimer's Disease: Morbid risk among first degree relatives approximately 50% by 90 years of age. *Archives of General Psychiatry, 44*, 405–408.

Morris, H. R., Janssen, J. C., Bandmann, O., Daniel, S. E., Rossor, M. N., Lees, A. J., et al. (1999). The tau gene A0 polymorphism in progressive supranuclear palsy and related neurodegenerative diseases. *Journal of Neurology and Neurosurgery Psychiatry, 66*, 665–667.

Morris, H. R., Perez-Tur, J., Janssen, J. C., Brown, J., Lees, A. J., Wood, N. J., et al. (1999). Mutation in the tau exon 10 splice site region in familial frontotemporal dementia. *Annals of Neurology, 45*, 270–271.

Mullan, M., Crawford, F., Axelman, K., Houlden, H., Lilius, L., Winblad, B., et al. (1992). A pathogenic mutation for probable Alzheimers disease in the APP gene at the N-terminus of β-amyloid. *Nature Genetics, 1*, 345–347.

Mullan, M., Houlden, H., Windelspecht, M., Fidani, L., Lombardi, C., Diaz, P., et al. (1992). A locus for familial early-onset Alzheimer's disease on the long arm of chromosome 14, proximal to the α-antichymostrypsin gene. *Nature Genetics, 2*, 340–342.

Mullan, M., Tsuji, S., Miki, T., Katsuya, T., Naruse, S., Kaneko, K., et al. (1993). Clinical comparison of Alzheimer's disease in pedigrees with codon 717 Val-Ile mutation in the amyloid precursor protein gene. *Neurobiology and Aging, 14*, 407–419.

Murrell, J., Farlow, M., Ghetti, B., & Benson, M. (1991). A mutation in the amyloid precursor protein associated with hereditary Alzheimer's disease. *Science, 254*, 97–99.

Myers, A., Holmas, P., Marshall, H., Kwon, J., Meyer, D., Ramic, R., et al. (2000). Susceptibility locus for Alzheimer's disease on chromosome 10. *Science, 290*, 2304–2305.

Myers, A., Wavrant De-Vrieze, F., Holmans, P., Hamshere, M., Crook, R., Compton, D., et al. (2002). A full genome screen for Alzheimer's disease: Stage two analysis. *American Journal of Medical Genetics, 114*, 235–244.

Myers, R. H., Schaffer, E. J., Wilson, P. W. F., D'Agostino, R., Ordovas, J. M., Espino, A., et al. (1996). Apolipoprotein E e4 association with dementia in the population-based study: the Frammington study. *Neurology, 46*, 673–677.

Nacharaju, P., Lewis, J., Easson, C., Yen, S., Hackett, J., Hutton, M., et al. (1999). Accel-

erated filament formation from tau protein with specific FTDP-17 missense mutations. *FEBS Letters, 447*, 195–199.
Nalbantoglu, J., Gilfix, B. M., Bertrand, P., Robitaille, Y., Gauthier, S., Rosenblatt, D. S., et al. (1994). Predictive value of apolipoprotein E genotyping in Alzheimer's disease. *Annals of Neurology, 36*, 889–895.
Namba, Y., Tomonaga, M., Kawasaki, H., Otomo, E., & Okeda, K. (1991). Apolipoprotein E immunoreactivity in cerebral amyloid deposits and neurofibrillary tangles in Alzheimer's disease and kuru plaque amyloid in Creutzfeldt-Jakob disease. *Brain Research, 541*, 163–166.
Nasreddine, Z. S., Loginov, M., Clark, L. N., Lamarche, J., Miller, B. L., Lamontagne, A., et al. (1999). From genotype to phenotype: A clinical pathological, and biochemical investigation of frontotemporal dementia and Parkinsonism (FTDP-17) caused by the P301L tau mutation. *Annals of Neurology, 45*, 704–715.
Neary, D., Snowden, J. S., Northen, B., & Goulding, P. (1988). Dementia of frontal lobe type. *Journal of Neurology and Neurosurgery Psychiatry, 51*, 353–361.
Nee, L. E., Eldridge, R., Sunderland, T., Thomas, C. B., Katz, D., Thompson, K. E., et al. (1997). Dementia of the Alzheimer's Type: Clinical and family study of 22 twin pairs. *Neurology, 37,* 359–363.
Oliva, R., Tolosa, E., Ezquerra, M., Molinuevo, J. L., Valledeoriola, F., Burguera, J., et al. (1998). Significant changes in the tau A0 and A3 alleles in progressive supranuclear palsy and improved genotyping by silver detection. *Archives of Neurology, 55*, 1122–1124.
Owen, M. J. (1994). Dementia. In P. McGuffin, M. J. Owen, M. C. O'Donovan, A. Thapar, & I. I. Gottesman (Eds.), *Seminars in psychiatric genetics* (pp. 192–218). London: Gaskell.
Payami, H., Montee, K. R., Kaye, J. A., Bird, T. D., Yu, C. E., Wijsman, E. M., et al. (1994). Alzheimer's disease, apolipoprotein E4, and gender. *Journal of the American Medical Association, 271*, 1316–1317.
Pericak-Vance, M. A., Bass, M. P., Yamaoka, L. H., Gaskell, P. C., Scott, W. K., Terwedow, H. A., et al. (1997). Complete genomic screen in late-onset familial Alzheimer disease. Evidence for a new locus on chromosome 12. *Journal of the American Medical Association, 278*, 1237–1241.
Pericak-Vance, M. A., Bass, M. L., Yamaoka, L. H., Gaskell, P. C., Scott, W. K., Terwedow, H. A., et al. (1998). Complete genomic screen in late onset familial Alzheimer's disease. *Neurobiology of Ageing, 19*, 39–42.
Pericak-Vance, M. A., Grubber, J., Bailey, L. R., Hedges, D., West, S., Santoro, L., et al. (2000). Identification of novel genes in late-onset Alzheimer's disease. *Experimental Gerontology, 35*, 1343–1352.
Pickering-Brown, S. M., Baker, M., Yen, S. H., Liu, W. K., Hasegawa, M., Cairns, N., et al. (2000). Pick's disease is associated with mutations in the tao gene. *Annals of Neurology, 48,* 859–867.
Poorkaj, P., Bird, T. D., Wijsman, E., Nemens, E., Garruto, R. M., Anderson, L., et al. (1998). Tau is a candidate gene for chromosome 17 frontotemporal dementia. *Annals of Neurology, 43*, 815–825.
Rebeck, G. W., Cheung, B. S., Growdon, W. B., Deng, A., Akuthota, P., Locascio, J., et al. (1999). Lack of independent associations of APOE lipoprotein e promoter and Intron 1 polymorphisms with Alzheimer's disease. *Neuroscience Letters, 272*, 155–158.
Rebeck, G. W., Reiter, J. S., Strickland, D. K., & Hyman, B. T. (1993). Apolipoprotein E in sporadic Alzheimers disease: Allelic variation and receptor interactions. *Neuron, 11*, 575–580.
Reed, L. A., Grabowski, T. J., Schmidt, M. L., Morris, J. C., Goate, A., Solodokin, A., et al. (1997). Autosomal dominant dementia with widespread neurofibrillary tangles. *Annals of Neurology, 42*, 564–572.
Rizzu, P., Van Swieten, J. C., Joosse, M., Hasegawa, M., Stevens, M., Tibben, A., et al. (1999). High prevalence of mutations in the microtubule-associated protein tau in a population study of frontotemporal dementia in the Netherlands. *American Journal of Human Genetics, 64*, 414–421.
Robakis, N. K., Wisniewski, H. M., Jenkins, E. C., Devine-Gage, E. A., Houck, G. E., Yao, X. L., et al. (1987). Chromosome 21q21 sublocalisation of gene encoding beta-amyloid

peptide in cerebral vessels and neuritic (senile) plaques of people with Alzheimer disease and Down syndrome. *Lancet, 1*, 384–385.

Roks, G., Dermaut, B., Heutink, P., Julliams, A., Backhovens, H., Van de Broeck, M. M., et al. (1999). Mutation screening of the tau gene in patients with early onset Alzheimer's disease. *Neuroscience Letters, 277*, 137–139.

Rossi, G., Giaccone, G., Giampaolo, L., Issuich, S., Puoti, G., Frigo, M., et al. (2000). Creutzfeldt-Jakob Disease with a novel for extra-repeat insertion mutation in the PrP Gene. *Neurology, 55*, 405–410.

Schellenberg, G. D., Bird, T. D., Wijsman, E. M., Moore, D. K., Boehnke, M., Bryant, E. M., et al. (1988). Absence of linkage chromosome 21q21 markers to familial Alzheimers disease. *Science, 241*, 1507–1510.

Schellenberg, G. D., Bird, T. D., Wijsman, E. M., Orr, H. T., Anderson, L., Nemens, E., et al. (1992). Genetic linkage evidence for a familial Alzheimers disease locus on chromosome 14. *Science, 259*, 668–671.

Schellenberg, G. D., D'Souza, I., & Poorkaj, P. (2000). The genetics of Alzheimer's disease. *Current Psychiatry Reports, 2*, 158–164.

Scheumer, D., et al. (1996). Secreted amyloid β-protein similar to that in the senile plaques of Alzheimer's disease is increased in vivo by the pre-senilin-1 and -2 and APP mutations linked to familial Alzheimer's disease. *Nature Medicine, 2*, 864–870.

Schmechel, D. E., Saunders, A. M., Strittmatter, W. J., Crain, B. J., Hulette, C. M., Joo, S. H., et al. (1993). Increased amyloid β-peptide deposition in cerebral cortex as a consequence of apolipoprotein E genotype in late-onset Alzheimer disease. *Proceedings of the National Academy of Sciences USA, 90*, 9649–9653.

Scott, W. K., Grubber, J. M., Conneally, P. M., Smaa, G. W., Hulette, C. M., Rosenburg, C. K., et al. (2000). Fine mapping of the chromosome 12 late-onset Alzheimer disease locus: Potential genetic and phenotypic heterogeneity. *American Journal of Human Genetics, 66*, 922–932.

Selkoe, D. J. (1999). Translating cell biology into therapeutic advances in Alzheimer's disease. *Nature, 399*, A23-A31.

Sergeant, N., Wattez, A., & Delacourte, A. (1999). Neurofibrillary degeneration in progressive supranuclear palsy and corticobasal degeneration: Tau pathologies with exclusively "exon 10" isoforms. *Journal of Neurochemistry, 72*, 1243–1249.

Sherrington, R., Rogaev, R., Liang, Y., Rogaeva, E. A., Levesque, G., Ikeda, M., et al. (1995). Cloning of a gene bearing mis-sense mutations in early onset familial Alzheimer's disease. *Nature, 375*, 754–760.

Snowden, J. S., Neary, D., & Mann, D. M. A. (1996). *Fronto-temporal lobar degeneration: Fronto-temporal dementia, progressive aphasia, semantic dementia.* New York: Churchill Livingstone.

Sperfeld, A. D., Collatz, M. B., Baier, H., Palmbach, M., Storch, A., Schwarz, J., et al. (1999). FTDP-17: An early-onset phenotype with Parkinsonism and epileptic seizures caused by a novel mutation. *Annals of Neurology, 46*, 708–715.

Spillantini, M. G., Bird, T. D., & Ghetti, B. (1998). Frontotemporal dementia and Parkinsonism linked to chromosome 17: A new group of taupathies. *Brain Pathology, 8*, 387–402.

Spillantini, M. G., Crowther, R. A., Kamphorst, W., Heutink, P., & van Swieten, J. C. (1998). Tau pathology in two Dutch families with mutations in the microtubule-binding region of tau. *American Journal of Pathology, 153*, 1359–1363.

Spillantini, M. G., Murrell, J. R., Goedert, M., Farlow, M. R., Klug, A., & Ghetti, B. (1998). Mutation in the tau gene in familial multiple system tauopathy with presenile dementia. *Proceedings of the National Academy of Sciences USA, 95*, 7737–7741.

Stevens, M., van Duijn, C. M., Kamphorst, W., de Knijff, P., Heutink, P., van Gool, W. A., et al. (1998). Familial aggregation in frontotemporal dementia. *Neurology, 50*, 1541–1545.

St George-Hyslop, P. H., Haines, J. L., Farrer, L. A., Polinsky, R., Van Broeckhoven, C., & Goate, A. (1990). Genetic linkage studies suggest that Alzheimers disease is not a single homogenous disorder. *Nature, 347*, 194–197.

St George-Hyslop, P., Haines, J., Rogaev, E., Mortilla, M., Vaula, G., Pericak-Vance, M., et al. (1992). Genetic evidence for a novel familial Alzheimer's disease locus on chromosome 14. *Nature Genetics, 2*, 330–334.

St George-Hyslop, P. H., Tanzi, R. E., Polinsky, R. J., Haines, J. L., Nee, L., Watkins, P. C., et al. (1987). The genetic defect causing familial Alzheimer's disease maps on chromosome 21. *Science, 235,* 885–890.

Stern, Y., Gurland, B., Tatemichi, T. K., Tang, M. X., Wilder, D., Mayeux, R. (1994). Influence of education and occupation on the incidence of Alzheimer's disease. *Journal of the American Medical Association, 271,* 1004–1010.

Strittmatter, W., Weisgraber, K., Huang, D. Y., Dong, L. M., Salvesen, G. S., Pericak-Vance, M., et al. (1993). Binding of human apolipoprotein E to synthetic amyloid beta-peptide: Isoform-specific effects and implications for late-onset Alzheimer disease. *Proceedings of the National Academy of Sciences USA, 90,* 8098–8102.

Tanzi, R. E., Gusella, J. F., Watkins, P. C., Bruns, G. A., St George-Hyslop, P., Van Keuren, M. L., et al. (1987). Amyloid beta protein gene: cDNA, mRNA distribution, and genetic linkage near the Alzheimer locus. *Science, 235,* 880–884.

Van Broeckhoven, C., Backhovens, H., Cruts, M., De Winter, G., Bruyland, M., Cras, P., et al. (1992). Mapping of a gene predisposing to early-onset Alzheimer's disease to chromosome 14q24.3. *Nature Genetics, 2,* 335–339.

van Duijn, C. (1996). Epidemiology of the dementias: Recent developments and new approaches. *Journal of Neurology, Neurosurgery and Psychiatry, 60,* 478–488.

van Duijn, C. M., Der Knijff, P., Curts, M., Wehnert, A., Havekes, L. M., Hofman, A., et al. (1994). Apolipoprotein E4 allele in a population-based study of early-onset Alzheimer's disease. *Nature Genetics, 7,* 74–78.

van Duijn, C. M., Havekes, L. M., Van Broeckhoven, C., De Kniff, P., & Hofman, A. (1995). Apolipoprotein E genotype and association between smoking and early on-set Alzheimer's disease. *British Medical Journal, 310,* 627–631.

van Duijn, C. M., Meijer, H., Witteman, J. C. M., et al. (1996). Estrogen apolipoprotein E and the risk of Alzheimer's disease [Abstract]. *Neurobiology and Aging, 17(Suppl.),* S79.

van Duijn, C., Tanja T., Haaxma, R. et al. (1992). Head trauma and the risk of Alzheimer's disease. *American Journal of Epidemiology, 135,* 775–782.

van Swieten, J. C., Stevens, M., Rosso, S. M., Rizzu, P., Joossee, M., de Koning, I., et al. (1999). Phenotypic variation in hereditary frontotemporal dementia with tau mutations. *Annals of Neurology, 46,* 617–626.

Vassar, R., Bennet, B. D., Babu-Khan, S., Kahn, S., Mendiaz, E. A., Denis, P., et al. (1999). B-secretase cleavage of Alzheimer's amyloid precursor protein by the transmembrane aspartic protease BACE. *Science, 286,* 735–741.

Warwick Daw, E., Payami, H., Nemens, E. J., Nochlin, D., Bird, T. D., Schellenberg, G. D., et al. (2000). The number of trait loci in late on-set Alzheimer's disease. *American Journal of Human Genetics, 66,* 196–204.

Zareparsi, S., Wirdefeldt, K., Burgess, C. E., Nutt, J., Kramer, P., Schalling, M., et al. (1999). Exclusion of dominant mutations within the FTDP-17 locus on chromosome 17 for Parkinson's disease. *Neuroscience Letters, 272,* 140–142.

Zhang, Z., Nadeau, P., Song, W., Donoviel, D., Yuan, M., Bernstein, A., et al. (2000). Presenilins are required for gamma-secretase cleavage of beta-APP and transmembrane cleavage of Notch-1. *Natural Cell Biology, 2,* 463–465.

Part IX

Conclusion

26

Behavioral Genomics

Robert Plomin, John C. DeFries, Ian W. Craig, and Peter McGuffin

The field of behavioral genetics has come a long way in the decade since the previous volume of this kind (Plomin & McClearn, 1993). Although a few chapters in the earlier volume mentioned the potential of molecular genetics, the only chapter to have a section on molecular genetics was the chapter on reading disability (DeFries & Gillis, 1993) because the 6p21 linkage with reading disability was just beginning to emerge at that time, the first quantitative trait locus (QTL) linkage found for a human behavioral disorder (Cardon et al., 1994). The concluding chapter of the volume noted that the accelerating pace of behavioral genetic research would propel the field into the next century, a momentum that would be fueled by the findings themselves, new methods, many large-scale ongoing projects, the use of genetic strategies by mainstream psychologists and, especially, molecular genetics (Plomin, 1993).

The most obvious change seen in the present volume is that molecular genetics has come on center stage in behavioral genetics much faster than anyone anticipated. Several of the chapters in this volume are almost entirely about molecular genetic research, including several genomewide hunts for genes. Although the focus now is on finding genes associated with behavior, most exciting are the first signs of the emergence of a behavioral genomics, a top-down analysis that considers how genes work at the level of the behavior of the whole organism. Behavioral genomics can add to the understanding of how genes work even when little is known about the molecular biology or neurobiology of the gene. Behavioral genomics is the future of behavioral genetics in the postgenomic era.

In the postgenomic era, all functional DNA variation will be known, and eventually many of these DNA variations will be found to be associated with complex traits including behavior. The key issue in the postgenomic era will be to understand how genes work—functional genomics. Functional genomics has come to mean molecular biology and proteomics, a bottom-up approach that identifies a polymorphism's DNA difference and traces its effect on protein structure and function in the cell and in neighboring cells. The bottom-up agenda aims eventually to trace pathways from cells through the brain and finally to behavior, but this is a long road to travel, especially for complex traits.

A top-down behavioral genomics approach is complementary and will eventually meet the bottom-up approach in the brain (see Figure 26.1.) Behavioral genomics focuses on the behavior of the whole organism—how genetic effects on behavior contribute to change and continuity in development (developmental genetics), how genetic effects contribute to comorbidity and heterogeneity between disorders (multivariate genetics), and how genetic effects interact and correlate with the environment (environmental genetics). These are issues central to quantitative genetic analyses that have gone beyond merely estimating heritability. Table 26.1 describes examples of these developmental, multivariate, and environmental issues from quantitative genetic research and lists questions they raise for behavioral genomics.

Specific DNA polymorphisms will provide much sharper scalpels to dissect these issues with greater precision. It is important that functional genomics be applied to behavior sooner rather than later because of the societal importance of behavioral dimensions and disorders such as mental

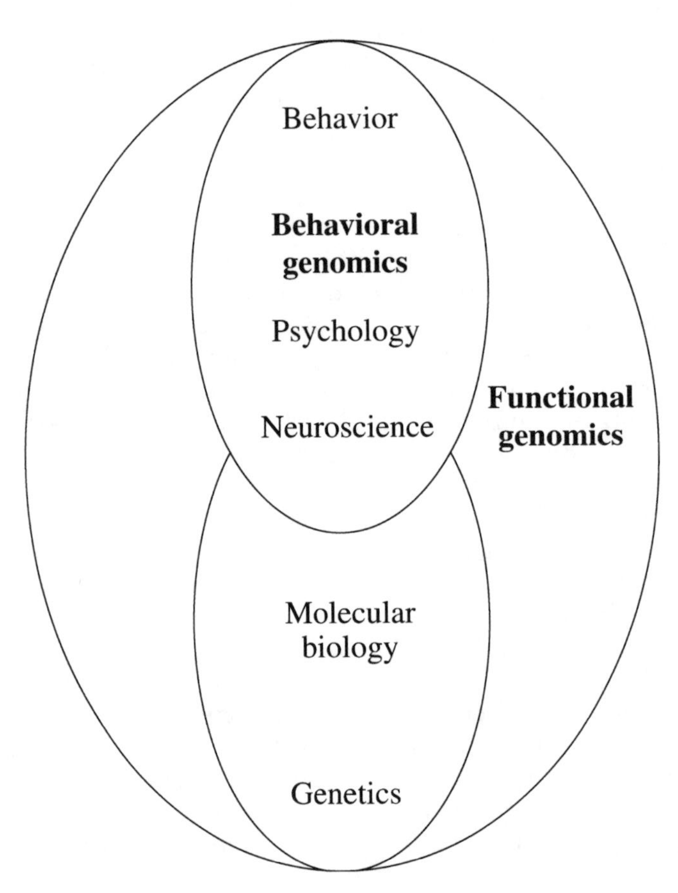

Figure 26.1. Top-down behavioral genomic approaches are complementary to bottom-up molecular biological approaches to functional genomics.

Table 26.1. Examples of Quantitative Genetic Findings and the Questions They Raise for Behavioral Genomics

Quantitative genetics findings	Behavioral genomics questions
Heritability/QTLs	
Heritability established for most traits	Can QTLs be identified?
Developmental issues	
Cross-sectional	
Heritability tends to increase with age	Does the strength of QTL associations increase with age?
Longitudinal	
From age to age, genetic influence shows much continuity and some change	To what extent do QTLs expressed early in life predict later behavior?
Multivariate issues	
Heterogeneity and comorbidity	
Some disorders consist of genetically different components (heterogeneity); some supposedly different disorders overlap genetically (comorbidity)	Can QTLs create a genetically based nosology?
Links between normal and "abnormal"	
Common disorders are generally the quantitative extreme of the same genetic factors responsible for heritability throughout the distribution	Are QTLs that are associated with disorders the same QTLs associated with normal variation?
Environmental issues	
Gene–environment interaction	
Heritability sometimes differs as a function of environment	Do QTL associations differ as a function of environment?
Gene–environment correlation	
Associations between environmental measures and behavior are often mediated genetically	To what extent do QTLs drive experience?

Note. QTL = quantitative trait locus.

health and illness and cognitive abilities and disabilities. The behavioral genomics level of analysis seems likely to yield quicker returns on investment in terms of diagnosis, treatment, and prevention than starting from the molecular biology of the cell. Behavioral genomics will lead to gene-assisted diagnoses and treatment programs as well as to early identification of potential problems that might lead to behavioral interventions that prevent the onset of problems before they create collateral damage. Moreover, the behavioral genomic level of analysis may be the most appropriate level of understanding for evolution because the functioning of the whole organism drives evolution. That is, behavior is often at the cutting edge of natural selection.

Behavioral genomics is likely to accelerate in the postgenomic era as behavioral scientists incorporate DNA polymorphisms in their research, capitalizing on the new scientific horizons for understanding behavior. Although it is difficult and expensive to find genes associated with complex traits, it is relatively easy and inexpensive to use these genes once they have been identified. New techniques will make it increasingly inexpensive and easy to genotype large numbers of subjects for many thousands of genes.

In this concluding chapter, we had expected to adjudicate disagreements among the chapters, but surprisingly few can be found. Although there are differences in emphasis and directions, the chapters are marked by a tolerance of different approaches that seems entirely appropriate given that the dust has not yet settled on the explosion that is the Human Genome Project. There is a humility in the face of the amazing advances that have been made in the past few years and a sense that much more is to come.

Rather than summarizing what has been said in the other chapters in this volume, we thought it would be more useful to extract from these chapters recommendations about what is needed to maximize the potential of the field at this exciting moment. This goal was the motivation for a meeting that we organized in March 2001 from which this volume evolved. The 3-day meeting was held at the Sanger Centre campus near Cambridge, England, and was funded by the Wellcome Trust. The 45 participants included 26 of the top behavioral genetic researchers, 8 of the most promising younger people in the field, molecular geneticists involved in the Human Genome Project, and several policymakers and officials from the Wellcome Trust. Nearly all of the senior authors of the chapters in this volume were participants. At the end of the meeting, we circulated a report to the participants outlining recommendations derived from the meeting. Their supportive responses to the report allow us to present the following five recommendations as a consensus view.

Recommendations to Foster Behavioral Genomics Research

Breaking the 1% QTL Barrier

A theme running through the present volume is that progress to date in identifying QTLs has been slower than expected. For example, although there are several promising leads, no clear-cut associations with schizophrenia and bipolar affective disorder have been identified (see Owen & O'Donovan, chapter 23, and Kalidindi & McGuffin, chapter 24). However, these were the first areas in which molecular genetic approaches were applied at a time when large pedigree linkage designs were in vogue. It is now generally accepted that such designs are only able to detect genes of major effect size, a strategy that has had some notable successes such as early-onset Alzheimer's disease and the epilepsies. Newer areas of application have been more successful for behavioral disorders because they have used designs such as sibpair QTL linkage, parent–child trios, and case control association designs that can detect genes of smaller effect size. In addition to the well-known apolipoprotein E QTL for late-onset dementia (see Williams, chapter 25), other successes include a well-replicated QTL linkage for reading disability on the short arm of chromosome 6 (see Willcutt et al., chapter 13), associations between hyperactivity and dopamine genes (see Thapar, chapter 22), and associations between anxiety and the serotonin transporter in several species in addition to the human species (see Lesch, chapter 20). Animal models are particularly powerful in identifying QTLs; dozens of QTL associations have been found and replicated in the most studied areas of learning and memory (see Wehner & Balogh, chapter 7), anxiety (see Flint, chapter 21), and behavioral responses to drugs (see Crabbe, chapter 16).

One reason for this slower-than-expected progress in identifying QTLs is that research to date has been underpowered for finding QTLs of small effect size (see Cardon, chapter 4, and Sham, chapter 3). The problem is that we do not know the distribution of effect sizes of QTLs for any complex trait in plant, animal, or human species. Not long ago, a 10% effect size was thought to be small, at least from the single-gene perspective in which the effect size was essentially 100%. However, for behavioral disorders and dimensions, a 10% effect size may turn out to be a very large effect. If QTLs account for 1% rather than 10% of the liability to disorders, this would explain why QTLs of smaller effect size, as in the case of associations between dopamine genes and hyperactivity, yield less consistent results than for apolipoprotein E and dementia where the QTL has a relatively large effect that accounts for more than 10% of the liability to dementia. A daunting target for molecular genetic research on complex traits such as behavior is to design research powerful enough to break the "1% QTL barrier," that is, to detect QTLs that account for 1% of the variance while providing protection against false-positive results in genome scans of thousands of genes. To break the 1% QTL barrier, samples of thousands of individuals are needed for research on disorders (comparing cases and controls) and on dimensions (assessing individual differences in a representative sample).

Clinically ascertained samples will continue to be of value, especially for rare disorders such as autism with frequencies less than 0.1% and for disorders such as schizophrenia and bipolar depression with frequencies around 1% or less. However, many behavioral disorders are much more common, with frequencies of 5% and even 10% such as reading disability and other learning disabilities, attention deficit hyperactivity disorder, depression, and alcohol abuse. Although clinically ascertained samples are likely to continue to be useful for common disorders, a complementary but more novel approach is to study these common disorders in the context of normal variation in a single extremely large sample. In addition to the cost-effectiveness of such a study as compared with undertaking separate studies for several common disorders and dimensions, such a design is synergistic in two ways. First, quantitative genetic analyses suggest that common disorders may be the quantitative extremes of the same genetic and environmental factors responsible for variation throughout the distribution. Second, many common disorders tend to co-occur or overlap with each other, and such a multivariate design will permit QTL analyses of comorbidity. Studying several common disorders in the context of normal dimensions of variation will facilitate finding QTL associations for protective as well as risk factors, advance our understanding of the genetic links between the normal and abnormal, investigate comorbidity and heterogeneity at the level of QTLs, begin to address gene–gene interactions (epistasis), and identify QTL–environment interactions and correlations. Achieving these objectives will have far-reaching implications for diagnosing, treating, and preventing disorders. A related point that emerged repeatedly throughout this volume is the need for better phenotypic descriptions of behavioral dimensions and disorders. A collaborative study focused on behavior would aid this objective as well. It also would provide a large representative sample that could be used as a standardized control group in other studies.

As described in Meade, Rawle, Doyle, and Greenaway (chapter 6), the proposed Medical Research Council and Wellcome Trust prospective study of medical illness and life events in 500,000 individuals could serve as a model for a collaborative study of behavioral disorders and dimensions. An ideal sample for the future of behavioral genetics in the postgenomic era would consist of twins for two reasons. First, twins make it possible to use quantitative genetic analyses that can chart the course for molecular genetic analyses. A clear message from this volume is that quantitative genetic analyses continue to be much needed to address issues such as those listed in Table 26.1. Second, fraternal twins provide an ideal sibling comparison for within-family QTL analyses, especially for developmental disorders that change with age because nontwin siblings differ in age. In addition, identical twins facilitate analyses of DNA expression and identification of genes involved in sensitivity to environmental influences, and female pairs of identical twins also provide a unique opportunity to explore the effects of X-chromosome inactivation.

Such a study could focus on late childhood when some interesting common disorders emerge such as learning disabilities and hyperactivity or in adolescence when disorders such as depression begin to emerge. A good argument could also be made for a study focused on adults to assess lifelong problems such as depression and alcohol abuse and the understudied adult concomitants of childhood disorders such as reading disability and hyperactivity. A long-term longitudinal study would be best to address issues of change and continuity during development. Obtaining data from nontwin siblings of twins as well as data from parents of the twins would also be valuable.

> For these reasons, our first recommendation is a collaborative study of a representative sample of 50,000 twin pairs (100,000 individuals) focused on common behavioral disorders as well as normal variation in behavioral dimensions.

Genes, Brain, and Behavior

The pace of discovery of genes associated with behavior will increase as systematic QTL scans are conducted using functional DNA variants in the brain that affect coding regions or regulation of gene expression. Identifying functional DNA variants in the brain associated with behavior will also facilitate functional genomic research that links genes, brain, and behavior. Of the approximately 40,000 genes, perhaps half are expressed in the brain, but a substantial proportion of these are housekeeping genes. In other words, there may be fewer than 10,000 genes expressed in the brain that are relevant specifically to brain function. Identifying functional DNA variants in these genes and DNA variants that regulate the expression of these genes is a high priority for behavioral genomic research because these genes are likely to be the source of much heritable influence on behavior. Not only will these DNA variants be useful for systematic genome scans for identifying genes associated with behavioral dimensions and disorders, but they also will make it possible to study integrated systems of gene pathways, an important step on the road to functional genomics.

Functional polymorphisms are especially important now because it is generally recognized that power for "indirect" association in which a DNA marker is in linkage disequilibrium with a QTL can drop off precipitously when the marker is more than a few thousand base pairs away from the QTL. For this reason, association studies have focused on candidate genes, especially functional polymorphisms in candidate genes, although only a few candidate genes have been investigated. Functional polymorphisms facilitate a much more powerful "direct" association strategy in which the hypothesis is tested that the marker *is* the QTL. The ultimate systematic approach to direct association is to screen all functional polymorphisms in the genome. New techniques, discussed by Craig and McClay (chapter 2), will soon make it feasible to genotype large numbers of markers for large samples.

Although there is much interest in general in the identification of functional DNA variants, an initiative focused on genes expressed in the brain would facilitate postgenomic research on the links among genes, brain, and behavior. One direction for such a program of research would be to focus on gene pathways in known neurotransmitter systems, identifying functional DNA variants in coding regions and regulatory elements of gene systems most likely to affect behavior. Research of this type has begun, but a concentrated initiative is needed. Another direction for research is the creation of a brain "transcriptome" map in which gene expression is mapped throughout the brain in mice (see Grant, chapter 8). Linking behavior-relevant functional DNA variants in the brain with differences in gene expression in brain regions represents a key bottom-up postgenomic goal for behavioral genomic research.

> Our second recommendation is to facilitate the identification of functional DNA variants of genes expressed in the brain and the investigation of their association with behavior.

Mouse Models of Behavior

It is no longer necessary to use mouse models to identify QTLs for behavior because of the potential power to identify QTL associations for human behavior directly. However, mouse models will be crucial as behavioral genetics enters the postgenomic era and the field moves beyond identifying genes associated with behavior toward understanding how these genes achieve their effect. Mouse models will increasingly contribute to functional genomic research on behavior, especially in terms of brain mechanisms that mediate genetic effects. In addition to their ability to manipulate the genome through breeding and through transgenics, mouse models also make it possible to control and to manipulate the environment, which will facilitate research on gene–environment interaction.

Although mouse models already play an important role in QTL and transgenic research in behavioral pharmacogenetics and in learning and memory, even in these areas there is consensus that the models are quite limited behaviorally (see Wehner & Balogh, chapter 7 and Crabbe, chapter 16). In other areas such as personality, psychopathology, and cognitive abilities and disabilities, few mouse models are available. What is especially needed are batteries of multiple tasks that can be used to assess a latent construct relatively free of test-specific factors. Greater integration of mouse and human behavioral genomic research also needs to be fostered. DNA is the ultimate arbitrator of the validity of a mouse model for human research. That is, it is not necessary to create mouse models of behavior that are phenotypically similar to human behavior. Mice do not get drunk the way humans do, yet mouse models of alcohol-related processes have made major contributions to behavioral genomics (see Crabbe, chapter 16). Although a mouse model of reading disability sounds absurd, brain processes related to reading in humans might well be illuminated by investigating gene–behavior pathways in mice. The ultimate test is whether the same genes affect the same brain processes in mouse and man.

> Our third recommendation is to foster the development of mouse models of behavior for use in functional genomic research on brain mechanisms that link genes and behavior.

Behavioral Bioinformatics

In the postgenomic era, behavioral genetic research will require integration of research on genomics, DNA variants, gene expression, proteomics, brain structure and function, and behavior in many species. Although bioinformatics resources are under development in most of these areas, integration of these resources from the perspective of behavioral genomics would greatly facilitate postgenomic research in the field.

> Our fourth recommendation is the establishment of behavioral bioinformatics facilities whose goal is to integrate resources needed for postgenomic research on behavior.

Postgraduate Training in Behavioral Genomics

Because behavior is a key aspect of function, demand for behavioral geneticists in the postgenomic era that focuses on gene function will increase. Although behavioral genetics has provided interdisciplinary training that combines genetics and behavior, both the demands and prospects for postgraduate behavioral genetic training in the postgenomic era are greater than ever. Increased support for postgraduate training in behavioral genomics is needed.

> Our fifth recommendation is to provide increased support for postgraduate training in behavioral genomics.

Behavioral Genetics in the Postgenomic Era

The future for behavioral genetics looks brighter than ever in the dawn of the postgenomic era. Behavioral genetics will be the major beneficiary of postgenomic developments that will facilitate investigation of complex traits influenced by many genes as well as by many environmental factors—first for finding genes associated with behavior and then for understanding the mechanisms by which those genes affect behavior at all levels of analysis from the cell to the brain to the whole organism. The most exciting prospect is the integration of quantitative genetics, molecular genetics, and functional genomics in a new focus on behavioral genomics. This integration augurs well for the future of behavioral genetics because behavioral disorders and dimensions are the ultimate complex traits and will be swept along in the wake of the Human Genome Project as it increasingly provides the tools needed for the genetic analysis of complex traits.

This integration is more than methodological and technological. Because DNA is the ultimate common denominator, genetic research on behavior is becoming integrated into the life sciences. Behavioral genetics will profit from as well as contribute to this integration. It will profit from the advances in molecular genetic and postgenomic research on common complex medical disorders such as diabetes, hypertension, and obesity. Behavioral genetics will contribute a quantitative genetic and QTL perspective that shifts the focus of common disorders to dimensions of normal variation in which common disorders are viewed as the quantitative extreme of the same genetic and environmental factors that create variation throughout the distribution. Behavioral genetics also will contribute to the statistical sophistication and the theoretical framework of quantitative genetics and QTL analysis, which will facilitate a more complete understanding of the genetic and environmental etiologies of complex traits in the postgenomic era.

References

Cardon, L. R., Smith, S. D., Fulker, D. W., Kimberling, W. J., Pennington, B. F., & DeFries, J. C. (1994). Quantitative trait locus for reading disability on chromosome 6. *Science, 266,* 276–279.

DeFries, J. C., & Gillis, J. J. (1993). Genetics and reading disability. In R. Plomin & G. E.

McClearn (Eds.), *Nature, nurture, and psychology* (pp. 121–145). Washington, DC: American Psychological Association.
Plomin, R. (1993). Nature and nurture: Perspective and prospective. In R. Plomin & G. E. McClearn (Eds.), *Nature, nurture, and psychology* (pp. 457–483). Washington, DC: American Psychological Association.
Plomin, R., & McClearn, G. E. (1993). *Nature, nurture, and psychology.* Washington, DC: American Psychological Association.

Glossary

Note. Italics indicate items that are cross-referenced in this glossary.

additive genetic variance Individual differences caused by the independent effects of *alleles* or *loci* that "add up." (For contrast, see *nonadditive genetic variance.*)

adoption studies A range of studies that use the separation of biological and social parentage brought about by adoption to assess the relative importance of genetic and environmental influences. Most commonly, the strategy involves a comparison of adoptees' resemblance to their biological parents who did not rear them and to their adoptive parents. May also involve the comparison of genetically related siblings and genetically unrelated (adoptive) siblings reared in the same family.

adoptive siblings Genetically unrelated children adopted by the same family and reared together.

affected sib pair (ASP) design A popular *linkage* design involving many pairs of siblings in which both members meet diagnostic criteria.

allele An alternative form of a *gene* at a *locus,* for example, A, B, and O for the ABO blood marker locus.

allele sharing Presence of 0, 1, or 2 of the same parents' *alleles* in two siblings.

allelic association An *association* between *allelic frequencies* and a *phenotype.* For example, the frequency of allele 4 of the apolipoprotein E gene is about 40 percent for individuals with Alzheimer's disease and 15 percent for control individuals who do not have the disorder.

allelic frequency Population frequency of an alternate form of a gene. For example, the frequency of the *Phenylketonuria* allele is about 1 percent. (For contrast, see *genotypic frequency.*)

alternative splicing *Introns* in genes can be spliced out in different ways, resulting in one gene producing several different RNAs and thus protein products.

amino acid One of the 20 building blocks of proteins, specified by a *codon* of *DNA.*

anticipation The severity of a disorder becomes greater or occurs at an earlier age in subsequent generations. In some disorders, this phenomenon is caused by the intergenerational expansion of DNA repeat sequences.

association The correlation between an *allele* and a *phenotype.* The correlation may be direct in which a *DNA marker* is functional and causes the phenotypic variation or indirect in which a DNA marker is near a functional *polymorphism* that causes the phenotypic variation. (For contrast, see *linkage.*)

assortative mating Nonrandom mating that results in similarity between spouses. Assortative mating can be negative ("opposites attract") but is usually positive.

assortment Independent assortment is Gregor Mendel's second law of heredity. It states that the inheritance of one *locus* is not affected by the inheritance of another locus. Exceptions to the law occur when genes are inherited close together on the same *chromosome*. Such *linkages* make it possible to map genes to chromosomes.

autosome Any *chromosome* other than the X or Y *sex chromosomes*. Humans have 22 pairs of autosomal chromosomes and 1 pair of sex chromosomes.

balanced polymorphism *Stabilizing selection* that maintains genetic variability, for example, by selecting against both *dominant* homozygotes and *recessive* homozygotes.

band (chromosomal) A chromosomal segment defined by staining characteristics.

base pair (bp) One step in the spiral staircase of the double helix of *DNA*, consisting of adenine bonded to thymine or cytosine bonded to guanine.

bioinformatics Information systems applied to biological phenomena such as *DNA sequence*.

candidate gene A *gene* presumed to be involved in a particular *phenotype*.

carrier An individual who is heterozygous at a given *locus* for a normal and a mutant recessive *allele* and who appears normal phenotypically.

centimorgan (cM) Measure of genetic distance on a *chromosome*. Two *loci* are 1 cM apart if there is a 1 percent chance of *recombination* due to crossover in a single generation. In humans, 1 cM corresponds to approximately 1 million *base pairs*.

chi-square test A statistical test used to determine the probability of obtaining the observed results by chance, under a specific hypothesis.

chromatid One member of one newly replicated *chromosome* in a chromosome pair, which may cross over with a chromatid from a homologous chromosome during *meiosis*.

chromosome A structure that is composed mainly of chromatin, which contains *DNA*, and resides in the *nucleus* of cells. The term in Latin for "colored body," because chromosomes stain differently from the rest of the cell. See also *autosome, sex chromosome*.

cloning The process of making copies of a specific piece of *DNA*, usually a *gene*. In genetics, cloning does not mean the process of making genetically identical copies of an entire organism.

codon A sequence of three *base pairs* that codes for a particular *amino acid* or the end of a chain.

coefficient of relationship (*r*) The proportion of *alleles* held in common by two related individuals.

complementary DNA (cDNA) Synthetic *DNA* reverse transcribed from a specific RNA through the action of the enzyme reverse transcriptase.

complex trait A trait influenced by multiple genes as well as environmental factors. Synonymous with multifactorial and quantitative trait.

concordance Presence of a particular condition in two family members, such as twins.

correlation An index of resemblance that ranges from .00, indicating no resemblance, to 1.00, indicating perfect resemblance.

crossover See *recombination.*

developmental genetic analysis Analysis of change and continuity of genetic and environmental parameters during development. Applied to longitudinal data, assesses genetic and environmental influences on age-to-age change and continuity.

DF extremes analysis An analysis of *familial* resemblance that takes advantage of quantitative scores of the relatives of *probands* rather than just assigning a dichotomous diagnosis to the relatives and assessing *concordance.* (For contrast, see *liability threshold model.*)

diathesis–stress A type of *genotype*–environment interaction in which individuals at genetic risk for a disorder (diathesis) are especially sensitive to the effects of risky (stress) environments.

dichotomous trait See *qualitative disorder.*

directional selection *Natural selection* operating against a particular *allele,* usually selection against a deleterious allele. See *balanced polymorphism, stabilizing selection.*

dizygotic (DZ) Fraternal or nonidentical twins; literally, "two *zygotes.*"

DNA (deoxyribonucleic acid) The double-stranded molecule that encodes genetic information. The two strands are held together by hydrogen bonds between two of the four bases, with adenine bonded to thymine and cytosine bonded to guanine.

DNA marker A *polymorphism* in *DNA* itself such as a *single-nucleotide polymorphism, restriction fragment-length polymorphism,* and a *simple-sequence repeat* polymorphism.

DNA sequence The order of *base pairs* on a single chain of the *DNA* double helix.

dNTP A mix of the four nucleotide bases A, C, G, and T added to a PCR reaction so that replication can occur.

dominant An *allele* that produces a particular *phenotype* when present in the heterozygous state.

effect size The proportion of individual differences for the trait in the population accounted for by a particular factor. For example, *heritability* estimates the effect size of genetic differences among individuals.

electrophoresis A method used to separate *DNA* fragments by size. When an electrical charge is applied to DNA fragments in a gel, smaller fragments travel farther.

epistasis Nonadditive interaction between *genes* at different *loci.* The effect of one gene depends on that of another. Compare with *dominance,* which refers to nonadditive effects between *alleles* at the same locus.

equal environments assumption In twin studies, the assumption that environments are similar for identical and fraternal twins.

exon *DNA sequence* transcribed in *messenger RNA* and translated into protein. Compare with *intron.*

expanded triplet repeat A repeating sequence of three *base pairs,* such as the C-G-G repeat responsible for *fragile X,* that increases in number of repeats over several generations.

F_1, F_2 The offspring in the first and second generations following mating between two inbred strains.

familial Among family members.

family study Assessing the resemblance between genetically related parents and offspring and between siblings living together. Resemblance can be due to heredity or to a family's *shared environment.*

first-degree relative See *genetic relatedness.*

fragile X Fragile sites are breaks in *chromosomes* that occur when chromosomes are stained or cultured. Fragile X is a fragile site on the X chromosome that is the second most important cause of mental retardation in males after Down syndrome and is due to an *expanded triplet repeat* (C-G-G).

full siblings Individuals who have both biological (birth) parents in common.

gamete Mature reproductive cell (sperm or ovum) that contains a haploid (half) set of *chromosomes.*

gametic imprinting See *genomic imprinting.*

gene The basic unit of inheritance. A sequence of *DNA* bases that codes for a particular product. Includes *DNA sequences* that regulate *transcription.* See *allele, locus.*

gene expression *Transcription* of *DNA* into RNA and *translation* into *amino acid* sequences.

gene frequency Can refer either to *allelic frequency* or *genotypic frequency.*

gene map Visual representation of the relative distances between *genes* or genetic *markers* on *chromosomes.*

gene targeting *Mutations* that are created in a specific *gene* and can then be transferred to an embryo.

genetic anticipation See *anticipation.*

genetic correlation A statistic from *multivariate genetic analysis* that indicates the extent to which genetic effects on one trait correlate with genetic effects on another trait, independent of the *heritability* of the two traits.

genetic counseling Conveys information about genetic risks and burdens and helps individuals come to terms with the information and to make their own decisions concerning actions.

genetic relatedness The extent or degree to which relatives have genes in common. *First-degree relatives* of the *proband* (parents and siblings) are 50 percent similar genetically. *Second-degree relatives* of the proband (grandparents, aunts, and uncles) are 25 percent similar genetically. *Third-degree relatives* of the proband (first cousins) are 12.5 percent similar genetically.

genome All the *DNA* of an organism for one member of each *chromosome* pair. The human genome contains about 3 billion DNA *base pairs.*

genomic imprinting The process by which an *allele* at a given *locus* is expressed differently, depending on whether it is inherited from the mother or the father.

genotype The genetic constitution of an individual, or the combination of *alleles* at a particular *locus.*

genotype–environment correlation Genetic influence on exposure to environment; experiences that are correlated with genetic propensities.

genotype–environment interaction Genetic sensitivity or susceptibility to environments. In *quantitative genetics,* this interaction is usually limited to statistical interactions such as genetic effects that differ in different environments.

genotypic frequency The frequency of *alleles* considered two at a time as they are inherited in individuals. The genotypic frequency of individuals with *Phenylketonuria* (homozygous for the *recessive* PKU allele) is .0001. The genotypic frequency of PKU *carriers* (who are heterozygous for the PKU allele) is 2 percent.

half siblings Individuals who have just one biological (birth) parent in common.

haplotype A set of closely linked genetic *marker*s present on one *chromosome* that tend to be inherited together (not easily separable by recombination). Some haplotypes may be in *linkage disequilibrium.*

Hardy–Weinberg equilibrium *Allelic* and *genotypic frequencies* remain the same generation after generation in the absence of forces such as *natural selection* that change these frequencies. If a two-allele *locus* is in Hardy–Weinberg equilibrium, the frequency of genotypes is $p^2 + 2pq + q^2$, where p and q are the frequencies of the two alleles.

heritability The proportion of phenotypic differences among individuals that can be attributed to genetic differences in a particular population. Broad-sense heritability involves all *additive* and *nonadditive genetic variances,* whereas narrow-sense heritability is limited to additive genetic variance.

heterosis See *hybrid vigor*.

heterozygosity The presence of different *alleles* at a given *locus* on both members of a *chromosome* pair at a given locus.

Human Genome Project An international research project to map each human gene and to completely sequence human *DNA.*

Huntington's disease A late-onset lethal human disease of nerve degeneration inherited as an autosomal dominant *phenotype.* The gene, encoding the protein huntingtin, has been cloned.

homozygosity The presence of the same *allele* at a given *locus* on both members of a *chromosome* pair at a given locus.

hybrid vigor The increase in viability and fertility that can occur during outbreeding, for example, when inbred strains are crossed. The increase in *heterozygosity* masks the effects of deleterious *recessive* alleles.

imprinting See *genomic imprinting.*

inbred strain study Comparing inbred strains, created by mating brothers and sisters for at least 20 generations. Differences between strains can be attributed to their genetic differences when the strains are reared in the same laboratory environment. Differences within strains estimate environmental influences, because all individuals within an inbred strain are virtually identical genetically.

inbreeding Mating between genetically related individuals.

inbreeding depression A reduction in viability and fertility that can occur following *inbreeding,* which makes it more likely that offspring will have the same *alleles* at any *locus;* therefore, deleterious *recessive* traits are more likely to be expressed.

inclusive fitness The reproductive fitness of an individual plus part of the fitness of kin that is genetically shared by the individual.

index case See *proband.*

innate Evolved capacities and constraints, not rigid hard-wiring that is impervious to experience.

instinct An *innate* behavioral tendency.

intron *DNA sequence* within a *gene* that is transcribed into *messenger RNA* but spliced out before *translation* into protein. Compare with *exon.*

kilobase (kb) 1,000 *base pairs* of *DNA.*

knockout Inactivation of a gene by *gene targeting.*

latent-class analysis A multivariate technique that clusters traits or symptoms into hypothesized underlying or latent classes.

liability threshold model A model that assumes that dichotomous disorders are due to underlying genetic liabilities that are distributed normally. The disorder appears only when a threshold of liability is exceeded.

lifetime expectancy See *morbidity risk estimate.*

linkage Close proximity of *loci* on a *chromosome.* Linkage is an exception to Gregor Mendel's second law of independent assortment, because closely linked loci are not inherited independently within families.

linkage analysis A technique that detects linkage between DNA markers and traits, used to map *genes* to *chromosomes.* See *linkage, DNA marker, mapping.*

linkage disequilibrium *Loci* close together on a *chromosome.* When the observed frequencies of *haplotypes* in a population do not agree with haplotype frequencies predicted by multiplying together the frequency of individual genetic *markers* in each haplotype.

linkage equilibrium *Loci* not close together on a *chromosome.* When the observed frequencies of *haplotypes* in a population agree with haplotype frequencies predicted by multiplying together the frequency of individual genetic *markers* in each haplotype.

locus (plural, **loci**) The site of a specific *gene* on a *chromosome.* The term is Latin for "place."

LOD score Log of the odds, a statistical term that indicates whether two *loci* are linked or unlinked. A LOD score of +3 or higher is commonly accepted as showing *linkage,* and a score of −2 excludes linkage.

mapping The study of the position of *genes* on *chromosomes.*

map unit See *centimorgan (cM).*

marker Any *phenotype* that detects a single-gene *polymorphism.* A *DNA marker* is a polymorphism in DNA itself.

meiosis Cell division that occurs during *gamete* formation and results in halving the number of *chromosomes* so that each gamete contains only one member of each chromosome pair.

messenger RNA (mRNA) Processed RNA that leaves the *nucleus* of the cell and serves as a template for protein synthesis in the cell body. Each set of three bases, called *codons,* specifies a certain protein in the sequence of *amino acids* that comprise the protein. The sequence of a strand of mRNA is based on the sequence of *complementary DNA.*

microsatellite repeat marker See *simple-sequence repeat (SSR) marker.*

mitosis Cell division that occurs in *somatic cells* in which a cell duplicates itself and its *DNA.*

model fitting In *quantitative genetics,* a method to test the goodness of fit between a model of environmental and *genetic relatedness* against observed data. Different models can be compared, and the best-fitting model is used to estimate genetic and environmental parameters.

molecular genetics The investigation of the effects of specific genes at the *DNA* level. In contrast to *quantitative genetics,* which investigates genetic and environmental components of *variance.*

monozygotic (MZ) Identical twins; literally, "one *zygote.*"

morbidity risk estimate An incidence figure that is an estimate of the risk of being affected.

multiple-gene trait See *polygenic trait.*

multivariate genetic analysis Quantitative genetic analysis of the covariance between traits.

mutagenesis Inducing *mutations* in the *genome,* randomly by the use of chemicals or targeted mutations such as gene *knock outs.*

mutation A heritable change in *DNA base pair* sequences.

natural selection The driving force in evolution in which individuals' *alleles* are spread on the basis of the relative number of their surviving and reproducing offspring.

nonadditive genetic variance Individual differences due to the effects of *alleles* (dominance) or *loci* (epistasis) that interact with other alleles or loci. (In contrast to *additive genetic variance.*)

nondisjunction Uneven division of members of a *chromosome* pair during *meiosis.*

nonshared environment Environmental influences that contribute to differences between family members.

nucleotide One of the building blocks of *DNA* and RNA. A nucleotide consists of a base (one of four chemicals: adenine, thymine, guanine, and cytosine) plus a molecule of sugar and one of phosphoric acid.

nucleus The part of the cell that contains *chromosomes.*

oligogenic A trait influenced by a few genes of major effect rather than a single gene or many genes of small effect. Compare with *polygenic.*

pedigree A family tree. Diagram depicting the genealogical history of a family, especially showing the inheritance of a particular condition in the family members.

penetrance The proportion of individuals with a specific *genotype* who manifest that genotype at the *phenotype* level.

phenocopy An environmentally induced *phenotype* that resembles the phenotype produced by a *polymorphism.*

phenotype The appearance of an individual that results from *genotype* and environment.

Phenylketonuria (PKU) A human metabolic disease caused by a *mutation* in a gene coding for a phenylalanine-processing enzyme (phenylalanine hydroxylase), which leads to accumulation of phenylalanine and mental retardation if not treated. Inherited as an autosomal *recessive* phenotype.

pleiotropy Multiple effects of a *gene.*

polygenic A trait influenced by many genes.

polymerase chain reaction (PCR) A method to amplify a particular *DNA sequence.*

polymorphism A *locus* with two or more *alleles.* The term is Latin for "multiple forms."

population genetics The study of *allelic* and *genotypic frequency* in populations and forces that change these frequencies, such as *natural selection.*

positional cloning Screening the entire *genome* by *linkage* analysis.

premutation Production of eggs or sperm with an unstable expanded number of repeats (up to 200 repeats for *fragile X*).

primer A short sequence of *DNA* used to initiate DNA replication in the *polymerase chain reaction.*

proband The index case from whom other family members are identified.

promoter A regulatory region a short distance upstream from the *transcription* start site of a *gene* that acts as the binding site for RNA polymerase. A region of DNA to which RNA polymerase binds to initiate *transcription.*

qualitative disorder An either-or trait, usually a diagnosis.

quantitative dimension Psychological and physical traits that are continuously distributed within a population, for example, general cognitive ability, height, and blood pressure.

quantitative genetics A theory of multiple-gene influences that, together with environmental variation, result in quantitative (continuous) distributions of *phenotypes.* Quantitative genetic methods, such as the twin and adoption methods for human analysis and inbred strain and selection methods for nonhuman analysis, estimate genetic and environmental contributions to phenotypic variance in a population.

quantitative trait loci (QTL) Genes of various *effect sizes* in multiple-gene systems that contribute to quantitative (continuous) variation in a *phenotype.*

random mating The mating of individuals in a population such that the union of individuals with the trait under study occurs according to the product rule of probability. (For contrast, see *assortative mating.*)

recessive *Alleles* that produce a particular *phenotype* only when present in the homozygous state.

recombinant inbred strains Inbred strains derived from brother–sister matings from an initial cross of two inbred progenitor strains. Called *recombinant* because, in the F_2 and subsequent generations, *chromosomes* from the progenitor strains recombine and exchange parts. Used to map genes.

recombination During *meiosis, chromosomes* exchange parts by crossing over of *chromatids.*

restriction enzyme Recognizes specific short *DNA sequences* and cuts DNA at that site.

restriction enzyme cleavage site The point in a *DNA sequence* where a *restriction* enzyme cuts the *DNA* strand. For example, one commonly used restriction enzyme, EcoRI, recognizes the sequence GAATTC and severs the DNA molecule between the G and A bases.

restriction fragment length polymorphism (RFLP) Variation in the length of *DNA* fragments generated after DNA is digested with a particular *restriction enzyme.* Caused by presence or absence of a particular sequence of DNA (restriction site) recognized by the restriction enzyme, which cuts the DNA at that site.

second-degree relative See *genetic relatedness.*

segregation The process by which two *alleles* at a *locus*, one from each parent, separate during heredity. This is Gregor Mendel's law of segregation, his first law of heredity.

selection study Breeding for a *phenotype* over several generations by selecting parents with high scores on the phenotype, mating them, and assessing their offspring to determine the response to selection. Bidirectional selection studies also select in the other direction, that is, for low scores.

selective placement In *adoption studies,* a method in which children are placed into adoptive families in which adoptive parents are similar to the children's biological parents.

sequence tagged site (STS) A short *DNA* segment that occurs only once in the human *genome* and whose exact location and order of bases are known. Because each is unique, STSs serve as landmarks on the physical map of the human genome.

sex chromosome One of the two *chromosomes* that specify an organism's genetic sex. Humans have two kinds of sex chromosomes, X and Y. Normal females possess two X chromosomes and normal males one X and one Y. Other chromosomes are called *autosomes.*

sex-linked trait See *X-linked trait.*

shared environment Environmental factors responsible for resemblance between family members.

simple-sequence repeat (SSR) marker *DNA markers* that consist of two, three, or four DNA bases that repeat several times and are distributed throughout the *genome* for unknown reasons. The best studied is the two-base (dinucleotide) repeat, C-A (cytosine followed by adenine). Also called *microsatellite repeat markers.*

single-nucleotide polymorphism (SNP) A variation in a single-nucleotide *base pair* that occurs in at least 1 percent of the population. SNPs occur in human *DNA* at a frequency of about one every 1,000 bases and are used as *DNA markers.* SNP is pronounced "snip."

sociobiology An extension of evolutionary theory that focuses on *inclusive fitness* and kin selection.

somatic cells All cells in the body except *gametes.*

stabilizing selection Selection that leads to a *balanced polymorphism,* for example, selection against both *dominant* homozygotes and *recessive* homozygotes.

standard deviation The square root of the *variance.*

stratification *Associations* due to chance population-specific *allelic frequencies* rather than to true *linkage disequilibrium.*

synteny *Loci* on the same *chromosome.* Synteny homology refers to similar ordering of loci in chromosomal regions in different species.

targeted gene knock out The introduction of a *mutation* in a *gene* by a designed alteration in a cloned *DNA sequence* that disrupts the gene's *transcription.* The altered gene is then introduced into the *genome,* and animals are bred to replace the normal *allele* with the *knock out.*

third-degree relative See *genetic relatedness.*

transcription The synthesis of an RNA molecule from *DNA* in the cell *nucleus.*

transgenic Containing foreign *DNA.* For example, *gene targeting* can be used to replace a gene with a nonfunctional substitute to knock out the gene's functioning.

translation Assembly of *amino acids* into peptide chains on the basis of information encoded in *messenger RNA.* Occurs on ribosomes in the cell cytoplasm.

transmission disequilibrium test (TDT) A within-family association design that involves at least two biological parents and one offspring that avoids the problem of population *stratification.*

triplet code See *codon.*

triplet repeat See *expanded triplet repeat.*

trisomy Having three copies of a particular *chromosome* due to *nondisjunction.*

twin study Comparing the resemblance of identical and fraternal twins to estimate genetic and environmental components of *variance.*

Type I error In statistics, the rejection of a true hypothesis. Also called false negative.

Type II error In statistics, the acceptance of a false hypothesis. Also called false positive.

variable expression A single genetic effect may result in variable manifestations in different individuals.

variance A measure of the variation around the central class of a distribution. The average squared deviation of the observations from their mean value.

X-linked trait A *phenotype* controlled by a *locus* on the X *chromosome.*

zygote The cell, or fertilized egg, resulting from the union of a sperm and an egg.

Author Index

Numbers in italics refer to listings in the reference sections.

Subject Index

About the Editors

Robert Plomin, PhD, is professor of behavioral genetics at the Institute of Psychiatry in London, where he is deputy director of the Social, Genetic and Developmental Psychiatry Research Centre at the Institute. The goal of the Research Centre is to bring together genetic and environmental research strategies to investigate behavioral development, a theme that characterizes his research. Plomin is currently conducting a study of all twins born in England during the period 1994–96, focusing on developmental delays in early childhood and their association with behavioral problems. After receiving his doctorate in psychology from the University of Texas, Austin, in 1974, he worked with John DeFries and Gerald E. McClearn at the Institute for Behavioral Genetics at the University of Colorado, Boulder. Together, they initiated several large longitudinal twin and adoption studies of behavioral development throughout the life span. From 1986 until 1994, he worked with McClearn at Pennsylvania State University. They launched a study of elderly twins reared apart and twins reared together to study aging, and they developed mouse models to identify genes in complex behavioral systems. Plomin's current interest is in harnessing the power of molecular genetics to identify genes for psychological traits. He has been president of the Behavior Genetics Association.

John C. DeFries, PhD, is professor of psychology and director of the Institute for Behavioral Genetics, University of Colorado, Boulder. After receiving his doctorate in agriculture (with specialty training in quantitative genetics) from the University of Illinois in 1961, he remained on the faculty of that institution for six years. In 1962, he began research on mouse behavioral genetics and, the following year, was a research fellow in genetics at the University of California, Berkeley, where he conducted research in the laboratory of Gerald E. McClearn. After returning to Illinois in 1964, DeFries initiated an extensive genetic analysis of open-field behavior in laboratory mice that included a classic bidirectional selection experiment with replicate selected and control lines. Three years later, he joined the Institute for Behavioral Genetics, which McClearn had founded in 1967. DeFries and Steven G. Vandenberg founded the journal *Behavior Genetics* in 1970; and DeFries and Robert Plomin founded the Colorado Adoption Project in 1975. For over two decades, DeFries's major research interest has concerned the genetics of reading disabilities, and he is currently director of the Colorado Learning Disabilities Research Center. He served as president of the Behavior Genetics Association from 1982 to 1983, receiving the association's Theodosius Dobzhansky Award for Outstanding Research in 1992, and he became a Fellow of the American Association for the Advancement of Science (Section J, Psychology) in 1994.

Ian W. Craig, PhD, is professor of molecular genetics and head of the molecular genetics group of the Social, Genetic and Developmental Psychiatry Research Centre at the Institute of Psychiatry and also professor in the Department of Biochemistry at Oxford University. His early work focused on identifying genes responsible for single-gene disorders and he is now involved in several projects aimed at finding genes for complex behavioral dimensions and disorders. He is especially interested in candidate gene strategies that integrate entire gene systems. Craig has chaired the Chromosome 12 Human Gene Mapping Committee of the Human Genome Organization.

Peter McGuffin, MB, ChB, PhD, is director of the Medical Research Council's Social, Genetic and Developmental Psychiatry Research Centre at the Institute of Psychiatry, Kings College, in London. He was previously professor and head of the Division of Psychological Medicine at the University of Wales College of Medicine, Cardiff, Wales. He graduated from Leeds University Medical School in 1972 and underwent a period of postgraduate training in internal medicine before specializing in psychiatry at the Bethlem Royal and Maudsley Hospitals, London. In 1979 he was awarded a Medical Research Council Fellowship to train in genetics at the Institute of Psychiatry in London and at Washington University Medical School, St. Louis, Missouri. During this time, he completed the work for his doctoral dissertation, which constituted one of the first genetic linkage studies on schizophrenia. He went on to carry out family and twin studies of depression and other psychiatric disorders, attempting to integrate the investigation of genetic and environmental influences. His current work continues with this general theme, at the same time incorporating molecular genetic techniques and their applications in the study of both normal and abnormal behaviors. McGuffin has been president of the International Society of Psychiatric Genetics since 1995 and is a founding Fellow of Britain's Academy of Medical Sciences.